Advances in Wildlife and Exotic Animals Anatomy

Advances in Wildlife and Exotic Animals Anatomy

Guest Editors

Matilde Lombardero Fernández
María del Mar Yllera Fernández

MDPI

Basel • Beijing • Wuhan • Barcelona • Belgrade • Novi Sad • Cluj • Manchester

Guest Editors
Matilde Lombardero Fernández
Department of Anatomy, Animal Production and Clinical Veterinary Sciences
University of Santiago de Compostela
Lugo
Spain

María del Mar Yllera Fernández
Department of Anatomy, Animal Production and Clinical Veterinary Sciences
University of Santiago de Compostela
Lugo
Spain

Editorial Office
MDPI AG
Grosspeteranlage 5
4052 Basel, Switzerland

This is a reprint of the Special Issue, published open access by the journal *Animals* (ISSN 2076-2615), freely accessible at: https://www.mdpi.com/journal/animals/special_issues/Advances_in_Wildlife_and_Exotic_Animals_Anatomy.

For citation purposes, cite each article independently as indicated on the article page online and as indicated below:

Lastname, A.A.; Lastname, B.B. Article Title. *Journal Name* **Year**, *Volume Number*, Page Range.

ISBN 978-3-7258-8231-1 (Hbk)
ISBN 978-3-7258-8232-8 (PDF)
https://doi.org/10.3390/books978-3-7258-8232-8

Cover image courtesy of María del Mar Yllera Fernández

Contents

About the Editors

Matilde Lombardero Fernández

Matilde Lombardero Fernández earned her Doctor in Veterinary Medicine (D.V.M.) title from the University of Santiago de Compostela (USC) in 1993 and obtained her Ph.D. in veterinary medicine from the same university in 1997. In 1998, she began her teaching career as Associate Professor at the Faculty of Veterinary Sciences in Lugo (USC) and, since 2010, has continued as Professor specializing in veterinary anatomy and histology. Her research interests focus on diverse fields, from the macroscopic to the microscopic morphology of various body systems in different animal species of veterinary interest (both domestic and exotic). Additionally, she investigates alternative preservation methods to formaldehyde (aiming to maintain its preservative properties while being less harmful to the health of teachers and students) for specimens used in veterinary anatomy teaching, such as preservation with saturated saline solution.

María del Mar Yllera Fernández

María del Mar Yllera Fernández earned her Doctor in Veterinary Medicine (D.V.M.) title from the Complutense University of Madrid (UCM) (1984) and her Ph.D. in veterinary medicine from the University of Santiago de Compostela (USC) (1988). In the same year as her Ph.D., she was hired as a University Assistant at the Faculty of Veterinary Sciences in Lugo (USC), obtaining a position as Associate Professor at the same faculty in 1994, where she taught the subjects of veterinary anatomy, embryology, and histology until her retirement in January 2023. Specialized in laboratory animals, she was the Responsible Veterinarian at the USC research animal facilities for 10 years and President of the USC Bioethics Committee. She has taught and organized various courses for personnel working with research animals and authored several publications on different aspects of animal experimentation (pain recognition, anatomy, animal welfare, etc.), as well as applied anatomy of animals of veterinary interest, including both domestic and exotic species.

Preface

The study of the anatomy of domestic animals is essential in the training of future veterinarians and serves as the basis for understanding the physiology of healthy animals, as well as how its alteration can lead to various pathologies. However, at present, the professional scope of veterinarians is diversifying, and in addition to domestic animals, wild and exotic animals are gaining prominence, particularly in relation to participation in conservation programs and the maintenance of biodiversity.

In the training of future veterinarians, there is a growing trend to progressively reduce the teaching load. Conversely, the variety of species that they may attend to in the clinics or conservation parks is increasing. Therefore, we hope that our Reprint will serve as a reference tool for these professionals, as it includes studies on the applied anatomy of various species of marine mammals, herbivores, carnivores, rodents, marsupials, birds, and reptiles, addressing a range of body structures.

Matilde Lombardero Fernández and María del Mar Yllera Fernández
Guest Editors

animals

MDPI

Editorial

Advances in Wildlife and Exotic Animals' Anatomy

María del Mar Yllera † and Matilde Lombardero *

Unit of Veterinary Anatomy and Embryology, Department of Anatomy, Animal Production and Clinical Veterinary Sciences, Faculty of Veterinary Sciences, Terra Campus, University of Santiago de Compostela, 27002 Lugo, Spain; mar.yllera@usc.es
* Correspondence: matilde.lombardero@usc.es; Tel.: +34-982-822-333
† Retired.

This Special Issue follows a previous one entitled "Advances in Animal Anatomy". In this editorial, we discuss the significant reduction in the number of hours allocated to the teaching of basic sciences in university curricula. Additionally, undergraduate veterinary programs have traditionally focused on domestic animals. However, many professionals now work with wild or exotic species in zoos, conservation reserves, or veterinary clinics, where these animals have become pets. Veterinarians, researchers, and specialized technicians often lack familiarity with these wild species and those known as new companion animals (NCAs). Thus, in a first contact, and in the absence of specific information about their anatomy, physiology, and pathologies, it is common to consider these animals similar to domestic species in these aspects. Nevertheless, most species have developed unique anatomical and physiological adaptations that can influence their behavior, pathologies, and specific needs.

In the introduction to this issue, we invited experts and researchers to submit articles detailing the particularities of wild species that could assist various professionals working with them. The response has been highly satisfactory.

This Special Issue compiles twenty-four applied anatomy research articles focused on wild species, some of which, due to their peculiar behavior and social nature, have also been introduced into homes, becoming new companion animals. Many of these studies employ modern imaging techniques, often complemented by traditional methods such as macroscopic anatomical dissections, histological preparations, and anatomical sections to identify organic structures (Figure 1). Knowledge of anatomy is fundamental to interpreting the obtained results. Accordingly, six articles utilized computed tomography, another six employed magnetic resonance imaging, one used arthroscopy (rhinoscopy), another conducted histochemical and immunohistochemical studies, and two manuscripts incorporated electron microscopy, in addition to traditional methods such as morphometry, cross-sections, and macroscopic dissections.

Sixteen manuscripts examined wild mammals, occasionally comparing their anatomy with that of domestic species, while four focused on birds, and the remaining four on reptiles.

This Special Issue includes four research articles on carnivorous mammals. Hadžiomerović et al. [Contribution 1] conducted a geometric and morphometric analysis of the auditory ossicles in the red fox (*Vulpes vulpes*). Their research indicated that the anatomical organization of the ossicles was similar, in terms of morphometric parameters, to that described in other carnivores. However, the small sample size of nine animals (seven females and two males) precluded the determination of sexual dimorphism parameters.

Figure 1. A word cloud featuring keywords from the articles included in this Special Issue.

The second research article on carnivorous mammals examines the nasal cavity of three large cats: the leopard (*Panthera pardus kotiya* L.), the lion (*Panthera leo leo* L.), and the cheetah (*Acinonyx jubatus jubatus* S.), comparing them to the domestic cat. Although there is some literature on the nasal cavity of the domestic cat using diagnostic imaging techniques such as computed tomography (CT), magnetic resonance imaging (MRI), and rhinoscopy, none is as comprehensive as the study conducted by Díaz Martínez et al. [Contribution 2]. Regarding large cats, there were no previous studies on the anatomy of their nasal cavity using imaging techniques. Transverse, parasagittal, and horizontal images obtained with CT and MRI, along with rhinoscopy images, allowed visualization of the arrangement of the bones, cartilage, and mucosa of the nasal cavity in each species. This information will assist veterinary clinicians in diagnosing pathologies affecting this region, which are quite common in cats.

Vélez García et al. [Contribution 3] investigated the brachial plexus of two procyonids, *Procyon cancrivorus* and *Nasua nasua*, focusing on both the origin and distribution of the nerves derived from it. Meticulous dissections of formalin-preserved specimens were performed, tracing each nerve from its origin to its termination along the thoracic limb to the hand. In both species, the brachial plexus originates from the C5-T2 nerve roots, although variations were observed as C5 and T2 were not constant. The course and distribution of the plexus-derived nerves were similar in all specimens. However, the authors identified specific differences that could be associated with the distinct behaviors of *P. cancrivorus* and *N. nasua*.

Kato et al. [Contribution 4] published the latest paper focusing on a carnivorous mammal, the raccoon (*Procyon lotor*), recently introduced in Japan. After collecting the carcasses, the researchers determined their gender and age, the latter by examining the teeth. Individuals under five months were excluded. Body weight and length were measured. A total of 206 animals were collected and divided into three age groups: juveniles, yearlings, and adults, separated by sex and region of origin. The correlation of body length with sex, age class, and region was tested using multiple regression. Craniometry was performed using digital calipers. The study examines the relationships between craniometrics and body measurements with sex, age class, body mass index, length, and region using a regression model. The researchers also compared the body length and craniometry of their specimens with those described for North American subspecies. Due to inconsistencies in the relationship, the authors suggested that hybridization occurred prior to the introduction of the raccoon in Japan.

This Special Issue includes three publications on rodents: one utilized imaging techniques to describe the anatomy of the porcupine's head, while the other two examined the digestive tract of the European beaver and the Brazilian agouti from different perspectives.

Morales-Bordon et al. [Contribution 5] combined anatomical sections with magnetic resonance imaging (MRI) to study the skull of the crested porcupine (*Hystrix cristata*) for the first time. Three carcasses were used in this research. After acquiring transverse plane MRI images, two specimens were frozen and cut into transverse slices from the olfactory bulb to the first cervical vertebra. The third specimen was used to obtain transverse, sagittal, and dorsal planes with MRI. By comparing the sections with the MRI images, the researchers were able to identify and describe in detail the cranial structures of the porcupine. This information is an invaluable tool for veterinarians, aiding in the identification of possible pathologies such as fractures, tumors, and abscesses, among others.

The European beaver is the largest rodent in Eurasia. As an herbivorous species, its diet varies with the seasons. Previous studies have indicated that the structure of the esophagus differs between species, and is related, among other factors, to the physicochemical properties of food. Martyniuk et al. [Contribution 6] examined the histology and ultrastructure of the esophagus in 18 adult animals and three fetuses. Their findings revealed that the mucosa, which lacks glands, was lined with stratified squamous keratinized epithelium in both adults and fetuses, and it was thicker in winter, when the animals consumed large quantities of woody plants. Immunohistochemical studies demonstrated a high rate of epithelial proliferation in winter, higher than in spring and summer. No seasonal differences were observed in the other layers of the esophagus.

Jones et al. [Contribution 7] conducted a similar study on the orange-rumped agouti (*Dasyprocta leporine*, Linnaeus, 1758). Observations of this species in the wild led to its classification as omnivorous due to its behavior. A gross anatomical, histologic, morphometric, and histometric analysis of the digestive tract was performed on six adult animals. Their esophagus was also lined with keratinized epithelium, an adaptation to roughage feeding, and esophageal glands were present. The simple stomach, with abundant gastric glands, was typical of carnivorous species, while the large development of the caecum and colon corresponded to a frugivorous and herbivorous diet. The histology of the different sections of the gastrointestinal tract also supported this conclusion.

Ticks can serve as vectors for multiple pathogens. They feed on blood and climb onto a host's hair coat from their habitat. Interestingly, they do not attack all animals equally; some species appear to be more attractive or sensitive to them than others. Lee et al. [Contribution 8] studied the skin characteristics of two deer species, the Korean water deer (*Hydropotes inermis argyropus*) and the roe deer (*Capreolus pygargus*), to understand why the former had a much lower tick infestation rate than the latter. They collected tissue samples from five individuals of each species, counted their number of ticks, and conducted microscopic studies of the hair and skin. As expected, their results confirmed a much lower infestation rate in the Korean water deer. The article also describes significant differences in hair properties, including diameter and density. The hair of the Korean water deer was thicker and its density much higher, which could hinder the movement and feeding of ticks. The skin of the Korean water deer had mainly primary follicles, while the skin of the roe deer was dominated by a pattern composed of primary and secondary follicles. The authors suggested that this difference could create environments less favorable for ticks in the former case.

The rostral epidural rete mirabile (*rete mirabile epidurale rostrale*) is a structure exclusively described in mammals of the order Artiodactyla. It comprises a network of small anastomosed arterial vessels located at the base of the brain on either side of the pituitary gland. Its primary function is to maintain a constant brain temperature. The arterial blood supplied to this *rete* may, in certain instances, be warmer than the organ itself. Situated within the cavernous sinus, whose blood has been cooled in the nasal cavity, the rete facilitates heat transfer from the arterial to the venous blood, thereby equalizing their temperatures. The presence of the rostral epidural rete mirabile is characteristic of nearly all animals within the Artiodactyla order. Zdun [Contribution 9] investigated its morphology in some wild Suiformes and hippopotamuses by injecting a red solution of the chemo-setting acrylic material Duracryl® Plus into both common carotid arteries and/or a blue solution into the bilateral external jugular veins. Some specimens were subjected to enzymatic digestion of soft tissues, while others were fixed in 5% formalin for dissection. They observed that the rostral branch to the rostral epidural rete mirabile is a single vessel in Suiformes, whereas in hippopotamuses, there were three to five vessels.

Three articles focusing on marsupials are included in this Special Issue. Zdun et al. [Contribution 10] described, for the first time, the arterial blood supply to the brain of the red-necked wallaby (*Notamacropus rufogriseus*). Using the same methods as in their previous manuscript, and incorporating an angio-CT examination, they obtained detailed information about the brain vascularization of this species. The internal carotid artery was identified as the main vessel supplying blood to the brain, with no carotid sinus present. This vessel entered the cranial cavity through the optic canal, and branched into smaller vessels. In the red-necked wallaby, Zdun et al. [Contribution 10] observed that the arterial circle of the brain was closed from the caudal side. The authors compared the cerebral anatomy of this species with that described in other mammalian groups.

Yllera et al. [Contributions 11,12] focused their studies on the sugar glider, a small Australian marsupial that is becoming increasingly popular as a companion animal. Given the unique reproductive characteristics of marsupials, they examined both the female [Yllera et al., Contribution 11] and male [Yllera et al., Contribution 12] genital apparatus through meticulous anatomical dissections, including some microscopic insights in males. The females exhibited duplicated ovaries, oviducts, uteri, and vaginas. During pregnancy, a third vagina, known as the birth canal, forms to facilitate the emergence of the offspring. In males, the reproductive system, similar to that of other marsupials, closely resembled that of placental mammals. It included a pair of testicles, two epididymides, two deferent ducts, the urethra, a bifid penis, and two types of accessory glands: the prostate and two pairs of bulbourethral glands. In this species, the genital, urinary, and digestive systems converge into a cavity called the cloaca, which communicates with the exterior. These articles provide detailed and valuable information to veterinary clinicians, aiding in accurate diagnosis, interpretation of modern imaging techniques such as MRI and CT, and successful surgical interventions in the genital tract when necessary.

Marine mammals constitute a diverse group of animals that have adapted to life in the sea or rely on the sea for sustenance. Many populations are vulnerable or endangered due to a long history of commercial exploitation, habitat degradation, or pollution. Currently, significant efforts are being made to halt their decline and promote their recovery. However, a comprehensive understanding of their habitat, behavior, feeding, reproduction, physiology, and anatomy is essential. Consequently, the study of marine mammals is one of the most active fields of research.

Three articles in this Special Issue focus on the anatomy of the Franciscana dolphin (*Pontoporia blainvillei*) and the South American fur seal (*Arctocephalus australis*). It is widely accepted that anatomy is crucial for understanding many physiological adaptations. Therefore, these data can be instrumental in developing effective conservation plans, and provide essential knowledge for clinical veterinarians who treat pathologies in marine mammals, both in the wild and in captivity.

Tostado-Marcos et al. [Contribution 13] conducted meticulous anatomical dissections of 11 adult and juvenile dolphins, examining their digestive tracts and comparing them with those of other marine and terrestrial carnivores. They identified several peculiarities in the tongue, teeth, stomach, and small intestine, suggesting that the Franciscana dolphin's digestive system has adapted to a diet consisting of small prey.

Martín-Orti et al. [Contribution 14] and Molpeceres-Diego et al. [Contribution 15] investigated the South American fur seal, focusing on the digestive and respiratory systems, respectively. Both studies primarily employed anatomical dissection, supplemented by radiological analysis of the main digestive viscera in the former, and by inflating the lungs with compressed air and creating a polyurethane cast of the entire bronchial tree in the latter. Both articles compared the anatomy of the fur seal with that of terrestrial carnivores, particularly the dog. The authors identified significant similarities between the two species, although the seal exhibited some specific characteristics. These differences were suggested to be adaptations to their feeding habits and marine environment, respectively.

Five manuscripts focused on seabirds, including two on Cory's Shearwater (*Calonectris borealis*) and three on the Atlantic puffin (*Fratercula arctica*), are included in this Special Issue. It is well known that significant anatomical differences exist between birds and mammals. Among birds, seabirds possess unique characteristics due to their adaptation to various environments: aquatic, terrestrial, and aerial. Some seabirds are classified as vulnerable species, necessitating conservation programs, while others are kept in zoos and wildlife reserves or are rescued from the wild when sick or injured, requiring veterinary care. Often, professionals (veterinarians, biologists) responsible for their care or recovery plans lack specific data on the species in question, and must adapt information from closely related species. These articles partially describe their anatomy and employ modern diagnostic imaging methods, providing knowledge that could be highly useful in aiding the conservation of seabirds.

Morales-Espino et al. investigated the coelomic cavity [Contribution 16] and the head [Contribution 17] of Cory's Shearwater using anatomical cross-sections, computed tomography (CT), and magnetic resonance imaging (MRI). Birds lack a diaphragm, resulting in a single body cavity: the coelomic cavity. The authors combined dissections of fresh cadavers with CT images and anatomical cross-sections of frozen specimens to identify and determine the spatial organization of the various organs within the body cavity. In their second study [Morales-Espino et al., Contribution 17], the authors focused on the anatomy of the bird's head. By combining transverse and dorsal anatomical sections of frozen specimens with sequential transverse CT images and MRI sequences in the transverse plane, they described the detailed anatomy of Cory's Shearwater's head, including the central nervous system and its associated structures.

Jaber et al. [Contribution 18] and Fumero-Hernández et al. [Contributions 19,20] conducted analogous studies utilizing cross-sectional anatomy and CT or MRI on the coelomic cavity [Jaber et al., Contribution 18], head [Fumero-Hernández et al., Contribution 19], and eyeball [Fumero-Hernández et al., Contribution 20] of the Atlantic puffin (*Fratercula arctica).* Additionally, Jaber et al. [Contribution 18] performed dissections on three specimens to visualize the organ locations within the coelomic cavity. These dissections enabled the precise identification and correlation of anatomical features observed in anatomical sections and CT images.

Birds possess one of the most acute senses of sight in the animal kingdom. Their eyes contain anatomical structures absent in mammals, such as the pecten and the scleral ring. Fumero-Hernández et al. [Contribution 20] obtained CT images in various planes of the puffin's head, concentrating on the ocular structures. The morphometric study results of these anatomical elements provided significant information that could aid in interpreting the behavioral patterns of this species.

Turning to reptiles, this Special Issue includes four studies. The first study by Kandyel et al. [Contribution 21] focuses on the Egyptian bridled skink (*Heremites vittatus*). Gross anatomy, electron microscopy, histochemistry, and immunohistochemistry were employed to examine the morphological characteristics of the tongue and laryngeal mound. The tongue of this species is triangular in shape, and is divided into three zones: foretongue, midtongue, and hindtongue. The authors identified nine subtypes of filiform papillae with a specific distribution on the dorsal surface of the tongue, while the ventral apical surface of the foretongue comprised conical papillae. Taste pores and taste buds were observed on the dorsal surface of the foretongue and midtongue, respectively. Histological studies revealed a keratinized layer only on the dorsal surface of the foretongue, whereas lingual glands were present only in the midtongue and hindtongue. The researchers noted a strong expression of cytokeratin throughout the tongue. The authors associated these tongue characteristics with the animal's feeding behavior.

González Rodríguez et al. [Contribution 22] utilized macroscopic cross-sections of frozen specimens in conjunction with modern imaging techniques (CT and MRI) to identify the primary structures of the rhinoceros iguana head (*Cyclura cornuta cornuta*), a reptile classified as vulnerable by the International Union for Conservation of Nature (IUCN). The limited literature available on this species primarily focuses on genetic and evolutionary studies or descriptions of pathological cases. Therefore, the anatomical information presented in this article, encompassing both soft tissues and bones, could be highly valuable for diagnosing pathologies such as fractures, skull malformations, or neoplasia. Additionally, it could serve as a useful tool for recognizing diseases resulting from an unbalanced diet, such as osteodystrophy.

Mohamad et al. [Contribution 23] examined the skulls of two reptile species, the loggerhead turtle (*Caretta caretta*) and the green iguana (*Iguana iguana*), using computed tomography (CT) reconstructive techniques such as maximum intensity projections (MIP) and volume rendering (VR). They studied five adult female turtles and four adult male iguanas, taking various measurements of their skulls. Transverse CT images were also obtained, and the data were transferred to a CT workstation to generate VR and MIP reconstructed images of the head. Three-dimensional reconstructions facilitated the visualization of cranial bones. Consequently, Mohamad et al. [Contribution 23] described the skull bones in both species, and highlighted the most significant anatomical differences between them. They concluded that these reconstruction techniques could be highly useful not only for diagnosing various pathological conditions, but also for teaching veterinary anatomy with realistic images.

The latest article on reptiles included in this Special Issue focused on the eyeball of the loggerhead turtle (*Caretta caretta*). This species spends most of its life at sea, necessitating certain adaptations, including well-developed nictitating membranes and sclerotic rings to protect the eye. Fumero-Hernández et al. [Contribution 24] conducted a morphometric study of the loggerhead turtle's eyeball using computed tomography. They studied 10 live animals and measured the length of their shells. The authors also took various measurements of different elements of the eyeball ($n = 20$), as well as the scleral ring, using oblique sagittal, transverse, and dorsal CT images of the skulls, and performed statistical analysis. Consequently, they obtained reference values for the eyeball and scleral ring. The authors suggested that these values could be related to the vision capabilities and habits of these animals. They acknowledged two limitations to their study: the low number of animals used and their varying sizes, and the inherent error associated with the manual contouring of computed tomography images.

We believe that this Special Issue has successfully met our objective of assisting various professionals (veterinarians, biologists, keepers, and technicians) who work with wild animals, both in their natural habitats and in captivity, by providing them with new anatomical knowledge and aiding in the interpretation of results obtained through modern diagnostic imaging techniques. We acknowledge that these contributions are on a limited number of species, but we hope they will positively impact on the care and welfare of wild animals. In this manner, we aim to contribute to the conservation of biodiversity.

Acknowledgments: We extend our gratitude to all the authors for submitting their research to this Special Issue. Similarly, we also thank the peer reviewers and academic editors for their expertise and insightful comments that contributed to the improvement of all manuscripts.

Conflicts of Interest: The authors declare no conflicts of interest.

List of Contributions:

1. Hadžiomerović, N.; Gundemir, O.; Tandir, F.; Avdić, R.; Katica, M. Geometric and Morphometric Analysis of the Auditory Ossicles in the Red Fox (*Vulpes vulpes*). *Animals* **2023**, *13*, 1230. https://doi.org/10.3390/ani13071230.
2. Díaz Martínez, E.; Arencibia Espinosa, A.; Soler Laguía, M.; Kilroy, D.; Martínez Gomariz, F.; Casas García, D.L.; Sánchez Collado, C.; Gil Cano, F.; Jaber, J.R.; Ramírez Zarzosa, G. An Anatomical Study Using Computed Tomography, Magnetic Resonance Imaging, and Rhinoscopy of the Nasal Cavity of Domestic Cat (*Felis silvestris catus* L.) and Big Cats: Lion (*Panthera leo leo* L.), Leopard (*Panthera pardus kotiya* L.), and Cheetah (*Acinonyx jubatus jubatus* S.). *Animals* **2024**, *14*, 1172. https://doi.org/10.3390/ani14081172.
3. Vélez García, J.F.; de Carvalho Barros, R.A.; Miglino, M.A. Origin and Distribution of the Brachial Plexus in Two Procyonids (*Procyon cancrivorus* and *Nasua nasua*, Carnivora). *Animals* **2023**, *13*, 210. https://doi.org/10.3390/ani13020210.
4. Kato, T.; Yamasaki, F.; Doi, K.; Kawamichi, M.; Hayama, S. Body Length and Craniometrics of Non-Native Raccoons in Two Regions in Middle Japan during Early Invasion Stages. *Animals* **2022**, *13*, 55. https://doi.org/10.3390/ani13010055.
5. Morales-Bordon, D.; Encinoso, M.; Arencibia, A.; Jaber, J.R. Cranial Investigations of Crested Porcupine (*Hystrix cristata*) by Anatomical Cross-Sections and Magnetic Resonance Imaging. *Animals* **2023**, *13*, 2551. https://doi.org/10.3390/ani13162551.
6. Martyniuk, K.; Ziółkowska, N.; Hanuszewska-Dominiak, M.; Szyryńska, N.; Lewczuk, B. Histology and Ultrastructure of the Esophagus in European Beaver (*Castor fiber*) Displays Features Adapted to Seasonal Changes in Diet. *Animals* **2023**, *13*, 635. https://doi.org/10.3390/ani13040635.
7. Jones, K.R.; John, R.E.; Sundaram, V. Morpho-Histological Studies of the Gastrointestinal Tract of the Orange-Rumped Agouti (*Dasyprocta leporina* Linnaeus, 1758), with Special Reference to Morphometry and Histometry. *Animals* **2022**, *12*, 2493. https://doi.org/10.3390/ani12192493.
8. Lee, S.-J.; Kim, K.-Y.; Kim, G.; Moon, S.; Park, Y.-C.; Cho, H.-S.; Oh, Y. Potential Tick Defense Associated with Skin and Hair Characteristics in Korean Water Deer (*Hydropotes inermis argyropus*). *Animals* **2024**, *14*, 185. https://doi.org/10.3390/ani14020185.
9. Zdun, M. Comparison of the Rostral Epidural Rete Mirabile and the Patterns of Its Blood Supply in Selected Suiformes and Hippopotamuses. *Animals* **2023**, *13*, 644. https://doi.org/10.3390/ani13040644.
10. Zdun, M.; Ruszkowski, J.J.; Gogulski, M.; Józefiak, A.; Hetman, M. Arterial Circle of the Brain of the Red-Necked Wallaby (*Notamacropus rufogriseus*). *Animals* **2022**, *12*, 2796. https://doi.org/10.3390/ani12202796.
11. Yllera, M.d.M.; Alonso-Peñarando, D.; Lombardero, M. Gross Anatomy of the Female Reproductive System of Sugar Gliders (*Petaurus breviceps*). *Animals* **2023**, *13*, 2377. https://doi.org/10.3390/ani13142377.

12. Yllera, M.d.M.; Alonso-Peñarando, D.; Lombardero, M. Anatomy of the Male Reproductive System of Sugar Gliders (*Petaurus breviceps*). *Animals* **2024**, *14*, 2748. https://doi.org/10.3390/ani14182748.
13. Tostado-Marcos, C.; Olocco Diz, M.J.; Martín-Orti, R.; Loureiro, J.-P.; Molpeceres-Diego, I.; Tendillo-Domínguez, E.; Pérez-Lloret, P.; Santos-Álvarez, I.; González-Soriano, J. Nature or Nurture: Is the Digestive System of the Pontoporia Blainvillei Influenced or Determined by Its Diet? *Animals* **2024**, *14*, 661. https://doi.org/10.3390/ani14050661.
14. Martín-Orti, R.; Tostado-Marcos, C.; Loureiro, J.-P.; Molpeceres-Diego, I.; Tendillo-Domínguez, E.; Santos-Álvarez, I.; Pérez-Lloret, P.; González-Soriano, J. The Digestive System of the Arctocephalus Australis in Comparison to the Dog as a Land-Carnivore Model. *Animals* **2022**, *12*, 1634. https://doi.org/10.3390/ani12131634.
15. Molpeceres-Diego, I.; Martín-Orti, R.; Loureiro, J.-P.; Tostado-Marcos, C.; Tendillo-Domínguez, E.; Santos-Álvarez, I.; Pérez-Lloret, P.; González-Soriano, J. The Respiratory System of the Arctocephalus Australis in Comparison to the Dog as a Land-Carnivore: Are There Adaptations to Marine Life? *Animals* **2023**, *13*, 199. https://doi.org/10.3390/ani13020199.
16. Morales Espino, A.; Déniz, S.; Fumero-Hernández, M.; Encinoso, M.; Calabuig, P.; Conde-Felipe, M.; Jaber, J.R. A Cadaveric Study Using Anatomical Cross-Section and Computed Tomography for the Coelomic Cavity in Juvenile Cory's Shearwater (Aves, *Procellariidae, Calonectris borealis*). *Animals* **2024**, *14*, 858. https://doi.org/10.3390/ani14060858.
17. Morales-Espino, A.; Déniz, S.; Paz-Oliva, P.; Roldán-Medina, N.; Encinoso, M.; Suárez-Cabrera, F.; Jaber, J.R. Cory's Shearwater (*Calonectris borealis*): Exploring Normal Head Anatomy through Cross-Sectional Anatomy, Computed Tomography and Magnetic Resonance Imaging. *Animals* **2024**, *14*, 1962. https://doi.org/10.3390/ani14131962.
18. Jaber, J.R.; Fumero-Hernández, M.; Corbera, J.A.; Morales, I.; Amador, M.; Ramírez Zarzosa, G.; Encinoso, M. Cross-Sectional Anatomy and Computed Tomography of the Coelomic Cavity in Juvenile Atlantic Puffins (Aves, *Alcidae, Fratercula arctica*). *Animals* **2023**, *13*, 2933. https://doi.org/10.3390/ani13182933.
19. Fumero-Hernández, M.; Encinoso, M.; Melian, A.; Nuez, H.A.; Salman, D.; Jaber, J.R. Cross Sectional Anatomy and Magnetic Resonance Imaging of the Juvenile Atlantic Puffin Head (Aves, Alcidae, *Fratercula arctica*). *Animals* **2023**, *13*, 3434. https://doi.org/10.3390/ani13223434.
20. Fumero-Hernández, M.; Encinoso, M.; Ramírez, A.S.; Morales, I.; Suárez Pérez, A.; Jaber, J.R. A Cadaveric Study Using Computed Tomography for Measuring the Ocular Bulb and Scleral Skeleton of the Atlantic Puffin (Aves, *Alcidae, Fratercula arctica*). *Animals* **2023**, *13*, 2418. https://doi.org/10.3390/ani13152418.
21. Kandyel, R.M.; Choudhary, O.P.; El-Nagar, S.H.; Miles, D.B.; Abumandour, M. Tongue of the Egyptian Endemic Bridled Skink (*Heremites vittatus*; Olivier, 1804): Gross, Electron Microscopy, Histochemistry, and Immunohistochemical Analysis. *Animals* **2023**, *13*, 3336. https://doi.org/10.3390/ani13213336.
22. González Rodríguez, E.; Encinoso Quintana, M.; Morales Bordon, D.; Garcés, J.G.; Artiles Nuez, H.; Jaber, J.R. Anatomical Description of Rhinoceros Iguana (*Cyclura cornuta cornuta*) Head by Computed Tomography, Magnetic Resonance Imaging and Gross-Sections. *Animals* **2023**, *13*, 955. https://doi.org/10.3390/ani13060955.
23. Mohamad, J.R.J.; González-Rodríguez, E.; Arencibia, A.; Déniz, S.; Carrascosa, C.; Encinoso, M. Anatomical Description of Loggerhead Turtle (*Caretta caretta*) and Green Iguana (*Iguana iguana*) Skull by Three-Dimensional Computed Tomography Reconstruction and Maximum Intensity Projection Images. *Animals* **2023**, *13*, 621. https://doi.org/10.3390/ani13040621.
24. Fumero-Hernández, M.; Encinoso Quintana, M.; Ramírez, A.S.; Morales Fariña, I.; Calabuig, P.; Jaber, J.R. Morphometric Study of the Eyeball of the Loggerhead Turtle (*Caretta caretta*) Using Computed Tomography (CT). *Animals* **2023**, *13*, 1016. https://doi.org/10.3390/ani13061016.

animals

MDPI

Article

Anatomy of the Male Reproductive System of Sugar Gliders (*Petaurus breviceps*)

María del Mar Yllera [1,†], Diana Alonso-Peñarando [2,3] and Matilde Lombardero [1,*]

1 Unit of Veterinary Anatomy and Embryology, Department of Anatomy, Animal Production and Clinical Veterinary Sciences, Faculty of Veterinary Sciences, Terra Campus, University of Santiago de Compostela, 27002 Lugo, Spain; mar.yllera@usc.es

2 Department of Animal Pathology, Faculty of Veterinary Sciences, Terra Campus, University of Santiago de Compostela, 27002 Lugo, Spain; diana.alonso.penarando@rai.usc.es

3 Veterinary Clinic Madivet, Calle Comercio, 5, Bargas, 45593 Toledo, Spain

* Correspondence: matilde.lombardero@usc.es; Tel.: +34-982-822-333

† Retired.

Simple Summary: Sugar gliders are exotic animals that have become pets in recent years. As a result, these small marsupials are increasingly coming to the veterinary practices. Marsupials are a group of mammals characterized by their peculiar way of reproduction. This translates into the distinct anatomy of their genital apparatus, both male and female. There are not many references concerning the anatomy of this particular species. The male reproductive system of the *Petaurus breviceps* follows the general pattern in marsupials, but each species has its own peculiarities that the veterinary clinician should know. It includes two testes, epididymis, and deferent ducts, as well as two types of accessory glands: the prostate and the bulbourethral glands, in addition to the male urethra and the penis, which is bifid at its end. The purpose of this study was to describe the male genital gross anatomy of the sugar gliders (supplemented with some microscopic observations), to provide detailed information to veterinary clinicians to facilitate accurate diagnosis and surgery.

Abstract: The present study provides a detailed macroscopic examination (with some microscopic insights) of the genital apparatus of seven adult and intact male sugar gliders, as well as one castrated individual. The scrotum is pendulous and attached to the ventral abdominal wall, situated in the caudal part of the abdomen and cranial to the cloacal opening. The testes are oval shaped with their long axes oriented vertically. The epididymides are attached along the caudomedial border of their respective testes. The head and tail of the epididymides are in close proximity to the poles of the testes but are not directly attached. The deferent ducts are positioned laterally to their ipsilateral ureter as they run near the dorsal surface of the urinary bladder. The ampulla of the deferent duct is absent. The penis is located post-scrotally, lacks insertion into the bony pelvis, and has a bifid distal end. The crura of the penis originate within the ischiocavernosus muscles, and there are two bulbs of the penis. When flaccid, the entire penis is concealed in the perineal region and externalizes through an orifice situated in the ventral part of the cloaca, traversing its floor towards the cloacal opening. The urethral external orifice is positioned at the point where the bifurcation of the free part of the penis begins. The prostate gland resides in the pelvic cavity and consists of two parts: a macroscopically visible body surrounding the urethra and a microscopically identifiable portion embedded within the walls of the duct. Sugar gliders possess two pairs of bulbous bulbourethral glands, located dorsally and laterally to the rectum, but lack vesicular glands.

Keywords: male genital anatomy; testis; epididymis; deferent duct; penis; corpus cavernosum; corpus spongiosum; bulbospongiosus muscle; ischiocavernosus muscle; prostate; bulbourethral glands

1. Introduction

The prevalence of novel companion animal species, particularly small marsupials such as sugar gliders, is increasing significantly. These animals now represent a growing segment of the patient population in veterinary practices. Consequently, veterinary professionals are encountering species with anatomical structures, physiological processes, and pathological conditions that diverge substantially from those of traditional domestic animals. However, the paucity of species-specific scientific literature and clinical data compels practitioners to extrapolate information from better-studied species. This extrapolation may potentially lead to complications and inaccuracies in diagnostic procedures, surgical interventions, and therapeutic regimens.

Sugar gliders (*Petaurus breviceps*) are nocturnal, arboreal marsupials indigenous to Australia and New Guinea [1]. These animals exhibit communal roosting behavior, sheltering during daylight hours in tree hollows [2]. In their natural habitat, sugar gliders form social groups characterized by scent-marking behaviors, with the dominant male's odor serving as a group identifier.

Their social nature makes them a lovable pet that can recognize their owners if they have been properly socialized from an early age [2,3], otherwise, the stress of isolation leads to self-mutilation which, in the case of males, often affects the penis, and amputation of the bifid penis may be necessary if it is severely damaged [4,5].

Sugar gliders are omnivorous, with a diet comprising tree sap, nectar, fruits, and insects [4]. Some populations have been observed to include avian eggs and nestlings in their diet [6]. Sexual dimorphism in this species is quite subtle, but males can be distinguished by the presence of scent glands on the forehead and chest [4,5], as well as their pendular scrotum and their larger size.

Marsupials distinguish themselves from other mammals through their unique reproductive processes, which involve distinct anatomical features of their genital apparatus. These differences are particularly pronounced in females, who possess double reproductive organs, but males also exhibit unique characteristics, primarily related to the penis and associated structures.

The physiology, reproduction, and reproductive anatomy and behavior of marsupials have been extensively studied, especially in wild species [7–25]. However, there is a notable lack of specific references to the internal anatomy of the reproductive system of *Petaurus breviceps*. This study aims to describe the male genital anatomy of the sugar gliders, providing detailed and valuable information to veterinary clinicians to facilitate accurate diagnosis and surgery.

2. Materials and Methods

The study sample comprised eight male sugar gliders (*Petaurus breviceps*), including seven intact adults and one neutered individual. The specimens were obtained through owner donations following natural mortality due to various pathologies.

Preservation methods varied among the specimens: four were frozen, one was preserved in formalin, one in 76° alcohol, and two were examined fresh. For one frozen specimen, the entire genital apparatus was extracted post-macroscopic examination and subsequently preserved in formalin for detailed analysis.

Dissections were performed using a Nikon AFX-II magnifying device with magnification ranging from 0.66 to 4.0× and 10× oculars. When necessary, ventral midline incisions were made to access body cavities. Photographic documentation was accomplished using a Coolpix S6 camera (Zoom Nikkor ED 5.8–17,4 mm 1:3.0–5.4) equipped with a macro lens.

For histological examination, tissue samples were processed using an automatic processor and embedded in paraffin. Sections of 6 μm thickness were cut using a microtome, deparaffinized, hydrated, and stained with hematoxylin-eosin. Following dehydration, slides were mounted in DPX (Sigma-Aldrich, St. Louis, MO, USA). Light microscopy was performed utilizing two distinct microscope setups. The first configuration employed a Nikon microscope coupled with an Olympus EP50 camera (Hamburg, Germany). The second setup utilized a Zeiss Axiophot microscope (Zeiss, Jena, Germany) equipped with a Zeiss AXIOCAM 208 color camera (Zeiss, Jena, Germany). Photomicrographs were acquired at various magnifications using these imaging systems. For low-magnification imaging, a Nikon magnifying loupe was utilized in conjunction with an Olympus EP50 camera.

3. Results

The anatomical structure of the male sugar glider's reproductive tract includes paired gonads, the testes (*testis*), paired gonadal duct systems (each comprises an epididymis (*epididymis*) and a deferent duct (*ductus deferens*)), as well as two accessory glands (the prostate (*prostata*) and a pair of bulbourethral glands (*glandula bulbourethralis*)), the male urethra (*urethra masculina*), and the penis (*penis*).

3.1. Testis

The testes (*testis*) are external in the *Petaurus breviceps*. They are ventral to the caudal part of the abdomen, cranial to the cloacal opening and the penis. Both gonads, together with the ipsilateral epididymis, are enclosed in a cutaneous pouch of skin known as the scrotum (scrotum). The scrotum is fur covered and pendulous, attached to the ventral abdominal wall by a well-developed pedicle measuring 0.5–0.7 cm in length (Figure 1).

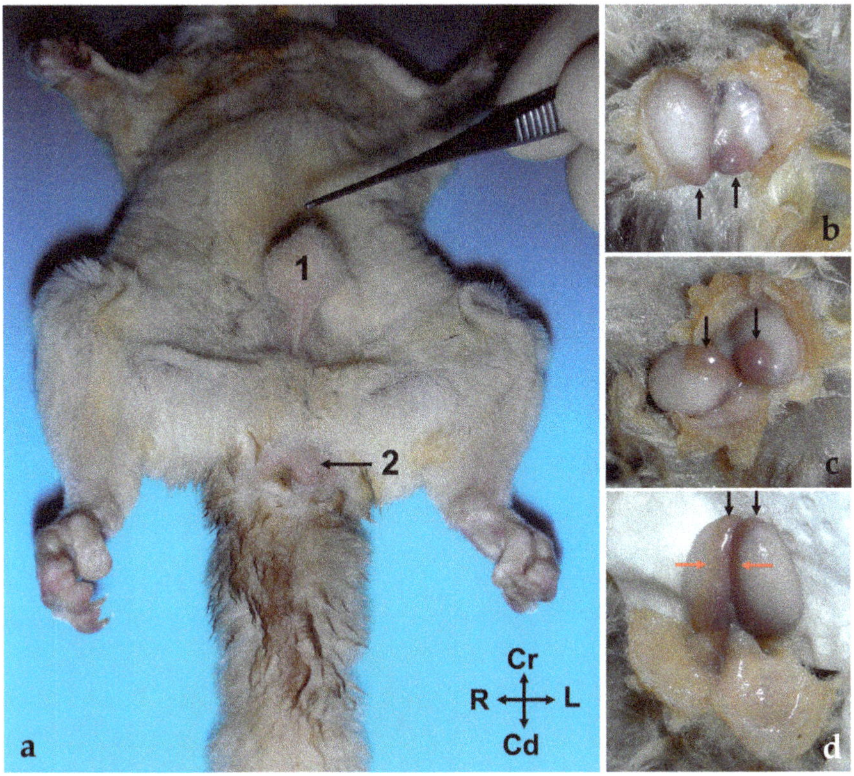

Figure 1. (**a**) Ventral view of the abdominal wall of a male sugar glider (fresh specimen); 1, the scrotum; 2, the cloacal orifice. (**b–d**) Images providing a detailed view of the testes and epididymis in their natural position. The scrotum has been incised for better visibility. The black arrows indicate the tail of the epididymis, while the red arrows point to the body of the epididymis. (**b**) Cranial view, (**c**) ventral view, and (**d**) caudal view. All photographs maintain the same orientation as indicated in image (**a**).

The testes are oval-shaped organs with their longitudinal axes oriented vertically. They possess two distinct poles: the head end (*extremitas capitata*) and the tail end (*extremitas caudata*), which are named based on their topographical relationship with the respective parts of the epididymis (head and tail). In the sugar glider, the head end of the testes is positioned dorsally (proximally), while the tail end is situated ventrally and distally. The testes are enveloped by a serous, unpigmented membrane known as the tunica vaginalis (Figure 1). In one specimen, an anomaly was observed, characterized by a prominent longitudinal sulcus on the medial surface of each testis, which was apparent in histological sections but not detectable macroscopically.

The efferent duct originates from the rete testis and transports spermatozoa towards the head of the epididymis. Histologically, it appears that there is a single efferent duct in the sugar glider.

3.2. Epididymis

The epididymis is a highly convoluted tubular structure that adheres to the caudomedial border of the ipsilateral testis. It is conventionally segmented into three distinct regions: the head, body, and tail. The head (*caput epididymidis*) and tail (*cauda epididymidis*) are situated in close proximity to the proximal and distal poles of the testis, respectively, though they are not directly attached. The body (corpus epididymidis) extends along the caudomedial border of the testis but remains unfused with it. Between the epididymis and testis lies a space known as the testicular bursa (*Bursa testicularis*), which is medially enclosed by the distal mesorchium (*mesorchium distale*) (Figure 2). The tail of the epididymis forms a distinctive rounded protuberance (Figures 1 and 2). Two ligaments are attached to the tail of the epididymis: the proper ligament of the testis (ligamentum testis proprium), which connects it to the gonad, and the ligament of the tail of the epididymis (ligamentum caudae epididymidis), which anchors it to the parietal layer of the peritoneal sac that surround both the epididymis and testicle.

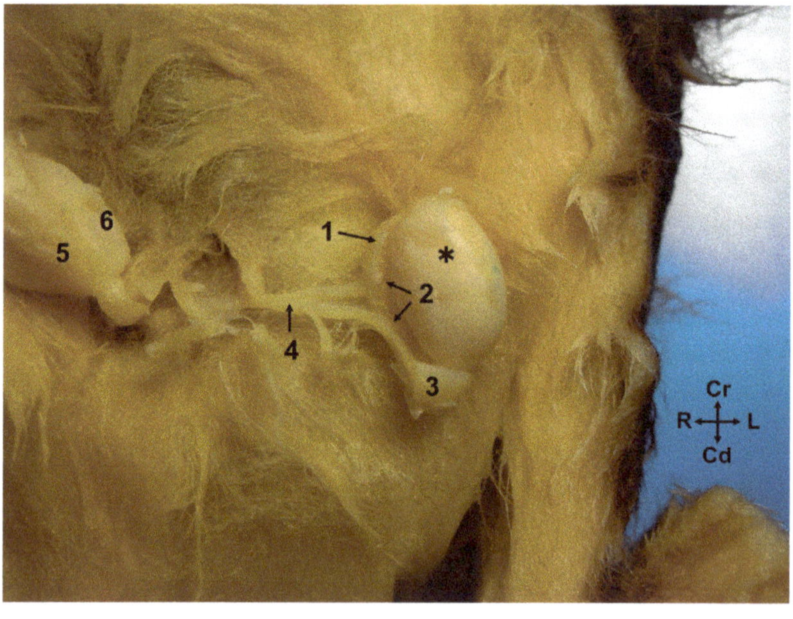

Figure 2. Ventral view of the abdominal wall of a male sugar glider. The *scrotum* has been incised (alcohol-fixed specimen). The left testis (∗) is in a displaced position, and the left tunica vaginalis parietalis has been excised. Left epididymis: 1, head; 2, bursa testicularis, a space situated between the body of the epididymis and the testis; 3, tail; 4, left deferent duct; 5, right epididymis; and 6, right testis.

3.3. Deferent Duct

The ductus deferens emerges from the tail of the epididymis and runs medially to the body of the epididymis towards the abdomen, enclosed within the testicular spermatic cord (Figure 3a). From the dorsal aspect of the pedicle, the ductus deferens courses subcutaneously in a caudolateral direction until reaching the superficial inguinal ring (*anulus inguinalis superficialis*), to pass through the canal inguinal (*canalis inguinalis*) to enter the abdominal cavity. It is noteworthy that in a minority of cases, anatomical variations may occur. In two out of seven intact male specimens examined, the deferent ducts were observed to cross each other as they entered the pedicle near the head of the epididymis, with the right duct passing to the left side and vice versa. These ducts subsequently recrossed before reaching the abdominal wall, ultimately returning to their original sides.

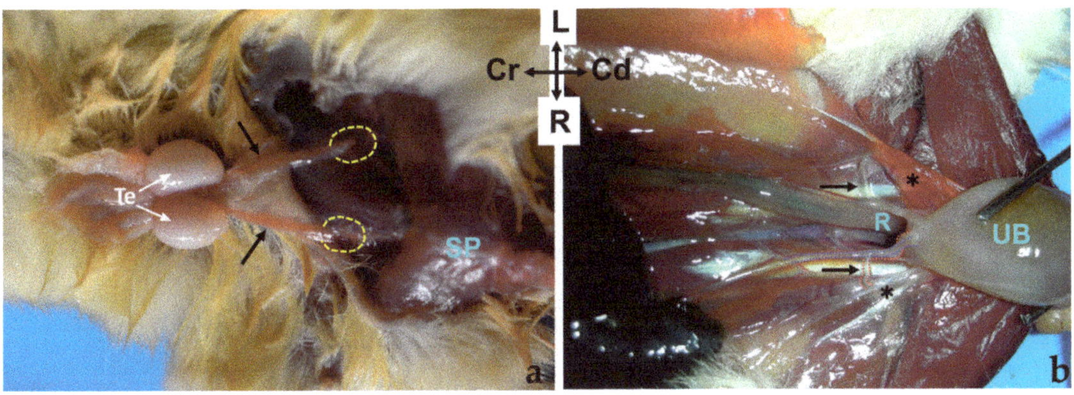

Figure 3. (**a**) Ventral view of a male sugar glider's abdominal wall. The extra-abdominal course of the deferent ducts is indicated by black arrows, leading towards the superficial inguinal ring, represented by a discontinuous oval shape; Te, testes (**b**) ventral view of a male sugar glider's abdominal cavity, with the digestive viscera displaced cranially. The intra-abdominal trajectory of the deferent ducts, again indicated by black arrows, leads towards the dorsal and cranial prostate. The urinary bladder (UB) has been reflected in a caudal direction to reveal its dorsal surface and the beginning of the prostate gland. The right deferent duct is demarcated by a red dotted line, while the terminal segment of the right ureter is indicated by a blue dotted line. Both males were fresh specimens. R, Rectum; SP, symphysis pubica; Te, testes; UB, urinary bladder; $*$, lateral vesical ligaments with round ligaments of the bladder. The orientation is the same in both images.

Upon entering the abdominal cavity, the ductus deferens curves medially to gain the dorsal surface of the urinary bladder neck, running lateral to the ipsilateral ureter (Figure 3b). As the ureters enter the bladder, the ductus deferens shift slightly caudally to enter into the cranial portion of the urethra, passing through the prostate. The ductus deferens maintains a consistent diameter throughout its course. Thus, the ampulla of the deferens duct (*ampulla ductus deferentis*) is absent.

3.4. Penis and Urethra

The penis (*penis*) is situated post-scrotally. When flaccid, the entire penis is concealed in the perineal region, extending to the caudal part of the pelvic cavity, with its body positioned ventrally to the rectum, forming an S-shaped curve known as the flexura sigmoidea (*flexura sigmoidea penis*). However, one intact male specimen exhibited a penis whose body extended to the cranial part of the pelvic cavity, with its flexura sigmoidea situated at the same level as the neck of the urinary bladder or the initial part of the urethra. When externalized, the penis emerges through an orifice located in the ventral part of the common cavity known as the cloaca, running along its floor. The cloaca is an internal cavity where the digestive system and urogenital apparatus terminate, communicating with the exterior through a single opening. Upon opening the cloacal orifice to inspect the short cloaca, two holes can be identified: one dorsal where the rectum empties and another ventral through which the bifid end of the penis is visible (Figure 4a). The mucosa of the cloacal floor invaginates to form a sac, known as the preputial sac, where the free part of the penis is located at rest (Figure 4b). At the deepest part of the preputial sac, the mucosa reflects onto the penis, demarcating the boundary between the penile body and the free part (pars libera penis), which comprises approximately half of the total penile length.

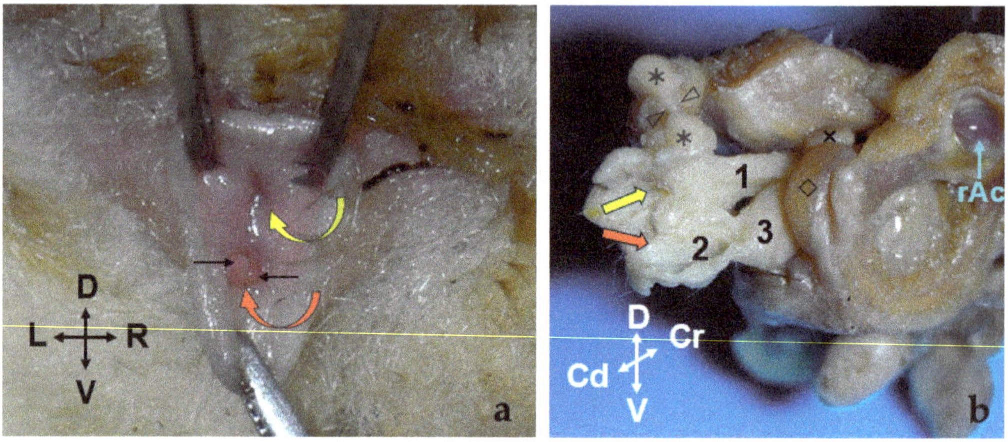

Figure 4. (**a**) Caudal view of a male sugar glider. The cloacal orifice is opened to reveal the opening of the rectum (yellow arrow) and the tips of the bifurcated penis (red arrow) positioned in the roof and floor of the cloaca, respectively. The forked end of the penis is visible protruding through the hole in the floor of the cloaca (black arrows). Fresh specimen. (**b**) Right lateral view of the caudal part of a male sugar glider. Dissection reveals the terminal sections of the digestive and urogenital systems, with the right cloacal wall removed. 1, Rectum; 2, preputial sac; 3, penis inside the preputial sac; ◇ right bulbospongiosus muscle; ×, right bulbourethral gland; ∗, paracloacal glands with one of its ducts indicated between the two arrowheads; r Ac, right acetabulum. Specimen fixed in 76° alcohol.

In *Petaurus breviceps*, the free part is bifurcated and caudally oriented (Figure 5). Each bifurcation of the penis exhibits a wider conical morphology near its terminus, with the exception of the castrated specimen. A "urethral groove" extends along the medial aspect of each bifurcation, spanning the entire length to the distal tip. Macroscopic examination reveals a pair of blood vessels visible beneath the penile mucosa, following the course of the aforementioned grooves, one on the dorsal wall and the other on the ventral wall.

These venous sinuses are discernible both macroscopically and microscopically (Figure 6). Histological analysis confirms the presence of a venous sinus in both the dorsal and ventral walls, corresponding to these observed vessels. The external urethral orifice (*ostium urethrae externum*) is situated in the bifurcation of the penis.

The penis originates in proximity to the caudal surface of the ischial arch (*arcus ischiadicus*). It runs caudally, ventral to the rectum, and terminates at the level of the cloaca. The organ starts with two structures known as the crura of the penis (*crura penis*), the dorsal portion of which is not firmly affixed to the bony pelvis. However, a thin fascia extends from the caudolateral wall of each crus to the ipsilateral ischial arch. The crura, resembling two columns, progress caudoventrally and converge at the midline to form the body of the penis (*corpus penis*). These columns, in conjunction with the two bulbs of the penis (*bulbi penis*), which are situated dorsolaterally to the body, constitute the root of the penis (radix penis).

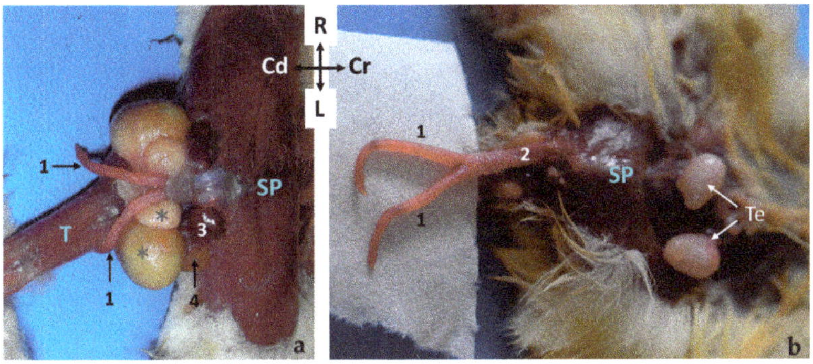

Figure 5. Ventral view of the caudal part of the body of a male sugar glider. Part of the penis is exteriorized through the cloacal orifice. Fresh specimen. (**a**) The free part of the penis is exteriorized through the cloacal orifice. (**b**) Displays both the body and the free part of the penis exteriorized through the cloacal orifice. 1, Bifurcated free part of the penis; 2, body of the penis; 3, left bulbospongiosus muscle; 4, left bulbourethral gland; ∗, left paracloacal glands; SP, symphysis pubica; T, tail; Te, testes. The orientation is the same in both images.

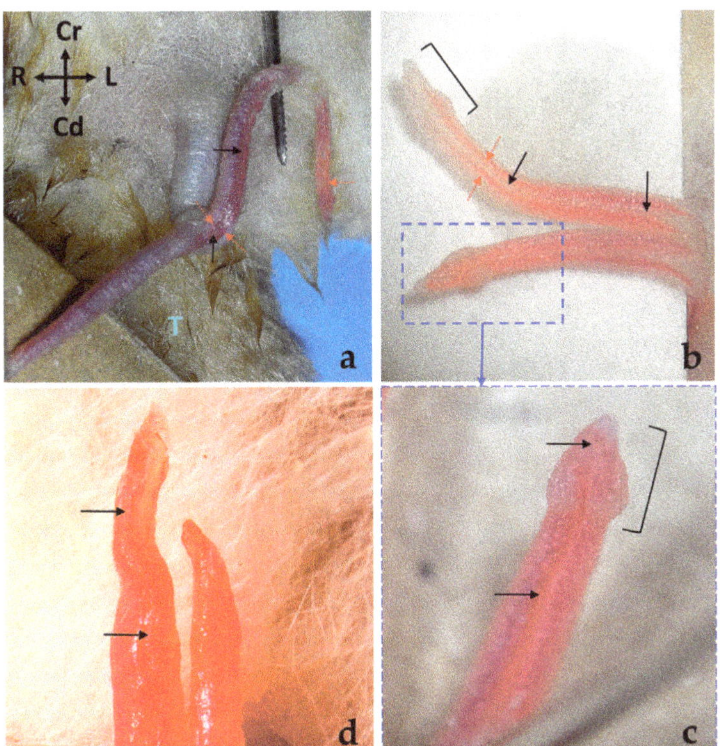

Figure 6. (**a**) Ventral view of the externalized penis of a male sugar glider in situ, exteriorized through the cloacal orifice. Fresh specimen. The left half of the free part has been displaced to observe its medial wall. (**b**) Medial wall of the bifurcated end of the penis, showing venous sinuses (red arrows). (**c**) Inset of (**b**) displaying the cone-shaped termination of the tip ( ]) of the bifurcated penis, with a prominent widening at its base. Note the urethral groove (black arrows) continues up to the tip. (**d**) Tips of the bifurcated penis from a castrated male, showing the absence of the widened conical termination.

The penis includes two types of erectile tissue named corpus cavernosum (*corpus cavernosum penis*) and corpus spongiosum (*corpus spongiosum penis*). Both the crura and the bulbs are made of erectile tissue; the crura are made up of the corpus cavernosum while the bulbs are paired, cranial, and lateral expansions of the corpus spongiosum that surrounds the penile urethra. From these bulbs, the erectile tissue extends to form a thin band that encircles the penile urethra. As the urethra transitions into the urethral grooves at the bifurcation of the penis, the corpus cavernosum adopts a "C" shape, positioned laterally to each groove. However, the corpus spongiosum continues to surround the groove, with a particular concentration along its lateral walls. Both types of erectile tissue reach the tip of the penis.

Two pyriform or globose paired muscles are closely associated with the penis of the sugar glider: the bulbospongiosus (*m. bulbospongiosus*) and ischiocavernosus muscles (*m. ischiocavernosus*). The ischiocavernosus muscles are situated caudodorsally to the crura (*crura penis*) and encompass the initial segment of the ipsilateral crus. The bulbospongiosus muscles are positioned dorsolaterally to the body of the penis (*corpus penis*), with one muscle on the right side and the other on the left side (Figures 7 and 8). These muscles are connected to the penis by pedicles that contain extensions of the corpus spongiosum, which surround the penile urethra and are referred to as the bulbs.

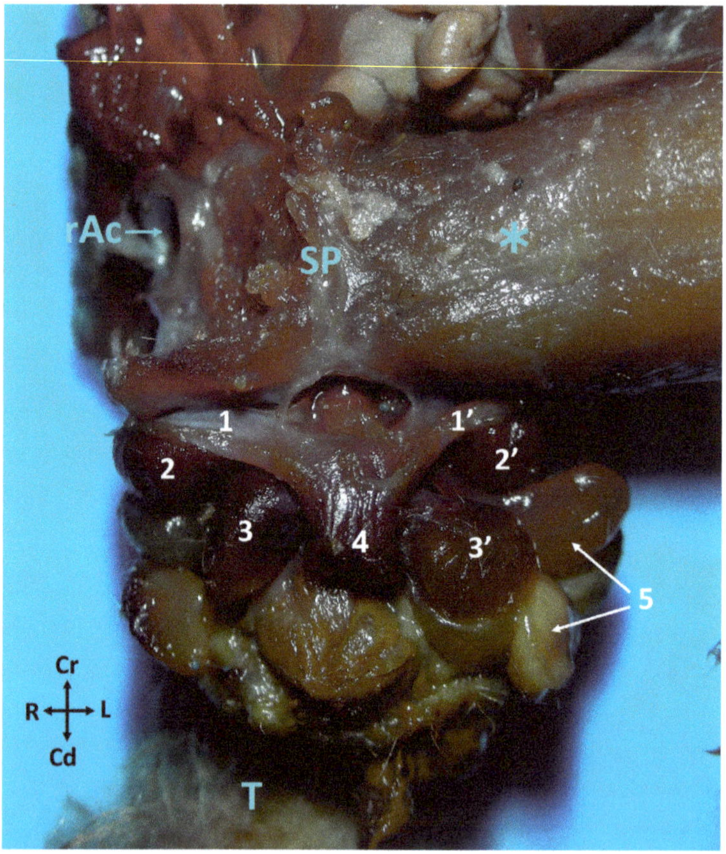

Figure 7. Ventral view of the caudal part of a male sugar glider to show the initial portion of the penis. The right leg has been removed. Fresh specimen. 1, 1′, Crura of the penis; 2, 2′, muscles ischiocavernosus; 3, 3′, muscles bulbospongiosus; 4, body of the penis; 5, left bulbourethral glands; SP, symphysis pubica; rAc, right acetabulum; T, tail; ∗ medial view of the left thigh.

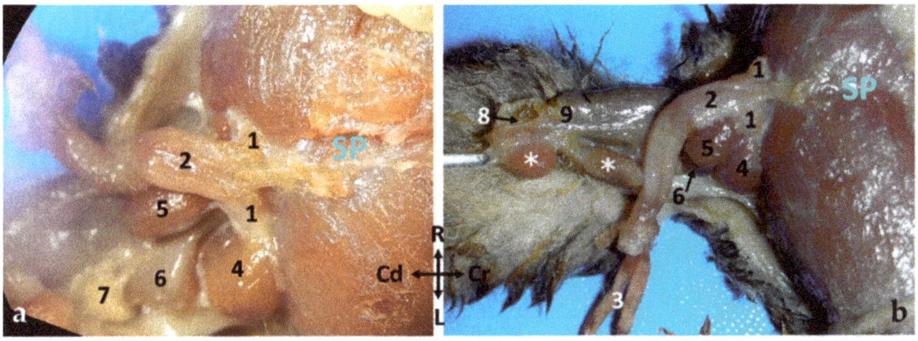

Figure 8. Examination of a fresh specimen. (**a**) Ventral view of a male sugar glider's penis in situ. The anatomical structures on its left side have been dissected. (**b**) Ventral left view of a male sugar glider's penis. The prepuce has been incised and the penis has been extracted from the preputial sac. 1, Crus penis; 2, corpus penis; 3, pars libera penis; 4, left ischiocavernosus muscle; 5, left bulbospongiosus muscle; 6, left bulbourethral gland I; 7, left bulbourethral gland II; 8, cloacal cavity; 9, rectum; ∗, paracloacal glands. SP, Symphysis pubica; both images have the same orientation.

The male urethra (*urethra masculina*) extends from the internal orifice at the urinary bladder neck to an external orifice (*ostium urethrae externum*) located at the caudal portion of the corpus, at the point where the bifurcation of the free part of the penis commences. It is composed of two segments: the pelvic part (*pars pelvina*) and the penile part (*pars penina*), which is incorporated within the penis. The entirety of the urethra, both pelvic and penile portions, is enveloped by the corpus spongiosum.

The body and free part of the penis are covered by a cutaneous mucosa consisting of keratinized epithelium, which also lines the walls of the urethral grooves. However, the bottom of the groove is lined with urothelium. Small cornified spicules may be present at the conical tip of the penis, except in the castrated specimen.

3.5. Accessory Glands

3.5.1. Prostate

The prostate of the sugar glider is situated within the pelvic cavity and comprises two distinct parts. First, there is the macroscopically visible body (*corpus prostatae*), which has a carrot-shaped appearance and surrounds the urethra. Second, there is the microscopically observed portion (*pars disseminata prostatae*), located within the walls of the duct. The compacted body envelops the entire pelvic urethra (*pars pelvina urethrae*). Macroscopically, the gland can be divided into three segment (Figure 9): the cranial segment, which is broad and rounded, followed by a constriction that separates it from the second segment, and the third segment, narrower in shape. Following the body of the prostate, the disseminated prostate accompanies the urethra until it is incorporated into the penis, becoming the penile urethra (Figure 9).

3.5.2. Bulbourethral Glands

In the sugar glider, two pairs of bulbous bulbourethral glands (*glandula bulbourethralis*) are observed within the pelvic cavity. These glands are situated dorsally and laterally to the rectum, respectively. They are closely associated with the crura of the penis. Structurally, they exhibit compactness and possess a smooth external surface (Figures 5 and 13). The excretory ducts of these glands run parallel between the bulbospongiosus and ischiocavernosus muscles, ultimately entering the urethra ventromedially just before it transitions into the penile urethra.

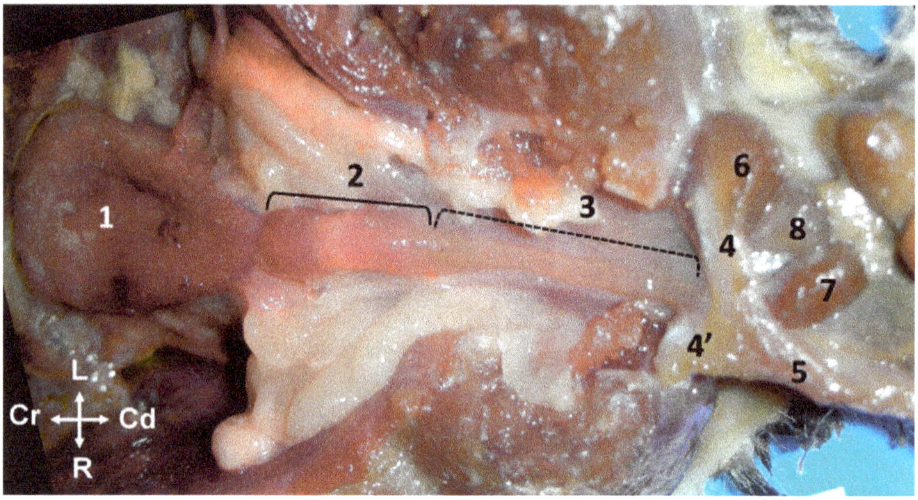

Figure 9. Ventral view of the prostate gland in situ. The floor of the pelvic cavity has been removed to observe the prostate and the urethra pelvina in the male sugar glider. Fresh specimen. 1, Bladder; 2, prostate; 3, urethra pelvina; 4, 4′, *Crura penis;* 5, *Corpus penis;* 6, left ischiocavernosus muscle; 7, left bulbospongiosus muscle; 8, left bulbourethral gland.

3.6. Microstructure

Histologically, the testis is a parenchymatous organ, comprising stroma and parenchyma. The stroma forms a capsule known as the tunica albuginea, which comprises dense irregular connective tissue and encapsulates the parenchyma. The parenchyma is composed of seminiferous tubules interspersed with sparse connective tissue and clusters of interstitial cells (Figure 10). The tunica albuginea is enveloped by a simple squamous epithelium, a serous membrane referred to as the tunica vaginalis lamina visceralis. This, in conjunction with the lamina parietalis, encloses the cavum vaginale.

The absence of septae testis resulted in no observable testicular lobules. The mediastinum testis, along with its rete testis, was situated at the proximal pole of the testis, approximately at a 45° angle from the major axis of the testicle, on the side opposite to the epididymis. Leydig cells, or interstitial cells, were abundant, highly acidophilic, and distributed in richly vascularized clusters. Additionally, a thin layer of myofibroblasts was observed surrounding the seminiferous tubules (Figure 10).

In relation to the parenchyma, in the seminiferous tubules, the nuclei of sustentacular cells, or Sertoli cells, were discernible in a basal location amidst the male germ cells. These germ cells exhibited a spectrum of developmental stages, from undifferentiated to mature stages, the latter being situated in an apical position, in closer proximity to the lumen. Within a microscopic field, seminiferous tubules exhibited various associations of sperm cell developmental stages. The majority of the parenchyma was constituted by convoluted seminiferous tubules, with straight seminiferous tubules being seldom observed, even in proximity to the mediastinum testis. These straight tubules were lined exclusively by Sertoli cells, or simple to columnar epithelium.

There seems to be a single efferent duct originating from the rete testis, which courses through the tunica albuginea. This duct, characterized by a mild sinuosity, is lined with an epithelium that varies from squamous-cuboidal to low columnar epithelium. Notably, certain cells within this lining display long apical membrane projections, known as stereocilia.

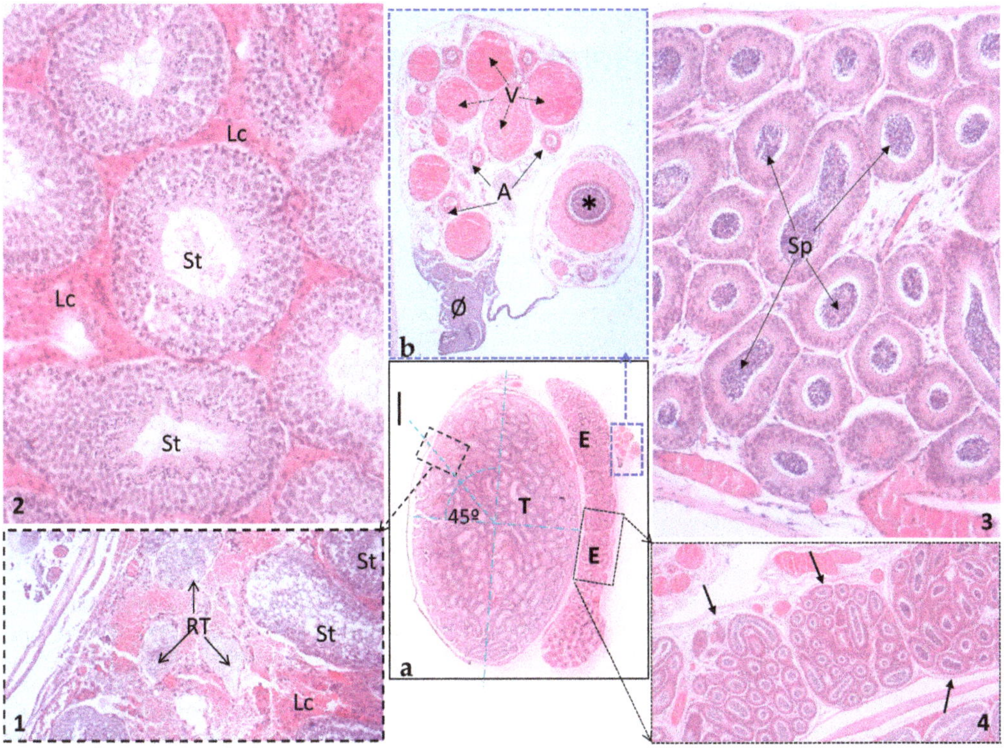

Figure 10. Testicle and epididymis. (**a**) A low-magnification section showing the testicle (T) and epididymis (E). Scale bar represents 1 mm. (1) An inset of (**a**) displays the rete testis, positioned at a 45° angle from the perpendicular axes (as depicted in (**a**)), from which the efferent duct transports spermatozoa to the epididymis. (2) Seminiferous tubules (St) teeming with numerous interstitial or Leydig cells (Lc). Note the distinct associations within the seminiferous tubules, exhibiting various stages of sperm cell development. (3) Histological section of the epididymis, revealing numerous cross-sections of the same epididymal duct filled with spermatozoa (Sp). (4) The black inset displays the characteristic lobuli epididymidis (black arrows) at a low magnification (objective 4×). (**b**) The blue inset in panel (**a**) provides a closer look at the spermatic cord and its components: ∗, deferent duct surrounded by a thick circular muscular layer; A, sections of the testicular artery; V, veins of the plexus pampiniformis; Ø, cremaster muscle (distal portion).

The epididymis consists of a single, highly convoluted tubule surrounded by loose connective tissue and encased in a dense connective tissue capsule. This tubular structure is arranged in a compact, zig-zag pattern to minimize its overall volume. The epididymis extends from its head (*caput*) through the body (*corpus*) to the tail (*cauda*), where it transitions into the vas deferens. When examined in longitudinal section, the microstructure of the epididymis reveals distinct lobuli epididymidis separated by thin transverse septa composed of dense connective tissue. Up to 15 lobuli epididymidis have been observed, with the largest and most bulky being the one from the tail (Figure 10). The ductus of the epididymis is lined by a pseudostratified columnar epithelium that displays varying morphology (Figure 10), with its thickness progressively increasing from the head towards the caudal portion. Based on these morphological variations in the pseudostratified epithelium, the epididymal duct can be divided into five distinct zones, each with its unique cellular characteristics. Similarly, the smooth muscle surrounding the ductus in a circular layer also increases in thickness from the head to the tail. This muscular layer is notably thin in the caput and progressively thickens to several cell layers thick in the tail, as a transition to the deferent duct.

Upon integration into the spermatic cord, the deferent duct lumen is significantly wider than that of the epididymis duct (Figure 10) and is lined with cuboidal to low columnar epithelium. The muscular layer is substantially thickened, appearing to consist exclusively of circularly oriented smooth muscle cells. It is accompanied by numerous sinuous arterial vessels and the veins of the pampiniform plexus, and the cremaster muscle.

Both vasa deferentia terminate at the roof of the prostatic urethra, forming a small elevation known as the seminal colliculus. It seems that there is no crista urethralis between the termination points of the ducts. The point of confluence of the vasa deferentia with the prostatic urethra is located in the cranial region of the prostate, but approaching the middle portion of the gland.

The penis is anchored by two lateral projections of dense connective tissue, which are introduced as an oblique sheet into the ischiocavernosus muscle, apparently without fixation to the ischial arch. These lamellar structures of dense connective tissue thicken, and blood vessels arrive at them that are incorporated into their inner part. In this way, these structures widen and gain volume in their medial part and incorporate the pelvic urethra, becoming penile urethra. This remains in a ventral position and the connective tissue structures merge in the midline and present large spaces filled with blood (caverns), forming the corpus cavernosum, in a dorsal position to the urethra. Thus, the penis is visualized as formed by two columns or longitudinal structures of dense connective tissue (of semilunar section) fused in their dorsal midline by a shared septum of dense connective tissue. Inside they present septa or trabeculae of dense connective tissue with radial arrangement from the urethra to the periphery and vice versa, which can be complete (when they cross the entire cavity) or incomplete and among which appear spaces filled with blood. They would form the corpus cavernosum of the penis (Figure 11).

At the level of the body of the penis, the urethra is a closed duct, lined by urothelium, which conducts urine and semen. In the proximal part of the penile urethra, the ducts of the bulbourethral glands join, potentially contributing more constituents to form the semen. The penile urethra runs in a ventral position to the corpus cavernosum and in its course it is surrounded by a multitude of blood vessels between its muscular wall, which form the corpus spongiosus. The penile urethra remains closed until the formation of a ventral groove that reaches it and opens it; this point is considered the external urethral orifice, just before the bifurcation of the free part of the penis (the part that is externalized during mating) (Figure 11). This groove becomes deeper and separates the two cavernous bodies, leading to the bifurcation of the penis. However, there is still an area that continues to be lined by transitional epithelium that transports semen to the tip of each part of the bifurcated penis. At this point, the penis begins to be externally covered by cutaneous mucosa all around its perimeter, originating the free parts of the penis, free of unions or adhesions to the prepuce. In the bifurcated free parts, a deep groove appears whose bottom is lined by urothelium, forming the urethral grooves. On both sides of the walls of the groove there are venous sinuses visible macroscopically and microscopically. And the cavernous body forms a concave or grooved structure, whose concave part protects the urothelium of the urethral groove, in a medial position, looking towards it contralateral.

At the terminal ends of each bifurcation of the free part of the penis, there is a conical widening formed by the corpus cavernosum that partially surrounds the urethral groove. This urethral groove does not end exactly at the vertex of the conical structure, but very close to it. Histologically, this conical dilation is formed by the cavernous body that surrounds the urethral groove, under whose urothelium the spongy body is arranged. This structure appears externally covered by cutaneous mucosa that presents small cornified spicules, without apparent orientation (Figure 11).

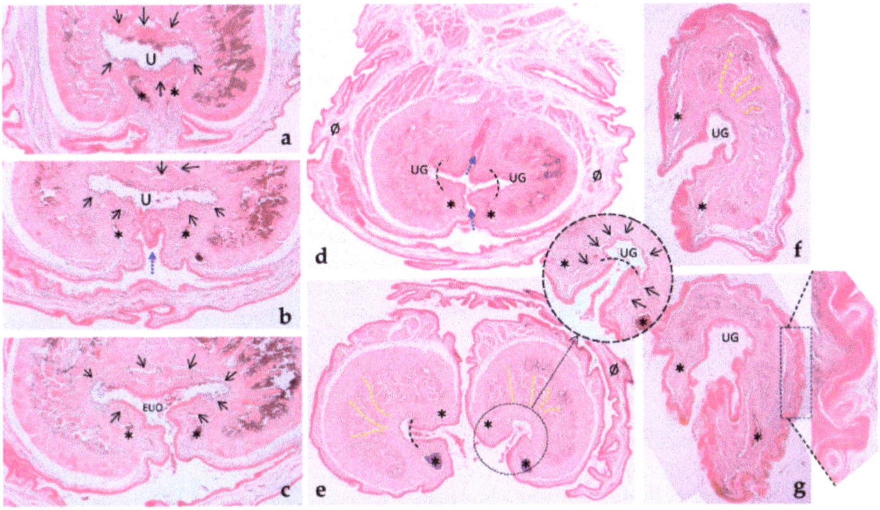

Figure 11. Penile histology. (**a**) The penile urethra is surrounded by the corpus spongiosum (black arrows) and the corpus cavernosum. (**b**) A ventral groove (blue arrow) deepens and almost reaches the penile urethra, just before the external urethral orifice. (**c**) The ventral groove reaches the urethra, which opens to the exterior, forming the external urethral orifice (EUO). (**d**) The groove continues dorsally (blue arrows) to form the bifurcated tip of the penis, and the continuation of the urethra creates two urethral grooves (UG) in a medial position of each part. Note the presence of two venous sinuses (∗) flanking the groove walls. The penis is protected by the mucosa of the prepuce (Ø). (**e**) The bifurcated penis is now separated from the prepuce, and each half is independent. The inset shows the urethral groove (UG), with urothelium at the groove's bottom (the limits are indicated by the dotted curved line) and cutaneous mucosa forming the groove walls. This entire structure is surrounded by the corpus spongiosum (black arrows) and two venous sinuses on either side of the groove (one dorsal and one ventral (∗)). In the corpus cavernosum, yellow lines indicate dense connective tissue septa with a radial arrangement between blood-filled spaces. (**f**) The termination of the bifurcated penile tip, at the beginning of the conical formation. The corpus cavernosum is clearly visible, suggesting the absence of a glans. (**g**) An oblique section of the penile termination reveals cornified spines on the surface (inset). In both (**f**,**g**), the corpus cavernosum is evident, further supporting the absence of a glans.

The prostate is a parenchymatous gland enveloped by a dense connective tissue capsule. Histologically, it can be divided into three distinct segments based on the glandular tissue types: the cranial (anterior) segment, the middle (central) segment, and the caudal (posterior) segment (Figure 12). Within this glandular tissue, simple branched tubular glands exhibit a radial arrangement. Their basal closed ends are in contact with the connective capsule, while their openings converge toward the urethra. These openings do not directly connect to the urethra but instead lead to single secretory ducts lined by cuboidal-columnar to transitional epithelium, which ultimately release their secretions into the urethra. Each glandular tubule is enveloped by a thin layer of smooth muscle. In the three sections of the prostate, the tubulo-alveolar glands have a different appearance (Figure 12). In the cranial portion, the glandular epithelium that covers the outermost region is formed by columnar cells, while towards the part closest to the urethra, the cells are more cubic. Inside the lumen appears acidophilic secretion in the form of individualized droplets of different diameters. Both in the middle and caudal parts, no differences are observed between the outer and inner parts. The glands in the middle part are internally lined by small basophilic cubic cells that release a basophilic secretion. In contrast, in the caudal part, the glandular epithelium is made up of large cells with clear cytoplasm that release a secretion forming large independent droplets of colorless appearance.

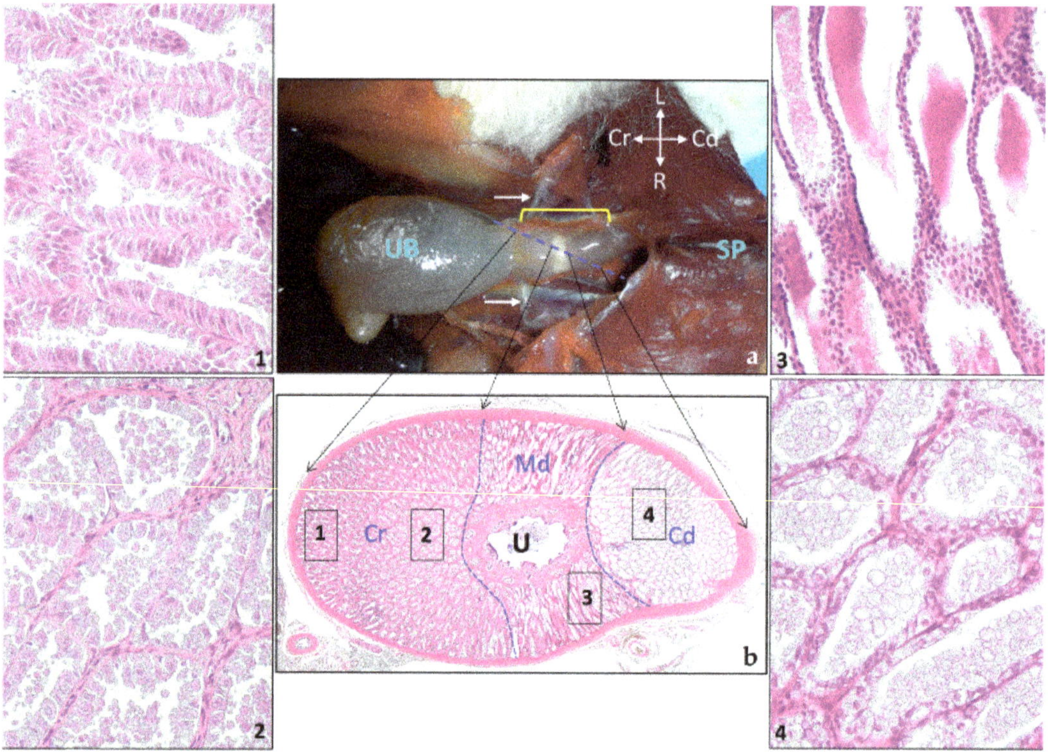

Figure 12. Prostate gland. (**a**) Macroscopic ventral view of the prostate gland in situ. The three parts of the prostate (yellow bracket) can be distinguished. White arrows, deferent ducts; SP, symphysis pubica; UB, urinary bladder. (**b**) Oblique histologic section of the prostate at the level of the blue dotted line in (**a**). The central image is a composition of three images made with the 4× objective and the numbers represent different areas of the cranial portion (Cr), middle (Md), and caudal part (Cd) easily distinguishable by their morphology. Photographs 1–4 represent the glandular tissue from the areas indicated in (**b**) to display their glandular epithelium and the secretion appearance (made with the 40× objective). U, Urethra.

The prostatic urethra receives the prostate secretion via numerous ducts and is encircled by erectile tissue containing small vessels that contribute to the formation of the corpus spongiosum (Figure 12). Toward the end of the prostate body, glandular tissues become visible within the thickness of the urethral walls, situated beneath the smooth muscle layer. This arrangement constitutes the disseminated prostate. In the final segments of the pelvic urethra, the glandular epithelium emerges from the muscular layer and is positioned around the urethra.

The two bulbourethral glands of the sugar glider have different histological characteristics, as well as the appearance of their secretion. Both are tubulo-alveolar exocrine glands surrounded by a capsule of dense connective tissue that emits septa of loose connective tissue among which the glandular epithelium is arranged, forming a single layer of cells that pour their secretion towards the light (Figure 13). In bulbourethral gland I (which is located closer to the pillars of the penis), the glandular cells are cubic, with clear cytoplasm, while in bulbourethral gland II, the secretory cells are low cubic and produce a very acidophilic secretion that is stored in the light of the alveoli, with granules of more intense staining. Both glands are externally surrounded by a double layer of striated muscle tissue, oriented perpendicular to each other. Their ducts convey their secretion into the penile urethra (Figure 13).

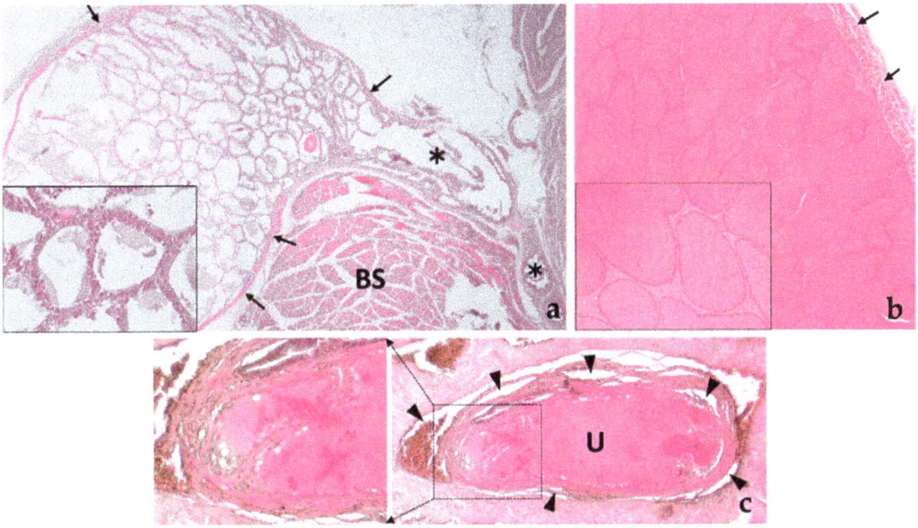

Figure 13. Histology of the bulbourethral glands. (**a**) Image showing bulbourethral gland I at low magnification to observe its tubulo-alveolar morphology and its secretion duct (∗). The entire gland is externally lined by striated muscle (black arrows) arranged in two perpendicular layers. The inset shows at higher magnification (40× objective) the morphology of the glandular epithelium made up of a single layer of cubic cells. BS, bulbospongiosus muscle. (**b**) Bulbourethral gland II, also of tubulo-alveolar nature, filled with acidophilic secretion. Two layers of striated muscle also appear surrounding the gland capsule. The inset shows at higher magnification (40× objective) the appearance of low cubic secretory cells arranged forming a single layer of cells lining the alveoli. The secretion, strongly acidophilic, presents some granules of more intense staining. (**c**) Section of the penile urethra showing the point of discharge of the ducts of the bulbourethral glands pouring their secretion into the lumen of the urethra. The inset shows in greater detail the discharge of the left duct. Note the presence of the spongy body surrounding the urethra (black arrowheads).

4. Discussion

The male reproductive system of the sugar glider is representative of marsupials and exhibits a greater similarity to the male genitalia of placental mammals as compared to the female counterpart [26]. This system fundamentally comprises the testes, which is the site of gamete formation, a duct system that transports the spermatozoa, and a collection of glands that are responsible for the secretion of seminal fluid [7]. Previous investigations into this system have been undertaken by Ness and Booth (2004) [27], Carboni and Tuly (2009) [4], and Girling (2013) [28].

The results obtained from our observations, at both macroscopic and microscopic levels, will be collectively discussed.

4.1. Testes

As in other marsupials, the scrotum of the sugar glider, with the testes and the epididymis inside, is always external and cranial to the cloacal opening [4], as well as to the penis [5,8,15,16,21,29,30]. Its pendular nature, characterized by a lengthy scrotal neck [27], follows the general pattern in marsupials, facilitating the cooling of the testes and epididymis by radiation and, in some species, also via the evaporation of sweat or saliva [7]. The cremaster muscle, which extends along the scrotal stalk, further aids in the thermoregulation of the testicle, drawing it nearer to the body in cold climates or distancing it in warmer conditions [7]. Only a few marsupial burrowing species, such as the *Notoryctes* genre (marsupial moles), lack this pendulous scrotum. Their testes are positioned prepenially, but they lie between the skin and the abdominal wall [7], just cranial to the pubic bones [31,32]. In the southern hairy-nosed wombat *(Lasiorhinus latifrons)* and the common wombat *(Vombatus ursinus)* [15,16], the scrotum is situated close to the body and is non-pendulous. A similar scenario is observed in the *Acrobates pygmaeus*, the smallest Australian glider [14].

Based on our findings, the testes of the *Petaurus breviceps* appear to conform the general marsupial norm; they possess an ellipsoid shape and they are permanently located within the scrotum [22]. Contrary to our observations in the sugar glider, the tunica vaginalis that encapsulates the testes appears to be to be melanized in numerous marsupials [8,12,33]. Nevertheless, the sugar glider is not the sole species where it is not of a dark hue. This has also been documented in the *Acrobates pygmaeus* [14], the common wombat (*Vombatus ursinus*) [15], the hairy-nosed wombat *(Lasiorhinus latifrons)* [16], and certain South American marsupials: Caenolestes [12], *Didelphis sp.* [30], and *Chironectes minimus* [23]. The functional or phylogenetic importance of this characteristic remains undetermined. Biggers (1966) [34] proposed a potential role in the regulation of testicular temperature, yet there appears to be no correlation between climate and tunic pigmentation as might be expected [22]. However, the findings of de Barros et al. (2013) [30] seem to emphasize this hypothesis, giving that the young and sexually inactive Didelphis sp. exhibited a whitish tunica vaginalis, while the adults displayed a high degree of dark pigmentation.

4.2. *Epididymis*

The epididymis is similar in shape to that of eutherians and other marsupials. However, in numerous marsupials, the head of the epididymis is not affixed to the testicular capsule as we observed in the *Petaurus breviceps*, but it is connected by the efferent ducts that transport the sperm from the testicle [35,36]. Contrary to placental mammals, a singular efferent duct has been documented in many marsupials, encompassing both Australian and South American species [8,11,35], albeit not universally [37]. Histologically, in our studied specimens, it appears that several straight ducts (lined with pale low-cuboidal epithelium) merged into a unique efferent duct that departs from the testis and reaches the head of the epididymis. The duct is lined by a dark, low ciliated columnar epithelium. Moreover, these ducts do not seem to follow a straight course, but rather appear coiled and are surrounded by a plexus of small blood vessels. Despite the requirement for additional histological studies to validate the distinct sections of the duct system between the testis and epididymis, certain parallels were observed with the descriptions provided by Woolley (1987) [38] in American marsupials. However, none of the descriptions corresponding to the various species entirely aligned with our findings in the sugar gliders.

In general, the position of the epididymis in relation to the testicle is not explicitly defined in the literature. Only de Barros et al. (2013) [30] have described it as lateral to the testis in certain Brazilian species *(Didelphis sp).* In contrast, the position of the epididymis in the sugar glider is caudomedial, similar to the arrangement found in domestic mammals with a pendulous scrotum [39].

The deferent duct establishes a connection between the epididymis and the prostatic portion of the urethra [39,40]. However, due to the migration of the ureters during embryonic development of the urogenital tract, the relative position of the ureter and the deferent duct varies between different groups of mammals. Specifically, in marsupials, the ureters are medial to the genital ducts (mesonephric duct in males), while in eutherians, they are lateral [41]. Consequently, in *Petaurus breviceps*, as is the case with other marsupials, the deferent duct is positioned laterally to the ipsilateral ureter.

4.3. *Deferent Duct*

In the majority of eutherian species, the final segment of the deferent duct, which lies on the bladder, displays a fusiform enlargement. This is known as the ampulla of the deferent duct, where the ampullary glands are located [39]. In contrast, the sugar glider, like other marsupials, maintains a consistent thickness throughout the length of the deferent duct [15]. Consequently, due to the absence of the ampulla of the deferent duct, ampullary glands are also absent [4,40]. However, exceptions may exist. For instance, a convoluted structure resembling an ampulla has been described in *Caenolestes obscurus* by Rodger in 1982 [12]. Histologically, we have observed that both vas deferens reach the roof of the prostatic urethra forming a small elevation known as the seminal colliculus. However, this elevation is not as pronounced as in domestic placental mammal species. The point of confluence of the vas deferens into the urethra is in the cranial part of the prostate but in its posterior part, very close to the middle part of the prostate. Our results contrast with those reported by Ward and Renfree (1988) [14] in *Acrobates pygmeus*, another Australian glider. They found that both vas deferens empty into the urethra at the level of the middle or central segment of the prostate.

4.4. *Penis*

The penis of the *Petaurus breviceps* is situated in a post-scrotal position, a characteristic shared by both South American and Australian marsupials [10,15,23,28].

Existing literature indicates that, during rest, the penis of the sugar glider is located inside the cloaca, specifically lying on its ventral wall [4,28,29,40]. However, our detailed dissections, revealed a different arrangement. The cloaca of this species is notably short, and the penis is actually positioned ventral to the rectum. During copulation, the free part of the penis is externalized through an orifice located in the caudoventral wall of the cavity. This anatomical configuration aligns with descriptions provided by Nogueira et al. in Brazilian didelphids [10] and the koala [25]. The cloaca, common to urogenital apparatus and the digestive system, is a common feature in marsupials. An exception to this is *Chironectes minimus,* where the digestive tract has its own separate opening, distinct from that of the urogenital tract [23].

The sigmoid flexure (*flexura sigmoidea penis)* that we saw in our specimens was previously described by other authors [22,28], and seems to be a consistent characteristic among marsupials [10,15,19,23–25].

According to the *Nomina Anatomica Veterinaria* (N.A.V.) [42], when at rest, the end of the penis is concealed by a fold of abdominal skin known as the prepuce *(preputium).* This fold's internal layer is attached to the organ. In the sugar glider, the penis, when flaccid, is hidden within the body, positioned ventral to the rectum, rendering the prepuce absent. However, the term 'prepuce' is often used in the literature to refer to the fold of the cloacal mucosa that attaches to the marsupial penis. Similarly, the sac housing the resting penis is called the preputial sac [10,15,19,25]. Precisely, in placental mammals, the attachment of this skinfold demarcates the separation between the body and the free part of the penis. In *Petaurus breviceps,* this boundary could potentially be identified by the location of the external urethral orifice and the bifurcation of the terminal part of the organ. However, as the urethra does not extend to the end of the penis, the structure of the eutherian free part of the penis cannot be recognized. Moreover, it is also not possible to identify the typical mammalian glans *(glans penis)*, which, according to the N.A.V. [42] is characterized by its position at the end of the free part and being formed exclusively by the urethra surrounded by some erectile tissue, the corpus spongiosum *(corpus spongiosum glandis).* Despite this, the terms 'free part' and 'glans' of the penis are frequently used in the referenced literature [7,22,24], even for species with a forked penis [10,15,17,21,25,30,43,44]. This fact could be attributed to the consideration of the glans penis as the terminal portion of the penis, where the corpus cavernosum expands and surrounds the corpus spongiosum, as suggested by Woolley and Webb (1977) [25].

In summary, knowing the distribution of the erectile tissue at the end of the penis is crucial to confirm the presence of a 'proper' glans penis in a species. To our knowledge, it is absent in most marsupials, as demonstrated by Warburton et al. (2019) [19] in their study on the red kangaroo. Histological studies conducted on Australian and New Guinea dasyurids, which possess slight or non-bifid penises [9,13,25,45] revealed that the corpus cavernosum consistently extends to the tip of the penis. Similar results have been obtained in studies on South American marsupials with bifurcated penises; the two types of erectile tissue extend to the tip of the penis [10,30,46]. This is a characteristic we also observed in the *Petaurus breviceps*. Other histological studies conducted on dasyuroides with a non-bifid penis by Woolley (1987) [45] revealed the presence of only a corpus cavernosum at the tip of the penis. Consequently, the existence of a true glans penis, as defined by the N.A.V. [42], cannot be confirmed in marsupials.

The distribution of the corpus cavernosum in the bifurcated free part of the penis in the bushy-tailed opossum (*Glironia vetusta*) [46] is similar to our observations in the sugar glider. In both instances, this erectile tissue forms two semilunar structures situated laterally to the urethral grooves. However, the arrangement of the corpus spongiosum differs; it surrounds the cavernosum in the bushy-tailed opossum, as it does in other Australian and South American marsupials, both with slightly or non-bifid free part of the penis [9,10,13,17,25,46], while in the sugar glider, it is positioned around the urethral groove.

In certain species with a forked penis, a dorsal projection known as the dorsal lobe, containing a corpus spongiosum, is present at the beginning of the free part [17]. This feature was not observed in *Petaurus breviceps*. Additionally, we were unable to identify the accessory corpora cavernosa, which forms an appendage of the penis as described by Woolley and Webb [25], and Woolley [17] in some Australian and New Guinean dasyurid marsupials with a non-bifid penis tip.

The bifid tip of the penis is a common feature in marsupials, as described in certain species of *Sminthopsis* [13], *Antechinus* [17], Brazilian *Didelphidae* [10], and the common wombat *(Vombatus ursinus)* [15]. However, it is not observed in *Dasyuroides byrnei, Dasycercus cristicauda* [45], *Dasykaluta, Pseudantechinus, Parantechinus* [17], or the grey kangaroo *(Macropus fuliginosus)* [19]. The bifurcated free part of the penis varies between species in terms of length, shape of the tip, the presence or absence of a diverticulum, and the positioning of the urethral grooves. In the sugar glider, the absence of a diverticulum, the considerable length of the free part, and the urethral grooves—located in the medial wall end very close to the tip of each half—are similar to the features observed in *Caluromys,* a South American marsupial. In contrast, other didelphidae *(Didelphis, Lutreolina, and Metachirus)* have a shorter free part with a diverticulum, and the grooves, also medial, do not extend to the end of the penis [10]. Among Australian marsupials, *Antechinus* species also exhibit long urethral grooves [17], akin to the sugar glider.

For all species that have a bifurcated penis [10], the urethra terminates at the caudal part of the body of the penis, between the body of the penis and the free part, as described in the literature for the sugar glider [3,5] and in line with our own observations. This specific anatomical feature is of considerable relevance, given that males do not urinate from the forked end of the penis, but from the proximal portion [27]. Consequently, the distal penis can be amputated in cases of penile necrosis, penile trauma, or paraphimosis. When an amputation procedure of the distal penis is required [47], the amputation does not compromise the survival of the individual. This is attributable to the fact that the urethra does not pass through the bifurcated sections of the penis but extends as a urethral groove along the medial surface of each half. This implies that even in the event of an amputation of one or both of the free, bifurcated parts of the penis, the male retains the ability to urinate without any complications, as the urethra remains intact and fully functional. Given that penile self-mutilation is relatively common in stressed captive males, the position of the urethral orifice enables the amputation of the terminal ends, when necessary, without causing damage or alterations to the urinary tract [3,27].

The same situation of the urethral opening has been previously described in the *Antechinus stuartii* [25]. The urethra opens into urethral grooves, the location of which appears to be species-dependent. Woolley et al. (2007) [13] conducted a morphological study of the penis in *Sminthopis* species (dasyuridae). They discovered that in species with a slightly bifurcated tip, the external urethral orifice is located dorsally on the free part of the penis, next to its end. The urethral grooves are either on the mesial aspect (in twelve species) or on the ventral aspect (in one species) of the bifid portion of the penis. However, in a previous study carried out in *Dasyuroides byrnei* and *Dasycercus cristicauda* (Dasyuridae), the same author [38] described the urethral grooves in the dorsal part of the penis tip. In *Antechinus*, six out of seven species with a forked penis exhibited the urethral grooves on the dorsal surface; in the seventh species, they were dorsally located in the cranial part of the free portion but they were present on the medial surface of the apices [17]. The author proposed that this positional change could be attributed to the curving of the apices towards the midline. In *Petaurus breviceps*, Carboni and Tully (2009) [4] and Johnson-Delaney and Lennox (2017) [48] identified a ventral groove in each part of the forked end of the penis. Our observations revealed a similar structure, but it was medially (mesially) positioned when the organ was placed in the physiological position, as seen in the Virginia opossum *(Didelphis virginiana)* [43], Brazilian *Didelphis sp.* [30], *Chironectes minimus* [23], and *Antechinus stuartii* [25]. The length of these urethral grooves varies among species: in the sugar glider, they extend almost to the tip of each part, similar to Chironectes minimus *Chironectes minimus* [23] and *Caluromys* [43], whereas in *Didelphis virginiana,* they terminate some distance from the tips [43].

The *Nomina Anatomica Veterinaria* (N.A.V.) [42] describes three parts in the penis of placental mammals: the root *(radix penis)*, the body *(corpus penis)*, and the free part *(pars libera penis)*. The root comprises two crura (*crura penis*) and the bulb of the penis *(bulbus penis)*. Each crus is essentially the proximal part of a corpus cavernosum anchored to the *archus ischiaticus* and enveloped by the ischiocavernosus muscle. The bulb of the penis represents the caudal expansion of the corpora spongiosum penis and is sheathed by the bulbospongiosus muscle [42].

In placental/domestic mammals, the corpus spongiosum *(corpus spongiosum penis)* commences at the pelvic outlet with a bilobed expansion, constituting the bulb of the penis *(bulbus penis)*. Subsequently, it narrows and encapsulates the segment of the urethra situated within the penis [39]. A distinctive feature of placental mammals is the presence of a singular penile bulb, contrasting with marsupials that exhibit dual expansions in the cranial region of the corpus spongiosum, situated on the right and left sides [9,10,13,19,21,25,30]. Regardless of the species, the penile bulb(s) are enveloped by the bulbospongiosus muscles.

The bulbospongiosi muscles, due to their form and location on either side of the penile body, and the slender, duct-like pedicles connecting them to the penile corpus, can be easily mistaken for bulbourethral glands, as noted by Young in his study of the koala [21].

The anatomical configuration and globular shape of the *bulbospongiosi* and *ischiocavernosus* muscles as observed in the *Petaurus breviceps,* have been previously documented in both South American and Australian marsupials [10,19,21]. In these species, the right and left *bulbospongiosi* muscles remain distinct from each other, unlike in placental mammals where they merge ventrally to form a single bulb. The significance of this anatomical difference remains ambiguous. Various hypotheses have been proposed, but none have been definitively validated. Young postulated in 1879 [21] that the double bulb might be essential to sustain the turgescence of the bifurcated glans, a characteristic feature of most marsupials. However, given that the corpus spongiosum does not bifurcate within the penile body, Warburton et al. (2019) [19] deem any functional differentiation between the right and left muscles improbable. According to these authors, the observed characteristics likely represent the bilateral development of the reproductive organs.

However, the anatomical differences between the bulbospongiosus and ischiocavernosus muscles in placental mammals and marsupials imply divergent functions. In placental mammals, the ischiocavernosus muscles contribute to the augmentation of blood pressure in the corpus cavernosus of the penis, facilitating erection [49]. The rhythmic contractions of the bulbospongiosus muscle aid in the propulsion of sperm through the urethra during ejaculation [39,43,50], enhance the erections of the glans penis [43,50], and the contractions of the ischiocavernosus muscle elevate the penis, assisting the male in achieving the intromission [43]. These contractions also induce or assist in penile movements, such as the "penile flips" observed in rats [50]. In marsupials, according to Warburton et al. (2019) [19] and Dixon (2021) [43], the function of both muscles could be construed as "mechanical pumps" that increase the pressure in the erectile tissues during the erection and ejaculation, However, these muscles are not believed to play a role in the eversion of the penis from the cloaca or in the execution of penile movements.

In our study, we identified the presence of horny or epidermal papillae, also referred to as spines, at the tip of the penis. This observation aligns with descriptions of the free portion of the penis in other marsupial species, both Australian [13,15,16,21,45,51] and American [10,30]. However, in this small species, given the size of these structures, we prefer to call them spicules, rather than spines. The morphology and arrangement of these penile spines vary significantly across species, with different shape and orientations. For instance, in the koala they are retroverted [51], whereas in the sugar glider they do not appear to possess a specific orientation. The function of these spicules in marsupials remains undetermined. In certain placental mammals, analogous structures are associated with the necessary vaginal stimulation for ovulation, facilitated by the release of hormones in ovulation-induced females. However, existing studies in marsupials have yet to yield definitive results [52]. Additional proposed functions of the spicules include inducing physical alterations in the female genital mucosa to facilitate the penetration of seminal substances responsible for triggering ovulation, providing tactile feedback to the male during copulation, or potentially playing a role in sexual selection [53]. However, none of these studies include marsupial species. One specimen examined in our study was a neutered male, which exhibited an absence of epidermal spicules at the tip of the penis. While no histological analysis was conducted to confirm their absence, no spicules were detectable under binocular loupe examination. This may be attributable to the fact that, as in other mammals, castration induces the regression of penile spines, given that they are secondary sexual characteristics and thus testosterone-dependent, as extensively documented in the male cat [54,55].

4.5. Accessory Glands

4.5.1. Prostate

According to the available literature, the prostate of the *Petaurus breviceps* is a large and disseminate gland [4,28,40,48]. The *Nomina Anatomica Veterinaria* [42] characterizes the prostate as a glandular structure composed of two distinct parts: a compact component termed as the "body" *(corpus prostatae)*, and a diffusely distributed part within the wall of the pelvic urethra *(pars disseminata prostatae)*. Given that the authors identify a large, macroscopic prostatic portion enveloping the urethra, the prostate of the sugar glider does not align with the disseminated part and should be referred to as the body. Furthermore, it is a compact structure. In terms of its anatomical structure, it comprises three macroscopically distinguishable regions: cranial, media, and caudal, referred as anterior, central, and posterior segments in the literature. However, the *Nomina Anatomica Veterinaria* [56] stipulates that the terms *'anterior' and 'posterior' (as well as 'superior' and 'inferior')* 'cannot be generally applied to quadrupeds because of the confusion arising from their meaning in human anatomy'. Consequently, the application of these terms is confined to certain structures of the head, excluding other parts of the body. Therefore, in the context of the sugar glider's prostate, being a quadruped animal, the part contiguous to the urinary bladder should be designated as the cranial part, and the part oriented towards the tail should be termed as the caudal part. The part situated between these two extremes could be interchangeably referred to as the central part or media.

In the sugar glider the prostate displays a morphology that is analogous to the shape of a carrot, a form that is predominantly observed among marsupials [15,16,23,24,30,57]. The constriction that segregates the initial two portions has been previously described in *Petaurus breviceps* by Johnson-Delaney (2021) [5]. A heart-shaped prostate has been reported in a few species, such as the *Acrobates pygmaeus* [14], bandicoots *(Isoodon macrourus* and *Parameles nasuta)* [24], and the koala *(Phascolarctos cinereus)* [21].

Histological studies of the prostate of marsupials have revealed that it is divided into several regions, each comprising different types of glands that produce a variety of secretion products, with some variation observed between family groups [20,24]. In the sugar glider prostate, we observed a transverse division into three zones: cranial (or anterior), media (or central), and caudal or posterior. To our knowledge, this tripartite transverse segmentation is common in marsupials [11,14–16,23,24,30,57]. However, despite its carrot-like shape, the distribution pattern of the three parts may vary depending on the species [24]; the limit of the cranial part is always transverse, but the former central and the caudal part could exhibit a different distribution, hence the nomenclature should be altered to internal/external depending on their distribution. Additionally, several exceptions have been documented with only two morphological and functional zones in the prostate gland: dorsal and ventral in the bandicoot *(Isoodon macrourus)* [24,58] or anterior and posterior in the honey possum (*Tarsipes rostratus)* [59].

4.5.2. Bulbourethral Glands

In the majority of placental mammal species, males possess a pair of bulbourethral glands, also known as Cowper's glands. These are situated adjacent to the rectum and discharge into the caudal section of the pelvic urethra. The literature reveals a variation in the number of bulbourethral gland pairs in marsupials, with some sources citing one, two, or three pairs [15,18,24,28,40]. According to our study, the sugar glider has two pairs, which aligns with the existing bibliography [4,5,40,48]. These glands positioned dorsally and laterally to the rectum, were initially characterized by Carboni and Tully (2009) [4] as multilobed. Other authors refer to them as lobulated as well [7,40]. However, our observations suggest a smooth external surface and a compact structure. Rodger and Huges (1973) [24] noted interspecies variation in gland shape, and even intraspecies variation between different gland pairs. The glands may exhibit a kidney or bulbous shape and may or may not be lobed. This observation is consistent with the findings of de Barros et al. (2013) [30] in Didelphis sp. Furthermore, these authors described three pairs of bulbourethral glands, each with a distinct glandular epithelium, but all encased within a double layer of striated muscle. In the same way, histological differences in the epithelium of the three gland pairs have also been reported in the hairy-nosed wombat (*Lasiorhinus latifrons*) [16] and the *Didelphis virginiana* [11].

In species that possess two pairs of bulbourethral glands, these can be situated dorsally to the pelvic urethra, as observed in the sugar glider, or alternatively, one pair may be dorsal and the other ventral, as in the case of *Chinorectes minimus* [23]. In instances where a third pair is present, two pairs are positioned dorsally and one ventrally to the urethra. This configuration is evident in the hairy-nosed wombat *(Lasiorhinus latifrons)* [16] and *Didelphis sp* [30,60]. The ducts of these glands discharge into the proximal part of the penile urethra [4], in close proximity to the crura penis. Each gland's duct enters the urethra on the ventral surface near the crus penis [40,48]. In the case of *Didelphis albiventris*, three pairs of bulbourethral glands are present: lateral, intermediate, and medial. The ducts of the first two pairs converge near the urethra to form a common duct, which, in conjunction with the duct of the medial gland, leads to a conical papilla ventrolateral to the initial section of the penile urethra [60]. However, in the sugar glider, our observations indicate that both ducts join the urethra in a caudal position, just cranial to the crura penis. The ducts initially run dorsally to the urethra before incorporating into the urethra, where they continue laterally, remaining independent and opening further cranially, without a papilla. To the best of our knowledge, this specific course of the bulbourethral ducts has not been previously documented in marsupial literature.

Although the histology of the bulbourethral glands of *Petaurus breviceps* has not been previously studied, the striated muscle envelope surrounding each gland has been previously described in other species [11,16,24,30,57].

According to previous studies, marsupials lack vesicular and coagulating glands [4,12,15,23,24,28,30,40,48,61,62]. Our findings corroborate the conclusions of these earlier investigations.

5. Conclusions

Sugar gliders (*Petaurus breviceps*) are becoming increasingly popular as marsupial pets due to their diminutive size and sociable disposition, which facilitates interaction with their owners. Consequently, these animals are more frequently presented as patients in veterinary clinics. However, veterinarians often face challenges in obtaining comprehensive information about their anatomical structure. Our study provides insight into the male reproductive system of *Petaurus breviceps*, that consists of two testes, epididymides, and deferent ducts, a single prostate, two bulbourethral glands, and a bifurcated, post-scrotal penis, a feature commonly misinterpreted as abnormal by pet owners. While not prevalent, certain pathologies such as penile fibrosarcoma and prostatitis have been documented in marsupials, including sugar gliders. In contrast, penile mutilation and necrosis are commonly observed in pet *Petaurus breviceps*. These conditions are typically stress-induced, often due to inadequate care in captivity, such as when these animals are kept in isolation or with an aggressive cage mate. In certain instances, surgery may be required to remove the damaged part or even the entire bifid portion of the penis. In such cases, a thorough understanding of the male genital tract's anatomy is essential for accurately diagnosing and effectively treating the pathological condition.

Author Contributions: Conceptualization, M.d.M.Y., D.A.-P. and M.L.; Writing—Original Draft Preparation, M.d.M.Y., D.A.-P. and M.L.; Dissections, M.d.M.Y., D.A.-P. and M.L.; Writing—Review and Editing, M.d.M.Y., D.A.-P. and M.L. All authors have read and agreed to the published version of the manuscript.

Funding: This research received no external funding.

Institutional Review Board Statement: Not applicable.

Informed Consent Statement: Not applicable.

Data Availability Statement: Data sharing is not applicable to this article as all data associated is available in the text.

Acknowledgments: The authors are thankful to Nuria Pérez for donating the corpses of some specimens used in this study.

Conflicts of Interest: The authors declare no conflicts of interest.

References

1. Brust, D.M. Sugar Gliders. *Exot. DVM A Pract. Resour. Clin.* **2009**, *11*, 32–41.
2. Raftery, A. Sugar Gliders (Petaurus Breviceps). *Companion Anim.* **2015**, *20*, 422–426. [CrossRef]
3. Kubiac, M. Sugar Gliders. In *Handbook of Exotic Pet Medicine*; Kubiac, M., Ed.; Willey-Blackwell: Chichester, UK, 2021 ISBN 9781119389958.
4. Carboni, D.; Tully, T.N. Manual of Exotic Pet Practice. In *Saunders*; Mitchell, M.A., Tully, T.N.J., Eds.; Elsevier: St. Louis, MO, USA 2009; ISBN 9781416001195.
5. Johnson-Delaney, C. Sugar Gliders. In *Ferrets, Rabbits, and Rodents: Clinical Medicine and Surgery*; Quesenberry, K.E., Orcutt, C.J. Eds.; Elsevier Inc.: St. Luis, MO, USA, 2021; pp. 385–400, ISBN 978-0-323-48435-0.
6. Stojanovic, D.; Webb, M.H.; Alderman, R.; Porfirio, L.L.; Heinsohn, R. Discovery of a Novel Predator Reveals Extreme but Highly Variable Mortality for an Endangered Migratory Bird. *Divers. Distrib.* **2014**, *20*, 1200–1207. [CrossRef]
7. Shaw, G. Reproduction. In *Marsupials*; Armati, P.J., Dickman, C.R., Hume, I.D., Eds.; Cambridge University Press: Cambridge UK, 2006; pp. 83–107, ISBN 9780511541889.

Cummins, J.M.; Temple-Smith, P.D.; Renfree, M.B. Reproduction in the Male Honey Possum (*Tarsipes rostratus*: Marsupialia): The Epididymis. *Am. J. Anat.* **1986**, *177*, 385–401. [CrossRef] [PubMed]

Woolley, P.A. Male Members of Some Endemic New Guinean Dasyurid Marsupials: They Help Resolve Relationships. *J. Mammal.* **2019**, *100*, 142–149. [CrossRef]

Nogueira, J.C.; Martinelli, P.M.; Costa, S.F.; Carvalho, G.A.; Câmara, B.G.O. The Penis Morphology of Didelphis, Lutreolina, Metachirus and Caluromys (Marsupialia, Didelphidae). *Mammalia* **1999**, *63*, 79–92. [CrossRef]

Chase, E.B. The Reproductive System of the Male Opossum, Didelphis Virginiana Kerr and Its Experimental Modification. *J. Morphol.* **1939**, *65*, 215–239. [CrossRef]

Rodger, J.C. The Testis and Its Excurrent Ducts in American Caenolestid and Didelphid Marsupials. *Am. J. Anat.* **1982**, *163*, 269–282. [CrossRef]

Woolley, P.A.; Westerman, M.; Krajewski, C. Interspecific Affinities within the Genus Sminthopsis (Dasyuromorphia: Dasyuridae) Based on Morphology of the Penis: Congruence with Other Anatomical and Molecular Data. *J. Mammal.* **2007**, *88*, 1381–1392. [CrossRef]

Ward, S.J.; Renfree, M.B. Reproduction in Males of the Feathertail Glider Acrobates Pygmaeus (Marsupialia). *J. Zool.* **1988**, *216*, 241–251. [CrossRef]

Hogan, L.A.; Janssen, T.; Johnston, S.D. Wombat Reproduction (Marsupialia; Vombatidae): An Update and Future Directions for the Development of Artificial Breeding Technology. *Reproduction* **2013**, *145*, R157–R173. [CrossRef] [PubMed]

Brooks, D.E.; Gaughwin, M.; Mann, T. Structural and Biochemical Characteristics of the Male Accessory Organs of Reproduction in the Hairy-Nosed Wombat (*Lasiorhinus latifrons*). *Proc. R. Soc. London. Ser. B Biol. Sci.* **1978**, *201*, 191–207. [CrossRef]

Woolley, P.A. Phallic Morphology of the Australian Species of Antechinus (Dasyuridae, Marsupialia): A New Taxonomic Tool? In *Carnivorous Marsupials*; Archer, M., Ed.; Royal Zoological Society of New South Wales: Sydney, New South Wales, Australia, 1982; pp. 767–781.

Taggart, D.A.; Shimmin, G.A.; Ratcliff, J.R.; Steele, V.R.; Dibben, R.; Dibben, J.; White, C.; Temple-Smith, P.D. Seasonal Changes in the Testis, Accessory Glands and Ejaculate Characteristics of the Southern Hairy-nosed Wombat, Lasiorhinus Latifrons (Marsupialia: Vombatidae). *J. Zool.* **2005**, *266*, 95–104. [CrossRef]

Warburton, N.M.; Bateman, P.W.; Fleming, P.A. Anatomy of the Cavernous Muscles of the Kangaroo Penis Highlights Marsupial–Placental Dichotomy. *J. Anat.* **2019**, *234*, 306–315. [CrossRef]

Dawson, T.J.; Finch, E.; Freedman, L.; Hume, I.D.; Renfree, M.B.; Temple-Smith, P.D. Morphology and Physiology of the Metatheria. In *Fauna of Australia*; Walton, D.W., Richardson, B.J., Eds.; Australian Government Publishing Service: Camberra, Australia, 1989; Volume 1b-Mammali, Chapter 17; pp. 1–102, ISBN 978 0 644 06056 1.

Young, A.H. The Male Generative Organs of the Koala (*Phascolarctos cinereus*). *J. Anat. Physiol.* **1879**, *13*, 305–317.

Tyndale-Biscoe, H.; Renfree, M. *Reproductive Physiology of Marsupials*; Cambridge University Press: Cambridge, UK, 1987; ISBN 9780521337922.

Nogueira, J.C.; Castro, A.C.S.; Câmara, E.V.C.; Câmara, B.G.O. Morphology of the Male Genital System of Chironectes Minimus and Comparison to Other Didelphid Marsupials. *J. Mammal.* **2004**, *85*, 834–841. [CrossRef]

Rodger, J.; Hughes, R. Studies of the Accessory Glands of Male Marsupials. *Aust. J. Zool.* **1973**, *21*, 303–320. [CrossRef]

Woolley, P.A.; Webb, S.J. The Penis of Dasyurid Marsupials. In *The Biology of Marsupials*; Macmillan Education UK: London, UK, 1977; pp. 307–323.

Yllera, M.d.M.; Alonso-Peñarando, D.; Lombardero, M. Gross Anatomy of the Female Reproductive System of Sugar Gliders (Petaurus Breviceps). *Animals* **2023**, *13*, 2377. [CrossRef] [PubMed]

Ness, R.D.; Booth, R. Sugar Gliders. In *Ferrets, Rabbits, and Rodents*; Elsevier: Amsterdam, The Netherlands, 2004; pp. 330–338, ISBN 9780721693774.

Girling, S.J. Basic Small Mammal Anatomy and Physiology. In *Veterinary Nursing of Exotics Pets*; Girling, S.J., Ed.; Wiley-Blackwell: Ames, IA, USA, 2013; pp. 2–26, ISBN 9780470659175.

Judah, V.; Nuttal, K. Sugar Gliders. In *Exotic Animal Care & Management*; Thomson Delmar Learning: New York, NY, USA, 2008; pp. 118–131.

De Barros, M.A.; Panattoni Martins, J.F.; Samoto, V.Y.; Oliveira, V.C.; Gonçalves, N.; Furlaneto Mançanares, C.A.; Vidane, A.; Carvalho, A.F.; Ambrósio, C.E.; Miglino, M.A. Marsupial Morphology of Reproduction: South America Opossum Male Model. *Microsc. Res. Tech.* **2013**, *76*, 388–397. [CrossRef]

Benshemesh, J. *Recovery Plan for Marsupial Moles Notoryctes Typhlops and N. Caurinus. 2005-2010*; Northern Territory Department of Infrastructure, Planning and Environment, Northern Territory: Alice Springs, Northern Territory, Australia, 2004.

Johnson, K.A.; Walton, D.W. Notoryctidae. In *Fauna of Australia*; Walton, D.W., Richardson, B., Eds.; Australian Government Publishing Service: Camberra, Australia, 1989; Chapter 23; pp. 1–24.

Taggart, D.A.; Temple-Smith, P.D. Comparative Studies of Epididymal Morphology and Sperm Distribution in Dasyurid Marsupials during the Breeding Season. *J. Zool.* **1994**, *232*, 365–381. [CrossRef]

Biggers, J.D. Reproduction in Male Marsupials. In *Proceedings of the Comparative Biology of Reproduction in Mammals*; Symposia of the Zoological Society of London; Rowlands, I.W., Ed.; Academic Press: London, UK, 1966; pp. 251–280.

Hughes, R.L. Comparative Morphology of Spermatozoa from Five Marsupial Families. *Aust. J. Zool.* **1965**, *13*, 533–543. [CrossRef]

36. Taggart, D.A.; Temple-Smith, P.D. Structural Features of the Epididymis in a Dasyurid Marsupial (*Antechinus stuartii*). *Cell Tissu Res.* **1989**, *258*. [CrossRef] [PubMed]
37. Gibson, R.J.; Leigh, C.M.; Breed, W.G. Unusual Macromorphology of the Ductuli Efferentes and Epididymis of the Koal (*Phascolarctos cinereus*). *Aust. J. Zool.* **2000**, *48*, 681. [CrossRef]
38. Woolley, P.A. The Seminiferous Tubules, Rete Testis and Efferent Ducts in Didelphid, Caenolestid and Microbiotheriid Marsupial In *Possums and Opossums*; Archer, M., Ed.; Royal Zoological Society of New South Wales: Mosman, Australia, 1987.
39. Dyce, K.M.; Sack, W.O.; Wensing, C.J.G. *Textbook of Veterinary Anatomy*, 4th ed.; Saunders-Elsevier: St. Louis, MO, USA, 201 ISBN 978-1-4160-6607-1.
40. Johnson-Delaney, C.A. Reproductive Medicine of Companion Marsupials. *Vet. Clin. N. Am. Exot. Anim. Pract.* **2002**, *5*, 537–55 [CrossRef] [PubMed]
41. Renfree, M.; Shaw, G. Reproduction in Monotremes and Marsupials. In *Encyclopedia of Life Sciences*; Wiley: Hoboken, N USA, 2001.
42. Schaller, O. *Illustrated Veterinary Anatomical Nomenclature*, 2nd ed.; Enke Verlag: Stuttgart, Germany, 2007; ISBN 978-3-8304-1069-
43. Dixon, A.F. Phallic Structure and Function. In *Mammalian Sexuality: The Act of Mating and the Evolution of Reproduction*; Cambridg University Press: Cambridge, UK, 2021; pp. 111–154.
44. Hershkovitz, P. *The South American Gracile Mouse Opossums, Genus Gracilinanus Gardner and Creighton, 1989 (Marmosidae, Marsup alia): A Taxonomic Review with Notes on General Morphology and Relationships/Philip Hershkovitz*; Field Museum of Natural Histor Chicago, IL, USA, 1992.
45. Woolley, P.A. Phallic Morphology of Dasyuroides-Byrnei and Dasycercus-Cristicauda (Marsupialia, Dasyuridae). *Aust. J. Zoo* **1987**, *35*, 535. [CrossRef]
46. Nogueira, J.C.; Ferreira da Silva, M.N.; Câmara, B.G.O. Morphology of the Male Genital System of the Bushy-Tailed Opossur Glironia Venusta Thomas, 1912 (Didelphimorphia, Didelphidae). *Mammalia* **1999**, *63*, 231–236. [CrossRef]
47. Jepson, L. Sugar Gliders. In *Exotic Animal Medicine: A Quick Reference Guide*; Elsevier: Amsterdam, The Netherlands, 201 pp. 231–257, ISBN 978-0-323-32849-4.
48. Johnson-Delaney, C.A.; Lennox, A.M. Reproductive Disorders of Marsupials. *Vet. Clin. N. Am. Exot. Anim. Pract.* **2017**, *2* 539–553. [CrossRef]
49. Schmidt, M.H.; Schmidt, H.S. The Ischiocavernosus and Bulbospongiosus Muscles in Mammalian Penile Rigidity. *Sleep* **1993**, *1* 171–183. [CrossRef]
50. Hart, B.; Melese-D'Hospital, P. Penile Mechanisms and the Role of the Striated Penile Muscles in Penile Reflexes. *Physiol. Beha* **1983**, *31*, 807–813. [CrossRef]
51. Esson, C.; Armati, P.J. Development of the Male Urogenital System of the Koala Phascolarctos Cinereus. *Anat. Embryol. (Berl* **1998**, *197*, 217–227. [CrossRef]
52. Hogan, L.A.; Phillips, C.J.C.; Lisle, A.; Keeley, T.; Horsup, A.B.; Janssen, T.; Johnston, S.D. Non-Invasive Methods of Oestru Detection in Captive Southern Hairy-Nosed Wombats (*Lasiorhinus latifrons*). *Anim. Reprod. Sci.* **2010**, *119*, 293–304. [CrossRef [PubMed]
53. Orr, T.J.; Brennan, P.L.R. All Features Great and Small—The Potential Roles of the Baculum and Penile Spines in Mammals. *Integ Comp. Biol.* **2016**, *56*, 635–643. [CrossRef] [PubMed]
54. Ryane, E. Spines Along the Feline Penis. In *Common Clinical Presentations in Dogs and Cats*; Wiley: Hoboken, NJ, USA, 201 pp. 733–735.
55. Little, S. Male Reproduction. In *The Cat: Medicine and Management*; Elsevier-Saunders: St. Louis, MO, USA, 2012; pp. 1184–119 ISBN 978-1-4377-0660-4.
56. International Committee on Veterinary Gross Anatomical Nomenclature. *Nomina Anatomica Veterinaria*, 6th ed.; Editoria Committee: Hanover, Germany; Ghent, Belgium, 2017.
57. Inns, R.W. Seasonal Changes in the Accessory Reproductive System and Plasma Testosterone Levels of the Male Tammar Wallab Macropus Eugenii, in the Wild. *Reproduction* **1982**, *66*, 675–680. [CrossRef] [PubMed]
58. Todhunter, R.; Gemmell, R.T. Seasonal Changes in the Reproductive Tract of the Male Marsupial Bandicoot, Isoodon Macrourus *J. Anat.* **1987**, *154*, 173–186. [PubMed]
59. Woolley, P.A.; Scarlett, G. Observations of the Reproductive Anatomy of Male Tarsipes Rostratus (Maesupialia:Tarsipedidae). I *Possums and Gliders*; Smith, A.P., Hume, I.D., Eds.; Australian Mammal Society: Sydney, Australia, 1984; pp. 445–450.
60. Nogueira, J.C. Anatomical Aspects and Biometry of the Male Genital System of the White-Belly Opossum Didelphis Albiventri Lund, 1841 during the Annual Cycle. *Mammalia* **1988**, *52*, 233–242. [CrossRef]
61. Tyndale-Biscoe, H.; Renfree, M. Male Anatomy and Spermatogenesis. In *Reproductive Physiology of Marsupials*; Cambridg University Press: Cambridge, UK, 1987; pp. 124–171, ISBN 9780521337922.
62. Rodger, J.C. Marsupials the Alternative Therians—From Gametes to Birth. *Theriogenology* **2020**, *150*, 405–411. [CrossRef]

Article

Cory's Shearwater (*Calonectris borealis*): Exploring Normal Head Anatomy through Cross-Sectional Anatomy, Computed Tomography and Magnetic Resonance Imaging

Alejandro Morales-Espino [1], Soraya Déniz [2], Pablo Paz-Oliva [3], Natalia Roldán-Medina [3], Mario Encinoso [2,*], Francisco Suárez-Cabrera [3] and Jose Raduan Jaber [3,*]

1 IVC Evidensia Los Tarahales, 35013 Las Palmas, Gran Canaria, Spain
2 Hospital Clínico Veterinario, Facultad de Veterinaria, Universidad de Las Palmas de Gran Canaria, Trasmontaña, Arucas, 35413 Las Palmas, Gran Canaria, Spain
3 Department of Morphology, Facultad de Veterinaria, Universidad de Las Palmas de Gran Canaria, Trasmontaña, Arucas, 35413 Las Palmas, Gran Canaria, Spain; natalia.roldan101@alu.ulpgc.es (N.R.-M.)
* Correspondence: mencinoso@gmail.com (M.E.); joseraduan.jaber@ulpgc.es (J.R.J.)

Simple Summary: This investigation describes the intricate anatomy of the central nervous system (CNS) and its related structures in Cory's shearwater, employing anatomical cross-sections and imaging methodologies like computed tomography (CT) and magnetic resonance imaging (MRI). This study represents the inaugural comprehensive portrayal of the head anatomy in Cory's shearwater, and the findings derived from this study promise to enrich future research in both the anatomical and pathological fields of closely related species.

Abstract: Cory's shearwater, or *Calonectris borealis*, stands out as a symbolic figure in the world of seabirds, playing a crucial role in marine ecosystems globally. Belonging to the Procellariidae family, it is singularized by its imposing wingspan and intricate migration patterns connecting it to various regions from the North Atlantic to the Pacific. Its role in the marine food chain, specialized diet and adaptation for nesting in the Canary Archipelago underscore its ecological importance. However, Cory's shearwater also faces important threats, such as the invasion of foreign species, highlighting the need for its conservation. Among the conservation issues, studies on its biology, the main threats it faces and its normal anatomy are essential to preserve marine biodiversity. Additionally, a variety of imaging techniques, such as computed tomography and magnetic resonance, facilitates the understanding of the bird's neuroanatomy and opens future research possibilities in comparative neuroscience. Moreover, this approach proves particularly relevant given the increasing attention these seabirds receive in environments such as zoos, rehabilitation centers and their natural habitat, where veterinarians play a crucial role in their care and well-being.

Keywords: computed tomography; magnetic resonance imaging; head; anatomical sections; seabirds; Cory's shearwater

1. Introduction

Cory's shearwater (*Calonectris borealis*) is a remarkable seabird known for its long-distance migrations and significant ecological role within the Canary Islands' ecosystem [1–4]. Belonging to the distinguished Procellariidae family, it stands out for intricate migration patterns linking it to various regions worldwide, from the North Atlantic to the Indian and Pacific Oceans, and it occupies marine and coastal habitats [1–4]. During their migratory journeys, Cory's shearwater populations trace routes covering significant distances, feeding and reproducing in a delicate balance with their marine environment. This migratory pattern reveals the complexity of their connection to oceanic ecosystems, where species survival and reproduction are intrinsically linked to the availability of food reserves [1,4]. As a vital component of the marine food chain, the shearwater deploys a specialized diet, focusing primarily on small fish, crustaceans and cephalopods. Its presence, therefore, contributes to the dynamics and stability of oceanic ecosystems, exerting a significant impact on the regulation of prey populations [1]. However, the importance of Cory's shearwater transcends its direct ecological function. These seabirds, through their behavioral patterns and distribution, emerge as valuable bioindicators of marine environmental health [3,4]. Changes in their migration, reproduction, or abundance can serve as early warning signals for potential environmental threats, such as food scarcity or contaminants in the ocean [1,5,6]. In this context, the protection of nesting habitats, sustainable fisheries management and the mitigation of anthropogenic threats emerge as essential elements in conservation efforts to ensure the ongoing well-being of this seabird species [4]. Among these conservation issues, studies on its biology, the main threats it faces and anatomical investigations are essential for preserving marine biodiversity, since they elucidate the underlying mechanisms of avian physiology, and for comprehending the evolutionary adaptations that have allowed birds to inhabit a diversity of ecological niches [7–29].

Despite its ecological importance, there is a notable gap in the research concerning the neuroanatomy and cranial characteristics of this species using modern imaging diagnostic techniques. However, the revolution brought about by modern imaging techniques has marked a milestone in exploring this doctrine. These imaging techniques have already been employed in research on the nervous system of birds [30–38], but to date, specific research on the CNS and associated structures of Cory's shearwater has not been conducted [32,38–40]. Therefore, this study aimed to describe the CNS and its associated structures through anatomical cross-sections and computed tomography and magnetic resonance imaging techniques.

2. Materials and Methods

2.1. Animals

In this research, we utilized a cohort of 10 juvenile Cory's shearwater specimens, characterized by an average mass of 0.520 kg (with a range of 0.480–0.820 kg) and an average length of 42 cm (with a range of 45–56 cm) from beak to tail base. These avian subjects were sourced from the Consejeria de Área de Medio Ambiente, Clima, Energía y Conocimiento of the Cabildo Insular de Gran Canaria, following strandings attributed to artificial lights, which have been recognized as a relevant threat to biodiversity conservation [3]. While most specimens were received post mortem, those initially alive but succumbing due to their weakened state were promptly preserved via freezing for subsequent CT and MRI procedures. It is important to emphasize that no animals were intentionally sacrificed or captured solely for scientific research.

2.2. CT Technique

For the CT assessment, we allowed our avian subjects to thaw at room temperature for 12 h. Sequential transverse CT images were obtained using a 16-slice helical CT scanner (Toshiba Astelion, Canon Medical System®, Tokyo, Japan). The birds were positioned symmetrically in dorsal recumbency on the stretcher with a craniocaudal entry. We used a standard protocol with the following parameters: 120 kVp, 80 mA, 512 × 512 acquisition matrix, 1809 × 858 field of view, a pitch of 0.94 and a gantry rotation of 1.5. The acquired images had a slice thickness of 0.6 mm. To enhance the visualization of various anatomical structures using CT, we employed different CT window settings by adjusting the window widths (WWs) and window levels (WLs): a bone window setting (WW = 1500; WL = 300), a soft tissue window setting (WW = 248; WL = 123) and a lung window setting (WW = 1400; WL = −500). No significant differences in CT density or anatomy were observed in the heads of the avian specimens utilized in this study.

2.3. MRI Technique

The ten Cory's shearwater specimens underwent imaging at Los Tarahales Veterinary Hospital (Las Palmas, Canary Islands, Spain). Imaging data were acquired using a Canon Vantage Elan 1.5 T imaging system (Toshiba, Vantage Elan, Japan), utilizing the following sequences: T1-weighted (T1W) sequences in the transverse plane (TR: 634 ms, TE: 10, FOV: 1809 × 829, slice thickness: 2 mm, matrix size: 192 × 160), T2-weighted (T2W) sequences in the transverse plane (TR: 4769 ms, TE: 120, FOV: 1809 × 829, slice thickness: 2 mm, matrix size: 192 × 224), T2W sequences in the dorsal plane (TR: 5271 ms, TE: 120, FOV: 1809 × 829, slice thickness: 2.5 mm, matrix size: 240 × 192) and T2W sequences in the sagittal plane (TR: 4450 ms, TE: 120, FOV: 1809 × 829, slice thickness: 2.9 mm, matrix size: 224 × 224). Additionally, we obtained enhanced spin-echo sequences in the dorsal, transverse and sagittal planes. The resulting MRI images exhibited a slice thickness of 2–3 mm.

2.4. Macroscopic Anatomical Sections

After imaging, the birds were preserved by freezing at −80 °C for 72 h. Following this preservation process, we sliced five specimens into transverse and dorsal sections, each 1 cm thick, utilizing an electric band saw. These sections underwent careful irrigation with water to eliminate any artefacts, such as feathers, which were removed using Adson forceps and identified, before being photographed on both sides to facilitate the precise identification of anatomical cross-sections and CT and MRI images. This dissection step greatly aided in achieving accurate identification and correlation with the CT and MRI images.

2.5. Anatomic Evaluation

To enhance the accurate identification of relevant structures within Cory's shearwater heads, we selected anatomical cross-sections that closely matched the CT and MR images. Extensive reference materials, including textbooks, bird anatomy literature and bone preparations from other seabird specimens, were consulted to aid this endeavor [30,39–44]. Furthermore, to ensure the meticulous interpretation of cranial structures, we relied on anatomical preparations provided by the Department of Anatomy at the Faculty of Veterinary Medicine, University of Las Palmas de Gran Canaria. These supplementary resources significantly enriched our understanding and accuracy in the anatomical interpretation.

3. Result

This investigation delved into the intricate anatomy of the Cory's shearwater CNS and associated structures. We present ten figures that mainly correspond to anatomical cross-sections and CT and T2W MR images (Figures 1–10). T2-weighted magnetic resonance imaging images were selected based on their relevance to identifying principal anatomical structures of the brain. The similarity of the sections was prioritized to achieve a better resemblance between the computed tomography and magnetic resonance imaging studies. In addition, we complemented the images with anatomical cross-sections for enhanced clarity and understanding. A pivotal representation of this synergy is depicted in Figure 1, where a sagittal view illustrates the transverse (A) and dorsal (B) orientations, aligning with the anatomical sections delineated in Figures 2–10. In this context, the transverse and dorsal section levels (I–VIII) were meticulously aligned, promoting a seamless correlation between the imaging data and the anatomical contexts. Finally, Figure 10 depicts a mid-sagittal anatomical section and the corresponding CT and T2W images, displaying the main structures that compose the head of this bird, especially those related to the CNS and its associated structures.

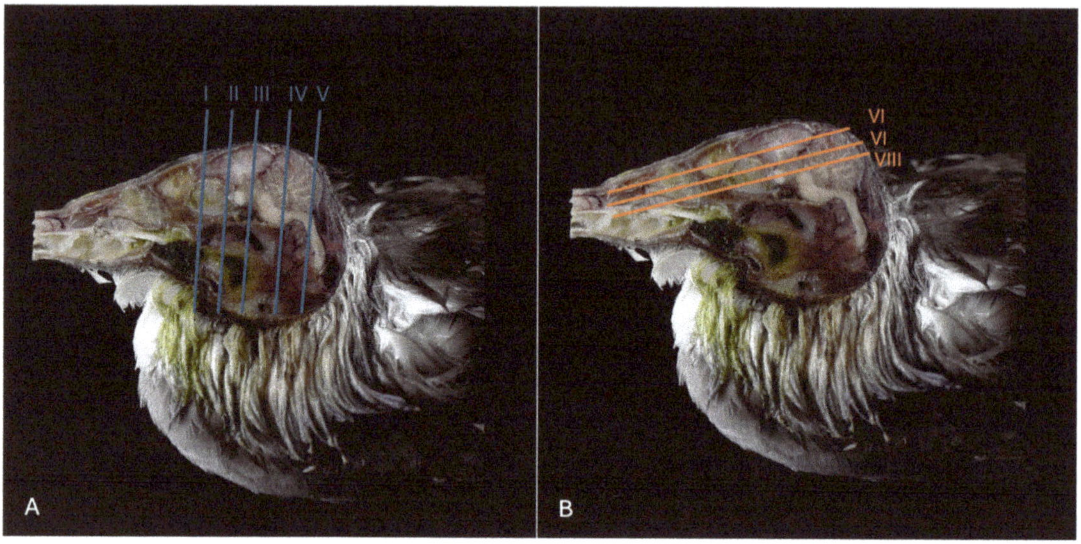

Figure 1. Sagittal cross-section images of the head of a Cory´s shearwater (*Calonectris Borealis*). The vertical (**A**) (labelled with blue) and horizontal (**B**) (labelled with orange) lines correspond to the approximate levels of the respective transverse and dorsal slices.

3.1. Anatomical Sections

The anatomical sections utilized in this study proved invaluable for distinguishing the various structures comprising the central nervous system and associated structures. Consequently, we were able to discern the *cerebrum* (*telencephalon*) and its telencephalic hemispheres, located under the frontal bone (labelled with number 16 in Figures 3A, 4A, 7A and 10A), delineated by the longitudinal cerebral fissure (Figures 2A, 3A, 4A, 7A and 8A). Furthermore, transverse cross-sections provided valuable insights into additional features. These included the identification of a slightly caudolaterally positioned rostral groove, known as the *vallecula telencephalic*, as well as the *hyperpallium,* which were prominently displayed on the dorsal surface of each hemisphere (labelled as numbers 23 and 21, respectively, in Figures 3A and 7A). Additionally, these sections revealed the presence of small, pointed olfactory bulbs at the rostral pole of each hemisphere (identified as number 11 in Figures 2A, 7A and 10A). The diencephalon was unmistakably recognized as a continuation rostrally from the *mesencephalon,* which represents the rostral limit of the brain stem and rests on the *basis cranii* (represented as numbers 57 and 51 in Figure 10A). The dorsal and sagittal anatomical images facilitated the identification of the course of the optic nerve penetrating the sclera (labelled as number 30 in Figures 9A and 10A). In addition, the dorsal sections display excellent detail of the optic chiasm located rostrally to the *hypophysis,* which rested on the *basis cranii* (identified as number 58 in Figure 10A). The sagittal section allowed us to identify the pineal gland, an important structure located near the dorsal surface of the diencephalon. It is typically small and appears as a flattened, oval-shaped structure identified ventrally to the parietal bone (labelled as number 65 in Figure 10A). It is a part of the avian endocrine system that plays a role in regulating various physiological processes, including the circadian rhythm and reproductive functions.

These sections also provided insights into other components of the Cory's shearwater brain. Among these were the *mesencephalon*, the notable large *corpus cerebelli*, the internal medullary body characterized by an internal white substance, and caudally, the small paired cerebellar hemispheres, which were identified in the transverse and dorsal sections (illustrated as numbers 45, 80 and 43, respectively, in Figures 5A, 6A, 7A, 8A, 9A and 10A). Additionally, these sections facilitated the identification of the ventral portion of the *rhomben-*

cephalon, encompassing different components such as the *pons* and the *medulla oblongata*, which rested on the basioccipital bone (illustrated as numbers 47 and 40, respectively, in Figures 5A, 6A and 10A).

Moreover, our examination facilitated a meticulous eye delineation, including various ocular components and enabling a deeper understanding of avian ocular anatomy. Noteworthy elements such as the cornea, sclera, retina, vitreous chamber and lens were precisely identified (displayed as numbers 5, 3, 8, 4, 6 and 1, respectively in Figures 2A, 3A, 7A, 8A, 9A and 10A). Simultaneously, our investigation uncovered many associated structures surrounding the avian eyeball, unveiling the intricate network of anatomical support and functionality. Furthermore, associated eyeball structures, including the interorbital septum and extraocular muscles, were among the structures distinguished, which help in maintaining the structural integrity and functionality of the avian eye (illustrated as numbers 9 and 10 as depicted in Figures 2A, 3A, 7A, 8A, 9A and 10A).

Regarding the morphology of the skull and its diverse bony structures, these sections offered crucial insights. Thus, the examination revealed detailed information about various components, including the nasal, frontal, parietal, temporal, otic, occipital (basioccipital and supraoccipital), pterygoid, interparietal and mandible/maxillary bones (illustrated as numbers 59, 12, 32, 79, 44, 28, 52, 67, 69, 14 and 66, respectively in Figures 2A, 3A, 4A, 5A, 6A, 7A, 8A, 9A and 10A). Additionally, the utilization of these sections proved indispensable in discerning the structure of both the roof of the oral cavity and the pharynx and the trachea (represented as numbers 13 and 17 in Figures 2A, 3A, 4A, 5A and 10A).

Furthermore, various head muscles, including the musculus tterygoideus pars ventralis, musculus tracheolateralis, musculus rectus capitus (comprising the musculus rectus dorsalis, musculus rectus ventralis and musculus rectus lateralis), musculus depressor mandibulae pars superficialis, musculus adductor mandibulae externus and constrictor colli were distinguished (labelled as numbers 19, 20, 41, 36, 37 and 48, respectively in Figures 2A, 3A, 4A, 5A, 6A, 8A and 10A).

3.2. Computed Tomography (CT)

In terms of the skull, the transverse, dorsal and sagittal CT scans provided clear delineation of the bones that encompasses it as the nasal, frontal, parietal, otic, occipital and mandible/maxilla bones, displaying high CT density with the soft tissue window settings (labelled as numbers 59, 12, 32, 44, 28, 52, 14 and 66, respectively, in Figures 2B, 3B, 4B, 5B, 6B, 7B, 8B, 9B and 10B). In addition, these CT images displayed various components of the encephalon, which presented a moderate or intermediate CT density and appeared grey. Therefore, the transverse CT image showed two telencephalic hemispheres, separated by the *fissura longitudinalis cerebri* (identified as number 24 in Figures 2B, 3B, 4B, 7B, 8B and 10B). These included the identification of the *vallecula telencephali*, prominently visible on the dorsal surface of each hemisphere (cited as number 23 in Figures 3B, 7B and 8B). Additionally, these sections depicted the presence of the pointed olfactory bulbs, which showed moderate attenuation (shown as number 11 in Figures 2B, 7B and 10B). Furthermore, other structures of the CNS displaying intermediate attenuation were identified, including the *hyperpallium*, the *cerebellum*, the optic lobe and the brain stem (depicted as numbers 21, 45, 31, 47 and 40, respectively, in Figures 3B, 4B, 5B, 6B, 7B, 8B and 10B).

Additionally, we identified diverse components of Cory´s shearwater eyeballs. Therefore, we distinguished the dorsal and ventral arch of the sclerotic ossifications and the lens, appearing as hyperattenuating structures just posterior to the iris (labelled as numbers 2 and 1, respectively, in Figures 2B, 3B, 8B and 9B). In contrast, other elements, such as the sclera or the vitreous chamber, were accurately pinpointed due to their moderate or low CT attenuation (represented as numbers 3 and 4, respectively, in Figures 2B, 3B, 8B and 9B).

3.3. Magnetic Resonance Imaging (MRI)

No discernible anatomical differences were observed in the examined shearwaters; however, it is noteworthy that compared to CT, CNS structures were better visualized.

Therefore, T2W MR images demonstrated precise alignment with the cranial structures observed in the cadaveric cross-sections, providing a comprehensive view of the CNS and its associated structures. Thus, the examination of Cory's shearwater brains in transverse, dorsal and sagittal planes allowed the observation of the two telencephalic hemispheres, revealing regions of a moderate-intensity signal (labelled as number 16 in Figures 2C, 3C, 4C, 7C, 8C and 10C). Notably, these imaging planes also facilitated the precise identification of the olfactory bulb, characterized by its small, tapering rostral structure, with moderate and uniform signal intensity (depicted as number 11 in Figures 2C, 7C and 10C). Moreover, the transverse T2W MR images displayed two curved dorsal structures with moderate signal intensity, corresponding with the *hyperpallium* (corresponding to number 21 in Figures 3C and 4C). The *diencephalon*, another constituent of the forebrain, exhibited limited differentiation from the adjacent *mesencephalon*, displaying signals of moderate to low intensity (identified as numbers 49 and 57, respectively, in Figure 10C). Moreover, notable features of the mesencephalon, such as the optic lobe, were perfectly discernible in the transverse and dorsal T2-weighted MRI images, showcasing similar signal characteristics (represented as number 31 in Figures 4C and 8C). Furthermore, other crucial components of the central nervous system, such as the *pons* and the *medulla oblongata,* were identifiable across transverse, dorsal and sagittal planes, and characterized by low-intensity signals (depicted as numbers 47 and 40, respectively, in Figures 5C, 6C and 10C). Adjacent to the brain stem, the *corpus cerebellum* and the small paired cerebellar hemispheres displayed indistinct areas of hypo- and moderate-intensity signals (shown as numbers 45 and 43, respectively, in Figures 5C, 6C, 7C, 8C, 9C and 10C).

Concerning CNS-associated structures, a variety of the *bulbus oculi* structures were depicted. Subsequently, the vitreous chamber consistently exhibited hyperintense signals across all planes examined (labelled as number 4 in Figures 2C, 3C, 8C, 9C and 10C). However, the cornea, sclera, lens and sclerotic ossifications displayed a hypointense signal in these MR images (corresponding to numbers 5, 3, 1 and 2, respectively, in Figures 2C, 3C and 8C). Furthermore, various head muscles, such as the *musculus pterygoideus pars ventralis, musculus tracheolateralis, musculus rectus capitus (musculus rectus dorsalis* plus *musculus rectus ventralis* plus *musculus rectus lateralis*), *musculus adductor mandibulae externus* and *constrictor colli* exhibited signals of intermediate intensity in the T2-weighted MR images (represented as numbers 19, 20, 41, 37 and 48, respectively, in Figures 2C, 3C, 4C, 8C and 10C).

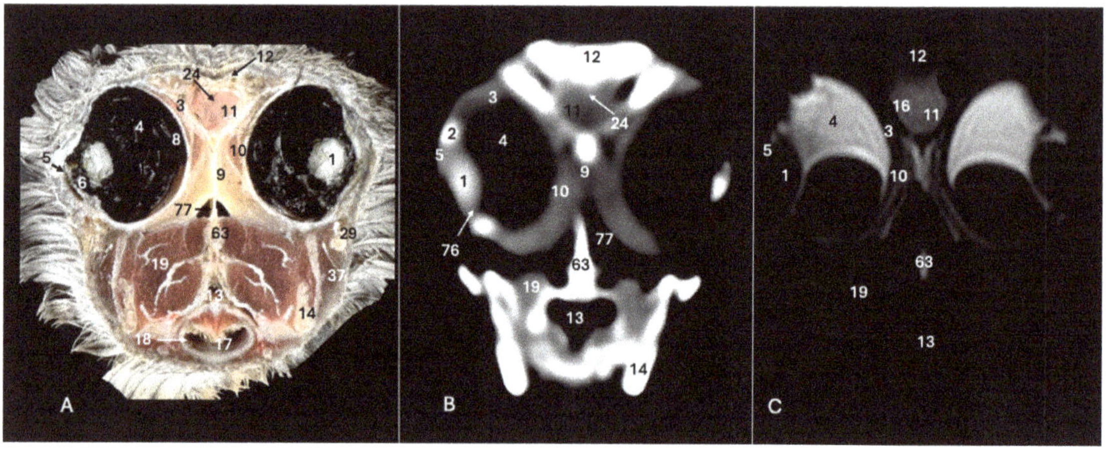

Figure 2. Anatomical cross-section (**A**), soft tissue CT window (**B**) and T2-weighted MR (**C**) transverse images of the Cory's shearwater head at the level of the eyes, corresponding to line I in Figure 1. 1: *Lent cristali*; 2: sclerotic ossification; 3: sclera; 4: *camera vitrea bulbi*; 5: cornea; 6: *camera anterior bulbi*;

9: *septum interorbitalis*; 10: extraocular muscles; 11: olfactory bulb; 12: *os frontale*; 13: pharynx; 14: mandible; 16: cerebral hemisphere; 17: trachea; 18: tracheal ring; 19: *musculus pterygoideus pars ventralis*; 24: *fissura longitudinalis cerebri*; 29: *processus postorbitalis*; 37: *musculus adductor mandibulae externus*; 63: parasphenoid rostrum; 76: ciliary body; 77: *choana*.

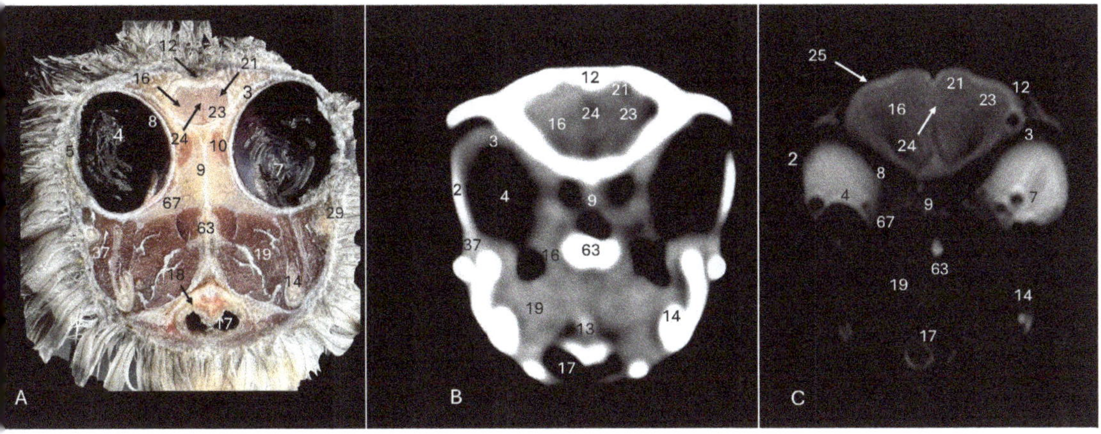

Figure 3. Anatomical cross-section (**A**), soft tissue CT window (**B**) and T2-weighted MR (**C**) transverse images of the Cory's shearwater head at the level of the *hyperpallium*, corresponding to line II in Figure 1. 2: Sclerotic ossification; 3: sclera; 4: *camera vitrea bulbi*; 5: cornea; 7: *pecten oculi*; 8: retina; 9: *septum interorbitalis*; 10: extraocular muscles; 12: *os frontale*; 13: pharynx; 14: mandible; 16: cerebral hemisphere; 17: trachea; 18: tracheal ring; 19: *musculus pterygoideus pars ventralis*; 21: *Hyperpallium*; 23: *vallecula telencephali*; 24: *fissura longitudinalis cerebri*; 25: meninges; 29: *processus postorbitalis*; 37: *musculus adductor mandibulae externus*; 63: parasphenoid rostrum; 67: *os pterygoideus*.

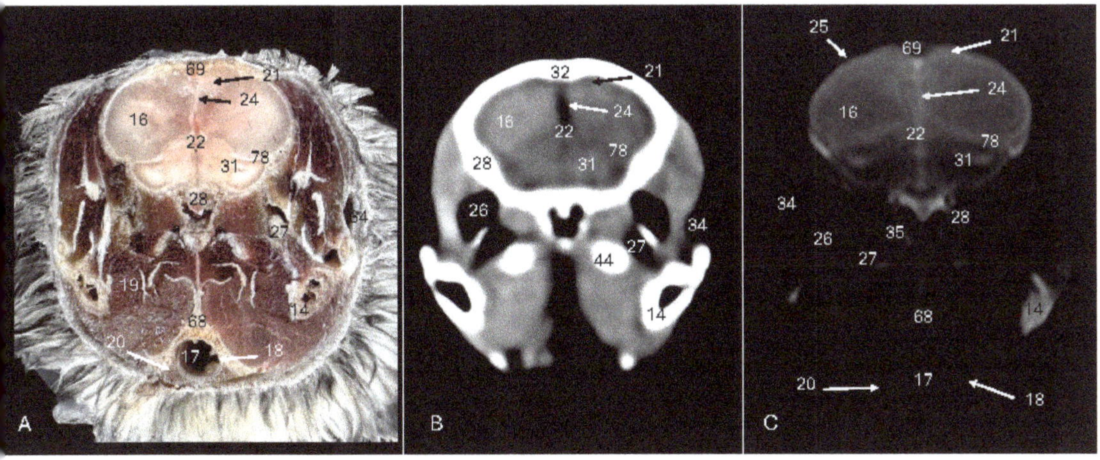

Figure 4. Anatomical cross-section (**A**), soft tissue CT window (**B**) and T2-weighted MR (**C**) transverse images of the Cory's shearwater head at the level of the optical lobes, corresponding to line III in Figure 1. 14: Mandible; 16: cerebral hemisphere; 17: trachea; 18: tracheal ring; 19: *musculus pterygoideus pars ventralis*. 20: *musculus tracheolateralis*; 21: *hyperpallium*; 22: 3rd ventricle; 24: *fissura longitudinalis cerebri*; 25: meninges; 26: middle ear; 27: tympanic bulla; 28: basioccipital bone; 31: optical lobe; 32: *os parietale*; 34: external acoustic canal; 35: inner ear; 44: otic bone; 68: cricoid cartilage; 69: interparietal bone; 78: tentorial process.

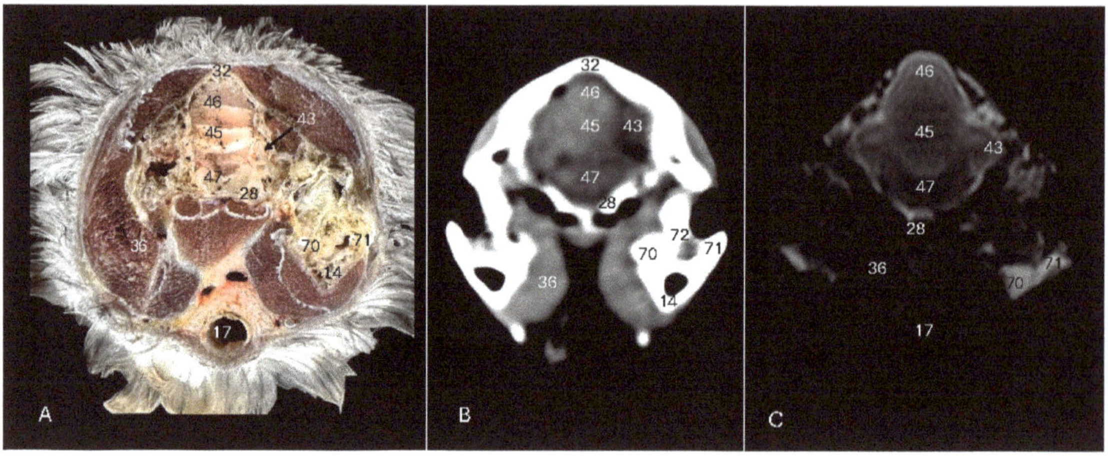

Figure 5. Anatomical cross-section (**A**), soft tissue CT window (**B**) and T2-weighted MR (**C**) transverse images of the Cory's shearwater head at the level of the *cerebellum,* corresponding to line IV in Figure 1. 14: Mandible; 17: trachea; 28: basioccipital bone; 32: *os parietale*; 36: *musculus depressor mandibulae pars superficialis* plus *musculus adductor mandibulae externus*; 43: cerebellar hemisphere; 45: *cerebellum* (body); 46: folia of *cerebellum*; 47: *pons*; 70: medial process of the mandible; 71: lateral process of the mandible; 72: quadrate articular fossa.

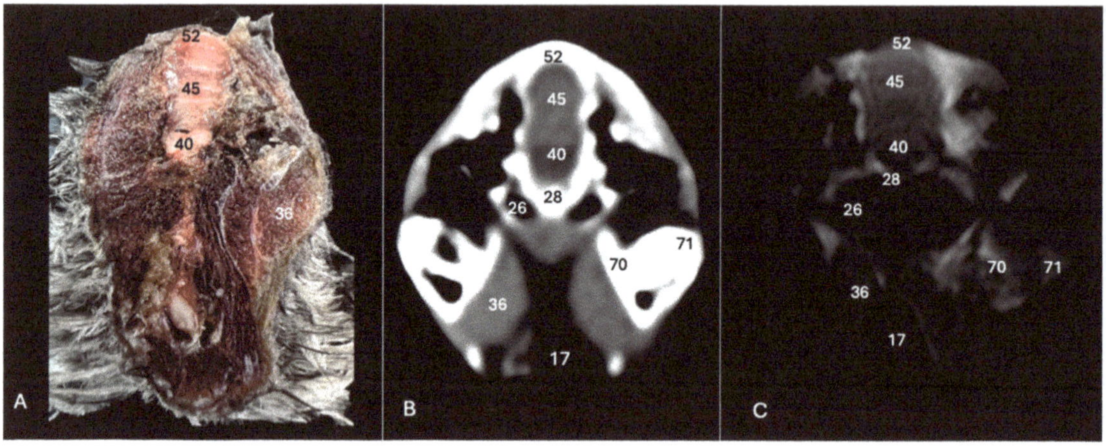

Figure 6. Anatomical cross-section (**A**), soft tissue CT window (**B**) and T2-weighted MR (**C**) transverse images of the Cory's shearwater head at the level of the caudal *cerebellum,* corresponding to line V in Figure 1. 17: Trachea; 26: middle ear; 28: basioccipital bone; 36: *depressor mandibulae pars superficialis* plus *musculus adductor mandibulae externus*; 45: *cerebellum* (body); 40: *medulla oblongata*; 52: *os supraoccipitale*; 70: medial process of the mandible; 71: lateral process of the mandible.

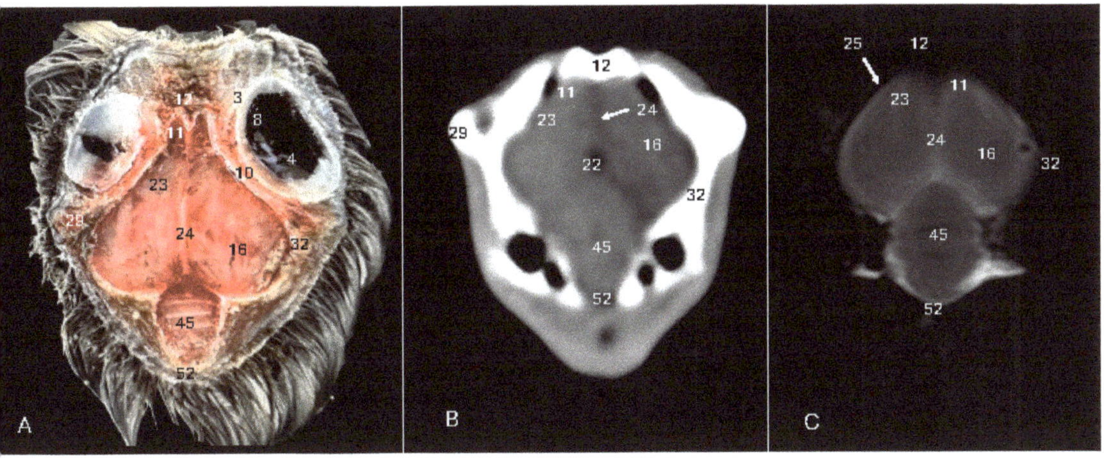

Figure 7. Anatomical cross-section (**A**), soft tissue CT window (**B**) and T2-weighted MR (**C**) dorsal images of the Cory's shearwater head at the level of the dorsal cerebral hemispheres, corresponding to line VI in Figure 1. 3: Sclera; 4: *camera vitrea bulbi*; 10: extraocular muscles; 11: olfactory bulb; 12: *os frontale*; 16: cerebral hemisphere; 22: 3rd ventricle; 23: *vallecula telencephali*; 24: *fissura longitudinalis cerebri*; 25: meninges; 29: *processus postorbitalis*; 32: *os parietale*; 45: *cerebellum* (body); 52: *os supraoccipitale*.

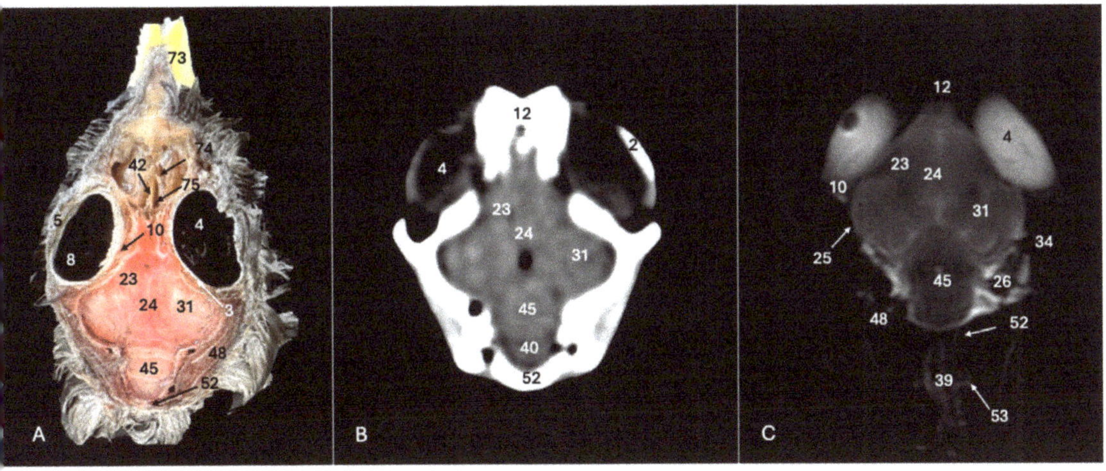

Figure 8. Anatomical cross-section (**A**), soft tissue CT window (**B**) and T2-weighted MR (**C**) dorsal images of the Cory's shearwater head at the level of the ventral cerebral hemisphere, corresponding to line VII in Figure 1. 2: Sclerotic ossification; 3: sclera; 4: *camera vitrea bulbi*; 5: cornea; 8: retina; 10: extraocular muscles; 12: *os frontale*; 23: *vallecula telencephali*; 24: *fissura longitudinalis cerebri*; 25: meninges; 26: middle ear; 31: optical lobe; 34: external acoustic canal; 39: *medulla spinalis*; 40: *medulla oblongata*; 42: nasal septum; 45: *cerebellum* (body); 48: *musculus constrictor colli*; 52: *os supraoccipitali*; 53: cervical vertebra; 73: beak; 74: *concha nasalis media*; 75: *concha nasalis caudalis*.

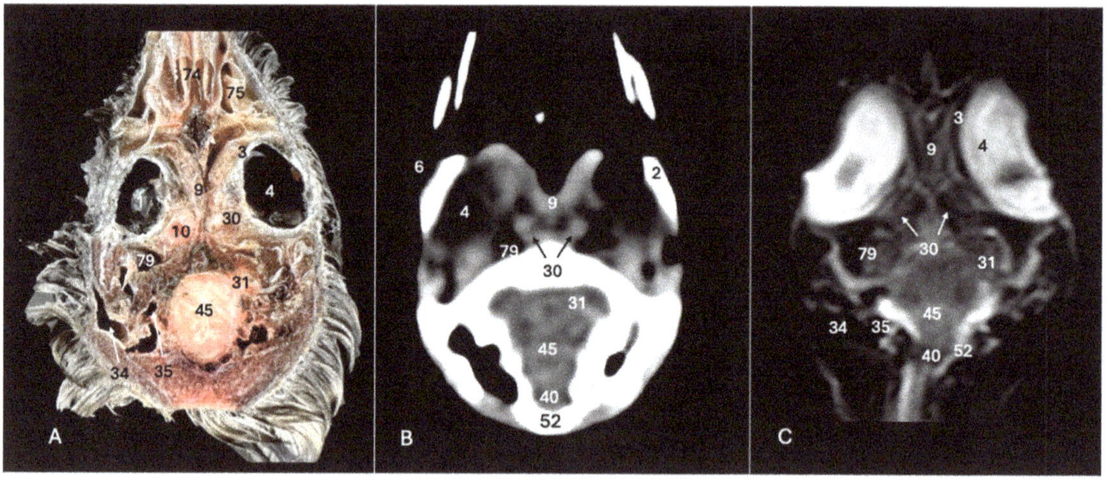

Figure 9. Anatomical cross-section (**A**), soft tissue CT window (**B**) and T2-weighted MR (**C**) dorsal images of the Cory's shearwater head at the level of the optic nerve, corresponding to line VIII in Figure 1. 1: *Lent cristali*; 2: sclerotic ossification; 3: sclera; 4: *camera vitrea bulbi*; 6: *camera anterior bulbi*; 9: *septum interorbitalis*; *10: extraocular muscle*; 30: optic nerve; 31: optical lobe; 34: external acoustic canal; 35: inner ear; 40: *medulla oblongata*; 45: *cerebellum* (body); 52: *os supraoccipitali*; 74: *concha nasalis media*; 75: *concha nasalis caudalis*; *79:* pneumatization of the temporal bone.

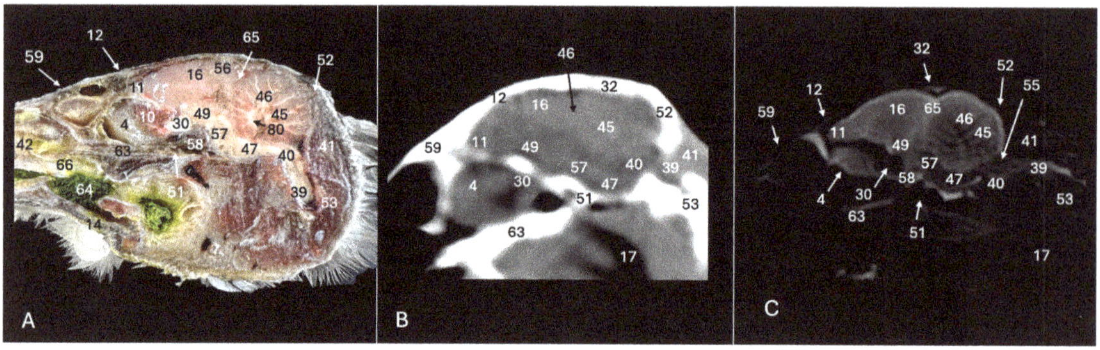

Figure 10. Anatomical cross-section (**A**), soft tissue CT window (**B**) and T2-weighted MR (**C**) sagittal images of the Cory's shearwater head at the level of the *fissura longitudinalis cerebri*. 4: *Camera vitrea bulbi*; 10: extraocular muscles; 11: olfactory bulb; 12: *os frontale*; 14: mandible; 16: cerebral hemisphere; 17: trachea; 30: optic nerve; 32: *os parietale*; 39: *medulla spinalis*; 40: *medulla oblongata*; 41: *musculus rectus* (*musculus rectus dorsalis* plus *musculus rectus ventralis* plus *musculus rectus lateralis*). 42: nasal septum; 45: *cerebellum* (body); 46: folia of *cerebellum*; 47: *pons*; 49: *diencephalon*; 51: *basis cranii*; 52: *os supraoccipitale*; 53: cervical vertebra; 55: fourth ventricle; 56: *area hipocampalis*; 57: *mesencephalon*; 58: hypophysis; 59: *os nasale*; 63: parasphenoid rostrum; 64: ingesta; 65: *glandula pinealis*; 66: maxilla; 80: internal white substance.

4. Discussion

To the authors' knowledge, this is the first description of a Cory's shearwater head using anatomical cross-sections and their corresponding CT and MR images. In avian medicine, conventional radiography has historically been a fundamental tool for diagnostic imaging, providing valuable information on musculoskeletal and respiratory processes and alterations in the coelomic cavity [31,36]. However, its utility is limited in the cranial region due to structure overlap and low resolution [30,31]. Despite modern imaging modalities presenting more promising options capable of offering precise anatomical and pathological information, their use remains challenging due to the high economic cost of these types of equipment, the need for animal sedation and possibly restraining the animal, and the duration of the examination in the case of MR studies [32–38]. Other imaging equipment, such as the micro-CT, offers images with higher spatial resolution and thinner slice thickness than those obtained with conventional CT scanners. Nonetheless, this equipment is not usually available in veterinary hospitals [45].

Throughout this study, we selected CT and MR images that better matched the anatomical cross-sections. The acquisition of adequate-resolution images of the central nervous system in this study was enhanced by the physical volumetric factor of the Cory's shearwater brain. In other words, animals with larger brain volumes tend to exhibit better resolution in the obtained images. This relationship between brain size and the quality of acquired images is particularly relevant when considering the strength of the equipment. In our study, we utilized a 1.5 Tesla magnet, which has certain limitations compared to previous research employing higher field magnets, specifically 3 and 4.7 Tesla magnets [30–36]. Our investigation has demonstrated a significant improvement in image resolution, especially when compared with other studies on seabirds using similar strength fields, such as Atlantic puffins, that present small head dimensions, leading to the more limited resolution of MR images [46]. Despite this, our results displayed images with similar resolutions to other studies on larger birds such as the red kite, common buzzard, or African grey parrot [30–36]. The significance of this finding lies in a deeper understanding of the factors influencing the quality of central nervous system images in animals. By acknowledging the relationship between brain sizes, the imaging technique utilized and the resolution obtained, we can optimize future studies to ensure greater precision in evaluating brain structures across different species.

As reported in recent studies [30,32,46], modern imaging diagnostic techniques showed adequate detail of the different regions of the avian brain. Here, both techniques allowed the observation of specific formations within the telencephalic hemispheres, such as the *vallecula telencephali*, the sagittal eminence observed as a prominence lateral to the *fissura longitudinalis cerebri*, the olfactory bulb and a prominent *hyperpallium*. This last formation is the homologue of the mammalian neocortex involved in visual capabilities and spatial cognition [47–49]. Further studies of the *hyperpallium* across multiple bird species, mainly birds of prey, should be performed to compare their development.

Concerning the Cory's shearwater mesencephalon, it was easily identified through CT and T2W MR images. Thus, the basioccipital bone was used as a landmark to visualize the optic lobes in the CT images, whereas their caudoventral situation related to the telencephalon and the identification of the tentorial process were pivotal to observing them with T2W images. These lobes were quite prominent mainly due to their visual, auditory and somatosensory functions [30,46,49]. As reported in other studies [30,50], the largest optic lobes have been identified in birds of prey. However, further investigations comparing these birds with seabirds are necessary since important developments of these structures, and optic nerves, as well as the optic chiasm, have also been observed on them [43,46].

Sagittal T2W images were essential to observing the large volume of the cerebellum, and its situation above the brain stem, which did not present a clear distinction between *pons* and *medulla oblongata,* as happens in mammals [30,43]. The combination of anatomical cross-sections and transverse T2W images displayed the cerebellar hemispheres. This finding contrasts with other studies using higher-strength field magnets, where these structures were not easily depicted [30].

As reported in other investigations performed on birds, felids and reptiles, the dorsal, transverse and sagittal anatomical sections offered intricate insights into the different elements and shapes of the avian eyeball [51–53]. It was characterized by its significant size relative to cranial volume and its lateral positioning, inherent to avian anatomy, showing a distinctive globose morphology with a subtle medial flattening. In addition, the incorporation of CT and T2W images confirmed the visualization of the vitreous chamber, the sclera, the extraocular muscles, the lens and the sclerotic ossifications. These ossifications were distinguishable in the CT images that depicted the dorsal and ventral arches. Reports of this structure have stated its association with the visual capabilities and activity patterns of seabird species [46].

In contrast to other studies performed on birds, the combination of CT and MR images was essential in the visualization of the shape and the different bones that comprise the bird skull. Employing a soft tissue CT window allowed us to distinguish its dome shape, and how it was robust and flattened compared to the skull of other seabird species [46]. These bones showed relevant pneumatization that could be related to their living and foraging ecology, mainly focused on small fishes and squids, which were captured by shallow immersions [2–4]. In addition, this CT window was quite helpful in displaying with adequate resolution the eyes and the nervous structures from the remaining tissues.

This approach has proven invaluable in evaluating the anatomical intricacies of the CNS and its related structures across a diverse range of wildlife species. From reptiles to rodents, terrestrial mammals to exotic avian species like the red kite, common buzzard and common kestrel, as well as domestic birds such as the pigeon and the African grey parrot, MRI has played a crucial role in identifying CNS components [46,52,54–57]. Unlike traditional imaging methods, modern imaging diagnostic techniques allow detailed images from various anatomical perspectives without specimen repositioning [46,51,52,54,55,58]. The unparalleled ability of MRI to discern between cranial bones and soft tissues makes it a superior tool for assessing the CNS and associated structures compared to computed tomography. However, we should highlight the limitations related to the use of carcasses, which were mainly observed in the MR images due to changes in the signal intensity of some tissues. Hence, it would be interesting to perform future investigations using adult and live specimens to assess these changes and the differences in the resolution and volume of the CNS and associated structures.

5. Conclusions

The application of these modern imaging techniques combined with anatomical cross-sections not only reveals the detailed anatomy of Cory's shearwater but also provides a platform for a better understanding of the CNS and its related structures, contributing to advancing knowledge in the fields of ornithology and comparative neuroscience.

Author Contributions: Conceptualization, J.R.J., S.D. and A.M.-E.; methodology, M.E., J.R.J., P.P.-O., N.R.-M. and A.M.-E.; investigation, J.R.J., A.M.-E., S.D., P.P.-O., N.R.-M., F.S.-C. and M.E.; resources, S.D. and F.S.-C.; writing—original draft preparation, J.R.J. and A.M.-E.; writing—review and editing, J.R.J., A.M.-E., S.D., P.P.-O., N.R.-M., F.S.-C. and M.E. All authors have read and agreed to the published version of the manuscript.

Funding: This research received no external funding.

Institutional Review Board Statement: In this study, we did not need ethical approval, since the birds arrived at our hospital to determine, through diagnostic imaging techniques, the cause of their stranding. From these studies, we obtained the information presented in our paper. As occurs with domestic mammals, informed consent from the owner allowed us to carry out this study.

Informed Consent Statement: Not applicable.

Data Availability Statement: The data supporting reported results can be found at "accedacris.ulpgc.es".

Acknowledgments: In loving memory of Alvaro Domingo Rodriguez Garcia and Maria Jose Oliva de la Nuez. We would also like to thank Ayesh Mohamad, Carmen Mingot, Concepción Mingot, Emilia Mingot, Nicolas Aquino, Marisa Mohamad and Jamal Jaber for their support and constructive comments.

Conflicts of Interest: Author Alejandro Morales-Espino was employed by the company IVC Evidensia Los Tarahales. The remaining authors declare that the research was conducted in the absence of any commercial or financial relationships that could be construed as a potential conflict of interest.

eferences

Cory's Shearwater—The Sound Approach. Available online: https://soundapproach.co.uk/species/corys-shearwater/ (accessed on 1 January 2024).

Seabird of the Month—Cory's Shearwater (Calonectris Borealis). Available online: https://www.birdlife.org/news/2021/05/31/seabird-month-corys-shearwater-calonectris-borealis/ (accessed on 1 January 2024).

Madroño, A.; González, C.; Atienza, J.C. (Eds.) Libro Rojo de las Aves de España. In *Dirección General para la Biodiversidad;* SEO-BirdLife: Madrid, Spain, 2004.

Migración y Ecología Espacial de las Poblaciones Españolas de Pardela Cenicienta. SEO BirdLife: Madrid, Spain, 2017. [CrossRef]

La Pardela Cenicienta Elegida "Ave del año 2013" | Natura | elmundo.es. Available online: https://www.elmundo.es/elmundo/2013/03/21/natura/1363873461.html (accessed on 1 January 2024).

Pardela Cenicienta, Especie Clave para la Conservación de los Mares-SEO/BirdLife. Available online: https://seo.org/pardela-cenicienta-especie-clave-para-la-conservacion-de-los-mares/ (accessed on 1 January 2024).

Lauridsen, H.; Hansen, K.; Wang, T.; Agger, P.; Andersen, J.L.; Knudsen, P.S.; Rasmussen, A.S.; Uhrenholt, L.; Pedersen, M. Inside Out: Modern Imaging Techniques to Reveal Animal Anatomy. *PLoS ONE* **2011**, *6*, e17879. [CrossRef] [PubMed]

Beuf, O.; Jaillon, F.; Saint-Jalmes, H. Small-Animal MRI: Signal-to-Noise Ratio Comparison at 7 and 1.5 T with Multiple-Animal Acquisition Strategies. *Magn. Reson. Mater. Phys.* **2006**, *19*, 202–208. [CrossRef] [PubMed]

Gómez-Díaz, E.; González-Solís, J. Geographic assignment of seabirds to their origin: Combining morphologic, genetic, and biogeochemical analyses. *Ecol. Appl.* **2007**, *17*, 1484–1498. [CrossRef] [PubMed]

Cory's Shearwater | Audubon Field Guide. Available online: https://www.audubon.org/field-guide/bird/corys-shearwater (accessed on 1 January 2024).

Navarro, J.; Kaliontzopoulou, A.; González-Solís, J. Sexual Dimorphism in Bill Morphology and Feeding Ecology in Cory's Shearwater (*Calonectris diomedea*). *Zoology* **2009**, *112*, 128–138. [CrossRef] [PubMed]

Paiva, V.H.; Guilford, T.; Meade, J.; Geraldes, P.; Ramos, J.A.; Garthe, S. Flight Dynamics of Cory's Shearwater Foraging in a Coastal Environment. *Zoology* **2010**, *113*, 47–56. [CrossRef] [PubMed]

Atchoi, E.; Mitkus, M.; Kus Vitta, P.; Machado, B.; Rocha, M.; Juliano, M.; Bried, J.L.; Rodríguez, A. Ontogenetic Exposure to Light Influences Seabird Vulnerability to Light Pollution. *J. Exp. Biol.* **2023**, *226*, jeb245126. [CrossRef] [PubMed]

González-Varo, J.P.; Rumeu, B.; Albrecht, J.; Arroyo, J.M.; Bueno, R.S.; Burgos, T.; da Silva, L.P.; Escribano-Ávila, G.; Farwig, N.; García, D.; et al. Limited Potential for Bird Migration to Disperse Plants to Cooler Latitudes. *Nature* **2021**, *595*, 75–79. [CrossRef] [PubMed]

Gearheart, G.; Kooyman, G.L.; Goetz, K.T.; McDonald, B.I. Migration Front of Post-Moult Emperor Penguins. *Polar Biol.* **2013**, *37*, 435–439. [CrossRef]

Bradarić, M.; Bouten, W.; Fijn, R.C.; Krijgsveld, K.L.; Shamoun-Baranes, J. Winds at Departure Shape Seasonal Patterns of Nocturnal Bird Migration over the North Sea. *J. Avian Biol.* **2020**, *51*, e02562. [CrossRef]

Manola, I.; Bradarić, M.; Groenland, R.; Fijn, R.; Bouten, W.; Shamoun-Baranes, J. Associations of Synoptic Weather Conditions with Nocturnal Bird Migration Over the North Sea. *Front. Ecol. Evol.* **2020**, *8*, 542438. [CrossRef]

Clairbaux, M.; Fort, J.; Mathewson, P.; Porter, W.; Strøm, H.; Grémillet, D. Climate Change Could Overturn Bird Migration: Transarctic Flights and High-Latitude Residency in a Sea Ice Free Arctic. *Sci. Rep.* **2019**, *9*, 17767. [CrossRef] [PubMed]

Piersma, T.; Gutiérrez, J.S. Bird Migration: Flying High to Avoid Overheating? *Curr. Biol.* **2021**, *31*, R955–R957. [CrossRef] [PubMed]

Galluzzo, F.G.; Cumbo, V.; Cammilleri, G.; Calabrese, V.; Pulvirenti, A.; Cicero, N.; Pantano, L.; Mascetti, A.; Lo Cascio, G.; Bacchi, E.; et al. Fatty Acids Composition of Stomach Oil of Scopoli's Shearwater (*Calonectris diomedea*) from Linosa's Colony. *Animals* **2022**, *12*, 1069. [CrossRef] [PubMed]

Amerson, A.B. Tick Distribution in the Central Pacific as Influenced by Sea Bird Movement. *J. Med. Entomol.* **1968**, *5*, 332–339. [CrossRef] [PubMed]

Gómez-Díaz, E.; González-Solís, J.; Peinado, M.A.; Page, R.D.M. Phylogeography of the Calonectris Shearwaters Using Molecular and Morphometric Data. *Mol. Phylogenet. Evol.* **2006**, *41*, 322–332. [CrossRef]

Ramírez, A.S.; Poveda, J.B.; Dijkman, R.; Poveda, C.; Suárez-Pérez, A.; Rosales, R.S.; Feberwee, A.; Szostak, M.P.; Ressel, L.; Viver, T.; et al. Mycoplasma Bradburyae Sp. Nov. Isolated from the Trachea of Sea Birds. *Syst. Appl. Microbiol.* **2023**, *46*, 126472. [CrossRef] [PubMed]

Chrastansky, A.; Callies, U.; Fleet, D.M. Estimation of the Impact of Prevailing Weather Conditions on the Occurrence of Oil-Contaminated Dead Birds on the German North Sea Coast. *Environ. Pollut.* **2009**, *157*, 194–198. [CrossRef] [PubMed]

Allen, J.R.; Thompson, A. PCBs and Organochlorine Pesticides in Shag (*Phalacrocorax dristotelis*) Eggs from the Central Irish Sea: A Preliminary Study. *Mar. Pollut. Bull.* **1996**, *32*, 890–892. [CrossRef]

Mironov, O.G. Hydrocarbon Pollution of the Sea and Its Influence on Marine Organisms. *Helgol. Mar. Res.* **1968**, *17*, 335–339. [CrossRef]

27. Camphuysen, C.J.; Heubeck, M. Marine Oil Pollution and Beached Bird Surveys: The Development of a Sensitive Monitoring Instrument. *Environ. Pollut.* **2001**, *112*, 443–461. [CrossRef] [PubMed]
28. Dunnet, G.M. Seabirds and North Sea Oil. *Philos. Trans. R Soc. Lond. B* **1987**, *316*, 513–524.
29. Stienen, E.W.M.; Courtens, W.; Van de walle, M.; Vanermen, N.; Verstraete, H. Long-Term Monitoring Study of Beached Seabirds Shows That Chronic Oil Pollution in the Southern North Sea Has Almost Halted. *Mar. Pollut. Bull.* **2017**, *115*, 194–200. [CrossRef] [PubMed]
30. Stańczyk, E.K.; Velasco Gallego, M.L.; Nowak, M.; Hatt, J.M.; Kircher, P.R.; Carrera, I. 3.0 Tesla Magnetic Resonance Imaging Anatomy of the Central Nervous System, Eye, and Inner Ear in Birds of Prey. *Vet. Radiol. Ultrasound* **2018**, *59*, 705–714. [CrossRef] [PubMed]
31. Vilaplana Grosso, F. Orthopedic Diagnostic Imaging in Exotic Pets. *Vet. Clin. Exot. Anim. Pract.* **2019**, *22*, 149–173. [CrossRef] [PubMed]
32. Fleming, G.J.; Lester, N.V.; Stevenson, R.; Silver, X.S. High field strength (4.7T) magnetic resonance imaging of hydrocephalus in an African Grey parrot (*Psittacus erithacus*). *Vet. Radiol. Ultrasound* **2003**, *44*, 542–545. [CrossRef] [PubMed]
33. de Francisco, O.N.; Feeney, D.; Armién, A.G.; Wuenschmann, A.; Redig, P.T. Correlation of Brain Magnetic Resonance Imaging of Spontaneously Lead Poisoned Bald Eagles (*Haliaeetus leucocephalus*) with Histological Lesions: A Pilot Study. *Res. Vet. Sci.* **2016**, *105*, 236–242. [CrossRef] [PubMed]
34. Delk, K.W.; Mejia-Fava, J.; Jiménez, D.A.; Kent, M.; Myrna, K.; Mayer, J.; Divers, S. Diagnostic Imaging of Peripheral Vestibular Disease in a Chinese Goose (*Anser cygnoides*). *J. Avian Med. Surg.* **2014**, *28*, 31–37. [CrossRef] [PubMed]
35. Wernick, M.B.; Dennler, M.; Beckmann, K.; Schybli, M.; Albini, S.; Hoop, R.K.; Steffen, F.; Kircher, P.; Hatt, J.M. Peripheral Nerve Sheath Tumor in a Subadult Golden Eagle (*Aquila chrysaetos*). *J. Avian Med. Surg.* **2014**, *28*, 57–63. [CrossRef] [PubMed]
36. Stauber, E.; Holmes, S.; DeGhetto, D.L.; Finch, N. Magnetic Resonance Imaging Is Superior to Radiography in Evaluating Spinal Cord Trauma in Three Bald Eagles (*Haliaeetus leucocephalus*). *J. Avian Med. Surg.* **2007**, *21*, 196–200. [CrossRef] [PubMed]
37. Behroozi, M.; Helluy, X.; Ströckens, F.; Gao, M.; Pusch, R.; Tabrik, S.; Tegenthoff, M.; Otto, T.; Axmacher, N.; Kumsta, R.; et al. Event-Related Functional MRI of Awake Behaving Pigeons at 7T. *Nat. Commun.* **2020**, *11*, 4715. [CrossRef] [PubMed]
38. Orije, J.E.M.J.; Van der Linden, A. A Brain for All Seasons: An in Vivo MRI Perspective on Songbirds. *J. Exp. Zool. A Ecol. Integr. Physiol.* **2022**, *337*, 967–984. [CrossRef] [PubMed]
39. Romagnano, A.; Shiroma, J.T.; Heard, D.J.; Johnson, R.D.; Schiering, M.R.; Mladinich, C.R.J. Magnetic resonance imaging of the brain and coelomic cavity of the domestic pigeon (*Columba livia domestica*). *Vet. Radiol. Ultrasound* **1996**, *37*, 431–440. [CrossRef]
40. Jirak, D.; Janacek, J.; Kear, B.P. A Combined MR and CT Study for Precise Quantitative Analysis of the Avian Brain OPEN. *Sci. Rep.* **2015**, *5*, 16002. [CrossRef] [PubMed]
41. Faillace, A.C.L.; Vieira, K.R.A.; Santana, M.I.S. Computed tomographic and gross anatomy of the head of the blue-fronted Amazon Parrot (*Amazona aestiva*). *Anat. Histol. Embryol.* **2021**, *50*, 192–205. [CrossRef]
42. Baumel, J.J.; Anthony, S.; King, J.E.; James, E. *Handbook of Avian Anatomy: Nomina Anatomica Avium*, 2nd ed.; Nuttall Ortinithological Club: Cambridge, MA, USA, 1993; pp. 318–467.
43. Hadden, P.W.; Ober, W.C.; Gerneke, D.A.; Thomas, D.; Scadeng, M.; McGhee, C.N.J.; Zhang, J. Micro-CT guided illustration of the head anatomy of penguins (Aves: Sphenisciformes: Spheniscidae). *J. Morphol.* **2022**, *283*, 827–851. [CrossRef] [PubMed]
44. Koenig, H.E.; Korbel, R.; Liebich, H.G.; Klupiec, C. *Avian Anatomy: Textbook and Colour Atlas*, 2nd ed.; 5m Books Ltd.: Sheffield, UK, 2016.
45. Banzato, T.; Hellebuyck, T.; Van Caelenberg, A.; Saunders, J.H.; Zotti, A. A review of diagnostic imaging of snakes and lizards. *Vet. Rec.* **2013**, *173*, 43–49. [CrossRef] [PubMed]
46. Fumero-Hernández, M.; Encinoso, M.; Melian, A.; Nuez, H.A.; Salman, D.; Jaber, J.R. Cross Sectional Anatomy and Magnetic Resonance Imaging of the Juvenile Atlantic Puffin Head (Aves, Alcidae, Fratercula Arctica). *Animals* **2023**, *13*, 3434. [CrossRef] [PubMed]
47. Medina, L.; Abellán, A.; Desfilis, E. Evolving Views on the Pallium. *Brain Behav. Evol.* **2021**, *96*, 181–199. [CrossRef] [PubMed]
48. Agarwal, A.; Sarel, A.; Derdikman, D.; Ulanovsky, N.; Gutfreund, Y. Spatial Coding in the Hippocampus and Hyperpallium of Flying Owls. *Proc. Natl. Acad. Sci. USA* **2023**, *120*, e2212418120. [CrossRef] [PubMed]
49. Evans, H.E. Avian anatomy. In *Handbook of Bird Biology*, 3rd ed.; Lovette, I.J., Fitzpatrick, J.W., Eds.; Wiley & Sons: Chester, UK, 2016; p. 219.
50. Portman, A.; Werner, S. The central nervous system. In *Biology and Comparative Physiology of Birds*; Marshall, A.J., Ed.; Academic Press: New York, NY, USA, 1961.
51. González Rodríguez, E.; Encinoso Quintana, M.; Morales Bordon, D.; Garcés, J.G.; Artiles Nuez, H.; Jaber, J.R. Anatomical Description of Rhinoceros Iguana (*Cyclura cornuta cornuta*) Head by Computed Tomography, Magnetic Resonance Imaging and Gross-Sections. *Animals* **2023**, *13*, 955. [CrossRef] [PubMed]
52. Arencibia, A.; Hidalgo, M.R.; Vázquez, J.M.; Contreras, S.; Ramírez, G.; Orós, J. Sectional Anatomic and Magnetic Resonance Imaging Features of the Head of Juvenile Loggerhead Sea Turtles (*Caretta caretta*). *Am. J. Vet. Res.* **2012**, *73*, 1119–1127. [CrossRef] [PubMed]

. Díaz Martínez, E.; Arencibia Espinosa, A.; Soler Laguía, M.; Kilroy, D.; Martínez Gomariz, F.; Casas García, D.L.; Sánchez Collado, C.; Gil Cano, F.; Jaber, J.R.; Ramírez Zarzosa, G. An Anatomical Study Using Computed Tomography, Magnetic Resonance Imaging, and Rhinoscopy of the Nasal Cavity of Domestic Cat (*Felis silvestris catus* L.) and Big Cats: Lion (*Panthera leo leo* L.), Leopard (*Panthera pardus kotiya* L.), and Cheetah (*Acinonyx jubatus jubatus* S.). *Animals* **2024**, *14*, 1172. [CrossRef] [PubMed]

. Morales-Bordon, D.; Encinoso, M.; Arencibia, A.; Jaber, J.R. Cranial Investigations of Crested Porcupine (*Hystrix cristata*) by Anatomical Cross-Sections and Magnetic Resonance Imaging. *Animals* **2023**, *13*, 2551. [CrossRef] [PubMed]

. Raduan Jaber, J.; Encinoso, M.; Morales, D.; Artiles, A.; Santana, M.; Blanco, D.; Arencibia, A. Anatomic Study of the Normal Bengal Tiger (*Panthera tigris tigris*) Brain and Associated Structures Using Low Field Magnetic Resonance Imaging. *Eur. J. Anat.* **2016**, *20*, 195–203.

. Arencibia, A.; Matos, J.; Encinoso, M.; Gil, F.; Artiles, A.; Martínez-Gomariz, F.; Vázquez, J.M. Computed Tomography and Magnetic Resonance Imaging Study of a Normal Tarsal Joint in a Bengal Tiger (*Panthera tigris*). *BMC Vet. Res.* **2019**, *15*, 126. [CrossRef] [PubMed]

. Capello, V. Diagnostic Imaging of Dental Disease in Pet Rabbits and Rodents. *Vet. Clin. Exot. Anim. Pract.* **2016**, *19*, 757–782. [CrossRef] [PubMed]

. Morales Espino, A.; Déniz, S.; Fumero-Hernández, M.; Encinoso, M.; Calabuig, P.; Conde-Felipe, M.; Jaber, J.R. A Cadaveric Study Using Anatomical Cross-Section and Computed Tomography for the Coelomic Cavity in Juvenile Cory's Shearwater (Aves, *Procellariidae, Calonectris borealis*). *Animals* **2024**, *14*, 858. [CrossRef] [PubMed]

animals

MDPI

Article

An Anatomical Study Using Computed Tomography, Magnetic Resonance Imaging, and Rhinoscopy of the Nasal Cavity of Domestic Cat (*Felis silvestris catus* L.) and Big Cats: Lion (*Panthera leo leo* L.), Leopard (*Panthera pardus kotiya* L.), and Cheetah (*Acinonyx jubatus jubatus* S.)

Elena Díaz Martínez [1], Alberto Arencibia Espinosa [2], Marta Soler Laguía [3], David Kilroy [4], Francisco Martínez Gomariz [1], Diego Luis Casas García [5], Cayetano Sánchez Collado [1], Francisco Gil Cano [1], José Raduán Jaber [2] and Gregorio Ramírez Zarzosa [1,*]

[1] Department of Anatomy and Comparative Pathological Anatomy, Veterinary Faculty, University of Murcia, 30100 Murcia, Spain; elena.diaz2@um.es (E.D.M.); f.gomariz@colvet.es (F.M.G.); scollado@um.es (C.S.C.); cano@um.es (F.G.C.)
[2] Department of Morphology, Anatomy and Embriology, Veterinary Faculty, University of Las Palmas de Gran Canaria, Transmontaña, Arucas, 35416 Las Palmas, Spain; alberto.arencibia@ulpgc.es (A.A.E.); joseraduan.jaber@ulpgc.es (J.R.J.)
[3] Department of Medicine and Surgery, Veterinary Faculty, University of Murcia, 30100 Murcia, Spain; mtasoler@um.es
[4] UCD Veterinary Science Centre, University College Dublin, D04 V1W8 Dublin, Ireland; david.kilroy@ucd.ie
[5] Centro Veterinario de Mínima Invasión Canarias CVMIC, 35019 Las Palmas de Gran Canaria, Spain; dlcasas@gmail.com
* Correspondence: grzar@um.es; Tel.: +34-868887546; Fax: +34-868884147

Simple Summary: The nasal cavity has been studied extensively in dogs. However, in cats, there are few studies of the specific anatomy of the nasal cavity, and as far as big cats are concerned, we have not currently found any studies. This work describes the anatomy of the nasal cavity of the domestic cat and three big cats (leopard, lion, and cheetah) using computed tomography, magnetic resonance, and rhinoscopy. The aim of this study is to highlight the specific anatomical characteristics of each of these species, as well as the differences between the domestic cat and its nondomestic congeners. This study highlights the advantages and disadvantages that each imaging technique provides for the study of the nasal cavity. This will help veterinary clinicians diagnose the pathologies that can affect this anatomical region and thereby improve their treatment. In addition, these big cats are in a vulnerable state, close to extinction, and better diagnosis and treatment of their pathologies would improve the survival of these species.

Abstract: The objective of this work was to study the normal anatomy of the nasal cavity of the three species of big cats (leopard, lion, and cheetah) compared to the domestic cat through the use of computed tomography, magnetic resonance imaging, and rhinoscopy. Computed tomography allowed us to clearly visualize the entire bony and cartilaginous framework that supports the nasal cavity. Magnetic resonance imaging permitted better visualization of the soft tissues of this cavity. On the other hand, rhinoscopy enabled the direct visualization of the mucosa of the vestibule and nasal cavity, which is very useful in the diagnosis of masses or foreign bodies. Furthermore, with this technique, it has been possible to observe several small orifices from the nasolacrimal duct, the pharyngeal auditory tube, and the lateral nasal gland. Computed tomography, magnetic resonance imaging, and rhinoscopy are useful tools in analysis of the anatomical characteristics of the nasal cavity in these species.

Keywords: CT; MRI; endoscopy; rhinoscopy; nasal cavity; anatomy; feline; big cats

1. Introduction

The nasal cavity is the first region of the respiratory tract. It begins with the nostrils, openings for air entry, and ends with the nasal choanae [1]. The nasal conchae are several delicate bony structures that are referred to as conchae in the rostral nasal cavity and turbinates in the caudal nasal cavity [2]. These conchae create spaces between them, the nasal meatuses (dorsal, middle, ventral, and common), through which inhaled and exhaled air flows [3]. Caudoventrally, the nasal cavity communicates with the nasopharynx through the choanae or nasopharyngeal openings [4]. The nasal cavity is divided into two halves by the nasal septum. Caudally, this septum is osseous and fuses with the cribriform plate of the ethmoid bone, and it becomes cartilaginous as it extends rostrally. In addition, the ventral part of the nasal septum is supported by the vomer bone. The cartilaginous nasal septum cannot be identified in radiographs, but it can be distinguished in computed tomography (CT) and magnetic resonance imaging (MRI). Moreover, it should be noted that cats have frontal sinuses, lateral maxillary recesses, and small sphenoidal sinuses [5].

The nostrils of carnivores are developed in a nasal plane, represented by the vertex of the nose. The vertex of the nose is usually moist; however, it lacks glands [6]. Most of the respiratory tract is covered with ciliated pseudostratified epithelium and a varying number of goblet cells containing mainly serous glands. In the caudal part of the nasal cavity (the olfactory region), the mucosa is specialized for olfaction and contains the olfactory nerve endings and glands. The submucosa of the respiratory region has several vascular plexuses. This vascularization warms the air and, by causing evaporation of the glandular secretions, adds moisture to the inhaled air [4]. Therefore, this complex architecture composed of conchae, turbinates, and meatuses serves to increase the surface area of the nasal passages, thus enhancing the main functions of olfaction, filtration, humidification, and warming of the nasal cavity [2].

The nasal cavity is the area that is most frequently affected by diseases of the upper respiratory tract. A complete knowledge of this region is of great importance, since pathologies in the nasal cavity of cats are common, among which neoplasms, infectious and non-infectious inflammatory diseases, trauma, dental diseases, and nasopharyngeal diseases predominate [2,7–9].

Since its inception, radiology has been the most widely used technique for the study of the nasal cavity. However, this technique has several drawbacks that limit and make the study of the nasal cavity difficult, and the main one is the overlapping of structures [9,10]. Nevertheless, CT and MRI are techniques that allow for a more reliable study of the nasal cavity and, therefore, its associated pathologies. Both techniques are more sensitive than radiology and more accurate in determining the extent of disease [9]. De Rycke et al. [11] reported that all the structures of the nasal cavity can be identified by both CT and MRI studies. The major advantages that CT can provide over MRI are the reduction in both the time and the cost to carry out the tomographic study [11]. In fact, CT is considered the best imaging technique for an initial evaluation of diseases of the nasal cavity [1]. In contrast to CT studies, MRI allows for the differentiation between soft tissues and fluids and the generation of multiple images without having to reposition the patient [5,9,11]. However, the use of CT or MRI in certain pathologies, such as neoplasms or chronic nasal diseases, does not allow for a clear definitive diagnosis, and it is necessary to take a tissue biopsy (performed by endoscopy) [12]. In this context, the rhinoscopy of the nasal cavity is particularly important. This imaging technique displays the internal structures of the nasal cavity, detects anatomical abnormalities such as fungal plaques and neoplasms, makes it possible to take biopsies, and aids the localization and extraction of foreign bodies [3,13]. Nevertheless, prior knowledge of the normal anatomy of the nasal cavity accelerates and simplifies an endoscopic study [14].

In addition, the leopard, the lion, and the cheetah are species that are threatened with extinction according to the Red List of Threatened Species of the International Union for Conservation of Nature (IUCN). Due to the declining population of these big cats, all three species are classified as vulnerable [15–17]. For this reason, it is important to conduct these anatomical studies and consider the respective pathologies of these species of large cats.

In the veterinary literature, there are few reports on the use of CT, MRI, and rhinoscopy in the study of the anatomy of the feline nasal cavity, since most of the published studies are based on the dog as the species of interest [1–3,14,18]. Only Conchou et al., Losonsky et al., and Gil Cano et al. [9,10,19] studied the specific anatomy of the feline nasal cavity using CT and MRI. However, to the best of the authors' knowledge, there are no publications that use rhinoscopy to study the anatomy of the feline nasal cavity. Most of the relevant published articles only studied certain specific clinical cases using CT, MRI, or rhinoscopy [13,20–29].

To date, there is no anatomical atlas of the morphology of the nasal cavity of big cats. Only a few studies have been published regarding the anatomy of the tiger head, brain, tarsal, and stifle joints using MRI [30–32]. There are just a few published studies on certain clinical cases that affect this region in large cats (nondomestic cats) [33–36]. In this context, our study provides a comprehensive anatomical atlas of the nasal cavity of the domestic cat, as well as of the leopard, lion, and cheetah, using CT, MRI, and rhinoscopy and highlights the interspecific felid structural differences. Advanced knowledge of the anatomy of the nasal cavity of these species through imaging techniques will provide veterinary clinicians with better diagnostic skills, as well as a better understanding of the respiratory physiology of each species, by knowing their anatomical differences.

2. Materials and Methods

2.1. Animals

A total of six felids were used in this study. Three big cat specimens, a lion (*Phantera leo leo*), a leopard (*Phantera pardus kotiya*), and a cheetah (*Acinonyx jubatus jubatus*), were obtained from the Terra Natura's natural park (Murcia, Spain). They were euthanized for reasons beyond our study. Before performing the necropsies, the common carotid arteries and external jugular veins were cannulated. After the necropsies, all feline heads were sectioned at the level of the C3-C4 intervertebral space. The heads were then fixed in an embalming solution (10% formaldehyde (1.5 L), glycerine (3 L), methyl alcohol (12.5 L), and phenol (1 L)) for 7 days under refrigeration. Finally, the three heads of the big cats were used for CT, MRI, and rhinoscopic studies. In addition, a crossbreed male cat cadaver (*Felis silvestris catus*), 3.8 kg and approximately 3.5 years old, euthanized for reasons not related to this study, was obtained from the Zoonoses Service of Murcia (Murcia, Spain). The cat head was used to obtain a sagittal section for the anatomical study (level sections) and to compare the anatomical structures of this region with the heads of the big cats. In this cat, the same pre-necropsy procedure was carried out as described above, but subsequently, the vasculature was injected with colored epoxy to better visualize the arterial and venous supply. Blue epoxy was used for the venous system and was injected via the external jugular veins, and red epoxy was used for the arterial system and was injected via the common carotid arteries (E20 plus BIODUR®, Biodur Products, Biodur, Heidelberg, Germany).

The CT and MRI study of the domestic cat were carried out on a pet cat with the prior consent of its owners (they were informed about the study and signed the consent form). It was a healthy 2-year-old male cat, weighing 3.5 kg. The rhinoscopic study was carried out on another cat, again after obtaining the informed consent of its owners (they were also informed about the procedure and signed the consent form). In this case, it was a 4-year-old male cat, 4 kg in weight, and without any health problems. For the CT and MRI study of the cat, the anesthetic protocol consisted of premedication with methadone (Semfortan® 10 mg/mL, Dechra, Northwich, UK) (0.1 mg/kg IV) and induction with propofol (Propofol-Lipuro 1%, Braun, Kronberg, Germany) (4–6 mg/kg IV), followed by volatile isoflurane (Forane, Abott, Chicago, IL, USA) (2–3%) in a flow of 0.8 L/min of O_2, and mechanical ventilation was used. For the rhinoscopic study, the following anesthetic protocol was used: premedication with medetomidine (Sedator® 1 mg/mL, Dechra) (0.025 mg/kg IM), methadone (Semfortan® 10 mg/mL, Dechra) (0.4 mg/kg IM), and midazolam (Dormazolam® 5 mg/mL) (0.1 mg/kg IM); induction with propofol (Propofol-

Lipuro 1%, Braun) (dose response IV), followed by volatile isoflurane (Forane, Abbott) (2%).

This study was supervised, and the research protocol was authorized by the Animal Ethical Committee of Veterinary Medicine of the University of Murcia, Spain (REGA ES300305440012 CEEA: 305/2017; extended on 25 July 2022 as project Type II).

2.2. Computed Tomography

CT was performed using a 16-row scanner (Toshiba Astelion, Toshiba Medical System, Madrid, Spain). Big cats and feline heads were placed in a ventral recumbency, with the mandibles parallel to the table using positional aids on the scanner. We used the following technical parameters for all four species: kVp 120, acquisition matrix 512 × 512, spiral pitch factor 0.625, and slice thickness 1.25 mm. A different mA was used depending on the species, ranging from 35 to 150 mAs.

The transverse original data were stored and transferred to the CT workstation. All images were analyzed using the DICOM viewer software (RadiAnt DICOM Viewer 2023.1; Medixant Co., Poznan, Poland). Images were assessed using a bone algorithm (window width, WW > 1500 UH; window level, WL: 400–500 UH).

2.3. Magnetic Resonance Imaging

MRI was acquired with a high-field (1.5 Tesla) scanner (General Electric Sigma Excite, Schenectady, NA, USA), and a human head coil with 8 channels was used. For the timing of data acquisition, the heads of the big cats and the domestic cat were placed in ventral recumbency, same as in the CT examinations.

For the MRI study of the feline nasal cavity, we used the MRI parameters shown in Table S1 (Supplementary Materials).

2.4. Rhinoscopy

A fixed rhinoscopy unit (Karl Storz®, Tuttlingen, Germany) with a camera processor (Karl Storz® image 1 hub, camera head Karl Storz® Image 1 H3 HD, a Storz power led 175) was used to obtain the rhinoscopic images of the nasal cavity of the big cats. A forward telescope (0° enlarged view, diameter 3 mm, length 14 cm, with incorporated fiber optic light transmission) or a forward-oblique telescope (30° enlarged view, diameter 2.7 mm, length 18 cm, with incorporated fiber optic light transmission) was employed for the nose depending on the width of the specimen's nasal cavity.

To carry out rhinoscopy on the domestic cat, a fixed rhinoscopy unit (Karl Storz®, Tuttlingen, Germany) with an endoscopy tower with an HD monitor (9619 NB—Karl Storz®), a camera unit (Telecam One-Chip 20212023—Karl Storz®), a video unit (Telecam SL II 202130 20—Karl Storz®), a 100 W xenon cold light source (XENON 100 SCB 201326 20—Karl Storz®), and a forward telescope (diameter 1.9 mm, 3 Fr working channels, 30° enlarged view) was used.

The rhinoscopic procedures were performed on the feline specimens following standard protocols. Each specimen was placed in ventral recumbency and facing the endoscopist. Initial image acquisition was of the external nose morphology, using an optic of 3 mm diameter and 0° vision angle. The rhinoscopic study of the four species was carried out by introducing the rhinoscope into the left nasal orifice. Following this, the nasal cavity was examined by introducing an optic of 2.7 mm and a 30° vision angle, protected by a 3 mm sheath, visualizing first the vestibule, then turning the optic to obtain a complete exposition of this part of the nasal cavity, observing the alar folds. After visualizing the nasal vestibule, the rhinoscope was introduced directly towards the common nasal meatus, which allowed us to observe structures such as the nasal conchae (cornetts), the ethmoturbinates, the maxillary recess, and the lateral nasal gland.

2.5. Anatomic Evaluation

The interpretation of the CT, MRI, and rhinoscopic study of the cats was carried out after an exhaustive consultation of various atlases of head anatomy [4,6,19,37–40]. We tried to accurately describe the anatomical structures of the nasal cavity, while always following the veterinary anatomical nomenclature.

3. Results

3.1. Computed Tomography and Magnetic Resonance Imaging

Eleven CT, as well as T1-weighted and T2-weighted MRI, figures of each of the feline species were selected based on their relevance to identify important anatomical nasal structures. The similarity of the section in the nasal cavity was prioritized for a greater resemblance between the CT and the MRI study. This is due to slight differences, which may appear within the same level section. The attenuation and signal intensity levels of the structures depending on whether a CT or MRI study was performed are identified in Table S2 (Supplementary Materials).

The anatomical level sections of the cat heads are shown in Figure 1. At the level of the nasal plane (Figure 2(A1–D3)), the development of the entire cartilaginous structure that forms the framework of the nose could be identified in the four studied feline species: dorsal and ventral lateral cartilage, lateral accessory cartilage, alar fold, and its corresponding alar groove, with an intermediate degree of attenuation and signal intensity in CT and MRI, respectively. In this context, it is worth highlighting the greater difficulty in identifying nasal cartilages on MRI compared to CT. On MRI, they are visualized as hyperintense lines with respect to the nasal wings that make up the opening of the nose. In contrast, it was observed that the conformation of the alar fold is more elongated and vertical in the lion and the cat, who have a much narrower space in the opening of the nose, than in the leopard and cheetah. Parallel folds were only identified in the leopard and the cheetah. The subnasal groove was visualized in the leopard by means of CT and MRI (Figure 2(A1–A3)) and less clearly in the cat by means of MRI (Figure 2(D3)).

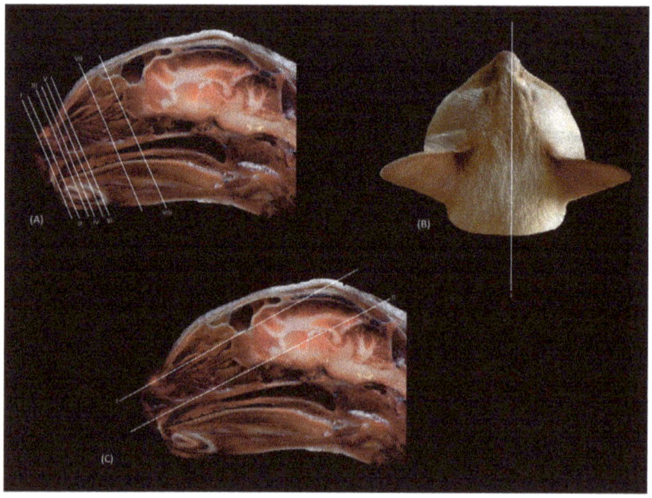

Figure 1. Approximate anatomical level sections of the cat head. Lines depict the transverse (**A**), parasagittal (**B**), and dorsal (**C**) planes. Each number represents the location for each CT and MR transverse (I to VIII), sagittal (I), and dorsal (I–II) images.

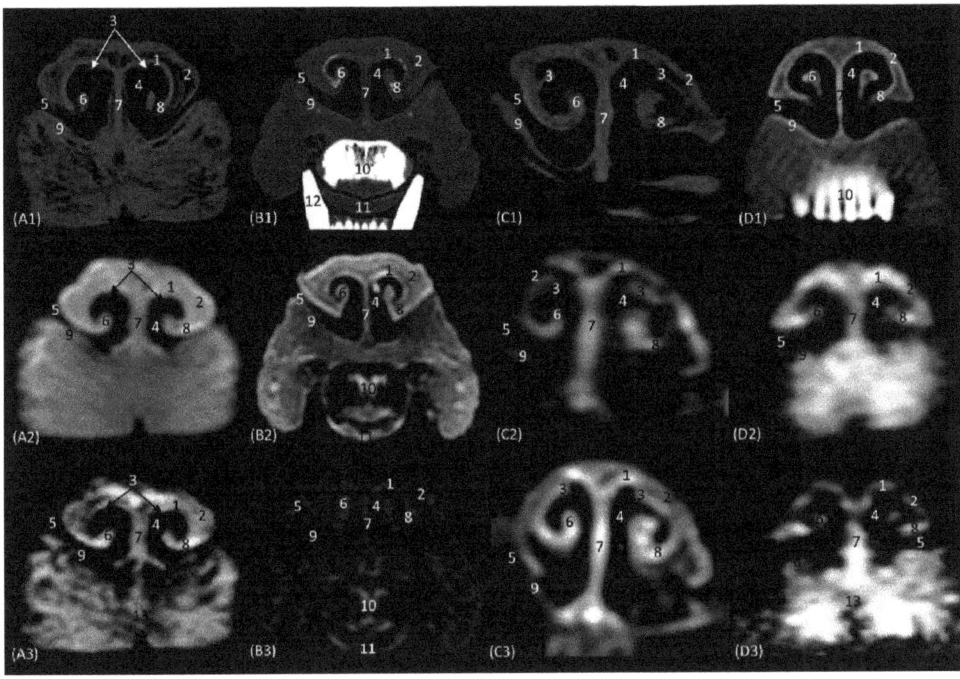

Figure 2. Representative transverse CT (**A1–D1**), T1-weighted MR (**A2–D2**), and T2-weighted MR (**A3–D3**) images at the level of the nasal orifices (nasal plane). Level section (**A**) I. Images are oriented so that the left side of the head is to the right and the dorsal is at the top. All views are rostral. (**A1–A3**): Leopard; (**B1–B3**): lion; (**C1–C3**): cheetah; and (**D1–D3**): cat. 1. Dorsal lateral nasal cartilage; 2. nasal wing; 3. parallel folds; 4. nasal orifice; 5. alar groove; 6. alar fold; 7. nasal septum: cartilage; 8. ventral lateral nasal cartilage; 9. lateral accessory nasal cartilage; 10. incisive bone and upper incisive teeth; 11. tongue; 12. lower canine tooth; 13. subnasal groove.

The next plane shows the nasal vestibule (Figure 3(A1–D3)). The ventral limit of the beginning of the nasal cavity is established at the level of the incisive bone in the four species, with the cortical and medullary cavity being well differentiated on CT. The root of the upper canine tooth is also identified, as well as the maxillary bone. Unlike the CT study, MRI permitted the observation of the incisive bone with a hypointense cortical and a hyperintense trabeculated medullary cavity due to the fat content of the bone marrow. This view shows how, in contrast to the other three species, the maxillary bone of the cheetah already begins to project dorsally as the scan sequence progresses. The incisive bone forms the hard palate at the level of the nasal vestibule. At this section level, we can more easily observe the structures making up the nasal cartilages of the four species, observing hyperintense lines on MRI with respect to the surrounding tissue. The straight fold in the four species at this level is still poorly developed. However, the alar fold is very evident in this section, being much more curved in the cheetah than in the three other felids, in which it develops more vertically. The basal fold is much more developed in the domestic cat than in the other species. On the other hand, in the domestic cat, it can be observed that, proportionally, the nasal wing and the alar fold have a greater depth, since at the same level of the nasal vestibule, these two structures can be identified only in domestic cats (Figure 3(D2,D3)). The nasal cavernous plexuses already begin to become evident in the middle of the nasal septum, being less noticeable at this level in the lion than in the rest of the species studied.

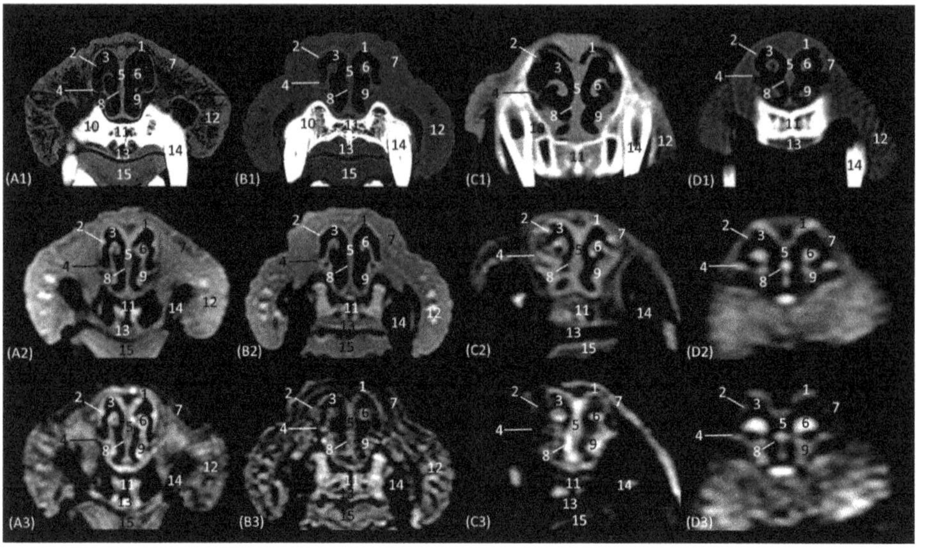

Figure 3. Representative transverse CT (**A1–D1**), T1-weighted MR (**A2–D2**), and T2-weighted MR (**A3–D3**) images at the level of the nasal vestibule. Level section (**A**) II. Transverse images are oriented so that the left side of the head is to the right, and the dorsal is at the top. All views are rostral. (**A1–A3**): Leopard; (**B1–B3**): lion; (**C1–C3**): cheetah; and (**D1–D3**): cat. 1. Dorsal lateral nasal cartilage; 2. straight fold; 3. nasal vestibule; 4. lateral accessory nasal cartilage; 5. nasal septum: cartilage; 6. alar fold; 7. nasal wing; 8. nasal cavernous plexuses; 9. basal fold; 10. maxillary bone; 11. incisive bone; 12. upper lip; 13. hard palate; 14. root of the upper canine tooth; 15. tongue.

The next section delves into the respiratory portion of the nasal cavity, since the nasal conchae are already beginning to be observed with intermediate attenuation and signal intensity in CT and MRI, respectively (Figure 4(A1–D3)). The bones that make up the floor, side walls, and roof of the nasal cavity are the incisive and vomer, maxillary, and nasal bones, respectively. At this section level, the maxillary bone can already be seen vertically, limiting the nasal cavity laterally and with the nasal bone marking its dorsal border. In the leopard and the lion, the most dorsal part is not visible, since the closure of the nasal bones extends more caudally in these two species. On MRI, both this dorsal limit and the junction between the maxillary bone and the nasal bone are difficult to differentiate, since both are hypointense. Within the nasal cavity, we can see the turbinates or conchae, the vomeronasal organ, and the nasal cavernous plexuses. The dorsal nasal concha of the big cats is identified as flat with only minimal extension into the nasal cavity. In the domestic cat, its development is proportionally much greater. The ventral nasal concha, with a dorsal and a ventral coil, is much more developed than the dorsal nasal concha, and it is slightly ossified at this level in all four species. This concha presents a similar morphology among the species of big cats, having a single start from the conchal crest of the maxillary bone while, at this level, a double start was identified in the domestic cat. These conchae create spaces in the nasal cavity: the dorsal, middle, ventral, and common meatuses, which are observed to be hypoattenuated in CT and hypointense in MRI. In the MRI study, they are not fully labelled to allow for better visualization of the rest of the structures. These spaces are described in all four species, but it is in the lion that they reach the greatest amplitude. At this level, the vomeronasal organ is very well identified; a hypoattenuated duct can be seen in the four species that we studied on CT, while on MRI, it appears slightly hypointense on T1-weighted and barely distinguishable on T2-weighted images. The vomer bone is located medially and has a groove for the support of the cartilage of the nasal septum. On CT, it is very well visualized in a hyperattenuated manner; however, on

MRI, a small external hypointense cortical line is displayed, and internally, a hyperintense area, with much greater signal intensity, is seen in the T2-weighted images. It is attached to the maxillary bone and is clearly identified on CT, but on T1 and T2-weighted MRI, this bone is hypointense and is viewed with difficulty. The morphology of this bone is a "Y" shape in the leopard and the lion and more vertical in cats, with a double Y shape and being centrally fused with a stool shape in cheetah. The nasal cavernous plexuses at this section level are well distinguished dorsally and laterally to the vomer bone and are less developed on the lateral walls of the nasal cavity.

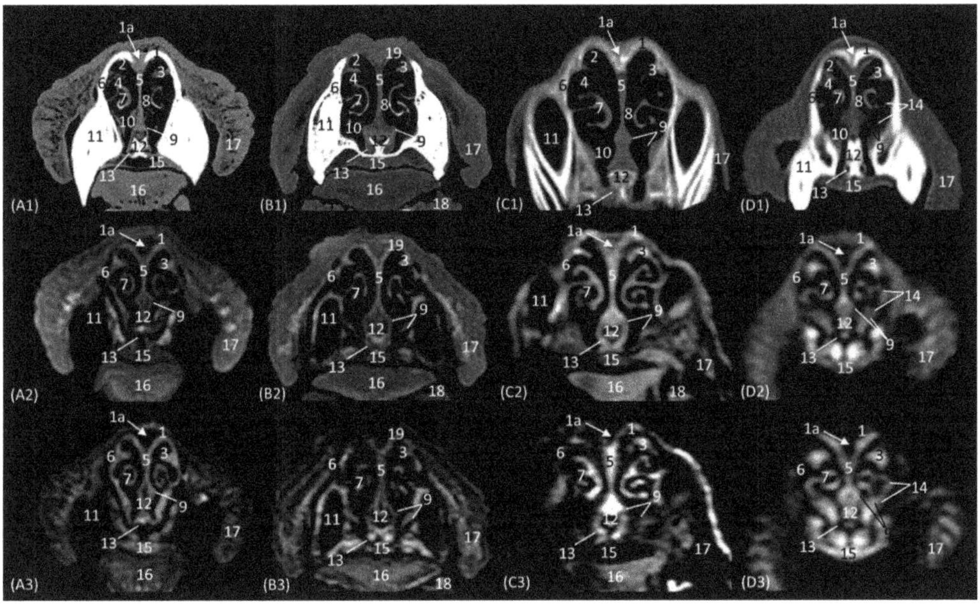

Figure 4. Representative transverse CT (**A1–D1**), T1-weighted MR (**A2–D2**), and T2-weighted MR (**A3–D3**) images at the level of the rostral portion of the respiratory part. Level section (**A**) III. Transverse images are oriented so that the left side of the head is to the right and the dorsal is at the top. All views are rostral. (**A1–A3**): Leopard; (**B1–B3**): lion; (**C1–C3**): cheetah; and (**D1–D3**): cat. 1. Nasal bone; 1a. nasal bone: nasal symphysis (arrow); 2. dorsal nasal meatus; 3. dorsal nasal concha: straight fold; 4. middle nasal meatus; 5. nasal septum: cartilage; 6. maxillary bone; 7. ventral nasal concha; 8. common nasal meatus; 9. nasal cavernous plexuses; 10. ventral nasal meatus; 11. root of the upper canine tooth; 12. vomer bone; 13. vomeronasal organ; 14. conchal crest of the maxilla; 15. hard palate; 16. tongue; 17. upper lip; 18. lower lip; 19. dorsal lateral nasal cartilage.

In the next section of the nasal cavity (Figure 5(A1–D3)), the conchae begin to appear slightly hyperattenuated on CT. The bones that make up the limits of the nasal cavity at this level are the palatine processes of the maxilla and the vomer bone at the base, the body of the maxillary bone laterally and dorsally, together with the nasal bone. It can be observed that the nasal bone is already closed dorsally in the leopard, while in the lion, it is not yet completely closed at this level. Within the nasal cavity, we can identify the dorsal nasal concha, the ventral nasal concha, and the third endoturbinate, as well as the nasal cavernous plexuses and the vomeronasal organ. At this level in CT, it is observed that the dorsal nasal concha is hyperattenuated in the leopard and the cheetah and slightly in the cat, but in the lion, the tissue continues to show an intermediate attenuation. With MRI, it is more difficult to identify this ossification, but a more hyperintense signal can be observed in the more cranial sections. This bony attachment of the dorsal nasal concha to the lateral wall of the nasal cavity is represented by the ethmoidal crest of the nasal bone.

At this level, in addition, the maxillary recess begins to be observable in the four species. The nasal cavernous plexuses continue to be identified in all four species, which is also evident in the ventrolateral wall, with less development in the lion and the cheetah. The third endoturbinate also begins to appear at this section level between the dorsal and the ventral nasal concha in the leopard, the cheetah, and the cat, but not yet in the lion. The vomeronasal organ in the leopard is a large-caliber duct, unlike in the cheetah and the domestic cat. It is topographically next to the long central stem of the vomer bone, which is concave and supports the base of the cartilage of the nasal septum, while in the lion, the concavity of the groove is formed without a bone stem from the maxillary bone.

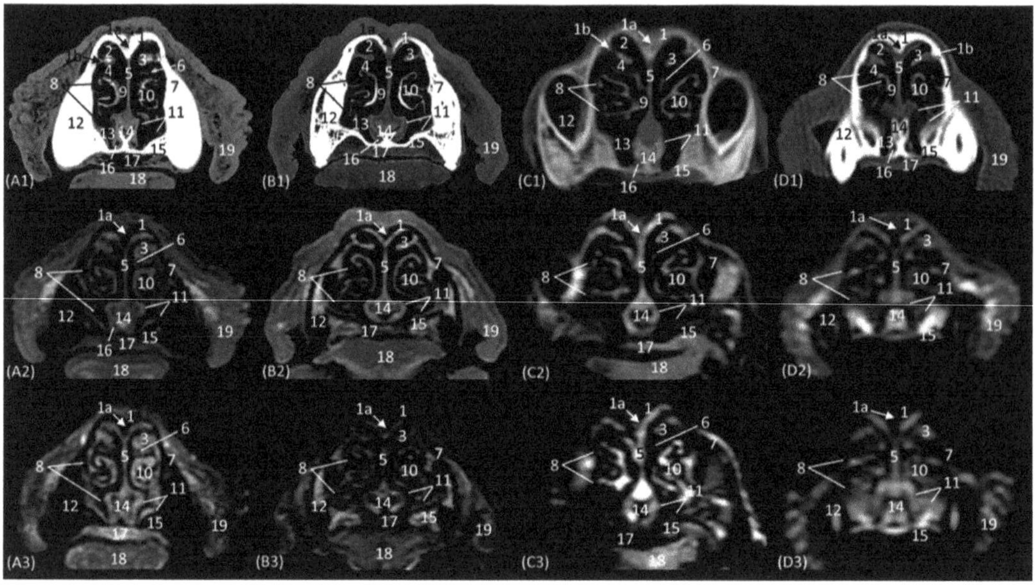

Figure 5. Representative transverse CT (**A1–D1**), T1-weighted MR (**A2–D2**), and T2-weighted MR (**A3–D3**) images at the level of the dorsal and ventral nasal conchae (rostral part). Level section (**A**) IV. Transverse images are oriented so that the left side of the head is to the right and the dorsal is at the top. All views are rostral. (**A1–A3**): Leopard; (**B1–B3**): lion; (**C1–C3**): cheetah; and (**D1–D3**): cat. 1. Nasal bone; 1a. nasal bone: nasal symphysis (arrow); 1b. nasal bone: ethmoidal crest (arrow); 2. dorsal nasal meatus; 3. dorsal nasal concha; 4. middle nasal meatus; 5. nasal septum: cartilage; 6. 3rd endoturbinate; 7. maxillary bone: body; 8. maxillary recess; 9. common nasal meatus; 10. ventral nasal concha; 11. nasal cavernous plexuses; 12. root of the upper canine tooth; 13. ventral nasal meatus; 14. vomer bone; 15. maxillary bone: palatine process; 16. vomeronasal organ; 17. hard palate; 18. tongue; 19. upper lip.

The next level is represented in Figure 6(A1–D3). Figure 6(C1) shows the section where the external plate of the frontal bone ventrally relates to the ethmoid bone and frontal sinus in the cheetah, since at this level in this species, the nasal bone is no longer sectioned. In the nasal cavity, it is worth highlighting the presence of the infraorbital canal. It can be identified both on CT and MRI in a rounded area that is hypoattenuated and hypointense, respectively, with respect to the maxillary bone. In this section, the third endorturbinate is observed, entering and intertwining between the dorsal and ventral nasal concha. It is evident again in these images that the nasal conchae and endoturbinates in the lion are much thinner than in the other three species, resulting in much wider meatuses. In this section, we can, again, see how the ventral nasal concha in the domestic cat has two bony starts from the maxilla, unlike the big cat species, which only have one.

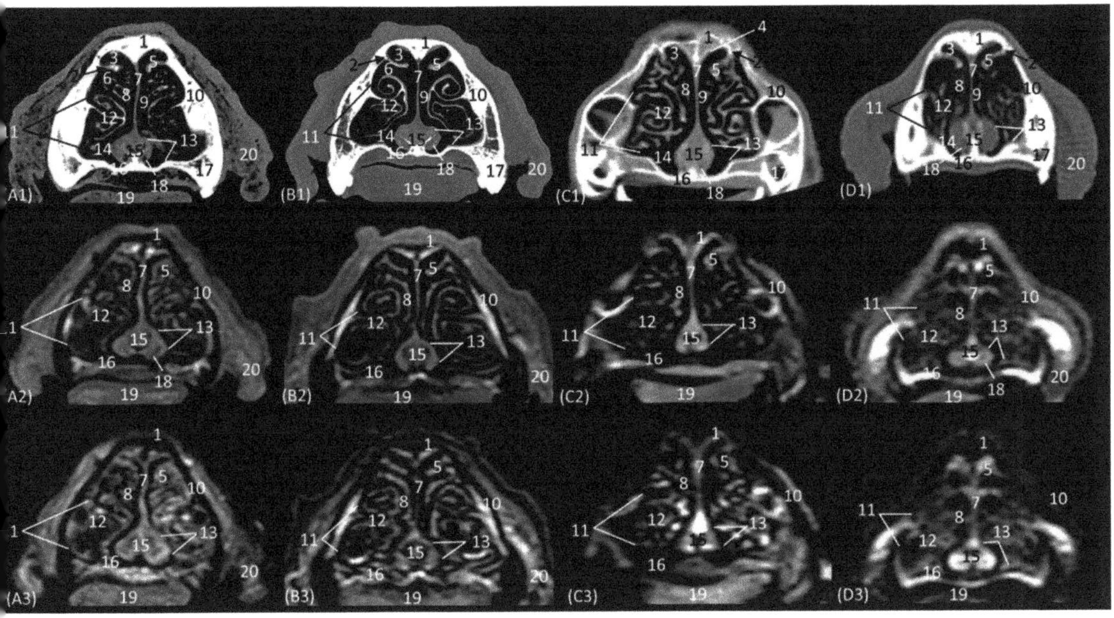

Figure 6. Representative transverse CT (**A1–D1**), T1-weighted MR (**A2–D2**), and T2-weighted MR (**A3–D3**) images at the level of the dorsal and ventral nasal conchae (middle part) and maxillary recess. Level section (**A**) V. Images are oriented so that the left side of the head is to the right and the dorsal is at the top. All views are rostral. (**A1–A3**): Leopard; (**B1–B3**): lion; (**C1–C3**): cheetah; and (**D1–D3**): cat. 1. Nasal bone; 2. nasal bone: ethmoidal crest (arrow); 3. dorsal nasal meatus; 4. ethmoid bone: tectorial plate; 5. dorsal nasal concha; 6. middle nasal meatus; 7. nasal septum: cartilage; 8. 3rd endoturbinate; 9. common nasal meatus; 10. maxillary bone: body; 11. maxillary recess; 12. ventral nasal concha; 13. nasal cavernous plexuses; 14. ventral nasal meatus; 15. vomer bone; 16. maxillary bone: palatine process; 17. infraorbital canal; 18. vomeronasal organ; 19. tongue; 20. upper lip.

A more caudal section of the nasal cavity is depicted in Figure 7(A1–D3). The bony framework that supports the nasal cavity at this level is composed dorsally of the nasal bone (only in the lion, as it extends more caudally), the external lamina of the frontal bone (squama), the maxillary bone (palatine process), and the vomer. In addition, the zygomatic bone is observed in the lion and in the domestic cat. At this level, the tectorial plate of the ethmoid bone can already be seen in the leopard and cheetah, although it is not yet visible in the lion, delimiting the frontal sinus ventrally, which also occurs in the domestic cat in a more rostral view (Figure 6(D2,D3)). In the lion, the frontal sinus develops much further caudally than in the other species. In addition, only in felids does the frontal sinus present a lateroventral opening towards the nasal cavity. Moreover, the bones of the roof (nasal and frontal bones) have a much greater thickness in lions than in the other species. The vomer bone in this plane already appears in its full extension: in large cats, it is composed of two long vertical arms and a short trunk, while in the domestic cat, it develops with a much longer trunk and shorter arms. In the nasal cavity, we observed the dorsal, middle, and ventral nasal conchae; the third endoturbinate; the vomeronasal organ; the cavernous plexuses; and the infraorbital canal. The vomeronasal organ was observed with a wide duct in the cheetah and a small duct in the leopard, and a duct was not observed in the lion or the cat. In this cross-section, the middle nasal concha is identified for the first time but is camouflaged by the extensive development of the third endoturbinate at this level. The ventral nasal concha shows ventral coiling and is dorsally flattened in the leopard. In the lion, the dorsal and ventral conchae are large but narrow, and in cheetahs and cats, they are smaller.

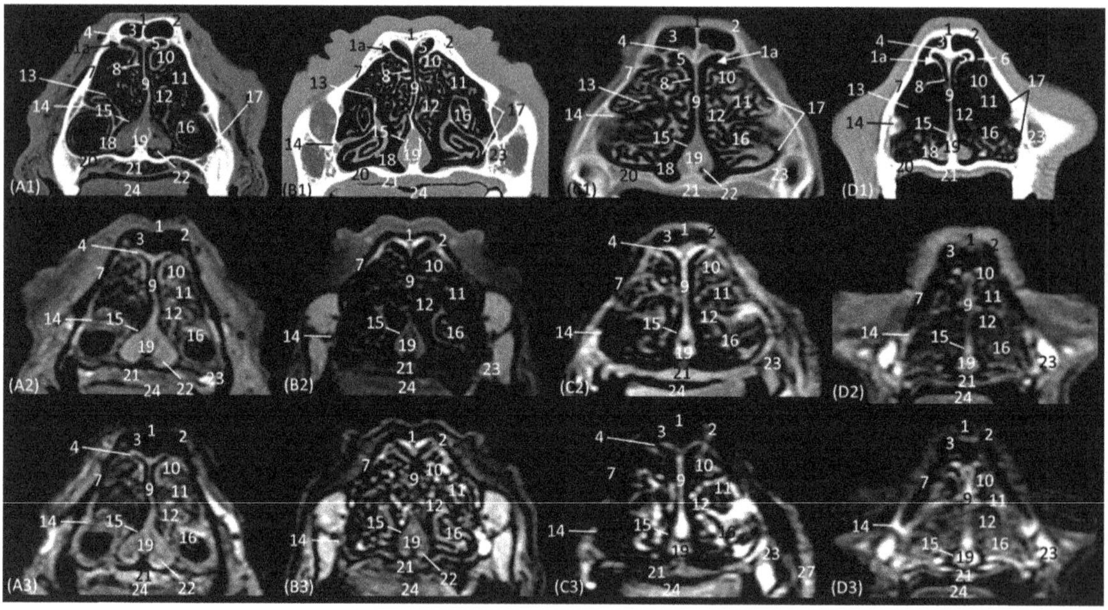

Figure 7. Representative transverse CT (**A1–D1**), T1-weighted MR (**A2–D2**), and T2-weighted MR (**A3–D3**) images at the level of the dorsal and ventral nasal conchae (caudal part). Level section (**A**) VI. Transverse images are oriented so that the left side of the head is to the right and the dorsal is at the top. All views are rostral. (**A1–A3**): Leopard; (**B1–B3**): lion; (**C1–C3**): cheetah; and (**D1–D3**): cat. 1. Nasal bone; 2. frontal bone: squama (external plate); 3. frontal sinus; 4. ethmoid bone: tectorial plate; 5. dorsal nasal meatus; 6. frontal sinus: opening; 7. maxillary bone: body; 8. common nasal meatus; 9. nasal septum: cartilage; 10. dorsal nasal concha; 11. middle nasal concha; 12. 3rd endoturbinate; 13. middle nasal meatus; 14. conchal crest of maxillary bone; 15. nasal cavernous plexuses; 16. ventral nasal concha; 17. maxillary recess; 18. ventral nasal meatus; 19. vomer bone; 20. maxillary bone: palatine process; 21. hard palate; 22. vomeronasal organ; 23. infraorbital canal; 24. tongue.

The next section level is shown in Figure 8(A1–D3), which allow the zygomatic bone to be identified in all species. The bony structures that make up the nasal cavity at this level is established dorsolaterally by the external and orbital surface of the frontal bone, the lacrimal bone, and the zygomatic bone; ventrally by the vomer bone, as well as the palatine bone; ventrally to the frontal sinus by the tectorial plate of the ethmoid bone; and medially by the cartilage of the nasal septum. This section also highlights the significant thickness of the frontal bone in lions, which is unlike the other feline species. It is observed to be hyperattenuated on CT, while on MRI, it is observed with a hypointense cortical with a hyperintense medullary cavity. Furthermore, at this level, the frontal sinus does not yet appear in the lion, while in the rest of the species, the bony septum of the frontal sinus can already be seen, and inside the nasal cavity, the first, second, and third ectoturbinates are already identifiable. The nasal septum is already ossified at this level, since it appears to be hyperattenuated on CT and hyperintense on MRI, although only very slightly in the lion. The bony component of the hard palate at this section is provided by the palatine bone. Within the nasal cavity at this level, we highlight the three ectoturbinates, the dorsal, middle, and ventral nasal conchae, the third endoturbinate, the vomeronasal organ, the nasal cavernous plexuses, and the lateral nasal gland. The ventral nasal concha is already identified as very small at this level. However, the third endoturbinate reaches its greatest development in this view. The tectorial plate, which occurred in a more rostral view in the cat, also joins more ventrally in the leopard and cheetah to leave the necessary space for the development of the ectoturbinates. From the dorsal to the ventral, the first ectoturbinate develops more laterally, the second more medially, and the third again more laterally. Also notable at this section is the identification of the lateral nasal gland in the four species in the maxillary recess, with an intermediate attenuation in CT and an intermediate signal intensity in MRI.

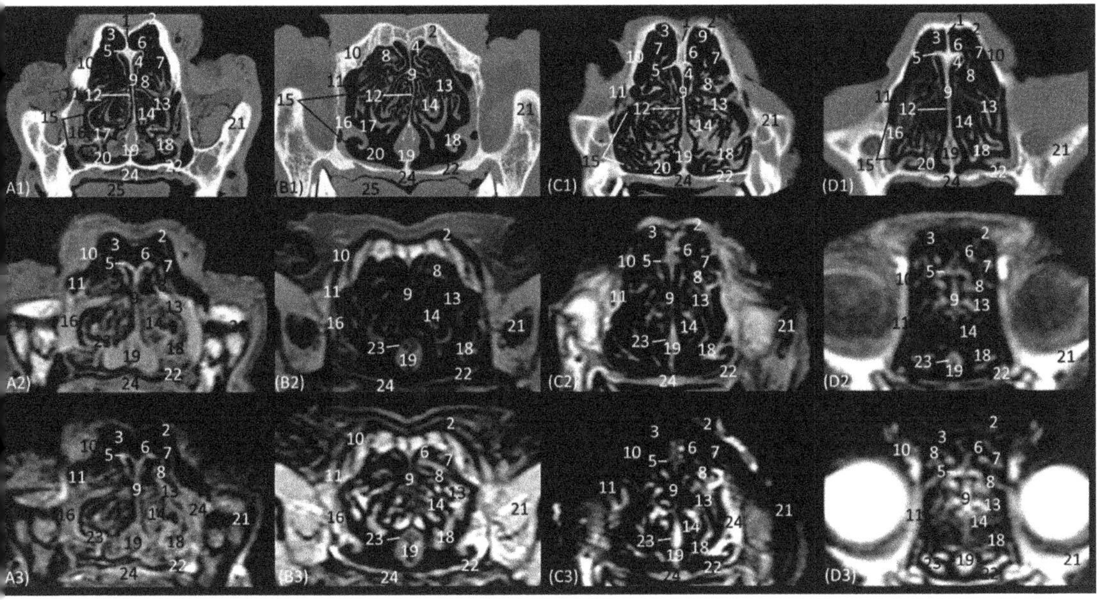

Figure 8. Representative transverse CT (**A1–D1**), T1-weighted MR (**A2–D2**), and T2-weighted MR (**A3–D3**) images at the level of the ethmoid labyrinth and lateral nasal gland. Level section (**A**) VII. Transverse images are oriented so that the left side of the head is to the right and the dorsal is at the top. All views are rostral. (**A1–A3**): Leopard; (**B1–B3**): lion; (**C1–C3**): cheetah; and (**D1–D3**): cat. 1. Frontal sinus: septum; 2. frontal bone: squama (external plate); 3. frontal sinus; 4. dorsal nasal meatus; 5. ethmoid bone: tectorial plate; 6. 2nd ectoturbinate; 7. 3rd ectoturbinate; 8. dorsal nasal concha; 9. nasal septum: cartilage/ethmoid bone (perpendicular plate); 10. frontal bone: orbital surface; 11. lacrimal bone: orbital surface; 12. common nasal meatus; 13. middle nasal concha; 14. 3rd endoturbinate; 15. maxillary recess; 16. lateral nasal gland; 17. middle nasal meatus; 18. ventral nasal concha; 19. vomer bone; 20. ventral nasal meatus; 21. zygomatic bone; 22. palatine bone; 23. nasal cavernous plexuses; 24. hard palate; 25. tongue.

The last transverse plane of the nasal cavity from this CT and MRI study is presented in Figure 9(A1–D3). The bony part of the nasal cavity at this level is represented dorsally by the external plate of the frontal bone; laterally by the orbital surface of the ethmoid bone and the wing of the presphenoid bone; and ventrally by the basal plate of the ethmoid bone, the vomer, and the palatine bone. In addition, ventral to the frontal sinus, the tectorial plate of the ethmoid bone was observed, as was, medially, the bony nasal septum. Ventrally to the nasal cavity, the choanae were observed. The vomer bone ends at this level, and it differs among the species. In the leopard and the lion, the vomer has an oval morphology, while in the cheetah, it is arrow-shaped, and in the domestic cat, it is smaller and rounded. The nasal septum, at this section level, can already be observed as ossified in the four species, hyperattenuated in CT, and in MRI, it has a higher signal intensity than in the more rostral sections. The nasal cavity of this section is made up of three ectoturbinates, the dorsal and middle nasal concha, and the third and fourth endoturbinates. The choanae were observed in all four felids and were delimited dorsally by the basal plate of the ethmoid bone and ventrally by the perpendicular and horizontal plate of the palatine bone. In this last plane, the frontal sinus was seen in the lion, along with the development of the three ectoturbinates, which at this level appeared to be larger in all four species. Furthermore, at this level, we can already identify the fourth endoturbinate in the most ventral part of the nasal cavity.

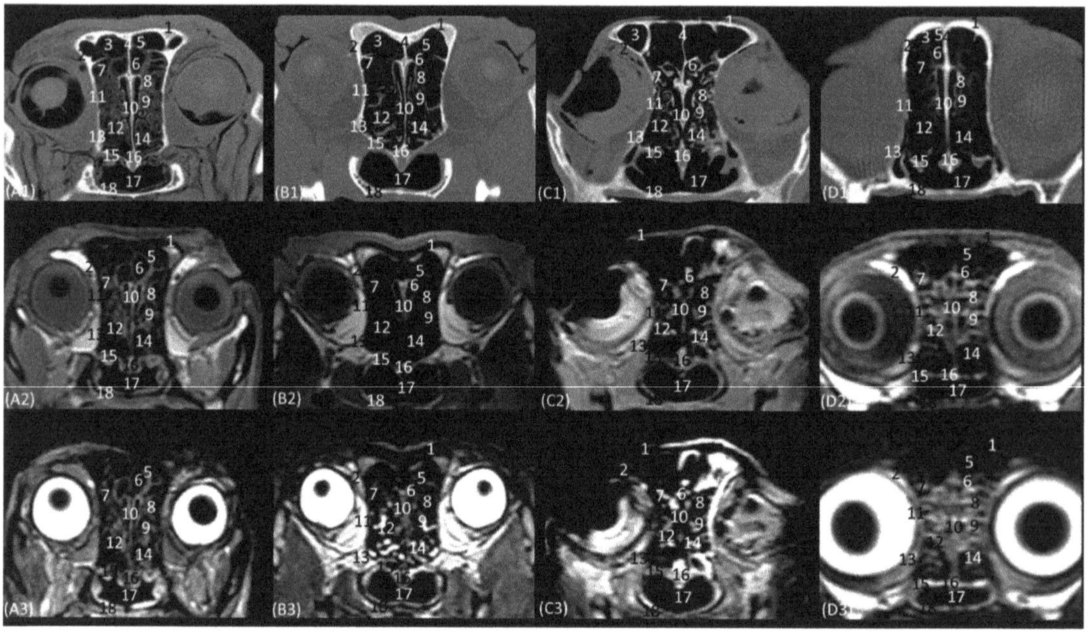

Figure 9. Representative transverse CT (**A1–D1**), T1-weighted MR (**A2–D2**), and T2-weighted MR (**A3–D3**) images at the level of the frontal sinuses and choana. Level section (**A**) VIII. Transverse images are oriented so that the left side of the head is to the right and the dorsal is at the top. All views are rostral. (**A1–A3**): Leopard; (**B1–B3**): lion; (**C1–C3**): cheetah; and (**D1–D3**): cat. 1. Frontal bone: squama (external plate); 2. frontal bone: orbital surface; 3. frontal sinus; 4. frontal sinus: septum; 5. 1st ectoturbinate. 6. 2nd ectoturbinate; 7. 3rd ectoturbinate; 8. dorsal nasal concha; 9. middle nasal concha; 10. nasal septum: ethmoid bone (perpendicular plate); 11. ethmoid bone: orbital surface; 12. 3rd endoturbinate; 13. presphenoid bone: wing; 14. 4th endoturbinate; 15. ethmoid bone: basal plate; 16. vomer bone; 17. choana; 18. palatine bone: horizontal and perpendicular plate.

Figure 10(A1–D3) represents a sagittal plane of the nasal cavity using CT and MRI. A parasagittal plane was selected to better identify the ethmoidal labyrinth. The entire skull and nasal cavity can be clearly seen. Thus, dorsally and from the rostral to the caudal, we can identify the nasal bone and the external plate of the frontal bone, with the frontal sinus caudally limiting the internal plate of the frontal bone and the tectorial plate of the ethmoid bone. At the caudal limit of the nasal cavity, we can see the cribriform plate of the ethmoid bone and the presphenoid bone with its sinus, and ventral to this, we can see the palatine and the vomer bones and the palatine processes of the maxillary and incisive bones. The rostral limit of the nasal cavity is determined ventrally by the body of the incisive bone, laterally by the nasal processes of the incisive bone, dorsally by the rostral end of the nasal bones, and medially by the cartilage of the nasal septum. The bones appeared to be hyperattenuated on CT, while on MRI, the thin hypointense cortex surrounding the hyperintense medullary bone was visualized throughout all the bones. Within the frontal sinus, the first, second, and third ectoturbinates were observed. In the nasal cavity, we can identify the straight, alar, and basal folds; the dorsal, middle, and ventral nasal conchae; the three ectoturbinates; the third and the fourth endoturbinate; the olfactory bulb; and the nasopharynx. In this view, we can see the beginnings and the course of the dorsal, middle, and ventral nasal conchae, with the latter remaining at more rostral levels and being caudally overlapped by the third endoturbinate, which is greatly expanded in all four species. The nasal conchae were hyperattenuated and slightly hyperintensified at the caudal level due to their bony base. It was also possible to identify more caudally the sphenoid sinus from which the fourth endoturbinate projects. In this plane, the beginning and the entire extension of the nasopharynx were also very well visualized, being observed to be hypoattenuated on CT and hypointense on MRI throughout its entire length. On the other hand, in this section, the straight, alar, and basal folds were clearly identified from dorsal to ventral, observed with attenuation and intermediate intensity.

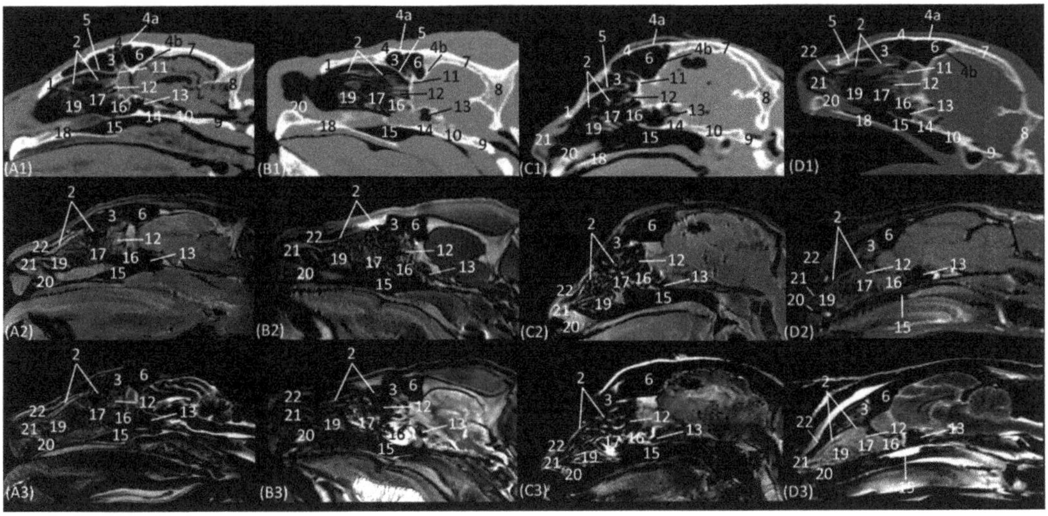

Figure 10. Representative parasagittal multiplanar reconstruction (MPR) CT (**A1–D1**), T1-weighted MR (**A2–D2**), and T2-weighted MR (**A3–D3**) images cutting off the nasal conchae, frontal, and sphenoidal sinuses. Level section (**B**) I. Parasagittal images are oriented so that the rostral part is to the left and the dorsal is at the top. All views are left lateral. (**A1–A3**): Leopard; (**B1–B3**): lion; (**C1–C3**): cheetah; and (**D1–D3**): cat. 1. Nasal bone; 2. dorsal nasal concha; 3. 1st, 2nd, and 3rd ectoturbinates; 4. frontal bone: 4a frontal bone: external plate; 4b. frontal bone: internal plate; 5. ethmoid bone: tectorial plate; 6. frontal sinus; 7. parietal bone; 8. occipital bone: squama; 9. occipital bone: basilar part; 10. basisphenoid bone; 11. ethmoid bone: cribiform plate; 12. middle nasal concha; 13. sphenoidal sinus; 14. presphenoid bone; 15. choana; 16. 4th endoturbinate; 17. 3rd endoturbinate; 18. incisive and maxillary bones: palatine process; 19. ventral nasal concha; 20; basal fold; 21. alar fold; 22. straight fold.

The horizontal or coronal view of the nasal cavity by CT and MRI is represented in Figure 11(A1–D3) and Figure 12(A1–D3). The first dorsal section (I) (Figure 11(A1–D3)) of the nasal cavity was delimited rostrally by the incisive bone, laterally by the maxillary bone, also partially involving the zygomatic and lacrimal bones, as well as the orbital surface of the frontal bone and its zygomatic process, and caudally by the cribriform plate of the ethmoid bone. The nasal septum can be seen as it begins rostrally, appearing as soft tissue due to its cartilaginous nature, and finally ends up ossifying more caudally. In the nasal cavity, we can highlight the alar fold, the ventral nasal concha, and the third endoturbinate. The alar fold can be observed in all four species, as it is part of the ventral nasal concha. The third endoturbinate embraces the ventral nasal concha and is well developed, occupying a large part of the nasal cavity. It was fixed caudally to the cribriform plate of the ethmoid bone. The frontal sinus was only identified in this plane on CT in the domestic cat and on MRI in the lion due to the slight inclination of some sections. The frontal sinus of the lion was divided by the bony septum of the frontal sinus.

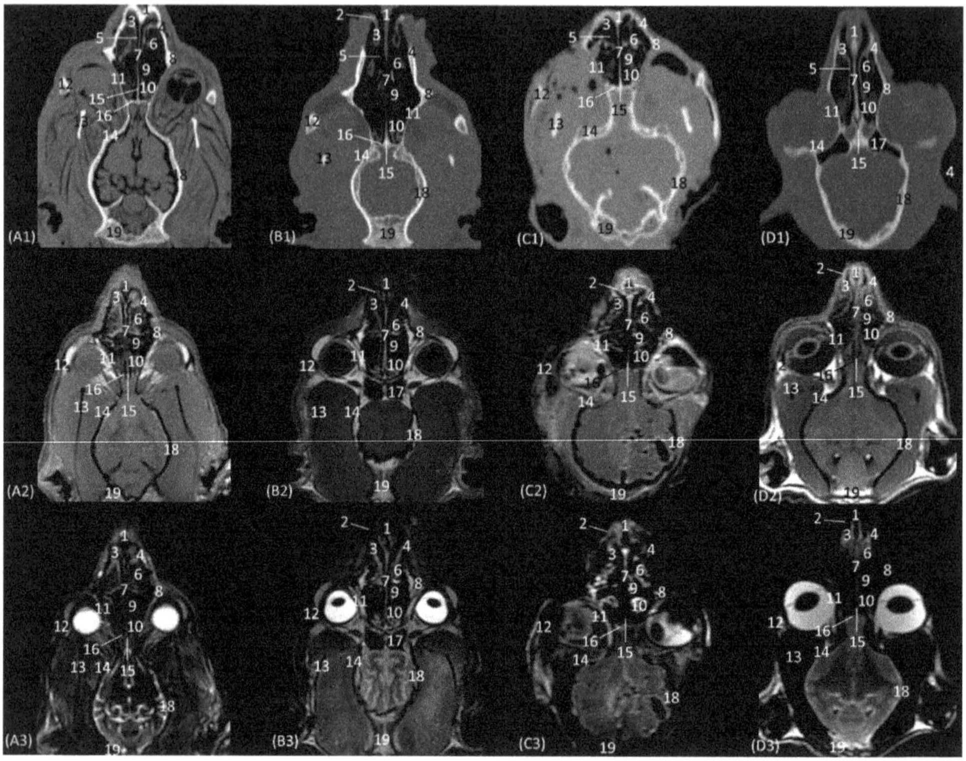

Figure 11. Representative dorsal multiplanar reconstruction (MPR) CT (**A1–D1**), T1-weighted MR (**A2–D2**), and T2-weighted MR (**A3–D3**) images at the level of the ventral nasal concha and perpendicular plate of the ethmoid bone. Level section (**C**) I. Dorsal images are oriented so that the rostral part is to the top. All views are dorsal. (**A1–A3**): Leopard; (**B1–B3**): lion; (**C1–C3**): cheetah; and (**D1–D3**): cat. 1. Incisive bone; 2. ventral lateral nasal cartilage; 3. alar fold; 4. maxillary bone; 5. common nasal meatus; 6. ventral nasal concha; 7. nasal septum: cartilage; 8. zygomatic bone: frontal process; 9. 3rd endoturbinate; 10. 4th endoturbinate; 11. frontal bone: orbital surface; 12. zygomatic bone; 13. mandible: ramus; 14. presphenoid bone: wing; 15. ethmoid bone: perpendicular plate; 16. ethmoid bone: cribriform plate; 17. frontal sinus; 18. temporal bone: squamous part; 19. occipital bone: squama.

Figure 12(A1–D3) represents a ventral horizontal view of the nasal cavity that sections the entire zygomatic arch, the temporal process of the zygomatic bone, as well as the zygomatic process of the temporal bone, although the latter is observed only in the leopard and the cheetah in the study by CT. The maxillary bone was left without rostral continuation in leopard, lion, and cat while in the cheetah, it continued with the incisive bone rostrally. In the nasal cavity, the alar fold, the ventral nasal concha, the third and the fourth endoturbinates, the nasal cavernous plexuses, and the lateral nasal gland were observed. The alar fold was still identified at this level, but in the lion, the basal fold was also visible. The lateral nasal gland was visible in this section in all four species in an area of soft tissue close to the lacrimal bone, which was identified as hypoattenuated on CT and hypointense on MRI with respect to the bone. On the other hand, the sphenoid sinus can be identified in the leopard and the cheetah, although it does not appear in the lion or the cat. Furthermore, at this level and caudal to the fourth endoturbinate, we can identify the sphenoid sinus with a thin septum that was hyperattenuated on CT and hyperintense on MRI in the leopard, cheetah, and cat, but it was not evident in the lion on both CT and MRI, nor in the cat on CT. Finally, unlike the other species, the nasal cavernous plexuses were highly developed in the cat and a little in the lion, and they were located in the most rostral part of the nasal septum.

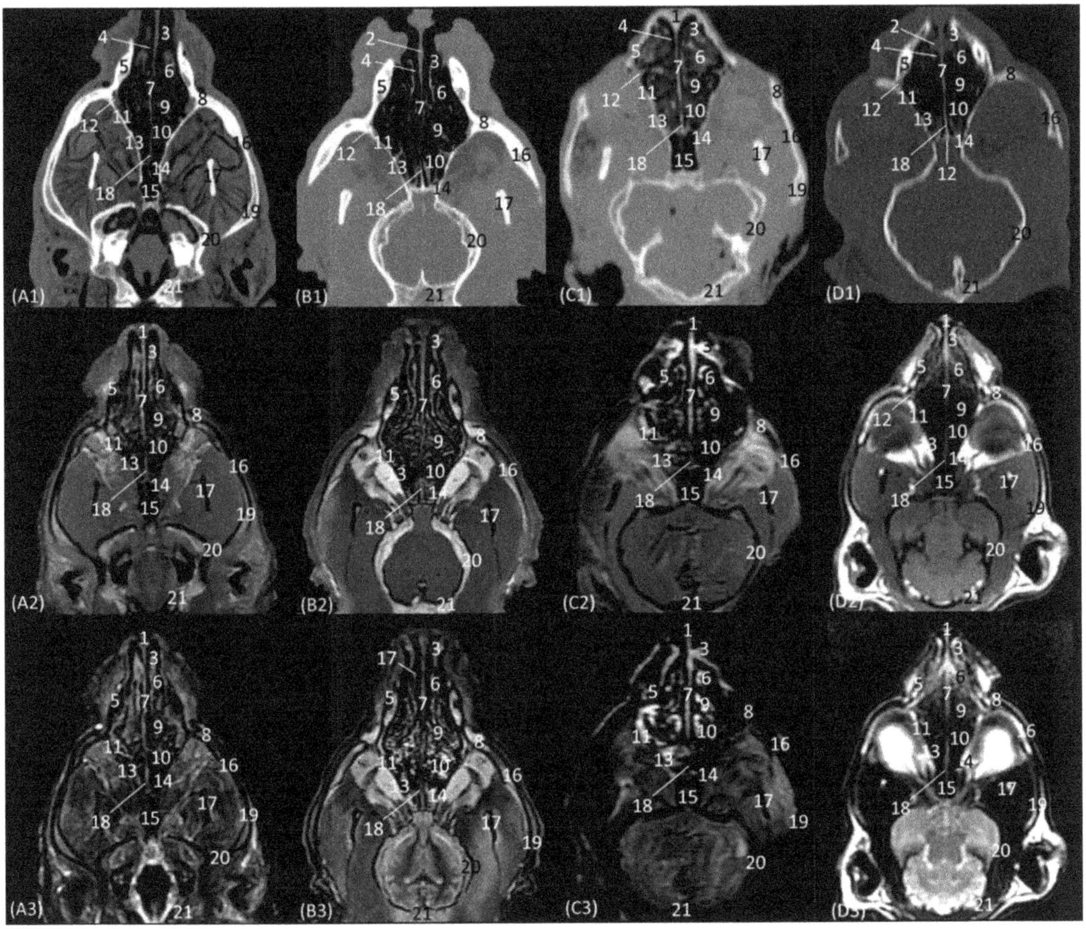

Figure 12. Representative dorsal multiplanar reconstruction (MPR) CT (**A1–D1**), T1-weighted MR (**A2–D2**), and T2-weighted MR (**A3–D3**) images at the level of the 3rd and 4th endoturbinates. Level section (**C**) II. Dorsal images are oriented so that the rostral part is to the top. All views are dorsal. (**A1–A3**): Leopard; (**B1–B3**): lion; (**C1–C3**): cheetah; and (**D1–D3**): cat. 1. Incisive bone; 2. nasal cavernous plexuses; 3. alar fold; 4. common nasal meatus; 5. maxillary bone; 6. ventral nasal concha; 7. nasal septum: cartilage; 8. zygomatic bone; 9. 3rd endoturbinate; 10. 4th endoturbinate; 11. lacrimal bone; 12. lateral nasal gland; 13. presphenoid bone: wing; 14. basisphenoid bone: wing; 15. sphenoidal sinus; 16. zygomatic bone: temporal process; 17. mandible: ramus; 18. ethmoid bone: perpendicular plate; 19. temporal bone: zygomatic process; 20. temporal bone: squamous part; 21. occipital bone: squama.

3.2. *Rhinoscopy*

The rhinoscopic study of the nasal cavity is seen in Figures 13–18. Figure 13 shows the nasal plane (the external nose) of the four felids. The vertex of the nose of the leopard and lion resemble each other in morphology, exhibiting an inverted triangle shape. However, in the cheetah, this shape is much more rectangular, and in cats, it is more elongated. The subnasal groove can be observed in the four species ventrally to the external nose. The alar fold was observed in all four species, and in the leopard and the lion, it extends lateromedially to the middle of the external nose, while in the cheetah, it runs almost the entire extent of the external nose; in cats, however, it ends more dorsally. The development of this fold also surrounds and shapes the nostrils. In the cheetah, this nasal orifice is larger than in the rest of the felids that we studied due to the conformation of its alar fold.

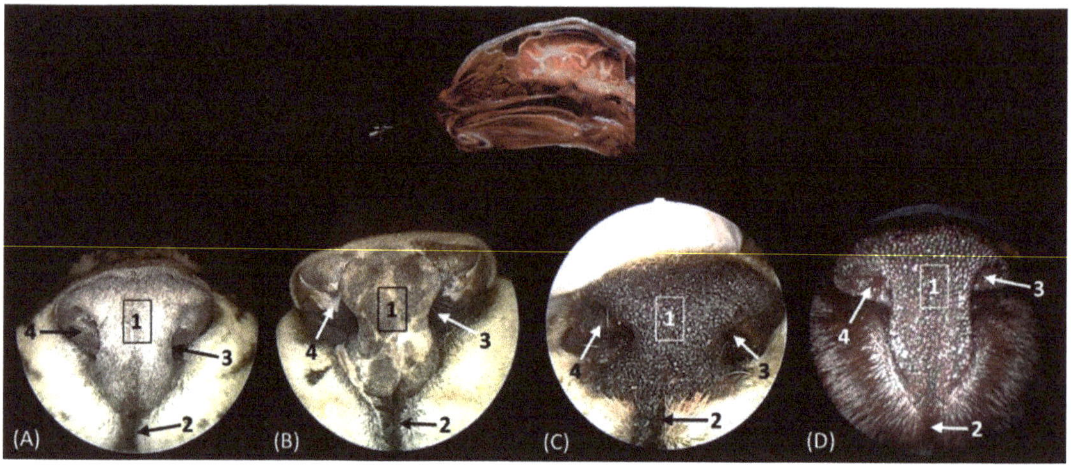

Figure 13. Endoscopic images of the nasal cavity at the level of the nasal orifices. The level of this endoscopic study is shown above in the sagittal anatomical section. Images are observed so that the right side of the head is to the left of the image. (**A**): Leopard; (**B**): lion; (**C**): cheetah; and (**D**): cat. 1. Tip of the nose; 2. subnasal groove; 3. nasal orifice; 4. alar fold.

Figure 14 shows the nasal vestibule, the most rostral part of the nasal cavity. In this and subsequent planes, the rhinoscope was introduced into the left nostril. In Figure 14 (A1–D1), the path of the alar fold was observed in all four species, forming the wing of the nose and ventrally forming the alar groove, which develops deeper in the leopard and the lion. As previously mentioned, the width of the nostril depends on the morphology of the alar fold, exhibiting a greater width in the cheetah. Figure 14(A2–D2) show the nasal vestibule in more detail. We can see the mucosa of the nasal vestibule and the mucosa of the nasal cavity, as well as the mucocutaneous junction between them. This transition is much more gradual in domestic cats than in big cats. The orifice of the nasolacrimal duct was identified close to this junction, on the floor of the vestibule. It was observed that this orifice is larger in cheetah than in the other species. Lateral to this opening, the end of the basal fold is seen.

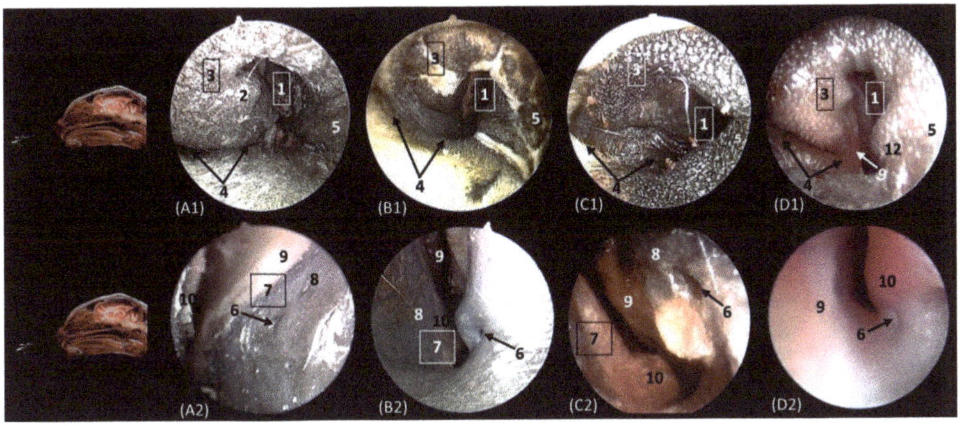

Figure 14. Endoscopic images of the nasal cavity at the level of the nasal vestibule and nasolacrimal duct opening. The levels of this endoscopic study are shown on the left in the sagittal anatomical sections. Images are observed so that the right side of the head is to the left of the image. (**A1**,**A2**): Leopard; (**B1**,**B2**): lion; (**C1**,**C2**): cheetah; and (**D1**,**D2**): cat. 1. Nasal orifice; 2. alar fold; 3. wing of the nose; 4. alar groove; 5. mobile rostral part of nasal septum; 6. nasolacrimal duct opening; 7. mucocutaneous junction of the nose; 8. nasal cavity: nasal vestibule; 9. nasal cavity: respiratory part; 10. basal fold.

Figure 15 delves into the respiratory portion of the nasal cavity. Three views were observed in the four species, from dorsal to ventral. In the most dorsal view (Figure 15(A1–D1)), the development of the dorsal nasal concha can be seen, starting from the lateral side of the nasal cavity. Ventral to it, the ventral nasal concha was observed, but between the two conchae, the third endoturbinate was also seen. The meatuses can also be observed between the conchae. The dorsal nasal meatus was visualized between the roof of the nasal cavity and the dorsal nasal concha. The middle nasal meatus was located between the dorsal nasal concha and the ventral nasal concha, while the ventral nasal meatus appeared between the ventral nasal concha and the floor of the nasal cavity. Finally, the common nasal meatus was identified between the medial surface of the nasal conchae and the nasal septum. These meatuses were described as narrower in the domestic cat than in the rest of the studied felids. Furthermore, it is worth highlighting how in the cheetah and the lion, the third endoturbinate presents a more sinuous, narrow, and flattened morphology than in the other species. In a more ventral view, the development of the maxillary recess was observed, located laterally to the ventral nasal concha (Figure 15(A1–D1)).

Figure 16A–C represent a more caudal view at the level of the ethmoidal labyrinth. Only views of the lion, cheetah, and cat are included, as due to the rigidity of the fixed structures of the nasal cavity of the leopard, access by rhinoscopy for a more caudal study of the nasal cavity is presented. In this section, the third and fourth endoturbinates were identified, as well as the opening of the sphenoid sinus, which is close to the floor of the nasal cavity. In the cat, the nasopharyngeal or choanal opening can also be identified ventrally, and in the cheetah, the pharyngeal orifice of the auditory tube can be identified on the lateral wall of the nasal cavity, ventral to the third and fourth endoturbinates.

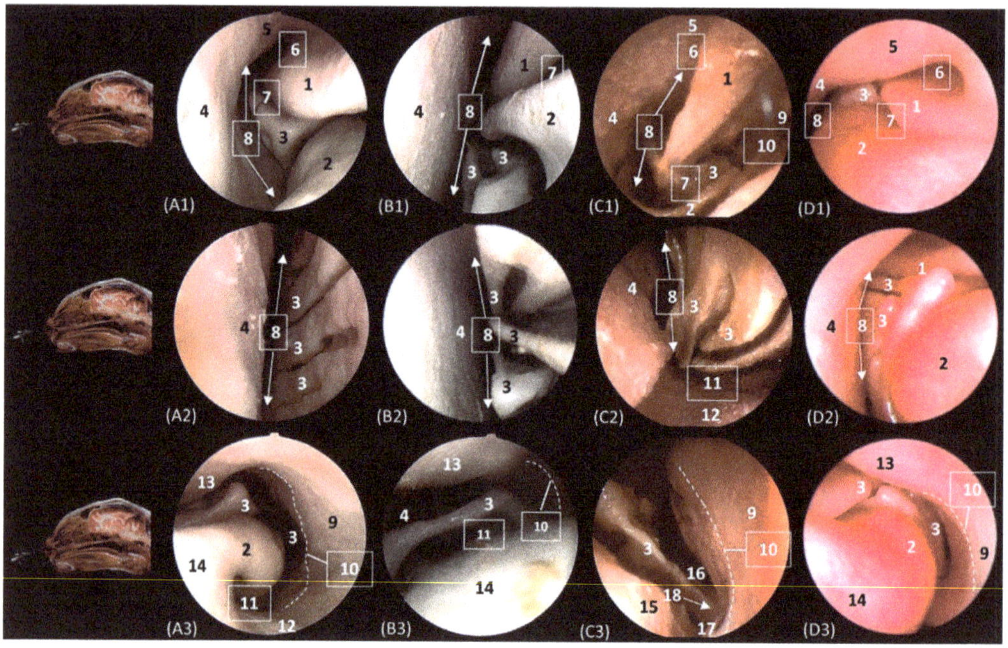

Figure 15. Endoscopic images of the nasal cavity at the level of the dorsal and ventral nasal conchae. The levels of this endoscopic study are shown on the left in the sagittal anatomical sections. Images are observed so that the right side of the head is to the left of the image. (**A1–A3**): Leopard; (**B1–B3**): lion; (**C1–C3**): cheetah; and (**D1–D3**): cat. 1. Dorsal nasal concha; 2. ventral nasal concha; 3. 3th endoturbinate; 4. nasal septum; 5. nasal cavity: roof; 6. dorsal nasal meatus: 7. middle nasal meatus; 8. common nasal meatus; 9. nasal cavity: lateral wall; 10. maxillary recess; 11. ventral nasal meatus; 12. nasal cavity: floor; 13. straight fold; 14. alar fold; 15. basal fold; 16. dorsal part of maxillary recess; 17. ventral part of maxillary recess; 18. beginning of ventral nasal concha in the conchal crest of the maxillary bone.

The pharyngeal orifice of the auditory tube can be identified in more detail in Figure 17C, located on the lateral wall. The leopard is the only species where the orifice of the lateral nasal gland was identified, being located close to the straight fold and, more specifically, in the oblique fold (Figure 17A).

Below, Figure 18A–C detail the nasopharyngeal or choanal opening in the lion, cheetah, and cat on the floor of the nasal cavity, showing the soft palate at this level. In this view, we can also see the pharyngeal opening of the auditory tube on the lateral wall of the nasal cavity.

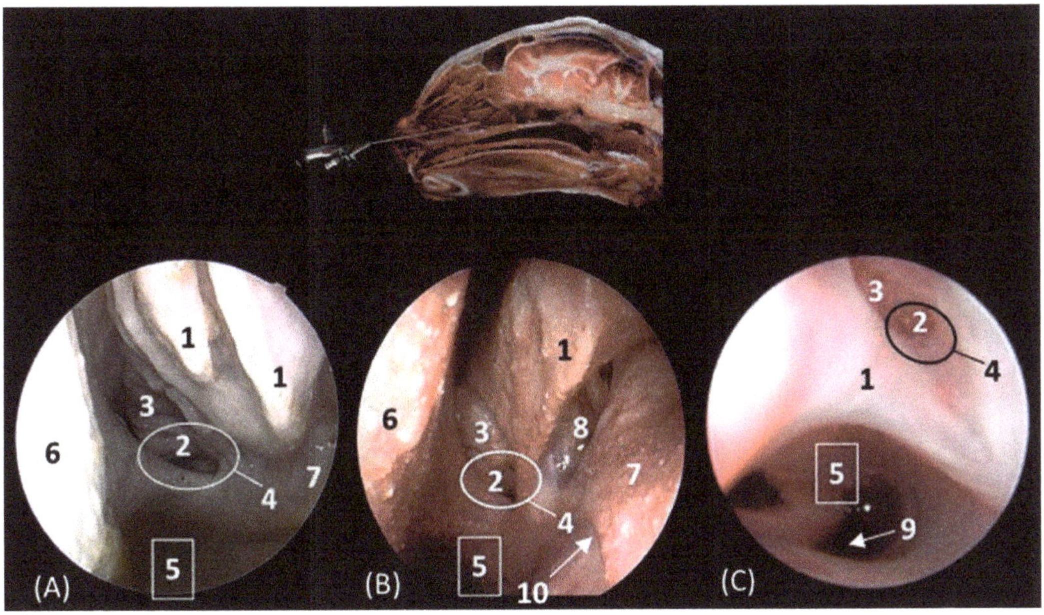

Figure 16. Endoscopic images of the nasal cavity at the level of the ethmoidal labyrinth. The level of this endoscopic study is shown above in the sagittal anatomical section. Images are observed so that the right side of the head is to the left of the image. (**A**): Lion; (**B**): cheetah; and (**C**): cat. 1. Ventral nasal concha; 2. 4th endoturbinate; 3. 3rd endoturbinate; 4. opening of the sphenoidal sinus; 5. ventral nasal meatus; 6. nasal septum; 7. nasal cavity: lateral wall; 8. lateral nasal gland; 9. choana; 10. pharyngeal orifice of the auditory tube.

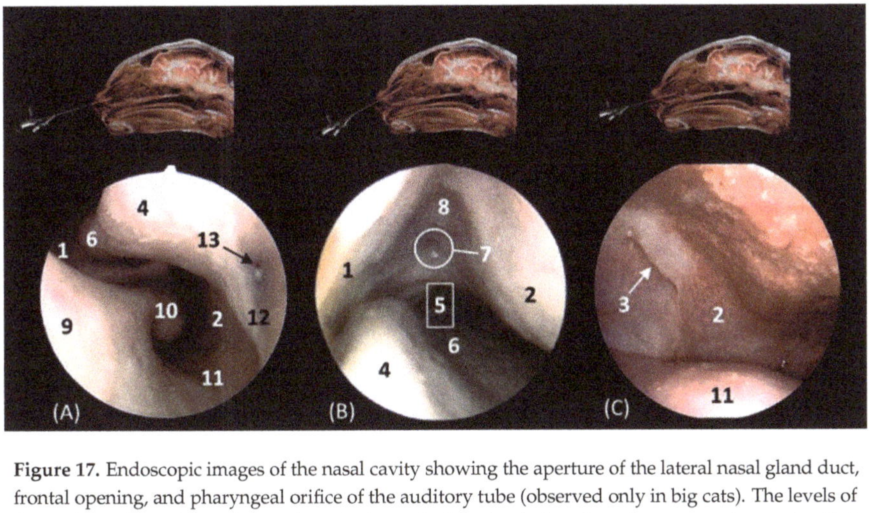

Figure 17. Endoscopic images of the nasal cavity showing the aperture of the lateral nasal gland duct, frontal opening, and pharyngeal orifice of the auditory tube (observed only in big cats). The levels of this endoscopic study are shown above in the sagittal anatomical sections. Images are observed so that the right side of the head is to the left of the image. (**A**): Leopard; (**B**): lion; and (**C**): cheetah. 1. Nasal septum; 2. nasal cavity: lateral wall; 3. pharyngeal orifice of the auditory tube; 4. straight fold; 5. dorsal nasal meatus; 6. dorsal nasal concha; 7. frontal opening; 8. nasal cavity: roof; 9. alar fold; 10. ventral nasal concha; 11. nasal cavity: floor; 12. oblique fold; 13. aperture of the lateral nasal gland duct.

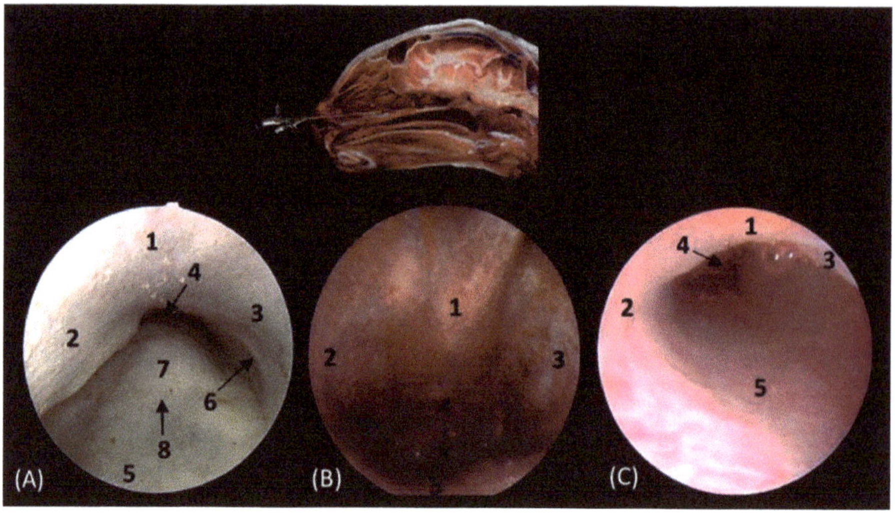

Figure 18. Endoscopic images at the level of the choana and nasopharynx. The level of this endoscopic study is shown above in the sagittal anatomical section. Images are observed so that the right side of the head is to the left of the image. (**A**): Lion; (**B**): cheetah; and (**C**): cat. 1. Presphenoid bone; 2. nasal septum; 3. nasal cavity: lateral wall; 4. choana; 5. nasal cavity: floor; 6. pharyngeal orifice of the auditory tube; 7. soft palate; 8. palatine aponeurosis.

4. Discussion

4.1. Anatomical, Comparative, and Functional Considerations

4.1.1. External Nose

In the nasal plane, the width of the nostril depends on the morphology of the alar fold. Thus, in the leopard and the cheetah, a smaller extension of this fold was observed in both CT and MRI, which means that on observation of the nasal plane of these felids by rhinoscopy, the lion and cat presented a narrower orifice than the leopard and cheetah did. The rostral nasal plane of felids was not described in any study using diagnostic imaging techniques and rhinoscopy published to date. For its study and for that of its cartilaginous structure, we have consulted anatomical texts [6,38,39,41] and used rhinoscopic vision.

4.1.2. Nasal Vestibule

In this plane, it worth nothing that, due to the more caudal projection of the alar fold in lions and cats, the incisive bone and teeth can already be visualized, while at the same level of the nasal section as in the leopard and cheetah, we can only see the alar fold. Parallel folds [6,38,39] are structures that are only visualized in a very rostral plane, and in our study, it was only possible to identify them with CT and MRI in the leopard and the cheetah.

The straight, alar, and basal nasal folds can be identified in the four species in our study, as Conchou et al. [9] do in the domestic cat. However, Losonsky et al. [10] only identified the alar fold in the cat. In addition, the same authors call the basal fold the "caudal portion of the alar fold".

4.1.3. Nasal Cavity

The nasal cavity of the domestic cat is similar to that of nondomestic cats; however, certain differences are worth highlighting. Unlike the big cats studied here, the domestic cat has two "conchal crests of the maxillary bone" for anchoring the ventral nasal concha to the internal wall of the maxillary bone. In the leopard, the lion, and the cheetah, we observed how the ventral nasal concha is attached to the lateral wall of the nasal cavity by only one crest. For its part, the dorsal nasal concha is attached to the lateral wall of the nasal cavity by the ethmoidal crest of the nasal bone. These findings were not described in any study to date.

In the lion, we observed that the internasal symphysis (medial union of the nasal bone) appeared to be much more caudal than in the rest of the feline species (Figures 4 and 5), so its mucosa and cartilaginous structure at this level (the dorsal lateral cartilage) must be quite rigid and resistant in the most rostral planes.

With respect to the nasal cavernous plexuses, it should be noted that they were neither described by Losonsky et al. [10], nor by Conchou et al. [9]. Schaller [38] described them in bovids as a set of dense venous networks in the submucosa, adjacent to both the nasal septum, as well as the lateral floor of the nasal cavity. Sandoval [39] also identified them without indicating a specific species and described their drainage into the sphenopalatine vein. We can identify these plexuses in our study next and laterally to the nasal septum. In the species of big cats that we studied, the plexuses were much more developed in the leopard than in the lion or the cheetah. We can relate this to the leopard's need to heat inhaled air, as they can withstand temperatures ranging between −30 and +70 °C [16]. In the lion and the cheetah, this is not necessary, as their habitats are not so extreme, and consequently, their plexuses are less developed [15,17].

Regarding the nasal conchae and the ecto- and endoturbinates, neither Losonsky et al. [10], nor Conchou et al. [9] described the middle nasal concha and the third and fourth endoturbinates. They mentioned ethmoturbinates in passing. Additionally, Losonsky et al. [10] described the dorsal nasal concha, where the second ectoturbinate was located in the frontal sinus. In our study, using both CT/MRI and rhinoscopy, in the lion and the cheetah, we observed that the nasal conchae and endoturbinates were more tightly coiled than in the other studied felids, thus leaving wider meatuses. This may also be related to the speed that these two felids can reach when running, since the cheetah is the fastest feline species [17].

Furthermore, another peculiarity that we observed in the lion throughout this study was the great thickness of the bones of the roof of its nasal cavity, mainly the frontal and nasal bones, as well as the great thickness of the muscles of its head. This may be related to the degree of resistance to blows and impact in the head region by large herbivores (buffalos, giraffes, zebras) when they defend themselves from attacks with their horns and by kicking for territory and food. It is also important to highlight the effect that captivity has on the lower development of the trabecular bone in felids [42].

The maxillary recess was only identified using imaging techniques with CT and MRI by De Rycke et al. [11] and with CT by Weeden et al. [1], which are both studies on the dog. In our study, the recess was visible in CT, MRI, and rhinocopy, limited medially by the ethmoidal labyrinth, laterally by the lacrimal and maxillary bone, and lateroventrally by the palatine bone.

The frontal sinus in cats, according to Barone [6] and Conchou et al. [9], has two parts (a rostral and a caudal part). However, Schaller, Sandoval, Vázquez et al. [38,39,43] described a single frontal sinus in felids. Thus, in our study, it can be seen how the tectorial lamina of the ethmoid bone covers and dorsally fixes the ectoturbinates within the frontal sinus, and it appeared to divide the frontal sinus, but in fact, it was a single sinus. In the domestic cat, the opening of the frontal sinus was identified using CT, which joined the frontal sinus with the nasal cavity and indirectly with the maxillary recess. Furthermore, using the rhinoscopic technique, in the lion, we also identified the opening of the sinus in the root of the nasal cavity. In addition, it should be noted that this opening only opens directly to the caudal maxillary sinus in equids, receiving the name of the frontomaxillary opening [38,39,43].

It should be noted that the lateral nasal gland had not yet been identified using diagnostic imaging techniques. In our study, with the help of Barone [6] and Sandoval [39], we were able to locate it using CT, MRI, and rhinoscopy. Rhinoscopy has allowed us to observe its location in the lion, the cheetah, and the cat, since it is camouflaged between the third endoturbinate and the maxillary recess. In addition, this also allowed us to locate the orifice of this gland, at the caudal limit of the nasal vestibule in the leopard.

4.1.4. Nasopharynx

Regarding the nasopharynx, the nasal cavity communicates with the nasal portion of the pharynx through the choanae or nasopharyngeal openings [43]. These openings have been identified in the four species of felids by CT and MRI; however, using rhinoscopy, we were only able to visualize them in the lion, cheetah, and cat due to the rigidity of the anatomical structures caused by fixation, preventing the passage of the rhinoscope to the most caudal regions. Close to the choanae, we identified on the lateral wall of the nasopharynx the pharyngeal orifice of the auditory tube in both the lion and the cheetah, which has not been identified to date in any rhinoscopic study. The nasopharynx has a more flattened lumen in the domestic cat than in the rest of the big cats that we studied, a fact that is very well illustrated in both the transverse and sagittal views of CT and MRI.

4.2. Clinical and Pathological Considerations

Neoplasms, inflammatory and non-inflammatory diseases, trauma, dental diseases, and nasopharyngeal diseases can cause clinical signs that are localized to the feline nasal cavity and paranasal sinuses. A detailed knowledge of the normal anatomy that is specific to the feline species facilitates the diagnosis and characterization of the lesions and abnormalities that can appear in the nasal cavity and paranasal sinuses [10]. Furthermore, most tumors of the nasal cavity are locally invasive and malignant in dogs and cats; however, some of them can be resected by surgery, and it is important to know the specific anatomy of the nasal cavity [1]. Weeden et al. [1] present in their study an anatomical study of the dog's nasal cavity using CT with a surgical approach. Regarding dogs, it is also worth highlighting the study carried out by De Rycke et al. [11], who developed a study of the nasal cavity of this species using CT, MRI, and anatomical sections.

In the domestic cat, there are various studies of the nasal cavity using diagnostic imaging techniques. Specifically, Conchou et al. [9] present an atlas of the nasal cavity and paranasal sinuses of the cat using MRI. However, they presented only transverse and horizontal views of the nasal cavity. Furthermore, in our study, we could visualize structures that Conchou et al. [9] did not identify (among them, the third and fourth endoturbinate, the maxillary recess, and the lateral nasal gland). Losonsky et al. [10] carried out a similar study, but using CT, highlighting the scarcity of sections that have been studied and the lack of definition of the tomographic images to identify the structures, which required them to present a schematic drawing to identify many of the anatomical features.

In relation to the study of the nasal cavity by rhinoscopy, we can highlight other studies which describe the procedure to carry out a rhinoscopic study in dogs and cats [3,14]. McCarthy et al. [18] also described the rhinoscopic procedure and also added a small anatomical description of the nasal cavity, but only attaching a few anatomical images, some of which were of poor quality.

As far as big cats are concerned, to date, there are no studies of the anatomy of the nasal cavity using imaging techniques. Snow et al. [30] presented an MRI study of the tiger head but did not delve into the nasal cavity. Jáber et al. [32] also performed a study of the Bengal tiger´s brain using MRI, but it does not describe the nasal cavity either. There are very few studies published to date of clinical cases that affect the nasal cavity in big cats. We can cite some relevant studies on clinical cases, such as the one carried out by Van der Weyden et al. [35], which proposed a clinical case of a meningioma of the nasal cavity of a Siberian tiger without the prior use of imaging techniques, which was diagnosed post mortem in the necropsy; or the work of Noguchi et al. [44], which described a clinical case of a cheetah involving excision of a nasopharyngeal nodule, diagnosed with CT and removed by rhinoscopy. Similarly, Ready et al. [36] described clinical cases of melanocytic neoplasms in tigers, African lions, and snow leopards that, among other locations, affect the nasal plane, and their study used radiographic and tomographic studies.

In this study, the advantages and disadvantages of studying the nasal cavity using CT, MRI, and rhinoscopy can be observed. To date, there are no such exhaustive studies of the nasal cavity that encompass both imaging techniques and rhinoscopy. CT allowed the bony limits and extensions of the nasal cavity to be identified much better than with MRI, since, in their entirety, they are completely hyperattenuated. With MRI, the bone was observed with different signal intensities depending on whether it was cortical or medullary, which in many cases makes it difficult to visualize the bone limits. The cortical layer, due to its low concentration of hydrogen atoms, was observed to be hypointense, while the medullary, as it contains bone marrow, which is rich in fat and therefore has a greater amount of hydrogen atoms, was observed to be hyperintense. In contrast, the MRI study provided better detail of the soft tissue, which allowed for clearer observation of the entire development of the turbinates, vomeronasal organ, lateral nasal gland, and nasal cavernous plexuses. Furthermore, with T2-weighted MRI, fluids and edema can be visualized as hyperintense, which can help in the interpretation of pathological conditions [9]. The rhinoscopic study, for its part, allowed for a direct visualization of the nasal cavity; however, it has the limitation that it cannot access all structures due to the restriction that rhinoscope access can pose in many areas. In our study, due to the rigidity after preservation with fixative solution of the leopard's head, it was impossible to access the more caudal levels. In this context, it is also worth highlighting the different colors of the mucosa of the nasal cavity of the domestic cat compared to the big cats, which was shown to have a much more vivid and pink color in the cat, "as it is a living animal". This is due to the effect of the fixative solution on the mucosa, since the rhinoscopic study of big cats was performed, unlike in the domestic cat, on fixed specimens.

The information provided in this article was intended to be key to the identification of pathological alterations in the nasal cavity. Diseases such as inflammatory rhinitis can cause hyperplasia of the nasal mucosa, as well as an increase in secretions [2]. Among the origins of rhinitis, we can highlight fungal conditions, for which the definitive diagnosis is taking biopsies through rhinoscopy [21]. Nasopharyngeal masses are also common in felids, usually of tumor origin, and their location and limits can be observed with considerable precision by both CT and MRI. Foreign bodies can also be identified in the nasal cavity, which are mainly diagnosed by CT and rhinoscopy [26]. Additionally, abscesses in the roots of the maxillary canine teeth can cause nasal discharge if the abscess extends to the nasal cavity [2]. The maxillary sinus is frequently affected when an abscess is generated due to infection of the caudal maxillary premolar teeth [1]. In this context, it is also worth noting that after having visualized the connection of the maxillary recess with the frontal sinus in the cat through the opening of the frontal sinus in our study, a condition at the level of these teeth could even develop into sinusitis or infection in this sinus.

All these pathologies that affect the nasal cavity, if they become chronic, can lead to the destruction of structures of the nasal cavity such as the turbinates [2]. Therefore, prior knowledge of nasal anatomy is necessary for the diagnosis of any nasal condition [9,10,14].

5. Conclusions

The combination of CT, MRI, and rhinoscopy techniques for examining the nasal cavity has allowed us to assess the benefits that each technique has for the study of this cavity.

CT is a technique of great value for the study of the bony nasal cavity, since the bones are visualized in their entire thickness in a hyperattenuated manner, which facilitates the identification of their limits. However, MRI presents certain difficulties for this, since it shows the bones with a hypointense cortical and a hyperintense medullary, which makes visualization difficult in many cases. In addition, MRI is very useful in visualizing soft tissue, which includes the mucosa of the turbinates or conchae, as well as structures such as the vomeronasal organ or the lateral nasal gland.

Rhinoscopy is a key technique in the direct visualization of the mucosa of the nasal vestibule and nasal cavity and allows for the differentiation of some structures such as the orifice of the nasolacrimal duct, the pharyngeal orifice of the auditory tube, or the orifice of the lateral nasal gland, which cannot be identified with any other imaging technique used.

In the external nose, the opening of the nostrils depends on the route and morphology of the alar fold, a fact that causes the lion and cat to have narrower nostrils.

At the level of the nasal cavity, it has been observed that the ventral nasal concha is attached to the lateral wall of the nasal cavity by a single conchal crest of the maxillary bone in big cats, while in domestic cats, it has two conchal crests. On the other hand, the attachment of the dorsal nasal concha to the lateral wall of the nasal cavity is through the ethmoidal crest of the nasal bone. In lions, we could highlight that the medial union of the nasal bone or internasal symphysis is much more caudal than in the rest of the feline species that we studied. Furthermore, the bones of the nasal cavity of the lion, as well as the muscles of this area, are much thicker in this species than in the others.

The nasal cavernous plexuses are more developed in leopards than in the rest of the species.

The nasal conchae, as well as the endoturbinates, present a narrower and flatter morphology in the lion and the cheetah, which means that the spaces between them, or the nasal meatuses, have a greater width.

With respect to the frontal sinus, in the domestic cat, it has been observed that the opening of the frontal sinus connects the frontal sinus with the nasal cavity and, indirectly, with the maxillary recess. This particularity has not been found in big cats.

The nasopharynx has a much narrower lumen in the domestic cat than in the big cats.

Finally, we can conclude by emphasizing that adequate knowledge of the specific anatomy of the nasal cavity of these species through imaging techniques allows us to more easily detect and diagnose alterations or pathologies that may affect this level.

Supplementary Materials: The following supporting information can be downloaded at https://www.mdpi.com/article/10.3390/ani14081172/s1: Table S1: MRI parameters used in this study; Table S2: Tissue CT density and signal intensity characteristics for this CT and MRI study.

Author Contributions: Conceptualization, E.D.M., G.R.Z. and A.A.E.; formal analysis, F.G.C. and A.A.E.; investigation, E.D.M., G.R.Z. and F.G.C.; methodology, M.S.L., C.S.C. and A.A.E.; resources, F.M.G., C.S.C., D.L.C.G. and J.R.J.; supervision, G.R.Z., M.S.L. and A.A.E.; validation, D.K., F.M.G., D.L.C.G. and J.R.J.; writing—original draft, E.D.M., G.R.Z., M.S.L. and A.A.E.; writing—review and editing, E.D.M., G.R.Z., M.S.L., D.K. and A.A.E. All authors have read and agreed to the published version of the manuscript.

Funding: This research received no external funding.

Institutional Review Board Statement: The animal study protocol was approved by The Animal Ethical Committee of Veterinary Medicine of Murcia University, who authorized the research protocol (REGA ES300305440012 CEEA: 305/2017; Extended on 25 July 2022 as project Type II).

Informed Consent Statement: Written informed consent has been obtained from the owners of the animals involved in this study.

Data Availability Statement: The information can be requested from dlcasas@gmail.com, almoradi@mivet.com, and ricardo.n.l@um.es.

Acknowledgments: We give our special thanks to Oscar Blázquez Pérez and Javier Lucas Mantione for the CT and MRI scans performed at Centro Veterinario de Diagnostico por Imagen del Levante, Ciudad Quesada, Rojales Alicante, Spain.

Conflicts of Interest: The authors declare no conflicts of interest.

References

1. Weeden, A.M.; Degner, D.A. Surgical approaches to the nasal cavity and sinuses. *Vet. Clin. N. Am. Small Anim. Pract.* **2016**, *46*, 719–733. [CrossRef] [PubMed]

Norris, A.M.; Laing, E.J. Diseases of the nose and sinuses. *Vet. Clin. N. Am. Small Anim. Pract.* **1985**, *15*, 865–890. [CrossRef] [PubMed]

Elie, M.; Sabo, M. Basics in canine and feline rhinoscopy. *Clin. Tech. Small Anim. Pract.* **2006**, *21*, 60–63. [CrossRef] [PubMed]

Nickel, R.; Schummer, A.; Seiferle, E. *The Viscera of the Domestic Mammals*, 2nd ed.; Verlag Paul Parey: Berlin, Germany, 1979; pp. 1–401.

Donald, E.; Thrall, D.E. *Textbook of Veterinary Diagnostic Radiology*, 7th ed.; Saunders: Durham, NC, USA, 2018; pp. 183–203.

Barone, R. *Anatomie Comparée des Mammifères Domestiques: Appareil Digestif, Appareil Respiratoire. Tome 3. Splanchnologie I*, 3rd ed.; Vigot: Paris, France, 2017; pp. 639–643.

Schoenborn, W.C.; Wisner, E.R.; Kass, P.P.; Dale, M. Retrospective assessment of computed tomographic imaging of feline sinonasal disease in 62 cats. *Vet. Radiol. Ultrasound* **2003**, *44*, 185–195. [CrossRef] [PubMed]

Henderson, S.M.; Bradley, K.; Day, M.; Tasker, S.; Caney, S.M.; Hotston Moore, A.; Gruffydd-Jones, T.J. Investigation of nasal disease in the cat--a retrospective study of 77 cases. *J. Feline Med. Surg.* **2004**, *6*, 245–257. [CrossRef]

Conchou, F.; Sautet, J.; Raharison, F.; Mogicato, G. Magnetic resonance imaging of normal nasal cavity and paranasal sinuses in cats. *Anat. Histol. Embryol.* **2011**, *41*, 60–67. [CrossRef]

10. Losonsky, J.M.; Abbott, L.C.; Kuriashkin, I.V. Computed tomography of the normal feline nasal cavity and paranasal sinuses. *Vet. Radiol. Ultrasound* **1997**, *38*, 251–258. [CrossRef] [PubMed]
11. De Rycke, L.M.; Saunders, J.H.; Gielen, I.M.; Van Bree, H.J.; Simoens, P.J. Magnetic resonance imaging, computed tomography, and cross-sectional views of the anatomy of normal nasal cavities and paranasal sinuses in mesaticephalic dogs. *Am. J. Vet. Res.* **2003**, *64*, 1093–1098. [CrossRef]
12. Johnson, L.R.; Clarke, H.E.; Bannasch, M.J.; De Cock, H.E. Correlation of rhinoscopic signs of inflammation with histologic findings in nasal biopsy specimens of cats with or without upper respiratory tract disease. *J. Am. Vet. Med. Assoc.* **2004**, *225*, 395–400. [CrossRef]
13. Kuehn, N.F. Chronic rhinitis in cats. *Clin. Tech. Small Anim. Pract.* **2006**, *21*, 69–75. [CrossRef]
14. Rawlings, C.A. Diagnostic rigid endoscopy: Otoscopy, rhinoscopy, and cystoscopy. *Vet. Clin. N. Am. Small Anim. Pract.* **2009**, *39*, 849–868. [CrossRef] [PubMed]
15. Bauer, H.; Packer, C.; Funston, P.F.; Henschel, P.; Nowell, K. *Panthera leo*. *The IUCN Red List of Threatened Species* 2016: e.T15951A115130419. Available online: http://dx.doi.org/10.2305/IUCN.UK.2016-3.RLTS.T15951A107265605.en (accessed on 10 April 2024).
16. Stein, A.B.; Athreya, V.; Gerngross, P.; Balme, G.; Henschel, P.; Karanth, U.; Miquelle, D.; Rostro-Garcia, S.; Kamler, J.F.; Laguardia, A.; et al. *Panthera pardus* (Amended Version of 2019 Assessment). *The IUCN Red List of Threatened Species* 2020: e.T15954A163991139. Available online: http://dx.doi.org/10.2305/IUCN.UK.2016-1.RLTS.T15954A50659089.en (accessed on 10 April 2024).
17. Durant, S.M.; Groom, R.; Ipavec, A.; Mitchell, N.; Khalatbari, L. *Acinonyx jubatus*. *The IUCN Red List of Threatened Species* 2022: e.T219A124366642. Available online: http://dx.doi.org/10.2305/IUCN.UK.2015-4.RLTS.T219A50649567.en (accessed on 10 April 2024).
18. McCarthy, T.C.; McDermaid, S.L. Rhinoscopy. *Vet. Clin. N. Am. Small Anim. Pract.* **1990**, *20*, 1265–1290. [CrossRef] [PubMed]
19. Gil Cano, F.; Latorre Reviriego, R.; Ramírez Zarzosa, G.; López Albors, O.; Ayala Florenciano, M.D.; Martínez Gomariz, F.; Sánchez Collado, C.; Vázquez Autón, J.M. *Atlas de la Anatomía del Gato*, 1st ed.; Multimédica Ediciones Veterinarias: Murcia, Spain, 2022; pp. 1–256.
20. Kudnig, S.T. Nasopharyngeal polyps in cats. *Clin. Tech. Small Anim. Pract.* **2002**, *17*, 174–177. [CrossRef] [PubMed]
21. Whitney, B.L.; Broussard, J.; Stefanacci, J.D. Four cats with fungal rhinitis. *J. Feline Med. Surg.* **2005**, *7*, 53–58. [CrossRef]
22. Schlueter, C.; Budras, K.D.; Ludewig, E.; Mayrhofer, E.; Koenig, H.E.; Walter, A.; Oechtering, G.U. Brachycephalic feline noses: CT and anatomical study of the relationship between head conformation and the nasolacrimal drainage system. *J. Feline Med. Surg.* **2009**, *11*, 891–900. [CrossRef] [PubMed]
23. Parker, V.J.; Morrison, J.A.; Yaeger, M.J. Olfactory neuroblastoma in a cat. *J. Feline Med. Surg.* **2010**, *12*, 867–871. [CrossRef] [PubMed]
24. Fischetti, A.J.; Gisselman, K.; Peterson, M.E. CT and MRI evaluation of skull bones and soft tissues in six cats with presumed acromegaly versus 12 unaffected cats. *Vet. Radiol. Ultrasound* **2012**, *53*, 535–539. [CrossRef] [PubMed]
25. Finck, M.; Ponce, F.; Guilbaud, L.; Chervier, C.; Floch, F.; Cadoré, J.L.; Chuzel, T.; Hugonnard, M. Computed tomography or rhinoscopy as the first-line procedure for suspected nasal tumor: A pilot study. *Can. Vet. J.* **2015**, *56*, 185–192. [PubMed]
26. Moreno-Aguado, B.; Carrera, I.; Holdsworth, A.; Agthe, P.; Maddox, T.W.; Trevail, T. CT findings in 20 dogs and six cats with confirmed nasal foreign bodies. *Vet. Radiol. Ultrasound* **2020**, *61*, 417–426. [CrossRef]
27. Sharma, D.; Pakravan, N.; Pritchard, J.C.; Hartmann, F.A.; Young, K.M. Mucoid Pseudomonas aeruginosa infection in a cat with severe chronic rhinosinusitis. *Vet. Clin. Pathol.* **2019**, *48*, 300–304. [CrossRef]
28. McEwan, S.A.; Sykes, J.E. Nasopharyngeal cryptococcosis in a cat: Interlaboratory variation in cryptococcal antigen assay test results. *JFMS Open Rep.* **2022**, *8*, 20551169221074624. [CrossRef]
29. Fawsitt, J.; Russell, O.; Alexander, A.; Peschard, A.L.; Wong, H.; Kortum, A. Clinical remission of feline sino-nasal aspergillosis despite evidence of persistent infection. *JFMS Open Rep.* **2023**, *9*, 20551169231201605. [CrossRef]
30. Snow, T.M.; Litster, A.L.; Gregory, R.J. Big cat scan: Magnetic resonance imaging of the tiger. *Australas. Radiol.* **2004**, *48*, 93–95. [CrossRef] [PubMed]

31. Arencibia, A.; Encinoso, M.; Jáber, J.R.; Morales, D.; Blanco, D.; Artiles, A.; Vázquez, J.M. Magnetic resonance imaging study in a normal Bengal tiger (*Panthera tigris*) stifle joint. *BMC Vet. Res.* **2015**, *11*, 192. [CrossRef]
32. Jáber, J.R.; Encinoso, M.; Morales, D.; Artiles, A.; Santana, M.; Blanco, D.; Arencibia, A. Anatomic study of the normal Bengal tiger (*Panthera tigris tigris*) brain and associated structures using low field magnetic resonance imaging. *Eur. J. Anat.* **2016**, *20*, 195–203.
33. Lara, V.M.; Benassi, J.C.; Bisetto, S.P.; De Oliveira, T.M.; Ferreira, H.L.; Júnior, J.P.A.; Carregaro, A.B. Molecular detection of infectious pathogens of the upper respiratory tract in captive nondomestic felids. *J. Zoo Wildl. Med.* **2017**, *48*, 529–531. [CrossRef] [PubMed]
34. Boudreaux, B.B.; LaRue, S.M.; Rademacher, N.; Neck, D.; Grasperge, B.; Wood, C.; Baker, D.G. Treatment of leiomyosarcoma in a tiger (*Panthera tigris*) with stereotactic radiotherapy. *Vet. Radiol. Ultrasound* **2019**, *60*, E33–E37. [CrossRef] [PubMed]
35. Van der Weyden, L.; Caldwell, P.; Steyrer, C.; O'Dell, N.; Henning, A. Sinonasal meningioma in a siberian tiger (*Panthera tigris altaica*). *Vet. Sci.* **2022**, *9*, 457. [CrossRef]
36. Ready, Z.C.; LoBato, D.; LaDouceur, E.; Garner, M.M.; Cushing, A.C. Melanocytic neoplasia in panthera species: Clinical presentations, pathologic findings and responses to treatment. *J. Zoo Wildl. Med.* **2023**, *53*, 844–854. [CrossRef]
37. Crouch, J.E. *Text-Atlas of Cat Anatomy*; Lea & Febiger: Philadelphia, PA, USA, 1969; pp. 1–416.
38. Schaller, O. *Illustrated Veterinary Anatomical Nomenclature*; Ferdinand Enke Verlag: Stuttgart, Germany, 1992; pp. 1–614.
39. Sandoval, J. *Tratado de Anatomía Veterinaria, Tomo III: Cabeza y Sistemas Viscerales*; Imprenta Sorles: León, Spain, 2000; pp. 1–281.
40. Liste Burillo, F. *Atlas Veterinario de Diagnóstico por Imagen*; Servet: Soria, Spain, 2009; pp. 146–198.
41. Hudson, L.C.; Hamilton, W.P. *Atlas of Feline Anatomy for Veterinarians*; W.B. Saunders Company: Philadelphia, PA, USA, 1993; pp. 1–287.
42. Chirchir, H.; Ruff, C.; Helgen, K.M.; Potts, R. Effects of reduced mobility on trabecular bone density in captive big cats. *R. Soc. Open Sci.* **2022**, *9*, 211345. [CrossRef]
43. Vázquez, J.M.; Gil, F.; Latorre, R.; Ramírez, G.; López Albors, O.; Ayala, M.D.; Martínez Gomariz, F.; Sánchez, C.; Orenes, M. *Manual de Prácticas de Anatomía Veterinaria: Sistemas Viscerales*; Diego Marín: Murcia, Spain, 2001; pp. 21–42.
44. Noguchi, S.; Toyota, K.; Ozaki, M.; Wada, Y.; Takami, Y.; Tanaka, M.; Kuwamura, M.; Shimada, T. Successful management of nasopharyngitis caused by Schizophyllum commune in a captive cheetah (*Acinonyx jubatus*). *J. Vet. Med. Sci.* **2023**, *85*, 1074–1076. [CrossRef] [PubMed]

animals

MDPI

Article

A Cadaveric Study Using Anatomical Cross-Section and Computed Tomography for the Coelomic Cavity in Juvenile Cory's Shearwater (Aves, *Procellariidae*, *Calonectris borealis*)

Alejandro Morales Espino [1], Soraya Déniz [2], Marcos Fumero-Hernández [2], Mario Encinoso [2,*], Pascual Calabuig [3], Magnolia Conde-Felipe [4] and José Raduan Jaber [5,*]

1 IVC Evidensia Los Tarahales, 35013 Las Palmas, Spain; alejandro.morales108@alu.ulpgc.es
2 Hospital Veterinario, Facultad de Veterinaria, Universidad de Las Palmas de Gran Canaria, Trasmontaña, Arucas, 35413 Las Palmas, Spain; soraya.deniz@ulpgc.es (S.D.); marcos.fumero101@alu.ulpgc.es (M.F.-H.)
3 Tafira Wildlife Rehabilitation Center (Cabildo de Gran Canaria), 35017 Las Palmas, Spain; pascual.calabuig@telefonica.net
4 Department of Pathology and Food Technology, Faculty of Veterinary Medicine, Universidad de Las Palmas de Gran Canaria, 35413 Las Palmas, Spain; magnolia.conde@ulpgc.es
5 Department of Morphology, Faculty of Veterinary Medicine, University of Las Palmas de Gran Canaria, Trasmontaña, Arucas, 35413 Las Palmas, Spain
* Correspondence: mencinoso@gmail.com (M.E.); joseraduan.jaber@ulpgc.es (J.R.J.)

Simple Summary: Cory's Shearwater (*Calonectris borealis*) is a fascinating subject to explore the life and complexities of marine birds. Hence, we believe that knowledge of its coelomic cavity could provide essential information for biologists, veterinarians, and researchers. In this paper, we describe its coelomic cavity using anatomical cross-sections combined with computed tomography images to better understand the complex organization that encloses the different components of this cavity.

Abstract: Birds play a crucial role in ecosystems, engaging in key functions such as pollination, pest control, and seed dispersal. The anatomical understanding of these species is essential to addressing emerging challenges, including climate change and habitat loss, which directly impact their survival. Detailed knowledge of avian anatomy is fundamental for research across various disciplines, ranging from ornithology to veterinary medicine. Therefore, this study aims to disseminate the understanding of avian anatomy and the application of computed tomography (CT) for visualizing the coelomic cavity in Cory's Shearwater (*Calonectris borealis*). Recent advances in comprehending the anatomical structures of this region in avian species are highlighted, with a specific focus on Cory's Shearwater as a study model. Various anatomical cross-sections and transverse CT images were described and analyzed in detail, offering a comprehensive insight into the coelomic cavity from different perspectives. The correlation between anatomical cross-sections and CT images is emphasized as crucial for a profound understanding of avian anatomy. This research contributes to the broader knowledge of avian anatomy, with potential implications for conservation efforts and veterinary practices.

Keywords: computed tomography; coelomic cavity; anatomical section; seabirds; Cory's Shearwater

1. Introduction

Cory's Shearwater (*Calonectris borealis*), also known as the common shearwater, stands as a fascinating subject of study in the avian field, providing a unique window to explore the life and complexities of marine birds [1,2]. With a wingspan that can reach remarkable dimensions, its wings play a crucial role in its long migratory flights [1,3]. Its plumage, characterized by white and black tones, provides suitable camouflage in its marine environment. The aerodynamic shape of its body and its sharp beak are testaments to evolutionary adaptations that enable it to survive in the rigors of the oceanic environment and distinctive migratory habits. Cory's Shearwater becomes an essential component of marine ecosystems, occupying vast areas ranging from the North Atlantic to the Indian and Pacific Oceans [1–4]. As a prominent member of the Procellariidae family, this species not only captures attention for its majestic presence but also for its vital role in the marine food chain [2,3]. Its specialized diet, focused on small fish, crustaceans, and cephalopods, positions it as a key component in the dynamics and balance of the oceans [1,2]. However, despite its ecological importance, Cory's Shearwater faces significant challenges, from environmental threats to the consequences of human activity in its habitats [2,4]. As a consequence, this bird has most recently been assessed for the IUCN Red List of Threatened Species as having the least concern [5].

The anatomical investigation of birds holds fundamental importance in the scientific field, mainly for biologists, veterinarians, and researchers, providing a detailed and better understanding of the structure and function of the biological systems of these species [6–8]. In addition, the anatomical study is essential for elucidating the underlying mechanisms of avian physiology, and for comprehending the evolutionary adaptations that have allowed birds to inhabit a diversity of ecological niches [6–8].

Conventional radiology stands out as the predominant auxiliary diagnostic technique in avian medicine, primarily owing to its widespread availability in veterinary centers, cost-effectiveness, and non-invasive nature; nonetheless, contemporary imaging modalities, such as magnetic resonance (MR) and computed tomography (CT), offer advancements in avian diagnostic capabilities [9–11]. These modern techniques enable the acquisition of high-resolution three-dimensional images with minimal distortion and even allow for simulations of dynamic processes within different organs, vascular structures, and nerves, all achieved through a minimally invasive approach. A plethora of investigations exist concerning the application of diagnostic imaging to exotic species [12–25]. However, to our knowledge, a dearth of studies address the anatomical characteristics of the avian coelom utilizing Cory's Shearwater as a model organism. Therefore, this study aimed to assess the coelomic cavity of Cory's Shearwater through anatomical cross-sections and CT image examination, establishing it as a potential anatomical model for pathological investigations involving other phylogenetically related avian species.

2. Materials and Methods

2.1. Animals

To perform this investigation, 8 juvenile Cory's (*Calonectris borealis*) cadavers were employed. The birds exhibited an average mass of 0.520 kg (ranging from 0.480 to 0.820 kg) and an average length, measured from the beak tip to the base of the tail, of 42 cm (ranging from 45 to 56 cm). The study group was provided by the Tafira Wildlife Rehabilitation Center (Consejeria de Área de Medio Ambiente, Clima, Energía y Conocimiento of the Cabildo de Gran Canaria, Spain) after a stranding event on the island coastline as a consequence of disorientation produced by light pollution. Six of our birds were quickly frozen for four days until the CT procedure, whereas the other two birds underwent dissection immediately after death to expose the coelomic cavity and visualize organ locations. This process facilitated the precise identification and correlation with the CT images.

2.2. CT Study

For the CT evaluation, we thawed our birds at room temperature for 12 h. We obtained transversal CT sequential images using a 16-slice helical CT scanner (Toshiba Astelion, Canon Medical System®, Tokyo, Japan). We positioned the birds symmetrically in dorsal recumbency on the stretcher with craniocaudal entry. We followed a standard protocol using the following parameters: 120 kVp, 80 mA, 512 × 512 acquisition matrix, 1809 × 858 field of view, a pitch of 0.94, and a gantry rotation of 1.5. The acquired images had a slice thickness of 0.6 mm. To elevate the identification of the different anatomical structures on CT, we used a variety of CT window settings by adjusting the window widths (WWs) and window levels (WLs): a bone window setting (WW = 1500; WL = 300), a soft tissue window setting (WW = 248; WL = 123), and a lung window setting (WW = 1400;

WL = −500). We did not perceive substantial differences in CT density or anatomy in the coelomic cavity of the animals used in this study. Moreover, we used the original data to create volume-rendered reconstructed images using a standard dicom 3D format (OsiriX MD, Geneva, Switzerland).

2.3. Anatomical Sections

After the CT study, the carcasses underwent freezing at −80 °C for 72 h. Thereafter, we sectioned three birds, generating parallel sections one centimeter thick, using an electric band saw. The obtained sections were carefully irrigated with water to remove artifacts (feathers), which were removed using Adson forceps and identified before being photographed on both sides.

2.4. Anatomic Evaluation

In pursuit of cross-sectional identification and labeling, concomitant with the associated CT images, we employed specific sources comprising textbooks and reports from scientific journals dedicated to bird anatomy [26–29]. Additionally, to achieve meticulous anatomical interpretation of coelomic structures, various anatomical preparations supplied by the Department of Morphology were utilized. These preparations were instrumental in augmenting our comprehension and precision of the organization of the coelomic cavity.

3. Results

A comprehensive array of illustrations delineating the anatomical intricacies of the Cory's Shearwater coelomic cavity is depicted through a series of figures (Figures 1–11). Figure 1 integrates diverse dissections, offering a comprehensive view of the principal formations included in the coelomic cavity. Figure 2 presents a multiplanar sagittal reconstruction (MPR) image, where the lines and their numerals (I–IX) denote an approximate level corresponding to subsequent anatomical and CT transverse planes. Figures 3–11 exhibit a set of four images for each case: (A) an anatomical cross-section, (B) a bone CT window, (C) a soft tissue CT window, and (D) a pulmonary CT window. These figures are sequentially arranged in a rostrocaudal progression, commencing from the lungs in Figure 3 and concluding at the cloaca levels in Figure 11. Figure 12 comprises four images: (A) an anatomical dissection and (B–D) dorsal MPR volume rendering images showcased at varying levels. This series of figures collectively provides a detailed and nuanced representation of the anatomical features within the coelomic cavity of Cory's Shearwater.

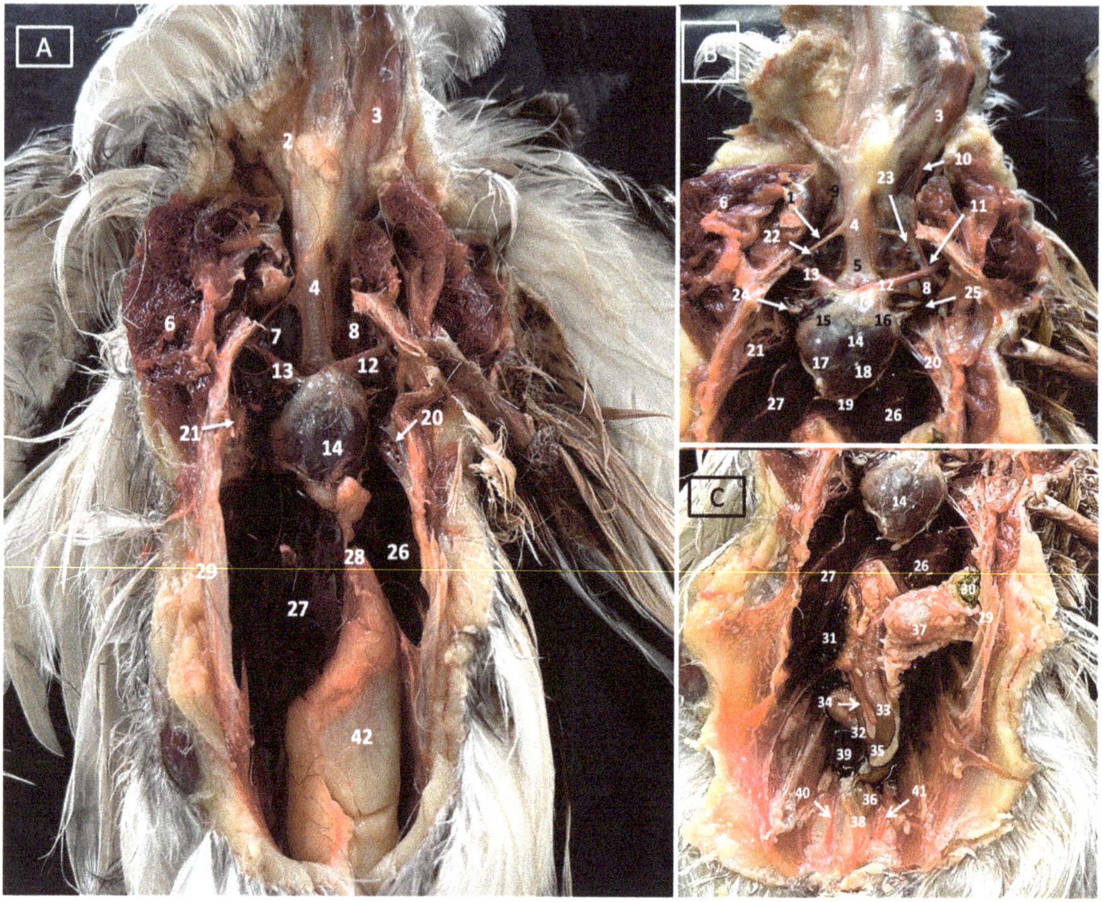

Figure 1. A gross dissection image illustrating the coelomic cavity of Cory's Shearwater (**A**), along with detailed views of its cardiovascular (**B**) and digestive (**C**) structures. Anatomical labels include 1: sternotracheal muscle. 2: esophagus. 3: longissimus colli muscle. 4: trachea. 5: syrinx. 6: pectoral muscle (thoracobrachialis muscle). 7: right cervical air sac. 8: left cervical air sac. 9: right jugular vein. 10: left jugular vein. 11: left subclavian artery. 12: left brachiocephalic trunk. 13: right brachiocephalic trunk. 14: heart. 15: right atrium. 16: left atrium. 17: right ventricle. 18: left ventricle. 19: pericardium. 20: left lung. 21: right lung. 22: right thyroid. 23: left thyroid. 24: right cranial cava. 25: left cranial cava. 26: left hepatic lobe. 27: right hepatic lobe. 28: hepatopericardial ligament. 29: ribs: 30: ingesta: 31: spleen. 32: descending duodenum. 33: ascending duodenum. 34: pancreas. 35: duodenal-jejunal flexure. 36: caecum. 37: ventriculus. 38: jejunoileum. 39: right kidney. 40: right ureter. 41: left ureter. 42: intestinal peritoneal cavity.

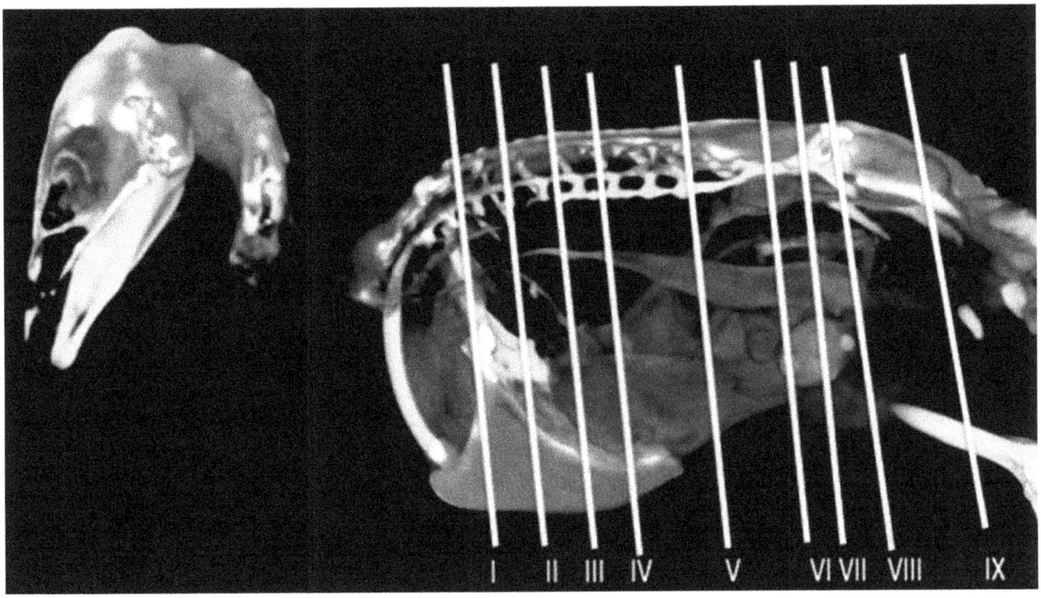

Figure 2. Sagittal MPR volume rendering image of a Cory's Shearwater specimen. The lines and corresponding numbers (I–IX) indicate the approximate positions of the subsequent transversal cross-sections and CT images.

3.1. Anatomical Dissections and Cross-Sections

In this study, we provide anatomical dissections (Figures 1A–C and 12A) and transversal cross-sections (shown in Figures 3A, 4A, 5A, 6A, 7A, 8A, 9A, 10A and 11A) of the coelomic cavity of the Cory's Shearwater. These figures play a crucial role in aiding the recognition of organs belonging to the respiratory, circulatory, digestive, and urinary systems inside this cavity. This distinct anatomical feature in Cory's Shearwater, observed through our meticulous examination, reflects the anatomical arrangement regarding the precise spatial disposition of vital organs within the coelomic cavity. Subsequently, we identified the heart that shows an oval conformation with a pointed apex located centrally inside this cavity and anterior to the liver (Figures 1A–C and 12A). Furthermore, we observed the different components of the heart, encompassing the atria and the ventricles. Notably, the anatomical images played a pivotal role in visualizing remarkable vascular formations, including the brachiocephalic trunks, the left subclavian vein, the left and right jugular veins, and the left and right cranial cava veins (depicted in Figures 1A,B, 3A, 4A and 12A, respectively). Moreover, both lungs were distinguishable, positioned in a craniodorsal location beneath the thoracic vertebrae and to the side of the ribs (Figures 1A,B and 12A). The anatomical study also facilitated the identification of the trachea (Figures 1A,B and 12A), coursing in a median position into the coelomic cavity before bifurcating into the right and left main bronchi. This bifurcation was clearly distinguished in the transverse anatomical sections (illustrated in Figures 3A and 4A). In specific dissected images, we successfully identified the *syrinx* (Figures 1B, 4A, and 12A). Dorsal to the *trachea*, we identify additional structures, including the *esophagus*, the longissimus colli muscle, and the sternotracheal muscle (Figures 1A,B, 3A, 4A, 5A, 6A and 12A, respectively). These images along the cross-section facilitated the observation of the different air sacs, including the cervical, thoracic, and abdominal air sacs (depicted in Figures 1A,B, 4A, 5A, 6A and 7A). These dissected images were helpful in distinguishing the thyroid gland (Figure 1B). It has an ovoid shape and is located at the base of the neck, just above the sternum bone. This gland consists of two lobes situated on either side of the trachea, connected by an isthmus.

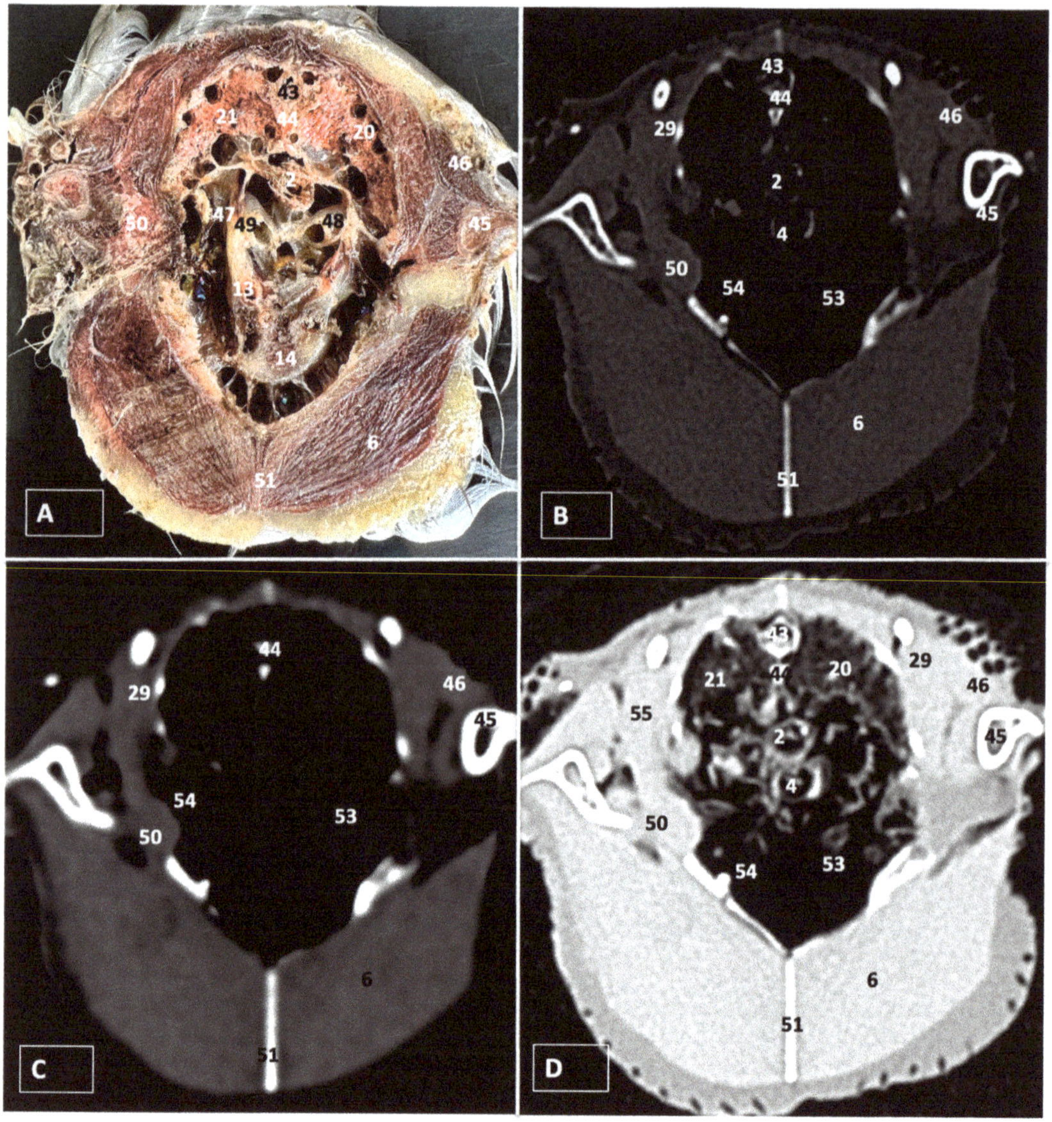

Figure 3. Transverse cross-section (**A**), bone (**B**), soft tissue (**C**), and pulmonary (**D**) CT images of the coelomic cavity of the Cory's Shearwater at the level of the lungs corresponding to line I in Figure 2. 43: spinal cord. 44: vertebra (vertebral body). 21: right lung. 20: left lung. 2: esophagus. 4: trachea. 45: humerus. 46: scapulohumeralis muscle. 47: common carotid artery. 48: left primary bronchus. 49: right primary bronchus. 13: right brachiocephalic trunk. 14: heart. 46: scapulohumeral caudal muscle. 6: pectoral muscle (thoracobrachialis muscle). 51: sternum. 29: ribs. 53: cranial thoracic air sac (left). 54: cranial thoracic air sac (right). 55: intercostal muscle. 50: scapulohumeral caudal muscle.

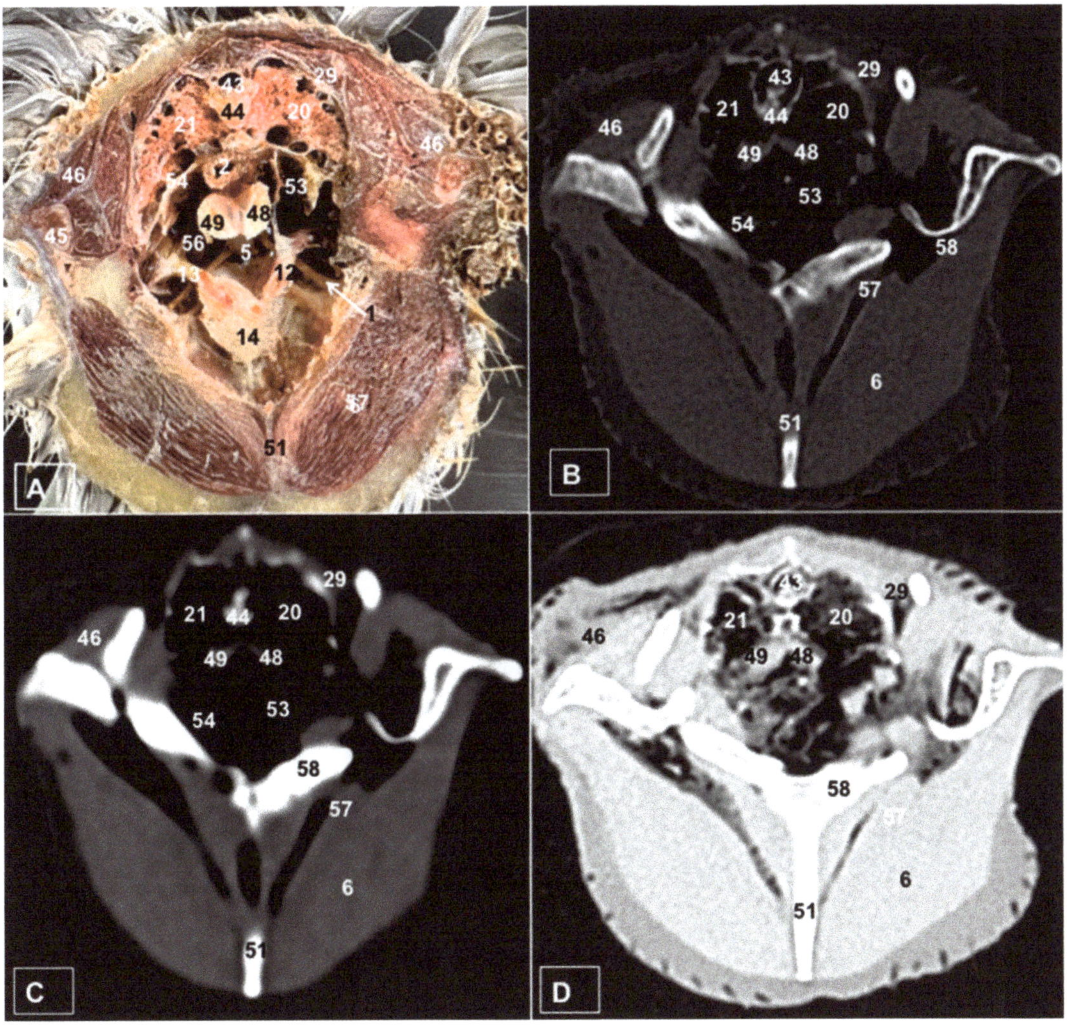

Figure 4. Transverse cross-section (**A**), bone (**B**), soft tissue (**C**), and pulmonary (**D**) CT images of the coelomic cavity of the Cory's Shearwater at the level of the main bronchi corresponding to line II in Figure 2. 43: spinal cord. 44: vertebra (vertebral body). 29: rib. 21: right lung. 20: left lung. 46: scapulohumeralis muscle. 45: humerus. 2: esophagus. 54: cranial thoracic air sac (right). 53: cranial thoracic air sac (left). 5: syrinx. 14: heart. 13: right brachiocephalic trunk. 12: left brachiocephalic trunk. 56: right subclavian artery. 1: sternotracheal muscle. 48: left primary bronchus. 49: right primary bronchus. 6: pectoral muscle (thoracobrachialis muscle). 57: supracoracoid muscle. 51: sternum. 58: coracoid bone.

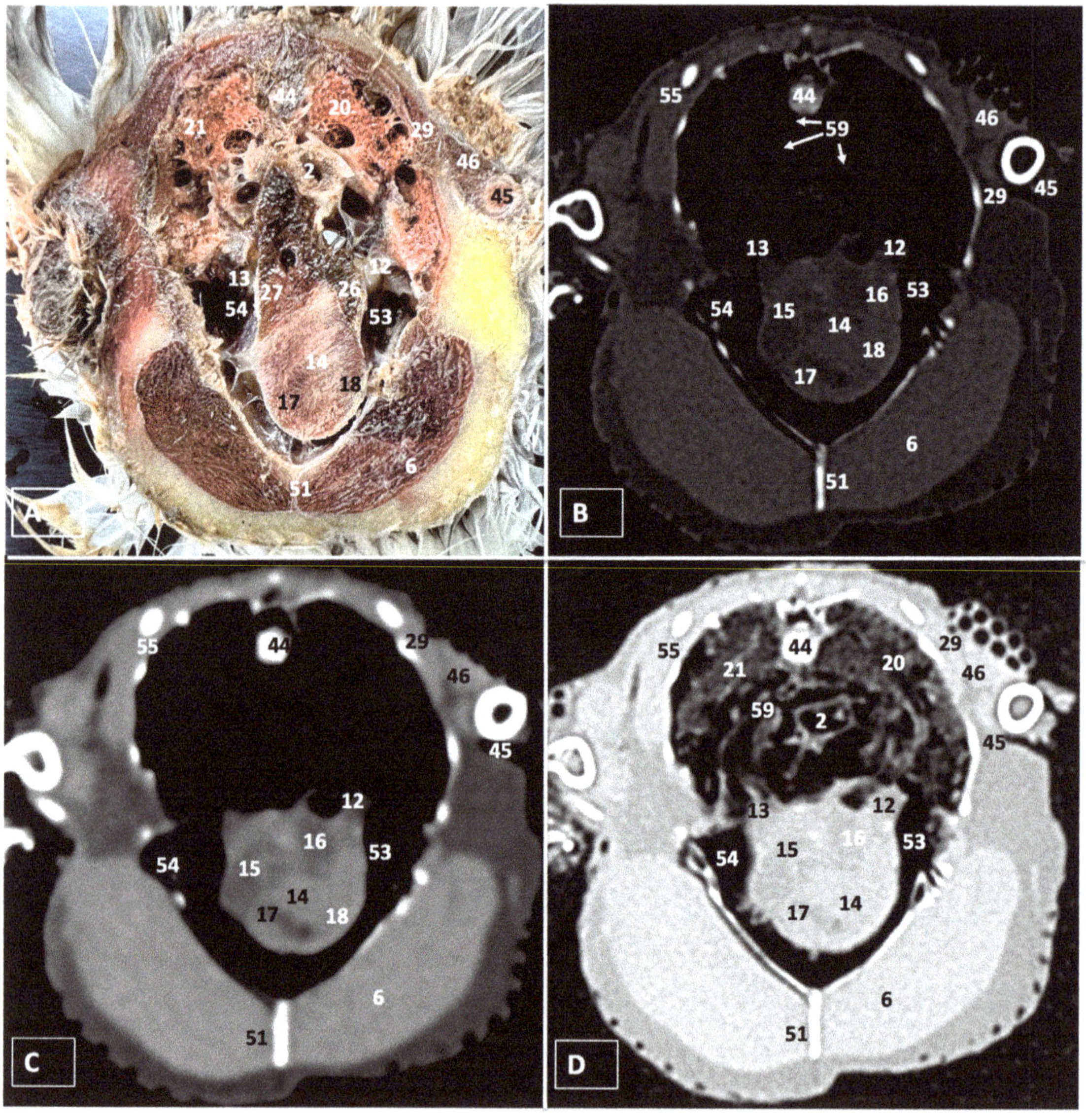

Figure 5. Transverse cross-section (**A**), bone (**B**), soft tissue (**C**), and pulmonary (**D**) CT images of the coelomic cavity of the Cory's Shearwater at the level of the heart, corresponding to line III in Figure 2. 44: vertebra (vertebral body). 21: right lung. 20: left lung. 2: esophagus. 29: ribs. 46: scapulohumeralis muscle. 45: humerus. 54: cranial thoracic air sac (right). 53: cranial thoracic air sac (left). 26: left hepatic lobe. 27: right hepatic lobe. 14: heart. 17: right ventricle. 18: left ventricle. 15: right atrium. 16: left atrium. 13: right brachiocephalic trunk. 12: left brachiocephalic trunk. 59: pulmonary vessels. 51: sternum. 6: pectoral muscle (thoracobrachialis muscle). 55: intercostal muscle.

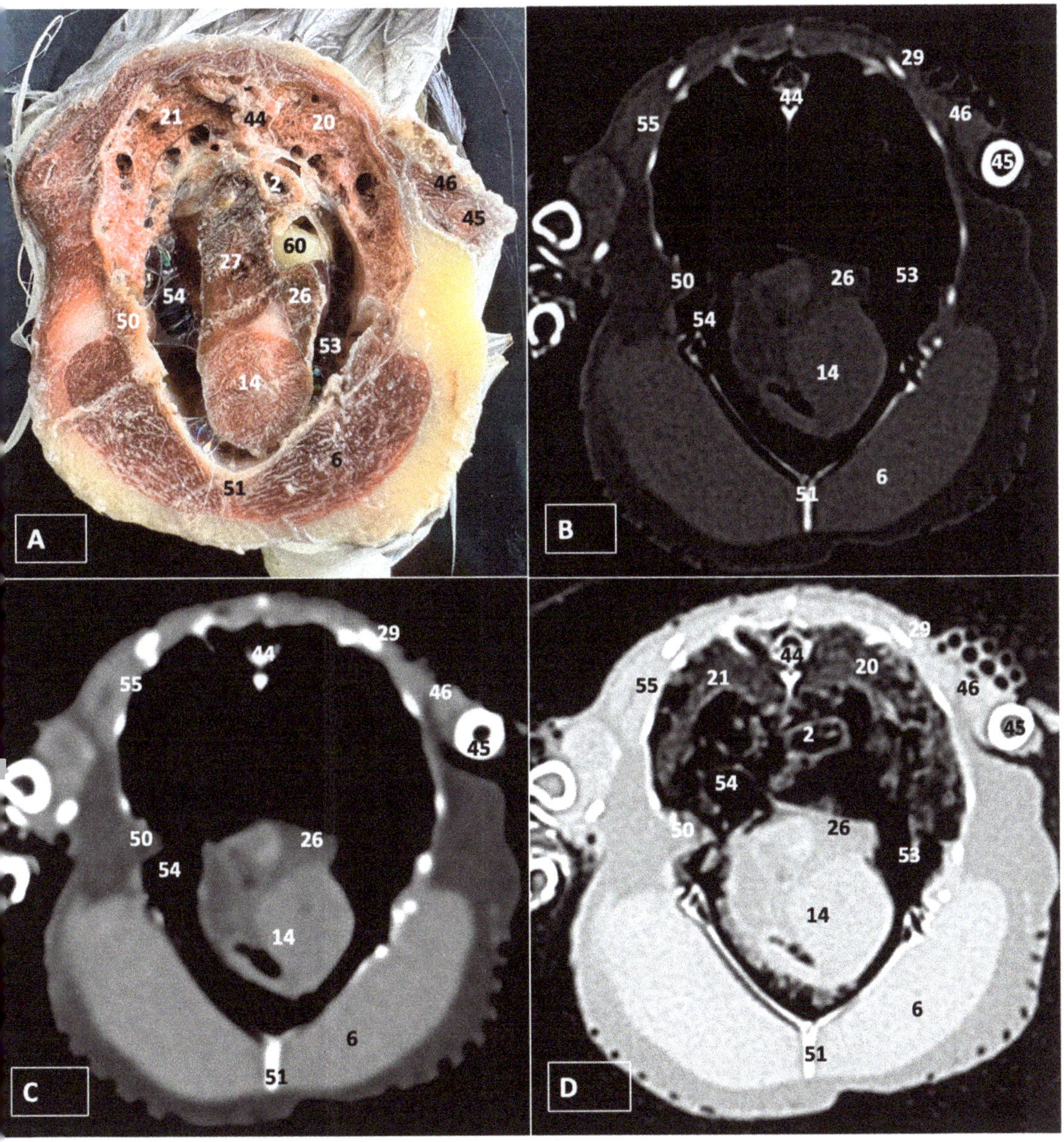

Figure 6. Transverse cross-section (**A**), bone (**B**), soft tissue (**C**), and pulmonary (**D**) CT images of the coelomic cavity of the Cory's Shearwater at the level of the hepatic lobes corresponding to line IV in Figure 2. 44: vertebra (vertebral body). 2: esophagus. 46: scapulohumeralis muscle. 45: humerus. 21: right lung. 20: left lung. 50: scapulohumeral caudal muscle. 27: right hepatic lobe. 26: left hepatic lobe. 60: proventriculus. 53: cranial thoracic air sac (left). 54: cranial thoracic air sac (right). 14: heart. 51: sternum. 6: pectoral muscle (thoracobrachialis muscle). 55: intercostal muscle.

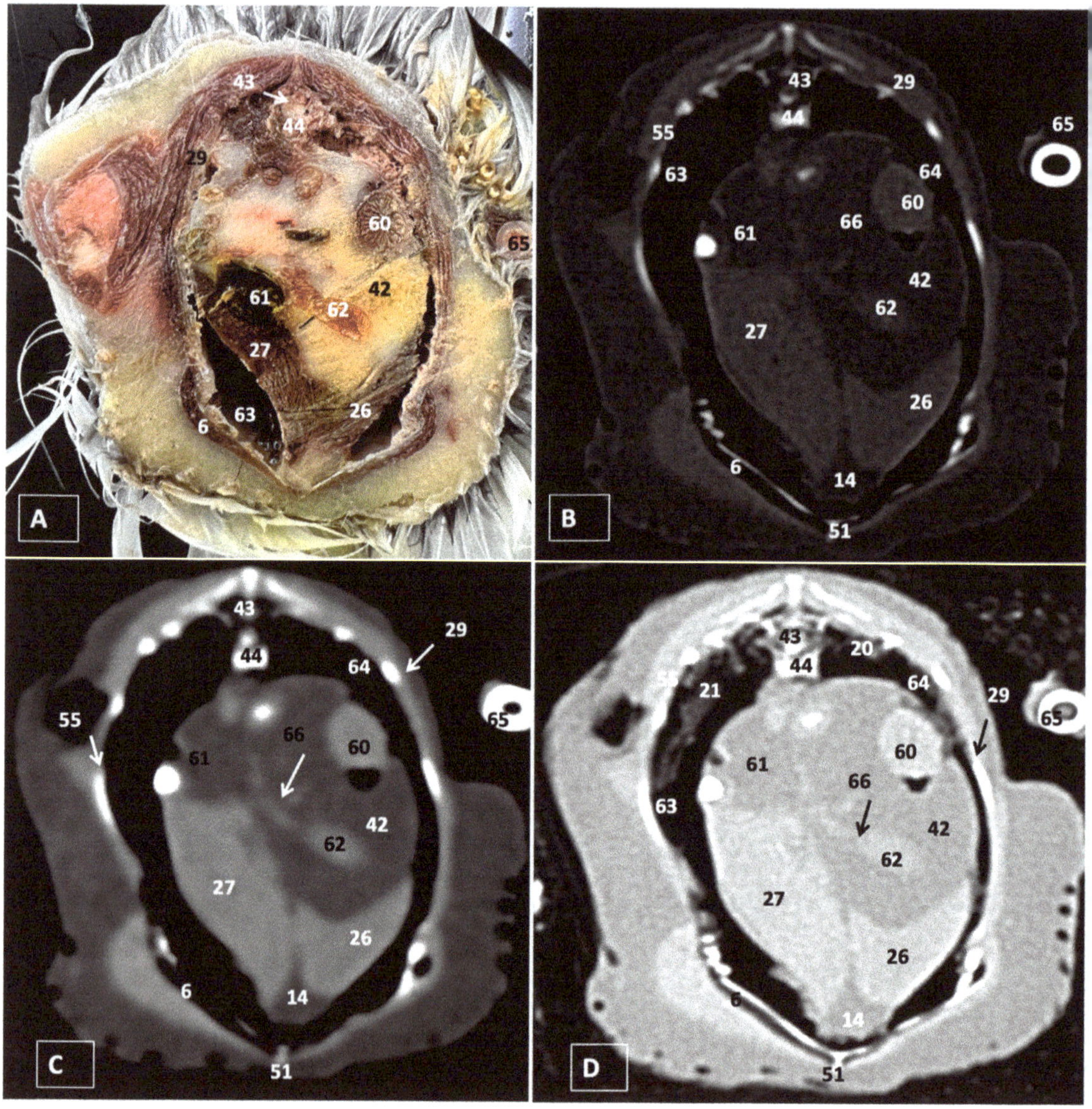

Figure 7. Transverse section (**A**), bone (**B**), soft tissue (**C**), and pulmonary (**D**) CT images of the coelomic cavity of the Cory's Shearwater at the level of the caudal thoracic air sac corresponding to line V in Figure 2. 44: spinal cord. 43: vertebra (vertebral body). 29: ribs. 65: femur. 21: right lung. 20: left lung. 27: right hepatic lobe. 66: common hepatoenteric duct. 26: left hepatic lobe. 14: heart. 61: gallbladder. 42: intestinal peritoneal cavity (full of fat). 63: caudal thoracic air sac (right). 64: caudal thoracic air sac (left). 60: proventriculus. 62: duodenum. 51: sternum. 6: pectoral muscle (abdominal portion). 55: intercostal muscle.

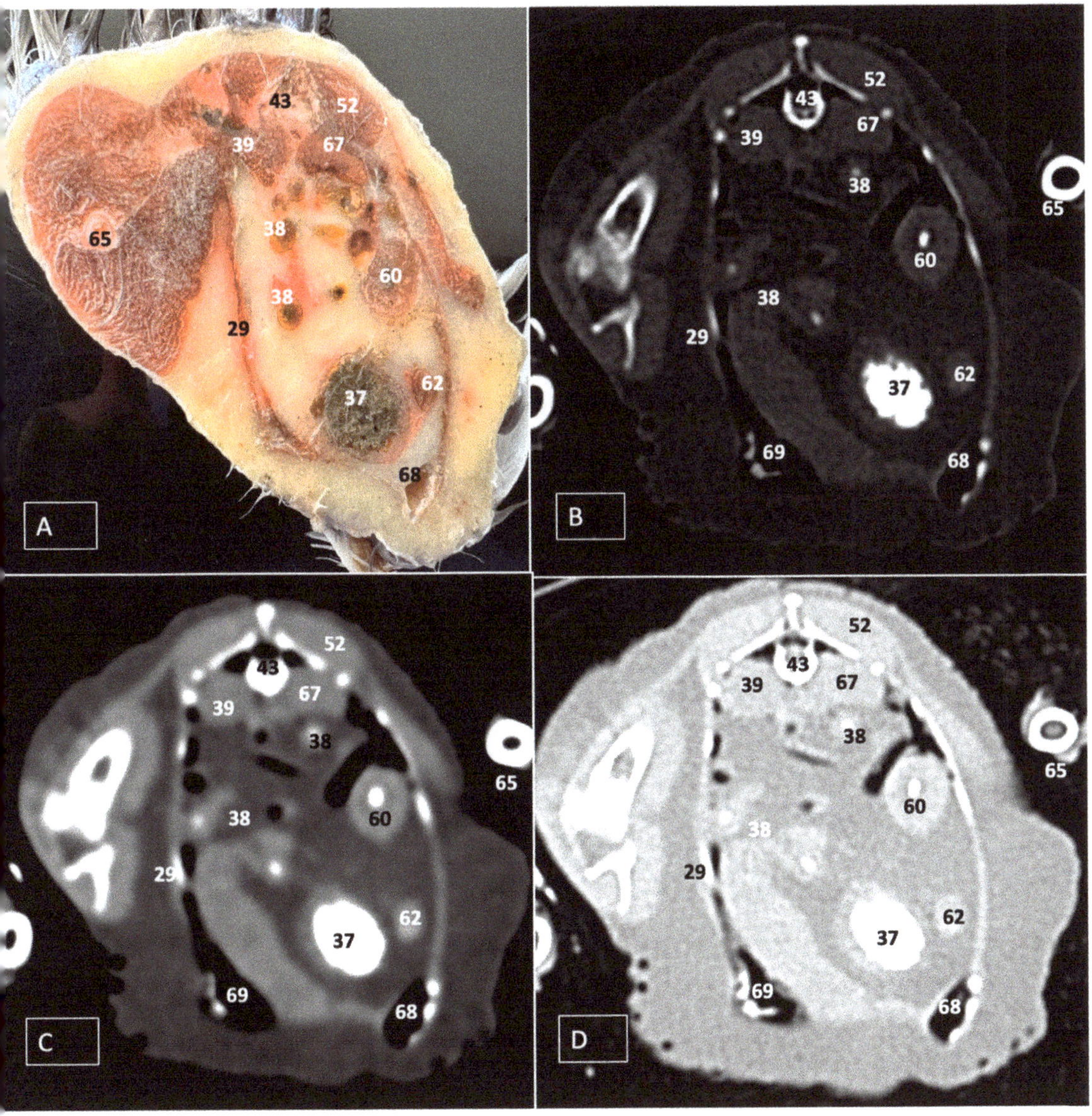

Figure 8. Transverse cross-section (**A**), bone (**B**), soft tissue (**C**), and pulmonary (**D**) CT images of the coelomic cavity of the Cory's Shearwater at the level of the kidneys, equivalent to line VI in Figure 2. 43: spinal cord. 39: right kidney. 67: left kidney. 38: jejunoileum. 60: proventriculus. 37: ventriculus. 62: duodenum. 69: right abdominal air sac. 68: left abdominal air sac. 29: ribs. 65: femur. 52: musculus levator caudae.

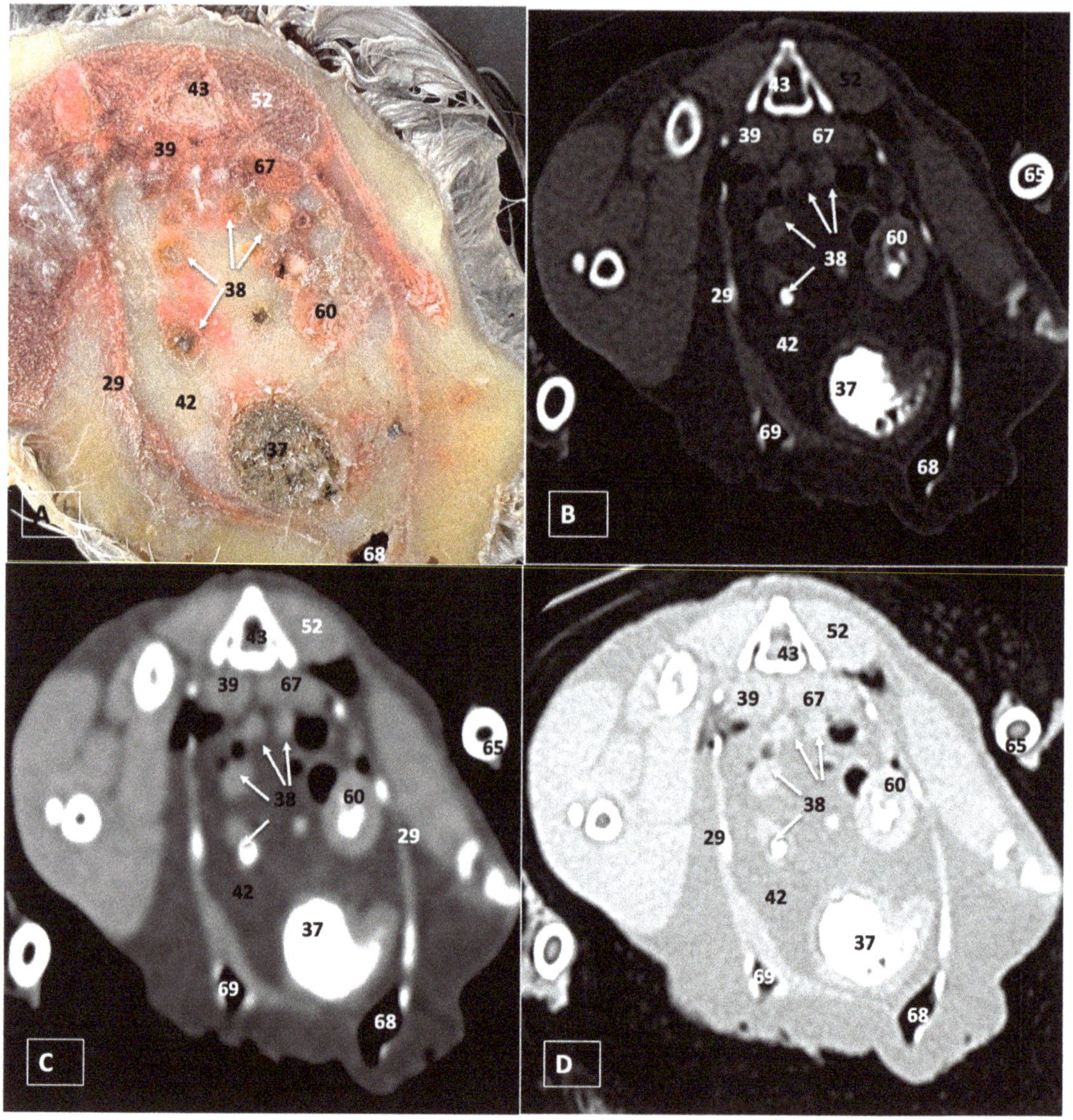

Figure 9. Transverse cross-section (**A**), bone (**B**), soft tissue (**C**), and pulmonary (**D**) CT images of the coelomic cavity of the Cory's Shearwater at the level of the ventriculus, equivalent to line VII in Figure 2. 43: spinal cord. 39: right kidney. 67: left kidney. 38: jejunoileum. 60: proventriculus. 37: ventriculus. 42: intestinal peritoneal cavity. 69: right abdominal air sac. 68: left abdominal air sac. 29: ribs. 65: femur. 52: musculus levator caudae.

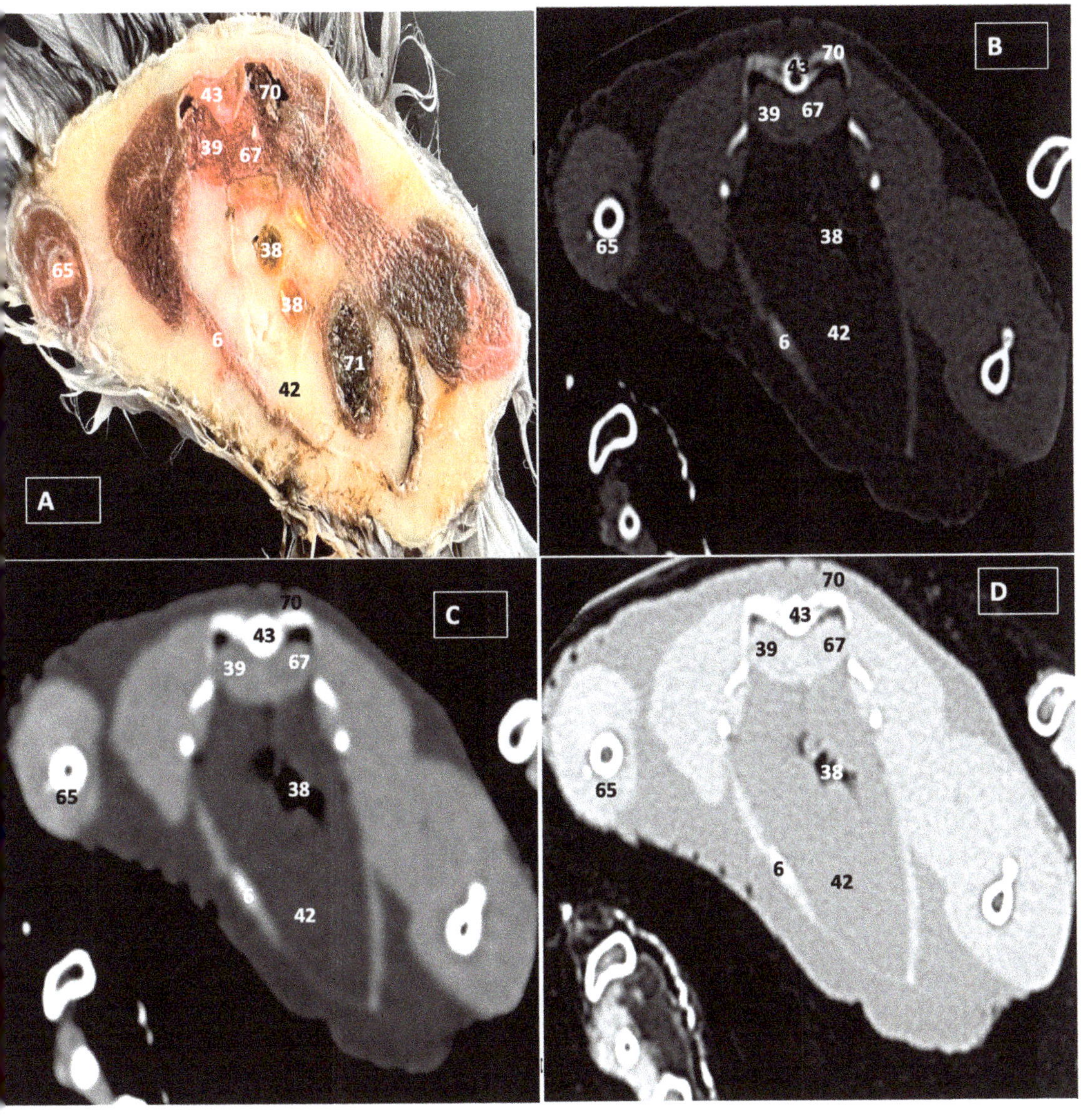

Figure 10. Transverse cross-section (**A**), bone (**B**), soft tissue (**C**), and pulmonary (**D**) CT images of the coelomic cavity of the Cory's Shearwater at the level of the caecum, equivalent to line VIII in Figure 1. 43: spinal cord. 70: synsacrum. 39: right kidney. 67: left kidney. 42: intestinal peritoneal cavity. 38: jejunoileum. 71: caecum. 65: femur. 6: pectoral muscle (abdominal portion).

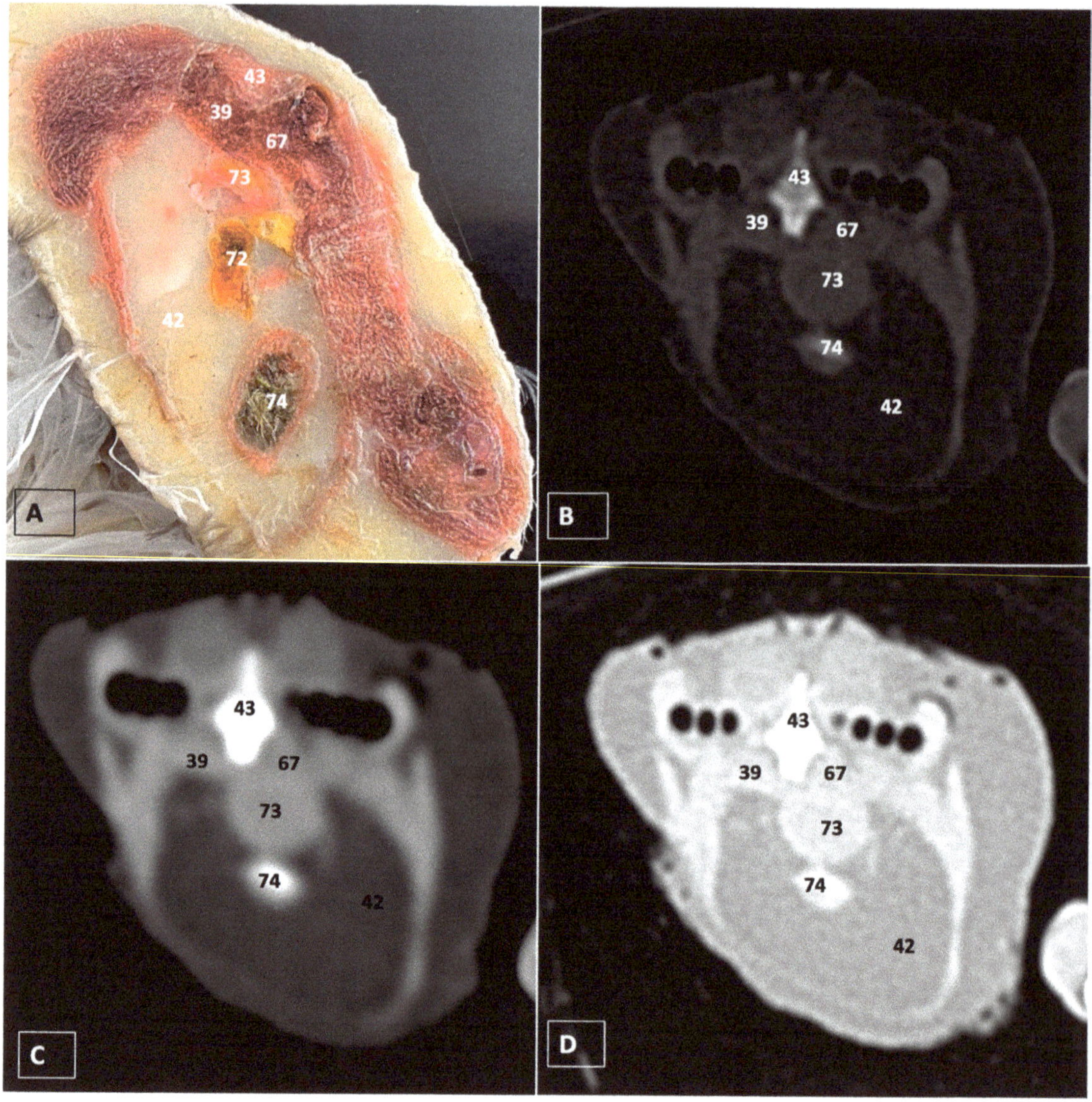

Figure 11. Transverse cross-section (**A**), bone (**B**), soft tissue (**C**), and pulmonary (**D**) CT images of the coelomic cavity of the Cory's Shearwater at the level of the oviduct and cloaca, equivalent to line IX in Figure 2. 43: spinal cord. 39: right kidney. 67: left kidney. 72: ileum. 42: intestinal peritoneal cavity (full of fat). 73: oviduct. 74: cloaca.

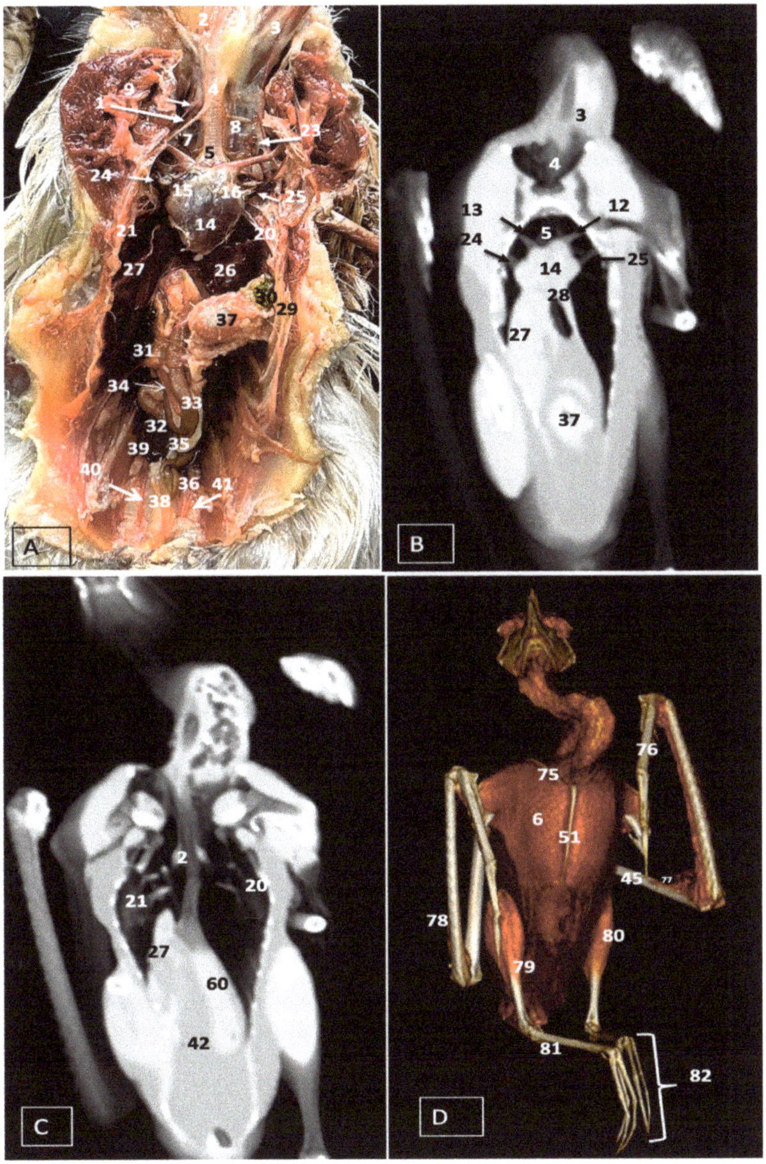

Figure 12. Anatomical dissection illustrating the coelomic cavity of Cory's Shearwater (**A**) and dorsal MPR volume rendering images in the pulmonary CT window of the coelomic cavity at level of the heart (**B**), the hepatic lobes (**C**), and a 3D CT reconstruction image (**D**). 1: sternotracheal muscle. 21: right lung. 20: left lung. 2: esophagus. 3: longissimus colli muscle. 4: trachea. 5: syrinx. 6: pectoral muscle (thoracobrachialis muscle). 7: right cervical air sac. 8: left cervical air sac. 9: jugular vein. 23: right thyroid. 24: right cranial cava. 25: left cranial cava. 12: left brachiocephalic trunk. 13: right brachiocephalic trunk. 14: heart. 15: right atrium. 16: left atrium. 26: left hepatic lobe. 27: right hepatic lobe. 60: proventriculus. 28: hepatopericardial ligament. 29: ribs: 30: ingesta: 31: spleen. 32: descending duodenum. 33 ascending duodenum. 34: pancreas. 35 duodenal-jejunal flexure. 36: caecum. 37: ventriculus. 38: jejunoileum. 39: right kidney. 40: right ureter. 41: left ureter.

42: intestinal peritoneal cavity. 75: clavicle. 76: metacarpal bone. 77: ventral collateral ligament + cranial cubital ligament. 45: humerus. 51: sternum. 78: ulna. 79: tibiotarsus. 80: fibularis longus muscle. 81: tarsometatarsus. 82: phalanges.

Moreover, these anatomical dissections and sections also enabled the examination of the Cory's Shearwater liver, comprising the right and left hepatic lobes (Figures 1A,C and 12A). The organ was conspicuously large, and its cranioventral segments of both hepatic lobes enveloped the heart (as illustrated in Figures 1A–C, 5A, 6A and 12A). Both lobes displayed comparable sizes, surpassing those of other organs within the coelomic cavity. The visceral surface of the organ is in close contact with the lungs. On the right, its visceral surface is in contact with a tubular structure that corresponds with the *duodenum* (Figure 7A). On the left visceral surface, it is associated with the glandular stomach (*proventriculus*) and with the muscular segment of the stomach referred to as the *ventriculus* (Figures 1C, 7A, 8A, 9A and 12A). Additionally, the medial border of the hepatic lobes exhibited the hepatopericardial ligament, establishing a connection between the liver and the heart apex (Figure 1A).

Additional elements of the digestive system, including the small and large bowel, were visible in these anatomical images. The dissected image located caudal to the liver provided a clearer depiction of the small intestine (*duodenum*) (Figures 1C, 7A, 8A, 9A, 10A, 11A and 12A). Within this region, the duodenum exhibited a distinct U-shaped configuration known as *ansa duodeni*, consisting of a descending segment, the *pars descendens*, and an ascending portion, the *pars ascendens* (Figures 1C and 12A). In the area proximal to the stomach and situated between the duodenal loops and these two segments, we noted the pancreas within the *mesoduodenum*, exhibiting a pale-pink coloration (Figures 1C and 12A). Further toward the caudal region, we were able to identify the terminal intestine (*caecum*) (Figures 1C and 12A). The *caecum* appears well developed and is an elongated pouch connected to the junction of the small and large intestines. In addition, it exhibited a unique coloration, adding to its anatomical distinctiveness.

Moreover, the dissected images facilitated the identification of the spleen, showing a flattened rectangular shape and brown to cherry red color (Figures 1C and 12A). It was medial to the proventriculus and ventriculus, close to the visceral surface of the liver. Furthermore, the employment of cross-sectional views proved indispensable in identifying both kidneys, which are positioned laterally to the spine and embedded dorsally within excavations of the synsacrum (Figures 1C, 8A, 9A, 10A, 11A and 12A). A noteworthy aspect revealed by these sections is the close association of the ventral surfaces of the kidneys with the paired abdominal air sacs, providing insight into the intricate anatomical relationships. Additionally, the ureters, a vital conduit for transporting urine from the kidneys to the cloaca for excretion, were distinctly observed to originate from the ventral surface of the kidneys (Figures 1C and 12A). In the posterior region of the coelomic cavity, we observed an excretory junction serving both the digestive and urogenital systems, referred to as the cloaca. This anatomical structure serves as a common passage for the digestive, urinary, and reproductive systems (illustrated in Figures 10A and 11A).

Furthermore, we observed various osseous structures pertaining to this cavity, encompassing the vertebrae, ribs, sternum, thoracic, and pelvic limbs, along with the pubis. These bony structures were intricately associated with different muscles, including distinct segments of the pectoral muscle (sternobrachialis, thoracic, and abdominal portions), as well as the supracoracoid, the scapulohumeralis, the scapulohumeral caudal, and the longissimus dorsi muscles (Figures 1A,B, 3A, 4A, 5A, 6A and 7A).

3.2. Computed Tomography Images

For CT examination, we selected those images that meticulously aligned with their anatomical sections (Figures 3B,D, 4B,D, 5B,D, 6B,D, 7B,D, 8B,D, 9B,D, 10B,D, 11B,D and 12B,C). Compared to the corresponding cross-section, these TC images offered enhanced morphological and tomographic insights into coelomic structures. Notably, CT images obtained with the pulmonary window setting (Figures 3B, 4B, 5B, 6B, 7B, 8B, 9B, 10B,

11B and 12B,C) showed remarkable visualization of different bones, related muscles, and soft tissues. These results were similar to those obtained with the soft tissue and bone window settings. As a result, we differentiated between several skeletal formations, which included the ulna, thoracic vertebrae, ribs, sternum, humerus, femur, metacarpal bones, and phalanges. In addition, we examined a variety of muscles and ligaments connected to these skeletal elements, such as the scapulohumeralis, scapulohumeral caudal, longissimus colli, longissimus dorsi, pectoral, intercostal muscles, ventral collateral and cranial cubital ligaments, and fibularis longus muscles (Figures 3D, 4D, 5D, 6D, 7D, 9D and 12D). More precisely, this window facilitated the identification of the trachea and its bifurcation in the left and right major bronchi, the syrinx, the pulmonary vessels, and the honeycomb-like pattern of the lung parenchyma (Figures 3D, 4D, 5D and 6D). This window also proved to be quite helpful in distinguishing the right and left primary bronchi entering into the ventromedial aspect of the lungs (Figure 4D). Other intrathoracic formations, including the heart and large vessels, such as the brachiocephalic trunks, were also identified (Figure 5D). Additionally, the walls of some air sacs, such as those of the cranial thoracic air sacs, could be distinguished (Figures 3D, 5D, 6D and 7D). Regarding the digestive components, this particular CT window skillfully outlined the *esophagus*, the hepatic lobes, the glandular and muscular stomach components, various intestinal segments, and the cloaca (Figures 3D, 5D, 6D, 7D, 8D and 9D). Additionally, the dorsal MPR volume rendering images distinguished the hepatopericardial ligament. This structure connects the liver to the pericardium, the membrane surrounding the heart (Figure 12B).

CT images acquired using the bone window and soft tissue settings (Figures 4B,C, 5B,C, 6B,C, 7B,C, 8B,C, 9B,C, 10B,C and 11B,C) demonstrated a remarkable distinction between bones and soft tissues. Interestingly, the different heart chambers and main arteries could be distinguished with these windows (as seen in Figure 5B,C). Moreover, various digestive structures, such as the right and left hepatic lobes, different intestinal segments, as well as the glandular and muscular portions of the cory's shearwater stomach, and the cloaca were also displayed (depicted in Figures 6B,C, 7B,C, 8B,C, 9B,C, 11B,C and 12B,C). These windows were helpful in the visualization of the thoracic and abdominal air sacs, which were joined by connective tissue with adjacent organs or muscles (as shown in Figures 3B,C,4B,C, 5B,C, 6B,C and 7B,C).

4. Discussion

In contrast to mammal species, birds do not present a structure separating the thoracic and abdominal cavities. As a result, these animals have a single cavity that contains all the viscera identified in mammals. To the best of the authors' knowledge, no reports have become available concerning the anatomy of Cory's Shearwater's coelomic cavity using advanced imaging diagnostic techniques. Nonetheless, these techniques have been applied to other bird species routinely seen by specialists in the field, such as the domestic pigeon (*Columba livia domestica*), toco toucans (*Ramphastos toco*), and gray parrot (*Psittacus erithacus*), demonstrating that they are essential for enabling normal anatomy and abnormalities in the internal organs of birds [20,23,24]. Traditionally, clinicians have used imaging modalities such as radiography [30] and ultrasonography (usually mode B) [16,27], which have been essential in obtaining images of the bony and the architecture of the internal organs that compose this region. Modern imaging modalities, such as CT and MRI, have become essential techniques in veterinary medicine because they provide early identification of focal and diffuse changes. In addition, they allow views of the body from several tomographic planes without repositioning the animal, thereby displaying images with excellent anatomic resolution in the absence of tissue superimposition, high contrast between different structures, and exceptional tissue differentiation that allows the evaluation of spatial relationships between organs that are not easily detected using radiography or ultrasonography. Despite all these advantages, there are also notable drawbacks, including the excessive cost of this equipment, the relatively high cost of the procedure, the need to sedate and possibly restrain the bird, and the longer examination time in the case of MRI. In consequence, their use is valuable only in those cases where it is specifically indicated or for remarkable birds such as breeding and falconry birds [8,20]. Due to these shortcomings, traditional diagnostic imaging techniques are still the most routinely used because they are fast, low-cost, and widely available in daily avian practice.

This investigation combined dissections and anatomical cross-sections, which showed essential information to help identify coelomic CT images using pulmonary, soft tissue, and bone window settings. Therefore, the utilization of dissections and anatomical cross-sections was immensely valuable in the depiction of the morphologic features of the Cory's Shearwater coelomic cavity. Thus, the dissected images played a pivotal role in distinguishing the spleen and some components of the endocrine system, such as the thyroid gland located in the angle between the subclavian and common carotid arteries. The identification of this gland helped in visualizing other main vessels, such as the jugular veins adjacent to both glands and converging at the venous angle to form the right and left cranial cava veins, as well as the subclavian arteries. However, due to the wide interval between slices employed in this study, some structures were not clearly discernible in the transverse and dorsal CT images using different windows. Comparable investigations conducted in other exotic species, such as rabbits, guinea pigs, iguanas, loggerhead turtles, and sea birds, have consistently demonstrated that this combination is essential to anatomical studies based on regions or divisions and in visualizing the relations between various organs [11,15,17,20,23,31,32].

In this research, the CT images obtained using pulmonary, soft tissue, and bone window settings were pivotal in displaying the main structures constituting the coelomic cavity. Thus, the use of the pulmonary window provided notable advantages in the visualization of the respiratory components and the related vascular structures. Therefore, the course of the trachea together with the esophagus on the right side of the neck, the distinction of the syrinx at the tracheal bifurcation and the course of the two brachiocephalic trunks, which continue as the subclavian arteries, were displayed using this window setting. Additionally, this window was essential in distinguishing the left and right cranial cava veins. Similar results have been described in other CT studies conducted on neonatal foals [30], reptiles [8,16], and a variety of avian species [11,20], including toco toucans and Atlantic puffins. Despite the fact that we did not administer contrast media, the combination of the different windows contributed to the visualization of the heart chambers and their associated vascular structures. We assumed that it could be due to the larger size of the animal, which provided better resolution in contrast to other studies performed on smaller birds [11,20], where cardiac chambers were not defined on CT. As in other reports performed on sea birds [11], this CT-based morphological evaluation has demonstrated the suitability of cadavers for studying diverse anatomical patterns. However, it is important to highlight the absence of blood flow in dead animals when comparing the results with those obtained from live specimens using contrast media, especially in small birds [11,17,18].

This comprehensive imaging also facilitated an adequate examination of Cory's Shearwater livers. Therefore, the precision afforded by the CT scans provided valuable insights about the liver. Thus, as observed in other avian species [18], the right lobe was larger than the left lobe. Moreover, the gallbladder was not only identified in the anatomical images but also in CT images. As explained before, it can be due to the larger size and volume of the bird, which could facilitate the visualization of these organs.

However, some limitations were observed, mainly due to work with fledglings that stranded as a consequence of artificial lights, which has been recognized as a relevant threat to biodiversity conservation. Birds can suffer from mass mortality episodes caused by artificial lights [33]. Thus, Cory's Shearwater fledglings are attracted and disorientated by artificial lights when they are leaving their nests for the first time and fly towards the sea [34]. As a result, we could not identify the gonads in these animals, which could be explained because they were sexually immature animals. Nevertheless, several avian species consequently do not show phenotypic sexual dimorphism, making it necessary to use endoscopy or DNA testing for definitive sex determination [28].

5. Conclusions

This study elucidates the spatial organization of the different organs comprising the coelomic cavity of this bird by using different CT windows, anatomical cross-sections, and dissections. Our results have demonstrated the efficacy of CT imaging to assess the avian anatomical organization through pulmonary, soft tissue, and bone windows without using a contrast medium. These images could be used as an initial anatomical reference in Procellariidae birds. However, further studies on adult specimens should be performed to detect possible differences with adult birds.

Author Contributions: Conceptualization, J.R.J., A.M.E., S.D. and M.E.; methodology, M.E., J.R.J., P.C., M.C.-F. and A.M.E.; investigation, J.R.J., A.M.E., M.F.-H., S.D. and M.E.; resources, P.C. and S.D.; writing—original draft preparation, J.R.J. and A.M.E.; writing—review and editing, J.R.J., A.M.E., P.C., S.D., M.F.-H., M.C.-F. and M.E. All authors have read and agreed to the published version of the manuscript.

Funding: This research received no external funding.

Institutional Review Board Statement: In this study, we did not need ethical approval, since the birds were coming to our hospital to rule out through CT any cause of stranding. From these studies, we obtained the information presented in our paper. As occurs in domestic mammals, just informed consent from the owner allowed us to carry out this study. Therefore, Tafira Wildlife Rehabilitation Center was informed that all animal identity information obtained from this study was treated as confidential to the extent permitted by law and only used for research or teaching purposes.

Informed Consent Statement: Informed consent was obtained from the owner of all subjects involved in the study.

Data Availability Statement: The data presented in this study are available in this article.

Acknowledgments: In loving memory of Alvaro Domingo Rodriguez Garcia. We would also like to thank Ayesh Mohamad, Carmen Mingot, Concepción Mingot, Emilia Mingot, Nicolas Aquino, Marisa Mohamad, Jamal Jaber, and Zaira Rivero for their support and constructive comments.

Conflicts of Interest: Author Alejandro Morales Espino was employed by the company IVC Evidensia Los Tarahales. The remaining authors declare that the research was conducted in the absence of any commercial or financial relationships that could be construed as a potential conflict of interest.

References

1. Pardela Cenicienta, Especie Clave para la Conservación de los Mares—SEO/BirdLife. Available online: https://seo.org/pardela-cenicienta-especie-clave-para-la-conservacion-de-los-mares/ (accessed on 1 January 2024).
2. Pardela Cenicienta Mediterránea—SEO/BirdLife. Available online: https://seo.org/ave/pardela-cenicienta-mediterranea/ (accessed on 1 January 2024).
3. (PDF) Pardela Cenicienta_Calonectris borealis borealis_. Available online: https://www.researchgate.net/publication/271271786_Pardela_Cenicienta_Calonectris_borealis_borealis (accessed on 1 January 2024).
4. MIGRACIÓN Y ECOLOGÍA ESPACIAL DE LAS POBLACIONES ESPAÑOLAS DE PARDELA CENICIENTA. Available online: https://doi.org/10.31170/0056 (accessed on 1 January 2024).
5. IUCN Red List of Threatened Species. Available online: https://www.iucnredlist.org/ (accessed on 1 February 2024).
6. Konishi, M.; Emlen, S.T.; Ricklefs, R.E.; Wingfield, J.C. Contributions of Bird Studies to Biology. *Science* **1989**, *246*, 465–472. Available online: https://www.science.org/doi/10.1126/science.2683069 (accessed on 13 January 2024). [CrossRef] [PubMed]
7. Current Therapy in Avian Medicine and Surgery | PDF | Veterinary Medicine | Pathology. Available online: https://es.scribd.com/document/384071318/Current-Therapy-in-Avian-Medicine-and-Surgery (accessed on 13 January 2024).
8. Diagnostic Imaging of Exotic Pets: Birds, Small Mammals, Reptiles | VetBooks. Available online: https://vetbooks.ir/diagnostic-imaging-of-exotic-pets-birds-small-mammals-reptiles/ (accessed on 13 January 2024).
9. Morales-Bordon, D.; Encinoso, M.; Arencibia, A.; Jaber, J.R. Cranial Investigations of Crested Porcupine (*Hystrix cristata*) by Anatomical Cross-Sections and Magnetic Resonance Imaging. *Animals* **2023**, *13*, 2551. Available online: https://www.mdpi.com/2076-2615/13/16/2551/htm (accessed on 13 January 2024). [CrossRef] [PubMed]
10. Fumero-Hernández, M.; Quintana, M.E.; Ramírez, A.S.; Fariña, I.M.; Calabuig, P.; Jaber, J.R. Morphometric Study of the Eyeball of the Loggerhead Turtle (*Caretta caretta*) Using Computed Tomography (CT). *Animals* **2023**, *13*, 1016. Available online: https://sciprofiles.com/publication/view/9aad60d6772e5b8f24e899e325335034 (accessed on 13 January 2024). [CrossRef] [PubMed]

11. Jaber, J.R.; Fumero-Hernández, M.; Corbera, J.A.; Morales, I.; Amador, M.; Ramírez Zarzosa, G.; Encinoso, M. Cross-Sectional Anatomy and Computed Tomography of the Coelomic Cavity in Juvenile Atlantic Puffins (Aves, *Alcidae, Fratercula arctica*). *Animals* **2023**, *13*, 2933. Available online: https://www.mdpi.com/2076-2615/13/18/2933/htm (accessed on 13 January 2024). [CrossRef] [PubMed]
12. Abraham, M.E.; Wakamatsu, N.; Lossie, G.A.; Karcher, D.M.; Heng, H.G.; Murakami, M. Computed tomographic and magnetic resonance imaging anatomy of the coelomic cavity in market-age commercial Pekin Ducks (*Anas platyrhynchos domesticus*). *Anat. Histol. Embryol.* **2023**, *52*, 843–861. Available online: https://pubmed.ncbi.nlm.nih.gov/37357377/ (accessed on 11 January 2024). [CrossRef]
13. Alonso-Farré, J.M.; Gonzalo-Orden, M.; Barreiro-Vázquez, J.D.; Ajenjo, J.M.; Barreiro-Lois, A.; Llarena-Reino, M.; Degollada, E. Cross-sectional anatomy, computed tomography and magnetic resonance imaging of the thoracic region of common dolphin (*Delphinus delphis*) and striped dolphin (*Stenella coeruleoalba*). *J. Vet. Med. Ser. C Anat. Histol. Embryol.* **2014**, *43*, 221–229. [CrossRef] [PubMed]
14. Ito, T.; Tanaka, T.; Kiyatake, I.; Izawa, T.; Furuya, M.; Sasai, K. Contrast-enhanced computed tomography and cross-sectional anatomy of the trunk in the brownbanded bamboo shark (*Chiloscyllium punctatum*). *J. Vet. Med. Ser. C Anat. Histol. Embryol.* **2023**, *52*, 437–447. [CrossRef] [PubMed]
15. Banzato, T.; Selleri, P.; Veladiano, I.A.; Martin, A.; Zanetti, E.; Zotti, A. Comparative evaluation of the cadaveric, radiographic and computed tomographic anatomy of the heads of green iguana (*Iguana iguana*), common tegu (*Tupinambis merianae*) and bearded dragon (*Pogona vitticeps*). *BMC Vet. Res.* **2012**, *8*, 53. [CrossRef] [PubMed]
16. Banzato, T.; Selleri, P.; Veladiano, I.A.; Zotti, A. Comparative evaluation of the cadaveric and computed tomographic features of the coelomic cavity in the green iguana (*Iguana iguana*), black and white tegu (*Tupinambis merianae*) and bearded dragon (*Pogona vitticeps*). *J. Vet. Med. Ser. C Anat. Histol. Embryol.* **2013**, *42*, 453–460. [CrossRef] [PubMed]
17. Veladiano, I.A.; Banzato, T.; Bellini, L.; Montani, A.; Catania, S.; Zotti, A. Computed tomographic anatomy of the heads of blue-and-gold macaws (*Ara ararauna*), African grey parrots (*Psittacus erithacus*), and monk parakeets (*Myiopsitta monachus*). *Am. J. Vet. Res.* **2016**, *77*, 1346–1356. [CrossRef] [PubMed]
18. Veladiano, I.A.; Banzato, T.; Bellini, L.; Montani, A.; Catania, S.; Zotti, A. Normal computed tomographic features and reference values for the coelomic cavity in pet parrots. *BMC Vet. Res.* **2016**, *12*, 182. Available online: https://pubmed.ncbi.nlm.nih.gov/27596377/ (accessed on 11 January 2024). [CrossRef] [PubMed]
19. dos Santos, G.J.; da Silva, J.P.; Hippólito, A.G.; Ferro, B.S.; Oliveira, E.L.R.; Okamoto, P.T.C.G.; Lourenço, M.L.G.; de Vasconcelos Machado, V.M.; Rahal, S.C.; Teixeira, C.R.; et al. Computed tomographic and radiographic morphometric study of cardiac and coelomic dimensions in captive blue-fronted Amazon parrots (*Amazona aestiva*, Linnaeus, 1758) with varying body condition scores. *Anat. Histol. Embryol.* **2020**, *49*, 299–306. Available online: https://onlinelibrary.wiley.com/doi/full/10.1111/ahe.12528 (accessed on 11 January 2024). [CrossRef] [PubMed]
20. da Silva, J.P.; Rahal, S.C.; Castiglioni, M.C.R.; Baldissera Gonçalves, R.A.; Doiche, D.P.; Moresco, A.; Mamprim, M.J.; Vulcano L.C. Radiography and computed tomography of the heart and lower respiratory tract in toco toucans (*Ramphastos toco*). *Anat Histol. Embryol.* **2020**, *49*, 541–549. Available online: https://onlinelibrary.wiley.com/doi/full/10.1111/ahe.12559 (accessed on 11 January 2024). [CrossRef] [PubMed]
21. Sisson: Anatomía de los Animales Domésticos v....—Google Académico. Available online: https://scholar.google.com/scholar_lookup?title=Anatom%C3%ADa+de+los+Animales+Dom%C3%A9sticos&author=Sisson,+S.&author=Getty,+R.&author=Grossman,+J.D.&publication_year=1982 (accessed on 13 January 2024).
22. König, H.; Liebich, H.; Bragulla, H. Anatomie und Propädeutik des Geflügels: Lehrbuch und Farbatlas für Studium und Praxis Available online: https://cir.nii.ac.jp/crid/1130000798308564608 (accessed on 13 January 2024).
23. Pepperberg, I.M.; Howell, K.S.; Banta, P.A.; Patterson, D.K.; Meister, M. Measurement of Grey Parrot (*Psittacus erithacus*) Trachea via Magnetic Resonance Imaging, Dissection, and Electron Beam Computed Tomography. *J. Morphol.* **1998**, *238*, 81–91. Available online: https://onlinelibrary.wiley.com/doi/10.1002/ (accessed on 13 January 2024). [CrossRef]
24. Romagnano, A.; Shiroma, J.T.; Heard, D.J.; Johnson, R.D.; Schiering, M.R.; Mladinich, C.R.J. MAGNETIC RESONANCE IMAGING OF THE BRAIN AND COELOMIC CAVITY OF THE DOMESTIC PIGEON (*Columba livia domestica*). *Vet. Radiol. Ultrasound* **1996**, *37*, 431–440. Available online: https://onlinelibrary.wiley.com/doi/full/10.1111/j.1740-8261.1996.tb01256.x (accessed on 13 January 2024). [CrossRef]
25. Tomographic Anatomy of the Golden Eagle (Aquila chrysaetos) on JSTOR. Available online: https://www.jstor.org/stable/20460266?casa_token=7aagwF4LiQ0AAAAA:tUlvpojf3tSoTAhSYAv_2dZYtJfM93_YoeG0-lHGvqc6FXdqyb41AuSL427u5pj1Fjt4EfF2xldnJ7GYD7ikHoQFcP3mpC5V4xJzT4p478Z9mntXPjA (accessed on 13 January 2024).
26. Sandoval, J. *Anatomía Veterinaria*; Imprenta Moderna: Logroño, Spain, 2003.
27. Dyce, K.M.; Aja Guardiola, S.; Morales Saavedra, J.L.; Palacios Martínez, J.R.; Wensing, C.J.G.; Sack, W.O. Anatomía Veterinaria Available online: https://books.google.com/books/about/Anatom%C3%ADa_veterinaria.html?hl=es&id=0AwwCgAAQBAJ (accessed on 10 December 2023).
28. König, H.E.; Korbel, R.; Liebich, H.G.; Bragulla, H.; Klupiec, C. *Avian Anatomy : Textbook and Colour Atlas*; 5m Publishing: Sheffield, UK, 2016; 340p.
29. MacLelland, J. Atlas en Color de Anatomía de las Aves. 1992. Available online: https://dialnet.unirioja.es/servlet/libro?codigo=157364 (accessed on 7 January 2024).

. Arencibia, A.; Corbera, J.A.; Ramírez, G.; Díaz-bertrana, M.L.; Pitti, L.; Morales, M.; Jaber, J.R. Anatomical Assessment of the Thorax in the Neonatal Foal Using Computed Tomography Angiography, Sectional Anatomy, and Gross Dissections. *Animals* **2020**, *10*, 1045. Available online: https://pubmed.ncbi.nlm.nih.gov/32560487/ (accessed on 1 February 2024). [CrossRef] [PubMed]
. Paré, J.A.; Veterinarian, S. BSAVA Manual of Psittacine Birds, 2nd ed. *Can. Vet. J.* **2008**, *49*, 268. Available online: https://www.ncbi.nlm.nih.gov/pmc/articles/PMC2249719/ (accessed on 1 February 2024).
. Fumero-Hernández, M.; Encinoso, M.; Melian, A.; Nuez, H.A.; Salman, D.; Jaber, J.R. Cross Sectional Anatomy and Magnetic Resonance Imaging of the Juvenile Atlantic Puffin Head (Aves, Alcidae, *Fratercula arctica*). *Animals* **2023**, *13*, 3434. [CrossRef] [PubMed]
. Rodríguez, A.; Rodríguez, B.; Negro, J. GPS tracking for mapping seabird mortality induced by light pollution. *Sci. Rep.* **2015**, *5*, 10670. [CrossRef] [PubMed]
. Imber, M.J. Behaviour of petrels in relation to the moon and artificial lights. *Notornis* **1975**, *22*, 302–306.

animals

MDPI

Article

Nature or Nurture: Is the Digestive System of the *Pontoporia blainvillei* Influenced or Determined by Its Diet?

Carlos Tostado-Marcos [1], María Julieta Olocco Diz [1], Rosario Martín-Orti [2], Juan-Pablo Loureiro [1], Ignacio Molpeceres-Diego [1,3], Enrique Tendillo-Domínguez [1], Pilar Pérez-Lloret [2], Inmaculada Santos-Álvarez [2,*] and Juncal González-Soriano [2,*]

1 Fundación Mundo Marino, Av. X 157, San Clemente del Tuyú B7105, Provincia de Buenos Aires, Argentina; tostadomarcoscarlos@gmail.com (C.T.-M.); mjoloccodiz@yahoo.com.ar (M.J.O.D.); juanploureiro@gmail.com (J.-P.L.); ignamolp11@gmail.com (I.M.-D.); enriquetd95@gmail.com (E.T.-D.)

2 Sección Departamental de Anatomía y Embriología (Veterinaria), Facultad de Veterinaria, Universidad Complutense, Avenida Puerta de Hierro s/n, 28040 Madrid, Spain; rosamart@ucm.es (R.M.-O.); pilper01@ucm.es (P.P.-L.)

3 Veterinary Histology and Pathology, Atlantic Center for Cetacean Research, University Institute of Animal Health and Food Safety (IUSA), Veterinary School, University of Las Palmas de Gran Canaria, 35413 Arucas, Spain

* Correspondence: inmasant@ucm.es (I.S.-Á.); juncalgs@vet.ucm.es (J.G.-S.)

Simple Summary: Anatomy is considered critical to understanding the size and shape of the different systems of animals belonging to the same group or individuals that are close in the phylogenetic scale. In the case of the Franciscana dolphin or *Pontoporia blainvillei*, the digestive apparatus shows some differential characteristics compared to other dolphins or other marine mammals. In the present work, the authors try to demonstrate whether these characteristics are conditioned by the animal's own diet or if they have more to do with certain phylogenetic adaptations. Considering that the Franciscana dolphin is an endangered species, any work that could facilitate better knowledge of these individuals has a high social, ecological, and scientific value because it will help in their care and conservation, and this is of utmost importance to implement effective management strategies for this species.

Abstract: The Franciscana (also known as the La Plata River Dolphin) is a small dolphin that lives in the coastal waters of Brazil, Uruguay, and Argentina. This species is considered the most endangered marine mammal in the western South Atlantic Ocean. Anatomic dissection of the digestive system of 19 animals of different ages, including 2 neonates, 12 juveniles, and 5 adults, was performed. Parameters related to length, breadth, weight, and diameter of the digestive viscera were considered in each case. Our results show that the Franciscana dolphin presents differential characteristics in relation to several parts of the digestive system, including, specifically, the tongue, the teeth, the stomach, and the small intestine. Thus, this paper add precious information to the actual knowledge of this vulnerable marine mammal species in order to improve conservation efforts.

Keywords: *Pontoporia blainvillei*; anatomy; digestive system; diet; adaptations

1. Introduction

The large group of marine mammals has developed numerous anatomical adaptations in response to their aquatic environment over time. Among them, cetaceans and sirenians stand out for presenting the highest degree of anatomical modifications, because they have achieved a complete adaptation to aquatic life. These two groups are distinguished, among other characteristics, by their feeding. Thus, while sirenians are the only herbivorous marine mammals, cetaceans are considered mostly carnivorous.

Cetaceans are divided into two main groups: those with teeth (Odontocetes) and those without teeth, or baleen cetaceans (Mysticetes). The Pontoporia dolphin (*Pontoporia blainvillei*), also known as the Franciscana dolphin, is classified in the suborder Odontocetes, which comprises six families and 71 species. Between them, the best known are the dolphins, but the odontocetes, or cetaceans with teeth, also include the killer whale, the sperm whale, the beaked whales, the beluga whale, the porpoises, and the narwhal. Many of them have undergone remarkable adaptations to their corresponding habitats and/or feedings, such as the loss of teeth in the upper arch (sperm whales), a small number of teeth (beaked whales), long beaks with small dorsal fins (river dolphins), or a rounded head and the absence of a beak and dorsal fin (narwhals and beluga whales). Many of these adaptations are related to their digestive system [1,2].

The Franciscana dolphin is endemic to the Atlantic coast of South America, and it represents one of the most emblematic species of the coastal waters of Argentina and Brazil [3]. Within its distribution range, up to five genetically distinct populations have been identified [4]. Like other cetaceans with coastal habits, the Franciscana is particularly vulnerable to the impact of human activities [5–7]. Habitat degradation, pollution, and, above all, bycatch have led to a high mortality rate of the species in recent years [8–12]. As a consequence, it has become the most endangered small cetacean in the southwest Atlantic Ocean [13,14], and has been classified as "Vulnerable" by the International Union for Conservation of Nature (IUCN) [15]. According to the Scientific Committee of the International Whaling Commission [16], an incidental population mortality of 1% per year is of concern in small cetacean populations, and a mortality of 2% may not be sustainable. For the Franciscana dolphin, annual mortality reaches between approximately 2% and 5% [8,11,17,18].

In Samborombón Bay (Argentina), where the species has been studied in greater detail, two distinct habitats have been observed: a brackish water estuary and an adjacent marine system. Studies have confirmed differences in the predominant prey species, but fish, especially juveniles less than 8 cm, as well as small loliginid squids constitute the preferred food in both territories [19]. In addition, evidence of predation by Franciscanas has been observed from a very early age, at approximately 2.5–3 months, when they initiate a transitional diet combining mother's milk with solid food. Weaning in Pontoporia is gradual, with early predation on crustaceans and fish, and the lactation period is estimated to last 6–7 months [19,20].

In light of the above, the authors wonders what is meant by "diet". In fact, what is diet? What does this term mean? What is meant when one talks about diet? The word *diet* comes from the Latin "*diaeta*", and this in turn derives from the Greek "*dayta*", which means regime of life. Diet, in nutrition, refers to the regular consumption of food and beverages. Therefore, a balanced diet is one that contains different types of foods in certain amounts and proportions so that requirements of calories, proteins, minerals, vitamins, and alternative nutrients are adequately met. But diet can also be defined as the total amount of food consumed by an animal or a group of animals. Moreover, diet can be considered the sum of the meals that an individual eats. This first statement includes certain parameters, such as food preferences, the prey's size, the degree of feed hardness, and daily feeding frequency, among others.

These are the parameters that were considered in carrying out this work. Thus, the main goal of the present study was to describe in detail the digestive system of the Franciscana dolphin in trying to determine whether these anatomical characteristics could be conditioned by several factors in relation to its diet or if the digestive system of this species is modified by phylogeny.

2. Materials and Methods

This anatomical study was carried out on 19 cadavers of the species *Pontoporia blainvillei*, made up of 5 adults, 12 juveniles, and 2 neonates. The individuals were classified into three groups according to age and body development according to data published by Arruda Ramos et al. [21]. Neonates have a body length of approximately 71 cm and no teeth. The vibrissae and umbilical cord can also be observed. Juveniles include individuals with a body length up to 130 cm and the presence of developing teeth (Figure 1). Adults have a body length greater than 130 cm and the presence of developed teeth.

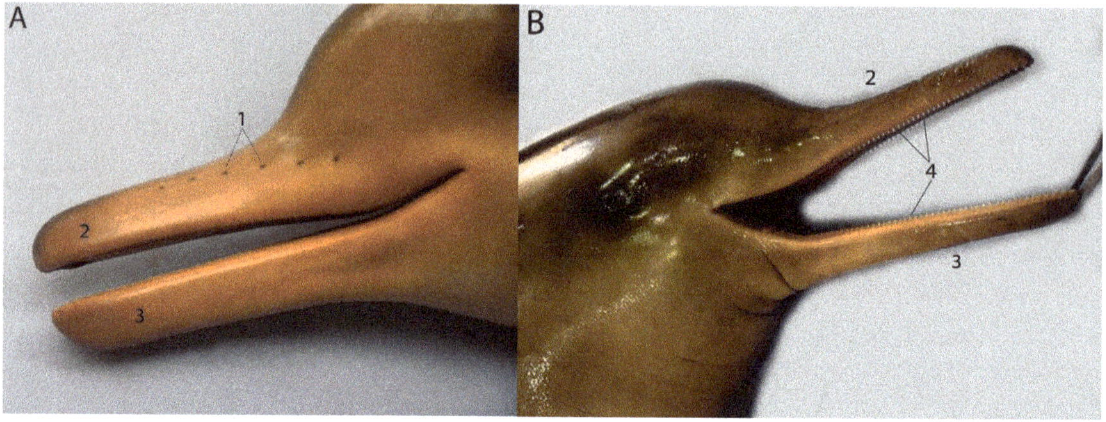

Figure 1. Snout of Pontoporia, external aspect: (**A**) neonate; (**B**) juvenile; (1) vibrissae; (2) maxilla; (3) mandible; (4) teeth.

In all cases, except one, they were found stranded dead on the beach. The dolphin cadavers, in all cases in good conditions, were transferred to the necropsy room of the rescue and rehabilitation center Fundación Mundo Marino, where anatomical dissections of each animal were carried out. Only one neonate was found stranded alive on the beach, because quite often, sick animals go to the shore to die. It later died during the rehabilitation process. All of them belonged to the group Area FMA sub-B (1), a population that inhabits the coast of the province of Buenos Aires. In no case were the animals frozen. Dissections were carried out over a period of less than 12 h from the time the cadavers were found. When it was necessary, the cadavers was kept under refrigeration (for less than 12 h). Practically in all cases, dissections were performed according to the "Documento Técnico sobre Protocolo Nacional de Actuación para Cetáceos Varamientos de Cetáceos" published by BEVACET [22], with the animals placed in right lateral recumbency. A few cases were placed in ventral recumbency and supine recumbency in order to have a complete visualization of the digestive system's topography. All dissections were carried out by veterinarians.

To keep the esophagus intact, the superficial musculature of the neck area was removed, exposing the trachea and esophagus. Then, a triangular incision along the inner sides of the jaws was made. The hyoid apparatus was cut, and the tongue was extracted through the intermandibular space. Then, the cardiorespiratory system was progressively separated.

To make the esophagus independent, its connections with the respiratory system (esophagus–trachea) were removed, and it was sectioned at the beginning, exactly at the level of the pharynx. To maintain the esophagus–stomach union, an eyelet-shaped cut was made around the esophageal hiatus. Finally, the entire digestive system was externalized by sectioning the most caudal part of the rectum, including the liver and the pancreas.

As mentioned before, the present study included different specimens—5 adults, 12 juveniles, and 2 neonates—each with obviously different degrees of development. Because of this, the corresponding arithmetic means of the data obtained were calculated separately to avoid deviations and erroneous data.

3. Results

In the Franciscana dolphin, cranial asymmetry (determined by the distance between the eye and the blowhole) was very slight, with a mean in juveniles of 9.2 cm on the right side and 7.8 cm on the left side; in neonates, it was 8 cm on the right side and 7 cm on the left side.

The snout of the Franciscana dolphin grows and lengthens as the animal ages until it reaches adulthood, as occurs in other river dolphin species (Figure 1). In neonate individuals, small vibrissae can be observed in the area of the snout (maxilla) that will disappear after a few days of life (Figure 2).

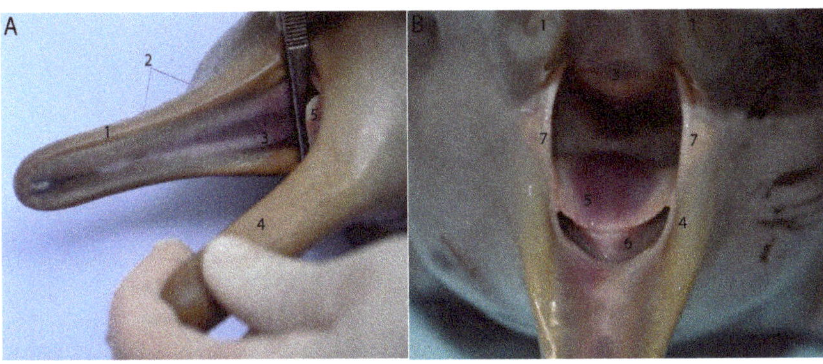

Figure 2. Roof and floor of the oral cavity of a neonate individual. Note the absence of teeth: (**A**) lateral view; (**B**) rostral view; (1) maxilla; (2) vibrissae; (3) hard palate; (4) mandible; (5) tongue; (6) musculature of the tongue; (7) corners of the mouth.

3.1. Oral Cavity

The buccal cavity is elongated, and it has a large buccal opening angle (Figure 3). This is due, in part, to the telescoping of the skull (a process of rostral elongation of the dolphin skull) and also the presence of folds at the level of the corners of the mouth (folds that join the palate and the jaw) (Figure 3). The distance between the two folds is related to the size of the isthmus of the fauces. This is important, as it will determine the size of the prey that the Franciscana dolphin is able to ingest. In neonates, the maw isthmus width was 2.7 cm, and in juveniles it was 3.2 cm.

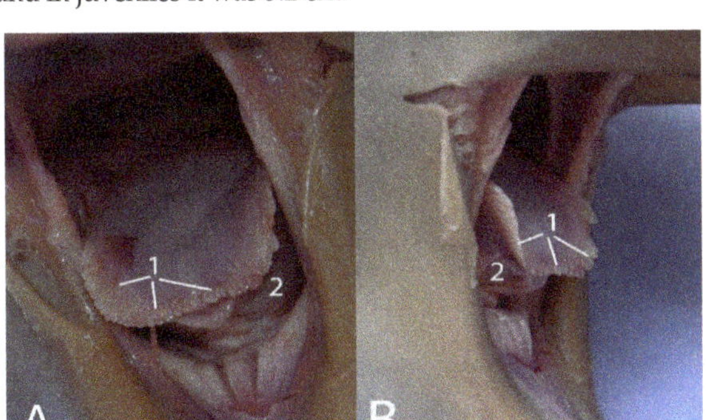

Figure 3. Tongue of lactating individuals: (**A**) rostral view; (**B**) lateral view; (1) anterolateral mechanical papillae; (2) tongue musculature.

3.1.1. Teeth

This species has a homodont dentition (Figure 1). The total number of teeth is high, with small variations among individuals. The total number ranges from 220 to 240. Normally, the maxilla has more teeth than the mandible (Table 1). Its morphology is conical, thin, and small in size. It has a saw-like appearance. They are very sharp teeth. Their feeding function is to press their prey, developing "ram feeding". There is a slight variation in the length of the teeth according to the area of the snout. There is growth as they advance from the most rostral part to the caudal, reaching a maximum size in the middle of the snout approximately, and then they begin to decrease, presenting the smallest size in the most caudal area of the snout, forming a plateau structure (Table 1).

Table 1. Morphometric data (number and length) of the teeth of the specimens of *Pontoporia blainvillei* used in the present study.

Age Group	ID	Maxilla (No)		Jaw (No)		Total (No)	Average (No)	Length (cm)
		Right	Left	Right	Left			R-M-C
A	M09	58	59	57	58	232	224	0.5-0.7-0.2
	M43	55	55	53	53	216		0.5-0.7-0.2
J	M20	56	56	51	52	215	218	0.4-0.5-0.2
	M21	52	54	55	53	214		0.4-0.6-0.2
	M60	57	55	53	54	219		0.4-0.5-0.2
	M65	57	57	55	55	224		0.3-0.5-0.2
	M71	52	52	53	52	209		0.5-0.6-0.4
	M78	59	58	56	56	229		0.4-0.5-0.2

A: adult; J: juvenile; ID: individual identification; R-M-C: rostral zone—middle zone—caudal zone of maxilla/mandible.

These animals are born without teeth. They do not have deciduous teeth, and they directly develop permanent teeth. They are monophyodont animals (teeth erupt only once). Eruption begins after approximately 2 months of lactation.

3.1.2. Tongue

The free portion of the tongue is small, short, thick, and somewhat mobile. It occupies most of the buccal cavity proper, and its limit is marked by the corner of the mouth (Figures 2 and 3). The average length is about 6 cm, with very little variation among the different age groups. There is also a slight difference in width between the base and the tip (Table 2). Neonate and juvenile individuals have anterolateral mechanical papillae arranged on both sides of the edge of the tongue apex in a single row. A clear decrease in the number of papillae was observed with age. As the individual develops, there is progressive atrophy/involution of the papillae until they disappear completely in adulthood because their functionality is attributed to the sucking process during lactation. These papillae are small, thin, and elongated. In lactating individuals, the average number of papillae was 50, with 32 in juveniles and 3 in adults.

3.2. *Esophagus*

The esophagus of the Franciscana is of no special anatomical interest, as its structure is similar to that of other mammals. It is a tubular organ that connects the pharynx to the stomach. It runs dorsal to the trachea and moves slightly to the left at the end of its course to cross the diaphragm at the esophageal hiatus, and it joins the stomach in the intrathoracic portion of the abdomen. The esophagus joins directly with the main stomach (Figures 4 and 5). The mean length was 17 cm in neonates, 25 cm in juveniles, and 31 cm in adults. It can be observed that there is a development of the organ according to body development. As for the diameter, it is very similar throughout its length. A slight variation was observed between sections. In adult specimens, the average was 5 cm in the cranial part, decreasing to 3.5 in the middle part and widening again in the caudal part near the entrance to the main stomach, with an average of 4 cm.

Table 2. Morphometric data of the tongue and the esophagus of the specimens of *Pontoporia blainvillei* used in the present study.

		Tongue					Esophagus				
Age Group	ID	Weight (g)	Length (cm)	Width (cm) Base	Width (cm) Vertex	Papillae m. (No)	Weight (g)	Length (cm)	Diameter (cm) cr	Diameter (cm) med	Diameter (cm) cd
A	M09	36	6	3.2	1.5	9	51	27	3.5	2.5	3
	M22	47	7	4	2	0	163	29	5	3.5	4.5
	M24	39	6.5	3.7	1.8	0	192	35	6.2	4.4	4.7
	M43	34	6	3.5	1.9	0	160	33	6	4	4.5
	M47	31	5	3.5	2	6	135	32	5	3	3.3
	mean	37	6.1	3.6	1.8	3.0	140	31	5.1	3.5	4.0
J	M12	28	4.3	2.5	1.9	37	19	19	2.4	---	1.5
	M16	35	5.7	3.6	2.1	27	77	26	3.5	---	2.8
	M20	31	5.7	3.5	1.7	29	92	26	4	3.2	3.6
	M21	27	5.5	3	1.4	15	95	24	2.8	2.4	2.6
	M23	28	5.5	2.8	2	38	44	20	3	1.6	2.5
	M28	33	6	3	2	19	63	24	3	2.5	2.8
	M29	27	6	2.8	2	25	38	23	2.8	2.2	2.5
	M45	27	6	2.9	2.4	32	67	29	3	2.4	2.6
	M60	25	6	2.8	2	41	40	24	2.2	---	1.8
	M71	28	6	2.5	2	35	59	27	2.8	---	3.2
	M78	34	6.7	3.2	2.4	30	116	31	4.3	3.7	4
	mean	29	5.8	3	2.0	30	65	25	3.1	2.6	2.7
N	M17	25	5.5	2.5	2	51	21	18	2	1.4	1.8
	M81	22	5	2.3	1.7	48	12	15	1.5	0.9	1.3
	mean	24	5.3	2.4	1.9	50	17	17	1.8	1.1	1.6

A: adult; J: juvenile; N: neonate; ID: individual identification; papillae m.: mechanical papillae; cr: cranial; med: medium; cd: caudal.

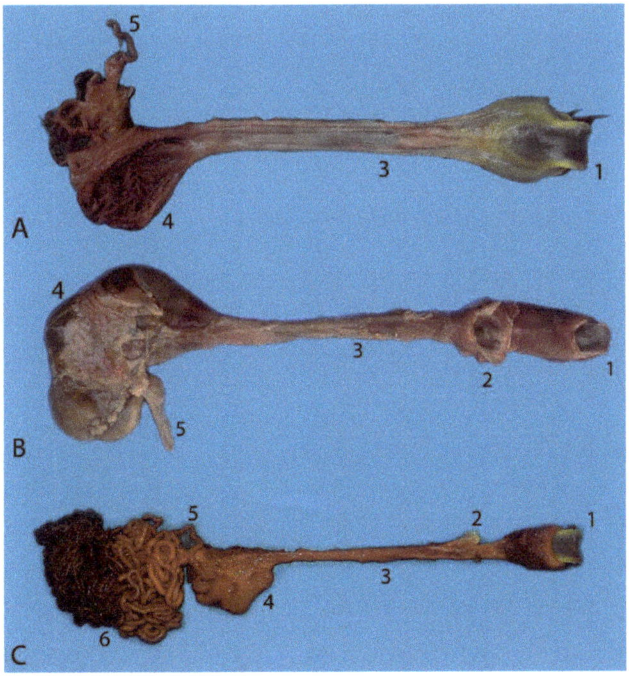

Figure 4. Digestive tract of a neonate individual: (**A**) opened; (**B**,**C**) closed. (1) Tongue; (2) goose beak (formed by epiglottic, arytenoid, and cricoid cartilages); (3) esophagus; (4) stomach; (5) duodenum; (6) jejunum.

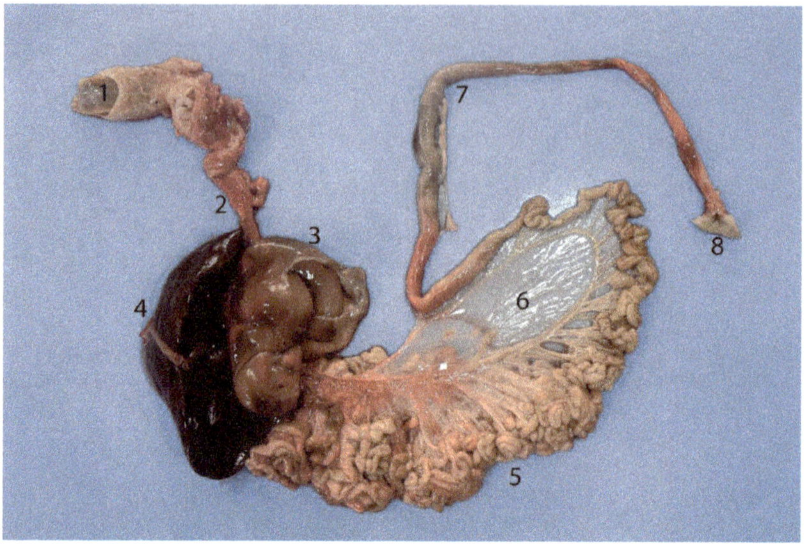

Figure 5. Complete digestive tract of a juvenile individual outside the abdominal cavity. (1) Tongue, (2) esophagus, (3) stomach, (4) liver, (5) small intestine, (6) mesojejunum, (7) colon, (8) anus.

3.3. Diaphragm

The diaphragm constitutes the cranial wall of the abdominal cavity and is formed by a muscular part and a tendinous or aponeurotic part. Its structure is similar to that observed in terrestrial mammals, but with some differences in this species and generally in cetaceans. It is highly developed, with a transverse/oblique orientation along a large part of the abdominal cavity. Its cranial insertion is ventral to the sternum, while the caudal part inserts near the cranial pole of the kidneys to the body of the thoracic vertebrae through the pillars of the diaphragm (Figure 6). This oblique arrangement of the diaphragm means that in these animals the intrathoracic portion of the abdomen is greater in length/depth.

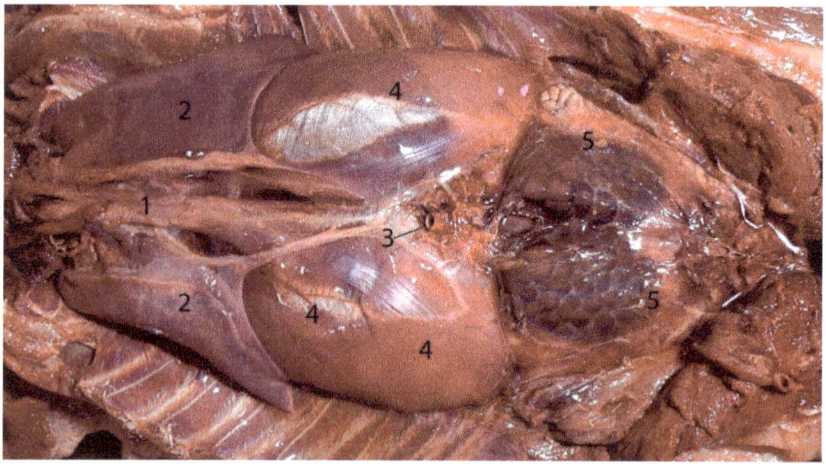

Figure 6. Juvenile animal positioned in prone position; dissection from dorsal to observe the topography with a dorso-ventral projection; cranial to the left, caudal to the right of the image: (1) esophagus; (2) lungs; (3) abdominal aorta; (4) diaphragm; (5) kidneys.

3.4. Stomach

Topographically, the stomach is located in the cranial part of the abdominal cavity, caudal to the liver and from ventral to dorsal. It is displaced to the left side with respect to the midline. More displaced to the right side is the liver, with which it maintains a close relationship. It is formed by two chambers: chamber 1, or the main stomach, and chamber 2, or the pyloric stomach. They also have a small U-shaped structure that serves as a junction between the two chambers, called the communication channel. The main stomach is located dorsally (Figure 7A), the communication channel is located ventral and to the left of the cavity, and the second chamber, or the pyloric stomach, is located ventral and slightly to the right to give way to the duodenum (Figure 7B). All of the compartments of the stomach are of the glandular type. The dimensions of the main and pyloric stomachs are given in Table 3.

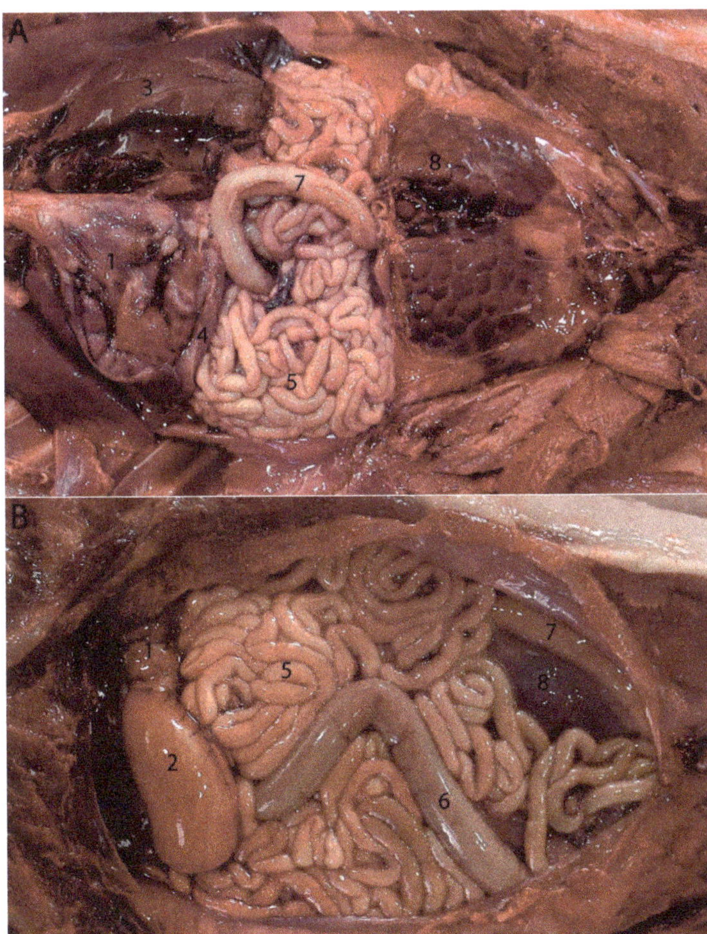

Figure 7. Abdominal cavity of juvenile individuals: (**A**) prone decubitus; (**B**) supine decubitus; (1) main stomach; (2) pyloric stomach; (3) liver (right lobe); (4) pancreas; (5) jejunum; (6) ascending colon; (7) descending colon (in image (**B**), it appears related to the lateral side of the left kidney, but the normal location is between both kidneys on the dorsal side of the abdomen. It was possibly displaced due to the manipulation of this individual); (8) kidney.

Table 3. Morphometric data of the main and pyloric stomachs of the specimens of *Pontoporia blainvillei* used in the present study.

Age Group	ID	Weight (g)	Capacity (mL)		Curvature MS (cm)		Length (cm)		Body Width (cm)		Fundus Width
		MS	MS	PS	Major	Minor	MS	PS	MS	PS	MS
A	M09	290	1150	105	39	13	15	9	9	5	7
	M22	489	1450	100	44	15	18	9.5	12	4	8
	M24	714	2100	245	46	17	21	10	14	5.5	9
	M43	575	1870	195	43	15	17	9.5	10	5	9
	M47	410	1600	170	40	14	16	9.4	10	4	7
	mean	495	1634	163	42	15	17	9	11	4.7	8
J	M12	96	140	16	29	9	10	5	8	3	6
	M16	281	800	90	36	16	19	10	14	4	8
	M20	380	1330	155	37	17	20	9	12	4	9
	M21	324	1200	120	38	13	17	11	11	4.5	7
	M23	136	320	40	27	6	10	6	7	3	5
	M28	327	820	60	38	15	14	8	9	4	7
	M29	136	400	65	30	9	11	5	9	3	6
	M45	265	790	105	31	11	18	6	10	4	8
	M60	110	290	25	28	8	11	6	7	4	6
	M71	230	600	75	26	11	15	5	10	4	7
	M78	385	940	105	32	13	22	6	15	4	10
	mean	243	694	78	32	12	15	7	10	4	7
N	M17	60	130	20	19	7	9	4	6	3.5	4
	M81	25	60	12	7	4	5	1.8	4	1.8	3
	mean	43	95	16	13	5.5	7	2.9	5	2.7	3.5

A: adult; J: juvenile; N: neonate; ID: individual identification; MS: main stomach; PS: pyloric stomach.

The differential characteristics shown by the Franciscana make it interesting to give a more detailed description of the stomach chambers (Figures 8–11):

- The main stomach or chamber 1 communicates directly with the esophagus. There is no pre-stomach in this species. The mucosa of the wall of the main stomach is of a glandular type, and ridges of the mucosa can be observed. The orifice of communication with the small U-shaped structure or communication channel is narrow, about 1 cm in diameter, and more than an orifice, it has the structure of a small channel. It is located in the fundus of the stomach (Figure 8).
- Communication channel: It is a small-sized, elongated, and thin compartment that serves as a communication between the main stomach and the pyloric stomach. The interior has a U-shaped structure, starting from ventral and, after bending, advancing dorsally (Figure 9);
- The pyloric stomach or chamber 2 is smaller than the main stomach. It connects with another smaller dilatation, which corresponds to the duodenal ampulla, through an orifice whose diameter varies between 0.3 and 0.5 cm. This communication does not appear to be a sphincter but rather an orifice that allows the passage of stomach contents in a continuous manner. In contrast, the pyloric sphincter (which marks the separation between the stomach and the small intestine) is located between the duodenal ampulla and duodenum (Figures 9 and 10).

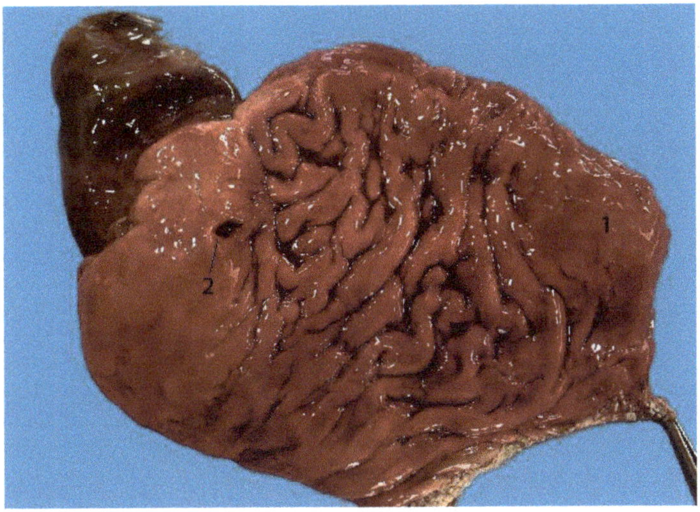

Figure 8. Image of the main stomach: (1) main stomach (mucosa); (2) orifice that communicates the main stomach with the communicating channel.

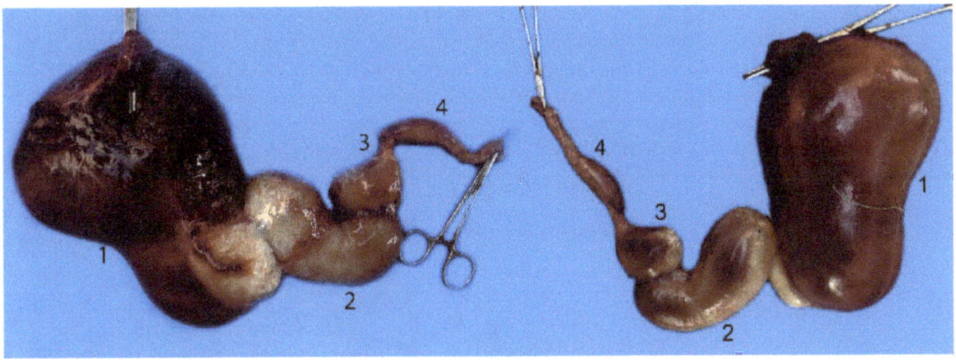

Figure 9. Images of stomach filled with water to obtain a better definition of the different chambers: (1) main stomach; (2) pyloric stomach; (3) duodenal ampulla; (4) duodenum.

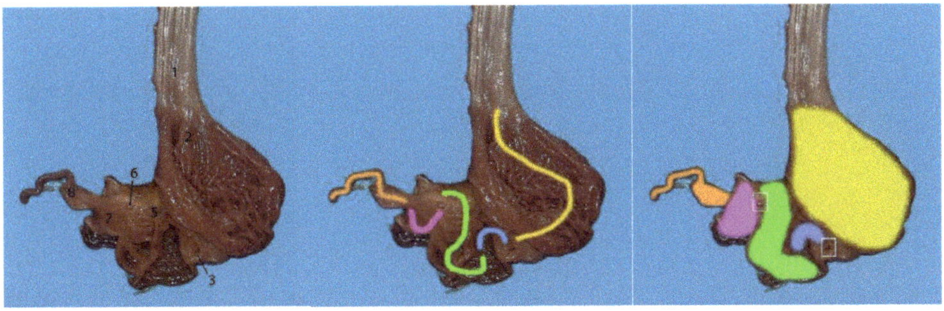

Figure 10. Stomach of a neonate individual. All of the chambers are opened: (1) esophagus; (2) main stomach (yellow); (3) orifice 1 (communication main stomach–communication channel) (white); (4) communication channel (blue); (5) pyloric stomach (green); (6) orifice 2 (communication pyloric stomach–duodenal ampulla) (white); (7) duodenal ampulla (violet); (8) duodenum (orange).

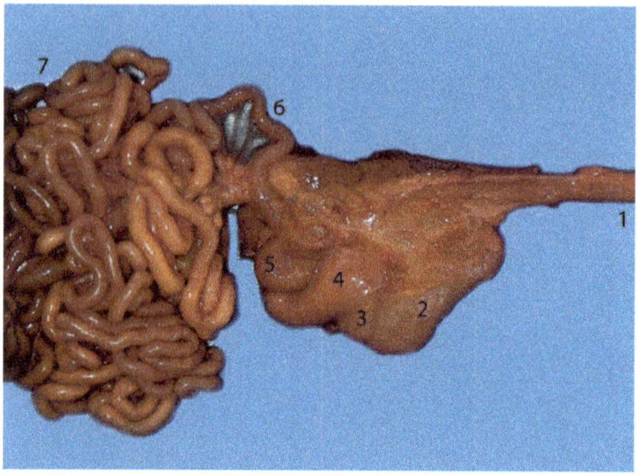

Figure 11. Stomach of a neonate individual: (1) esophagus; (2) main stomach; (3) communication channel; (4) pyloric stomach; (5) duodenal ampulla; (6) duodenum; (7) jejunum.

3.5. Small Intestine

The small intestine of the Franciscana is large in size, occupying a large part of the abdominal cavity. It consists of three sections: the duodenum, jejunum, and ileum. It is complex to macroscopically determine the change between sections due to the morphological similarity between them. The diameter is quite regular throughout (Table 4), as well as the thickness of the mucosa. The criterion used to differentiate between the duodenum and the jejunum is the location of the first jejunal loop. Likewise, the change from the jejunum to the ileum is determined by the last jejunal loop. The ileum presents a straight course, without curvatures or loops. The ratio between the length of the small intestine and the body length is spectacularly high, being approximately 27:1 in juveniles.

Table 4. Morphometric data of the small intestine of the specimens of *Pontoporia blainvillei* used in the present study.

Age Group	ID	Weight (g)			Length (cm)				Diameter (cm)					
		D	J	I	D (cm)	AD (cm)	J (m)	I (cm)	Duodenum cr	Duodenum cd	AD	Jejunum cr	Jejunum cd	Ileum
A	M09	15	1110	---	9	5	35	---	1.7	1.3	4	1.1	0.9	---
	M22	21	1200	7	10	5	37	9	2	1.5	4	1.2	1	1
	M24	24	1720	9	11	5	38	10	3	1.7	5	1.5	1.4	1
	M43	15	1040	6	14	4.5	37	11	2.6	1.4	4	1.5	1.3	1.1
	M47	10	930	5	14	5	36	8	3.2	1.5	4	1.4	1.2	1
	mean	17	1200	6.8	11.6	4.9	37	9.5	2.5	1.5	4.2	1.3	1.2	1.0
J	M12	6.5	200	---	8	3	16	---	2	1	3	0.5	0.4	---
	M16	14	850	---	12	4	34	---	1.6	1.1	3	0.8	0.7	---
	M20	11	1130	5	9	3	32	8	2.5	1.5	4	1.1	0.8	1
	M21	13	800	2.7	11	4	31	7	2	1	4	0.8	0.6	1
	M23	6.4	530	1.4	7	3	23	5	1.4	1	3.5	0.9	0.7	0.6
	M28	12.6	1000	2.5	11	5	32	4	2	1.2	4	1	0.8	0.9
	M29	6,2	485	1.2	9	4	22	4	2	1	4	0.9	0.7	0.6
	M45	18	750	3.5	10	5	31	5	2.2	1.2	4	0.8	0.7	0.7
	M60	5	310	2	8	---	21	6	---	---	---	0,7	0.6	1
	M71	22	400	---	10	5	25	---	2.3	1.4	5	0.8	0.7	---
	M78	20	640	---	9	7.5	30	---	2	1.2	5.5	0.9	0.7	---
	mean	12.2	672	2.7	9.5	4.4	27	5.9	2.0	1.2	4.0	0.8	0.7	0.9
N	M17	3	180	1.8	7	2.5	15	4	0.8	0.5	2	0.4	0.4	0.5
	M81	1	105	1	6	1.4	13	4	0.6	0.4	1.2	0.4	0.4	0.4
	mean	2	142	1.4	6.5	2.0	14	4	0.7	0.5	1.6	0.4	0.4	0.45

A: adult; J: juvenile; N: neonate; ID: individual identification; D: duodenum; AD: duodenal ampulla; J: jejunum; I: ileum; cr: cranial; cd: caudal.

3.5.1. Duodenum and Pancreas

As mentioned above, the beginning of the duodenum presents a dilatation called the duodenal ampulla. Afterwards, it narrows until it reaches a size and lumen similar to the jejunum (Table 4). The communication between the pyloric stomach and the duodenal ampulla is through an orifice. That is, the communication between the stomach and the small intestine is not delimited by the pyloric sphincter. However, this sphincter appears between the duodenal ampulla and the next section of the duodenum (Figures 9–14). The pancreas presents two lobes, and it is intimately related to the duodenum and stomach, as is the case in other species. The study of the pancreas was not possible in all individuals, given its high enzymatic activity and rapid post-mortem autolysis.

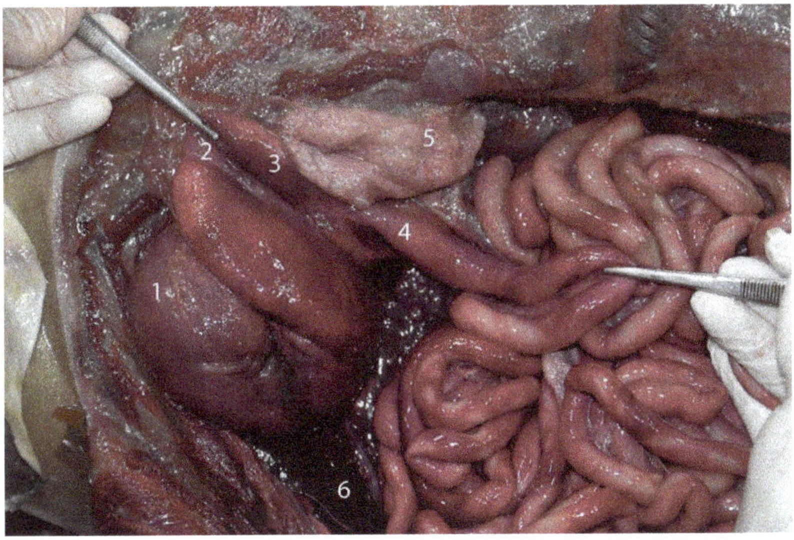

Figure 12. Cranial abdominal cavity of an adult individual; cranial (left)–caudal (right). By removing the stomach and pancreas, we can visualize the duodenum, located on the right side of the dorsal part of the cranial abdomen: (1) main stomach; (2) communication channel; (3) pyloric stomach; (4) duodenum; (5) pancreas (right lobe); (6) liver.

Figure 13. Initial portion of the duodenum opened. (1) Hepatopancreatic duct, (2) duodenum.

In the duodenum ends a duct that is common to the liver and pancreas and is called the hepatopancreatic duct. This duct runs along the wall of the duodenum from its beginning to its end inside the duodenum (Figure 13).

3.5.2. Jejunum

The jejunum is located caudal to the liver and stomach, occupying the entire dorsal and ventral region of the abdominal cavity. It is the longest section of the small intestine. Its length increases with the age of the animal until it reaches the adult stage (Table 4). In addition, it is worth noting a peculiar characteristic of the jejunum of this species that has not been previously described in the literature. It presents an arrangement within the abdominal cavity in the form of "discs", with different levels of organization depending on the degree of development of the individual. A clear difference between age groups has been observed. Thus, neonate individuals presented a low level of organization, with scarce formation of these discs. As the animal develops, the same occurs with the discs. In juvenile individuals, a certain degree of organization was observed, with the appearance of some jejunal discs. Meanwhile, adult individuals presented a high level of organization of jejunal discs. A higher organization of the jejunal discs can be observed in the cranial part of the jejunum than in the caudal part (Figure 15).

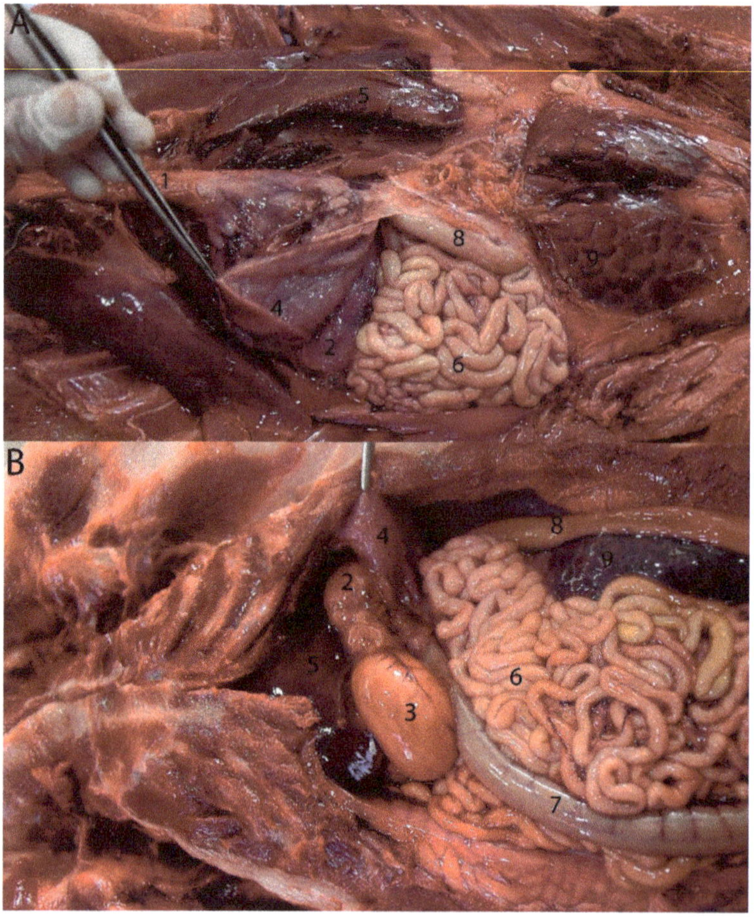

Figure 14. *Cont.*

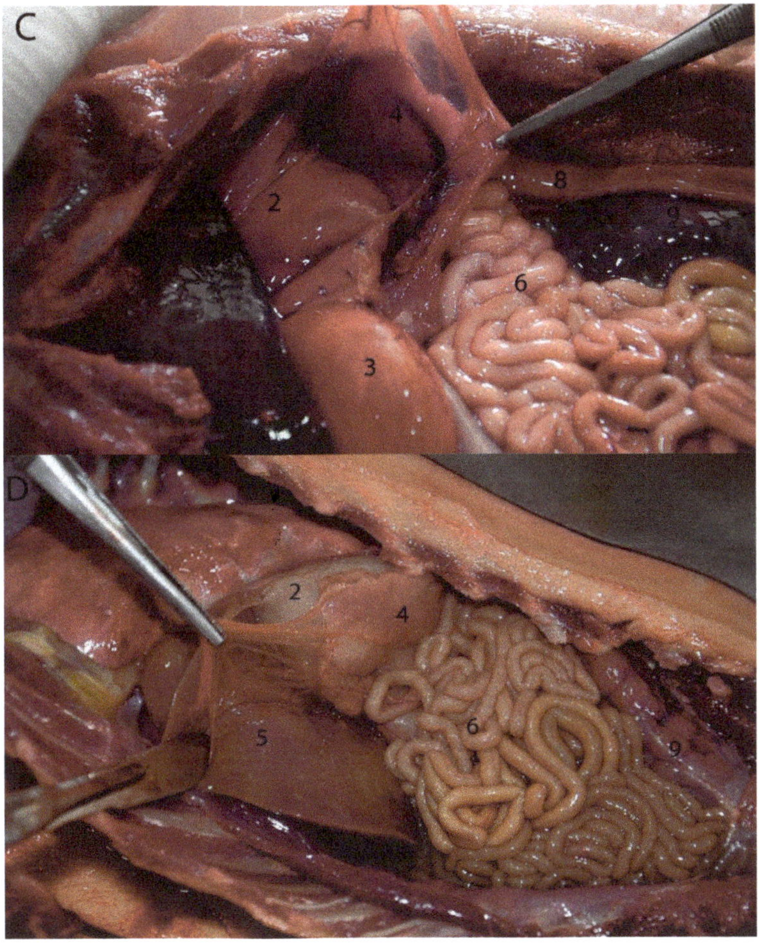

Figure 14. Juvenile individual: (**A**) prone decubitus; (**B**,**C**) supine decubitus; (**D**) neonate individual in supine decubitus. (1) Esophagus; (2) main stomach; (3) pyloric stomach; (4) pancreas; (5) liver; (6) jejunum; (7) ascending colon; (8) descending colon; (9) left kidney.

3.5.3. Ileum

The ileum is the final section of the small intestine; it is short and of a similar diameter to the jejunum (1, 0.9, and 0.4 cm in adult, juvenile, and neonate individuals, respectively) (Table 4) (Figure 16). Its beginning is determined by the last jejunal loop. Its end is marked by the beginning of the colon (with its mucosal ridges), as well as the last ileal blood vessel (Figures 16 and 17). The mean length was 9.5, 5.9, and 4 cm in adult, juvenile and neonate individuals, respectively.

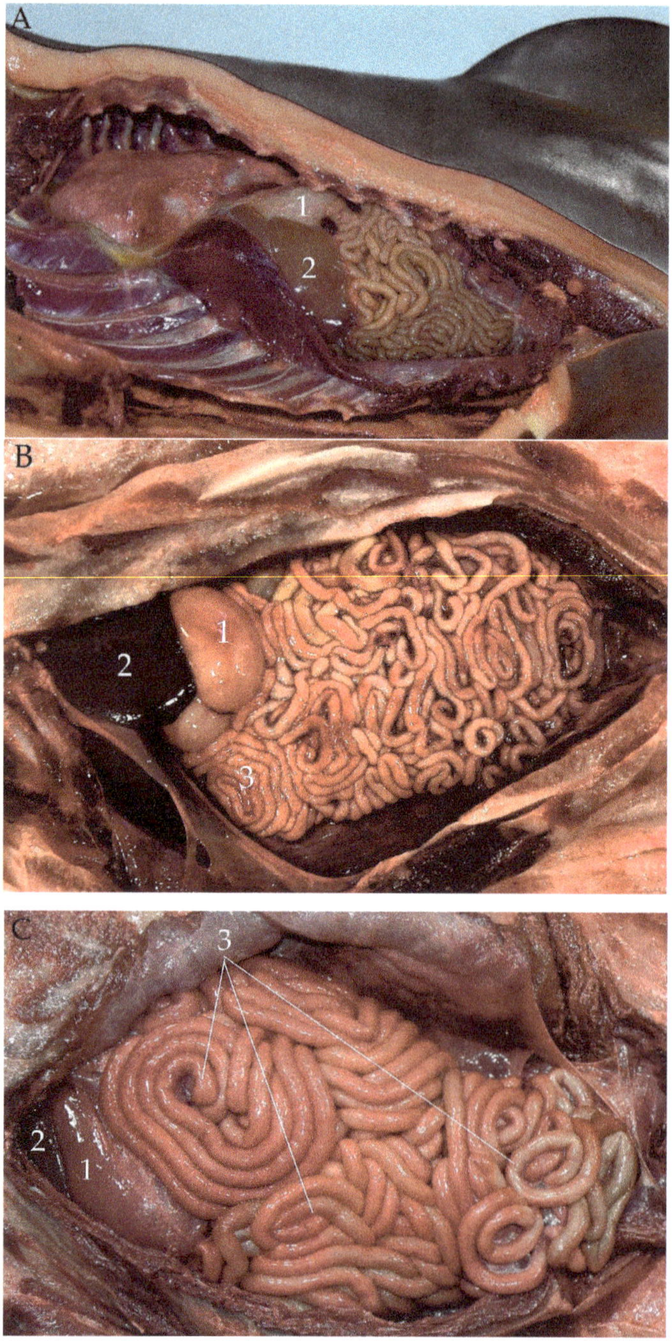

Figure 15. (**A**) Neonate individual in right lateral decubitus; cranial (left)–caudal (right) (without jejunal discs): (1) main stomach; (2) liver; (3) jejunum. (**B**) Juvenile individual in supine decubitus; ventral view of abdomen; cranial (left)–caudal (right): (1) main stomach; (2) liver; (3) jejunal discs. (**C**) Adult individual in right lateral decubitus–supine; cranial (left)–caudal (right): (1) stomach; (2) liver; (3) jejunal discs.

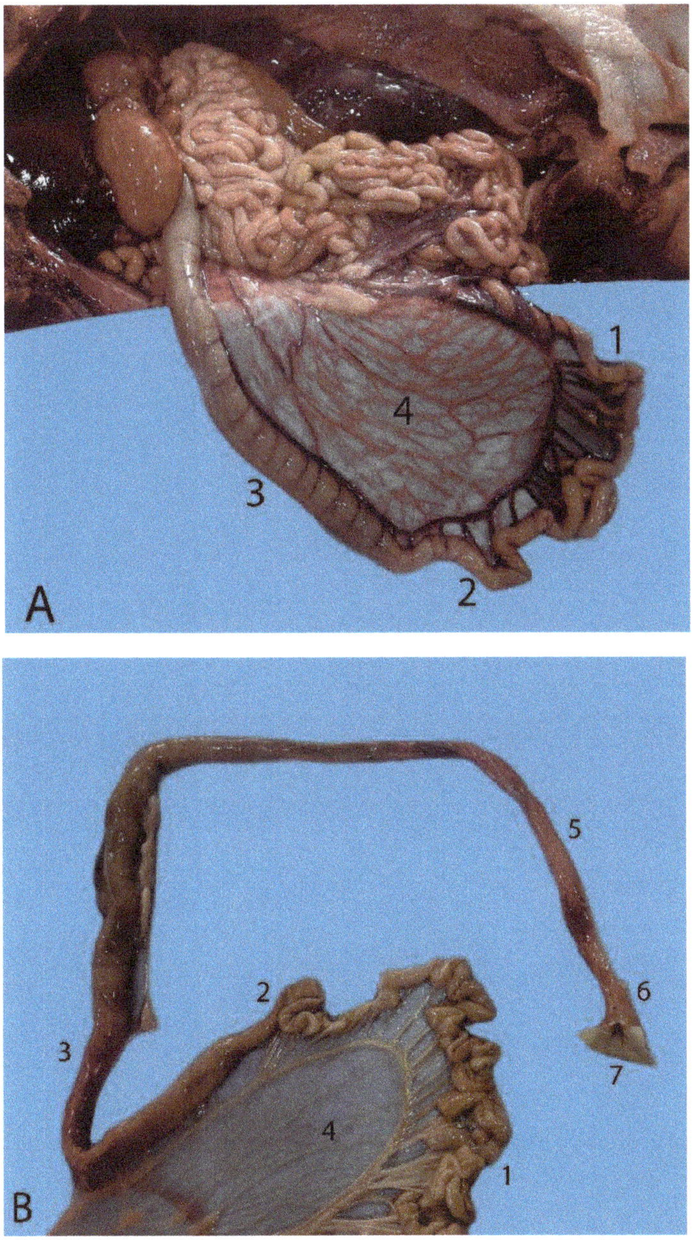

Figure 16. Intestine of a juvenile individual: (**A**) transition between small and large intestine; (**B**) view of the extended large intestine; (1) jejunum; (2) ileum; (3) ascending colon; (4) mesojejunum; (5) descending colon; (6) rectum; (7) anus.

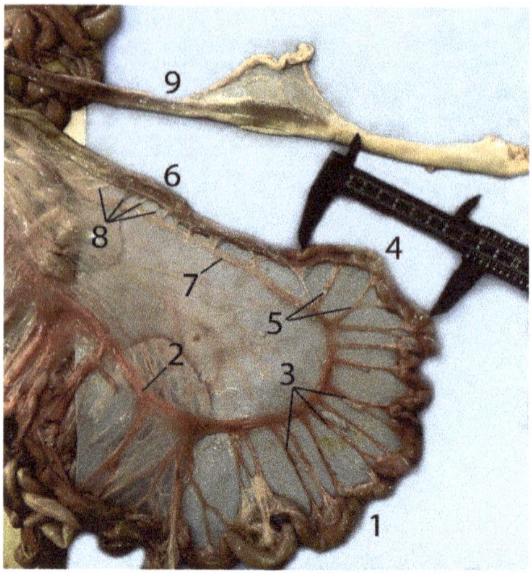

Figure 17. Intestine of an adult individual. The scale marks the ileum: (1) jejunum; (2) cranial mesenteric artery; (3) jejunal arteries; (4) ileum; (5) ileal arteries; (6) ascending colon; (7) caudal mesenteric artery; (8) colic arteries; (9) descending colon.

3.6. Large Intestine

The Franciscana dolphin does not have a cecum, which makes it difficult to differentiate the change to the large intestine. No histochemical techniques were performed to demonstrate histological differences between the two sections. Thus, to differentiate the end of the ileum and the beginning of the colon, we based ourselves on the following criteria:

- Diameter: macroscopically, a slight change in size can be observed at the transition from the small to the large intestine (Tables 4 and 5). They clearly mark the beginning of the colon and make it thicker.
- Wall thickness: on palpation, a change in the thickness of the wall between the small and the large intestine can be noted: ileum (thinner) and colon (thicker).

Table 5. Morphometric data of the large intestine of the specimens of *Pontoporia blainvillei* used in the present study.

Age Group	ID	Colon				Straight Length
		Weight (g)	Length cm	Diameter (cm) cr	Diameter (cm) cd	cm
A	M09	37	62	1.7	1.5	3
	M22	49	64	1.7	1.4	4
	M24	60	81	2.2	1.6	5
	M43	41	67	2	1.3	4
	M47	54	76	1.8	1.2	4
	mean	17	70	1.9	1.4	4
J	M12	28	45	1	0.7	3
	M16	49	51	1.3	0.8	3.5
	M20	37	55	1.1	1.6	4
	M21	60	68	1.2	1.4	2
	M23	42	48	1.6	1.3	3
	M28	50	65	1.8	1.3	4
	M29	43	60	1.6	1.3	4.5
	M45	42	63	1.6	1.4	4
	M60	30	51	1.4	1.1	3

Table 5. *Cont.*

Age Group	ID	Colon Weight (g)	Length cm	Diameter (cm) cr	cd	Straight Length cm
	M65	36	52	1.4	1.2	3
	M71	30	61	1.6	1.4	4
	M78	45	71	1.7	1.4	5
	mean	41	58	1.4	1.2	3.6
N	M17	17	44	0.8	0.6	2.5
	M81	8	36	0.6	0.5	1.5
	mean	13	40	0.7	0.6	2

A: adult; J: juvenile; N: neonate; ID: individual identification; cr: cranial; cd: caudal.

3.6.1. Colon

The colon is divided into three sections: the ascending, transverse, and descending colon (Figure 18). The ascending colon begins in the caudal part of the abdominal cavity, and it has an ascending course located in the ventral part of the right side. Once it has reached the cranial area, where it is related to the stomach, liver, and pancreas, there is a change of direction and it runs from ventral to dorsal in an obliquely crosswise plane. This section is short, corresponding to the transverse colon. From there, the descending colon runs along the dorsal midline caudal to the abdominal cavity. It passes between the two kidneys (Figure 18D) and descends again progressively towards the ventral pelvic area until it empties into the rectum and then into the anus.

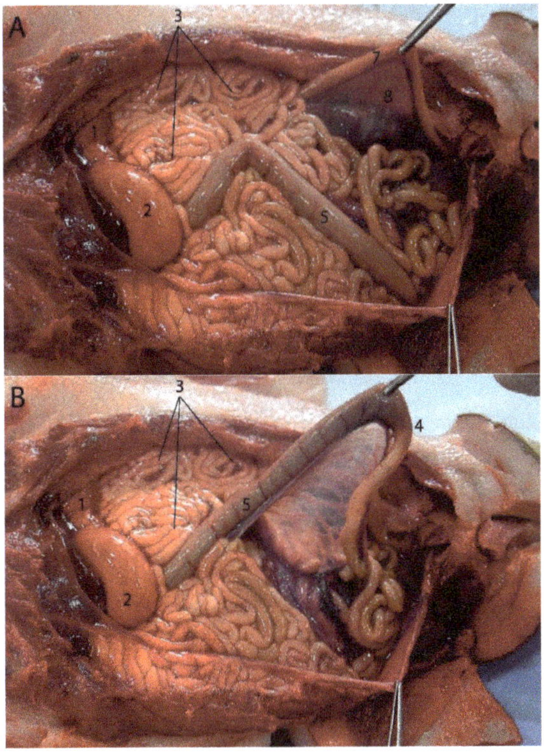

Figure 18. *Cont.*

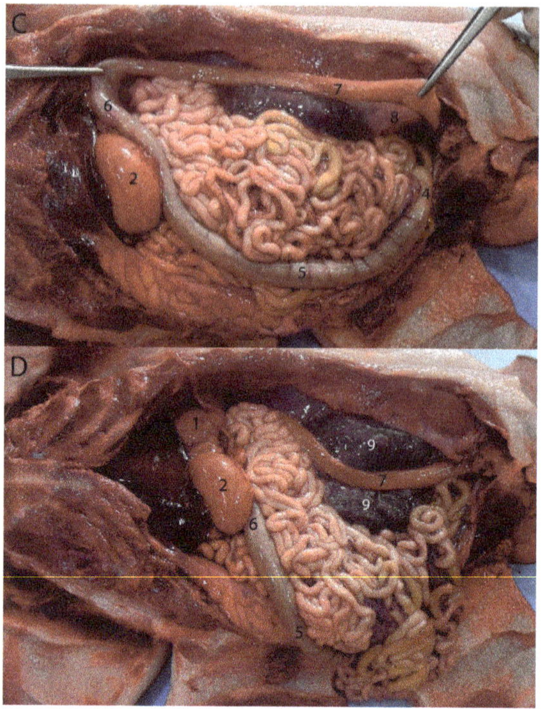

Figure 18. Abdominal cavity of a juvenile individual. Different planes in supine decubitus: (**A**) ventral view of the abdominal cavity; (**B**) the ascending colon is lifted to see the transition between small and large intestine; (**C**) the colon has been displaced in order to see all their parts; (**D**) the jejunum has been retired to see the course of the descending colon, between both kidneys; (1) main stomach; (2) pyloric stomach; (3) jejunal discs; (4) transition from small to large intestine (ileum–colon); (5) ascending colon; (6) transverse colon; (7) descending colon; (8) mesocolon; (9) kidneys.

In its first half, the colon presents a mucosa lined with Peyer's plaques/lymphatic tissue associated with the intestine. In addition, in its interior, two thick folds or ridges of the mucosa stand out (Figures 19 and 20). The last section of the colon, corresponding to the descending colon, has a smooth mucosa without folds. These characteristics, together with the coloration and thickness, are unique to the colon compared to the other parts of the intestine. No tapeworms coli or haustras were observed. The morphometric data of the colon are shown in Table 5.

3.6.2. Rectum

In certain cases, a macroscopic change in the mucosal lining epithelium was observed in the most distal part of the large intestine. It could be compatible with the change from colon to rectum (Figure 21). In other individuals with a more advanced degree of decomposition, this difference could not be observed, possibly due to the degradation and detachment of this rougher-looking mucosa. No histochemical techniques were performed to demonstrate histological differences between the two sections.

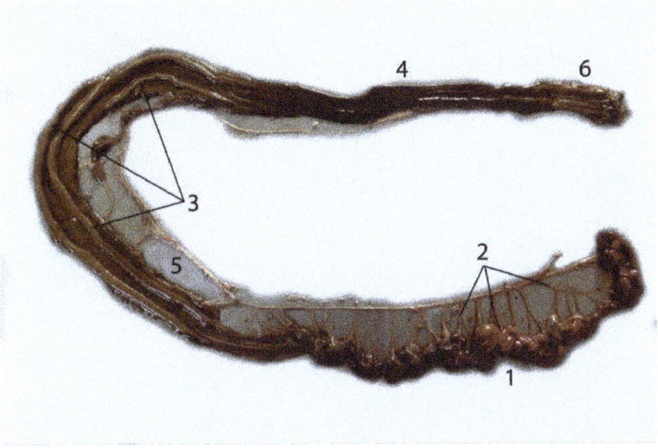

Figure 19. Large intestine, opened to expose the mucosa: (1) jejunum; (2) jejunal arteries; (3) folds of the colon mucosa; (4) smooth mucosa of the final stretch of the colon; (5) mesocolon; (6) rectum.

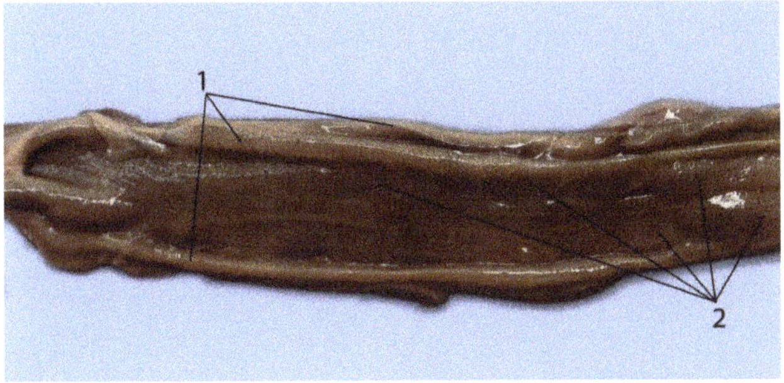

Figure 20. Initial section of the colon: (1) folds of the colon mucosa; (2) associated lymphatic tissue/Peyer's patches.

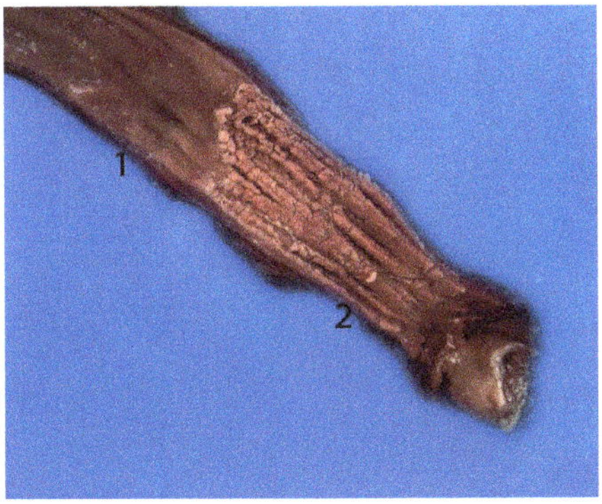

Figure 21. Final portion of the large intestine (opened) and anus; note the macroscopic change of the mucosa: (1) colon; (2) rectum.

3.7. Liver

The liver is formed by two lobes divided by a shallow but well-defined fissure (Figure 22). It is located in the intrathoracic portion of the abdominal cavity. The left lobe is smaller than the right lobe (Figure 22) (Table 6) because the left side of the cavity is mostly occupied by the stomach. This lobe is located ventrally, while the right lobe is located on the right side from ventral to dorsal. On its diaphragmatic side, it presents folds that join the diaphragm, corresponding to the coronary and falciform ligaments of other mammals (Figures 22 and 23). On the visceral side, it presents the hepatogastric ligament that joins it to the stomach (Figure 23). The ratio of liver to body mass ranged from 1 to 2.9% (Table 6). There was one exception, which was the infant individual that was found stranded alive and transferred to the rescue center for rehabilitation. The result was 7.5%; associated with this data, generalized jaundice was observed in the animal's mucous membranes.

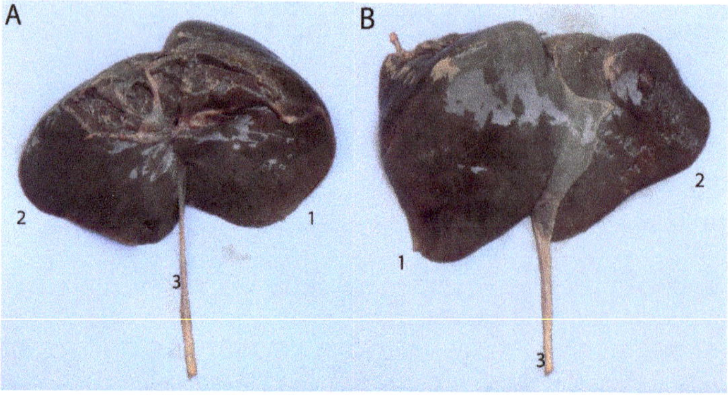

Figure 22. Liver: (**A**) visceral surface; (**B**) diaphragmatic surface; (1) right lobe; (2) left lobe; (3) round ligament of the liver.

Table 6. Measures of the liver weight of the specimens of *Pontoporia blainvillei* used in the present study.

Age Group	ID	Liver Weight (g) Total	Right Lobe	Left Lobe	Body Weight (kg)	Ratio Liver/Body Weight (%)
A	M09	455	277	178	20.6	2.2
	M22	510	326	184	42.3	1.2
	M24	1240	730	510	61.4	2
	M43	700	415	285	38.5	1.8
	M47	665	400	65	27	2.4
	mean	714	430	244	38	1.9
J	M12	220	130	90	9.2	2.4
	M16	196	140	56	18.3	1
	M20	526	332	194	22.2	2.4
	M21	506	357	149	23.8	2.1
	M23	277	174	103	10.6	2.6
	M28	440	287	153	16.2	2.7
	M29	327	228	99	12.5	2.6
	M45	380	240	140	14.7	2.5
	M60	245	140	105	10	2.4
	M65	250	150	100	14.5	1.7
	M71	240	140	100	13.2	2.4
	M78	545	365	180	19.2	2.8
	mean	366	235	131	15	2.4
N	M17	310	135	175	4.1	7.5
	M81	115	70	45	3.9	2.9
	mean	115	70	45	4.0	2.9

A: adult; J: juvenile; N: neonate; ID: individual identification; cr: cranial; cd: caudal. M17: individual with hepatomegaly, not included in the mean liver weight.

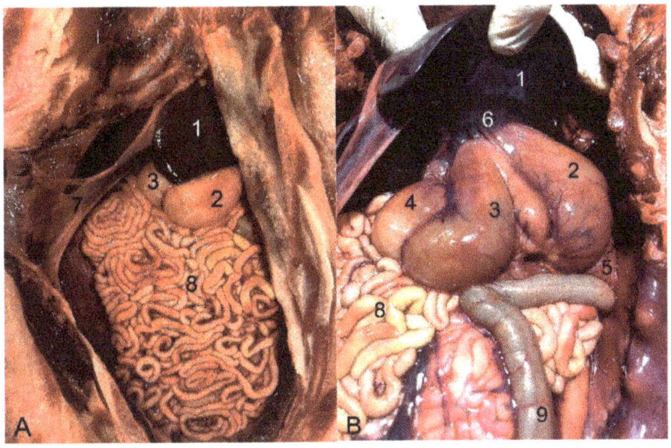

Figure 23. Abdominal cavity of a juvenile individual, supine decubitus: (**A**) ventral view of the abdominal cavity; (**B**) the stomach is being exposed to show the different chambers, and the jejunum has been displaced to visualize the course of the descending colon; (1) left lobe of the liver; (2) main stomach; (3) pyloric stomach; (4) duodenal ampulla; (5) pancreas; (6) lesser omentum; (7) round ligament; (8) jejunum; (9) ascending colon.

3.8. Spleen

It is associated with the gastrointestinal tract due to its topography, but not functionally. For this reason, it will not be detailed in this study. In several individuals, this organ was not found, possibly due to the degree of decomposition of the animals, its proximity to the pancreas with rapid autodigestion, and its small size (Figure 24). In the two neonate individuals, it could be located on the left side of the abdomen, attached by the gastrosplenic ligament to the main stomach, also in close relation to the pancreas. It is small in size and oval- or button-shaped. Small accessory spleens were also observed (Figure 24).

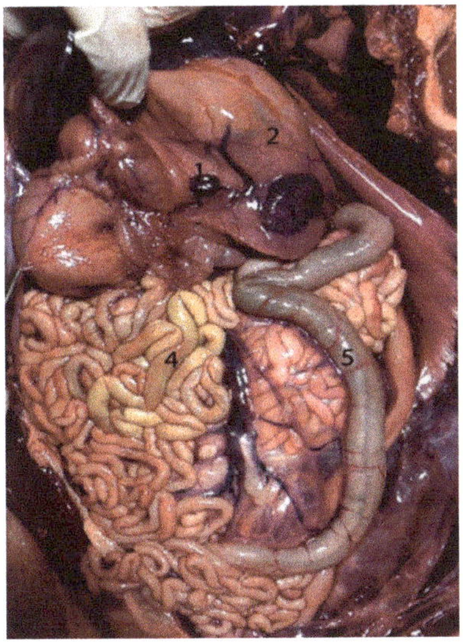

Figure 24. Infant individual. Abdominal cavity. Supine decubitus: (1) spleen and accessory spleen; (2) main stomach; (3) liver; (4) jejunum; (5) ascending colon.

4. Discussion

Among the described organs of the digestive system of the Franciscana dolphin, there are some that share the typical characteristics of the Delfinidae, and these will not be discussed in this paper. However, the focus will be on those that present particular characteristics of this species related to its diet.

First of all, the low skull asymmetry seen in these animals is directly related to its small prey size. The existence of a relationship between prey size and skull asymmetry has been previously described: animals with more asymmetrical skulls are able to capture larger prey than similarly-sized, more symmetrical animals [23,24]. Secondly, because what ultimately determines prey size is the diameter of the maw isthmus and the pharyngeal passage [25,26], the width of the maw isthmus in this study was consistent with the size of the prey consumed: small-sized or juvenile fishes, small marine cephalopods (67% = 11–13 cm mean length), and slightly larger squids [27–32].

One might think that the elongated and thin snout of the Pontoporia is a common feature in river species and that it could be related to freshwater life. However, this hypothesis does not hold true for Sotalia, which is also a river dolphin. The authors hypothesize that the Pontoporia, the Amazon dolphin, and the Ganges and Yangtze dolphins probably share a common ancestor, contrary to the Sotalia dolphin, which adapted to life in freshwater much later than the others. In summary, this snout morphology would have more justification from an ontological than an adaptive point of view.

These animals have conical teeth of different lengths, as do other estuarine dolphins, such as the Ganges dolphin (*Platanea platinids*), which are ideal for catching fish and shrimp at an early age [33]. The number of teeth in the upper arch of the Franciscana is similar to that of the lower arch, with 110 pieces on average vs. 108 in juveniles and 113 vs. 110.5 in adults, which differs from those species whose diet is based exclusively on squid and that lack teeth in the upper arch, as explained by Cozzi et al. and Martin Diaz [26,34].

Unlike the tongue of most dolphins [35–37], the tongue tip of the Franciscana is barely mobile [38] and very short, even differing from other freshwater dolphins [38]. This is probably compensated by the presence of the anterolateral papillae (anterolateral fimbriae) consisting of compact outgrows placed along a single row developed mainly in infants [35,36,38]. According to Ferrando et al. [39], these latter papillae may help to create a tight seal between the tongue and the roof of the oral cavity and therefore help with suction feeding, as the period of development of these papillae is coincident with the lactation period [40]. Even as we have described in the Franciscana, these papillae tend to be reduced or disappear with age, although they may persist in some individuals [26]. From the age of two months, this species begins to consume solid food and its feeding becomes mixed, and then these papillae assist in holding the prey against the palate while water flows through the papillae [33,34,36,39,41] until they begin to involute due to disuse and completely lose functionality in adulthood.

In the present work, the average number of marginal papillae was 3 in adults, 30 in juveniles, and 50 in neonates. These data are consistent with those of Guimaraes et al. [41], who obtained an average number of marginal papillae of 53 in immature and 36.7 in mature dolphins and only 3.75 on average in adults (being nonexistent in some specimens), but they differ from the 13 to 15 projections counted in juveniles by Yamasaki et al. [38], which was probably measured in older juveniles.

A multiple-chambered stomach is unusual in carnivores. However, all cetaceans are carnivores and have the presence of a multichambered stomach. The Cetartiodactyla shows a tendency towards plurilocular stomachs, which can be found and characterized by very different architectures in Suidae, Tayassuidae, Camelidae, Tragulidae, Pecora, Hippopotamidae, and Cetacea [26,28]. After all, whales are closely related to artiodactyls, which also have multichambered stomachs. While their multiple chambers may relate to the mechanical and enzymatic breakdown of an herbivorous diet (e.g., separation of food to be regurgitated and re-chewed as cud), it is unclear what functions multiple chambers play in the carnivorous cetaceans [24]. Perhaps, in a sense, they compensate for the absence of chewing by grinding and compressing in the first chamber before absorption takes place in the subsequent compartments [26].

While all dolphins share the multicameral anatomy of their stomachs, Mead [42] describes three morphological appearances of the stomach, which he calls the ziphiid stomach (one main stomach, one pyloric stomach), the derived stomach type I (two main stomachs, one pyloric stomach), and the derived stomach type II (two main stomachs, two pyloric stomachs). The Franciscana shows macroscopic features similar to those of a zyphid-type dolphin, with a main stomach, a communicating channel, and a pyloric stomach, as also described by Yamasaki et al. [43].

In the Franciscana, there is no pre-stomach present, as in all known zyphids and contrary to most eutherian mammals (Figure 25). This feature is also thought to be shared with the probably extinct Chinese river dolphin (*Lipotes vexillifer)*, but not so with the Amazon river dolphin (*Inia geoffrensis)*. Marigo and Groch [44] reconfirm that some odontocete species lack an anterior or pre-stomach chamber and, instead, the main chamber is divided into two or three compartments separated by septa [34]. Yet, this is not the case in Franciscana. Storage and mechanical degradation functions have been attributed to the anterior or pre-stomach chamber [25,44,45], which encourages us to think that in the face of the continuous consumption of small prey, this possibility of storage and this reinforcement in digestion for the chemical and physical degradation of the prey are not necessary. With this, the authors say that the storage function is probably being replaced by the increase in feeding frequency (continuous feeder), and, on the other hand, the mechanical action can be performed in the muscular chamber of the main stomach. This structure of the stomach with one less compartment could agree with its antiquity and primitiveness with respect to other cetacean species more evolved or new in the phylogenetic tree. It may also be compensated by the greater length of its small intestine, where enzymatic degradation processes occur.

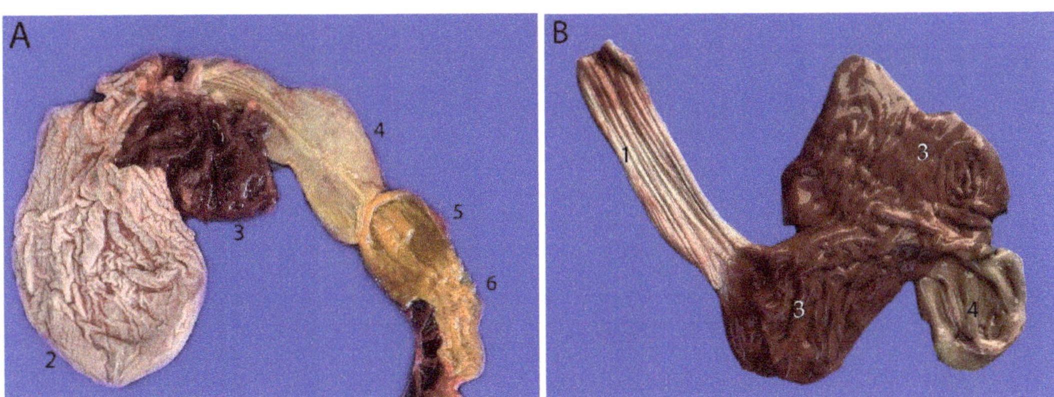

Figure 25. Comparative images of the stomachs of two different dolphins. The esophagus and the fore- and main stomachs are opened to show their mucous: (**A**) multichambered stomach of a *Stenella frontalis* (from IUSA-ULPGC); (**B**) bicameral stomach of a *Pontoporia blainvillei*; (1) esophagus; (2) forestomach; (3) main stomach; (4) pyloric stomach; (5) duodenal ampulla; (6) duodenum.

The stomach values recorded are consistent with the values published by Kamiya and Yamasaki [46]. After the animal has reached a body weight of 20 kg, the relative weight of the stomach at adulthood contributes only approximately 1.5% of the body weight and does not show considerable changes during body growth. While the relative size of the stomach to the live weight of the Franciscana is consistent with values for a strict carnivore of that size [47], the same is not true for the relative size of the intestines.

The small intestine of the Franciscana is similar to that of other carnivores. Yet, the total length of the gut in the Franciscana is remarkably high. The reason for this is the extraordinary length of the jejunum, corroborating what Yamasaki et al. described [20]. According to Crespo (2002), this species feeds mostly near the bottom on fish belonging to several families; it is listed as a predominantly teutophagous odontocete [48]. The reason for the long intestine in Franciscana and the influence of the types of food are not known. Büker, however, gives an interpretation of the lengthening of the intestinal tract in aquatic mammals: during diving, blood supply to the intestines is reduced, and a longer gut could compensate for the necessary assimilation of nutrients [49]. The authors would like to add the possibility of the development of a relatively longer small intestine that allows more time for the food to be in contact with the enzymes that will degrade it, as well as a greater absorption surface in response to a lower-quality diet compared to that of other dolphins, comprising bottom-dwelling juvenile teleosts, squid, and shrimp [31,50–54].

Among the toothed whales, there is considerable variation in the ratio between the length of the small intestine and the body length, ranging from 5/1 in bottlenose whales to 14/1 in some species of dolphin [55]. Baleen whales tend to show lower ratios (fin whales, 4/1; little piked and humpback whales, 5.5/1). These ratios do not necessarily correlate with either the diet or size of the species [56], with the pontoporie ratio being 24–37.7/1.

The small intestine of Franciscana occupies the greater part of the lower abdominal cavity. The length of the small intestine examined by Yamasaki et al. [20] is similar to the values found in this work. However, for this same author, the ratio varied considerably from specimen to specimen, and it seems to show no correlation to body length. Yet, it might have a close relationship with the lapse of time after death rather than individual variation. Also, the length varies in fixed and unfixed states. The length shortened about 8.5% after fixation in 10% formalin solution.

As explained by Cozzi et al. [26] and Marigo and Groch [44] for other species, for the Franciscana, it has been observed that pancreatic juices flow into the duodenum via the pancreatic duct. In most odontocetes, this process is individual, but in some cases, it joins the hepatic duct, thus flowing into what is called an hepatopancreatic duct similar, in this aspect, to equines [34,57], a continuous consumer.

Regarding the arrangement of the jejunal loops in adults, there is a hypothesis that this is in relation to the need to maximize efficiency when taking advantage of the available space. In this way, a longer intestine can be accommodated as time goes by in almost the same abdominal cavity, making space more efficient. And, even as the animals advance in age, their diets become of better quality due to learning how to capture prey and the ecological niches they occupy [51]. This could be consistent with a greater prey metabolism, and therefore less gas production, which allows for a more orderly arrangement of the loops. However, there is no literature to support this hypothesis.

As mentioned in the results, there is not a particular distinction between the small and large intestines, with no caecum or vermiform appendix present. According to the results of this study, the authors agree with Cozzi et al. [26], as this apparently simplified disposition results from the diet, which includes mostly proteins of animal origin (generally fish and cephalopods). The exception to the rule is the Ganges River Dolphin, belonging to the family Platinistidae, which possesses a cecum.

The mucosa of the colon shows one or two longitudinal folds that increase its inner surface and disappear as they move caudally. The Gangetic dolphin, on the other hand, has numerous longitudinal folds that begin in the distal colon and continue to the rectum; in the Amazon river dolphin, the folds appear along the entire colon [28,58,59] Although this fact may be disadvantageous with respect to having circular folds for them to widen the absorption area, the extraordinarily long intestines may compensate for the lack of these type of folds [20,60].

There are differences between our results and previously published data on the total length of the large intestine of the Franciscana dolphin. Apart from the absence of cecum and the similar thickness of both intestines at the junction point, these variances could be related to the criteria used to define the beginning and end of the large intestine, as there are no clear macroscopic signs. In the study by Mead and Brownel, the values obtained were 40–48 cm [61], which were within the range of those described by Yamakasi et al., 25–58 cm [20], and the results of the present study were slightly higher, with 62–81 cm for adults, 45–71 cm for juveniles, and 36–44 cm in infants.

The peculiar morphological characteristics of the gross intestine of dolphins, and specifically the shortage of any bowel-like structure in the gut, could suggest minimal or no storage capacity of undigested residue [62]. In fact, evacuated feces are generally rather liquid and may contain unprocessed materials, especially when shrimps are part of the diet [26]. However, the observation made in some of the studied animals that were under human care was of a more consistent fecal matter than that of other dolphins. This may be related to the greater length of the intestines and therefore the greater surface area of water absorption of the feces, which is consistent with less watery feces.

Regarding the relationship between liver weight and body weight, the average did not include the neonate with hepatomegaly that was found stranded alive and transferred to the rescue center for rehabilitation. In this specimen, the liver weight was 7.5% of the body weight. In association with this high percentage, generalized jaundice was observed in the mucous membranes of this animal. Interestingly, this condition has been observed in many of the neonates that were undergoing rehabilitation at the rescue center Fundación Mundo Marino. This finding is not described in the literature. It is currently being investigated by several Latin American organizations collaborating in the conservation of marine fauna. In the rest of the individuals, the relative values were 2.43%, which coincides with those already described by other authors, where the weight of the liver averages 2–3% of the total body weight in dolphins [21,26,63]. However, these values are higher than the relative values of other cetaceans, with the exception of estuarine dolphins [63], which makes the authors think of a higher metabolic rate in this type of dolphin. These higher metabolic rate data also coincide with the hypothesis expressed by Helm (1983) that the higher the metabolic rate, the longer the relative length of the intestines [64].

The liver of the Franciscana, like the liver of cetaceans and like that of many terrestrial species, lacks a gallbladder, and the bile produced in the hepatic stroma is carried into the duodenum by a common hepatic duct, as mentioned above. The absence of an extra-hepatic storage organ for the bile is possibly related to the continuous ingestion of food and the consequently frequent presence of food in the proximal intestine [26], as is the case in continuous consumers.

5. Conclusions

The anatomical characteristics of the digestive system of the Franciscana dolphin are similar to those present in other marine mammals. However, our results show that this dolphin shows several peculiarities, especially in relation to the tongue, the teeth, the stomach, and the small intestine.

Although further studies are needed to confirm the results of the present paper, it could be assessed that the digestive system of the Franciscana dolphin is adapted to a small prey diet, considering the size of its oral cavity, the scarce development of its tongue, and the absence of a pre-stomach that allows for the shredding of larger prey. These animals are continuous consumers. This fact is consistent with the absence of a forestomach and a gall bladder, as well as the presence of a hepatopancreatic duct similar to that of other continuous consumers. The composition of crustaceans and squid in their normal diet could explain the length of their intestine, as it would allow the food to be inside the digestive tract for a longer time to facilitate its enzymatical degradation and absorption.

The importance of this work is clear due to the following reasons. First of all, it is important for anatomists, as there are very few previous descriptions of the anatomy of the digestive system of this dolphin. Then, it is important for physiologists and nutritionists, as anatomy is the basis of a better comprehension of other basic sciences. It is not possible to explain any physiological process without knowing the anatomy of an organ, apparatus, or system. For example, anatomy, digestion, and nutrition are closely related. Once again, the knowledge of the digestive system's anatomy is critical to understanding certain processes related to the absorption of nutrients, the balance between proteins, fats, and carbohydrates, and metabolism. Therefore, our work supports the idea of anatomy in constant adaptation to nature or external factors to facilitate other processes. And, finally, this paper is also important because the species of interest is in danger, and any knowledge or data about these animals are of relevance and will increase their chances of survival.

In summary, it seems that the protein-based diet of the *Pontoporia blainvillei* could determine the anatomy of its digestive system, although certain phylogenetic adaptations should not be ruled out. Of course, further studies are needed to clarify this hypothesis.

Author Contributions: Conceptualization, C.T.-M., R.M.-O. and J.G.-S.; methodology, C.T.-M., J.G.-S. and R.M.-O.; data collection and validation, C.T.-M., I.M.-D., J.-P.L. and E.T.-D.; formal analysis, I.S.-Á., R.M.-O., M.J.O.D. and J.G.-S.; resources, C.T.-M., M.J.O.D., I.M.-D. and J.-P.L.; writing—original draft preparation, C.T.-M., I.M.-D., M.J.O.D. and P.P.-L.; writing—review and editing, J.G.-S., R.M.-O. and I.S.-Á.; supervision, C.T.-M., R.M.-O., I.S.-Á. and J.G.-S.; project administration, J.-P.L., I.M.-D., C.T.-M., E.T.-D. and M.J.O.D.; funding acquisition, J.-P.L. All authors have read and agreed to the published version of the manuscript.

Funding: This research received no external funding.

Institutional Review Board Statement: Ethical review and approval were waived for this study by the Fundación Mundo Marino, which authorized the use of several cadavers for an anatomical study.

Informed Consent Statement: Not applicable.

Data Availability Statement: All data concerning this research are include in the Tables of the present manuscript.

Acknowledgments: The authors thank the Fundación Mundo Marino for the use of its facilities and equipment to carry out this work.

Conflicts of Interest: The authors declare no conflicts of interest.

References

1. Langer, P. Evidence from the digestive tract on phylogenetic relationships in ungulates and whales. *J. Zool. Syst. Evol. Res.* **2001** *39*, 77–90. [CrossRef]
2. de Muizon, C.; Lambert, O.; Bianucci, G. River dolphins: Evolution. In *Encyclopedia of Marine Mammals*, 3rd ed.; Würsig, B. Thewissen, J.G.M., Kovacs, K.M., Eds.; Elsevier: San Diego, CA, USA, 2017.
3. Crespo, E.A. Franciscana Dolphin. Pontoporia blainvillei. In *Encyclopedia of Marine Mammals*, 2nd ed.; Perrin, W.F., Würsig, B. Thewissen, J.G.M., Eds.; Academic Press: San Diego, CA, USA, 2009; pp. 463–469. [CrossRef]
4. Gariboldi, M.C.; Tunis, J.I.; Dejean, C.B.; Failla, M.; Vitullo, A.D.; Negri, M.F.; Cappozzo, H.L. Population Genetics of Franciscana Dolphins (*Pontoporia blainvillei*): Introducing a New Population from the Southern Edge of Their Distribution. *PLoS ONE* **2015**, *10*, e0132854. [CrossRef] [PubMed]
5. Berggren, P.; Wade, P.R.; Carlström, J.; Read, A.J. Potential limits to anthropogenic mortality for harbor porpoises in the Baltic region. *Biol. Conserv.* **2002**, *103*, 313–322. [CrossRef]
6. Bordino, P.; Wells, R.S.; Stamper, M.A. Satellite tracking of Franciscana dolphins Pontoporia blainvillei in Argentina: Preliminary information on ranging, diving and social patterns. In Proceedings of the IWC Committee Meeting, Santiago, Chile, 23–27 June 2008.
7. Di Beneditto, A.P.M. Interactions between gillnet fisheries and small cetaceans in Northern Rio de Janeiro, Brazil (2001–2002). *Lat. Am. J. Aquat. Mamm.* **2003**, *2*, 79–86. [CrossRef]
8. Cappozzo, H.L.; Negri, M.F.; Pérez, F.H.; Albareda, D.; Monzón, F.; Corcuera, J.F. Incidental mortality of Franciscana Dolphin (*Pontoporia blainvillei*) in Argentina. *Lat. Am. J. Aquat. Mamm.* **2007**, *6*, 127–137. [CrossRef]
9. Crespo, E.A.; Corcuera, J.F.; Cazorla, A.L. Interactions between marine mammals and fisheries in some fishing areas of the coast of Argentina. *Rep. Int. Whal. Comm. Spec. Issue* **1994**, *15*, 269–282.
10. Secchi, E.R.; Zerbini, A.N.; Bassoi, M.; Dalla Rosa, L.; Moller, L.M.; Roccha Campos, C.C. Mortality of franciscanas Pontoporia blainvillei in coastal gillneting in southern Brazil (1994–1995). *Rep. Int. Whal. Comm.* **1997**, *47*, 653–658.
11. Negri, M.F.; Denuncio, P.; Panebianco, M.V.; Cappozzo, H.L. Bycatch of franciscana dolphins Pontoporia blainvillei and the dynamic of artisanal fisheries in the species' southernmost area of distribution. *Braz. J. Oceanogr.* **2012**, *60*, 151–160. [CrossRef]
12. Bordino, P.; Mackay, A.I.; Werner, T.B.; Northridge, S.P.; Read, A.J. Franciscana bycatch is not reduced by acoustically reflective or physically stiffened gillnets. *Endanger. Species Res.* **2013**, *21*, 1–12. [CrossRef]

Secchi, E.R.; Danilewicz, D.; Ott, P.H. Applying the phylogeographic concept to identify franciscana dolphin stocks: Implications to meet management objectives. *J. Cetacean Res. Manag.* **2003**, *5*, 61–68. [CrossRef]

Secchi, E.R. Review on the threats and conservation status of Franciscana Pontoporia blainvillei (Cetacea, Pontoporiidae). In *Biology, Evolution and Conservation of River Dolphins within South America and Asia*; Ruiz Garcia, M., Shostell, J.M., Eds.; Nova Science Publishers Inc.: New York, NY, USA, 2010; pp. 323–339.

Reeves, R.R.; Dalebout, M.; Jefferson, T.A.; Karczmarski, L.; Laidre, K.; O'Corry-Crowe, G.; Rojas-Bracho, L.; Secchi, E.; Slooten, E.; Smith, B.D.; et al. *Pontoporia blainvillei*; 2012. Database: IUCN Red List of Threatened Species. Gland, Switzerland and Cambridge. Available online: http://www.iucnredlist.org/details/17978/0 (accessed on 20 May 2023).

Donovan, G.P.; Bjørge, A. Dall's porpoise, Phocoenoides dalli-Introductory remarks. In *Biology of the Phocoenids*; Bjørge, A., Donovan, G.P., Eds.; International Whaling Commission: Cambridge, UK, 1995; Volume 16, pp. 378–380.

Crespo, E.A.; Pedraza, S.N.; Grandi, M.F.; Dans, S.L.; Garaffo, G. Abundance and distribution of endangered Franciscana dolphins in Argentine waters and conservation implications. *Mar. Mamm. Sci.* **2010**, *26*, 17–35. [CrossRef]

Secchi, E.R.; Ott, P.H.; Crespo, E.A.; Kinas, P.; Pedraza, S.N.; Bordino, P. A first estimate of franciscana (*Pontoporia blainvillei*) abundance off southern Brazil. *J. Cetacean Res. Manag.* **2001**, *3*, 95–100. [CrossRef]

Rodríguez, D.; Rivero, L.; Bastida, R. Feeding ecology of the Franciscana (*Pontoporia blainvillei*) in marine and estuarine waters of Argentina. *Lat. Am. J. Aquat. Mamm.* **2002**, *1*, 77–94. [CrossRef]

Yamasaki, F.; Tahahashi, K.; Kamiya, T. Digestive tract of La Plata dolphin, *Pontoporia blainvillei* II. Small and large intestines. *Okajimas Folia Anat. Jpn.* **1975**, *52*, 1–26. [CrossRef]

Arruda Ramos, R.M.; Madeira Di Beneditto, A.P.; Wille Lim, N.R. Growth parameters of *Pontoporia blainvillei* and *Sotalia fluviatilis* (Cetacea) in northern Rio de Janeiro, Brazil. *Aquat. Mamm.* **2000**, *26*, 65–75.

BEVACET (Spanish Database of Cetacean Strandings). *Technical Document on National Action Protocol for Cetacean Strandings*; Ministry for Ecological Transition and the Demographic Challenge, Government of Spain: Madrid, Spain, 2022; Available online: http://bevacet.uv.es/ (accessed on 20 May 2023).

MacLeod, C.D.; Reidenberg, J.S.; Weller, M.; Santos, M.B.; Herman, J.; Goold, J.; Pierce, G.J. Breaking symmetry: The marine environment, prey size, and the evolution of asymmetry in cetacean skulls. *Anat. Rec. (Hoboken)* **2007**, *290*, 539–545. [CrossRef] [PubMed]

Reidenberg, J.S. Anatomical adaptations of aquatic mammals. *Anat. Rec. (Hoboken)* **2007**, *290*, 507–513. [CrossRef]

Harrison, R.J.; Johnson, F.R.; Young, B.A. The oesophagus and stomach of dolphins (Tursiops, Delphinus, Stenella). *J. Zool.* **1970**, *160*, 377–390. [CrossRef]

Cozzi, B.; Huggenberger, S.; Oelschläger, H. *Anatomy of Dolphins: Insights into Body Structure and Function*; Academic Press: London, UK, 2016; Available online: https://www.goodreads.com/book/show/28407064-anatomy-of-dolphins (accessed on 18 April 2023).

Carwardine, M. *Whales, Dolphins, and Porpoises*; Dorling Kindersley: New York, NY, USA, 1995.

Zachos, F.E. *Handbook of Zoology Mammalia VIII*; Schmidt-Rhaesa, A., Ed.; Zoologisches Museum/Zoological Museum Hamburg, Leibniz Institute for the Analysis of Biodiversity Change: Hamburg, Germany, 2020.

Ott, P.H. Estudo da Ecologia Alimentar de *Pontoporia blainvillei* (Gervais & D.Orbigny, 1844) (Cetacea, Pontoporiidae) no Litoral Norte do Rio Grande do Sul, sul do Brasil. Bachelor's Dissertation, Federal University of Rio Grande do Sul, Porto Alegre, Brazil, 1995; 69p.

Bassoi, M. Feeding Ecology of Franciscana Dolphin, *Pontoporia blainvillei* (Cetacea: Pontoporiidae), and Oceanographic Processes on the Southern Brazilian Coast. Ph.D. Thesis, School of Ocean & Earth Sciences, University of Southampton, Southampton, UK, 2005.

Cremer, M.J.; Pinheiro, P.C.; Simões-Lopes, P.C. Prey consumed by Guiana dolphin *Sotalia guianensis* (Cetacea, Delphinidae) and franciscana dolphin *Pontoporia blainvillei* (Cetacea, Pontoporiidae) in an estuarine environment in southern Brazil. *Iheringia Ser. Zool.* **2012**, *102*, 131–137. [CrossRef]

Paso-Viola, M.N.; Denuncio, P.; Negri, M.F.; Rodríguez, D.; Bastida, R.; Cappozzo, H.L. Diet composition of franciscana dolphin *Pontoporia blainvillei* from southern Buenos Aires, Argentina and its interaction with fisheries. *Rev. Biol. Mar. Oceanogr.* **2014**, *49*, 393–400. [CrossRef]

Shil, S.K.; Zahangir, M.M.; Das, B.C.; Rahman, M.M.; Yadav, S.K.; Kibria, M.M.; Siddiki, A.Z. Macro and microanatomy of some organs of a juvenile male Ganges River dolphin (*Platanista gangetica* spp. gangetica). *Anat. Histol. Embryol.* **2023**, *52*, 180–189. [CrossRef] [PubMed]

Martín Díaz, N. Comparative Anatomy of the Suborder Odontoceti. Bechelor's Thesis, Veterinary Medicine, University of Extremadura, Badajoz, Spain, 2018. Available online: http://hdl.handle.net/10662/8526 (accessed on 28 March 2023).

Sokolov, V.E.; Volkova, O.V. Structure of the dolphin's tongue. In *Morphology and Ecology Ofmarine Mammals: Seals, Dolphins, and Porpoises*; Chapskii, K.K., Sokolov, V.E., Eds.; John Wiley and Sons: New York, NY, USA, 1973; pp. 119–127.

Kastelein, R.A.; Dubbeldam, J.L. Marginal papillae on the tongue of the harbour porpoise (*Phocoena phocoena*), bottlenosedolphin (*Tursiops truncatus*), and Commerson's dolphin (*Cephalo-rhynchus commersonii*). *Aquat. Mamm.* **1990**, *15*, 158–170.

Werth, A.J. Adaptations of the cetacean hyolingual apparatus for aquatic feeding and thermoregulation. *Anat. Rec. (Hoboken)* **2007**, *290*, 546–568. [CrossRef] [PubMed]

38. Yamasaki, F.; Satomi, H.; Kamiya, T. The tongue of Franciscana (*La Plata dolphin*), *Pontoporia blainvillei*. *Okajimas Folia Anat. Jpn* **1976**, *53*, 77–92. [CrossRef] [PubMed]
39. Ferrando, T.; Caresano, F.; Ferrando, S.; Gallus, L.; Wurtz, M.; Tagliafierro, G. The tongue morphology and lingual gland histochemistry of Ligurian Sea odontocetes. *Mar. Mamm. Sci.* **2010**, *26*, 588–601. [CrossRef]
40. Kasuya, T. Some information on the growth of the Ganges dolphin with a comment on the Indus dolphin. *Sci. Rep. Whales Res. Inst.* **1972**, *24*, 87–108.
41. Guimaraes, J.P.; Mari, R.B.; Marigo, J.; Rosas, F.C.W.; Watanabe, I. Gross and microscopic observations on the lingual structure of the franciscana (*Pontoporia blainvillei*-Gervais and D'orbigny, 1844). *Microsc. Res. Tech.* **2012**, *75*, 737–742. [CrossRef]
42. Mead, J.G. Stomach anatomy and use in defining systemic relationships of the cetacean family ziphiidae (Beaked Whales). *Anat. Rec.* **2007**, *290*, 581–595. [CrossRef]
43. Yamasaki, F.; Takahashi, K.; Kamiya, T. Digestive tract of La Plata dolphin, Pontoporia blainvillei. I. Oesophagus and stomach. *Okajimas Folia Anat. Jpn.* **1974**, *51*, 29–51. [CrossRef]
44. Marigo, J.; Groch, K.R. Cetacea (golfinhos e Baleias). In *Tratado de Animais Selvagens: Medicina Veterinária*, 2nd ed.; Cubas, Z.S., Silva, J.C.R., Catão-Dias, J.L., Eds.; Grupo Editorial Nacional: Rio de Janeiro, Brazil, 2014; Chapter 43; pp. 917–935.
45. Reynolds III, J.E.; Rommel, S.A. Anatomical dissection: Thorax and abdomen. In *Encyclopedia of Marine Mammals*, 2nd ed.; Perrin, W.F., Würsig, B., Thewissen, J.G.M., Eds.; Academic Press: San Diego, CA, USA, 2009; pp. 29–36.
46. Kamiya, T.; Yamasaki, F. Organ weights of *Pontoporia blainvillei* and *Platanista gangetica* (Platanistidae). *Sci. Rep. Whales Res. Inst.* **1974**, *26*, 265–270.
47. Dwight, D. Allometric relationships in lions vs. domestic cats. *Evolution* **1962**, *16*, 505–514.
48. Crespo, E.A. Franciscana Dolphin: Pontoporia blainvillei. In *Encyclopedia of Marine Mammals*, 3rd ed.; Würsig, B., Thewissen, J.G.M., Kovacs, K.M., Eds.; Academic Press: San Diego, CA, USA, 2022; pp. 388–392; ISBN 9780128043271.
49. Büker, M. *Literaturübersicht zur Ernährung, Verdauungsanatomie und -physiologie von Meeressäugern*; Weißensee Verlag: Berlin, Germany, 2008; ISBN 978-3-89998-150-6.
50. Tellechea, J.S.; Perez, W.; Olsson, D.; Lima, M.; Norbis, W. Feeding habits of franciscana dolphins (*Pontoporia blainvillei*): Echolocation or passive listening? *Aquat. Mamm.* **2017**, *43*, 440–448. [CrossRef]
51. Bassoi, M.; Secchi, E.; Danilewicz, D.; Moreno, I.; Santos, R.; Shepherd, J. Intrapopulation variation in the diet of franciscana dolphin (*Pontoporia blainvillei*) off southern Brazil. *JMBA* **2021**, *101*, 621–637. [CrossRef]
52. Denuncio, P.; Viola, M.N.P.; Machovsky-Capuska, G.E.; Raubenheimer, D.; Blasina, G.; Machado, R.; Polizzi, P.; Gerpe, M.; Cappozzo, H.L.; Rodriguez, D.H. Population variance in prey, diets and their macronutrient composition in an endangered marine predator, the Franciscana dolphin. *J. Sea Res.* **2017**, *129*, 70–79. [CrossRef]
53. Troina, G.; Botta, S.; Secchi, E.R.; Dehairs, F. Ontogenetic and sexual characterization of the feeding habits of franciscanas, *Pontoporia blainvillei*, based on tooth dentin carbon and nitrogen stable isotopes. *Mar. Mamm. Sci.* **2016**, *32*, 1115–1137. [CrossRef]
54. Henning, B.; de Sa Carvalho, B.; Pires, M.M.; Bassoi, M.; Marigo, J.; Bertozzi, C.; Araújo, M.S. Geographical and intrapopulation variation in the diet of a threatened marine predator, *Pontoporia blainvillei* (Cetacea). *Biotropica* **2017**, *50*, 157–168. [CrossRef]
55. Slijper Arruda, E.J. *Whales*; Basic Books: New York, NY, USA, 1962.
56. Stevens, C.E.; Hume, I.E. *Comparative Physiology of the Vertebrate Digestive System*; Cambridge University Press: New York, NY, USA, 1995.
57. Slijper, E.J.; Harrison, R.J. *Whales*, 2nd ed.; Cornell University Press: Ithaca, NY, USA, 1979.
58. Yamasaki, F.; Kito, K. A Morphological note on the intestine of the boutu with emphasis on its length and ileo-colic transition compared with other platanistids. *Sci. Rep. Whales* **1984**, *35*, 165–172.
59. Takahashi, K.; Yamasaki, F. Digestive tract of ganges dolphin, *Platanista gangetica*: II. small and large intestines. *Okajimas Folia Anat. Jpn.* **1972**, *48*, 427–451. [CrossRef] [PubMed]
60. Gaskin, D.E. Form and function in the digestive tract and associated organs in Cetacea, with a consideration of metabolic rates and specific energy budgets. In *Oceanography and Marine Biology: An Annual Review*; Barnes, H.B., Ed.; Aberdeen University Press: Aberdeen, UK, 1978; Volume 16, pp. 313–345. ISBN 0-900015-44-6.
61. Mead, J.G.; Brownell, R.L., Jr. Order Cetacea. In *Mammal Species of the World*, 3rd ed.; Wilson, D.E., Reeder, D.M., Eds.; The Johns Hopkins University Press: Baltimore, MD, USA, 2005; pp. 723–743.
62. Simpson, J.G.; Gardner, M.B. Comparative microscopic anatomy of selected marine mammals. In *Mammals of the Sea. Biology and Medicine*; Ridgway, S.H., Ed.; Charles C. Thomas: Springfield, IL, USA, 1972; pp. 298–418.
63. Yamasaki, F.; Takahashi, K.; Kamiya, T. The Liver of Franciscana (*La Plata Dolphin*), *Pontoporia blainvillei*. *Okajimas Folia Anat. Jpn* **1976**, *52*, 313–330. [CrossRef]
64. Helm, R.C. Intestinal length of three California pinniped species. *J. Zool. Lond.* **1983**, *199*, 297–304. [CrossRef]

animals

MDPI

ommunication

›otential Tick Defense Associated with Skin and Hair Characteristics in Korean Water Deer (*Hydropotes nermis argyropus*)

ang-Joon Lee [1,†], Ki-Yoon Kim [2,3,†], Gyurae Kim [1], Subin Moon [1], Yung-Chul Park [2], Ho-Seong Cho [4,*] nd Yeonsu Oh [1,*]

1 College of Veterinary Medicine and Institute of Veterinary Science, Kangwon National University, Chuncheon 24341, Republic of Korea; sjoon516@kangwon.ac.kr (S.-J.L.)
2 Division of Forest Science, College of Forest & Environmental Sciences, Kangwon National University, Chuncheon 24341, Republic of Korea
3 College of Veterinary Medicine, Chungbuk National University, Cheongju 28644, Republic of Korea
4 College of Veterinary Medicine and Bio-Safety Research Institute, Jeonbuk National University, Iksan 54596, Republic of Korea
* Correspondence: hscho@jbnu.ac.kr (H.-S.C.); yeonoh@kangwon.ac.kr (Y.O.); Tel.: +82-63-850-0960 (H.-S.C.); +82-33-250-8792 (Y.O.)
† These authors equally contributed to this work.

Simple Summary: Ticks, key vectors for multiple pathogens in veterinary medicine, lack the ability to fly. To feed on blood, they must climb onto a host's hair coat from their natural habitat. Their life cycle is closely linked to wildlife, with the Cervidae family serving as primary hosts for adult ticks. In South Korea, the Korean water deer and roe deer are prominent members of the Cervidae family. Notably, the Korean water deer exhibit a lower rate of tick infestation compared to the roe deer. This study aims to understand the reasons behind this difference by conducting an in-depth comparative analysis of the skin and hair characteristics of both species. We found significant variations in certain factors that may hinder ticks from reaching their feeding sites on Korean water deer, potentially acting as a natural defense mechanism against ticks. To our knowledge, this research is the first to focus on comparing the skin and hair attributes of Korean water deer and roe deer in the context of tick bite prevention.

Abstract: The Korean water deer (WD), a predominant wildlife species in South Korea, is listed as vulnerable by the IUCN Red List. Despite belonging to the same family, Cervidae, WD show significantly fewer adult ixodid tick infestations compared to roe deer (RD). Ticks, which cannot fly, engage in questing behavior in natural environments to latch onto hosts. They detect signals like body temperature and host skin chemicals to navigate through the hair coat to the preferred epidermis. In light of this, we performed an extensive comparative study of the skin tissue and hair characteristics of both deer species, focusing on elements contributing to the reduced tick bite incidence in WD. Remarkably, WD exhibited more prominent blood vessels, sebaceous glands, and sweat glands, which are crucial for skin barrier functions ($p < 0.005$). Moreover, WD had irregular scale patterns on their hair cuticles and possessed hair that was significantly stiffer and 2.83 times thicker than that of RD ($p < 0.001$). These characteristics potentially impede ticks from reaching the epidermis hair in WD and RD in the context of tick bite prevention. Further investigations in this area could enhance our understanding of tick–host dynamics and contribute to developing preventive measures against tick-borne diseases in other deer species.

Keywords: Korean water deer; skin and hair characteristics; comparative pathology; tick bites

1. Introduction

Ticks are recognized as significant vectors for a variety of protozoan, bacterial, and viral pathogens, impacting both public health and veterinary medicine more than any other group of blood-feeding arthropods [1]. Diseases such as severe fever with thrombocytopenia syndrome (SFTS), Lyme disease, tick-borne viral encephalitis, and Q fever are notable examples of tick-borne diseases affecting human and animals [1]. The recent global rise in tick-borne diseases is attributed to factors like climate change, land use modifications, expansion of host populations, and increased outdoor activities [2–4]. Ticks' life cycles heavily depend on wildlife, which serve as both hosts and reservoirs, playing a pivotal role in the maintenance and transmission of tick-borne pathogens [5,6]. The Cervidae family, in particular, is a crucial host group for adult ticks [7].

Ticks commonly engage in a behavior known as "questing", where they wait atop vegetation to latch onto a passing host [1]. They use sensory systems, such as Haller's organs on their forelegs' tarsi, to detect approaching hosts [1]. Once on a host, ticks seek optimal feeding sites, guided by stimuli like odor, body temperature, and humidity [8–11]. Factors like skin tissue properties, hair coat, and skin thickness can influence tick resistance. Furthermore, skin tissue with a hair coat can act as physical barriers that affect tick resistance, including the density of the fur coat and the skin thickness [12–14].

In the Republic of Korea, significant research on ticks, with over 143 articles published between 1966 and 2022, has identified prominent species in the Ixodidae and Argasidae families [15]. Notably, *Haemaphysalis longicornis*, *H. flava*, and *Ixodes nipponensis* are widespread. Among these, *H. longicornis* is the most prevalent (69.7%), followed by *H. flava* (22.2%), *I. persulcatus* (6.1%), *I. nipponensis* (1.8%), and *H. japonica* (0.2%) [16].

Korean water deer (WD; *Hydropotes inermis argyropus*), a member of the family Cervidae and a dominant species in the local ecosystem, are widely distributed across the entire Korean Peninsula, yet are classified as vulnerable by the International Union for Conservation of Nature (IUCN) [17]. Despite sharing several characteristics with other deer species, we found that the incidence of tick bites in WD is significantly lower than that in roe deer (RD; *Capreolus pygargus*) [18].

In this study, we explore the hypothesis that variations in skin tissue and hair coat between WD and RD may account for differences in tick bite incidence. We conducted a comparative analysis of the skin tissue and hair attributes of both deer species to understand these disparities better.

2. Materials and Methods

2.1. Tick Collection

During the active tick season, from July to October, we captured five individuals each of WD (Korean water deer) and RD (roe deer) from agricultural areas. We thoroughly inspected and collected ticks from the hair coats and body surfaces of these animals. The collected ticks were preserved in tick collection tubes for subsequent counting.

2.2. Sample Collection and Hair Analysis

Formalin-fixed skin tissues from two RD and two WD, selected from the captured animals, were collected from the Wildlife Research Lab at Kangwon National University. The selection was made based on the proximity of capture dates. Samples included tissues from four areas: head; neck; back; and axilla. From each tissue sample, ten guard hairs were randomly selected and extracted using forceps, leaving the remaining tissue for histopathological studies.

2.3. Hair Morphology Examination

We documented macroscopic features like hair type (straight or curled) and color. Hair shafts were then mounted on glass slides and secured with clear tape for measurements of diameter and length using a digital caliper and a light microscope equipped with CellSens standard 1.16 software (Olympus, Tokyo, Japan) [19]. The medulla structure was also examined. For cuticle analysis, we applied the gelatin casting method, leaving gelatin imprints of the hair scales for microscopic examination. The classification of cuticle and medulla patterns was based on established criteria [20,21]. Briefly, 20% pig native gelatin (Merck, Darmstadt, Germany) was prepared in boiling water. A thin film of gelatin was placed onto a clean glass slide. The hairs were placed superficially on the gelatin film and left overnight at room temperature. Subsequently, hairs were removed, leaving imprints of the scales on the gelatin cast. Images were acquired at ×400 using a BX53 light microscope (Olympus, Tokyo, Japan). Among the two subdivisions of guard hair, namely, the hair shield and the hair shaft, we selected the hair shaft, which is closer to the epidermis.

2.4. Histopathological Analysis

For each animal, we prepared four horizontal and four longitudinal sections from the skin tissues. The samples underwent a series of treatments, including dehydration, clearing in xylene, and embedding in paraffin blocks. Sections of 4 μm thickness were stained with hematoxylin and eosin (H&E) for histopathological comparison between the species, focusing on adnexa distribution, epidermal and dermal thickness, and follicle spacing. The analysis utilized NIH ImageJ software (version 1.54, Bethesda, MD, USA).

2.5. Statistical Analysis

Summary statistics were calculated for all groups to assess the overall quality of the data, including normality. The statistical relationship of hair measurements and histopathological results between RD and WD was determined with Student's *t*-test, followed by the Mann–Whitney U test and Wilcoxon rank-sum test to calculate the post-power of the findings using IBM SPSS Statistics (version 26, Armonk, NY, USA). Differences were considered significant if $p < 0.05$.

3. Results

3.1. Variation in Tick Infestation between Roe Deer and Korean Water Deer

The study revealed a notably lower prevalence of adult ixodid tick infestations and blood feeding activities in WD compared to RD during the active tick season (July to October), with significant differences ($p = 0.0079, 0.0317$) (Table 1). The tick species identified include *Haemophysalis longicornis, H. flava, H. japonica,* and *Ixodes nipponensis* from the Ixodidae family. The observed variability in standard deviation and error, despite statistical significance, was attributed to the absence of tick infestations in August. The anomaly coincided with unusually dry weather and higher temperatures, averaging 1.5 °C above normal, marking it as the hottest year recorded [22].

Table 1. Comparison of average counts of adult and blood-feeding ixodid ticks in Korean water deer and roe deer.

Species	No. of Adult Ixodid Ticks	No. of Blood-Feeding Ixodid Ticks
WD ($n = 5$)	25.2 ± 10.69	9.8 ± 4.18
RD ($n = 5$)	231.8 ± 81.81 **	85.2 ± 40.08 *

Values are presented as mean ± SE; *, significant difference compared to Korean water deer, $p < 0.05$; **, highly significant difference compared to Korean water deer, $p < 0.01$.

3.2. Morphological Distrinction in Hair of Roe Deer and Korean Water Deer

Upon macroscopic examination, significant differences were noted in hair thickness and color between the species. RD featured yellow-brown coats, while WD had grayish coats. Distinct differences in hair cuticles, including scale position, scale margin structure, margin distance, and pattern, were observed between RD and WD [21]. WD had thicker and more irregular hair patterns, whereas RD displayed uniform scale margins (Figure 1A,B). In terms of medulla structure, both species showed similar characteristics, with other features being indistinguishable (Figure 1C,D).

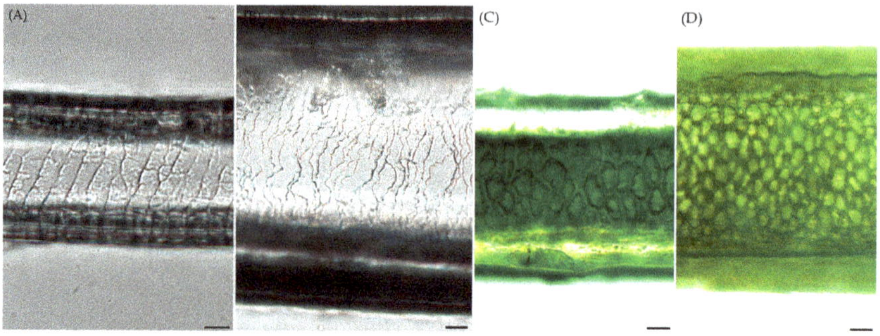

Figure 1. Hair cuticle and medulla characteristics. (**A**) Roe deer (*Capreolus pygargus*) hair cuticles exhibited transversal scale alignment, smooth scale edges, moderate distance between scales, and a consistent wave-like pattern. Scale bar = 20 μm. (**B**) Korean water deer (*Hydropotes inermis argyropus*) hair cuticles featured transversal scale alignment, slightly wavy scale edges, close proximity between scales, and an irregular wave pattern. Scale bar = 20 μm. (**C**) Roe deer (*Capreolus pygargus*) hair medullas were composed of multiple cells, a semi-filled lattice structure, a steady pattern, and scalloped edges. Scale bar = 50 μm. (**D**) Korean water deer (*Hydropotes inermis argyropus*) hair medullas were similar in structure with multiple cells, a semi-filled lattice, consistent pattern, and scalloped edges. Scale bar = 50 μm.

3.3. Comparison of Hair Diameter and Length in Roe Deer and Korean Water Deer

We measured hair diameter and length from 20 hair shafts across four body parts (head, neck, back, and axilla) in both species (Table 2). RD exhibited significantly thinner hair diameters across all body parts compared to WD ($p < 0.001$), with WD hair being approximately 2.83 times thicker overall ($p < 0.001$). RD had shorter hair lengths in the head and back regions than WD ($p < 0.001$ and $p = 0.020$, respectively), but overall, hair lengths were not significantly different between the species ($p = 0.437$).

Table 2. Analysis of hair diameter and length differences between roe deer and Korean water deer.

Parameter	Body Part	RD Hair (n = 20)	WD Hair (n = 20)
Diameter (μm) (Mean ± SD)	Head	136.49 ± 17.92 ***	216.65 ± 36.08
	Neck	172.45 ± 39.22 ***	436.63 ± 145.57
	Back	145.79 ± 17.74 ***	629.24 ± 26.93
	Axilla	118.88 ± 15.05 ***	326.86 ± 43.65
	Total	144.36 ± 3.47 ***	409.20 ± 167.56
Length (mm) (Mean ± SD)	Head	15.02 ± 3.04 ***	9.15 ± 1.49
	Neck	19.13 ± 3.55	20.56 ± 3.04
	Back	20.33 ± 3.50 *	24.03 ± 3.80
	Axilla	25.92 ± 4.56	25.59 ± 3.02
	Total	19.87 ± 5.24	19.98 ± 6.49

*, statistical significance between roe deer and Korean water deer, $p < 0.05$; ***, $p < 0.001$.

3.4. Skin Pathology Comparison between Roe Deer and Korean Water Deer

Our comparative pathological analysis focused on skin layers, blood vessels, and skin appendages, including sebaceous and sweat glands and hair follicles (Figure 2). Notable differences were observed in the sizes of skin appendages and hair follicle types, as well as the dermal blood vessel diameters. WD had larger skin appendages and simpler primary hair follicle pattens compared to RD' compound primary with secondary follicles. Although the blood vessel distribution in the dermis was similar, WD had significantly larger blood vessel diameters. No notable differences were found in epidermal thickness. Quantitative measurements of sebaceous glands, sweat glands, and blood vessels showed significant enlargement in WD ($p < 0.01$) (Table 3), but no significant differences in epidermal thickness or distance between hair follicles' midlines were found between the species (p = 0.403, 0.996).

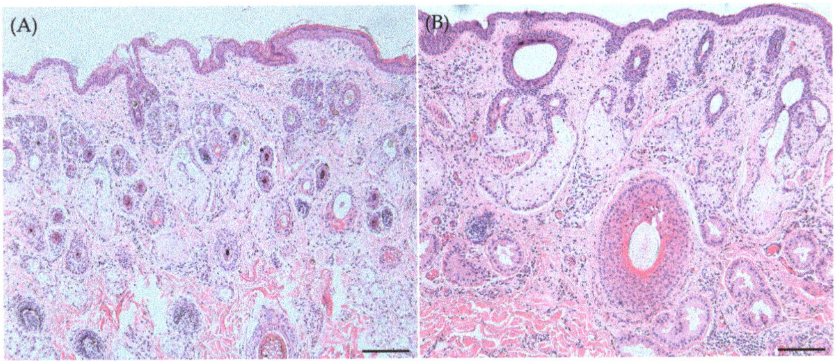

Figure 2. Cross-section of skin. (**A**) Roe deer (*Capreolus pygargus*) and (**B**) Korean water deer (*Hydropotes inermis argyropus*). In WD, the skin appendages were larger in size. The hair follicles in RD showed a compound primary with secondary follicle structure, more intricate compared to the basic primary follicle pattern seen in WD. WD had broader blood vessel diameters. The epidermal thickness did not show any notable variation. Scale bar = 200 μm. H&E staining.

Table 3. Comparison of skin appendage distribution in roe deer and Korean water deer.

Parameter	Tissue	RD	WD
Area (1000 μm^2) (Mean ± SD)	Sebaceous gland (n = 24)	111.11 ± 20.12 ***	314.52 ± 22.84
	Blood vessel (n = 24)	9.25 ± 1.29 ***	42.26 ± 6.53
	Sweat gland (n = 24)	203.24 ± 33.73 **	310.49 ± 29.61
Distance (μm) (Mean ± SD)	Epidermis (n = 64)	62.23 ± 22.11	65.00 ± 12.48
	Primary hair follicle (n = 64)	390.08 ± 151.33	389.94 ± 119.05

**, statistical significance between roe deer and Korean water deer, $p < 0.01$; ***, $p < 0.001$.

4. Discussion

The premise of this investigation was to determine whether the lower incidence of adult tick bites in Korean water deer (*Hydropotes inermis argyropus*, WD) as compared to roe deer (*Capreolus pygargus*, RD) could be linked to distinct variations in their dermal or fur attributes. Our study approached this by examining the integumentary systems of both species, with a focus on characteristics such as hair and skin tissue.

Contrary to flying blood-feeding arthropods, ticks rely on direct contact with their hosts. This interaction is influenced by various phenotypic aspects like hair length, skin thickness, and coat texture. Prior studies, predominantly in cattle breeds, have established correlations between these characteristics and tick prevalence, noting that animals with finer, shorter coats and lighter colors generally demonstrate reduced tick burdens. Studies in cattle have shown a positive correlation between hair length and coat thickness with tick infestation in zebu, Holstein–Gyr [13], Ngun, Bonsmara, and Hereford cattle [23]. In other words, cattle with shorter hairs and smoother coats generally exhibit lower tick counts compared to those with longer hairs and woolier coats [13,24]. Also, lighter-colored coats have been linked to lower tick infection rates than darker-colored coats, suggesting that hair color can impact tick resistance [14,25].

Our research extends these findings to deer species, evaluating specific parameters such as hair color, cuticle structure, and follicle density [26]. In our analysis, WD and RD differed markedly in hair properties, including color, cuticular pattern, diameter, and density. While no significant variance in hair length was observed, WD's hair was markedly thicker. Interestingly, hair density emerged as a crucial factor. WD's hair spacing was significantly narrower, potentially hindering tick mobility and feeding site localization. The hair spacing in WD was over eight times narrower (26.64 μm) compared to that of RD (235.61 μm), which may cause adult ticks to spend more time finding feeding sites. Feeding sites differ throughout the ticks' life stages; however, adult ticks primarily feed on the head and neck [7]. Furthermore, the microclimate created by these dense hairs could unfavorably affect tick survival [27]. Non-blood-feeding ticks can endure for about 40 h at 33% relative humidity [28], thus leading to prolonged exposure to the microclimate without rehydrating. The larger gaps between the hairs on RD, more than eight times wider than those of WD, could influence tick questing and penetration, potentially causing ticks to prefer RD as hosts.

Our comparative pathological examination of skin tissues revealed notable distinctions. WD possessed larger blood vessels and more prominent sweat and sebaceous glands. The role of skin appendages involves protection from the environment, temperature regulation, and maintenance of fluid balance, and the enhanced secretory output is attributed to the enlarged glands [29]. These physiological functions can influence the activity of blood-feeding arthropods [1,9,10], with WD primarily featuring simple primary follicles, while RD display compound primary with secondary follicle patterns. These factors, integral to skin health and homeostasis, could potentially alter the microenvironment, making it less hospitable for ticks. The difference in hair follicle structure and composition between the two species was also significant. WD predominantly featured stiffer primary hairs, while RD had softer, more pliable coats; such variations may impact the overall suitability of the host for ticks, considering factors like temperature and humidity regulation [13,14,23–25,30].

An emerging area of interest in vector biology is the role of the microbial community in attracting blood-feeding arthropods. The differential chemical signatures arising from microbial activity on the skin may influence host attractiveness to ticks [31]. Our study underscores the necessity of further research in this domain, particularly in understanding how host-specific microbiota and emitted semiochemicals, such as amino acids and fatty acids from the sweat and sebaceous glands, affect vector behavior [31].

In summary, this research provides a comprehensive comparison of the skin and fur characteristics of WD and RD, suggesting that the unique features of WD, including skin appendages, hair properties, and potential microbial influences, may confer a natural resistance to tick infestation. To the best of our knowledge, this is the first focused study comparing the skin and hair features of WD and RD in relation to tick bite prevention. This study not only advances our understanding of host–vector interactions in these species, but also lays the groundwork for exploring novel tick prevention strategies on other deer species and, potentially, in broader contexts.

5. Conclusions

This study aimed to elucidate the factors contributing to the lower tick infestation rates observed in Korean water deer (*Hydropotes inermis argyropus*, WD) compared to roe deer (*Capreolus pygargus*, RD), both belonging to the family Cervidae. Our hypothesis centered on the unique dermal and fur traits of WD. Through detailed comparative analysis, we identified several distinctive features in WD, including larger skin appendages, irregular cuticle patterns, increased hair thickness, and higher hair density. These findings illuminate the role of specific skin and fur characteristics in reducing tick prevalence in WD. The insights gained from this study are valuable for understanding the mechanisms of host defenses against tick infestation and could inform the development of preventive measures against tick-borne diseases in deer and, potentially, other species.

Author Contributions: Conceptualization, Y.O. and H.-S.C.; methodology, S.-J.L.; software, K.-Y.K.; validation, G.K., S.M. and Y.-C.P.; formal analysis, Y.O.; investigation, K.-Y.K.; resources, K.-Y.K.; data curation, H.-S.C.; writing—original draft preparation, S.-J.L. and K.-Y.K.; writing—review and editing, Y.O. and H.-S.C.; visualization, S.-J.L.; supervision, Y.O.; project administration, Y.O.; funding acquisition, Y.O. All authors have read and agreed to the published version of the manuscript.

Funding: This study was supported by 'The Government-wide R&D to Advance Infectious Disease Prevention and Control', Republic of Korea (grant number: HG23C1623).

Institutional Review Board Statement: The animal study protocol was approved by the Institutional Animal Care and Use Committee of Kangwon National University (KW-230925-1 and 10 October 2023).

Informed Consent Statement: Not applicable.

Data Availability Statement: The data presented in this study are available in article.

Conflicts of Interest: The authors declare no conflicts of interest.

eferences

Nicholson, W.L.; Sonenshine, D.E.; Noden, B.H.; Brown, R.N. Ticks (*Ixodida*). In *Medical and Veterinary Entomology*, 3rd ed.; Mullen, G.R., Durden, L.A., Eds.; Academic Press: Cambridge, MA, USA, 2019; pp. 603–672.

Ogden, N.H.; Ben Beard, C.; Ginsberg, H.S.; Tsao, J.I. Possible effects of climate change on ixodid ticks and the pathogens they transmit: Predictions and observations. *J. Med. Entomol.* **2021**, *58*, 1536–1545. [CrossRef] [PubMed]

Diuk-Wasser, M.A.; VanAcker, M.C.; Fernandez, M.P. Impact of land use changes and habitat fragmentation on the eco-epidemiology of tick-borne diseases. *J. Med. Entomol.* **2021**, *58*, 1546–1564. [CrossRef] [PubMed]

Beltrame, A.; Laroche, M.; Degani, M.; Perandin, F.; Bisoffi, Z.; Raoult, D.; Parola, P. Tick-borne pathogens in removed ticks Veneto, northeastern Italy: A cross-sectional investigation. *Travel Med. Infect. Dis.* **2018**, *26*, 58–61. [CrossRef] [PubMed]

Tsao, J.I.; Hamer, S.A.; Han, S.; Sidge, J.L.; Hickling, G.J. The contribution of wildlife hosts to the rise of ticks and tick-borne diseases in North America. *J. Med. Entomol.* **2021**, *58*, 1565–1587. [CrossRef] [PubMed]

Orkun, Ö.; Çakmak, A. Molecular identification of tick-borne bacteria in wild animals and their ticks in Central Anatolia, Turkey. *Comp. Immunol. Microbiol. Infect. Dis.* **2019**, *63*, 58–65. [CrossRef]

Kiffner, C.; Loedige, C.; Alings, M.; Vor, T.; Ruehe, F. Attachment site selection of ticks on roe deer, *Capreolus capreolus*. *Exp. Appl. Acarol.* **2011**, *53*, 79–94. [CrossRef]

Oorebeek, M.; Sharrad, R.; Kleindorfer, S. What attracts larval *Ixodes hirsti* (Acari: Ixodidae) to their host? *Parasitol. Res.* **2009**, *104*, 623–628. [CrossRef]

Carr, A.L.; Salgado, V.L. Ticks home in on body heat: A new understanding of Haller's organ and repellent action. *PLoS ONE* **2019**, *14*, e0221659. [CrossRef]

10. O'kelly, J.C.; Spiers, W.G. Observations on body temperature of the host and resistance to the tick *Boophilus microplus* (Acari: Ixodidae). *J. Med. Entomol.* **1983**, *20*, 498–505. [CrossRef]
11. Haarløv, N. Life cycle and distribution pattern of *Lipoptena cervi* (L.)(Dipt., Hippobosc.) on Danish deer. *Oikos* **1964**, *15*, 93–129. [CrossRef]
12. De Castro, J.J.; Capstick, P.B.; Nokoe, S.; Kiara, H.; Rinkanya, F.; Slade, R.; Okello, O.; Bennun, L. Towards the selection of cattle for tick resistance in Africa. *Exp. Appl. Acarol.* **1991**, *12*, 219–227. [CrossRef] [PubMed]
13. Verissimo, C.J.; Nicolau, C.V.J.; Cardoso, V.L.; Pinheiro, M.G. Haircoat characteristics and tick infestation on gyr (zebu) and crossbred (holdstein x gyr) cattle. *Arch. Zootec.* **2002**, *51*, 389–392.
14. Martinez, M.L.; Machado, M.A.; Nascimento, C.S.; Silva, M.V.G.B.; Teodoro, R.L.; Furlong, J.; Prata, M.C.A.; Campos, A.L.; Guimarães, M.F.M.; Azevedo, A.L.S.; et al. Association of BoLA-DRB3. 2 alleles with tick (*Boophilus microplus*) resistance in cattle. *Genet. Mol. Res.* **2006**, *5*, 513–524. [PubMed]
15. Alkathiri, B.; Lee, S.H. Review of ticks (families: *Ixodidae* and *Argasidae*) in the Republic of Korea. *J. Biomed. Transl. Res.* **2022**, *23*, 91–108.
16. Kim, H.C.; Han, S.H.; Chong, S.T.; Klein, T.A.; Choi, C.Y.; Nam, H.Y.; Chae, Y.H.; Lee, H.; Ko, S.; Kang, J.G.; et al. Ticks collected from selected mammalian hosts surveyed in the Republic of Korea during 2008–2009. *Korean J. Parasitol.* **2011**, *49*, 331. [CrossRef] [PubMed]
17. Lee, S.K.; Lee, E.J. Internationally vulnerable Korean water deer (*Hydropotes inermis argyropus*) can act as an ecological filter by endozoochory. *Glob. Ecol. Conserv.* **2020**, *24*, e01368. [CrossRef]
18. Kim, K.Y. Spatio-Temporal Distribution of Ectoparasites in Relation to Habitat use Characteristics of Mammalian Hosts in Cultivated Land and Nearby Habitats in Gangwon-do. Ph.D. Thesis, Kangwon National University, Chuncheon, Republic of Korea, December 2022.
19. Crocker, E.J. A new technique for the rapid simultaneous examination of medullae and cuticular patterns of hairs. *Microscope* **1998**, *46*, 169–173.
20. Cornally, A.; Lawton, C. A guide to the identification of Irish mammal hair. In *Irish Wildlife Manuals*; No. 92.; Marnell, F., Jeffrey, R., Eds.; National Parks and Wildlife Service, Department of the Arts, Heritage, Regional, Rural and Gaeltacht Affairs: Dublin, Ireland, 2016; pp. 1–40. ISSN 1393-6670.
21. De Marinis, A.M.; Asprea, A. Hair identification key of wild and domestic ungulates from southern Europe. *Wildl. Biol.* **2006**, *12*, 305–320. [CrossRef]

22. Korea Meteorological Administration. Available online: http://www.climate.go.kr/home/bbs_new/view.php?code=9 &bname=monthweather_new&vcode=302&skind=&sword=&category1=&category2= (accessed on 8 November 2023).
23. Spickett, A.M.; De Klerk, D.; Enslin, C.B.; Scholtz, M.M. Resistance of Nguni, Bonsmara and Hereford cattle to ticks in a bushveld region of South Africa. *Onderstepoort J. Vet. Res.* **1989**, *56*, 245–250.
24. Marufu, M.C.; Qokweni, L.; Chimonyo, M.; Dzama, K. Relationships between tick counts and coat characteristics in Nguni and Bonsmara cattle reared on semiarid rangelands in South Africa. *Ticks Tick Borne Dis.* **2011**, *2*, 172–177. [CrossRef]
25. Gasparin, G.; Miyata, M.; Coutinho, L.L.; Martinez, M.L.; Teodoro, R.L.; Furlong, J.; Silva, M.V.G.B.; Regitano, L.C.A. Mapping of quantitative trait loci controlling tick [*Riphicephalus* (*Boophilus*) *microplus*] resistance on bovine chromosomes 5, 7 and 14. *Anim. Genet.* **2007**, *38*, 453–459. [CrossRef] [PubMed]
26. Tabor, A.E.; Ali, A.; Rehman, G.; Rocha Garcia, G.; Zangirolamo, A.F.; Malardo, T.; Jonsson, N.N. Cattle tick *Rhipicephalus microplus*-host interface: A review of resistant and susceptible host responses. *Front. Cell. Infect. Microbiol.* **2017**, *7*, 506. [CrossRef] [PubMed]
27. Lee, E.; Choi, T.Y.; Woo, D.; Min, M.S.; Sugita, S.; Lee, H. Species identification key of Korean mammal hair. *J. Vet. Med. Sci.* **2014**, *76*, 667–675. [CrossRef] [PubMed]
28. Boehnke, D.; Gebhardt, R.; Petney, T.; Norra, S. On the complexity of measuring forests microclimate and interpreting its relevance in habitat ecology: The example of *Ixodes ricinus* ticks. *Parasit. Vectors* **2017**, *10*, 1–14. [CrossRef]
29. Lai-Cheong, J.E.; McGrath, J.A. Structure and function of skin, hair and nails. *Medicine* **2013**, *41*, 317–320. [CrossRef]
30. Lehane, M.J. Host-insect interactions. In *The Biology of Blood-Sucking in Insects*, 2nd ed.; Lehane, M.J., Ed.; Cambridge University Press: Cambridge, MA, USA, 2005; pp. 116–149.
31. Verhulst, N.O.; Qiu, Y.T.; Beijleveld, H.; Maliepaard, C.; Knights, D.; Schulz, S.; Berg-Lyons, D.; Lauber, L.C.; Verduijn, W.; Haasnoot, G.W.; et al. Composition of human skin microbiota affects attractiveness to malaria mosquitoes. *PLoS ONE* **2011**, *6*, e28991. [CrossRef]

animals

MDPI

rticle

Cross Sectional Anatomy and Magnetic Resonance Imaging of he Juvenile Atlantic Puffin Head (Aves, Alcidae, *ratercula arctica*)

Marcos Fumero-Hernández [1], Mario Encinoso [1,*], Ayose Melian [2], Himar Artiles Nuez [3], Doaa Salman [4] nd José Raduan Jaber [5,*]

1 Veterinary Hospital, Faculty of Veterinary Medicine, University of Las Palmas de Gran Canaria, Trasmontaña, Arucas, 35413 Las Palmas, Spain; marcos.fumero101@alu.ulpgc.es
2 Myofauna Servicios Veterinarios, Camino Lomo Grande, Arucas, 35411 Las Palmas, Spain; consulta@myofauna.com
3 IVC Evidensia Los Tarahales, 35013 Islas Canarias, Spain; administracion@hvtarahales.es
4 Department of Animal Medicine, Faculty of Veterinary Medicine, Sohag University, Sohag 82524, Egypt; doaasalman2020@gmail.com
5 Department of Morphology, Faculty of Veterinary Medicine, University of Las Palmas de Gran Canaria, Trasmontaña, Arucas, 35413 Las Palmas, Spain
* Correspondence: mencinoso@gmail.com (M.E.); joseraduan.jaber@ulpgc.es (J.R.J.)

Simple Summary: In this study, we described the anatomy of the central nervous system (CNS) and associated structures of the puffin using anatomical cross-sections and advanced imaging techniques such as magnetic resonance imaging (MRI). To the best of the authors' knowledge, this is the first description of the head in the Atlantic puffin. The results obtained in this study could contribute to future anatomical and pathological studies on related species.

Abstract: The Atlantic puffin is a medium-sized seabird with black and white plumage and orange feet. It is distributed mainly along the northern Atlantic Ocean, and due, among other reasons, to human activities, it is in a threatened situation and classified as a vulnerable species according to the International Union of Conservation of Nature (IUCN). In this study, we used a total of 20 carcasses of juvenile Atlantic puffins to perform MRI, as well as anatomical cross-sections. Thus, an adequate description of the head was made, providing valuable information that could be helpful as a diagnostic tool for veterinary clinicians, who increasingly treat these birds in zoos, rehabilitation centers, and even in the wild.

Keywords: anatomical cross-section; magnetic resonance imaging; head; seabirds; Atlantic puffin

1. Introduction

Fratercula arctica, commonly called Atlantic puffin or simply puffin, is a species of seabird belonging to the order *Charadriiformes* and family *Alcidae*. Its scientific name is related to the aspect of its black and white plumage (from the Latin *fratercula*; friar) and its distribution fundamentally throughout the North Atlantic Ocean [1,2]. These seabirds are medium-sized birds that spend most of their lives in pelagic and offshore habitats, as they only come ashore during the reproductive period to nest on cliffs and rocky areas [1,3–5]. Males have larger body dimensions than females and are heavier, but their colors are similar (more vivid in the male during mating season). This seabird has a monogamous mating system, reaching sexual maturity around five years of age (range 3 to 6 years) and incubating one egg per clutch [3,6–13]. It feeds mainly on fish, including variable amounts of polychaetes and crustaceans [14,15]. It is the only species of the three that constitute the genus *Fratercula* (along with the horned puffin (*Fraterculacorniculata*) and the tufted puffin (*Fratercula cirrhata*) that is included in the IUCN list of threatened species, being classified as vulnerable [1]. Among the threats these animals face are entrapment in fishing nets [16,17], pollution [7], predation by introduced species [18–20], or reduced food availability [21]. Another factor that has shown a tremendous negative impact on puffin populations is climate change, characterized by strong and increasingly common sea storms that can lead to massive strandings [22–25].

Puffins are essential indicators of the state of marine ecosystems and their changes over time due to their position in the trophic chains because they can act as bioaccumulator organisms [26–28]. Therefore, they play a helpful role in the transfer of phosphorus and nitrogen from oceanic waters to the continents [29]. In addition to its biological importance, this charismatic species from the North Atlantic Ocean has become an ecotouristic attraction, and many companies in North America and Europe have developed this type of activity near puffin colonies [3]. For all these reasons, private and governmental associations allocate part of their funds to puffin conservation projects [30,31].

It is considered that birds have a level of brain expansion close to that of mammals. Despite this, research into the neuroanatomy of this taxonomic group has been scarce. However, the irruption of modern imaging techniques such as magnetic resonance imaging (MRI) or computed tomography (CT) has enabled the development of this doctrine [32].

In avian medicine, one of the more used imaging diagnostic tools has been conventional radiography, which provides valuable information on musculoskeletal and respiratory processes and alterations of the coelomic cavity (gas, effusions, or masses). However, it is of little use in the head due to the structure overlapping and low resolution [33,34]. In this case, MRI and cross-sectional imaging modalities could provide adequate anatomical and pathological information, avoiding the main drawbacks of radiography [35–39]. It should be noted that, as in human medicine, MRI is the technique of choice for the nervous system study [34], which has already been used in the study of the avian nervous system [35–41]. Nonetheless, to the authors' knowledge, no research has been carried out on the nervous system of the Atlantic puffin. Therefore, this study aimed to describe the head of this bird using anatomical cross-sections and MRI, so the results obtained could serve as a basis not only for future research on the puffin but also on other birds or phylogenetically related animals.

2. Materials and Methods

2.1. Animals

We conducted this study using a cohort of 20 Atlantic puffins, exhibiting a weight ranging from 0.185 to 0.310 kg and a size range spanning 16 cm to 24 cm (measured from beak to tail). These specimens were sourced from the Consejeria de Área de Medio Ambiente, Clima, Energía y Conocimiento of the Cabildo Insular de Gran Canaria. They had perished as a result of mass strandings associated with episodes of severe open sea storms. Most of the specimens were received already deceased, and those that were initially alive but succumbed due to their weakened condition were promptly preserved via freezing for subsequent MRI procedures. These specimens were made available as part of the investigative efforts aimed at elucidating the potential causative factors behind the substantial stranding events. Additionally, the responsible authority granted explicit permission for their inclusion in this study. It is essential to clarify that no animals were intentionally sacrificed or captured for the sole purpose of scientific research.

2.2. MRI Technique

The twenty puffins were imaged at Los Tarahales Veterinary Hospital (Las Palmas, Canary Islands, Spain). Imaging data were acquired using a Canon Vantage Elan 1.5 T imaging system, employing the following sequences: T1-weighted (T1W) sequences in the transverse plane (TR: 634 ms, TE: 10, FOV: 1809 × 829, slice thickness: 2 mm, matrix size: 192 × 160), T2-weighted (T2W) sequences in the transverse plane (TR: 4769 ms, TE: 120, FOV: 1809 × 829, slice thickness: 2 mm, matrix size: 192 × 224), T2W sequences in the dorsal plane (TR: 5271 ms, TE: 120, FOV: 1809 × 829, slice thickness: 2.5 mm, matrix size: 240 × 192), and T2W sequences in the sagittal plane (TR: 4450 ms, TE: 120, FOV: 1809 × 829, slice thickness: 2.9 mm, matrix size: 224 × 224). Additionally, enhanced spin-echo sequences were obtained in the dorsal, transverse, and sagittal planes. The resulting MRI images exhibited a slice thickness ranging from 2.7 to 3.5 mm.

2.3. Anatomical Sections

Following the completion of the scanning procedure, seven out of the twenty juvenile puffins were placed in a dorsal decubitus position within expanded polystyrene containers and promptly subjected to deep freezing at −80 °C for a duration of 72 h. Subsequently, serial sections of half a centimeter thick were made using an electric band saw to obtain sequential transverse, sagittal, and dorsal anatomical cross sections. These slices were immediately cleaned with water, numbered, and photographed on both sides.

2.4. Anatomic Evaluation

Those anatomical cross-sections that better matched the MR images were selected to facilitate the identification of relevant structures of the Atlantic puffin head. In addition, it was necessary to consult textbooks, relevant references on bird anatomy, as well as bone preparations from other seabird specimens [32,33,42–45].

3. Results

In this study, we present a total of 10 figures, which correspond to anatomical sections and T2W MR images. T2W MR images were used because they exhibited higher resolution and contrast. Figure 1 is a sagittal image that shows transverse (A) and dorsal (B) lines, displaying the cross-section levels depicted by roman numerals (II–VIII), which approximately matched the anatomical sections (Figures 2–9). These figures are displayed in two images: (A) anatomical cross-section and (B) T2W MR image. In addition, Figure 10 shows a sagittal anatomical section and the corresponding T2W image, identifying the main structures that compose the head of the Atlantic puffin, especially those related to the central nervous system.

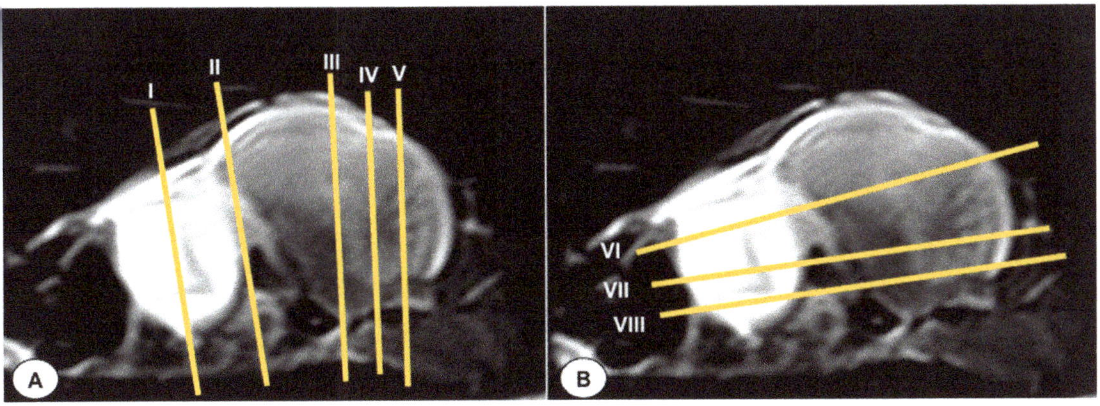

Figure 1. T2-weighted MR sagittal image of the head of an Atlantic puffin (*Fratercula arctica*). The vertical (labeled as (**A**)) and horizontal (labeled as (**B**)) lines correspond to the approximate level of the respective transverse and dorsal slices.

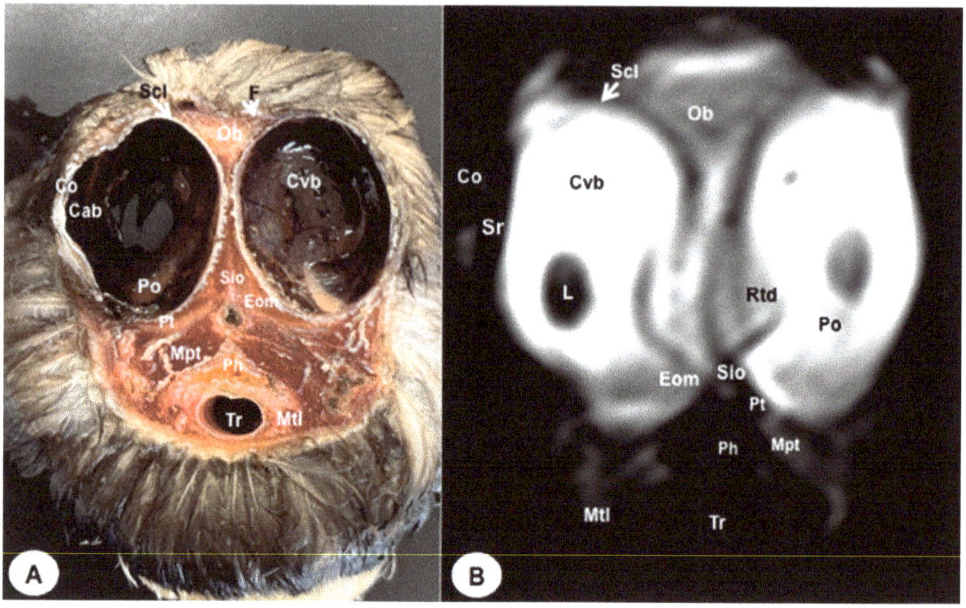

Figure 2. Transverse cross-section (**A**) and T2W MR (**B**) images of the Atlantic puffin head at the level of the *bulbus oculi* corresponding to the line I in Figure 1. Scl: sclera; Sr: sclerotic ring; Cvb: *camera vitrea bulbi*; Co: cornea; Cab: *camera anterior bulbi*: Po: *pecten oculi*; Rtd: retinal detachment; Sio: *septum interorbitalis*; Eom: extraocularmuscle; Ob: olfactory bulb; F: *Os frontale*; Ph: pharinx; Tr: trachea; Pt: *Os pterygoideus*; Mpt: *Musculus pterygoideus pars ventralis*; Mtl: *Musculus tracheolateralis*.

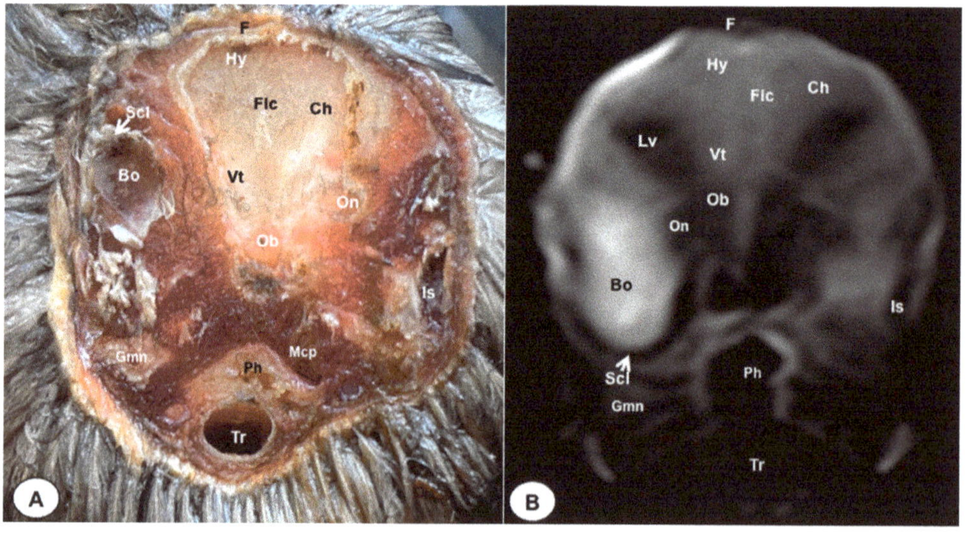

Figure 3. Transverse cross-section (**A**) and T2W MR (**B**) images of the Atlantic puffin head at the level of the olfactory bulb corresponding to line II in Figure 1. F: *Os frontale*; Hy: *hyperpallium*; Ch: cerebral hemisphere; Lv: lateral ventricles; Vt: *Vallecula telencephali*; Flc: *fissura longitudinalis cerebri*; Ob: olfatory bulb; On: optic nerve; Scl: sclera; Bo: *bulbus oculi*;Is: *Sinus infraorbitalis*; Gmn: *Glandula membranae nictitantis*; Ph: pharynx; Mcp: *Musculus constrictor pharyngis*; Tr: trachea.

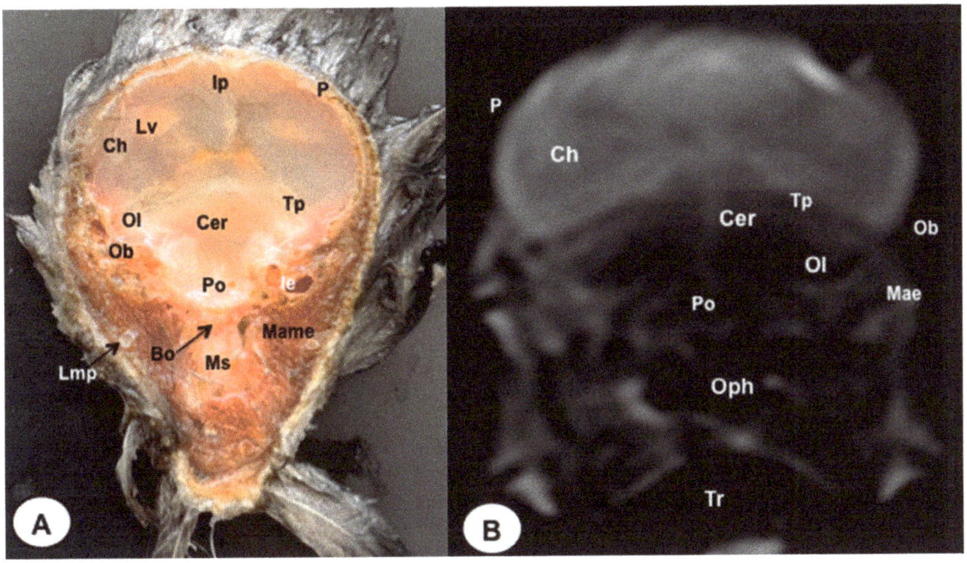

Figure 4. Transverse cross-section (**A**) and T2W MR (**B**) images of the Atlantic puffin head at the level of the optic lobe corresponding to line III in Figure 1. P: *Os parietale*; Ip: interparietal bone; Lv: lateral ventricle;Tp: tentorial process; Bo: *Os basioccipitale*; Ob: otic bones; Ch: cerebral hemisphere; Ol: optic lobe; Cer: *cerebellum*(body); Po: *pons*; Ms: *medulla spinalis*; Mae: *Meatus acusticus externus*; Ie: inner ear; Lmp: Lateral mandibular process; Mame: *Musculus adductor mandibulae externus*; Oph: oropharynx; Tr: trachea.

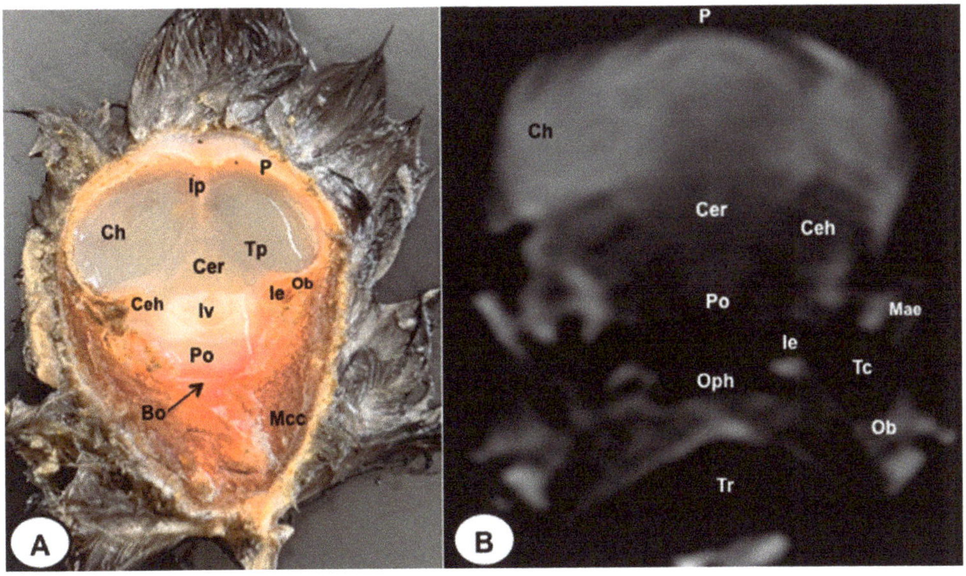

Figure 5. Transverse cross-section (**A**) and T2W MR (**B**) images of the Atlantic puffin head at the level of the pons corresponding to line IV in Figure 1. P: *Os parietale*; Ip: interparietal bone; Tp: tentorial process; Bo: *Os basioccipitale*; Ob: otic bones; Ch: cerebral hemisphere; Ceh: cerebellar hemisphere; Cer: *cerebellum* (body); Iv: fourth ventricle; Po: *pons*; Mae: *Meatus acusticus externus*; Ie: inner ear; Mcc: *Musculus cucularis capitis*; Oph: oropharynx; Tr: trachea.

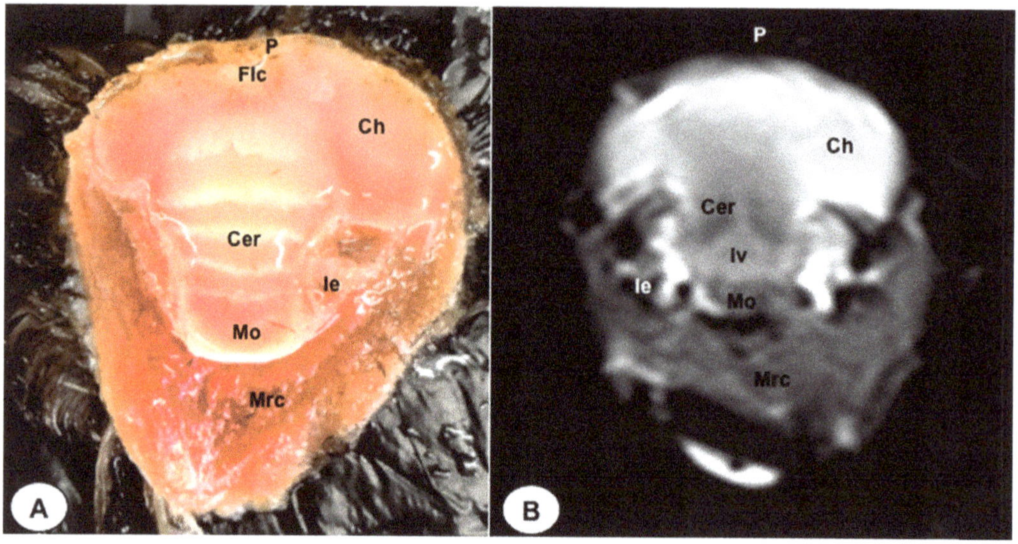

Figure 6. Dorsal cross-section (**A**) and T2W MR (**B**) images of the Atlantic puffin head at the level of the *medulla oblongata* corresponding to line V in Figure 1. P: *Os parietale*; Flc: *fissura longitudinalis cerebri*; Ch: cerebral hemisphere; Cer: *cerebellum*(body); Mo: *Medulla oblongata*; Ie: inner ear; Mrc: *Musculus rectus capitus (Musculus rectus dorsalis + Musculus rectus ventralis + Musculus rectus lateralis)*.

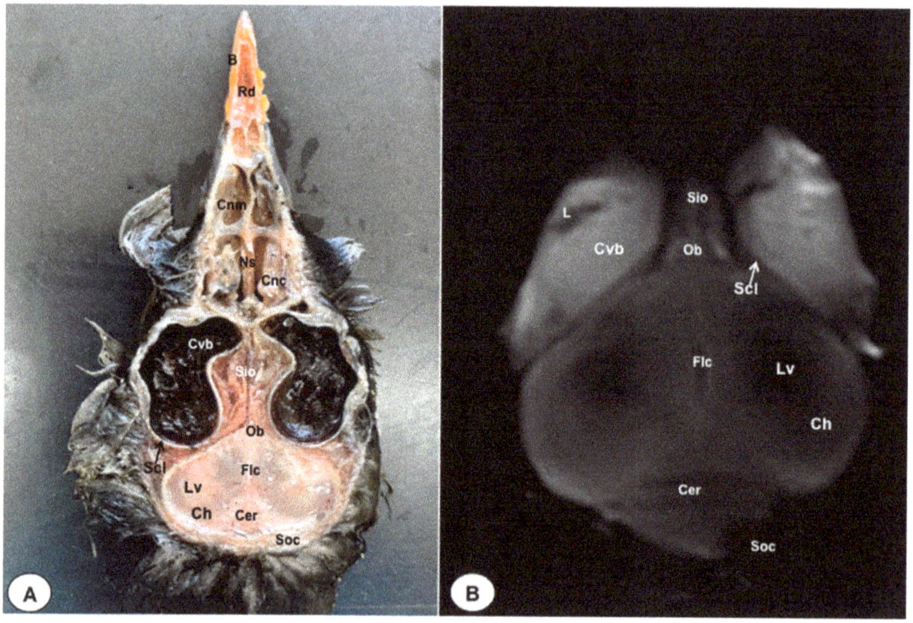

Figure 7. Dorsal cross-section (**A**) and T2W MR (**B**) images of the Atlantic puffin head at the level of the nasal cavity corresponding to line VI in Figure 1. B: beak; Rd: rostral diverticulum; Cnm: *Concha nasalis media*; Cnc: *concha nasalis caudalis*; Ns: nasal septum; Cvb: *camera vitrea bulbi*; Sio: *septum interorbitalis*; Ob: olfatory bulb; Flc: *fissura longitudinalis cerebri*; Ch: cerebral hemisphere; Lv: lateral ventricle; Cer: *cerebellum*; Scl: sclera; Soc: *Os supraoccipitale*.

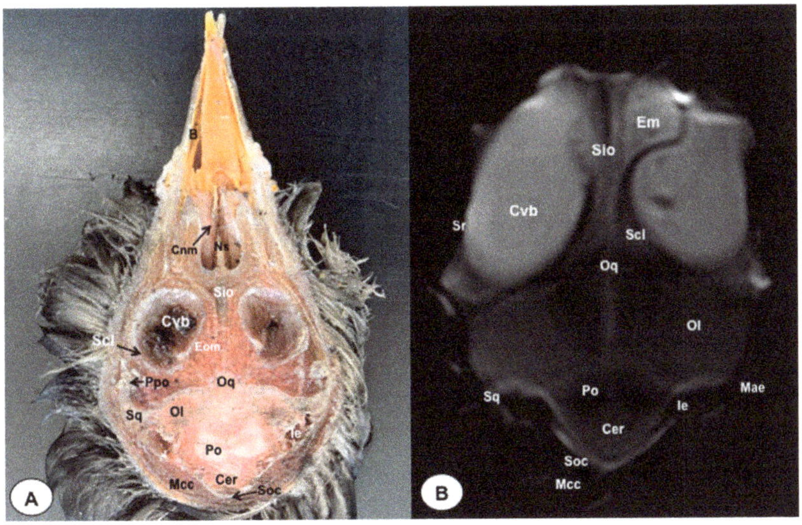

Figure 8. Dorsal cross-section (**A**) and T2W MR (**B**) images of the Atlantic puffin head at the level of optic lobes corresponding to line VII in Figure 1. Sr: sclerotic ring; Scl: sclera; Cvb: *camera vitrea bulbi*; Eom: extraocular muscles; Sio: *septum interorbitalis*; Po: *pons*; Ol: optic lobe; Oq: *chiasma opticum*; Ce: *cerebellum*; Ie: inner ear; Mae: *Meatus acusticus externus*; Sq: *Os squamosum*; Ppo: *Processus postorbitalis*; Soc: *Os supraoccipitale*; Mcc: *Musculus constrictor colli*; Em: *Musculus ethmomandibularis*; Ns: Nasal septum; Cnm: *Concha nasalis media*; B: Beak.

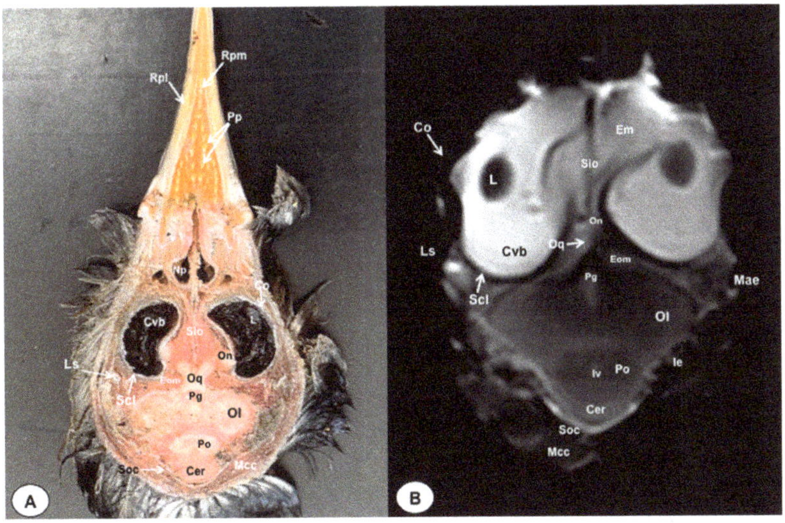

Figure 9. Dorsal cross-section (**A**) and T2W MR (**B**) images of the Atlantic puffin head at the level of the optic chiasm corresponding to line VIII in Figure 1. Cvb: *camera vitrea bulbi*; Sio: *septum interorbitalis*; On: optic nerve; Oq: *chiasma opticum*; Co: cornea; L: lens; Scl: sclera; Eom: extraocular muscle; Ls: *Os laterosphenoidale*; Soc: *Os supraoccipitale*; Po: *pons*; Cer: *cerebellum*; Ol: optic lobe; Pg: pituitary gland; Iv: fourth ventricle; Ie: Inner ear; Mae: *Meatus acusticus externus*; Mcc: *Musculus constrictor colli*; Np: nasopharynx; Em: *Musculus ethmomandibularis*; Cho: *Choanas*; Pp: *Papillae palatinae*; Rpl: *Ruga palatina lateralis*; Rpm: *Ruga palatina mediana*.

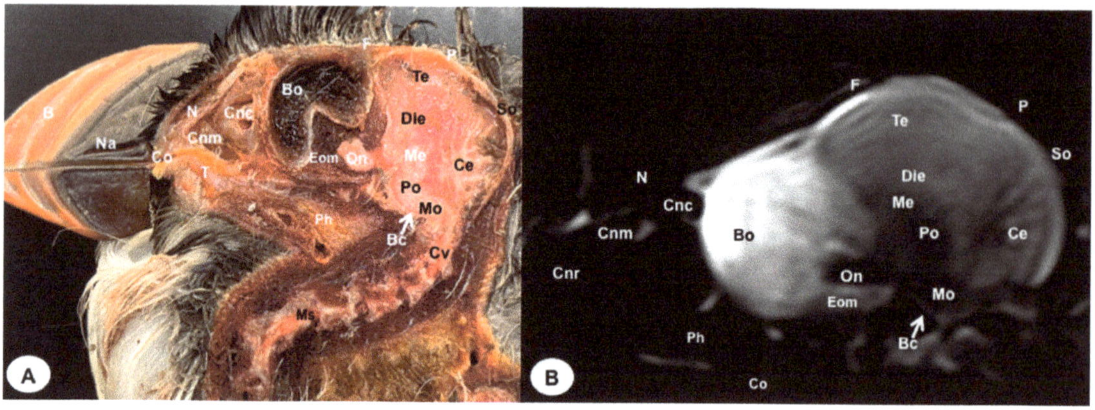

Figure 10. Sagittal cross-section (**A**) and T2W MR (**B**) images of the Atlantic puffin head at the level of the optic nerve. B: beak; Na: *naris*; Co: *cavum oris*; N: *Os nasale*; T: tongue; Cnm: *Concha nasalis media*; Cnc: *concha nasalis caudalis*; Bo: *bulbus oculi*; On: optic nerve; Ph: pharinx; F: *Os frontale*; Te: *telencephalon*; Die: *diencephalon*; Ce: *cerebellum*; Me: *mesencephalon*; Eom: extraocular muscles; Po: pons; Mo: *medulla oblongata*; Ms: *medulla spinalis*; Bc: *basis cranii*; P: *Os parietale*; So: *Os supraoccipitale*; Cv: Cervical vertebra.

3.1. Anatomical Sections

The anatomical sections presented in this investigation proved to be valuable for the discernment of various structures constituting the central nervous system and its related components, all of which were annotated in accordance with the *Nomina Anatomica Avium*. Therefore, we observed the avian brain with the telencephalic hemispheres containing the lateral ventricles (see Figures 4A and 7A). Both hemispheres were separated by the *Fissura longitudinalis cerebri* (Figures 6A and 7A). In addition, transverse cross sections were quite helpful in distinguishing a slightly caudolateral disposed rostral groove, the *Vallecula telencephali*, which was presented on the dorsal surface of each hemisphere, as well as a little pointed olfactory bulb situated at the rostral end of each hemisphere (see Figures 2A, 3A and 7A). The *Diencephalon* was identified as a rostral continuation to the *Mesencephalon* and represents the rostral limit of the brain stem (Figure 10A). The dorsal and sagittal anatomical sections allowed the visualization of some components of the *Hypothalamus*, such as the optic chiasm and the optic nerve penetrating the sclera (see Figures 8A–10A). Further, these sections allowed us to identify other components of the Atlantic puffin brain, such as the dorsal part of the *Mesencephalon*, the large unpaired median *Corpus cerebelli*, the internal medullary body with an internal white substance, and the little paired cerebellar hemispheres (Figures 4A–10A). Moreover, these sections allowed the identification of the ventral part of the *Rhombencephalon*, including the *pons* and the *Medulla oblongata*, which did not present an obvious demarcation (see Figures 4A–10A).

In addition, dorsal, transverse, and sagittal anatomical sections yielded fundamental insights regarding the avian eyeball, which exhibits a relatively substantial size in proportion to the cranial volume and a lateral positioning characteristic of avian anatomy, assuming a globose morphology with slight medial flattening, as depicted in Figure 2A. Furthermore, the delineation of various ocular components, including the cornea, sclera, retina, vitreous chamber, *Pecten oculi*, and lens, was achieved with precision (see Figures 2A, 3A and 7A–10A). Concurrently, our examination unveiled associated structures about the eyeball, encompassing the interorbital septum, extraocular muscles, and the infraorbital sinus that provides structural support to the eyeball, as illustrated in Figures 2A, 3A and 7A–10A. Moreover, these sections provided essential information about the skull shape and different bony structures comprising the skull, including the nasal, the frontal, the parietal, the pterygoid, and the otic and occipital bones (see Figures 2A–10A).

Regarding the nasal cavity, it was found to be bilaterally situated on either side of the median nasal septum. The nares were dorsally positioned at the base of the avian beak, as depicted in Figure 10A. Within the nasal cavity, three nasal concha were discerned in a rostrocaudal sequence: the rostral nasal concha, middle nasal concha, and caudal nasal concha, with particular emphasis on the enhanced development of the middle concha relative to the others (see Figures 7A, 8A and 10A). Moreover, the dorsal, sagittal, and transverse sections were essential to distinguish the roof of the oral cavity and the pharynx (see Figures 2A, 3A, 9A and 10A). Thus, we could observe how the roof was covered by a non-glandular keratinized mucosa forming the transverse ridges (*Rugae palatinae*), which were covered by numerous *Papillae* (see Figure 9A).

3.2. Magnetic Resonance Imaging (MRI)

No discernible anatomical distinctions were evident in the examined juvenile puffins. T2-weighted magnetic resonance (T2W MR) images demonstrated a precise alignment with the cranial structures observed in the cadaveric cross sections, furnishing comprehensive insights into the central nervous system (CNS) and its associated structures. Consequently, various constituents of the puffin head's CNS, the *Bulbus oculi*, and their related elements were adequately differentiated. Thus, in the transverse and dorsal planes, the two telencephalic hemispheres were homogeneous, displaying regions of moderate and hypointense signals corresponding to the cerebral hemispheres and lateral ventricles, respectively (see Figures 3B–7B and 10B). Notably, these two planes facilitated the identification of the olfactory bulb, which was a small rostrally tapering structure with moderate and homogeneous signal intensity (see Figures 2B, 3B and 7B). The *Hyperpallium*, distinguished by its curved dorsal contour and moderate signal intensity, was another distinct structure identified (see Figure 3B). Conversely, the *Diencephalon*, another forebrain component, displayed limited differentiation from the adjacent *Mesencephalon*, manifesting moderate to low-intensity signals (see Figure 10B). Additionally, prominent features of the *Mesencephalon*, such as the optic lobe, were exclusively discerned in the dorsal T2W MR images, showcasing analogous signal characteristics (see Figures 4B, 8B and 9B). Other essential components of the CNS, including the *Pons* and the *Medulla oblongata*, were identifiable in the transverse, dorsal, and sagittal planes, characterized by low-intensity signals (see Figures 4B–6B and 8B–10B). Adjacent to the brain stem, the *Corpus cerebelli* and the small paired cerebellar hemispheres exhibited poorly defined regions of hypo-and moderate intense signal (see Figures 4B–10B).

Regarding the *Bulbus oculi*, the vitreous chamber consistently exhibited hyperintense signals across all planes employed (see Figures 2B, 3B and 7B–10B). However, the cornea, sclera, lens, scleral skeleton, and *Pecten oculi* displayed a hypointense signal in the T2W MR images (see Figures 2B, 3B and 7B–10B). The optic nerve showed an accurate visualization using MRI, presenting a hypointense signal, and was surrounded by hyperintense cerebrospinal fluid (see Figures 9B and 10B). Furthermore, the oral cavity, the pharynx, and the trachea were visualized with low-intensity signals in the T2W images (see Figures 2B–5B). Moreover, diverse muscles of the head, including the *Musculus Pterygoideus pars Ventralis, Musculus Tracheolateralis, Musculus Rectus Capitus (Musculus Rectus Dorsalis + Musculus Rectus Ventralis + Musculus Rectus Lateralis)* and *Musculus Constrictor colli* were displayed with intermediate intensity signal in the T2W MR images (see Figures 2B, 6B, 8B and 9B). In addition, various skull bones, including the nasal, the frontal, the parietal, the pterygoid, and the otic and occipital bones, could be identified in the T2W MR images (see Figures 2B–10B).

4. Discussion

Atlantic puffins can reach latitudes close to the Canary Islands, but their presence on its coasts is uncommon [1]. In this investigation, the specimens employed were sourced from a substantial influx that happened along the coastlines of the Canary Islands in early 2023. The probable cause of this stranding event was attributed to the intense open-sea storms, which resulted in the disorientation and debilitation of the puffins, thereby hampering their foraging abilities. It is noteworthy that severe storms have previously inflicted a considerable adverse impact on puffin populations, culminating in widespread stranding incidents [22–25].

As a consequence, the avian specimens gathered for our research predominantly comprised juvenile individuals. It posed a constraining element encountered during the course of this study, as the diminutive dimensions of the head (measuring less than 5 cm from the beak to the occipital bone) rendered the acquisition of central nervous system images, with a brain length of less than 1.5 cm, particularly challenging. This challenge was further exacerbated when imaging the rostral aspect of the head that directly pertains to the beak. Furthermore, concerning image resolution, which is evidently more limited than in other species, such as dogs or cats, the physical-volumetric factor of the brains of these animals restricted the achievable resolution even when using a 1.5 T high-field magnet. Other studies performed on the avian brain used 3 T and 4.7 T magnets showed better resolution since they were performed in larger birds, including the red kite, the common buzzard, or the African grey parrot [33,39]. Thus, a higher field magnet and bigger specimens provided better resolution.

Some authors have reported that utilization of various coil types can enhance the signal-to-noise ratio and contrast, both in 1.5 T [46] and higher field intensity equipment [47]. This observation should be considered in future similar studies to improve image resolution. In addition, potential tissue changes associated with post-mortem phenomena should be taken into consideration, as they could also negatively impact the obtained image. However, conducting this type of in vivo study typically involves the use of anesthesia, thereby posing a higher risk to the animals undergoing this procedure [48,49]. The use of anatomical cross sections in different planes allowed the depiction of the normal anatomy of the Atlantic puffin brain and its associated structures with excellent detail, complementing the employ of MRI and enabling the acquisition of valuable anatomical information despite the aforementioned limitations.

In our study, we used magnetic resonance imaging to examine the head of the Atlantic puffin. Analogous anatomical investigations employing advanced diagnostic imaging methodologies have been conducted in this species, yielding encouraging outcomes [50]. To the best of the author's knowledge, this is the first description of the head of this species using a high-field strength magnet. This technique has already provided essential information in the assessment of the anatomical knowledge of the head and associated structures in other wildlife species, such as reptiles [51,52], rodents [53,54], terrestrial mammals [55,56], as well as different avian species, including red kite (*Milvus milvus*), common buzzard (*Buteo buteo*), and common kestrel (*Falco tinnunculus*) [33], domestic pigeon (*Columba livia domestica*) [41] or African grey parrot (*Psittacus erithacus*) [39], among others. In contrast to conventional imaging procedures, MRI can be used to obtain images via various anatomic planes without repositioning specimens [51,57–59]. This technique constitutes a high-value tool to evaluate the central nervous system (CNS) and its associated structures because it provides adequate differentiation between the cranial bones and soft tissue structures compared to other modern imaging techniques. Thus, we use sagittal, transverse, and dorsal MRI planes. These dorsal images were especially helpful in identifying different components of the CNS and soft tissue structures. Nonetheless, it is important to highlight that images could not be accurately interpreted without a thorough knowledge of the tomographic or planimetric anatomy of the subject species.

As previously mentioned in other species, a high field strength magnet was adequate to visualize the avian brain [33]. Hence, we could visualize the *Telencephalon* and the telencephalic hemispheres that were lissencephalic, the olfactory bulb, as well as a dorsal eminence corresponding to the *Hyperpallium*, whose curved dorsal contour was better visualized in the T2W MR images. The identification of this specific structure in this species and the dimensions of the sclerotic ring [60] suggested its strong visual specialization that could be related to the bird's feeding patterns. In contrast to the anatomical cross sections, the dorsal and transverse T2W MR images were quite helpful in distinguishing the lateral ventricles, which were located in the medial and occipital regions of the cerebral hemispheres until projected laterally towards the olfactory bulb. However, studies performed on birds with higher-resolution equipment did not clearly label the extension of the lateral ventricles [33]. Therefore, we assumed this finding could be attributed to post-mortem changes affecting the ventricular system.

In relation to the *Mesencephalon*, the optic lobes were discernible in both transverse and dorsal cross-sectional views, as well as in corresponding T2-weighted magnetic resonance (MR) images. These structures were situated in lateral proximity to the *Telencephalon*, and their differentiation from the telencephalic hemispheres was evident due to the presence of the tentorial process, which resulted in a distinct demarcation between these regions. Notably, these lobes exhibited substantial volume, primarily attributed to their processing of a significant portion of visual information [45,61].

Similar to terrestrial mammals, the cerebellum of the Atlantic puffin lies above the midsection of the *Mesencephalon*, the pons, and the *Medulla oblongata*, with no discernible clear demarcations between these regions. Examination using transverse cross-sections and T2-weighted MR images revealed the presence of two small cerebellar hemispheres and a larger unpaired central structure. These cerebellar structures were separated by transverse fissures, facilitating a clear distinction between them. Furthermore, these fissures exhibited marked differentiation between the gray and white matter within the cerebellum.

The T2W MR images combined with anatomical cross-sections allowed adequate differentiation of soft tissue structures. Therefore, different ocular components, including the lens, the vitreous and the anterior chambers, the optic nerve, and the sclera, were distinguished. Thus, they appeared with high signal intensity in T2-weighted MR images. However, other formations, such as the *Pecten oculi,* showed low signal intensity and resolution because we used cadaveric specimens, and the intravenous contrast use was not feasible. In vivo studies involving such substances [59] have demonstrated their utility in visualizing these structures in greater detail since their connective tissue scaffold encloses a compacted capillary network.

Despite the low resolution of the bony structures, the skull showed a dome shape, with large orbits separated by a thin interorbital septum and modification of the facial bones to form the beak. The skull bones visualized did not show important pneumatization as happens in other aquatic species. This lower pneumatization of the skull of the puffin [42] could be related to their living and eating habits since, as we have mentioned, they are excellent divers, and therefore, the pneumatization of the skull could be a problem when diving in oceanic waters. In addition to these findings, the ethmoidal labyrinth was not described, probably due to the poor development of the sense of smell, also related to the small rostrally tapering of the olfactory bulb.

5. Conclusions

This study presents an initial characterization of the cranial anatomy of the Atlantic puffin utilizing transverse, sagittal, and dorsal magnetic resonance imaging in conjunction with anatomical cross-sections. Magnetic resonance imaging has demonstrated its suitability as a tool, providing comprehensive anatomical insights into the various components comprising the puffin's cranial region. Despite the high economic cost associated with this equipment, its availability in routine clinical settings remains challenging. Nevertheless, MRI offers valuable information that may serve as a reference in research investigations and clinical evaluations of neurological disorders in seabirds. Subsequent investigations involving adult puffins are essential to facilitate a comparative analysis of cephalic anatomical structures and to explore potential age-related variations in these structures.

Author Contributions: Conceptualization, J.R.J., M.F.-H. and M.E.; methodology, M.E., H.A.N., A.M. and M.F.-H.; investigation, J.R.J., M.F.-H., D.S. and M.E.; resources, A.M. and H.A.N.; writing—original draft preparation, J.R.J. and M.F.-H.; writing—review and editing, J.R.J., M.F.-H., A.M., D.S. and M.E. All authors have read and agreed to the published version of the manuscript.

Funding: This research received no external funding.

Institutional Review Board Statement: In this study, we did not need ethical approval since the puffins were coming to our hospital to rule out using CT any cause of stranding. From these studies, we obtained the information presented in our paper. As happens in domestic mammals, just informed consent from the owner allowed us to carry out this study. Therefore, El Cabildo Insular de Gran Canaria was informed that all animal identity information obtained from this study was treated as confidential to the extent permitted by law and only used for research or teaching purposes.

Informed Consent Statement: Written informed consent has been obtained from El Cabildo Insular de Gran Canaria to publish this paper.

Data Availability Statement: The information is available at "https://accedacris.ulpgc.es", accessed on 25 June 2023.

Acknowledgments: In loving memory of Alvaro Domingo Rodriguez Garcia. We also thank Concepción Mingot, Carmen Mingot, Emilia Mingot, Nicolas Aquino, Ayesh Mohamad, Nicolasa Rodríguez, Marisa Mohamad, and Jamal Jaber for their support and constructive comments. Moreover, we also thank the Consejeria de Área de Medio Ambiente, Clima, Energía y Conocimiento of the Cabildo Insular de Gran Canaria for providing the animals of this study.

Conflicts of Interest: The authors have read the journal's guideline, and despite author A.M. is employed in Myofauna Servicios Veterinarios, this company does not finance the project or provided the animals. Therefore, we declare no conflict of interest.

References

1. IUCN. The IUCN Red List of Threatened Species. Version 2022-2. Available online: https://www.iucnredlist.org (accessed on 22 April 2023).
2. Jobling, J.A. *A Dictionary of Scientific Bird Names*, 1st ed.; Oxford University Press: Oxford, UK, 1991; p. 164.
3. Lowther, P.E.; Diamond, A.W.; Kress, S.W.; Robertson, G.J.; Russell, K. *The Birds of North America*; The Birds of North America Inc.: Philadelphia, PA, USA, 2002; pp. 1–23.
4. Harris, M.; Daunt, F.; Newell, M.; Phillips, R.; Wanless, S. Wintering areas of adult Atlantic Puffins *Fratercula arctica* from a North Sea colony as revealed by geolocation technology. *Mar. Biol.* **2010**, *157*, 827–836. [CrossRef]
5. Nettleship, D.N.; Kirwan, G.M.; Christie, D.A.; de Juana, E. *Handbook of the Birds of the World Alive*; Lynx Edicions: Barcelona Spain, 2014.
6. Burnham, K.K.; Burnham, J.L.; Johnson, J.A. Morphological measurements on Atlantic puffin *Fratercula arctica naumanni* in high Arctic Greenland. *Polar Res.* **2020**, *39*. [CrossRef]
7. Harris, M.P.; Wanless, S. *The Puffin*; Poyser Monographs; Bloomsbury Publishing: London, UK, 2011.
8. Painter, J. Fratercula arctica. Available online: https://animaldiversity.org/accounts/Fratercula_arctica/ (accessed on 25 April 2023).
9. Breton, A.; Diamond, A.; Kress, S. Encounter, survival and movement probabilities from an Atlantic puffin (*Fratercula arctica*) metapopulation. *Ecol. Monogr.* **2006**, *76*, 133–149. [CrossRef]
10. Durant, J.; Tycho, A.; Nills, C. Ocean climate prior to breeding affects the duration of the nestling period in the Atlantic puffin *Biol. Lett.* **2006**, *2*, 122–128. [CrossRef] [PubMed]
11. Rodway, M. Relationship between wing length and body mass in Atlantic puffin chicks. *J. Field Ornithol.* **1997**, *14*, 338–347.
12. Cramp, S. *The Birds of the Western Palearctic, full edition*; Oxford University Press: Oxford, UK, 1985.
13. Boag, D.; Alexander, M. *The Puffin*; Blandford Press: London, UK, 1995.
14. Harris, M.P.; Hislop, J.R.G. The food of young Puffins *Fratercula arctica*. *J. Zool.* **1978**, *185*, 213–236. [CrossRef]
15. Falk, K.; Jensen, J.K.; Kampp, K. Winter diet of Atlantic puffins (*Fratercula arctica*) in the northeast Atlantic. *Col. Waterbirds* **1992**, *15*, 230–235. [CrossRef]
16. Tasker, M.L.; Camphuysen, C.J.; Cooper, J.; Garthe, S.; Montevecchi, W.A.; Blaber, S.J.M. The impacts of fishing on marine birds. *J. Mar. Sci.* **2000**, *57*, 531–547. [CrossRef]
17. Rogan, E.; Mackey, M. Megafauna by catch in drift nets for albacore tuna (*Thunnus alalunga*) in the NE Atlantic. *Fish. Res.* **2007**, *86*, 6–14. [CrossRef]
18. Barrett, R.T. Atlantic Puffin *Fraterculaarctica* chick growth in relation to food load composition. *Seabird* **2015**, *28*, 17–29. [CrossRef]
19. Stempniewicz, L.; Jensen, J.K. Puffin harvesting and survival at Nólsoy, The Faeroes. *Ornis Svec.* **2007**, *17*, 95–99. [CrossRef]
20. Mitchell, P.I.; Newton, S.F.; Ratcliffe, N.; Dunn, T.E. *Seabird Populations of Britain and Ireland*, 1st ed.; Christopher Helm: London, UK, 2004.
21. Breton, A.R.; Diamond, A.W. Annual survival of adult Atlantic Puffins *Fratercula arctica* is positively correlated with Herring *Clupea harengus* availability. *Ibis* **2014**, *156*, 35–47. [CrossRef]

2. Durant, J.; Anker-Nilssen, T.; Stenseth, N.C. Trophic interactions under climate fluctuations: The Atlantic puffin as an example. *Proc. R. Soc. B Biol. Sci.* **2003**, *270*, 1461–1466. [CrossRef] [PubMed]
3. Sandvik, H.; Erikstad, K.E.; Barrett, R.T.; Yoccoz, N.G. The effect of climate on adult survival in five species of North Atlantic seabirds. *J. Anim. Ecol.* **2005**, *74*, 817–831. [CrossRef]
4. Melillo, J.M.; Richmond, T.T.C.; Yohe, G.W. Extreme Weather. In *Highlights of Climate Change Impacts in the United States: The Third National Climate Assessment*; U.S. Global Change Research Program: Washington, DC, USA, 2014.
5. Harris, M.; Elkins, N. An unprecedented wreck of Puffins in eastern Scotland in March and April 2013. *Scott. Birds* **2013**, *33*, 157–159.
6. Kovacs, C.; Meyers, R. Anatomy and histochemistry of flight muscles of wing-propelled diving bird, the *Atlantic puffin, Fraterculaarctica*. *J. Morphol.* **2000**, *244*, 109–125. [CrossRef]
7. Schreiber, E.; Burger, J. *Biology of Marine Birds*, 1st ed.; CRC Press: Boca Raton, FL, USA, 2001.
8. Moen, S. Morphologic and genetic variation among breeding colonies of the Atlantic puffin. *Auk* **1991**, *108*, 755–763.
9. Otero, X.L.; De La Peña-Lastra, S.; Pérez-Alberti, A.; Osorio Ferreira, T.; Huerta-Díaz, M.A. Seabird colonies as important global drivers in the nitrogen and phosphorus cycles. *Nat. Commun.* **2018**, *9*, 246. [CrossRef]
0. UK Government. Available online: https://www.gov.uk/government/news/englands-treasured-island-seabird-populations-to-be-protected-with-new-government-funding (accessed on 20 April 2023).
1. The National Audubon Society. Project Puffin. Available online: https://projectpuffin.audubon.org/about (accessed on 20 April 2023).
2. Smith, N.A.; Balanoff, A.M.; Ksepka, D.T. Symposium on 'Evolving approaches for studying the anatomy of the avian brain': Introduction. *J. Anat.* **2016**, *229*, 171–172. [CrossRef]
3. Stańczyk, E.K.; Velasco Gallego, M.L.; Nowak, M.; Hatt, J.M.; Kircher, P.R.; Carrera, I. 3.0 Tesla magnetic resonance imaging anatomy of the central nervous system, eye, and inner ear in birds of prey. *Vet. Radiol. Ultrasound* **2018**, *59*, 705–714. [CrossRef]
4. Grosso, F.V. Orthopedic diagnostic imaging in exotic pets. *Vet. Clin. N. Am. Exot. Anim. Pract.* **2019**, *22*, 149–173. [CrossRef]
5. Stauber, E.; Holmes, S.; DeGhetto, D.L.; Finch, N. Magnetic resonance imaging is superior to radiography in evaluating spinal cord trauma in three bald eagles (*Haliaeetus leucocephalus*). *J. Avian Med. Surg.* **2007**, *21*, 196–200. [CrossRef]
6. Wernick, M.B.; Dennler, M.; Beckmann, K.; Schybli, M.; Albini, S.; Hoop, R.K.; Steffen, F.; Kircher, P.; Hatt, J.M. Peripheral nerve sheath tumor in a subadult golden eagle (*Aquila chrysaetos*). *J. Avian Med. Surg.* **2014**, *28*, 57–63. [CrossRef]
7. Delk, K.W.; Mejia-Fava, J.; Jiménez, D.A.; Kent, M.; Myrna, K.; Mayer, J.; Divers, S. Diagnostic imaging of peripheral vestibular disease in a Chinese goose (*Ansercygnoides*). *J. Avian Med. Surg.* **2014**, *28*, 31–37. [CrossRef]
8. de Francisco, O.N.; Feeney, D.; Armién, A.G.; Wuenschmann, A.; Redig, P.T. Correlation of brain magnetic resonance imaging of spontaneously lead poisoned bald eagles (*Haliaeetus leucocephalus*) with histological lesions: A pilot study. *Res. Vet. Sci.* **2016**, *105*, 236–242. [CrossRef] [PubMed]
9. Fleming, G.J.; Lester, N.V.; Stevenson, R.; Silver, X.S. High field strength (4.7T) magnetic resonance imaging of hydrocephalus in an African Grey parrot (*Psittacus erithacus*). *Vet. Radiol. Ultrasound* **2003**, *44*, 542–545. [CrossRef]
0. Jirak, D.; Janacek, J.; Kear, B.P. A combined MR and CT study for precise quantitative analysis of the avian brain. *Sci. Rep.* **2015**, *5*, 1–7. [CrossRef]
1. Romagnano, A.; Shiroma, J.T.; Heard, D.J.; Johnson, R.D.; Schiering, M.R.; Mladinich, M.S. Magnetic resonance imaging of the brain and coelomic cavity of the domestic pigeon (*Columba livia domestica*). *Vet. Radiol. Ultrasound* **1996**, *37*, 431–440. [CrossRef]
2. Faillace, A.C.L.; Vieira, K.R.A.; Santana, M.I.S. Computed tomographic and gross anatomy of the head of the blue-fronted Amazon parrot (*Amazona aestiva*). *Anat. Histol. Embryol.* **2021**, *50*, 192–205. [CrossRef]
3. Baumel, J.J.; Anthony, S.; King, J.E.; James, E. *Handbook of Avian Anatomy: Nomina Anatomica Avium*, 2nd ed.; Nutttall Ortinithological Club: Cambridge, MA, USA, 1993; pp. 318–467.
4. Hadden, P.W.; Ober, W.C.; Gerneke, D.A.; Thomas, D.; Scadeng, M.; McGhee, C.N.J.; Zhang, J. Micro-CT guided illustration of the head anatomy of penguins (Aves: Sphenisciformes: Spheniscidae). *J. Morphol.* **2022**, *283*, 827–851. [CrossRef] [PubMed]
5. Koenig, H.E.; Korbel, R.; Liebich, H.G.; Klupiec, C. *Avian Anatomy: Textbook and Colour Atlas*, 2nd ed.; 5m Books Ltd: Sheffield, UK, 2016.
6. Yoshioka, H.; Ueno, T.; Tanaka, T.; Shindo, M.; Itai, Y. High-resolution MR imaging of triangular fibrocartilage complex (TFCC): Comparison of microscopy coils and a conventional small surface coil. *Skelet. Radiol.* **2003**, *32*, 575–581. [CrossRef] [PubMed]
7. Bittersohl, B.; Huang, T.; Schneider, E.; Blazar, P.; Winalski, C.; Lang, P.; Yoshioka, H. High-resolution MRI of the triangular fibrocartilage complex (TFCC) at 3T: Comparison of surface coil and volume coil. *J. Magn. Reson. Imaging* **2007**, *26*, 701–707. [CrossRef] [PubMed]
8. Ronaldson, H.L.; Monticelli, P.; Cuff, A.R.; Michel, K.B.; d'Ovidio, D.; Adami, C. Anesthesia and anesthetic-related complications of 8 elegant-crested tinamous (*Eudromia elegans*) undergoing experimental surgery. *J. Avian Med. Surg.* **2020**, *34*, 17–25. [CrossRef]
9. González, M.S.; Adami, C. Psittacine Sedation and Anesthesia. *Vet. Clin. N. Am.* **2022**, *25*, 113–134.
0. Jaber, J.R.; Fumero-Hernandez, M.; Corbera, J.A.; Morales, I.; Amador, M.; Ramírez, G.; Encinoso, M. Cross-Sectional Anatomy and Computed Tomography of the Coelomic Cavity in Juvenile Atlantic Puffins (Aves, *Alcidae, Fratercula arctica*). *Animals* **2023**, *13*, 2933. [CrossRef] [PubMed]

51. González-Rodríguez, E.; Encinoso Quintana, M.; Morales Bordon, D.; Garcés, J.G.; ArtilesNuez, H.; Jaber, J.R. Anatomical Description of Rhinoceros Iguana (*Cyclura cornuta cornuta*) Head by Computed Tomography, Magnetic Resonance Imaging and Gross-Sections. *Animals* **2023**, *13*, 955. [CrossRef] [PubMed]
52. Arencibia, A.; Hidalgo, M.R.; Vázquez, J.M.; Contreras, S.; Ramírez, G.; Orós, J. Sectional anatomic and magnetic resonance imaging features of the head of juvenile loggerhead sea turtles (*Caretta caretta*). *Am. J. Vet. Res.* **2012**, *73*, 1119–1127. [CrossRef]
53. Capello, V. Diagnostic Imaging of Dental Disease in Pet Rabbits and Rodents. *Veter. Clin. N. Am. Exot. Anim. Pract.* **2016**, *19*, 757–782. [CrossRef]
54. Morales-Bordon, D.; Encinoso, M.; Arencibia, A.; Jaber, J.R. Cranial Investigations of Crested Porcupine (*Hystrix cristata*) by Anatomical Cross-Sections and Magnetic Resonance Imaging. *Animals* **2023**, *13*, 2551. [CrossRef]
55. Jaber Mohamad, J.R.; Encinoso Quintana, M.Ó.; Morales, D.; ArtilesVizcaíno, A.; Santana, M.; Blanco Sucino, D.; Arencibia Espinosa, A. Anatomic study of the normal Bengal tiger (*Panthera tigris tigris*) brain and associated structures using low field magnetic resonance imaging. *Eur. J. Anat.* **2016**, *20*, 195–203.
56. Arencibia, A.; Matos, J.; Encinoso, M.; Gil, F.; Artiles, A.; Martínez-Gomariz, F.; Vázquez, J.M. Computed tomography and magnetic resonance imaging study of a normal tarsal joint in a Bengal tiger (*Panthera tigris*). *BMC Vet. Res.* **2019**, *15*, 126. [CrossRef]
57. Wyneken, J. Computed tomography and magnetic resonance imaging anatomy of reptiles. In *Reptile Medicine and Surgery*, 2nd ed.; Mader, D.R., Ed.; Elsevier: Amsterdam, The Netherlands, 2006; pp. 1088–1096.
58. Valente, A.L.S.; Cuenca, R.; Zamora, M.A.; Parga, M.L.; Lavin, S.; Alegre, F.; Marco, I. Sectional anatomic and magnetic resonance imaging features of coelomic structures of loggerhead sea turtles. *Am. J. Vet. Res.* **2006**, *67*, 1347–1353. [CrossRef] [PubMed]
59. Xiao, Y.D.; Paudel, R.; Liu, J.; Ma, C.; Zhang, Z.S.; Zhou, S.K. MRI contrast agents: Classification and application. *Int. J. Mol. Med.* **2016**, *38*, 1319–1326. [CrossRef] [PubMed]
60. Fumero-Hernández, M.; Encinoso, M.; Ramírez, A.S.; Morales, I.; Suárez Pérez, A.; Jaber, J.R. A Cadaveric Study Using Computed Tomography for Measuring the Ocular Bulb and Scleral Skeleton of the Atlantic Puffin (Aves, *Alcidae, Fratercula arctica*). *Animals* **2023**, *13*, 2418. [CrossRef] [PubMed]
61. Evans, H.E. Avian anatomy. In *Hand-Book of Bird Biology*, 3rd ed.; Lovette, I.J., Fitzpatrick, J.W., Eds.; Wiley & Sons: Chester, UK, 2016; p. 219.

animals

MDPI

rticle

ongue of the Egyptian Endemic Bridled Skink (*Heremites ittatus*; Olivier, 1804): Gross, Electron Microscopy, Histochemistry, and Immunohistochemical Analysis

amadan M. Kandyel [1,2], Om Prakash Choudhary [3,*], Sahar H. El-Nagar [4], Donald B. Miles [5] nd Mohamed Abumandour [6]

1 Department of Zoology, Faculty of Science, Tanta University, Tanta 31527, Egypt
2 Department of Biology, Faculty of Arts and Sciences, Najran University, Najran 61441, Saudi Arabia
3 Department of Veterinary Anatomy, College of Veterinary Science, Guru Angad Dev Veterinary and Animal Sciences University (GADVASU), Rampura Phul, Bathinda 151103, Punjab, India
4 Department of Animal Wealth Development, Faculty of Veterinary Medicine, Kafrelsheikh University, Kafr El-Sheikh 33516, Egypt
5 Department of Biological Sciences, Ohio Center for Ecological and Evolutionary Studies, Ohio University, Athens, OH 45701, USA
6 Department of Anatomy and Embryology, Faculty of Veterinary Medicine, Alexandria University, Alexandria 21321, Egypt
* Correspondence: om.choudhary@gadvasu.in or dr.om.choudhary@gmail.com

Simple Summary: The present study describes the morphological characteristics of the tongue and laryngeal mound of *Heremites vittatus.* In SEM analysis, the complicated papillary system showed a special filiform system (nine subtypes) in addition to the conical system on the ventral surface of the foretongue apex. The dorsal surface of the laryngeal mound represented 18 longitudinal folds. Histologically, the keratinized layer was present in the foretongue but was absent in the mid- and hindtongue. Significant histochemical signals with positive glandular AB and PAS-positive reactions were observed. Immunohistochemistry showed strong cytokeratin immunopositivity in all parts of the tongue.

Abstract: The present study used light and scanning electron microscopy to describe the integrative morphological description of the tongue and laryngeal mound of *Heremites vittatus*, an endemic lizard of Saharan Africa. Additionally, ultrastructure, histology, histochemistry, and immunohistochemical approaches were used to characterize the lingual apparatus adaptations. In the present study, *Heremites vittatus* consisted of a complex lingual papillary system in which the ventral apical surface of the foretongue comprised conical papillae. The dorsal surface consisted of different filiform papillary (*papillae filiformes*) types: the anterior section had two types (bifid and pointed), and the posterior section had four types (triangular, trifid, quadrifid, and pentafid) papillae. The dorsal midtongue surface exhibits scale-like, serrated filiform papillae with anterior gland openings. The hindtongue consisted of two overlapping filiform papillae: scale-like, board-serrated papillae on the median portion and finger-like papillae on the wings. The dorsal surface of the laryngeal mound had 18 longitudinal folds with glandular openings. Histologically, the foretongue was covered by a slightly keratinized layer that was absent in the mid- and hindtongue. The lingual glands were absent from the foretongue but present in the interpapillary space in the mid- and hindtongues. We observed a few rounded taste buds in the conical papilla epithelium. Histochemical analysis revealed strong glandular Alcian Blue (AB)-positive and Periodic Acid–Schiff (PAS)-positive reactions. Immunohistochemistry showed strong cytokeratin immunopositivity in all parts of the tongue. In conclusion, the obtained data about the lingual characterizations have been consistent with the active foraging behavior of the species and its environmental conditions.

Keywords: lingual papillae; laryngeal mound; immunohistochemical; SEM; *Heremites vittatus*

1. Introduction

The feeding strategy of a species is a core behavior that significantly influences survival and reproductive success [1,2]. The tongue is considered a key innovation in the evolution of the terrestrial lifestyle as it is involved in prey processing and the passage of food from the oral cavity to the gut [3,4]. The tongue serves various functions, including prey capture, drinking, respiration, and protective behavior [5,6].

Reptile tongues are distinguishable from other vertebrates by the morphological and physiological attributes of individual species that correlate with feeding behavior (e.g., prey capture, prey processing) and lingual functions [3,7,8], in which diet, nutritional habits, and food manipulation all play a role in the lingual surface variances between reptiles [9]. Prior research relied on light microscopy to describe the tongue surface structure of different reptile species [10,11], which revealed a wide range of morphological characteristics of the tongue, particularly lingual shape, papillary presence, and distribution. The presence or absence of the different papillary types on the dorsal lingual surface of the different reptilian species indicated the feeding behaviors and nature of the available food particles [12]. Meanwhile, the author has been aware of the similarities and variances in the lingual epithelial structure of many reptile species.

Heremites (=Mabuya) vittatus (*Bridled Mabuya* or *Bridled Skink*) is a skink species found in North Africa and the Middle East. *Heremites vittatus* is associated with open-sandy or stony soil with little grass or heavy vegetation. This species can commonly be observed near water, especially in the wetland region of Egypt. The diet of *Heremites vittatus* consists of insects [13].

There is scanty information regarding the microscopic features of the tongue and laryngeal mound of *Heremites vittatus*. Understanding more about the oropharyngeal cavity anatomy may aid in keeping these species from becoming endangered by describing the functional relationship of the oral cavity with their eating, predation, and respiration. Therefore, the primary purpose of our observational study is to characterize and provide complete gross and ultrastructure characterizations of the tongue with the laryngeal mound of *Heremites vittatus*. This study also uses different histological techniques (ordinary histological, histochemical, and immunohistochemical) to illustrate the lingual adaptations with the different feeding habits and available feeding particles.

2. Material and Methods

2.1. Sample Collection

Ten adults (six males and four females) of *Heremites vittatus* were captured alive in the Abo-Rawash district, Giza, Egypt. The collected reptile species were kept in the animal keeping facility at the Department of Zoology, Faculty of Science, Tanta University, Egypt. The housing of *Heremites vittatus* includes a reptile's cage with a temperature range of 18–24 °C and 55–70% humidity; crickets, mealworms, cockroaches, earthworms, superworms, hornworms, and other prey animals can all be found in diets. All the lizards were put to death by decapitation after an anesthetic overdose with ketamine hydrochloride (Parke-Davis, Berlin, Germany), 325 µg/g body weight [14], carried out by a licensed zoologist in conformity with local, national, and international ethical standards, and examined for any anatomical anomalies. The samples that were being evaluated were brought to the lab's facilities in ice boxes.

This study was carried out according to the Institutional Animal Care and Use Committee (IACUC) protocols of Laboratory Animals, Faculty of Science, Tanta University, and it was approved by Egyptian laws. The experiments were carried out following the ethical provisions established by the 2010/63/EU directive for animal experiments and were prepared according to the guidelines for the use and care of animals of the IACUC Tanta University, under research project number (IACUC-SCI-TU-0309).

2.2. Morphological Examinations

Heads were decapitated, and tongues and laryngeal entrances were carefully removed and submerged in 10% neutral buffered formal saline. After fixation, three tongues and laryngeal entrances were examined grossly using a stereomicroscope (Olympus VM VMF 2×, Eyepiece 10× Stereo Microscope, Tokyo, Japan). Then, the photographs of the tongue were taken with a digital camera a digital camera (Canon IXY 325, Tokyo, Japan). In this study, all the anatomical terminologies have been used according to Nomina Anatomia Veterinaria (NAV) [15].

2.3. Scanning Electron Microscopy (SEM) Analysis

The tongues of four individuals of *Heremites vittatus* were prepared for SEM techniques [16]. Each tongue with its laryngeal entrance was first rinsed in 2% glutaraldehyde (Mfcd00007025) dissolved in a 0.1 M phosphate buffer solution at pH 7.4 (Mfcd00131855). The tongues were then cut and fixed in 2% formaldehyde and 1.25% glutaraldehyde in a 0.1 M sodium cacodylate buffer (97068-100 ML-F), pH 7.2 at 4 °C. Samples were washed in 0.1 M sodium cacodylate containing 5% sucrose, processed by tannic acid (Mfcd00066397), dehydrated in ascending grades of ethanol (50%, 70, 80, 90, 95, and 100% ethanol, 15 min each), then held and processed for 2 h at room temperature [16]. Finally, the samples were dried in carbon dioxide (critical point drying), attached to stubs with colloidal carbon, and coated with gold–palladium in a sputtering device. SEM analysis was carried out with the JEOL JSM 6510 lv SEM unit at the Laboratory of Electron Microscopy at the Faculty of Agriculture, Mansoura University, Egypt. SEM image analysis was carried out at the Department of Veterinary Anatomy, College of Veterinary Science, Rampura Phul, Punjab, India.

2.4. Histological and Mucin Histochemical Examinations

Three tongue specimens (0.5 cm^3) from the three different lingual regions (fore, mid, and hindtongue) of *Heremites vittatus* were collected and immersed in 10% neutral buffered formal saline for 24 h. The tongue samples were thoroughly transferred to 70% alcohol after 48 h. The tongue samples were dehydrated with increasing gradations of ethanol (70–100%), cleared in xylene, impregnated, and embedded in paraffin wax. Histological sections of 5–6 μm were cut using a Leica rotatory microtome (RM 20352035; Leica Microsystems, Wetzlar, Germany) and mounted on glass slides. Paraffin sections were organized and stained with Hematoxylin and Eosin Stain (AB245880) according to the Suvarna et al. [17] protocol. Extra-paraffin sections were stained with AB (pH 2.5, Mucin Stain) (ab150662) [18] and PAS (ab150680) [19] to localize the acidic and neutral mucopolysaccharides, respectively. Moreover, the collagen fibers were examined using Masson's trichrome stain method (AB150686) [20]. The stained sections were examined with a BX50/BXFLA microscope (Olympus, Tokyo, Japan). From the micrographs, the height, width, and gland diameter of the papillae on different lingual parts were determined using ImageJ software (1.53t) [21].

2.5. Immunohistochemical Analysis of the Dorsal Mucosal Lingual Keratin

Lingual tissues were incubated overnight with the primary Anti-Pan Cytokeratin antibody [AE1/AE3] that was found as Invitrogen Anti-Pan Cytokeratin Monoclonal (AE1/AE3), eBioscience™, Catalog # 50-9003-82. After washing in a Phosphate-buffered saline (PBS), the sections were incubated for 30 min with Rabbit Anti-Mouse secondary antibody and then for 30 min with ABC, diluted to 1:200 [22]. After three PBS washes, immerse for 5–10 min in freshly prepared 0.1% 3, 3-diaminobenzidine tetrahydrochloride (DAB). Sections were counterstained in Mayer's Haematoxylin for 2 min, dehydrated, cleared in xylene, and mounted. The primary antibody was replaced by PBS for the negative control test. Finally, sections were examined and photographed under the bright field light microscope (Olympus BX 50 compound microscope).

2.6. Negative Image Analysis Using CMEIAS Color Segmentation (Supplementary Image)

The negative image of Figure 8 was performed using CMEIAS color segmentation and improved computing technology. Firstly, the image was opened with CMEIAS Color Segmentation, then selected "Process" from the menu items and clicked "Negative Image" [23].

2.7. Digital Coloring of Scanning and Transmission Electron Microscopic Images

We digitally colored the scanned electron microscopy images using the Photo Filter 6.3.2 program to recognize the different lingual papillary regions, as per the method described by Kandyel et al. [24].

3. Results

3.1. Gross Morphological Observations of the Tongue and Laryngeal Entrance of the Heremites Vittatus

The tongue is triangular and broadens posteriorly, with a pointed, black-pigmented tip. The dorsoventral flattened tongue was classified into three lingual regions: the foretongue, the midtongue, and the hindtongue, with two lateral lingual wings encircling the laryngeal mound and its glottis (Figure 1/FT, MT, HT, LM). The average length of the tongue was measured at about 1.164 ± 0.13 cm. The three lingual regions had an average length of (0.270 ± 0.031, 0.392 ± 0.024, and 0.49 ± 0.41 cm) and an average width of (3.61 ± 0.15, 4.81 ± 0.74, and 5.61 ± 0.89 cm) of the foretongue, midtongue, and hindtongue, respectively. The ventral surface of the tongue was attached to the oral cavity floor by the lingual frenulum.

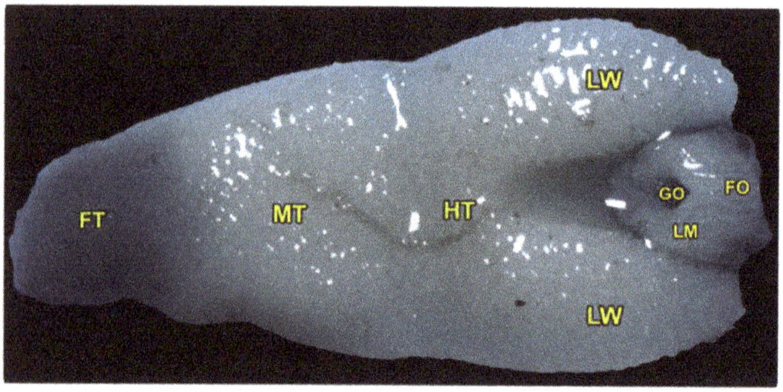

Figure 1. Gross morphology of the tongue of *Heremites vittatus*, showing the foretongue (FT), midtongue (MT), hindtongue (HT), the lingual wings (LW), the laryngeal mound (LM), the median glottic opening (GO), and laryngeal folds (FO).

The pharyngeal cavity was occupied by the elongated laryngeal mound that carried a median longitudinal glottis opening. The laryngeal mound was situated directly posterior to the lingual root and surrounded by the two lateral lingual wings (Figure 1/LM, GO, LW). Moreover, our morphometric findings showed that the average total length of the laryngeal mound and glottic opening were 0.289 ± 0.032 and 0.120 ± 0.014 cm, respectively, while the average total width of the laryngeal mound and glottic opening were 1.93 ± 0.124 and 0.31 ± 0.012 cm, respectively. Additionally, the slight and non-significant increase in its equatorial diameter (0.208 ± 0.02 cm) was greater than the axial one (0.1038 ± 0.003 cm).

Our description depends mainly on the location of the lingual frenulum, as shown in (Figure 1) to describe the three distinct lingual parts (the foretongue, the midtongue, and the hindtongue), in which the foretongue begins from the rostral part of the lingual tip until the beginning of the lingual frenulum. The midtongue begins from the beginning of the lingual frenulum until the terminal portion of the lingual frenulum, and the hindtongue begins from the terminal portion of the lingual frenulum.

3.2. Scanning Electron Microscopy Observations of the Tongue and Laryngeal Entrance

The ventral surface of the foretongue's apex has numerous conical papillae (*papillae conicae*) of hexagonal and pentagonal determinations with narrow space in-between (Figure 2/FT, VS, CP); additionally, with high magnifications, each conical papillae revealed the irregular ventral surface appearance (Figure 2E). Whereas the dorsal surface of the anterior apical part had two types of posteriorly directed filiform papillae (*papillae filiformes*): bifid and pointed papillae with a corrugated surface (Figure 3A–E/FT, DS, FP, BFP, black star, OFP).

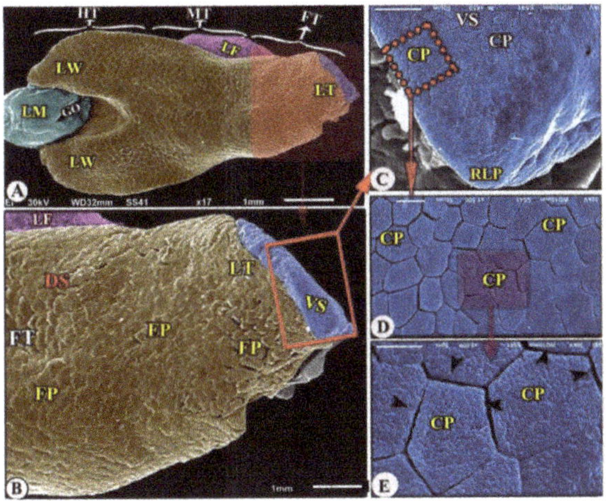

Figure 2. Scanning electron microscopic image (views (**A–E**)) of the *Heremites vittatus* tongue showing the foretongue (FT) with its lingual part (LT) with its anterior round tip (RLP), midtongue (MT) with lingual frenulum (LF), and hindtongue (HT). The dorsal surface (DS) possessed *papillae filiformes* (FP). The ventral surface (VS) possessed *papillae conicae* (CP) that separated narrow spaces (black arrowheads). The lingual wings (LW), the laryngeal mound (LM), the median glottic opening (GO) were clear.

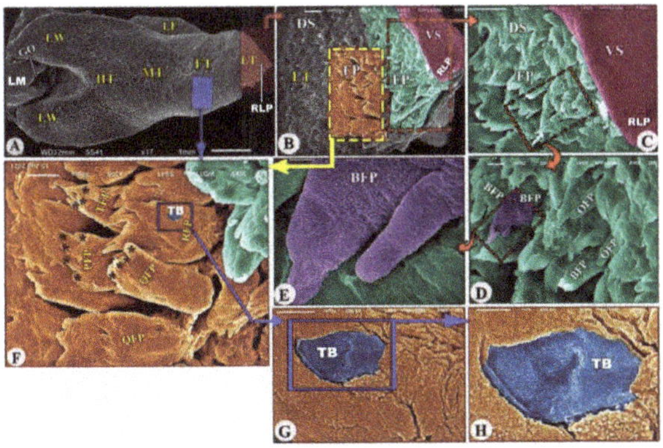

Figure 3. Scanning electron microscopic image (views (**A–H**)) of the *Heremites vittatus* tongue showing the foretongue (FT) with its lingual part (LT) with its anterior round tip (RLP), midtongue (MT) with lingual frenulum (LF), and hindtongue (HT) with its lingual wings (LW). The ventral (VS) and dorsal (DS) surfaces possessed filiform papillae (FP), bifid filiform papillae (BFP), pointed filiform papillae (OFP), trifid filiform papillae (TFP), quadrifid filiform papillae (QFP), pentafid filiform papillae (PFP) with processes (black star), triangular filiform papillae (RFP), and taste buds (TB). The laryngeal mound (LM) with the median glottic opening (GO) and laryngeal folds (FO).

The dorsal surface of the posterior part of the foretongue possessed four types of posteriorly directed filiform papillae (*papillae filiformes*): the triangular, trifid, quadrifid, and pentafid papillae (Figure 3A,F/RFP, TFP, QFP, PFP, black star); additionally, with high magnification, there were a small number of taste buds on the dorsal papillary surface and a corrugated papillary surface at the basal part (Figure 3G–H/TB).

The ventral surface of the midtongue was attached ventrally to the oral cavity floor by the thick lingual frenulum (Figure 4A,B/MT, LF). The dorsal surface of the midtongue possessed numerous posteriorly directed scale-like serrated filiform papillae (Figure 4A–F/MT, SFP) that overlapped each other; additionally, with high SEM magnification, the dorsal papillary surface possessed numerous microridges, small openings of the anterior lingual salivary glands, and taste buds (Figure 4G–I/Mr, red arrowheads, TB).

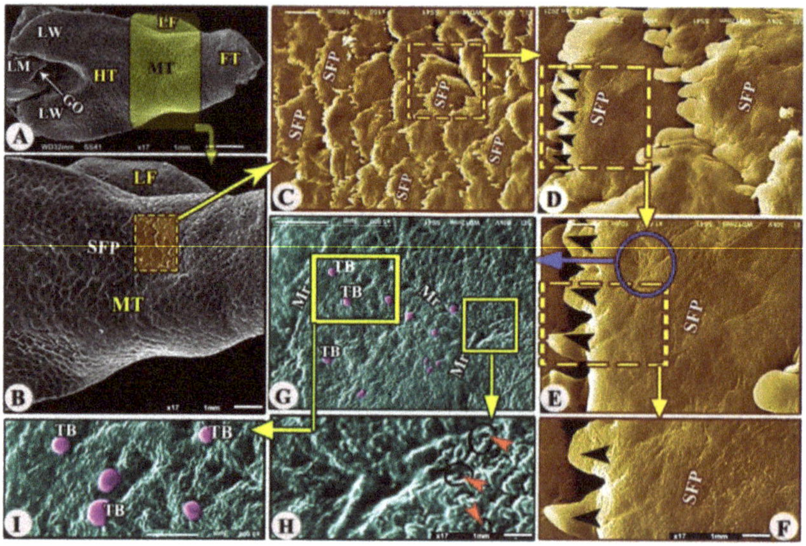

Figure 4. Scanning electron microscopic image (views (**A–I**) of the *Heremites vittatus* tongue showing the foretongue (FT), midtongue (MT), lingual frenulum (LF), and hindtongue (HT) with its wings (LW). The dorsal surface of the midtongue possessed overlapped scale-like serrated filiform papillae (SFP) with a serrated apex (black arrowheads). With a high magnification of the papillary surface, there were taste buds (TB), microridges (Mr), and anterior salivary gland openings (red arrowheads).

The dorsal surface of the hindtongue had three portions: the median and the two lateral lingual wings (Figure 5A/HT and LW). Its filiform papillae were divided into two types that overlap each other: the scale-like board serrated posteriorly directed filiform papillae on the dorsal surface of the median portion of the lingual root (Figure 5A–C/SRFP) and the finger-like projected filiform papillae on the dorsal surface of the two lingual wings (Figure 5A,D,E/FFP), while the medioposteriorly directed finger-like projected filiform papillae were observed on the dorsal surface of the two lingual wings (Figure 5A,D,E/FFP). With high magnification, the dorsal surface of these papillae possessed a small opening of the posterior salivary glands on their dorsal surface (Figure 5C,F).

The laryngeal entrance region was formed by the laryngeal mound with its median glottic opening (Figure 5A,G/LM, GO). The dorsal surface of the laryngeal mound carried about 18 longitudinal folds that had a corrugated surface (Figure 5A,G,H/LM, FO); additionally, with high magnification, these folds carried numerous small openings of the laryngeal salivary glands (Figure 5I/FO, red arrowheads).

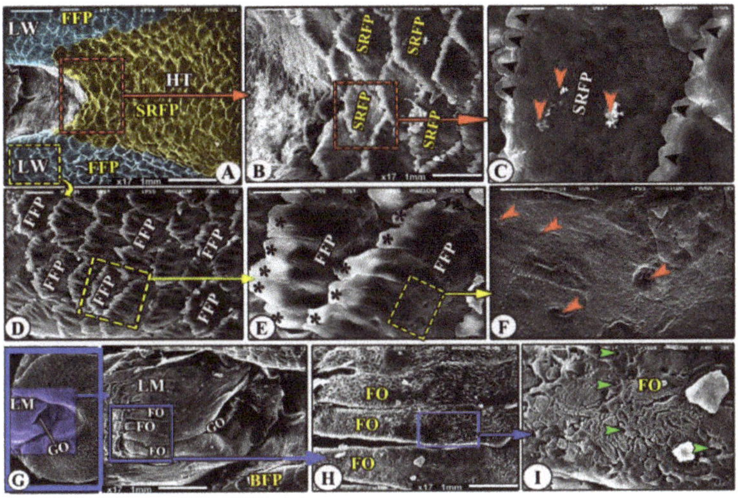

Figure 5. Scanning electron microscopic image of the *Heremites vittatus* tongue. (Views (**A–F**) show the dorsal surface of the hindtongue (HT) that carried scale-like board serrated papillae (SRFP) with a serrated apex (black arrowheads) in the median part, whereas the wings (LW) had finger-like projected papillae (FFP) with processes (black stars). The high magnification of the papillary surface showed numerous posterior salivary gland openings (red arrowheads). (Views (**G–I**)) showing the laryngeal mound (LM) and median glottic opening. (GO), with high magnification, its dorsal surface possessed 18 longitudinal laryngeal folds (FO) and numerous small openings of the laryngeal salivary glands (green arrowheads).

3.3. Histological, Histochemical, and Immunohistochemical Studies Examinations

Histologically, the numerous posteriorly directed lingual filiform papillae were dispersed along the dorsal surface with various morphological characteristics. The lingual papillary cores were formed by the connective tissue that was located above numerous muscle fiber bundles. These well-developed muscle bundles were organized into transverse and longitudinal courses (Figure 6A–C/CTC, MB). The foretongue was covered by a stratified, slightly keratinized squamous epithelium that grew out to form the pointed filiform papilla (Figure 6A–C/PFP, KE). Deep interpapillary notches separated the papillae, and their cores were filled with the loose collagenous connective tissue of the submucosal layer (Figure 6A–C/PFP, IPS). There were no lingual glands or sensory structures observed.

The midtongue was covered by stratified non-keratinized squamous epithelium, which formed the pointed filiform papilla (Figure 6D–F/PFP, NK). Several alveolar-like lingual glands were identified at higher magnification in the interpapillary space regions (Figure 6E/LG). Furthermore, the papillae were supported by a layer of well-developed muscle fibers arranged in different patterns (Figure 6E,F/MB). On the surface of the hindtongue, two different-shaped filiform papillae were detected; each one was covered by a stratified, non-keratinized squamous epithelium (Figure 6H–J/PFP, CFP, NK). The pointed filiform papillae were distributed over the anterior part of the hindtongue, similar to those observed at the midtongue, while the conical filiform papillae covered the posterior part (Figure 6I,J/CFP). Lingual glands were situated in the interpapillary spaces and in the depth of the underlying muscular tissue (Figure 6H–J/LG). A few rounded-shaped taste buds were observed in the epithelium of the conical filiform papillae (Figure 6I/blue arrowhead).

The collagen fibers were intertwined and filled the core of each lingual papilla as well as between the muscle fiber bundles beneath (Figure 7/PFP, yellow arrowheads). The histochemical investigations revealed that the lingual glands displayed strong AB and PAS positive reactions, in which the blue color indicated positive AB reactivity, whereas the red one indicated PAS reactions (Figures 8 and 9). Immunohistochemistry showed strong cytokeratin immunopositivity in all parts of the tongue (Figures 8, 9 and 10A,a,B,b).

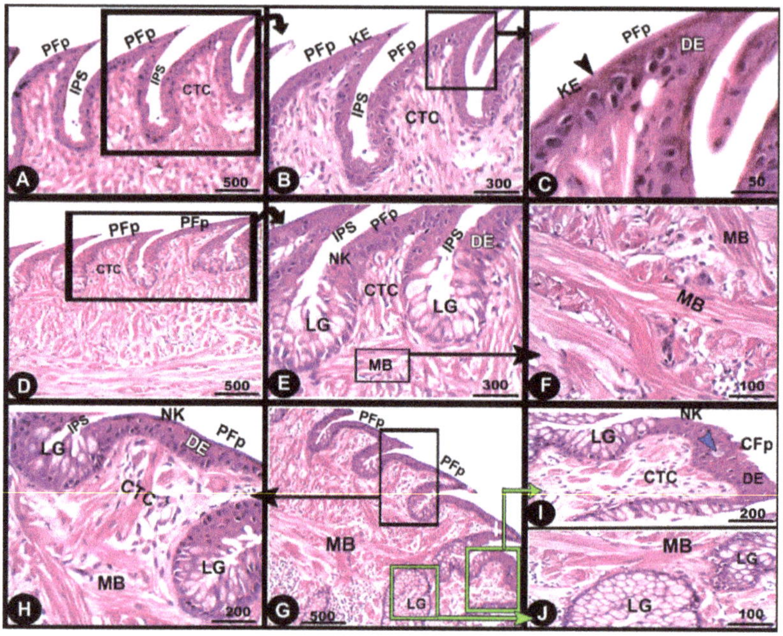

Figure 6. Micrograph image (views (**A–J**)) of the dorsal surface of the foretongue (views (**A–C**)), midtongue (views (**D–F**)), and hindtongue (views (**H–J**)) of the *Heremites vittatus* tongue. The pointed filiform papillae (PFP) and conical papillae (CFP), interpapillary space (IPS), connective tissue core (CTC), muscle bundles (MB), keratinized (KE) or non-keratinized (NK) dorsal squamous epithelium (DE), keratinized layer (black arrowhead), taste buds (blue arrowhead), and lingual glands (LG). H&E stain.

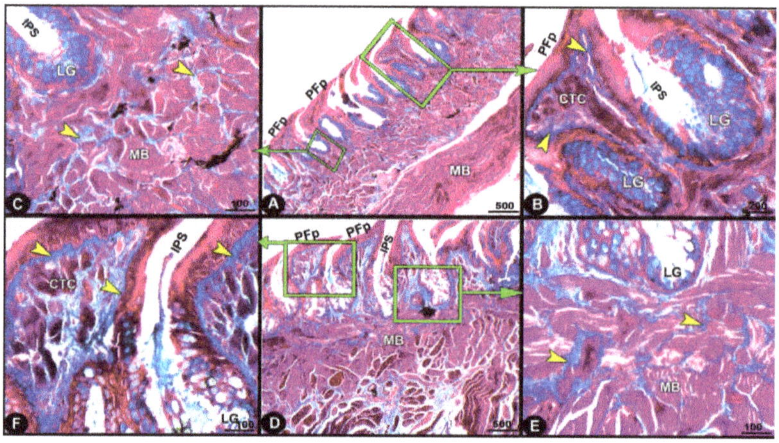

Figure 7. Micrograph image (views (**A–F**)) of the *Heremites vittatus* tongue showing the pointed filiform papillae (PFP), interpapillary space (IPS), connective tissue core (CTC), muscle bundles (MB), collagen fibers (yellow arrowheads) that were intertwined and filled the core of each papilla, and glands (LG). Masson's trichrome stain.

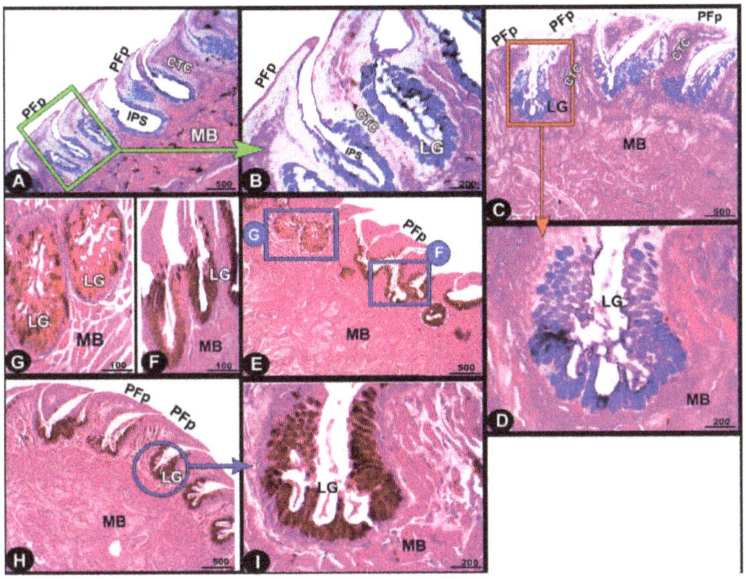

Figure 8. Histochemical micrograph image (views (**A–I**)) of the *Heremites vittatus* tongue showing the interpapillary space (IPS), the pointed filiform papillae (PFP), connective tissue core (CTC), muscle bundles (MB), and glands (LG). Note: the glands displayed strong AB and PAS-positive reactions, in which the blue color indicates positive AB reactivity while the red color indicates PAS reactions. PAS and AB stain.

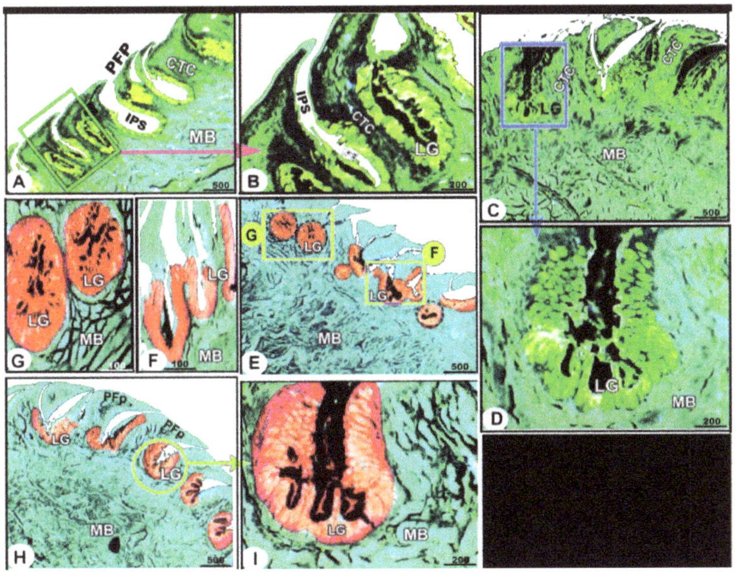

Figure 9. Histochemical micrograph image (views (**A–I**)) of the *Heremites vittatus* tongue showing the negative image of Figure 8 to clarify the pointed filiform papillae (PFP), interpapillary space (IPS), connective tissue core (CTC), muscle bundles (MB), and glands (LG). Note: the glands displayed strong AB and PAS-positive reactions, in which the blue color indicates positive AB reactivity while the red color indicates PAS reactions. PAS and AB stain.

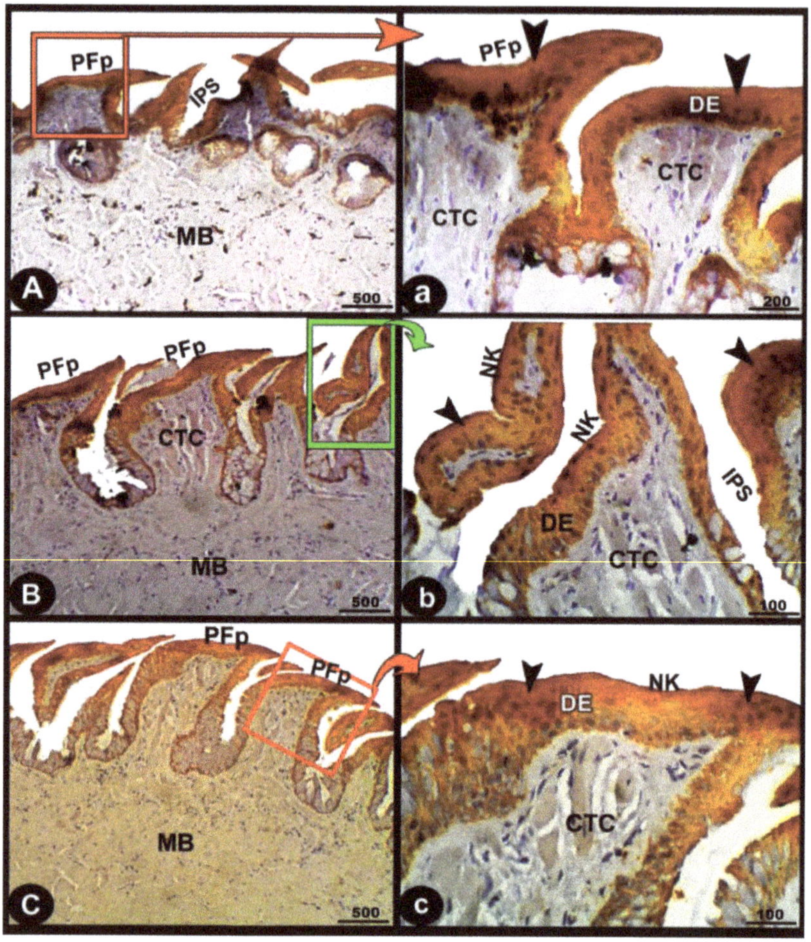

Figure 10. Immunohistological micrographs for cytokeratin of the dorsal surface of the foretongue (views (**A**,**a**)), midtongue (views (**B**,**b**)), and hindtongue (views (**C**,**c**)) of the *Heremites vittatus* tongue. The pointed filiform papillae (PFP), interpapillary space (IPS), connective tissue core (CTC), muscle bundles (MB), keratinized (KE) or non-keratinized (NK) squamous epithelium (DE), keratinized layer (black arrowheads), and glands (LG). Note: in immunohistochemistry analysis, there was a strong cytokeratin immunopositivity in all parts of the tongue (black arrowheads).

Our statistical analysis of the height, width, and gland diameter of papillae on different parts of the tongue revealed that the papillae height and width significantly increased from the fore- to midtongue and non-significantly decreased towards the hindtongue. While the gland diameter gradually and significantly increases from the foretongue towards the hindtongue (Figure 11).

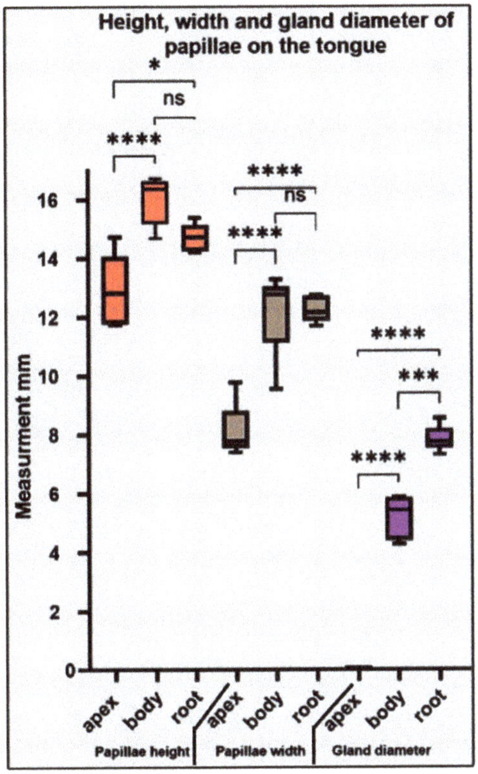

Figure 11. Represents the height, width, and gland diameter of the papillae on different parts of the *Heremites vittatus* tongue. Values have been represented as the mean ± SEM ($n = 5$). *, ***, and **** denote statistical significance with $p < 0.05$, $p < 0.001$, and $p < 0.0001$, respectively.

4. Discussion

The present study revealed fascinating insights into how the examined lizards have adapted to survive and thrive in their Egyptian ecosystem. For example, some reptiles have developed specialized jaws and teeth to efficiently capture and consume their preferred prey, while others have adapted to survive on a diet that may include plants, insects, or even other reptiles [1]. Anatomically, these different feeding strategies influenced the lingual structure and its framework, and these appeared in the currently examined reptile species, *Heremites vittatus*. The studied species resembled most reptile species, and they depended on a prompt bite to capture their prey by using their jaws, whereas most lizard species use an actionable searching strategy [25].

Dietary specialization is thought to have two effects on lingual morphology, in which these adaptations were observed on the foretongue surface, which comes into direct contact with various food particles during feeding [26], and the lingual papillae, which aid in the movement of food particles towards the esophagus [12,25]. According to prior published data on the reptile species, the *Heremites vittatus* tongue was categorized into three regions: foretongue, midtongue, and hindtongue, with two lateral lingual wings that encircled the laryngeal mound and glottis [25,27,28]. Furthermore, the thick lingual frenulum attached the midtongue ventrally to the oral cavity floor, which was already described in most reptile species [25], including Bosc's fringe-toed lizard and Sinai fan-fingered gecko. Also, Abbate et al. [29] reported that the blue-tongued lizard tongue was classified into three regions but with different nominations as follows: the lingual tip, foretongue, and hindtongue, while

Cizek and Hamouzova [28] gave other lingual classifications of the sand lizard *L. agilis* as bifid apex, body, and bifurcated root. The present gross and SEM observations revealed that the tongue of *Heremites vittatus* has a triangle shape that represents the typical lingual shape in most reptilian species, similar to that described [25,27,30] in Bosc's fringe-toed lizard *A. boskianus*, Sinai fan-fingered gecko *P. guttatus*, Leopard Gecko *E. macularius*, and Green Iguana *I. iguana*, respectively. Meanwhile, other lizards have unusual tongue shapes, such as the blue-tongued lizard's V-shaped tongue [29].

The tongue has an essential biological role in all vertebrates, particularly in reptile species, as it is related to numerous functions. For example, the tongue of the *Agama stellio* has been used for hunting [6,25], and the tongues of the *Oplurus cuvieri* and *Iguana iguana* have been used for transportation and swallowing food particles [25,30,31]. According to the previously published data, numerous appearances of the lingual tip in different reptile species correlate with the available food particles [7,25]. The current work revealed the presence of a pointed, black-pigmented tip in the *Heremites vittatus* tongue. Meanwhile, the most common lingual tip appearance in most reptilian species is the bifurcated tip, as in *Anoliscaro linensis* [32], *Oplurus cuvieri* [31], *Psammophis sibilans* [7], the green iguana [30], Bosc's fringe-toed lizard [25], and *Pogona vitticeps* [33], while a slightly bifurcated tip was observed in the tongues of the Sinai fan-fingered gecko [25] and *Gekko japonicus* [34]. Meanwhile, the round tip was reported in blue-tongue skinks [29]; additionally, the pointed or round tip in the current observations may assist in increasing the speed of protraction movement [7,25,29].

The current study verified previously published data indicating most reptile species used their projecting foretongue region to touch and taste the substrate, such as food or water [25] in Bosc's fringe-toed lizard and Sinai fan-fingered gecko, while Cooper [35] indicated that tongue-flicking is the basic squamate behavior. Moreover, the current study confirmed that *Heremites vittatus* depended on the "actively hunting" feeding technique described in Bosc's fringe-toed lizard [25], which predominantly utilizes chemical notification. In contrast, the Gecko species using a "sit-and-wait" feeding technique, similar to that described in the Sinai fan-fingered gecko [25], were thought to depend primarily on visual signals. In harmonization with Cooper [36] and proven by Baeckens et al. [37], who concluded that the lingual morphology and Jacobson's organ were believed to mirror these features as well as the reptilian species utilizing chemical signals to communicate with hetero- or conspecifics, moreover, the described broad feature of the hindtongue region, similar to that observed in most reptilian species, Gewily and Mahmoud [25], may provide a wide surface accessible for loading water during drinking or bearing prey back to the oral cavity.

The lingual papillary system reveals distinct feeding strategies and lifestyles in various vertebrate species, including the currently examined *Heremites vittatus*, blue-tongue skink (*Tiliqua scincoides*), green iguana (*Iguana iguana*), Bosc's fringe-toed lizard Acanthodactylus boskianus, Sinai fan-fingered gecko Ptyodactylus guttatus, Egyptian fruit bat (*Rousettus aegyptiacus*), and common quail (*Coturnix coturnix*) [8,25,26,29,30]. The unique "actively hunting" feeding strategies and lifestyles of *Heremites vittatus* contribute to their unique adaptations and survival in various environments. The examined *Heremites vittatus* tongue has a unique papillary system with distinct regional specifics with different directions, sizes, and distributions. They include a special filiform system on its dorsal surface and a conical system on the ventral surface of the foretongue region, which is suitable for its rapid prey capture using the "actively hunting" strategy. According to current SEM observations, the presence of papillary variations, including their sizes, shapes, directions, and distributions on the dorsal and ventral lingual surfaces, plays a mechanical role in the captured food handling; additionally, the interpapillary space acts as a lingual pouch to pick up a large quantity of the food particles contacting the lingual surface during its protrusion to gather airborne and substrate-fixed chemical particles, then retracts and transports them to the vomeronasal organ [7,38].

Our study on the *Heremites vittatus* tongue revealed a unique papillary distribution, with numerous conical papillae on the ventral surface, two types of posteriorly directed filiform papillae on the anterior foretongue's dorsal surface, and four types of posteriorly directed filiform papillae on the posterior foretongue part. The midtongue region had numerous scale-like serrated filiform papillae, while the hindtongue region had scale-like board-serrated filiform. Finger-like projected filiform papillae were also observed on the dorsal surface of the two lingual wings. The description of the different papillary subtypes on the reptile tongue was recorded previously, in which the three papillary subtypes were observed on the dorsal surface of the Bosc's fringe-toed lizard tongue, which named the conical flattened and conical round papillae on the fore- and midtongue and the long papillae on the hindtongue, while the Sinai fan-fingered gecko tongue consisted of the rounded cylindrical, conical papillae on the foretongue and midtongue regions, whereas the tall filiform papillae on the hindtongue [25]. Also, the three papillary subtypes with different nominations were seen on the dorsal lingual surface of the *Gekko japonicus* [34]: dome-shaped papillae on the apex, flat fan-shaped papillae on the body, and long, scale-like papillae on the lateral part of the apex and body. Iwasaki [34] reported that the gecko tongue had three papillary subtypes that were named: dome-like, flat leaf-like, and broad scale-like papillae on the fore-, mid-, and hindtongue, respectively.

Meanwhile, in lizards, there were broad scale-like papillae on the foretongue, transverse (parallel folds), broad short (folds), and leaf-like papillae on the midtongue, in addition to the presence of broad, flat, and fan-shaped papillae on the hindtongue [11]. Meanwhile, the two papillary subtypes with different nominations were reported by Abbate and Latella [29] in the blue-tongued lizard tongue, which named the foliate-like papillae on the median part of the dorsal surface of the fore- and hindtongue regions and the cylindrical papillae on the lateral part of the same regions, with the complete absence of these papillae from the lingual tip. In addition, Abbate and Latella [30] described the presence of the conical and cylindrical papillae on the dorsal surface of the fore- and hindtongue, respectively, with the complete absence of these papillae from the lingual tip of the green iguana. However, the reptile tongue in some turtles and snakes lacked lingual papillae [7]. In general, the absence of lingual papillae on the entire dorsal lingual surface of turtles [39] indicates that the presence of lingual papillae correlates with a species' adaptation to its habitat, diet, or feeding behavior.

The current work revealed that the dorsal papillary surface of the *Heremites vittatus* tongue possessed a small number of taste pores on the foretongue and taste buds on the midtongue. The Sinai fan-fingered gecko tongue has numerous taste buds along its two-thirds lingual region, while the Bosc's fringe-toed lizard tongue has a small number on its foretongue but an increase in its midtongue [25]. Previous studies [29,40] found that gekkotan and blue-tongued lizard tongues primarily lack taste buds, suggesting it may be a phylogenetic, functional, or adaptive feature. This view disagrees with that mentioned by Bayoumi and Abd-Elhameed [11], who observed taste buds on the apical and lateral surfaces of the dome-shaped papillae on the lingual tip of *P. guttatus*, while they were completely absent from the papillae of *A. boskianus* tongue; meanwhile, Taha [41] observed the presence of a few taste buds in the epithelium of the three lingual parts of *Trachylepis vittata*. Schwenk [5] supposed that in all iguanids, the taste buds were abundant on the papillary surface of the foretongue region and played an important role in testing the prey's palatability when touching its tongue surface during the capturing process. The tongue of some lizards is utilized as a chemoreceptor organ to follow the pheromone trails of prey, as described in Bosc's fringe-toed lizard [25].

The current histological study agrees with most published data that the dorsal lingual epithelium was formed from a stratified squamous type [11,25]. The previously published data showed that the common variations among the reptiles were focused on the presence, absence, or amount of keratinized material on the dorsal lingual surface, which is closely related to the feeding mechanism and the feeding habits of the different reptile species. The foretongue of *Heremites vittatus* is covered by a slightly keratinized layer that extends to the filiform papilla, whereas this keratinized layer completely disappears from the dorsal surface of the mid- and hindtongue. Meanwhile, Gewily and Mahmoud [25] described the presence of the keratinized layer over the dorsal surface of the midtongue of Bosc's fringe-toed lizard and the ventral surface of the foretongue of both Bosc's fringe-toed lizard and the Sinai fan-fingered gecko. However, the lingual keratinization was described only on the lingual tip of the scincine lizard [42], but the presence of the keratinization on the dorsal and ventral lingual surfaces was described in *Psammophis* sibilans, with those on the dorsal surface reaching three times that present on the ventral surface [7]. In the white spot gecko, there is a thin keratinized layer on the dorsal surface of the foretongue region. In contrast, this keratinized layer is absent in the Egyptian spiny tail lizard, *U. aegypticus* [11]. Furthermore, the complete absence of keratinization from the dorsal lingual surface is a popular characteristic in some reptile species [4,10,27,43]. The *Heremites vittatus* tongue's findings show that keratinization is absent from the mid- and hindtongue's dorsal lingual surfaces, but it compensates with a complex papillary system on its dorsal and ventral lingual surfaces, containing numerous papillary types and microridges, and numerous conical papillae on the ventral surface of the foretongue region. Functionally, the numerous non-keratinized papillary types on the dorsal lingual surface may facilitate the slipping process of caught prey towards the esophagus. Moreover, it was described that the presence of microridges on the lingual surface helps in the spreading of the mucosal film (secreted by the lingual glands) over the lingual surface to reduce friction forces [44,45].

The lingual glands In the *Heremites vittatus* tongue were absent in the foretongue, whereas these glands were observed in the interpapillary space of the mid- and hindtongue. In addition, these lingual glands were observed in the depth of the underlying muscular tissue. The lingual glands were observed between the papillary bases and were numerous in the hindtongue [45,46] in adult *C. ocellatus* and *E. schneideri*, respectively. The lingual gland's secretion acts as a lubricant material on the dorsal lingual surface to facilitate the caught food particle's movements, transportation, and swallowing [25,46].

The current histochemical investigations of the examined *Heremites vittatus* tongue revealed that the collagen fibers were intertwined and filled the core of each lingual papilla and between the muscle fiber bundles beneath. The sheet of collagen connective tissue was observed just beneath the epithelial layer, as reported in *C. ocellatus* [45] and *E. schneideri* [46]. The obtained findings confirmed that muscle contraction will lead to shortening or curving of the papillae due to the presence of the collagenous sheet. Furthermore, the current findings agree with Winokur [39] that the lingual muscles attached to the lingual glands will help in squeezing the glands to secrete their secretion on the dorsal lingual surface. The current histochemical investigations of the *Heremites vittatus* tongue revealed that the lingual glands displayed strong positive reactions when stained with AB and PAS. The AB blue color indicates the acidic mucin, while the reddish–purple color refers to the neutral mucin secretions when stained with PAS stain. Meanwhile, in the *P. vitticeps* lizard [47], it was reported that the tubular serous lingual glands of the foretongue region gave PAS-negative secretory cells, but the tubulo-alveolar mucous lingual glands were PAS-positive, revealing the presence of mucopolysaccharide. Meanwhile, the lingual glands gave an AB-PAS-positive reaction [45] in adult *Chalcides ocellatus.*

The present findings represent the first data on the cytokeratin expression of the *Heremites vittatus* tongue. As expected, we observed that the cytokeratin expression showed strong immunopositivity in all parts of the tongue.

5. Conclusions

Conclusively, in *Heremites vittatus*, the complicated papillary system included filiform papillae on the dorsal surface (nine subtypes) as well as conical papillae on the ventral surface of the foretongue region, which was ideal for capturing prey quickly using the "actively hunting" method. Histologically, the foretongue was covered by a slightly keratinized layer that disappeared entirely from the mid- and hindtongue regions. The lingual glands exhibit strong positive AB and PAS reactions. Immunohistochemistry showed strong cytokeratin immunopositivity in all parts of the tongue.

Author Contributions: Main manuscript text—writing, R.M.K., D.B.M., M.A, and O.P.C.; prepared the samples and performed the scanning electron microscopy and light microscopy, R.M.K., O.P.C. and M.A.; statistical analysis, S.H.E.-N.; gross morphology and morphometry, D.B.M., M.A. and R.M.K.; prepared figures, M.A. and R.M.K.; analyzed the scanning electron microscopy images, O.P.C. and R.M.K.; final proofing/editing of the manuscript, O.P.C. and M.A. All authors have read and agreed to the published version of the manuscript.

Funding: This research received no external funding.

Institutional Review Board Statement: The experiments were carried out following the ethical provisions established by the 2010/63/EU directive for animal experiments and were approved by the Institutional Animal Care and Use Committee (IACUC), Tanta University, under research project number (IACUC-SCI-TU-0309, approval date: 10 February 2023). The animals were kept according to Egyptian law. The present investigation was prepared according to ethical permission from the Faculty of Veterinary Medicine, Alexandria University, and approved by the Institutional Animal Care and Use Committee (ALEXU-IACUC, approval date: 1 February 2023).

Informed Consent Statement: Not applicable.

Data Availability Statement: The datasets used and/or analyzed during the current study are available from the corresponding author on reasonable request.

Conflicts of Interest: The authors declare no conflict of interest.

References

1. Montuelle, S.J.; Herrel, A.; Libourel, P.A.; Reveret, L.; Bels, V.L. Locomotor–feeding coupling during prey capture in a lizard (*Gerrhosaurus major*): Effects of prehension mode. *J. Exp. Biol.* **2009**, *6*, 768–777. [CrossRef]
2. Bels, V.; Paindavoine, A.S.; Zghikh, L.; Paulet, E.; Pallandre, J.; Montuelle, S. Feeding in Lizards: Form-Function and Complex Multifunctional System. In *Feeding in Vertebrates. Fascinating Life Sciences*; Bels, V., Whishaw, I., Eds.; Springer: Cham, Switzerland, 2019; pp. 469–525.
3. Iwasaki, S. Evolution of the structure and function of the vertebrate tongue. *J. Anat.* **2002**, *201*, 1–13. [CrossRef]
4. Herrel, A.; Canbek, M.; Ozelmas, U.; Uyanoglu, M.; Karakaya, M. Comparative functional analysis of the hyolingual anatomy in lacertid lizards. *Anat. Rec.* **2005**, *284*, 561–573. [CrossRef]
5. Schwenk, K. Of tongues and noses: Chemoreception in lizards and snakes. *Trends Ecol. Evol.* **1995**, *10*, 7–12. [CrossRef]
6. Herrel, A.; Timmermans, J.P.; De Vree, F. Tongue flicking in agamid lizards: Morphology, kinematics, and muscle activity patterns. *Anat. Rec.* **1998**, *252*, 102–116. [CrossRef]
7. El-Mansi, A.A.; Al-Kahtani, M.A.; Abumandour, M.M.A.; Ahmed, A.E. Structural and Functional Characterization of the Tongue and Digestive Tract of *Psammophis sibilans* (Squamata, Lamprophiidae): Adaptive Strategies for Foraging and Feeding Behaviors. *Microsc. Microanal.* **2020**, *26*, 524–541. [CrossRef]
8. Abumandour, M.M.A.; Farrag, F.A.; El-Mansi, A.; Lashen, S.E.; Shukry, M.; Kassab, M.A.; Hamoda, H.S. Posthatching developmental studies on the tongue and laryngeal entrance of the common quail (*Coturnix coturnix*, Linnaeus, 1758) in different five age-stages. *Microsc. Res. Tech.* **2021**, *84*, 1649–1672. [CrossRef]
9. Mohammed, M.B.H. Structure and function of the tongue and hyoid apparatus in *Acanthodactylus boskianus* (Lacertidae; Reptilia). *J. Egypt. Ger. Soc. Zool.* **1992**, *7*, 65–89.
10. Abbate, F.; Guerrera, M.C.; Montalbano, G.; Zichichi, R.; Germana, A.; Ciriaco, E. Morphology of the lingual dorsal surface and oral taste buds in Italian lizard (*Podarcis sicula*). *Anat. Histol. Embryol.* **2010**, *39*, 167–171. [CrossRef]
11. Bayoumi, S.S.; Abd-Elhameed, A.A.; Mohamed, E.M. Comparative studies on the dorsal lingual surface of two Egyptian squamate reptiles with two different feeding habits. *Egypt. J. Exp. Biol. (Zool.)* **2011**, *7*, 203–211.
12. Abumandour, M.M.A.; Madkour, N.; Morsy, K.; Haddad, S.; Abourashed, N.M.; El Basyouny, H.A.; Kandyel, R.M. Lingual papillary system of the Egyptian mouse-tailed bat (*Rhinopoma hardwickii*): New insights to its feeding strategies using scanning electron microscope, and immunohistochemical analysis of keratin. *Microsc. Res. Tech.* **2022**, *85*, 2192–2205. [CrossRef]
13. Bahmani, Z.; Rastegar-Pouyani, N.; Rastegar-Pouyani, E. Systematics of the Genus *Trachylepis Fitzinger*, 1843 (Sauria: Scincidae) in Iran: A review. *Iran. J. Anim. Biosyst.* **2016**, *12*, 211–224.
14. Verderame, M.; Chianese, T.; Rosati, L.; Scudiero, R. Molecular and Histological Effects of Glyphosate on Testicular Tissue of the Lizard Podarcis siculus. *Int. J. Mol. Sci.* **2022**, *23*, 4850. [CrossRef]

15. International Committee on Veterinary Gross Anatomical Nomenclature. *Nomina Anatomica Veterinaria*, 6th ed.; Revised Version Editorial Committee: Hanover, Germany; Ghent, Belgium; Sapporo, Japan; Columbia, MO, USA; Rio de Janeiro, Brazil, 2017; Available online: https://www.vetmed.uni-leipzig.de/fileadmin/Fakult%C3%A4t_VMF/Institut_Veterin%C3%A4r-Anatomisches/Dokumente/NAV_6th-Edition-2017.pdf (accessed on 20 July 2023).
16. Abumandour, M.M.; Kandyel, R.M. Age-related ultrastructural features of the tongue of the rock pigeon *Columba livia* dakhlae in different three age stages (young, mature, and adult) captured from Egypt. *Microsc. Res. Tech.* **2020**, *83*, 118–132. [CrossRef]
17. Suvarna, K.S.; Layton, C.; Bancroft, J.D. *Bancroft's Theory and Practice of Histological Techniques*, 8th ed.; Elsevier Health Sciences: Amsterdam, The Netherlands, 2019.
18. Bancroft, J.D.; Cook, H.; Turner, D. *Manual of Histological Techniques and Their Diagnostic Application*, 2nd ed.; Churchill Livingstone: London, UK, 1996.
19. Schumacher, U.; Duku, M.; Katoh, M.; Jörns, J.; Krause, W.J. Histochemical similarities of mucins produced by Brunner's glands and pyloric glands: A comparative study. *Anat. Rec. Part A Discov. Mol. Cell. Evol. Biol. Off. Publ. Am. Assoc. Anat.* **2004**, *278*, 540–550. [CrossRef]
20. Masson, P. Some histological methods: Trichrome staining and their preliminary technique. *J. Tech. Methods Bull. Int. Assoc. Med. Mus.* **1929**, *12*, 75–90.
21. Schneider, C.A.; Rasband, W.S.; Eliceiri, K.W. NIH Image to ImageJ: 25 years of image analysis. *Nat. Methods* **2012**, *9*, 671–675. [CrossRef]
22. Kandyel, R.M.; Elwan, M.M.; Abumandour, M.M.; El Nahass, E.E. Comparative ultrastructural-functional characterizations of the skin in three reptile species; *Chalcides ocellatus, Uromastyx aegyptia aegyptia,* and *Psammophis schokari aegyptia* (Forskal, 1775) Adaptive strategies to their habitat. *Microsc. Res. Tech.* **2021**, *84*, 2104–2118. [CrossRef]
23. Abd-Elkareem, M.; Abou Khalil, N.S.; Sayed, A.E.-D.H. Cytoprotective effect of Nigella sativa seed on 4-nonylphenol-induced renal damage in the African catfish (*Clarias gariepinus*). *Chemosphere* **2020**, *259*, 127379. [CrossRef]
24. Kandyel, R.M.; El Basyouny, H.A.; El Nahas, E.E.; Madkour, F.; Haddad, S.; Massoud, D.; Morsy, K.; Madkour, N.; Abumandour, M.M.A. A histological and immunohistochemical study on the parabronchial epithelium of the domestic fowl's (*Gallus gallus domesticus*) lung with special reference to its scanning and transmission electron microscopic characteristics. *Microsc. Res. Tech.* **2022**, *85*, 1108–1119. [CrossRef]
25. Gewily, D.I.; Mahmoud, F.A.; Saber, S.A.; ElSalkh, B.A.; El-Dahshan, A.A.; Abumandour, M.M.A.; Kandyel, R.M.; Gadel-Rab, A.G. Ultrastructural comparison between the tongue of two reptilian species endemic in Egyptian fauna; Bosc's fringe-toed lizard *Acanthodactylus boskianus* and Sinai fan-fingered gecko *Ptyodactylus guttatus*. *Microsc. Res. Tech.* **2021**, *84*, 1977–1991. [CrossRef]
26. Kandyel, R.M.; Abumandour, M.M.A.; Mahmoud, S.F.; Shukry, M.; Madkour, N.F.; El-Mansi, A.; Farrag, F.A. Structural and Functional adaptation of the lingual papillae of the Egyptian fruit bat (*Rousettus aegyptiacus*): Specific Adaptive feeding Strategies. *Folia Morphol.* **2021**, *81*, 400–411. [CrossRef]
27. Abbate, F.; Guerrera, M.C.; Levanti, M.; Laurà, R.; Montalbano, G.; Cavallaro, M.; Germanà, A. The tongue of Leopard Gecko (*Eublepharis macularius*): LM, SEM and confocal laser study. *Anat. Histol. Embryol.* **2020**, *49*, 51–59. [CrossRef]
28. Cizek, P.; Hamouzova, P.; Kvapil, P.; Kyllar, M. Light and scanning electron microscopy of the tongue of the sand lizard (*Lacerta agilis*). *Folia Morphol.* **2019**, *78*, 101–106. [CrossRef]
29. Abbate, F.; Latella, G.; Montalbano, G.; Guerrera, M.C.; Germana, G.P.; Levanti, M.B. The lingual dorsal surface of the blue-tongue skink (*Tiliqua scincoides*). *Anat. Histol. Embryol.* **2009**, *38*, 348–350. [CrossRef]
30. Abbate, F.; Latella, G.; Montalbano, G.; Guerrera, M.C.; Levanti, M.B.; Ciriaco, E. Scanning electron microscopical study of the lingual epithelium of green iguana (*Iguana iguana*). *Anat. Histol. Embryol.* **2008**, *37*, 314–316. [CrossRef]
31. Delheusy, V.; Toubeau, G.; Bels, V.L. Tongue structure and function in *Oplurus cuvieri* (Reptilia: Iguanidae). *Anat. Rec.* **1994**, *238*, 263–276. [CrossRef]
32. Rabinowitz, T.; Tandler, B. Ultrastructure of lingual salivary glands in the American chameleon: *Anoliscaro linensis*. *Anat. Rec.* **1991**, *229*, 489–494. [CrossRef]
33. Schaerlaeken, V.; Meyers, J.J.; Herrel, A. Modulation of prey capture kinematics and the role of lingual sensory feedback in the lizard *Pogonavitticeps*. *Zoology* **2007**, *110*, 127–138. [CrossRef]
34. Iwasaki, S. Fine structure of the dorsal lingual epithelium of the lizard, *Gekko japonicus* (Lacertilia, Gekkonidae). *Am. J. Anat.* **1990**, *187*, 12–20. [CrossRef]
35. Cooper, W.E. Chemical discrimination by tongue-flicking in lizards: A review with hypotheses on its origin and its ecological and phylogenetic relationships. *J. Chem. Ecol.* **1994**, *20*, 439–487. [CrossRef]
36. Cooper, W.E. Prey chemical discrimination indicated by tongue-flicking in the eublepharid gecko *Coleonyx variegates*. *J. Exp. Zool.* **1998**, *281*, 21–25. [CrossRef]
37. Baeckens, S.; Herrel, A.; Broeckhoven, C.; Vasilopoulou-Kampitsi, M.; Huyghe, K.; Goyens, J.; Van Damme, R. Evolutionary morphology of the lizard chemosensory system. *Sci. Rep.* **2017**, *7*, 101–141. [CrossRef]
38. Filoramo, N.I.; Schwenk, K. The mechanism of chemical delivery to the vomeronasal organs in squamate reptiles: A comparative morphological approach. *J. Exp. Zool.* **2009**, *34*, 20–34. [CrossRef]
39. Winokur, R.M. The buccopharyngeal mucosa of the turtles (*Testudines*). *J. Morphol.* **1988**, *196*, 33–52. [CrossRef]

Jamniczky, H.A.; Russell, A.P.; Johnson, M.K.; Montuelle, S.J.; Bels, V.L. Morphology and histology of the tongue and oral chamber of *Eublepharis macularius* (Squamata: Gekkonidae), with special reference to the foretongue and its role in fluid uptake and transport. *Evol. Biol.* **2009**, *36*, 397–406. [CrossRef]

Taha, A.M. Comparative Anatomical, Histological and Histochemistrical study of tongue in two species of insectivorous vertebrates. *Aust. J. Basic Appl. Sci.* **2013**, *7*, 401–410.

Wassif, E.T. The fine structure of the dorsal lingual epithelium of the scincine lizard *Chalcideso cellatus Forscal* (*Scincidae, Sauria, Reptilia*) I. Histogenesis of the lingual epithelium. *Egypt. J. Biol.* **2001**, *3*, 12–19.

Yang, C.; Wang, L. Histological and morphological observations on tongue of *Scincella tsinlingensis* (Reptilia, Squamata, Scincidae). *Micron* **2016**, *80*, 24–33. [CrossRef]

Iwasaki, S.; Kobayashi, K. Fine structure of the dorsal lingual epithelium of the Japanese lizard, *Takydromus takydromoides*. *Acta Anat.* **1992**, *67*, 214–225.

Wassif, E.T. Ultrastructure of the lingual epithelium of adult scincine lizard *Chalcides ocellatus*. *Egypt. J. Biol.* **2002**, *4*, 76–86.

Wassif, E.T.; El-Hawary, M.S. Scanning Electron Microscopy of the Dorsal Lingual Epithelium of the Golden Lizard; *Eumeces schneideri* (Reptilia: Scincidae). *J. Egypt. Ger. Soc. Zool.* **1998**, *26*, 11–30.

Zghikh, L.N.; Vangysel, E.; Nonclercq, D.; Legrand, A.; Blairon, B.; Berri, C.; Bordeau, T.; Rémy, C.; Burtéa, C.; Montuelle, S.J. Morphology and fibre-type distribution in the tongue of the Pogona vitticeps lizard (*Iguania, Agamidae*). *J. Anat.* **2014**, *225*, 377–389. [CrossRef] [PubMed]

animals

MDPI

Article

Cross-Sectional Anatomy and Computed Tomography of the Coelomic Cavity in Juvenile Atlantic Puffins (Aves, *Alcidae*, *Fratercula arctica*)

José Raduan Jaber [1,*], Marcos Fumero-Hernández [2,*], Juan Alberto Corbera [2], Inmaculada Morales [2], Manuel Amador [3], Gregorio Ramírez Zarzosa [4] and Mario Encinoso [2]

[1] Department of Morphology Faculty of Veterinary Medicine, University of Las Palmas de Gran Canaria, Trasmontaña, 35413 Las Palmas, Spain
[2] Veterinary Clinical Hospital, Faculty of Veterinary Medicine, University of Las Palmas de Gran Canaria, Trasmontaña, 35413 Las Palmas, Spain; juan.corbera@ulpgc.es (J.A.C.); inmaculada.morales@ulpgc.es (I.M.); mencinoso@gmail.com (M.E.)
[3] Dirección Insular de Medio Ambiente, Consejeria de Área de Medio Ambiente, Clima, Energía y Conocimiento of the Cabildo de Gran Canaria, 35002 Las Palmas, Spain; mcamador@grancanaria.com
[4] Department of Anatomy and Compared Pathological Anatomy, Veterinary Faculty, Campus de Espinardo, University of Murcia, 30100 Murcia, Spain; grzar@um.es
* Correspondence: joseraduan.jaber@ulpgc.es (J.R.J.); marcos.fumero101@alu.ulpgc.es (M.F.-H.)

Simple Summary: Birds constitute an important group within the new companion animals. Since their popularity has been increasing, it is more common to find them in homes, zoos and wildlife centres around the world, evidencing the need for clinicians, biologists and researchers to have a deep knowledge of their anatomy, pathology and physiology to provide adequate medical care.

Abstract: In birds, unlike mammals, there is no complete separation between the thoracic and abdominal cavities. Instead, they have the coelomic cavity where most main organs are found. Therefore, an adequate knowledge of the anatomy of the coelomic cavity is of great importance for veterinarians, biologists and the scientific community. This study aimed to evaluate the coelomic cavity anatomy in the Atlantic puffin (*Fratercula arctica*) using anatomical sections and computed tomography images.

Keywords: coelomic cavity; computed tomography; anatomical sections; seabirds; puffin

1. Introduction

Birds constitute an enormously diverse taxonomic group, including thousands of species worldwide [1]. These animals have historically caused the fascination of humans, who have adopted them as companion animals. The possession of birds as pets is a growing trend, and their presence in zoos, wildlife and veterinary centres is also increasingly common [2]. For all these reasons, knowledge of the anatomy and pathophysiology of these animals has become one of the objectives of current veterinary medicine.

Imaging diagnosis is an essential tool for bird clinicians since many psittacine species exhibit a special ability to mask disease signs [3]. Conventional radiology is the method most widely used as an auxiliary diagnostic technique in avian medicine due to its low cost, non-invasive nature and the fact that it is found in most veterinary centres [4–6]. However, modern imaging techniques such as magnetic resonance (MR) and computed tomography (CT) make it possible to obtain three-dimensional images with little distortion and even simulations of dynamic processes of different organs and vascular structures in a minimally invasive way [7–9]. Consequently, its use in anatomical studies has spread, and even smaller versions, such as micro-CT, have been created for animal research [10–21]. These anatomical studies are essential not only from the clinical point of view but also for endangered species conservation, enabling adequate knowledge of their anatomical characteristics and behaviour patterns [21–23].

A diverse range of studies can be found regarding the use of diagnostic imaging in the study of the coelomic cavity of birds; however, as far as we know, there are no studies that deal with the anatomy of the avian coelom using the Atlantic puffin as a model [24–27]. The Atlantic puffin (*Fratercula arctica*) is a medium-sized marine bird of the Alcidae family with characteristic black and white colouration. It lives mainly in the North Atlantic Ocean, where it faces a multitude of threats, many of them linked to human activity. All this has led to it being classified as a vulnerable species on the list of threatened species of the International Union for Conservation of Nature (IUCN) [28,29].

The aim of this paper was to study the coelomic cavity of the Atlantic puffin using anatomical sections and CT images, which could serve as an anatomical model for pathologic studies of other phylogenetically related birds.

2. Materials and Methods

2.1. Animals

In the present study, a total of 20 juvenile Atlantic puffin (*Fratercula arctica*) carcasses were utilised. The avian specimens displayed an average mass of 0.251 kg (with a range of 0.185–0.310 kg) and an average length, measured from the beak tip to the base of the tail, of 19.85 cm (with a range of 16–24 cm). The study group comprised young animals that were provided by the Consejeria de Área de Medio Ambiente, Clima, Energía y Conocimiento of the Cabildo de Gran Canaria following a massive stranding event on the island's coastline. Subsequently, all enrolled puffins for this study were sectioned and scanned at the Veterinary Hospital of Las Palmas de Gran Canaria University (Canary Islands, Spain).

2.2. CT Technique

For the CT examinations, the carcasses were thawed at room temperature for 12 h. Sequential transverse CT slices were acquired using a 16-slice helical CT scanner (Toshiba Astelion, Canon Medical System®, Tokyo, Japan). The animals were symmetrically positioned in dorsal recumbency on the stretcher, with craniocaudal entry. A standard clinical protocol was followed, employing the following parameters: 120 kVp, 80 mA, 512 × 512 acquisition matrix, 1809 × 858 field of view, pitch of 0.94 and a gantry rotation of 1.5. The obtained images had a slice thickness of 0.6 mm. To enhance the assessment of the coelomic structures on CT, three different CT Window settings were applied by adjusting the window widths (WWs) and window levels (WLs): a bone window setting (WW = 1500; WL = 300), a soft tissue window setting (WW = 248; WL = 123) and a lung window setting (WW = 1400; WL = −500). No significant variations in CT density or anatomy were detected within the coelomic cavity of the birds used in this investigation. Finally, all these images were uploaded to an image viewer (OsiriX MD, Apple, Cupertino, CA, USA) to facilitate data manipulation and analysis.

2.3. Anatomical Sections

Following the acquisition of CT images, the imaged cadavers were frozen at −80 °C for 72 h. Subsequently, eight specimens were sectioned in the transverse plane. To accomplish this, parallel sections, each one centimetre thick, were generated using an electric band saw. Finally, after carefully moistening the sections obtained with water, removing the artefacts (feathers and sand) with Adson forceps and identifying them, they were photographed on both sides.

For further examination and identification of internal structures in anatomical cross-sections and CT images, three puffins were dissected to expose the coelomic cavity and visualise the location of the organs. The dissection facilitated the accurate identification and correlation of anatomical features observed in anatomical sections and CT images.

2.4. Anatomic Identification

For the purpose of identifying and labelling the cross-section along with the corresponding CT images, we used reference materials, such as textbooks and articles from scientific journals focused on avian anatomy [30–32]. Furthermore, to ensure precise anatomical interpretation of the coelomic structures, we also used different anatomical preparations provided by the Department of Anatomy of the Faculty of Veterinary Medicine, University of Las Palmas de Gran Canaria. These additional resources proved beneficial in enhancing our understanding and accuracy in interpreting the anatomical features within the coelomic cavity.

3. Results

Different figures revealing the anatomical structures of the Atlantic puffin coelomic cavity are presented (Figures 1–12). Figure 1 comprises a compilation of various dissections displaying the main structures within the coelomic cavity. Figure 2 shows a sagittal multiplanar reconstruction (MPR) volume rendering, wherein each line and corresponding number (I–IX) represents approximately the level of the following anatomical and CT transverse planes. Figures 3–11 consist of four images for each case: (A) an anatomical cross-section, (B) a pulmonary CT window, (C) a soft tissue CT window and (D) a bone CT window. These images are presented in a rostrocaudal progression, starting from the lungs (Figure 3) to the cloaca levels (Figure 11). Finally, Figure 12 is composed of three images: (A) an anatomical dissection and (B,C) dorsal MPR volume rendering images in the pulmonary CT window at different levels.

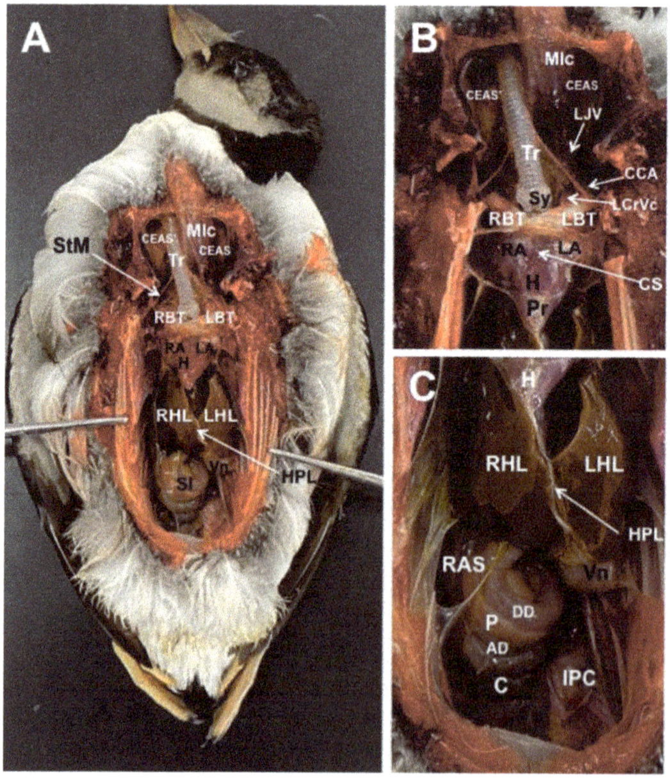

Figure 1. Gross dissection image of the coelomic cavity of the *Fratercula arctica* (**A**), and images of the cardiovascular (**B**) and digestive (**C**) structures. StM: sternotracheal muscle. Mlc: longissimus colli

muscle. Tr: trachea. Sy: syrinx. CEAS': Right Cervical Air Sac. CEAS: Left Cervical Air Sac. LBT: Left Brachiocephalic Trunk. RBT: Right Brachiocephalic Trunk. LJV: left jugular vein. CCA: common carotid artery. LCrVc: left cranial vena cava. H: heart. RA: Right Atrium. LA: Left Atrium. Pr: pericardium. CS: coronary sinus. LHL: left hepatic lobe. RHL: right hepatic lobe. HPL: Hepatopericardial Ligament. DD: Descending Duodenum. AD: Ascending Duodenum. P: pancreas. C: caecum. RAS: right abdominal air sac. Vn: ventriculus. SI: small intestine. IPC: Intestinal Peritoneal Cavity.

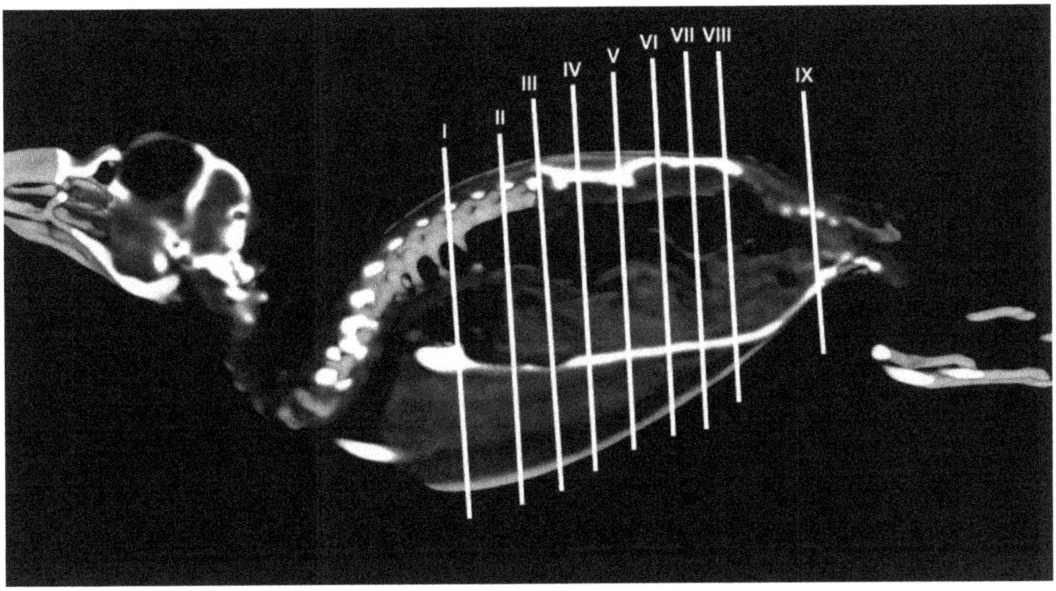

Figure 2. Sagital MPR volume rendering image of the body of an Atlantic puffin. The lines and numbers (I–IX) represent the approximate level of the following transverse cross-sections and CT images.

3.1. Anatomical Dissections and Cross-Sections

We present anatomical dissections (Figures 1A–C and 12A) and transverse (Figures 3A–11A) cross-sections of the coelomic cavity. All these figures were essential in facilitating the identification of the organs of the respiratory, circulatory, digestive and urinary systems within this cavity. Consequently, we identified the puffin's heart, which exhibits an oval shape with a sharply pointed apex located along the central axis of the coelomic cavity and cranial to the liver (Figure 1A–C). Moreover, the heart chambers, including the Right and Left Atrium and the right and left ventricles, were observed. Additionally, these images were crucial in visualising significant blood vessels, such as the Left and Right Brachiocephalic Trunks, the common carotid artery, the left cranial vena cava and the left jugular vein (Figure 1A,B). The right and left lungs were visually discernible in a craniodorsal location, ventral to the thoracic vertebra and lateral to the ribs. The anatomical dissections and transverse cross-sections also aided in distinguishing the trachea (Figure 1A,B). This structure coursed in a median position into the coelomic cavity until bifurcating into the right and left main bronchi (Figures 3A–5A). In contrast, we could identify the syrinx in specific dissected images (Figures 1B and 12A). Dorsal to the trachea, we visualised other formations, such as the oesophagus, the longissimus colli and the sternotracheal muscles (Figures 1A,B and 12A).

The dissected and transverse images played a crucial role in observing the walls of the air sac. As a result, the cross-sections allowed the observation of the topographic distribution of the different air sacs, including the clavicular, the cervical, the thoracic and the abdominal air sacs (Figures 1A,B, 3A–10A and 12A). These anatomical preparations also facilitated the observation of the puffin's liver, which consists of the right and the left hepatic lobes, whose cranioventral segments surrounded the heart (Figures 1A,C and 12A). Both lobes exhibited similar sizes and were larger than other organs within the coelomic cavity. At the level of the visceral surface of the right lobe of the liver, a membranous structure corresponding to the gallbladder was distinguishable (Figure 12A). Moreover, the medial border of the hepatic lobes showed the hepatopericardial ligament, which connects the liver with the heart apex (Figure 1C). Caudally and closely related to the visceral surface of the left hepatic lobe, we identified the muscular portion of the puffin's stomach, known as the *ventriculus* (Figures 1A,C and 12A). Its identification was facilitated by the presence of a thick muscular layer, as well as by the visualisation of two muscles, the *crassus caudodorsalis* and the *crassus cranioventralis*, which delineate the dorsal and ventral borders of the *ventriculus* (Figures 8A and 9A). In addition, these sections also allowed us to observe its glandular part, the *proventriculus*, which shared a similar location at the left side of the coelomic cavity after the junction with the oesophagus and the *ventriculus* (Figures 6A and 7A). Other digestive system components, such as the small and large intestine, were also visible in these anatomic sections. The small intestine (*duodenum*) was better identified in the dissected image caudal to the liver (Figures 1A,C and 12A). It showed a U-shaped *ansa duodeni* comprising a descending segment, *the pars descendens*, and an ascending portion, the *pars ascendens* (Figures 1C and 12A). More caudally, we could distinguish the terminal intestine (*caecum*), which was well developed in the puffin, exhibiting a distinctive colouration (Figures 1C and 12A). Between these two segments and close to the duodenal loops, we observed the pancreas that lies within the mesoduodenum, displaying a pale yellow colouration (Figures 1C, 8A, 10A and 12A). In addition, the utilisation of transverse sections was crucial in discerning the presence of paired kidneys situated laterally to the spine and embedded dorsally in excavations of the synsacrum (Figures 8A–10A). Interestingly, these transverse sections allowed us to visualise the ventral surface of the right and left kidneys in close contact with the paired abdominal air sacs (Figures 8A and 9A). Furthermore, the caudal pole of the kidney housed the ureters, as depicted in Figures 9A and 10A. These structures were distinguished emerging from the ventral aspect of the kidneys. Concerning the genital organs, we distinguished a round single yellow structure that closely resembled the ovary (Figure 6A).

In the caudal part of the coelomic cavity, we could identify an excretory passage for the digestive and urogenital systems known as the cloaca, which was surrounded by cloacae muscles (*musculus levator cloacae and musculus trans* versus *cloacae*) that allows the cloaca to become larger for different functions, including egg laying, copulation and defecation (Figure 11A). In addition, various bony structures related to the coelomic cavity were observed, including the vertebrae, ribs, sternum, thoracic and pelvic limbs, as well as the lateral parts of the pubis. These bone formations were associated with specific muscles, such as the different parts of the Pectoral Muscle (sternobrachialis, thoracic and abdominal portions), the Supracoracoid Muscle, the scapulohumeralis muscle, the scapulohumeral caudal muscle and the Longissimus Dorsi Muscle (Figures 3A–12A).

3.2. Computed Tomography Images

The CT images that closely corresponded to their anatomical sections were selected (Figures 3B–D–11B–D and 12B,C). These CT images provided additional morphological and tomographic details regarding the coelomic structures when compared with the corresponding cross-sections. The CT images acquired using the pulmonary window setting (Figures 3B–11B and 12B,C) provided excellent visualisation of various bones and their associated muscles, which were also seen with the soft tissue and bone window settings. Consequently, we observed different bony structures, such as the thoracic vertebrae, ribs, sternum, humerus, femur and pubis, as well as various muscles associated with these bony structures, including the scapulohumeralis, the scapulohumeral caudal, the longissimus colli, the longissimus dorsi, the pectoral, the supracoracoid and the intercostal muscles. In addition, the essential components of the respiratory tract, such as the trachea, the syrinx, the tracheal bifurcation, the course of the right and left main bronchi, the pulmonary vessels as well as the pulmonary parenchyma, could also be distinguished due to their higher attenuation (Figures 3B–7B, 12B and 12C). Interestingly, the lung parenchyma was well defined, showing a honeycomb-like pattern in all planes, exhibiting an irregular aspect in all puffins (Figures 3B–7B). Moreover, the right and left primary bronchi were visualised entering into the ventromedial aspect of the right and left lungs (Figures 4B, 5B and 12C). Furthermore, other intrathoracic structures, including the heart and large vessels, such as the two brachiocephalic trunks, were well visualised (Figures 4B and 5B). In addition, the walls of the clavicular, thoracic and abdominal air sacs could be distinguished (Figures 3B–10B). Concerning the digestive structures, this specific CT window effectively delineated the oesophagus, the right and left hepatic lobe, the glandular and the muscular parts of the puffin stomach, different intestinal segments and the cloaca (Figures 5B–11B). In addition, the dorsal CT images were helpful in observing the bifurcation of the caecum in the right and left caecum (Figure 12C).

The CT images obtained using the soft tissue windows (Figures 3C–Figure 11C) also provided satisfactory identification of the bony and muscle structures that comprise the coelomic cavity. In addition, this CT window allowed the identification of the oesophagus located ventrally to the vertebral body (Figures 3C–5C). Additionally, adequate visualisation of the heart was distinguished with this CT window. Therefore, we could observe the right ventricle surrounded by the right and left hepatic lobe (Figures 5C–7C), as well as the main arteries, such as the Right and Left Brachiocephalic Trunks, which appeared with intermediate attenuation (Figure 4C). Moreover, the lumen of the proventriculus and the ventricular folds of the puffin stomach were seen with excellent detail in the left portion of the coelomic cavity (Figures 6C–9C).

The CT images acquired with the use of the bone window setting (Figures 3D–11D) demonstrated excellent differentiation between the bones and the soft tissues within the coelomic cavity. In contrast, the heart chambers and the main arteries exhibited an intermediate CT attenuation (Figures 3D–7D). Additionally, various digestive structures, including the oesophagus, the hepatic lobes, different intestinal sections and the cloaca, as well as the glandular and the muscular part of the puffin stomach, were also observed with intermediate attenuation (Figures 3D–11D). Interestingly, this CT window enabled the observation of the left abdominal air sac covering the ventricular surface, which adjoins the left ventral hepatic peritoneal cavity on the left side of the ventriculus (Figure 8D). In addition, the crassus cranioventralis was discernible by its anatomical location ventral to the ventricle, as well as by its round shape and intermediate attenuation (Figure 8D).

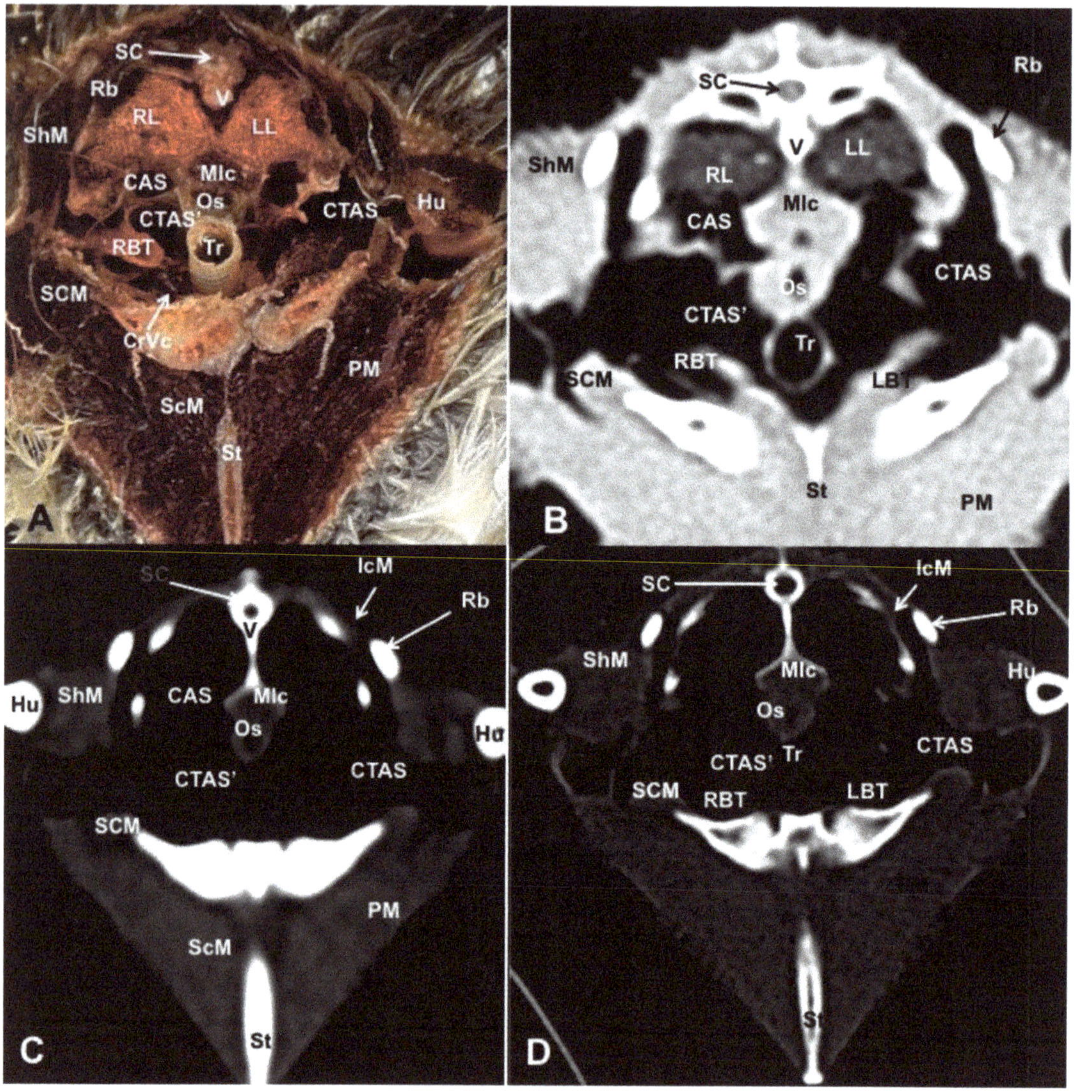

Figure 3. Transverse cross-sectional image (**A**), and pulmonary (**B**), soft tissue (**C**) and bone (**D**) CT images of the coelomic cavity of the *Fratercula arctica* at the level of the trachea corresponding to line I in Figure 1. SC: spinal cord. V: vertebra (vertebral body). ShM: scapulohumeralis muscle. SCM: scapulohumeral caudal muscle. RL: right lung. LL: left lung. Mlc: longissimus colli muscle. Os: oesophagus. Tr: trachea. CTAS: cranial thoracic air sac (left). CTAS′: cranial thoracic air sac (right). CAS: Clavicular Air Sac. LBT: Left Brachiocephalic Trunk. RBT: Right Brachiocephalic Trunk. CrVc: cranial cava vein. St: sternum. PM: Pectoral Muscle (thoracobrachialis muscle). ScM: Supracoracoid Muscle. IcM: intercostal muscle. Hu: humerus. Rb: ribs.

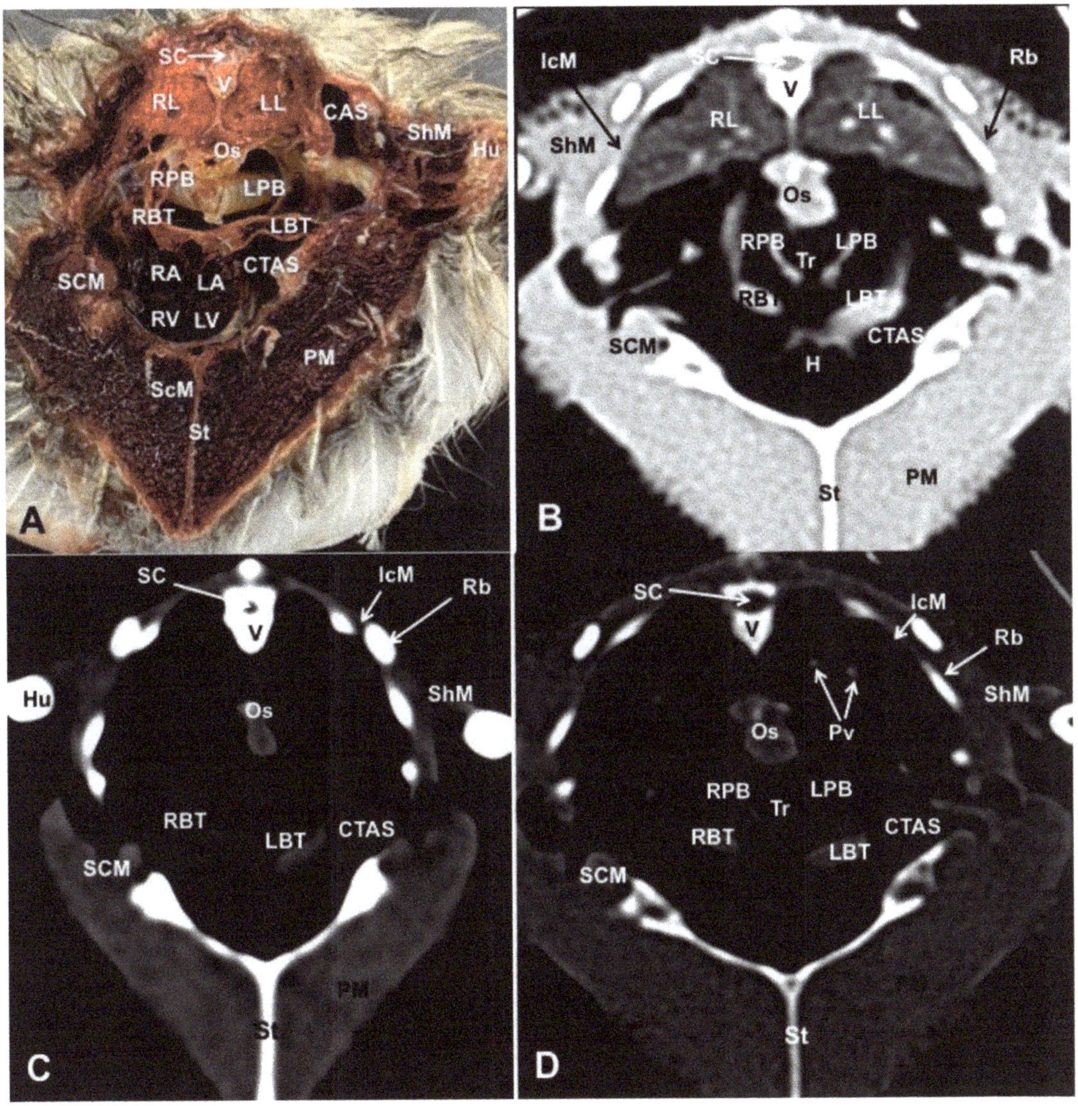

Figure 4. Transverse cross-sectional image (**A**), and pulmonary (**B**), soft tissue (**C**) and bone (**D**) CT images of the coelomic cavity of the *Fratercula arctica* at the level of the heart corresponding to line II in Figure 1. SC: spinal cord. V: vertebra (vertebral body). ShM: scapulohumeralis muscle. SCM: scapulohumeral caudal muscle. RL: right lung. LL: left lung. Pv: pulmonary vessels. Os: oesophagus. Tr: trachea. LPB: Left Primary Bronchus. RPB: Right Primary Bronchus. CTAS: cranial thoracic air sac (left). CAS: Clavicular Air Sac. LBT: Left Brachiocephalic Trunk. RBT: Right Brachiocephalic Trunk. H: heart. RA: Right Atrium. LA: Left Atrium. RV: right ventricle. LV: Left Ventricle. St: sternum. PM: Pectoral Muscle (thoracobrachialis muscle). ScM: Supracoracoid Muscle. IcM: intercostal muscle. Hu: humerus. Rb: ribs.

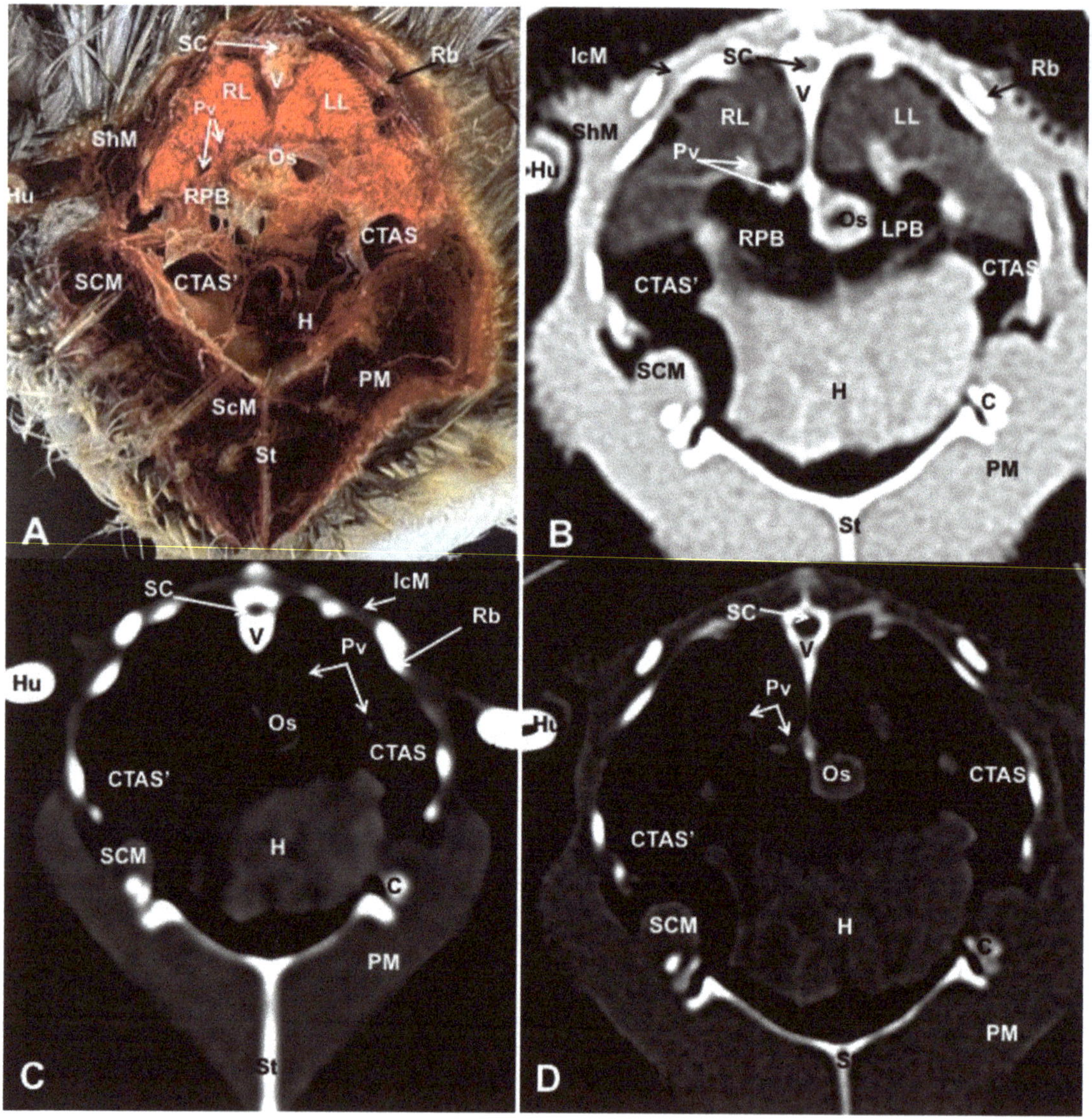

Figure 5. Transverse cross-sectional image (**A**), and pulmonary (**B**), soft tissue (**C**) and bone (**D**) CT images of the coelomic cavity of the *Fratercula* arctica at the level of the cranial thoracic air sacs corresponding to line III in Figure 1. SC: spinal cord. V: vertebra (vertebral body). ShM: scapulohumeralis muscle. SCM: scapulohumeral caudal muscle. RL: right lung. LL: left lung. Pv: pulmonary vessels. Os: oesophagus. LPB: Left Primary Bronchus. RPB: Right Primary Bronchus. CTAS': cranial thoracic air sac (right). CTAS: cranial thoracic air sac (left). H: heart. St: sternum. C: coracoid bone. PM: Pectoral Muscle. ScM: Supracoracoid Muscle. IcM: intercostal muscle. Hu: humerus. Rb: ribs.

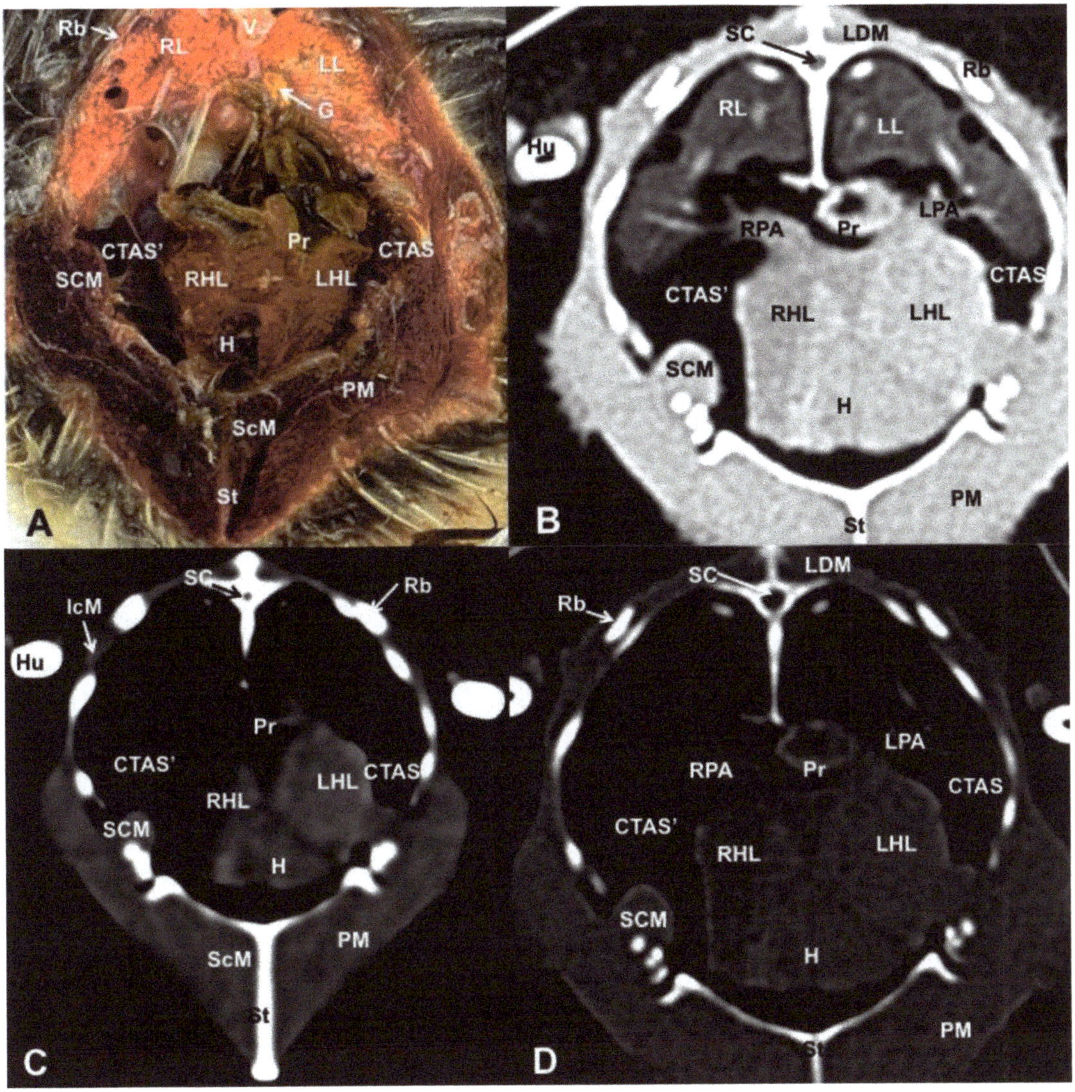

Figure 6. Transverse cross-sectional image (**A**), and pulmonary (**B**), soft tissue (**C**) and bone (**D**) CT images of the coelomic cavity of the *Fratercula arctica* at the level of the liver (hepatic lobes) corresponding to line IV in Figure 1. SC: spinal cord. V: vertebra (vertebral body). LDM: Longissimus Dorsi Muscle. RL: right lung. LL: left lung. RPA: right pulmonary artery. LPA: left pulmonary artery. G: gonad. CTAS: cranial thoracic air sac (left). CTAS': cranial thoracic air sac (right). Pr: proventriculus. RHL: right hepatic lobe. LHL: left hepatic lobe. H: heart (ventricle). SCM: scapulohumeral caudal muscle. St: sternum. ScM: Supracoracoid Muscle. PM: Pectoral Muscle. IcM: intercostal muscle. Hu: humerus. Rb: ribs.

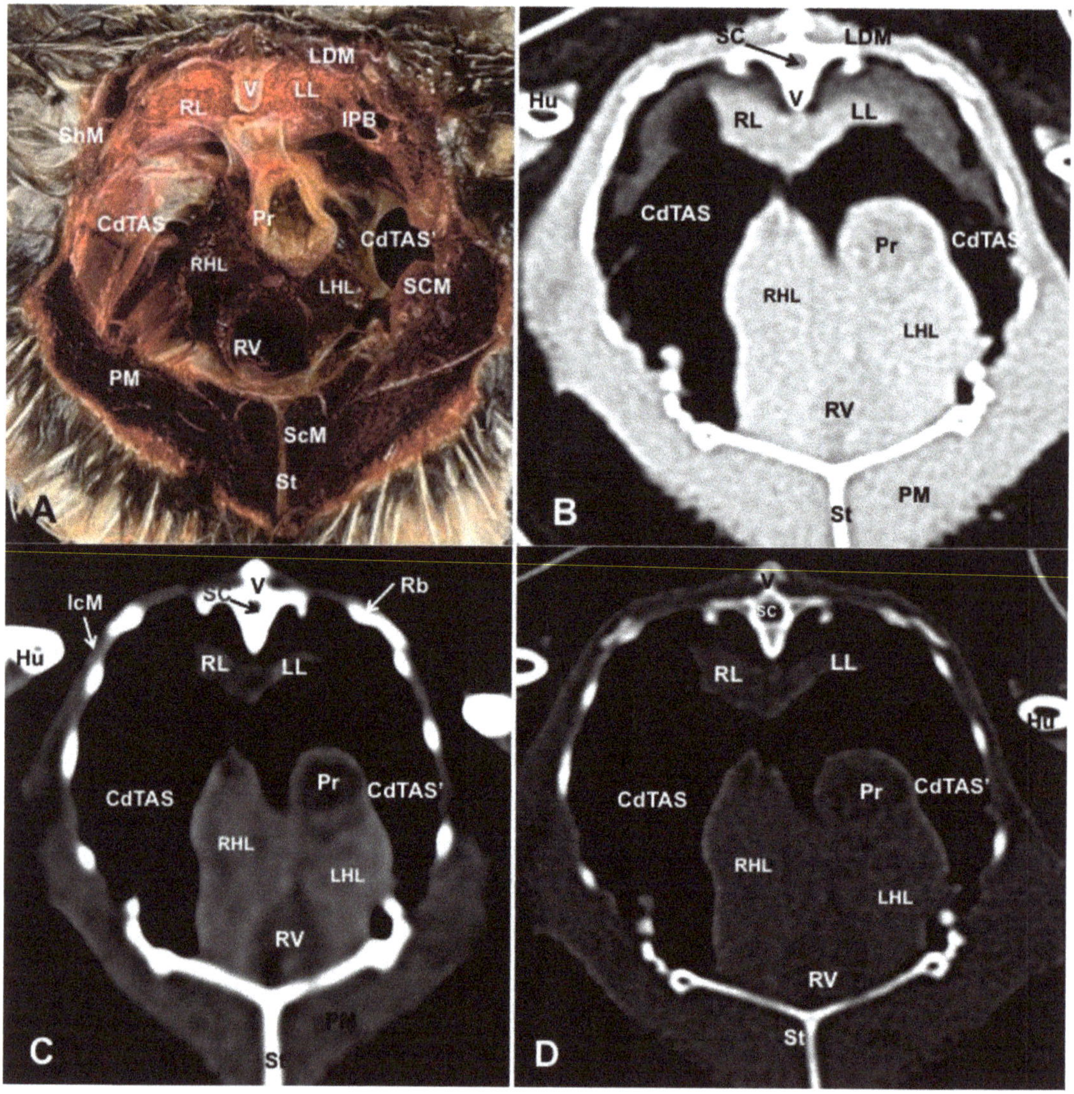

Figure 7. Transverse cross-sectional image (**A**), and pulmonary (**B**), soft tissue (**C**) and bone (**D**) CT images of the coelomic cavity of the *Fratercula arctica* at the level of the caudal thoracic air sac corresponding to line V in Figure 1. SC: spinal cord. V: vertebra (vertebral body and spinal process). LDM: Longissimus Dorsi Muscle. ShM: scapulohumeral muscle. RL: right lung. LL: left lung. CdTAS: caudal thoracic air sac (right). CdTAS′: caudal thoracic air sac (left). Pr: proventriculus. RHL: right hepatic lobe. LHL: left hepatic lobe. RV: right ventricle. St: sternum. ScM: Supracoracoid Muscle. PM: Pectoral Muscle. IcM: intercostal muscle. Hu: humerus. Rb: ribs.

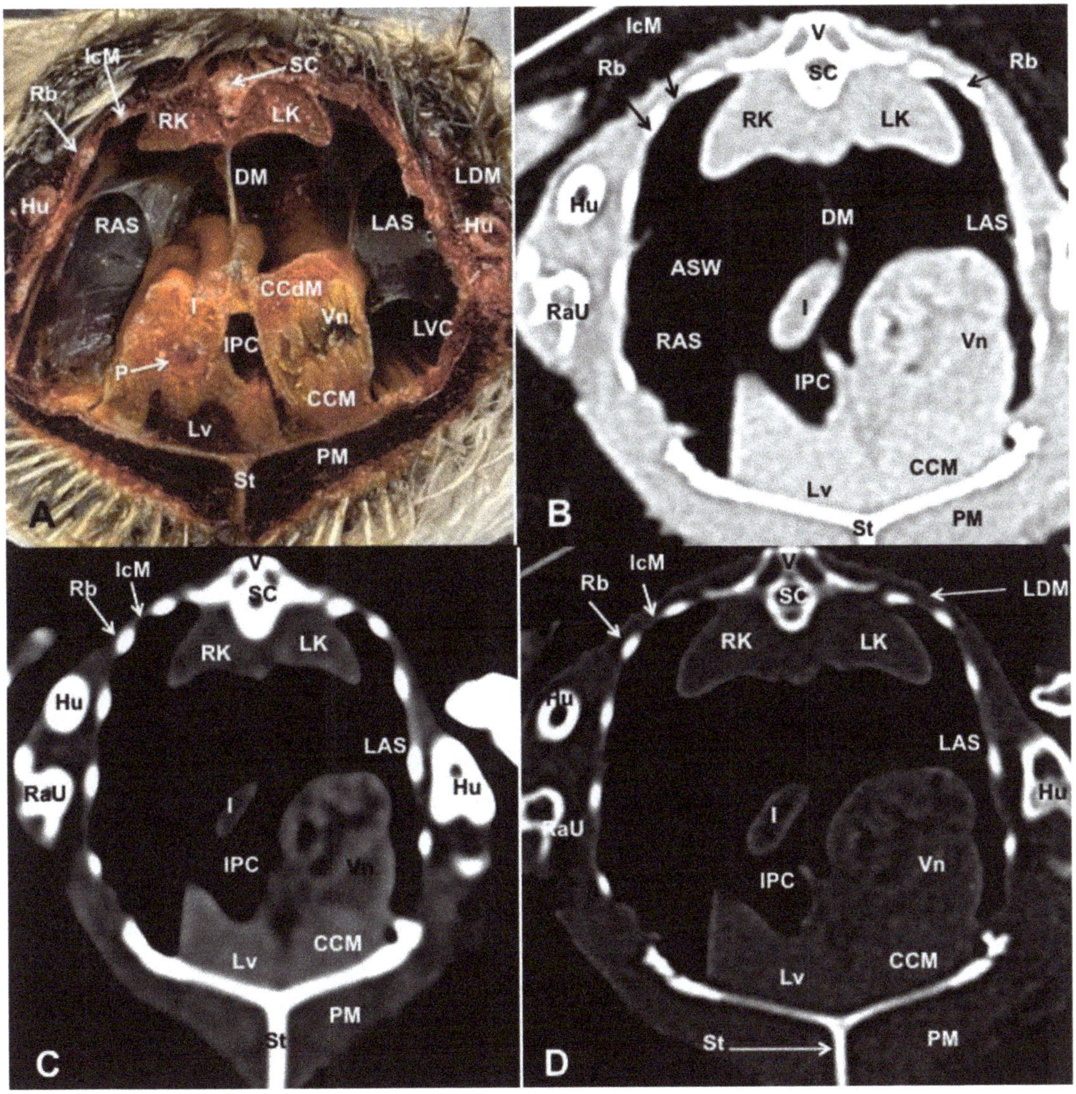

Figure 8. Transverse cross-sectional image (**A**), and pulmonary (**B**), soft tissue (**C**) and bone (**D**) CT images of the coelomic cavity of the *Fratercula arctica* at the level of the kidneys corresponding to line VI in Figure 1. SC: spinal cord. V: vertebra (spinal process). RK: right kidney. LK: left kidney. DM: Dorsal Mesenterium. LAS: left abdominal sac. RAS: right abdominal Sac. ASW: air sac wall. I: intestine. P: pancreas. Vn: ventriculus. CCdM: crassus caudodorsalis muscle. CCM: crassus cranioventralis muscle. IPC: Intestinal Peritoneal Cavity. LVC: left ventral hepatic peritoneal cavity. Lv: liver. St: sternum. PM: Pectoral Muscle. IcM: intercostal muscle. Hu: humerus. RaU: Radius and Ulna. Rb: ribs. LDM: Longissimus Dorsi Muscle.

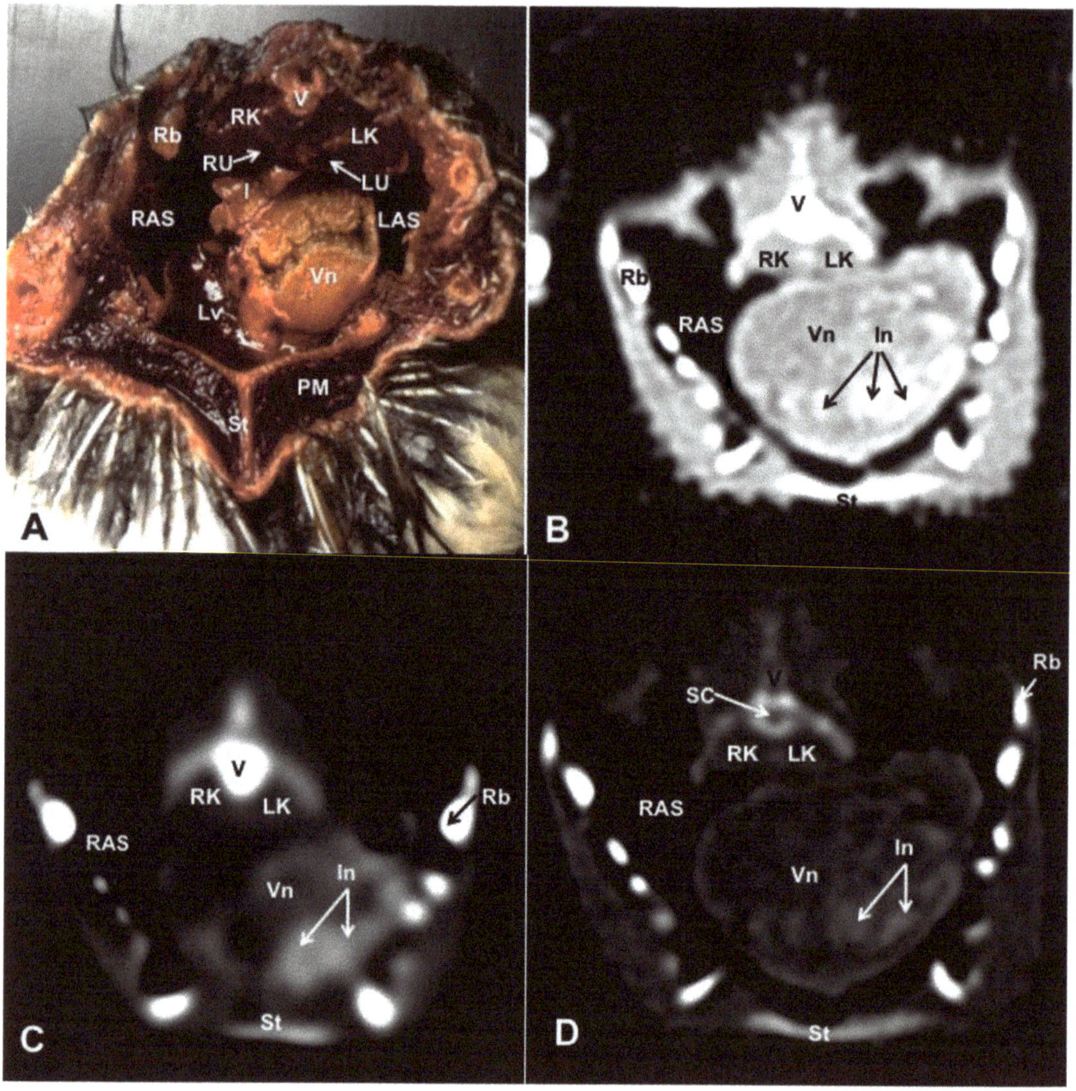

Figure 9. Transverse cross-sectional image (**A**), and pulmonary (**B**), soft tissue (**C**) and bone (**D**) CT images of the coelomic cavity of the *Fratercula arctica* at the level of the ventriculus corresponding to line VII in Figure 1. SC: spinal cord. V: vertebra (vertebral body and spinal process). RK: right kidney. LK: left kidney. RU: right ureter. LU: left ureter. Lv: liver. RAS: right abdominal air sac. LAS: left abdominal air sac. I: intestine. Vn: ventriculus. In: ingesta. St: sternum. PM: Pectoral Muscle (thoracic part). Rb: ribs.

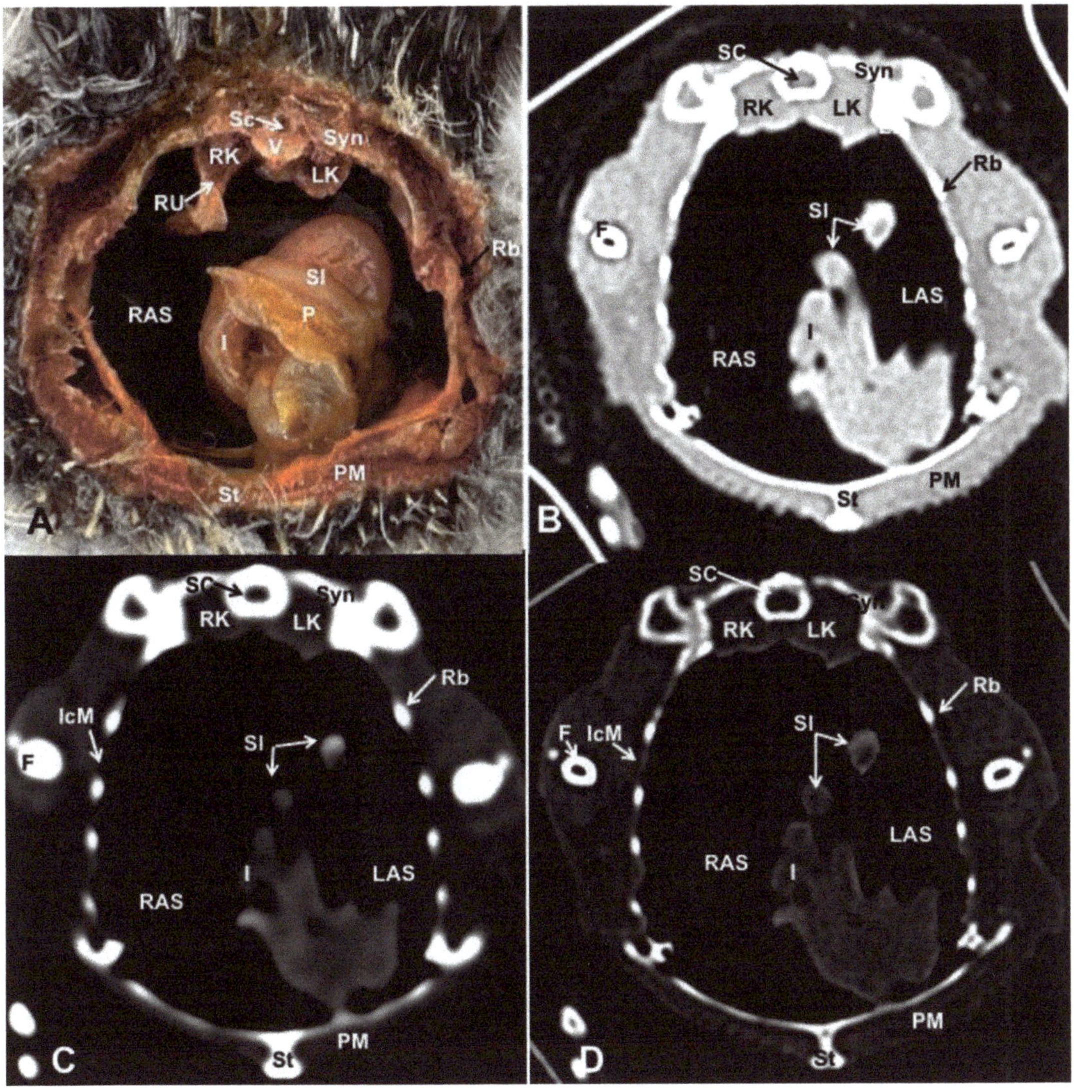

Figure 10. Transverse cross-sectional image (**A**), and pulmonary (**B**), soft tissue (**C**) and bone (**D**) CT images of the coelomic cavity of the *Fratercula arctica* at the level of the caudal end of the right and left kidneys corresponding to line VIII in Figure 1. SC: spinal cord. V: vertebra (vertebral body). Syn: synsacrum. F: femur. RK: right kidney. LK: left kidney. RU: right ureter. RAS: right abdominal air sac. LAS: left abdominal air sac. P: pancreas. I: intestine. SI: small intestine (duodenum). St: sternum. PM: Pectoral Muscle (thoracic part). IcM: intercostal muscle. Rb: ribs.

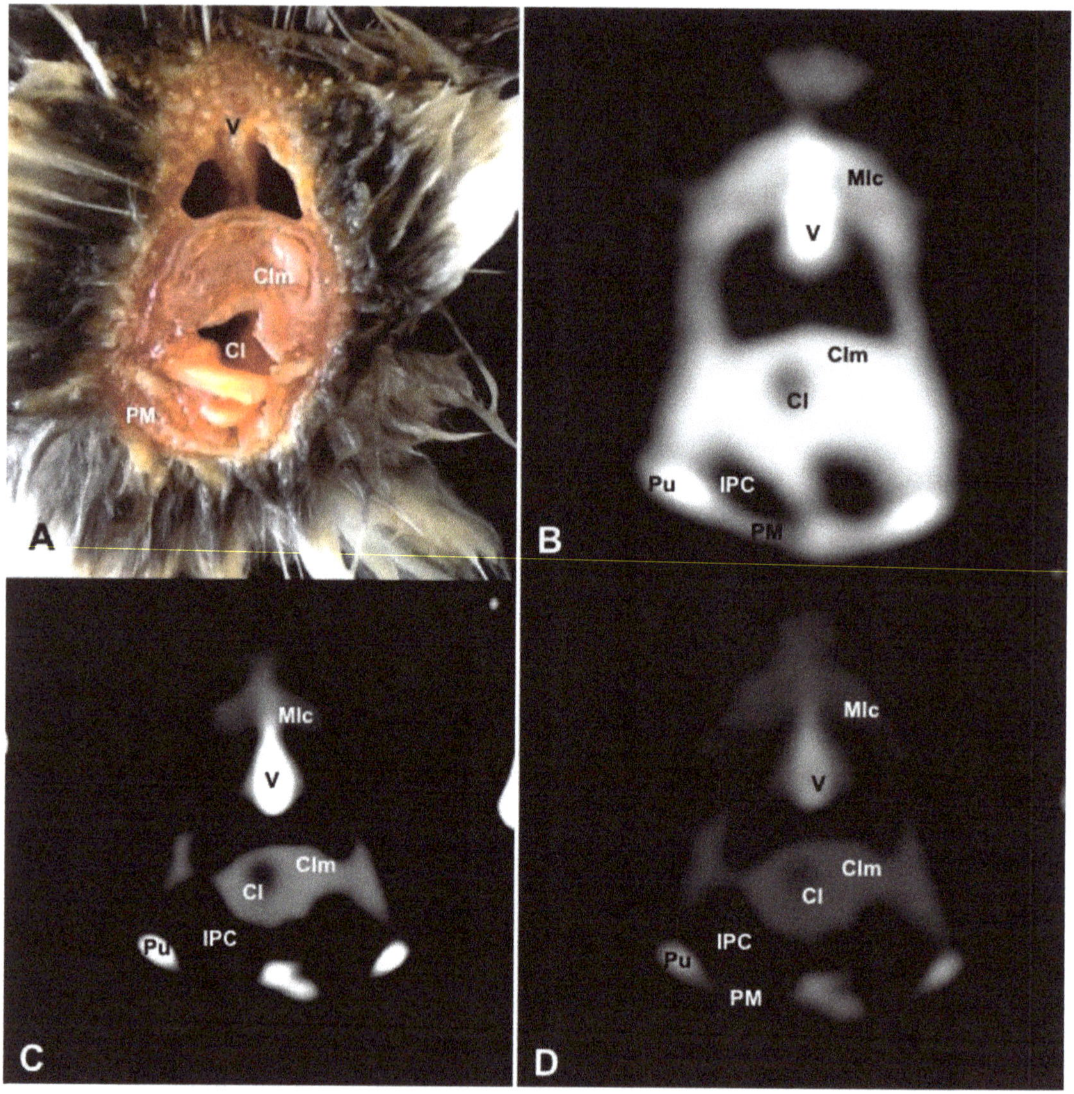

Figure 11. Transverse cross-sectional image (**A**), and pulmonary (**B**), soft tissue (**C**) and bone (**D**) CT images of the coelomic cavity of the *Fratercula arctica* at the level of the cloaca corresponding to line IX in Figure 1. V: vertebra. Clm: cloacal muscles (musculus levator cloacae + musculus transversus cloacae). Cl: cloaca. IPC: Intestinal Peritoneal Cavity. Pu: pubis. PM: Pectoral Muscle (abdominal portion). Mlc: musculus levator caudae.

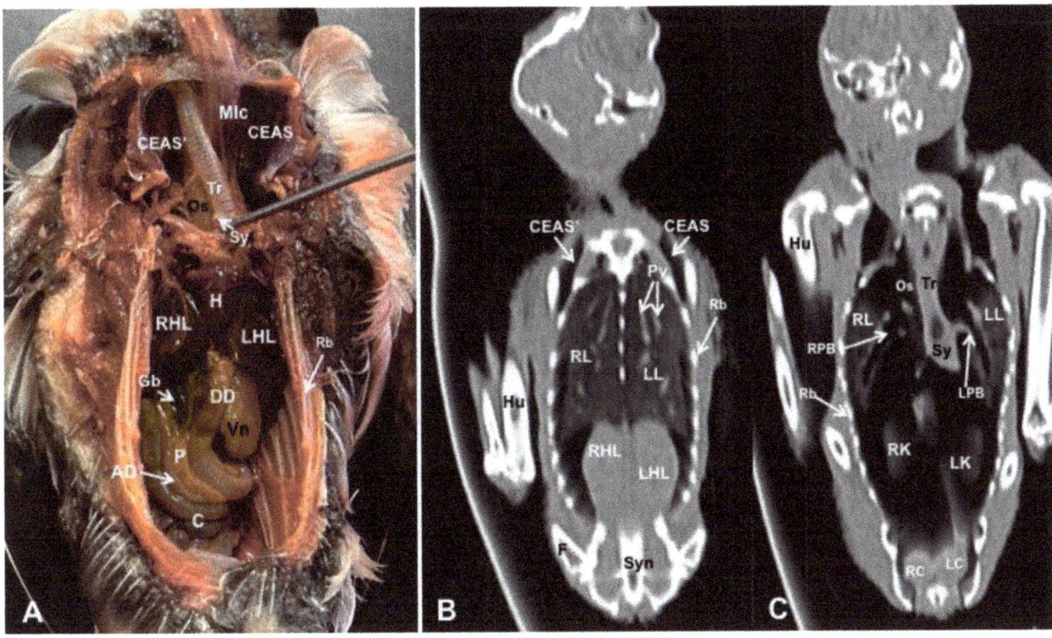

Figure 12. Gross dissection image of the coelomic cavity of the *Fratercula arctica* (**A**), and dorsal MPR volume rendering images in pulmonary CT window of the coelomic cavity at level of hepatic lobes (**B**) and kidneys (**C**). Mlc: longissimus colli muscle. CEAS': Right Cervical Air Sac. CEAS: Left Cervical Air Sac. Tr: trachea. Sy: syrinx. Os: oesophagus. LL: left lung. RL: right lung. Pv: pulmonary vessels. LPB: Left Primary Bronchus (intrapulmonary portion). RPB: Right Primary Bronchus (intrapulmonary portion). H: heart. LHL: left hepatic lobe. RHL: right hepatic lobe. Gb: gallbladder. DD: Descending Duodenum. AD: Ascending Duodenum. Vn: ventriculus. P: pancreas. C: caecum. RC: right caecum. LC: left caecum. Hu: humerus. F: femur. Syn: synsacrum. Rb: rib.

4. Discussion

Birds possess a complex anatomy, very different from mammals, as they lack a distinct separation between the thorax and the abdomen, resulting in the coelomic cavity. Consequently, this cavity has a complex organisation that encloses organs of the circulatory system and the respiratory, digestive and urogenital apparatus [24,25,30–34].

The conventional approach to studying animal anatomy, which involves dissections for visualising organs and internal structures, has undergone a profound transformation with the irruption of modern imaging techniques such as MRI and CT, which are the most advanced diagnostic imaging modalities available for identifying normal anatomy and abnormalities in the internal organs of live animals [9]. In contrast to conventional imaging methods, such as traditional radiographs or ultrasonography, these modern modalities are essential techniques for enabling better morphologic characterisation of the coelomic cavity by providing views of body sections from multiple tomographic planes without repositioning the animal, thereby delivering images with exceptional anatomic resolution in the absence of tissue superimposition, high contrast between different structures, and excellent tissue differentiation that permits the assessment of spatial relationships between organs that are not detected using radiography or ultrasonography. Disadvantages of these procedures include their high cost, the need to sedate and possibly restrain the bird, and the longer examination time in the case of MRI [4–6,21,25–27]. Due to these shortcomings, the traditional imaging procedures are still the most routinely used because they are fast and low-cost and are widely available in daily avian practice [5,26].

In the present study, coelomic CT images were assessed using pulmonary, soft tissue and bone window settings. The employment of the pulmonary window setting yielded notable benefits in the differentiation of the respiratory tract and diverse vascular structures. As a result, the lung parenchyma, with its honeycomb pattern related to the end-on view of the parabronchi; the course of the two pulmonary arteries supplying each lung; and the paired brachiocephalic trunks were well distinguished with this window setting (Figures 4B–6B). Moreover, this CT window and the different planes were quite effective in visualising the extrapulmonary and intrapulmonary portions of the primary bronchus into each lung (Figure 12C). Comparable findings have been reported in previous CT investigations performed on other species, such as neonatal foals [35], reptiles [36] and different avian species [25–27]. Meanwhile, the application of the soft tissue window provided additional insights into puffin cardiac topography and its associated blood vessels (Figures 4C–7C). However, the combination of the different CT window settings was not helpful for the visualisation of the four cardiac chambers, which was due to the absence of contrast media administration. Although there are limited studies on the use of contrast media in birds [25–27], its application has proven to enhance and facilitate the visualisation of cardiovascular structures and air sacs in live specimens [25,26]. Despite the constraint posed in our study by the use of cadaveric specimens, precluding the administration of intravenous contrast medium, the pulmonary CT window facilitated visualisation of the wall of the thoracic and abdominal air sacs (Figures 7B–10B). Nevertheless, studies performed on other bird species described how the air sac walls are not precisely distinguished in CT or radiographs [25,26]. Similar to previous studies [25], the combination of the aforementioned pulmonary CT window with the bone window setting revealed a distinct visualisation of the ventriculus of the stomach, whose content exhibited an amalgamation of air and dense materials (Figures 8C,D and 9C,D). In addition, this particular CT window was quite helpful in distinguishing the muscular layer of the puffin ventriculus, and the relation between the bird heart and the right and left hepatic lobes (Figure 7C,D). In contrast to other avian species, such as parrots [25], whose right lobe is generally larger than the left, the puffin liver presented lobes of similar size. In addition, they showed the gallbladder, which was only distinguishable in the dissected images. This fact could be due to the lack of contrast administration that could impede the visualisation of other organs, including the spleen, the ureters or the pancreas.

In avian species, the sexes are separated, consisting of either male or female genital organs. Nevertheless, several avian species consequently do not show phenotypic sexual dimorphism, making it necessary to employ endoscopy or DNA testing for definitive sex determination [32]. In this study, we identified a single round structure adjacent to the caudal margin of the left lung that closely resembled the left ovary. However, the small size of the puffins suggested that they were juvenile specimens and, therefore, sexually immature animals, which could explain the challenge encountered in distinguishing the gonads in these animals. As stated in other reports described in birds [25], the gonads are visible only in sexually mature individuals of large species.

This CT-based anatomical study has confirmed the suitability of cadavers for evaluating diverse anatomical patterns. However, it is essential to consider the absence of blood flow in dead animals when comparing the results with those obtained from live specimens using contrast media [6,17,18,25]. Results from the present study demonstrated that the utilisation of frozen anatomical sections and dissections was beneficial in identifying various coelomic structures observed on transverse CT images and ensured accurate correlations.

5. Conclusions

The current investigation elucidates the spatial arrangement of the different organs within the coelomic cavity of puffins, achieved through the integration of diverse CT window settings in conjunction with anatomical cross-sections. The outcomes underscore the efficacy of CT imaging for the comprehensive study of avian anatomical structure, obtaining high-quality images that could serve as a reference to evaluate Alcidae birds with pathologies at the coelom level. Notably, it is imperative to emphasise that the use of CT as a diagnostic modality in live birds requires the administration of anaesthesia and carries potential complications and mortality risks. Consequently, the utilisation of this technique in live avians demands judicious consideration and rigorous justification by the attending clinician.

Author Contributions: Conceptualization, J.R.J., M.F.-H., J.A.C., I.M., M.A., G.R.Z. and M.E.; methodology, J.R.J., M.F.-H., J.A.C., I.M., M.A., G.R.Z. and M.E.; investigation, J.R.J., M.F.-H., J.A.C., I.M., M.A., G.R.Z. and M.E.; resources, J.R.J., M.F.-H., J.A.C., I.M., M.A., G.R.Z. and M.E.; writing—review and editing, J.R.J., M.F.-H., J.A.C., I.M., M.A., G.R.Z. and M.E. All authors have read and agreed to the published version of the manuscript.

Funding: This research received no external funding.

Institutional Review Board Statement: In this study, we did not need ethical approval, since the puffins were coming to our hospital to rule out through CT any cause of stranding. From these studies, we obtained the information presented in our paper. As occurs in domestic mammals, just informed consent from the owner allowed us to carry out this study. Therefore, El Cabildo Insular de Gran Canaria was informed that all animal identity information obtained from this study was treated as confidential to the extent permitted by law and only used for research or teaching purposes.

Informed Consent Statement: Written informed consent has been obtained from El Cabildo Insular de Gran Canaria to publish this paper.

Data Availability Statement: The information is available at "https://accedacris.ulpgc.es", accessed on 25 June 2023.

Acknowledgments: In loving memory of Alvaro Domingo Rodriguez Garcia. We also thank Concepción Mingot, Carmen Mingot, Emilia Mingot, Nicolas Aquino, Ayesh Mohamad, Marisa Mohamad and Jamal Jaber for their support and constructive comments. Moreover, we also thank the Consejeria de Área de Medio Ambiente, Clima, Energía y Conocimiento of the Cabildo Insular de Gran Canaria for providing the animals of this study.

Conflicts of Interest: The authors declare no conflict of interest.

References

1. Barrowclough, G.F.; Cracraft, J.; Klicka, J.; Zink, R.M. How Many Kinds of Birds Are There and Why Does It Matter? *PLoS ONE* **2016**, *11*, e0166307. [CrossRef] [PubMed]
2. American Veterinary Medical Association. Available online: https://www.avma.org/resources-tools/reports-statistics/us-pet-ownership-statistis (accessed on 1 June 2023).
3. Raftery, A. *BSAVA Manual of Psittacine Birds*, 2nd ed.; British Small Animal Veterinary Association: Gloucester, UK, 2005; pp. 35–49.
4. Krautwald-Junghanns, M.E.; Pees, M.; Reese, S.; Tully, T. *Diagnostic Imaging of Exotic Pets*, 1st ed.; Schlütersche: Hannover, Germany, 2011.
5. McMillan, M.C. *Imaging Techniques*; Wingers Publishing: Lake Worth, FL, USA, 1994; pp. 246–326.
6. Zeeland, Y.R.A.; Schoemaker, N.J.; Hsu, E.W. *Current Therapy in Avian Medicine and Surgery*; Elsevier: St. Louis, MO, USA, 2016; pp. 531–549.
7. Haacke, E.M.; Brown, R.W.; Thompson, M.R.; Vankatesan, R. *Magnetic Resonance Imaging*; John Wiley & Sons: New York, NY, USA, 1999; pp. 331–380, 669–702.
8. Beuf, O.; Jaillon, F.; Saint-Jalmes, H. Small-animal MRI: Signal-to-noise ratio comparison at 7 and 15 T with multiple-animal acquisition strategies. *Magn. Reson. Mater. Phys.* **2006**, *19*, 202–208. [CrossRef] [PubMed]
9. Lauridsen, H.; Hansen, K.; Wang, T.; Agger, P.; Andersen, J.L.; Knudsen, P.S.; Rasmussen, A.S.; Uhrenholt, L.; Pedersen, M. Inside Out: Modern Imaging Techniques to Reveal Animal Anatomy. *PLoS ONE* **2011**, *6*, e17879. [CrossRef] [PubMed]
10. Dinley, J.; Hawkins, L.; Paterson, G.; Ball, A.D.; Sinclair, I.; Sinnett-Jones, P.; Lanham, S. Micro-computed X-ray tomography: A new non-destructive method of assessing sectional, fly-through and 3D imaging of a soft-bodied marine worm. *J. Microsc.* **2010**, *238*, 123–133. [CrossRef] [PubMed]
11. Ziegler, A.; Faber, A.; Mueller, S.; Bartolomaeus, T. Systematic comparison and reconstruction of sea urchin (*Echinoidea*) internal anatomy: A novel approach using magnetic resonance imaging. *BMC Biol.* **2008**, *6*, 33. [CrossRef]
12. Rieppel, O.; Gauthier, J.; Maisano, J. Comparative morphology of the dermal palate in squamate reptiles, with comments on phylogenetic implications. *Zool. J. Linn. Soc.* **2008**, *152*, 131–152. [CrossRef]

13. Digital Fish Library. A Collaborative Project at the University of California, San Diego Funded by the National Science Foundation's (NSF) Biological Infrastructure Program. Available online: www.digitalfishlibrary.org (accessed on 8 June 2023).
14. Webb, E.; Yuan, M.; Lemoine, N.; Wang, Y. Imaging in animal models. *Integr. Cancer Sci. Ther.* **2016**, *3*, 428–431. [CrossRef]
15. Schambach, S.J.; Bag, S.; Schilling, L.; Groden, C.; Brockmann, M.A. Application of micro-CT in small animal imaging. *Methods* **2010**, *50*, 2–13. [CrossRef]
16. Farrag, M.; Pukale, D.D.; Leipzig, N.D. Micro-computed tomography utility for estimation of intraparenchymal spinal cord cystic lesions in small animals. *Neural Regen. Res.* **2021**, *16*, 2293–2298.
17. Delk, K.W.; Mejia-Fava, J.; Jiménez, D.A.; Kent, M.; Myrna, K.; Mayer, J.; Divers, S. Diagnostic imaging of peripheral vestibular disease in a Chinese goose (*Anser cygnoides*). *J. Avian Med. Surg.* **2014**, *28*, 31–37. [CrossRef]
18. Fleming, G.J.; Lester, N.V.; Stevenson, R.; Silver, X.S. High field strength (4.7 T) magnetic resonance imaging of hydrocephalus in an African Grey parrot (*Psittacus erithacus*). *Vet. Radiol. Ultrasound* **2003**, *44*, 542–545. [CrossRef] [PubMed]
19. Jirak, D.; Janacek, J.; Kear, B.P. A combined MR and CT study for precise quantitative analysis of the avian brain. *Sci. Rep.* **2015** *5*, 16002. [CrossRef] [PubMed]
20. Romagnano, A.; Shiroma, J.T.; Heard, D.J.; Johnson, R.D.; Schiering, M.R.; Mladinich, M.S. Magnetic resonance imaging of the brain and coelomic cavity of the domestic pigeon (*Columba livia domestica*). *Vet. Radiol. Ultrasound* **1996**, *37*, 431–440. [CrossRef]
21. Fumero-Hernández, M.; Encinoso, M.; Ramírez, A.S.; Morales, I.; Suárez Pérez, A.; Jaber, J.R. A Cadaveric Study Using Computed Tomography for Measuring the Ocular Bulb and Scleral Skeleton of the Atlantic Puffin (Aves, Alcidae, Fratercula arctica). *Animals* **2023**, *13*, 2418. [CrossRef] [PubMed]
22. Hall, M.I. The anatomical relationships between the avian eye, orbit and sclerotic ring: Implications for inferring activity patterns in extinct birds. *J. Anat.* **2008**, *212*, 781–794. [CrossRef]
23. Andrade, S.B.D.; Araujo, N.L.L.C.D.; Raposo, A.C.S.; Muramoto, C.; Oriá, A.P. Morphometric descriptive report of scleral ossicle rings, by ultrasound and computed tomography, in three Testudines specimens. *Ciênc. Rural.* **2023**, *53*, e20210423. [CrossRef]
24. Orosz, S.E.; Toal, R.L. Tomographic anatomy of the golden eagle (*Aquila chrysaetos*). *J. Zoo Wildl. Med.* **1992**, *23*, 39–46.
25. Veladiano, I.A.; Banzato, T.; Bellini, L.; Montani, A.; Catania, S.; Zotti, A. Normal computed tomographic features and reference values for the coelomic cavity in pet parrots. *BMC Vet. Res.* **2016**, *12*, 182. [CrossRef]
26. Da Silva, J.P.; Rahal, S.C.; Castiglioni, M.C.R.; Baldissera Gonçalves, R.A.; Doiche, D.P.; Moresco, A.; Mamprim, M.J.; Vulcano, L.C Radiography and computed tomography of the heart and lower respiratory tract in toco toucans (*Ramphastos toco*). *Anat. Histol. Embryol.* **2020**, *49*, 541–549. [CrossRef]
27. Pepperberg, I.M.; Howell, K.S.; Banta, P.A.; Patterson, D.K.; Meisser, M. Measurement of grey parrot (*Psittacus erithacus*) trachea via magnetic resonance imaging, dissection, and electron beam computed tomography. *J. Morphol.* **1998**, *238*, 81–91. [CrossRef]
28. IUCN. The IUCN Red List of Threatened Species. Version 2022-2. Available online: https://www.iucnredlist.org (accessed on 2 June 2023).
29. Painter, J. *Fratercula arctica*. Available online: https://animaldiversity.org/accounts/Fratercula_arctica/ (accessed on 30 May 2023).
30. Sisson, S.; Getty, R.; Grossman, J.D. *Anatomía de los Animales Domésticos*, 5th ed.; Elsevier Masson: Issy les Moulineaux, France, 1982.
31. MacLelland, J. *Atlas en Color de Anatomía de las Aves*; McGraw-Hill Interamericana: Madrid, Spain, 1992.
32. König, E.H.; Liebich, H.G.; Bragulla, H. *Anatomie und Propädeutik des Geflügels*; Schattauer: Stuttgart, Germany, 2001.
33. Isaac, I.; Richardson, J.; Liuti, T.; Longo, M. Safety of intravenous iodinated contrast medium injection in rabbits undergoing conscious computed tomography. *Vet. Rec. Open* **2022**, *9*, e31. [CrossRef] [PubMed]
34. Seamon, A.B.; Hofmeister, E.H.; Divers, S.J. Outcome following inhalation anesthesia in birds at a veterinary referral hospital: 352 cases (2004–2014). *J. Am. Vet. Med. Assoc.* **2017**, *251*, 814–817. [CrossRef] [PubMed]
35. Arencibia, A.; Corbera, J.A.; Ramírez, G.; Díaz-Bertrana, M.L.; Pitti, L.; Morales, M.; Jaber, J.R. Anatomical Assessment of the Thorax in the Neonatal Foal Using Computed Tomography Angiography, Sectional Anatomy, and Gross Dissections. *Animals* **2020**, *10*, 1045. [CrossRef] [PubMed]
36. Banzato, T.; Selleri, P.; Veladiano, I.A.; Zotti, A. Comparative evaluation of the cadaveric and computed tomographic features of the coelomic cavity in the green iguana (*Iguana iguana*), black and white tegu (*Tupinambis merianae*) and bearded dragon (*Pogona vitticeps*). *Anat. Histol. Embryol.* **2013**, *42*, 453–460. [CrossRef]

animals

MDPI

rticle

Cranial Investigations of Crested Porcupine (*Hystrix cristata*) by Anatomical Cross-Sections and Magnetic Resonance Imaging

aniel Morales-Bordon [1], Mario Encinoso [2,*], Alberto Arencibia [3] and José Raduan Jaber [3,*]

[1] Departamento de Patología Animal, Producción Animal, Bromatología y Tecnología de los Alimentos, Facultad de Veterinaria, Universidad de Las Palmas de Gran Canaria, Trasmontaña, 35413 Arucas, Las Palmas, Spain; daniel.morales@ulpgc.es
[2] Hospital Clínico Veterinario, Facultad de Veterinaria, Universidad de Las Palmas de Gran Canaria, Trasmontaña, 35413 Arucas, Las Palmas, Spain
[3] Departamento de Morfologia, Facultad de Veterinaria, Universidad de Las Palmas de Gran Canaria, Trasmontaña, 35413 Arucas, Las Palmas, Spain; alberto.arencibia@ulpgc.es
* Correspondence: mencinoso@gmail.com (M.E.); joseraduan.jaber@ulpgc.es (J.R.J.)

Simple Summary: Advanced imaging diagnostic techniques, such as magnetic resonance imaging with anatomical sections, were used to evaluate the head of the crested porcupine (*Hystrix cristata*). These techniques were very helpful to delineate the main formations that compose the central nervous system (CNS), as well as its associated structures. To the authors' knowledge, the present study is the first to describe this area using anatomical sections and magnetic resonance imaging (MRI) in crested porcupines.

Abstract: This paper aimed to describe an atlas of the crested porcupine (*Hystrix cristata*) head by applying advanced imaging techniques such as MRI. Furthermore, by combining the images acquired through these techniques with anatomical sections, we obtained an adequate description of the structures that form the CNS and associated structures of this species. This anatomical information could serve as a valuable diagnostic tool for the clinical evaluation of different pathological processes in porcupines, such as abscesses, skull malformations, fractures, and neoplasia.

Keywords: magnetic resonance imaging; anatomical sections; rodents; head anatomy; CNS; crested porcupine

1. Introduction

In recent years, advanced diagnostic imaging techniques have facilitated the visualization of several diseases in captive and wildlife medicine. Traditionally, standard radiography was the choice more frequently used by clinicians, biologists, and researchers [1]. Nevertheless, the results of previous studies have demonstrated that computed tomography (CT) and magnetic resonance imaging (MRI) provide more information with which to improve diagnostic accuracy, prognosis, treatment of diseases, and anatomic knowledge [2–4]. These modern techniques avoid the superimposition of adjacent anatomical structures and delineate the anatomic detail of specific tissue densities more finely, which improving the capacity for interpretation [2]. These benefits provide important value for the anatomic investigations of specific regions, providing essential knowledge of domestic and exotic mammal species [2–5]. Among these species, we highlight the crested porcupine (*Hystrix cristata*), which is one of the most well-known members of the family Hystricidae and is included in the IUCN red list as least concern since they are regionally or locally threatened, mainly due to the fact that they are hunted illegally for food and killed because they are considered as an agricultural pest [6,7]. Consequently, these animals require appropriate conservation policies in regional and local contexts. It is a species of rodent native to Italy and Sicily, as well as a broad central strip ranging from Senegal west to Somalia and east to Kenya and Tanzania, but very little is known regarding its geographic variability in Africa [6–8]. These animals are found in forests, rocky areas, mountains, croplands, and sandhill deserts, and protect themselves in caves, aardvark holes, and burrows that they dig themselves [6,7].

Porcupines are characterized for presenting strong and pointed quills covering the tail, sides, and top of the body that can be raised into a crest and used for defensive purposes. Concerning their heads, they are large and robust with enlarged infraorbital foramina so that portions of the masseter extend through it and arise from the frontal side surface of the snout. In addition, the nasal cavities are enlarged [9]. Interestingly, Hystrix show prominent pockets in the skull, maxillary, lacrimal, and turbinate bones, which could be used for attachment of the masticatory muscles.

Some valuable literature on the anatomical, physiological, and pathological study of pet rabbits and rodents is already broadly available [10–16]. Among these, we highlighted those studies that deal with brain and head MRI anatomy, which are available at reduced resolution from low field-strength and anatomical cross-sections, and more recently, with a high field-strength magnet in excised and fixed rabbit brain for research and clinical guidance in animals with diverse pathological processes [14,15]. To the best of the authors' knowledge, only a few publications have been conducted on porcupines, which have been focused on their biology, morphometric geographic variability, and some clinical conditions such as upper respiratory tract disease [6,8,9,13]. However, no description of the brain has been reported in this specific rodent. Therefore, this study aimed to describe the normal anatomy of the CNS of the crested porcupine and its associated structures, using specific anatomical sections that better matched the images obtained by MRI. The combination of MRI and macroscopic anatomical sections could provide helpful information for anatomic teaching and clinical practice.

2. Materials and Methods

2.1. Animals

Three carcasses of adult (two males and one female) crested porcupines (*Hystrix cristata*) from the zoological park "Rancho Texas Lanzarote Park" (Lanzarote, Canary Islands, Spain) were collected. None of the porcupines used in this study had a history of central nervous system disease, and no abnormalities were detected during imaging examinations. Prior consent was obtained from the responsible person at the zoological park to include the porcupines in this study.

2.2. Anatomic Evaluation

After the MR images were acquired, the two scanned carcasses were frozen at −80 °C for 48 h before performing anatomical cross-sections. After that, the two frozen carcasses were sectioned using an electric band saw to obtain sequential anatomical cross-sections. Contiguous 1 cm transverse slices were obtained starting at the olfactory bulb and extending to the first cervical vertebra region. These slices were intentionally thicker than those of the MRI to retain the integrity and position of the anatomic formations in the slices. The obtained sections were cleaned with water, numbered, and photographed on the cranial and caudal surfaces. Later, we selected those anatomical sections that better correlated with the MRI images to identify the structures of central nervous system and associated structures of the crested porcupine. Moreover, we also used anatomical textbooks and relevant references describing the anatomy of cats, rabbits, and different rodent species to facilitate accurate anatomic identification of these structures [14,15,17–21].

2.3. MRI Technique

A magnetic resonance imaging study was conducted on individual crested porcupine using a 1.5-Tesla magnet (Toshiba, Vantage Elan, Japan) while the animal was positioned in a ventral recumbent position. A standard MRI protocol was employed to generate spin-echo (SE) T1-weighted and T2-weighted images in sagittal, transverse, and dorsal planes. SE T1-weighted transverse images were acquired with the following parameters: echo time (TE) of 10 ms, repetition time (TR) of 800 ms, acquisition matrix size of 536 × 384, and a slice thickness of 4.5 mm with 4 mm spacing between slices. For SE T2-weighted transverse images, the TE was set to 120 ms, TR to 10,541 ms, acquisition matrix size to 624 × 448, and slice thickness to 3 mm with 3 mm interslice spacing. SE T2-weighted sagittal images were obtained with a TE of 120 ms, TR of 7529 ms, acquisition matrix size of 512 × 804, and a slice thickness of 2.8 mm with 2 mm interslice spacing. Lastly, SE T2-weighted dorsal images were acquired using a TE of 120 ms, TR of 8282 ms, acquisition matrix size of 468 × 512, and a slice thickness of 3.4 mm with 3 mm interslice spacing. A medical imaging viewer (OsiriX MD, Geneva, Switzerland) was utilized to assess the images of the study.

3. Results

In this study, we present different anatomical cross-sections and T2W images of the crested porcupine, which were displayed in a rostral-to-caudal progression from the level of the ethmoturbinates and eyeballs to the caudal end of the brain stem. Therefore, Figure 1 represents a T2W sagittal image in which each line and number (I–VI) represents approximately the level of the following anatomical and MRI transverse images. Transverse anatomical sections and MR images of the porcupine revealing the relevant anatomical structures of the head were displayed (Figures 2–7). These figures were composed of two images: (A) macroscopic anatomical section and (B) T2W MR image. Additionally, two sagittal and two dorsal T2W MR images were presented to depict the relevant structures of the porcupine central nervous system (Figures 8 and 9).

3.1. *Anatomical Cross-Sections*

The anatomical sections obtained in this study allowed us to visualize the different structures belonging to the central nervous system and its associated structures, which were labeled according to the International Committee on Veterinary Gross Anatomical Nomenclature. Therefore, we identified the main components of the brain (the prosencephalon, mesencephalon, and rhombencephalon). Thus, the two telencephalic hemispheres surrounded by the cerebral cortex and separated by the longitudinal cerebral fissure were identified (Figures 5A, 6A and 7A). Both hemispheres were connected by fibers of white matter known as the corpus callosum (Figures 4A, 5A, 6A and 7A). Each cerebral hemisphere contained a lateral ventricle (Figures 4A, 5A, 6A and 7A). Ventrally, we distinguished a component of the basal ganglia, the nucleus caudatus (Figures 4A and 5A). Thus, we identified different parts of it, such as the head and the tail. Other structures which we observed were the septal nuclei, which were circumscribed by two parallel vertical lines through the most inferior and medial aspect of each lateral ventricle (Figure 4A). More caudally, the diencephalon enclosing the third ventricle was identified, as well as more ventrally specific components of the hypothalamus, such as the optic chiasm (Figures 4A and 5A). Additionally, these sections were quite helpful in showing the caudal parts of the thalamus. Hence, the lateral eminence on the caudodorsal surface of the thalamus, known as the lateral geniculate body, was distinguished, whereas caudoventrally, we identified the medial geniculate body of the thalamus (Figure 5A). Moreover, the dorsal part of the mesencephalon with the caudal and rostral colliculus and its ventral part with the cerebral peduncles were also shown (Figures 6A and 7A). These anatomical sections were helpful in identifying the vermis and the cerebellar peduncles, which connected the cerebellum to the adjacent brain stem and the cerebrum (Figure 7A). The ventral part of the cerebellum with the lingula, covering part of the fourth ventricle, could also be identified (Figure 7A). These sections were also helpful in distinguishing the medulla oblongata and the decussation of the pyramids. In addition, different bony structures comprising the neurocranium were observed, such as the frontal, the temporal (with its scamous, petrous, and tympanic parts), the sphenoid, and the occipital bones (Figures 2A, 3A, 4A, 5A, 6A and 7A). Furthermore, these sections showed different air-filled spaces, such as the frontal and the sphenoidal sinuses (Figures 2A, 3A, 4A, 5A, 6A and 7A), and structures associated with the nasal cavity, including the ethmoturbinates and the vomer, could be distinguished (Figures 2A and 3A). Main sensory organs such as the eyeball and its associated structures were also depicted. Consequently, we identified the retina, the vitreous chamber, and the optic nerve, which were surrounded by extraocular muscles. Among these, we distinguished the dorsal and ventral rectus muscles (Figure 2A). Moreover, the main components of the auditory system, such as the external auditory canal, the tympanic cavity, and the inner ear, were visualized (Figures 5A and 6A). Also, we identified the relevant muscles related to masticatory function, such as the temporalis, the medial and lateral pterygoid muscles, the masseter, the digastric muscle, and other important muscles, including the buccinator muscle and the longissimus capitis of the head (Figures 2A, 3A, 4A, 5A, 6A and 7A).

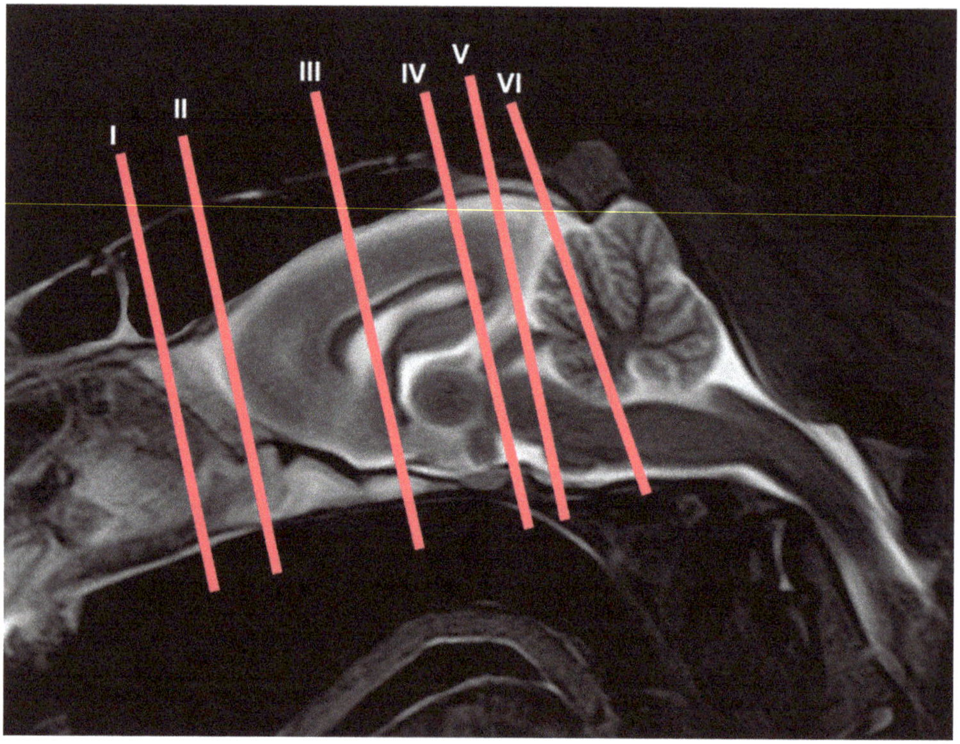

Figure 1. Sagittal T2W MR image of the head of a crested porcupine. The lines and numbers (I–VI) represent the approximate levels of the following transverse cross-sections and MR images.

3.2. Magnetic Resonance Imaging (MRI)

No significant anatomic differences were identified subjectively in the three porcupines which were imaged. Most anatomic structures distinguished on T2-weighted images of the cadaver specimens matched adequately with structures identified in the corresponding anatomical cross-sections. Hence, the central nervous system structures of the porcupine head, the eyeball's structures (vitreous humour and lens), and the masticatory muscles showed an accurate visualization using T2W MR images. Nonetheless, the bones that comprised the neurocranium, such as the frontal, the parietal, the temporal, the occipital, and the sphenoid bones, were identified with a hypointense signal (Figures 2B, 3B, 4B, 5B, 6B, 7B, 8 and 9).

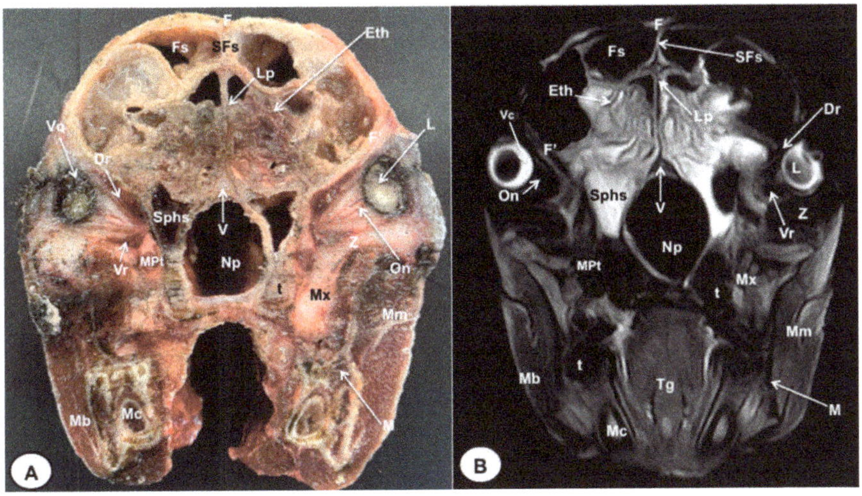

Figure 2. Transverse cross-section (**A**) and T2W MR (**B**) images of the crested porcupine head at the level of the ethmoturbinates corresponding to line I in Figure 1. F: frontal bone; F′: orbital plate of frontal bone; Fs: frontal sinuses; SFs: septum of frontal sinuses; Eth: ethmoturbinates; Lp: *Lamina perpendicularis ossis ethmoidei;* Sphs: sphenoid sinus; V: vomer; Np: nasopharynx; On: optic nerve; Vc: vitreous chamber; L: lens; Dr: *Musculus rectus dorsalis;* Vr: *Musculus rectus ventralis;* MPt: *Musculus pterygoideus medialis;* Z: zygomatic bone; Mx: maxillary bone; t: tooth; M: mandible; Mc: mandibular canal; Mm: *Musculus masseter;* Mb: *Musculus buccinator;* Tg: tongue.

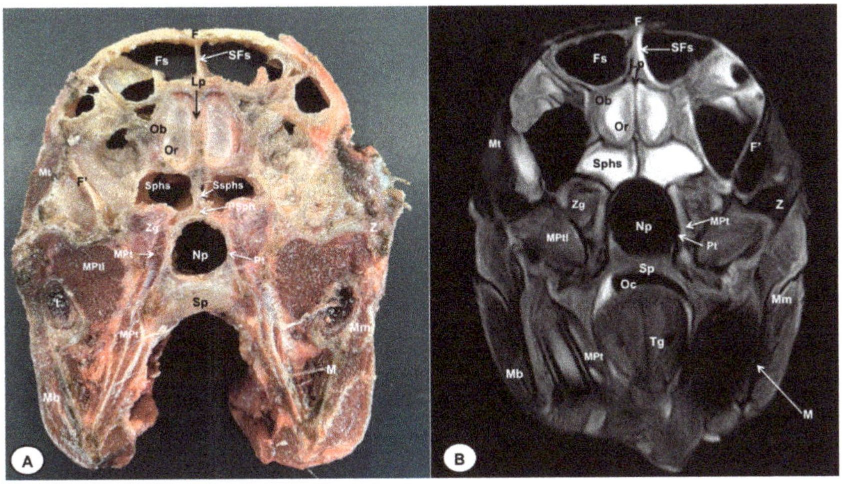

Figure 3. Transverse cross-section (**A**) and T2W MR (**B**) images of the crested porcupine head at the level of the olfactory bulb corresponding to line II in Figure 1. F: frontal bone; F′: orbital plate of frontal bone; Fs: frontal sinuses; SFs: septum of frontal sinuses; Ob: olfactory bulb; Or: olfactory recess; Lp: *Lamina perpendicularis ossis ethmoidei;* Sphs: sphenoid sinus; Ssphs: septum of sphenoidal sinuses; PSph: presphenoid bone; Np: nasopharynx; Pt: pterygoid bone; Sp: soft palate; Oc: oral cavity; MPt: *Musculus pterygoideus medialis;* MPtl: *Musculus pterygoideus lateralis;* Mt: *Musculus temporalis;* Zg: zygomatic glands; Z: zygomatic bone; t: tooth; M: mandible; Mm: *Musculus masseter;* Mb: *Musculus buccinator;* Tg: tongue.

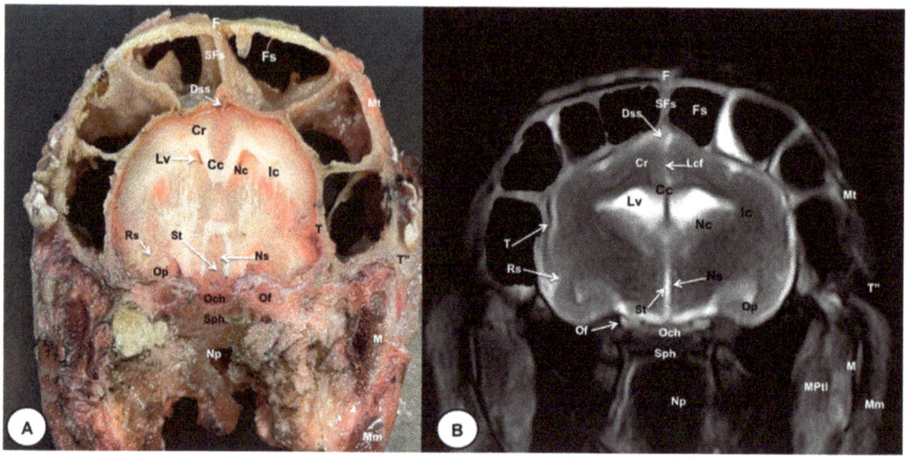

Figure 4. Transverse cross-section (**A**) and T2W MR (**B**) images of the crested porcupine head at the level of the optic chiasm, corresponding to line III in Figure 1. F: frontal bone; Fs: frontal sinuses; SFs: septum of frontal sinuses; Dss: dorsal sagittal sinus; Lcf: longitudinal cerebral fissure; Cc: corpus callosus; Cr: *Corona radiata*; Lv: lateral ventricle; Nc: *Nucleus caudatus;* Ic: internal capsule; Rs: rhinal sulcus; Op: olfactory peduncle; St: septum of telencephalon (*Septum telencephali*); Ns: *Nuclei septi*; Och: optic chiasm; Of: orbital fissure; T: temporal bone (squamous part); T″: zygomatic process of temporal bone; Sph: sphenoid bone; Np: nasopharynx; Mt: *Musculus temporalis*; MPtl: *Musculus pterygoideus lateralis*; M: mandible. Mm: *Musculus masseter*.

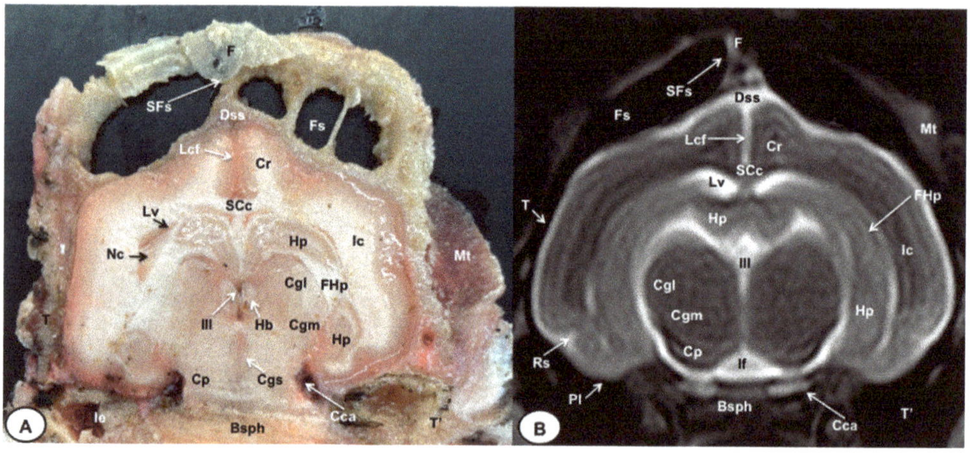

Figure 5. Transverse cross-section (**A**) and T2W MR (**B**) images of the crested porcupine head at the level of the caudal part of the diencephalon, corresponding to line IV in Figure 1. F: frontal bone; Fs: frontal sinuses; SFs: septum of frontal sinuses; Dss: dorsal sagittal sinus; Lcf: longitudinal cerebral fissure; SCc: splenium of corpus callosum; Cr: *Corona radiata*; Lv: lateral ventricle; Hp: *Hippocampus*; FHp. fimbria of hippocampus; Nc: *Nucleus caudatus* (tail); Ic: internal capsule; III: third ventricle; Hb: habenula; Cgl: *Corpus geniculatum laterale* (lateral geniculate body); Cgm: *Corpus geniculatum mediale* (medial geniculate body); Cgs: central grey substance; Cp: cerebral peduncle; If: interpeduncular fossa; Rs: rhinal sulcus; Pl: piriform lobe; T: temporal bone squama; T′: tympanic and petrous parts of temporal bone; Ie: inner ear. Cca: caudal communicating artery; Bsph: basisphenoid bone; Mt: *Musculus temporalis*.

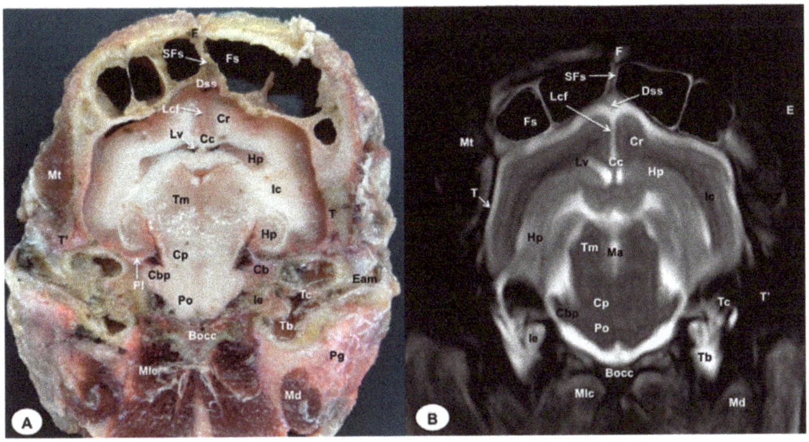

Figure 6. Transverse cross-section (**A**) and T2W MR (**B**) images of the crested porcupine head at the level of the caudal part of the hippocampus, corresponding to line V in Figure 1. F: frontal bone; Fs: frontal sinuses; SFs: septum of frontal sinuses; Dss: dorsal sagittal sinus; Lcf: longitudinal cerebral fissure; Cc: corpus callosus; Cr: *Corona radiata*; Lv: lateral ventricle; Hp: *Hippocampus*; Ic: internal capsule; Tm: tectum of mesencephalon; Ma: mesencephalic aqueduct; Po: Pons; Pl: piriform lobe; Cb: cerebellum; Cbp: cerebellar peduncle (lateral); Cp: cerebral peduncle; T: temporal bone squama; T′: tympanic and petrous parts of temporal bone; Bocc: basioccipital bone; Mt: *Musculus temporalis*; Mlc: *Musculus longissimus capitis*; Md: *Musculus digastricus*; Pg: parotid gland; Ie: inner ear (cochlea); Tc: tympanic cavity; Tb: tympanic bulla; E: ear (external part).

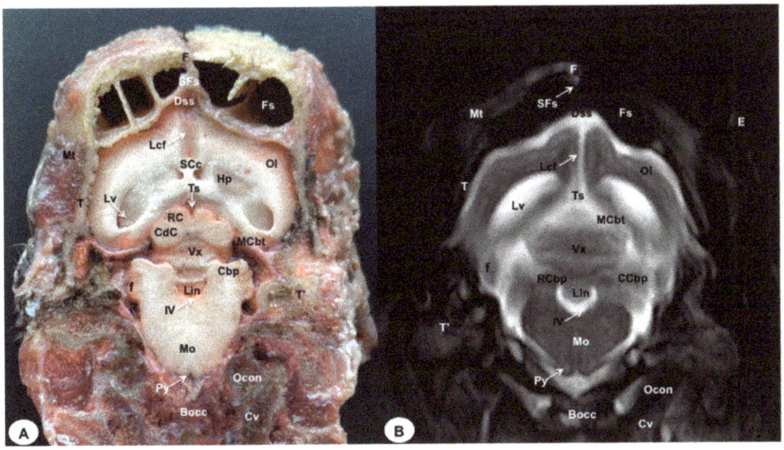

Figure 7. Transverse cross-section (**A**) and T2W MR (**B**) images of the crested porcupine head at the level of the cerebellum, corresponding to line VI in Figure 1. F: frontal bone; Fs: frontal sinuses; SFs: septum of frontal sinuses; T: temporal bone squama; Mt: *Musculus temporalis*; T′: tympanic and petrous parts of temporal bone; Dss: dorsal sagittal sinus; Lcf: longitudinal cerebral fissure; Lv: lateral ventricle; SCc: splenium of corpus callosum; Hp: hippocampus; Ts: transverse sinus; MCbt: membranous cerebellar tentorium; RC: rostral colliculus; CdC: caudal colliculus; Cbp: cerebellar peduncle; Vx: vermis of cerebellum; RCbp: rostral cerebellar peduncle; CCbp: caudal cerebellar peduncle; Lin: lingula; IV: fourth ventricle; Mo: *medulla oblongata*; Py: pyramids of the medulla oblongata; Cv: first cervical vertebra; Ocon: occipital condyle; Bocc: basioccipital bone; E: external ear.

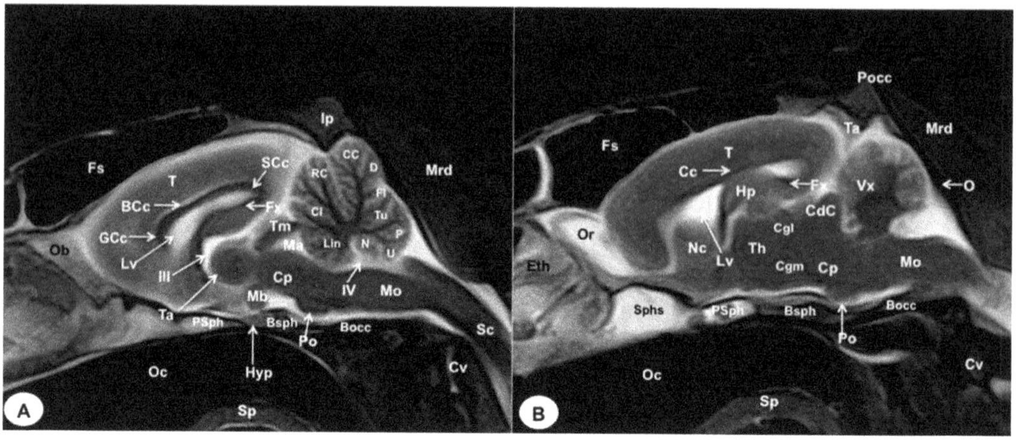

Figure 8. Sagittal (**A**) and parasagittal (**B**) T2W MR images of the crested porcupine head. Fs: frontal sinuses; Ip: interparietal bone; Pocc: *Protuberantia occipitalis externa;* Eth: ethmoturbinates; Ob: olfactory bulb; Or: olfactory recess; T: telencephalon; Lv: lateral ventricle; Cc: *corpus callosum;* GCc: genu of corpus callosum; BCc: body of corpus callosum; SCc: splenium of corpus callosum; Hp: *hippocampus;* Fx: fornix; Nc: *Nucleus caudatus;* III: third ventricle; Ta: interthalamic adhesion; Th: *thalamus;* Cgl: *Corpus geniculatum laterale* (lateral geniculate body); Cgm: *Corpus geniculatum mediale* (medial geniculate body); Mb: mamillary body; Hyp: *Hypophysis;* Ta: tentorial process; Tm: tectum of mesencephalon; Ma: mesencephalic aqueduct; CdC: caudal colliculus; Cp: cerebral peduncle; Vx: vermis of cerebellum; Lin: lingula; Cl: central lobe of cerebellum; RC: rostral culmen; CC: caudal culmen; D: declive; Fl: folium; Tu: tuber; P: pyramid; U: uvula; N: nodule; IV: fourth ventricle; Po: Pons; Mo: *Medulla oblongata;* Sc: spinal cord; Cv: cervical vertebra; Mrd: *Musculus rectus dorsalis;* O: occipital bone; Sphs: sphenoid sinus; PSph: presphenoid bone; Bsph: basisphenoid bone; Bocc: basioccipital bone; Oc: oral cavity; Sp: soft palate.

In the transverse planes of the encephalon, identifiable structures of the brain were more hyperintense than the white matter, which was more hypointense in T2W sequences (Figures 3B, 4B, 5B, 6B and 7B). Moreover, the two sagittal and dorsal (Figures 8 and 9) images and the different transverse T2W images were essential to depicting the components that comprise the ventricular system, which displayed a hyperintense signal. Hence, the lateral ventricles and the dorsal and ventral parts of the third ventricle were displayed (Figures 4B, 5B, 6B, 7B, 8 and 9), and among these, we observed interthalamic adhesion, limited laterally by the right and left sides of the thalamus (Figure 8A). In addition, transverse, sagittal, and parasagittal T2W images displayed with adequate detail the dorsal and ventral parts of the hippocampus (Figures 5B, 6B, 7B, 8B and 9). In addition, the tectum of mesencephalon (tectum mesencephali) with the caudal colliculus and the fourth ventricle was visualized in excellent detail (Figure 8A,B and Figure 9B). Other essential components of the CNS, such as the vermis of the cerebellum with its dorsal and ventral lobes, were distinguished in the sagittal T2W images (Figure 8A). Hence, the moderate contrast between grey and white matter was helpful in distinguishing the different lobes. Thus, the dorsal lobes of the cerebellum (the rostral and ventral culmen, the declive, the folium, the tuber, and the pyramid), as well as its ventral lobes (the lingula, the nodulus, and the uvula), were identified. In addition to these findings, we also observed the rostral and caudal cerebellar peduncles (Figures 7B and 9A). This technique also facilitated an adequate resolution to be achieved with which to identify the muscles involved in the masticatory function, which have already been mentioned in the anatomical sections.

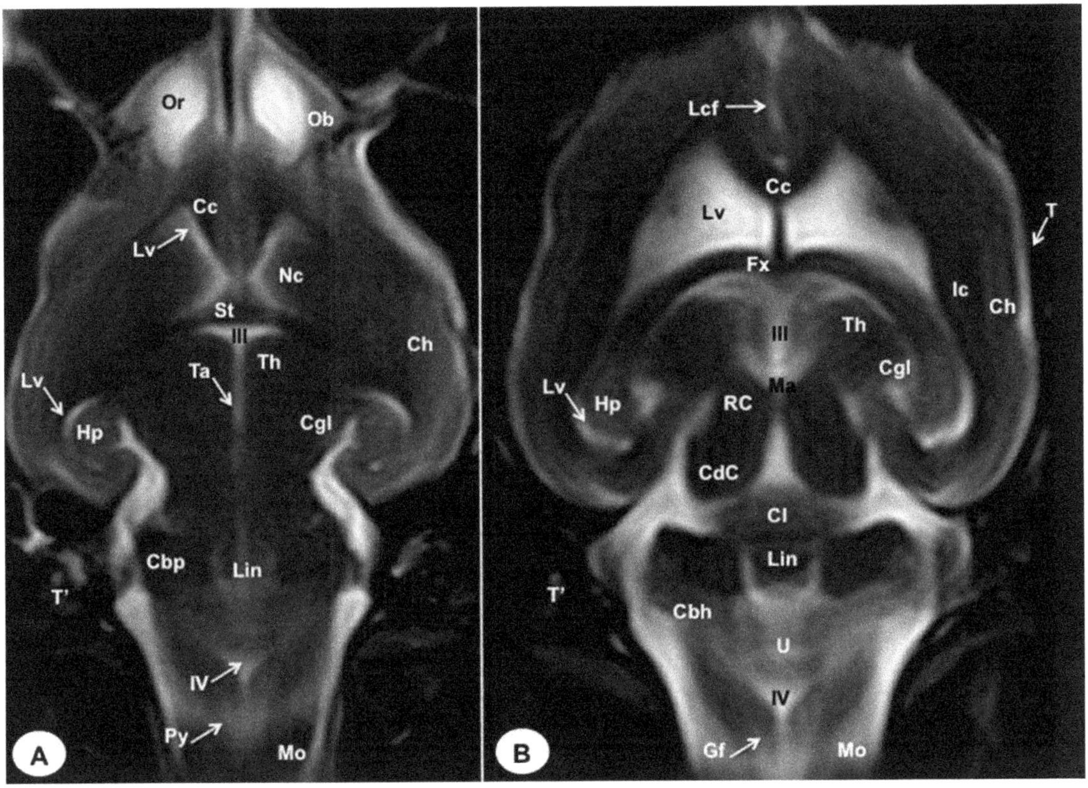

Figure 9. T2W dorsal MRI images of the crested porcupine head at the level of the olfactory bulb (**A**) and corpus callosum (**B**). Ob: olfactory bulb; Or: olfactory recess; Lcf: longitudinal cerebral fissure; Cc: corpus callosum; Lv: lateral ventricle; Nc: *Nucleus caudatus*; St: septum of telencephalon (*Septum telencephali*); Fx: *fornix*; III: third ventricle; Ic: internal capsule; Ch: cerebral hemisphere; Hp: *hippocampus*; Th. *thalamus*; Ta: interthalamic adhesion; Cgl: *corpus geniculatum laterale* (lateral geniculate body); Ma: mesencephalic aqueduct; RC: rostral colliculus; CdC: caudal colliculus; Cl: central lobe of cerebellum; Lin: *lingula*; Cbp: cerebellar peduncle; Cbh: cerebellar hemisphere; IV: fourth ventricle; Mo: *medulla oblongata*; Py: pyramids of the medulla oblongata; Gf: Gracile fasciculus; T: temporal bone squama; T′: tympanic and petrous parts of temporal bone.

4. Discussion

In recent years, modern diagnostic imaging techniques such as computed tomography and magnetic resonance imaging have become quite fashionable in captive and free-ranging animals to diagnose and treat diseases, since their availability for clinical use has dramatically increased. However, the costs and risk of complications due to general anesthesia [11,12] are limiting factors to the use of these techniques in exotic animals. In contrast with traditional imaging methods such as radiography and ultrasound, which are widespread among veterinarians [10], these procedures can provide images of the different structures in various planes without repositioning the animal [12,16,22,23]. CT provides more detailed anatomical information regarding the skull bones and dentition compared to MRI [16]. However, certain clinical conditions affecting rodents may involve secondary soft tissue structures that are better visualized using MRI. In the case of the CNS, magnetic resonance imaging is considered the gold standard imaging modality for humans and animals, as CT tends to depict the CNS structures as a homogeneous formation without a clear distinction between its different components [24]. Furthermore, MRI allows the identification of structures deep into the bone and air and is less limited by operator experience [3,11,12,14]. Therefore, these advanced imaging techniques enhance anatomic identification and lesion detection, allowing assessment, detailed prognosis, diagnosis of underlying lesions, and treatment choice to be carried out [11,12]. In the case of rodent species, these procedures have been used to evaluate a range of head processes not well visualized through conventional imaging techniques. Among these, we included dental disease and its associated problems, deformities, and osteomyelitis; the extension of the infection process to different bone cavities of the skull such as the nasal cavity, the paranasal sinuses, and the tympanic bullae; as well as fractures and different types of neoplasia [10–12]. However, it is important to consider their cost, the need for sedation, and the time required for data analysis before employing these methods. Despite these disadvantages, their application in the study of endangered species is justified due to the valuable information they can provide with minimal risk to the animal [3,22,23]. In this study, we successfully yielded high-resolution images of the central nervous system and its related structures using state-of-the-art diagnostic imaging techniques such as MRI. These images exhibited exceptional clarity and complemented the anatomical cross-section images. To the best of the authors' knowledge, this represents the first comprehensive description of the CNS and associated structures of the crested porcupine utilizing contemporary diagnostic imaging techniques, such as MRI, and their correlation with the anatomical cross-section images. As a result, our investigation has provided valuable insights into the intricate anatomic details of the crested porcupine's head, which hold potential diagnostic, experimental, and educational significance.

In our study, the utilization of anatomical cross-sections has proven to be immensely valuable in accurately characterizing the morphologic features of the CNS in the crested porcupine, as well as the structures related to the eyeball and the auditory system. These cross-sectional views played a crucial role in visualizing significant components of the diencephalon, brainstem, and cerebellum, including the habenula, the rostral and caudal colliculi, and the cerebellar peduncles. However, due to the wide interval between slices employed in this study, some of these structures were not clearly discernible in the transverse MR views. Comparable investigations conducted in other exotic species, such as rabbits, guinea pigs, iguanas, and loggerhead turtles, have consistently demonstrated that this combination is essential to comparing the relative positions and sizes of cephalic or other anatomical structures [3,14,25–27].

The MR images obtained in the transverse and sagittal planes were acquired without repositioning the head. The high-field magnet facilitated an adequate evaluation of the crested porcupine's head. Thus, the different planes utilized herein made structure identification more evident. Additionally, subjective image analysis and objective measurements revealed a relatively large dorsal *metencephalon* and a smaller *telencephalon* compared to dogs of a similar size and weight. One report described similar findings in size when they studied the rabbit brain with a high field-strength magnet [15]. Nonetheless, further studies are needed to confirm these findings. Moreover, the present study demonstrated that MRI was effective in visualizing the bones of the porcupine's head. Nonetheless, it is important to note, as described in other reports, that the skull bones and teeth could be identified due to their hypointense signal [24,27]. In contrast, excellent discrimination of the main components of the *encephalon*, including the *rhinencephalon*, *telencephalon*, *diencephalon*, *mesencephalon*, *metencephalon*, and *myelencephalon*, was achieved with the T2W images. These images have proven to be useful for anatomical and clinical studies of various exotic species [3,12,22–31]. Therefore, the transverse T2W images revealed the olfactory bulb and its recess, which was hyperattenuated compared to the *telencephalon*. Similar enhancement was observed in rabbits, particularly in the sagittal plane [14,15]. Additionally, the sagittal T2W images were essential for visualizing the different lobes of the cerebellar vermis, which displayed a hyperintense/isointense signal when compared with the hypointense cerebellar white matter. Our results confirmed those obtained in rodents and rabbits, which have demonstrated that MRI is an essential technique for evaluating the CNS and its associated soft tissue structures [2,11,12,14–17,27,31,32]. However, other structures, such as the nasal cavity and paranasal sinuses, were not evaluated in our study because we were focused on those structures better visualized by MRI.

As mentioned in previous studies [25], the use of cadavers in our research limited the administration of an intravenous contrast medium, which could have improved the resolution of MR images. This limitation, as well as the small number of specimens, should be considered in further studies. Nevertheless, both MRI and macroscopic anatomical sections provided adequate information for anatomic evaluation in teaching and clinical settings. In contrast to other studies conducted on rodents and rabbits [15,30–32], we employed anatomical sections and conducted a comprehensive anatomical description, including the location and intensity of the different structures comprising the central nervous system of the crested porcupine. The sagittal images, compared with those displayed in the transverse plane, facilitated a better assessment of the topographic anatomical structures in the median plane, primarily involving the intracranial cavity and the central nervous system. Similar findings have been reported in other MRI studies performed on exotic species [3,25]. All the images obtained in this study could serve as initial reference materials to support pathological investigations of crested porcupine heads.

5. Conclusions

This investigation is the first description of a crested porcupine's head using transverse, sagittal, and dorsal MR images in combination with anatomical cross-sections. The images obtained in this study were quite helpful in providing essential references for the different bone and soft tissue structures comprising the CNS and sensory organs of the crested porcupine. Therefore, the information obtained in this study could be adequate for the anatomic and clinical evaluation of numerous pathologic processes involving the heads of these animals, such as abscesses, metabolic bone diseases, fractures, inflammation, and neoplasia. Moreover, the MR images obtained in different spatial planes could facilitate our understanding of anatomic organization for our students, since these procedures allow for the visualization of structures without overlapping, eliminating the difficulties of visualizing specific anatomic structures. Nevertheless, the high cost and accessibility of this equipment do not facilitate its use on porcupines or rodents in daily veterinary practice.

Author Contributions: Conceptualization, J.R.J., A.A. and M.E.; methodology, J.R.J., D.M.-B. and M.E.; formal analysis, J.R.J., D.M.-B. and M.E.; investigation, J.R.J., A.A., D.M.-B. and M.E.; resources, D.M.-B.; writing—original draft preparation, J.R.J. and D.M.-B.; writing—review and editing, J.R.J., D.M.-B. and M.E. All authors have read and agreed to the published version of the manuscript.

Funding: This research received no external fundings.

Institutional Review Board Statement: As happens in domestic mammals, informed consent from the owner allowed us to carry out this study. Therefore, Rancho Texas Lanzarote Park was informed that all animal identity information obtained from this study would be treated as confidential to the extent permitted by law, and only used for research or teaching purposes.

Informed Consent Statement: Informed consent was obtained from all subjects involved in the study.

Data Availability Statement: The information is available at "https://accedacris.ulpgc.es".

Acknowledgments: In loving memory of Alvaro Domingo and Honorio Rodriguez Garcia. We also thank Ayesh Mohamad, Carmen Mingot, Emilia Mingot, Concha Mingot, Nicolas Aquino, Marisa Mohamad, and Jamal Jaber for their support and constructive comments, and Rancho Texas Lanzarote Park for providing the animals included in this study.

Conflicts of Interest: The authors declare no conflict of interest.

References

1. Banzato, T.; Russo, E.; Di Toma, A.; Palmisano, G.; Zotti, A. Anatomic imaging of the Boa constrictor head: A comparison between radiography, computed tomography and cadaver anatomy. *Am. J. Vet. Res.* **2011**, *72*, 1592–1599. [CrossRef] [PubMed]
2. Van Caelenberg, A.I.; De Rycke, L.M.; Hermans, K.; Verhaert, L.; Van Bree, H.J.; Gielen, I.M. Comparison of radiography and CT to identify changes in the skulls of four rabbits with dental disease. *J. Vet. Dent.* **2011**, *28*, 172–181. [CrossRef] [PubMed]
3. González Rodríguez, E.; Encinoso Quintana, M.; Morales Bordon, D.; Garcés, J.G.; Artiles Nuez, H.; Jaber, J.R. Anatomical Description of Rhinoceros Iguana (*Cyclura cornuta cornuta*) Head by Computed Tomography, Magnetic Resonance Imaging and Gross-Sections. *Animals* **2023**, *13*, 955. [CrossRef] [PubMed]
4. Arencibia, A.; Corbera, J.A.; Ramírez, G.; Díaz-Bertrana, M.L.; Pitti, L.; Morales, M.; Jaber, J.R. Anatomical Assessment of the Thorax in the Neonatal Foal Using Computed Tomography Angiography, Sectional Anatomy, and Gross Dissections. *Animals* **2020**, *10*, 1045. [CrossRef]
5. Zafra, R.; Carrascosa, C.; Rivero, M.; Peña, S.; Fernández, T.; Suarez-Bonnet, A.; Jaber, J.R. Analysis of equine cervical spine using three-dimensional computed tomographic reconstruction. *J. Appl. Anim. Res.* **2012**, *40*, 108–111. [CrossRef]
6. Amori, G.; De Smet, K. *Hystrix cristata*. In *IUCN Red List of Threatened Species*. 2016, p. e.T10746A22232484. Available online: https://www.iucnredlist.org/especies/10746/22232484 (accessed on 12 June 2023).
7. Wilson, D.E.; Reeder, D.M. (Eds.) Species Hystrix (Hystrix) cristata. In *Mammal Species of the World: A Taxonomic and Geographic Reference*, 3rd ed.; Johns Hopkins University Press: Baltimore, MA, USA, 2005; p. 1543.
8. Angelici, F.M.; Colangelo, P.; Gippoliti, S. Out of Europe: Investigating *Hystrix cristata* (Rodentia: Hystricidae) skull morphometric geographic variability in Africa. *Biogeographia* **2021**, *36*, a001. [CrossRef]
9. Storch, G. Porcupines. In *Grzimek1s Encyclopedia of Mammals*; Grzimek, B., Ed.; McGraw-Hill: New York, NY, USA, 1990; pp 300–307.
10. Capello, V.; Cauduro, A. Comparison of Diagnostic Consistency and Diagnostic Accuracy Between Survey Radiography and Computed Tomography of the Skull in 30 Rabbits with Dental Disease. *J. Exot. Pet Med.* **2016**, *25*, 115–127. [CrossRef]
11. Capello, V. Diagnostic Imaging of Dental Disease in Pet Rabbits and Rodents. *Veter Clin. N. Am. Exot. Anim. Pract.* **2016**, *19*, 757–782. [CrossRef]
12. Głodek, J.; Adamiak, Z.; Przeworski, A. Magnetic Resonance Imaging of Reptiles, Rodents, and Lagomorphs for Clinical Diagnosis and Animal Research. *Comp. Med.* **2016**, *66*, 216–219.
13. Fornazari, F.; Guimaraes, F.F.; Teixeira, C.R.; Langoni, H. Isolation of Staphylococcus epidermidis from inflamed upper respiratory tract of an orange-spined hairy dwarf porcupine (*Sphiggurus villosus*). *J. Venom. Anim. Toxins Incl. Trop. Dis.* **2012**, *18*, 455–458. [CrossRef]
14. Van Caelenberg, A.I.; De Rycke, L.M.; Hermans, K.; Verhaert, L.; van Bree, H.J.; Gielen, I.M. Low-field magnetic resonance imaging and cross-sectional anatomy of the rabbit head. *Vet. J.* **2011**, *188*, 83–91. [CrossRef]
15. Müllhaupt, D.; Augsburger, H.; Schwarz, A.; Fischer, G.; Kircher, P.; Hatt, J.M.; Ohlerth, S. Magnetic resonance imaging anatomy of the rabbit brain at 3 T. *Acta Vet. Scand.* **2015**, *57*, 47. [CrossRef]
16. Capello, V.; Lennox, A. Advanced diagnostic imaging and surgical treatment of an odontogenic retromasseteric abscess in a guinea pig. *J. Small Anim. Pract.* **2015**, *56*, 134–137. [CrossRef] [PubMed]
17. Muñoz-Moreno, E.; Arbat-Plana, A.; Batalle, D.; Soria, G.; Illa, M.; Prats-Galino, A.; Eixarch, E.; Gratacos, E. A Magnetic Resonance Image Based Atlas of the Rabbit Brain for Automatic Parcellation. *PLoS ONE* **2013**, *8*, e67418. [CrossRef]
18. Shek, J.W.; Wen, G.Y.; Wisniewski, H.M. *Atlas of the Rabbit Brain and Spinal Cord*; Karger: Basel, Switzerland, 1986.
19. Popesko, P.; Rajtova, V.; Horak, J. *Anatomy of Small Laboratory Animals*; Wolfe Publishing Ltd.: London, UK, 1990; pp. 14–53.
20. Gray-Edwards, H.L.; Salibi, N.; Josephson, E.M.; Hudson, J.A.; Cox, N.R.; Randle, A.N.; McCurdy, V.J.; Bradbury, A.M.; Wilson, D.U.; Beyers, R.J.; et al. High resolution MRI anatomy of the cat brain at 3Tesla. *J. Neurosci. Methods* **2014**, *227*, 10–17. [CrossRef] [PubMed]
21. International Committee on Veterinary Gross Anatomical Nomenclature. *Nomina Anatomica Veterinaria*; Ithaca: New York, NY, USA, 2012.
22. Arencibia, A.; Encinoso, M.; Jaber, J.R.; Morales, D.; Blanco, D.; Artiles, A.; Vázquez, J.M. Magnetic resonance imaging study in a normal Bengal tiger (*Panthera tigris*) stifle joint. *BMC Vet. Res.* **2015**, *11*, 192. [CrossRef]
23. Jaber, J.R.; Encinoso, M.; Morales, D.; Artiles, A.; Santana, M.; Blanco, D.; Arencibia, A. Anatomic study of the normal Bengal tiger (*Panthera tigris tigris*) brain and associated structures using low field magnetic resonance imaging. *Eur. J. Anat.* **2016**, *20*, 195–203.
24. Ben Khalifa, A.; Belkhiria, J.; Hamdi, H.; Chandoul, W.; Mattoussi, A. Computed tomography and magnetic resonance imaging of the brain and associated structures of the one humped camel (*Camelus dromedarius*): A comparative study. *J. New Sci.* **2019**, *68*, 4221–4231.
25. Arencibia, A.; Hidalgo, M.; Vázquez, J.M.; Contreras, S.; Ramírez, G.; Oros, J. Sectional anatomic and magnetic resonante imaging features of the head of juvenile loggerhead sea turtles (*Caretta careta*). *Am. J. Vet. Res.* **2012**, *73*, 1119–1127. [CrossRef]
26. Valente, A.L.; Cuenca, R.; Zamora, M.A.; Parga, M.L.; Lavin, S.; Alegre, F.; Marco, I. Sectional anatomic and magnetic resonance imaging of coelomic structures of loggerhead sea turtles. *Am. J. Vet. Res.* **2006**, *67*, 1347–1353. [CrossRef]
27. Mahdy, M.A.A. Correlation between computed tomography, magnetic resonance imaging and cross-sectional anatomy of the head of the guinea pig (*Cavia porcellus*, Linnaeus 1758). *Anat. Histol. Embryol.* **2022**, *51*, 51–61. [CrossRef] [PubMed]

. Arencibia, A.; Matos, J.; Encinoso, M.; Gil, F.; Artiles, A.; Martínez-Gomariz, F.; Vázquez, J.M. Computed tomography and magnetic resonance imaging study of a normal tarsal joint in a Bengal tiger (*Panthera tigris*). *BMC Vet. Res.* **2019**, *15*, 126. [CrossRef] [PubMed]
. Snow, T.M.; Litster, A.L.; Gregory, R.J. Big cat scan: Magnetic resonance imaging of the tiger. *Australas. Radiol.* **2004**, *48*, 93–95. [CrossRef]
. Encinoso, M.; Oros, J.; Ramírez, G.; Jaber, J.R.; Artiles, A.; Arencibia, A. Anatomic Study of the Elbow Joint in a Bengal Tiger (*Panthera tigris* tigris) Using Magnetic Resonance Imaging andGross Dissections. *Animals* **2019**, *9*, 1058. [CrossRef]
. Encinoso, M.; Morales, D.; Déniz, S.; Guerra, J.V.; Jaber, J.R. Computed Tomography and Magnetic Resonance Imaging of a Rhinosinusitis Secondary to a Dental Abscess in a Crested Porcupine (*Hystrix cristata*). *Slov. Vet. Res.* **2023**, *60*, 37–43. [CrossRef]
. Chuang, N.; Mori, S.; Yamamoto, A.; Jiang, H.; Ye, X.; Xu, X.; Richards, L.J.; Nathans, J.; Miller, M.I.; Toga, A.W.; et al. An MRI-based atlas and database of the developing mouse brain. *Neuroimage* **2011**, *54*, 80–89. [CrossRef] [PubMed]

animals

MDPI

Article

A Cadaveric Study Using Computed Tomography for Measuring the Ocular Bulb and Scleral Skeleton of the Atlantic Puffin (Aves, *Alcidae, Fratercula arctica*)

Marcos Fumero-Hernández [1], Mario Encinoso [1,*], Ana Sofia Ramírez [2], Inmaculada Morales [1], Alejandro Suárez Pérez [2] and José Raduan Jaber [3,*]

[1] Veterinary Hospital, Faculty of Veterinary Medicine, University of Las Palmas de Gran Canaria, Trasmontaña, 35413 Arucas, Las Palmas, Spain; marcos.fumero101@alu.ulpgc.es (M.F.-H.); inmaculada.morales@ulpgc.es (I.M.)

[2] Department of Pathology and Food Technology, Faculty of Veterinary Medicine, Universidad de Las Palmas de Gran Canaria, Trasmontaña, 35413 Arucas, Las Palmas, Spain; anasofia.ramirez@ulpgc.es (A.S.R.); alejandro.suarezperez@ulpgc.es (A.S.P.)

[3] Department of Morphology, Faculty of Veterinary Medicine, University of Las Palmas de Gran Canaria, Trasmontaña, 35413 Arucas, Las Palmas, Spain

* Correspondence: mencinoso@gmail.com (M.E.); joseraduan.jaber@ulpgc.es (J.R.J.)

Simple Summary: Advanced diagnostic imaging techniques, such as CT, can provide helpful information on the specific structures of the head, such as the ocular bulb, due to their high spatial resolution, avoidance of overlapping structures, and fast imaging acquisition. An adequate knowledge of the eye bird anatomy is essential for clinicians, biologists, and researchers to understand many aspects concerning its biology.

Abstract: Imaging diagnosis plays a fundamental role in avian medicine. However, there are few publications regarding its use in ophthalmology. Seabirds, in particular, present a peculiar ecology since their lives take place in very diverse environments: the aquatic, the terrestrial, and the aerial. This fact implies a series of adaptations at a visual level that are necessary for adequate interaction with the environment. Therefore, knowledge of eye particularities is of great importance for the scientific community since it allows us to deepen our understanding of the ocular anatomy and biology of these animals, which are increasingly present in veterinary and wildlife centers. In our study, we performed a morphometric analysis of the ocular bulb and its internal structures in the puffin (*Fratercula arctica*) using advanced imaging techniques such as CT.

Keywords: computed tomography; imaging; ocular morphometry; orbit; sclerotic ring; birds

1. Introduction

The Atlantic puffin (*Fratercula arctica*) is a member of the Alcidae family and one of three species in the Fratercula genus, alongside the horned puffin (*Fratercula corniculata*) and the tufted puffin (*Fratercula cirrhata*), which are typically found in the North Pacific [1]. These medium-sized alcids have a wingspan of approximately 50 cm and can weigh up to half a kilogram. They possess dense black plumage on their head, neck, and dorsum, with white patches on their chest, ventrum, and around the eyes. The puffin's bill is large and exhibits orange tones, which fade to less intense colors after the breeding season. Notably, there is no distinct sexual dimorphism, although males are generally slightly larger than females and their colors become more conspicuous during the mating season [2–6].

The Atlantic puffin has a wide distribution, inhabiting the entire northern Atlantic Ocean region, from northwestern Greenland, Newfoundland, and Maine (USA) in the west to northwestern Russia and the Canary Islands (Spain) in the east [1,7]. These birds spend the majority of their lives in the ocean, returning to land solely for reproduction.

They primarily nest on isolated islands and cliff sides, ranging from Brittany and the Bay of Fundy (Canada) to the Arctic sea ice on both sides of the Atlantic. Puffins return to their colonies in March to nest during April and May and then leave in July to return to the open sea [5,8,9]. They are monogamous birds, typically selecting the same partner and nesting location in successive breeding periods. Upon returning to the breeding area, males primarily engage in nest construction, which involves shallow burrows, rocky ledges, and crevices, sometimes, incorporating plant material [5,10,11]. Females, after laying a single egg per season, dedicate themselves to incubating it and subsequently feeding the chick [12–14]. Chicks are primarily fed with fish, particularly herring (*Clupea harengus*; 12–85%), hake (*Urophycis* spp., *Merluccius* spp.; 12–87%), or sand eel (*Ammodytes* spp.; 0–71%). Adult puffins are also piscivorous, with varying amounts of polychaetes and crustaceans in their diet [5,14,15].

The Atlantic puffin is currently listed as a threatened species on the International Union for Conservation of Nature (IUCN) Red List, categorized as vulnerable, due to a significant population decline over most of its range [1]. Climate change, linked to ocean temperature variations that impact plankton blooms and disrupt the food chain, is one of the main factors contributing to their vulnerability [16,17]. Puffins are also sensitive to extreme weather events, such as marine storms, which have been increasing in frequency and intensity and are associated with mass mortality [18,19]. Human activities, including hunting for consumption and oil spills or other pollutants, also negatively impact puffin populations [20,21]. Therefore, the conservation of puffins and other seabirds is a globally significant task. Efforts aimed at their conservation require a deeper understanding of their anatomy and biology as they relate to their behavior and the threats they face.

Numerous studies have emphasized the importance of vision in the interaction of these animals with their environment and their overall survival [22–24]. Bird vision ranks among the sharpest in the animal kingdom. Avian eyes possess unique anatomical features, such as the pecten and sclerotic ring, found only in birds and certain reptiles. These adaptations enable birds to effectively interact and navigate across diverse environments, encompassing aquatic, terrestrial, and aerial realms [25–27]. Consequently, comprehending the ocular structures and functionality of the Atlantic puffin holds significant importance for veterinarians, biologists, and the wider scientific community.

Studying internal animal anatomy can be highly intricate. While dissection has traditionally been the conventional approach, contemporary diagnostic imaging techniques such as magnetic resonance imaging (MRI) and computed tomography (CT) offer minimally invasive and efficient means to obtain precise anatomical information [28]. Although primarily employed in human medicine, smaller versions of these imaging modalities, such as micro-computed tomography, have been developed to minimize image distortion, proving helpful for investigating the anatomy of small animals [29–31].

In the case of puffins, the existing literature predominantly focuses on various aspects of their biology and ecology [9,14,21,32] or approaches ophthalmology from a clinical standpoint [33]. However, conducting anatomical and morphobiometric analyses of the puffin's eye and associated structures can yield valuable insights into animal behavior patterns. This has been demonstrated in studies exploring the fossils of extinct marine reptiles belonging to the *Ichthyosaur* order [34] and non-passerine birds [35]. Similar investigations have also been conducted on live animals, encompassing both domestic and wild species [36–38].

Therefore, the objective of this study was to employ non-invasive examinations, such as computed tomography, to measure the size and characteristics of the puffin's eye and its associated structures.

2. Materials and Methods

2.1. Animals

In order to conduct this study, a total of 29 Atlantic puffin carcasses were obtained by the Consejeria de Área de Medio Ambiente, Clima, Energía y Conocimiento of the

Cabildo de Gran Canaria (Gran Canaria, Canary Islands, Spain). The weight of the animals ranged from 0.185 kg to 0.251 kg, with a mean weight of 0.251 kg. Since the physical examination alone did not provide sufficient information to determine the precise cause of the stranded individuals, modern non-invasive imaging techniques were employed to assess their appendicular skeleton, identify potential metabolic bone diseases, and exclude the presence of foreign bodies and internal organ injuries. Prior consent was obtained from the responsible person at the Cabildo de Gran Canaria to include the puffins in this study. The scanning procedures were carried out at the Veterinary Hospital of Las Palmas de Gran Canaria University.

2.2. CT Technique

CT examinations of the skull were conducted on the thawed carcasses of these specimens after a 12-h defrosting period at room temperature. Sequential slices were acquired using a 16-slice helical CT scanner (Toshiba Astelion, Canon Medical System®, Tokyo, Japan). The animals were positioned symmetrically in a prone position on the stretcher, employing a craniocaudal entry. A standard clinical protocol was employed, utilizing parameters of 120 kVp, 80 mA, a 512 × 512 acquisition matrix, an 1809 × 858 field of view, a pitch of 0.94, and a gantry rotation of 1.5 s. The resulting images had a thickness of 0.6 mm. CT scans were obtained in the dorsal, transverse, and sagittal planes with both bone and soft tissue windows. Subsequently, all the acquired images were imported into an image viewer (OsiriX MD, Apple, Cupertino, CA, USA) to facilitate data manipulation and measurements of the puffin ocular bulb and its associated structures.

2.3. Measurements

We conducted measurements on the head length, width, and orbit depth. Additionally, measurements of both eyes ($n = 58$) were analyzed by two observers using oblique sagittal, transverse, and dorsal CT images obtained from all the skulls. A soft tissue attenuation window was utilized for image analysis. The measurement methodology followed a protocol previously described in studies on the loggerhead turtle [39], as well as in dogs and cats [36–38], with some modifications based on studies conducted on non-passerine birds [35]. The parameters and specific measurements taken are described as follows:

(A) Measurements in the transverse plane relative to the ocular bulb:

- Lens diameter: This parameter refers to the maximum distance between the lateral and medial edges of the lens, also known as the equatorial diameter (Figure 1A);
- Internal diameter of the sclerotic ring, which represents the maximum distance between the inner lateromedial edges of the ring close to the cornea (Figure 1A);
- External diameter of the sclerotic ring, which corresponds to the maximum distance between the outer lateromedial edges of the ring close to the sclera (Figure 1A);
- Thickness of the sclerotic ring, defined as the distance between the internal diameter of the sclerotic ring and the external diameter, measured in the dorsal arch (Figure 2).

(B) Transverse plane concerning the puffin's body

Height of the ocular bulb, which corresponds to the distance between the dorsal part of the *os frontale* and the *os quadratojugale*, indicating the vertical dimension of the ocular bulb (Figure 5B).

- Attenuation of the sclerotic ring and lens: This measurement was taken in the dorsal area and expressed in Hounsfield Units, providing information about the radiodensity of the sclerotic ring, lens, and vitreous humor (Figure 3).

(C) Dorsal plane

- Width of the ocular bulb: This measurement represents the lateromedial length of the ocular bulb, extending from the *os lacrimale* to the inner face of the *os frontale* (Figure 5A);
- Length of the ocular bulb: This measurement refers to the distance between the most rostral and caudal portions of the ocular bulb, spanning from the *septum interorbitale* to the *processus antorbitalis* (Figure 5A).

(D) Three-dimensional reconstruction

- Head length: This measurement is obtained from the centre point of the junction between the beak joint to the *prominentia cerebellaris*, providing the length of the head (Figure 4A);
- Depth of the orbit: This measurement represents the distance between the midpoint of the orbital diameter and the *foramen opticum*, indicating the depth of the orbit (Figure 4B).

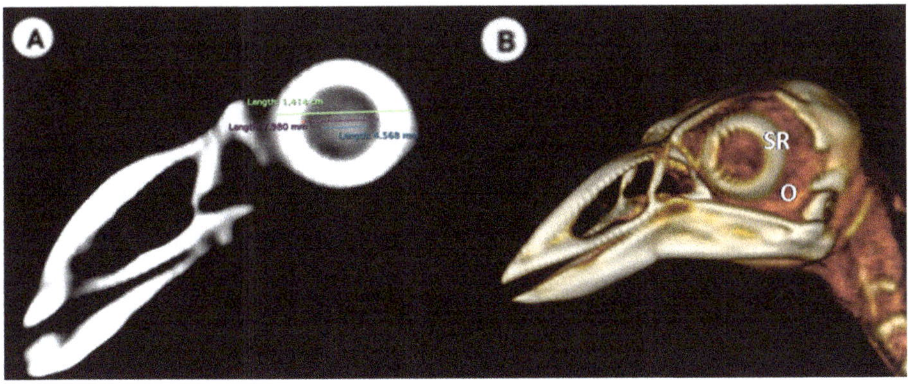

Figure 1. (**A**) Transverse multiplanar reconstruction (MPR) related to the ocular bulb of *F. Arctica* with measurements of the lens and the internal and external diameters of the sclerotic ring. (**B**) Volume rendering image of the *F. Arctica* skull with the sclerotic ring (SR) and orbit (O).

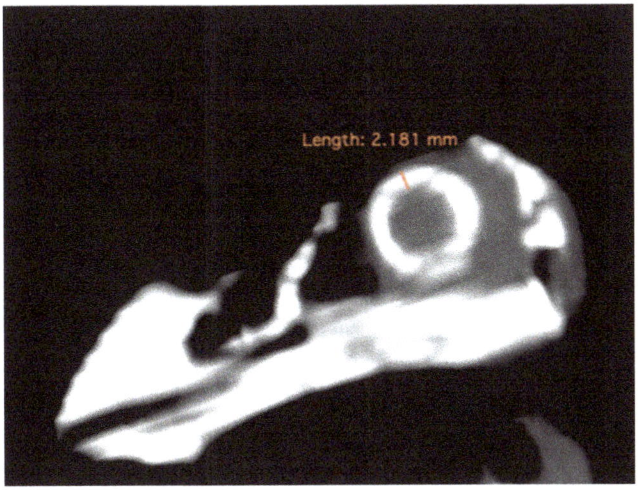

Figure 2. MPR related to the ocular bulb of *F. Arctica* with a measure of the sclerotic ring thickness.

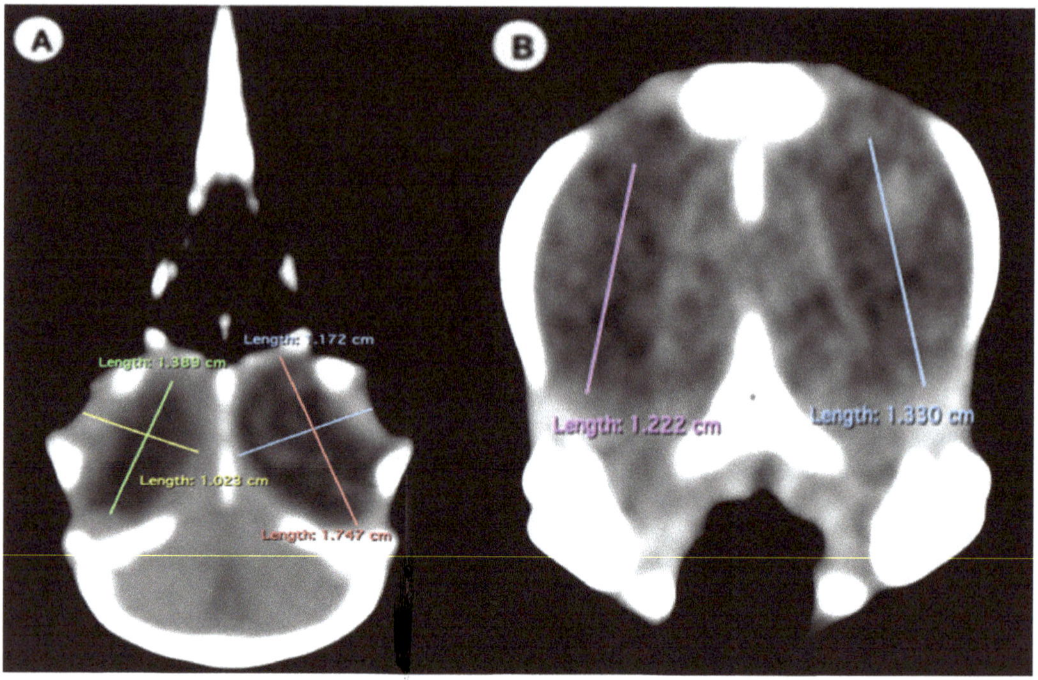

Figure 3. Transverse CT image in the soft tissue window of the *F. Arctica* head. Regions of interest (ROI) indicate the areas used to measure mean attenuation in Hounsfield Units (HU) of the lens (yellow circle), sclerotic ring (purple circle), and vitreous humor (green circle).

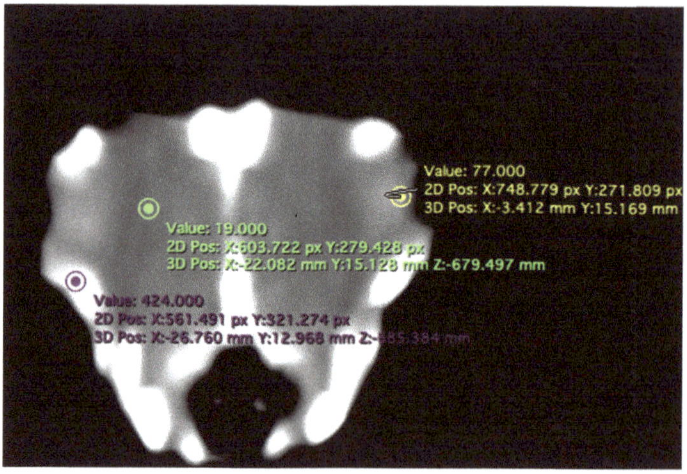

Figure 4. Volume rendering image of the *F. Arctica* skull with measurements of head length (**A**) and orbit depth (**B**). Image (**B**) has been modified by removing the cranial vault in order to measure the depth of the orbital basin.

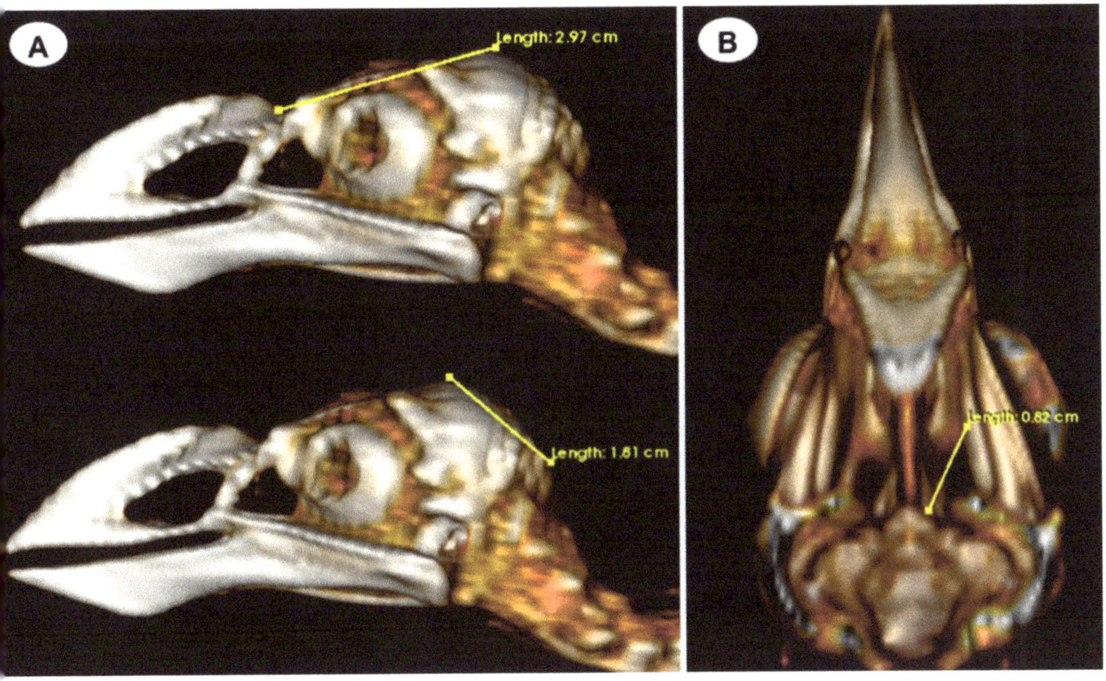

Figure 5. Dorsal multiplanar reconstruction (MPR) and transverse CT images in soft tissue windows of the head with measurements of the length, width (**A**), and height (**B**) of the ocular bulb of the *F. Arctica* head.

2.4. Statistical Analysis

The statistical analysis was performed using commercially available software (SPSS 19, Statistical Package for the Social Sciences, Chicago, IL, USA). Descriptive statistics, including the mean, median, range, and standard deviation (SD), were calculated for each measurement. The Shapiro-Wilk test was used to assess the normal distribution of quantitative data. The Mann-Whitney U test was employed to compare measurements between the right and left eyes. Furthermore, Spearman correlation was utilized to analyze the relationship between different ocular bulb variables and head length, as well as other quantitative variables. The statistical significance level was set at $p < 0.05$.

3. Results

In the computed tomography images, the sclerotic rings are visible from the anterior view of the respective eyes as circular structures with continuous morphology, distinct from the surrounding elements of the skull. The ellipsoid-shaped ocular bulbs exhibit well-defined soft tissue margins that likely correspond to the sclera. Various components of the eye, including the lens, vitreous and aqueous humors, and the anterior and posterior chambers, are also well distinguished in the images (Figures 1–5). From these images, we acquired the different measurements corresponding to both eyes (Table 1).

Table 1 presents the measurements of internal structures for the left, right, and both eyes of all 29 puffins included in the study. The mean, median, and standard deviation of the head length were 5.10 cm, 5.07 cm, and 0.13 cm, respectively, with a range of 4.8 to 5.32 cm. The average lens diameter for all eyes was 3.32 mm, with a range of 2.7 to 4.2 mm. The average internal diameter of the sclerotic ring was 0.63 cm, ranging from 0.59 to 0.7 cm. Additionally, the average external diameter of the sclerotic ring was 1.44 cm, ranging from 1.38 to 1.49 cm. The sclerotic ring thickness had a mean value of 2.16 mm, with a range of

2.04 to 2.29 mm. Regarding the ocular bulbs, they had an average height of 1.53 cm (range: 1.41–1.61 cm), an average width of 1.28 cm (range: 1.1–1.42 cm), and an average length of 1.61 cm (range: 1.46–1.71 cm). Additionally, the orbit depth had a mean of 8.19 mm (range: 7.9–8.5). Some variables exhibited a non-normal distribution. The Mann–Whitney U test indicated no significant differences between the measurements of the right and left eyes. Likewise, the same analysis demonstrated no statistically significant differences between the right and left eyes when considering all variables collectively.

Table 1. Measurements of the right and left eyes of the Atlantic puffin.

	Right Eye				Left Eye				Both Eyes			
	Mean	Median	Range	SD	Mean	Median	Range	SD	Mean	Median	Range	SD
Lens diameter (mm)	3.34	3.3	2.7–4.2	0.36	3.30	3.2	2.8–3.9	0.29	3.32	3.3	2.7–4.2	0.32
Internal diameter of the sclerotic ring (cm)	0.64	0.63	0.59–0.69	0.03	0.63	0.63	0.59–0.7	0.03	0.63	0.63	0.59–0.7	0.03
External diameter of the sclerotic ring (cm)	1.44	1.44	1.38–1.48	0.03	1.43	1.43	1.38–1.49	0.03	1.44	1.44	1.38–1.49	0.03
Sclerotic ring thickness (mm)	2.15	2.14	2.04–2.28	0.07	2.18	2.195	2.08–2.29	0.06	2.16	2.16	2.04–2.29	0.06
Ocular bulb height (cm)	1.54	1.52	1.47–1.61	0.04	1.52	1.52	1.41–1.61	0.05	1.53	1.52	1.41–1.61	0.05
Ocular bulb width (cm)	1.29	1.29	1.1–1.42	0.09	1.27	1.28	1.1–1.39	0.08	1.28	1.29	1.1–1.42	0.08
Ocular bulb length (cm)	1.62	1.64	1.46–1.7	0.07	1.60	1.6	1.48–1.71	0.07	1.61	1.62	1.46–1.71	0.07
Orbit depth (mm)	8.16	8.2	7.9–8.4	0.16	8.23	8.3	7.9–8.5	0.18	8.19	8.2	7.9–8.5	0.18
Lens attenuation (HU)	70.14	71	64–79	2.95	68.93	69	64–74	2.72	69.53	69.5	64–79	2.88
Vitreous humor attenuation (HU)	35.17	35	31–41	3.06	35.59	36	31–45	3.76	35.38	35.5	31–45	3.40
Sclerotic ring attenuation (dorsal arch) (HU)	721.21	725	687–752	18.39	719.79	723	687–745	17.64	720.50	723	687–752	17.87

The Spearman correlation analysis conducted to examine the relationship between eye measurements and head length did not reveal any statistically significant findings. However, significant correlations were observed in other aspects. There was a strong correlation between ocular bulb height and width (rho = 0.661; $p < 0.001$), indicating a substantial relationship between these variables. Additionally, moderate correlations were observed between ocular bulb height and length (rho = 0.403; $p = 0.002$), as well as between ocular bulb width and length (rho = 0.449; $p = 0.002$), suggesting meaningful associations between these measurements. Moreover, a good correlation was found between the external and internal diameters of the sclerotic ring (rho = 0.431; $p < 0.001$), indicating a relationship between these variables.

Correlations involving the internal diameter of the sclerotic ring showed statistical significance with ocular bulb height (rho = 0.601; $p < 0.001$), length (rho = 0.496; $p < 0.001$), but not with width (rho = 0.203; $p = 0.127$). In a similar way, the external diameter of the sclerotic ring exhibited significant correlations with ocular bulb height (rho = 0.688; $p < 0.001$) and width (rho = 0.642; $p < 0.001$), while in this case no significant correlation was found with ocular bulb length (rho = −0.059; $p = 0.662$). Significant correlations were observed when comparing the orbit depth with ocular bulb measurements: height (rho = −0.484; $p < 0.001$) and width (rho = −0.294; $p = 0.025$). However, no significant correlation was observed with ocular bulb length (rho = −0.114; $p = 0.395$).

In terms of attenuation, the regions of interest (ROI) for measuring the attenuations of the sclerotic ring, vitreous humor, and lens were indicated by the green, yellow, and purple circles, respectively (Figure 3). It is noteworthy that the vitreous chamber exhibited hypoattenuation compared to the sclerotic ring. Conversely, the lens displayed hyperattenuation in relation to the surrounding aqueous humor. The mean lens attenuation was determined to be 69.53 Hounsfield units (HU), with a range of 64–79 HU. The vitreous humor exhibited an average attenuation of 35.38 HU (range: 31–45 HU), while the sclerotic ring displayed an attenuation of 720.50 HU (range: 687–752 HU).

4. Discussion

In this study, we conducted computed tomography to assess the ocular bulb and associated hard tissues of the Atlantic puffin. To the best of the authors' knowledge, this is the first report describing measurements of Atlantic puffin eyes using a CT scan. This technique has proven quite useful in the evaluation of the morphological knowledge of the normal eye and associated structures in different species, such as dogs [36,37], cats [38], and reptiles [39,40]. CT facilitates the view of body sections from different tomographic planes, offering images with high anatomic resolution without tissue overlapping, adequate contrast between different structures, and excellent tissue differentiation [41,42]. Consequently, modern imaging techniques like CT have significantly improved the image quality, enabling exceptional evaluation of eye dimensions and different structures of the puffin ocular bulb, including the sclerotic ring, vitreous chamber, and lens. This anatomical knowledge is crucial for understanding the biology and ophthalmology of seabird species. Thus, the tissue structures related to the ocular bulb, including the sclerotic ring and the orbit, play a vital role in inferring the size and shape of the ocular bulb, which have been linked to the bird's activity patterns [35]. It is crucial to emphasize that mammals lack a sclerotic ring, and therefore, the sole bone correlation lies in the morphology of the orbit itself [35]. Nevertheless, studies on primates have indicated that as body size increases, the volume of the primate orbit increases at a higher rate compared to the volume of the eye. Consequently, in larger bodies, the dimensions of the orbit may not accurately predict the size or shape of the eye [43–46]. Our study does not compare body size with eye measurements. Instead, we focused on the correlation between head length and eye measurements, as previous studies on birds have established this association [35]. However, we did not observe any significant correlation between the eye measurement and head length in Atlantic puffins.

The morphometric analysis of the internal diameter of the sclerotic ring provides valuable insights into the relative size of the dilated pupil and the cornea, as this bone correlates well with the cornea. This relationship enables the estimation of the light-capturing capacity of the eye and consequently deduces the behavior pattern, whether diurnal or nocturnal [25,35]. However, it is important to note that this correlation may not be reliable if analyzed in isolation. In bird species that do not exhibit distinct activity patterns strongly skewed towards day or night, it becomes necessary to combine these measurements with those of the orbit depth to obtain more accurate and reliable results because just the dimensions of the sclerotic ring alone have not separated nocturnal and diurnal birds well [35].

Our CT studies have also revealed a significant and positive statistical correlation between the internal and external diameters of the sclerotic ring and the width, height, and length of the ocular bulb. These findings suggest a potential relationship between these parameters, visual acuity, and the diurnal pattern observed in the Atlantic puffin. Similar associations have been documented in extinct and extant bird species [35] and sea turtles [39]. However, further investigations focusing on the influence of water pressure during diving or air pressure on the Atlantic puffin's ocular bulb sclerotic ring should be performed to deepen our understanding in this area.

The assessment of orbit depth, from the midpoint of the orbital diameter to the optic foramen, has been previously investigated by other researchers in extinct and extant birds [35]. In those studies, the measurements were performed directly on the carcasses rather than indirectly, as in our study, where three-dimensional reconstruction techniques and specialized computer programs were employed. In our animals, significant correlations were observed when comparing the orbit depth with ocular bulb height and width, suggesting relevant animal visual acuity and sensitivity to low light. Close findings were observed in extinct aquatic reptiles such as *Ichthyosaurus* and *Ophtalmosaurus* with large eyes and similar feeding habits to Atlantic puffins [34], which spend part of the day feeding on small, fast-moving preys at depths ranging from 30–60 m [1,2]. However, it is essential to acknowledge that these measurements may vary and could be influenced by operator-dependent factors during the contouring of CT images, as previously reported in other studies [36,38].

The quality of the acquired tomographic images presented here could be influenced primarily by two factors: the utilization of conventional CT instead of micro-CT and the small size of the birds included in the study. Previous studies have demonstrated that micro-CT offers high contrast and superior image quality for small animals and tissues [47]. However, it is important to highlight that, despite using a conventional CT scanner, we obtained the essential information required to fulfill the study objectives.

5. Conclusions

In this study, the CT images obtained in different planes provided relevant information about the morphometric characteristics of the ocular bulb and the sclerotic ring in the Atlantic puffin. The reference values established included the presumed normal diameters of the ocular bulb and lens, as well as various measurements of the sclerotic ring, which may be associated with the visual capabilities and activity patterns of these seabird species. However, it is essential to note that further investigations involving live animals are necessary to evaluate potential differences compared to data obtained from carcasses. Additionally, it is essential to consider the inherent operational error associated with the manual contouring of computed tomography images.

Author Contributions: Conceptualization, J.R.J.M., M.F.-H. and M.E.Q.; methodology, M.E.Q. and M.F.-H.; investigation, J.R.J.M., M.F.-H., I.M., A.S.R. and M.E.Q.; resources, A.S.P. writing—original draft preparation, J.R.J.M. and M.F.-H.; writing—review and editing, J.R.J.M., M.F.-H., I.M., A.S.R., A.S.P. and M.E.Q. All authors have read and agreed to the published version of the manuscript.

Funding: This research received no external funding.

Institutional Review Board Statement: In this study, we did not need ethical approval since the puffins were coming to our hospital to rule out by CT any cause of stranding. From these studies, we obtained the information presented in our paper. As happens in domestic mammals, just informed consent from the owner allowed us to carry out this study. Therefore, El Cabildo Insular de Gran Canaria was informed that all animal identity information obtained from this study was treated as confidential to the extent permitted by law and only used for research or teaching purposes.

Informed Consent Statement: Written informed consent has been obtained from El Cabildo Insular de Gran Canaria to publish this paper.

Data Availability Statement: The information is available at "https://accedacris.ulpgc.es", accessed on 25 June 2023.

Acknowledgments: In loving memory of Alvaro Domingo Rodriguez Garcia. We also thank Marisa Mohamad, Ayesh Mohamad, Carmen Mingot, Emilia Mingot, Concepción Mingot, Nicolas Aquino and Jamal Jaber for their support and constructive comments, and El Cabildo Insular de Gran Canaria to provide the animals for this study.

Conflicts of Interest: The authors declare no conflict of interest.

eferences

BirdLife International. *Fratercula arctica.* In *The IUCN Red List of Threatened Species*; 2018; p. e.T22694927A132581443. Available online: https://www.iucnredlist.org/especies/22694927/132581443 (accessed on 12 July 2023).

Myers, P.; Espinosa, R.; Parr, C.S.; Jones, T.; Hammond, G.S.; Dewey, T.A. The Animal Diversity Web (Online). 2023. Available online: https://animaldiversity.org (accessed on 25 June 2023).

Breton, A.; Diamond, A.; Kress, S. Encounter, survival and movement probabilities from an Atlantic puffin (*Fratercula arctica*) metapopulation. *Ecol. Monogr.* **2006**, *76*, 133–149. [CrossRef]

Durant, J.; Tycho, A.; Nills, C. Ocean climate prior to breeding affects the duration of the nestling period in the Atlantic puffin. *Biol. Lett.* **2006**, *2*, 122–128. [CrossRef]

Lowther, P.E.; Diamond, A.W.; Kress, S.W.; Robertson, G.J.; Russell, K. *The Birds of North America*; The Birds of North America, Inc.: Philadelphia, PA, USA, 2002; pp. 1–23.

Rodway, M. Relationship between wing length and body mass in atlantic puffin chicks. *J. Field Ornithol.* **1997**, *14*, 338–347.

Burnham, K.K.; Burnham, J.L.; Johnson, J.A.; Huffman, A. Migratory movements of Atlantic puffins Fratercula arctica naumanni from high Arctic Greenland. *PLoS ONE* **2021**, *16*, e0252055. [CrossRef]

Harris, M.; Daunt, F.; Newell, M.; Phillips, R.; Wanless, S. Wintering areas of adult Atlantic Puffins Fratercula arctica from a North Sea colony as revealed by geolocation technology. *Mar. Biol.* **2010**, *157*, 827–836. [CrossRef]

Guilford, T.; Freeman, R.; Boyle, D.; Dean, B.; Kirk, H.; Phillips, R.; Perrins, C. A Dispersive Migration in the Atlantic Puffin and Its Implications for Migratory Navigation. *PLoS ONE* **2011**, *6*, e21336. [CrossRef] [PubMed]

10. Cramp, S. *The Birds of the Western Palearctic*, Full ed.; Oxford University Press: Oxford, UK, 1985.
11. Boag, D.; Alexander, M. *The Puffin*; Blandford Press: London, UK, 1995.
12. Creelman, E.; Storey, A.E. Sex differences in reproductive behavior of Atlantic Puffins. *Condor* **1991**, *93*, 390–398. [CrossRef]
13. Anker-Nilssen, T. The breeding performance of Puffins *Fratercula arctica* on Røst, northern Norway in 1979–1985. *Fauna Norv.* **1987**, *10*, 21–38.
14. Harris, M.P.; Hislop, J.R.G. The food of young Puffins *Fratercula arctica*. *J. Zool.* **1978**, *185*, 213–236. [CrossRef]
15. Falk, K.; Jensen, J.K.; Kampp, K. Winter diet of Atlantic puffins (*Fratercula arctica*) in the northeast Atlantic. *Col. Waterbirds* **1992**, *15*, 230–235. [CrossRef]
16. Durant, J.; Anker-Nilssen, T.; Stenseth, N.C. Trophic interactions under climate fluctuations: The Atlantic puffin as an example. *Proc. R. Soc. Lond. B Biol. Sci.* **2003**, *270*, 1461–1466. [CrossRef] [PubMed]
17. Sandvik, H.; Erikstad, K.E.; Barrett, R.T.; Yoccoz, N.G. The effect of climate on adult survival in five species of North Atlantic seabirds. *J. Anim. Ecol.* **2005**, *74*, 817–831. [CrossRef]
18. Melillo, J.M.; Richmond, T.T.C.; Yohe, G.W. Extreme Weather. In *Highlights of Climate Change Impacts in the United States: The Third National Climate Assessment*; U.S. Global Change Research Program; Cambridge University Press: Cambridge, UK, 2014.
19. Wilhelm, S.I.; Mailhiot, J.; Arany, J.; Chardine, J.W.; Robertson, G.J.; Ryan, P.C. Update and trends of three important seabird populations in the western North Atlantic using a geographic information system approach. *Mar. Ornithol.* **2015**, *43*, 211–222.
20. Thorup, S.H.; Jensen, J.K.; Petersen, K.T.; Kasper, D.B. *Færøsk Trækfugleatlas: The Faroese Bird Migration Atlas*; Faroe University Press: Tórshavn, Denmark, 2014.
21. Harris, M.P.; Wanless, S. *The Puffin*; Poyser Monographs; Bloomsbury Publishing: London, UK, 2011.
22. Martin, G.R.; Wanless, S. The visual fields of Common Guillemots *Uria aalge* and Atlantic Puffins *Fratercula arctica*: Foraging, vigilance and collision vulnerability. *Ibis* **2015**, *157*, 798–807. [CrossRef]
23. Martin, G.R.; White, C.R.; Butler, P.J. Vision and the foraging technique of Great Cormorants *Phalacrocorax carbo*: Pursuit or flush-foraging? *Ibis* **2008**, *150*, 39–48. [CrossRef]
24. Regular, P.M.; Hedd, A.; Montevecchi, W.A. Fishing in the dark: A pursuit-diving seabird modifies foraging behaviour in response to nocturnal light levels. *PLoS ONE* **2011**, *6*, e26763. [CrossRef]
25. Franz-Odendaal, T.A. Skeletons of the Eye: An Evolutionary and Developmental Perspective. *Anat. Rec.* **2020**, *303*, 100–109. [CrossRef]
26. Bartol, S.M.; Musick, J.A.; Ochs, A.L. Visual acuity thresholds of juvenile loggerhead sea turtles (*Caretta caretta*): An electrophysiological approach. *J. Comp. Physiol. A Neuroethol. Sens. Neural Behav. Physiol.* **2002**, *187*, 953–960. [CrossRef]
27. Waldvogel, J.A. The bird's eye view. *Am. Sci.* **1990**, *78*, 342–353.
28. Lauridsen, H.; Hansen, K.; Wang, T.; Agger, P.; Andersen, J.L.; Knudsen, P.S.; Rasmussen, A.S.; Uhrenholt, L.; Pedersen, M. Inside Out: Modern Imaging Techniques to Reveal Animal Anatomy. *PLoS ONE* **2011**, *6*, e17879. [CrossRef]
29. Webb, E.; Yuan, M.; Lemoine, N.; Wang, Y. Imaging in animal models. *Integr. Cancer Sci. Ther.* **2016**, *3*, 428–431. [CrossRef]
30. Schambach, S.J.; Bag, S.; Schilling, L.; Groden, C.; Brockmann, M.A. Application of micro-CT in small animal imaging. *Methods* **2010**, *50*, 2–13. [CrossRef] [PubMed]
31. Farrag, M.; Pukale, D.D.; Leipzig, N.D. Micro-computed tomography utility for estimation of intraparenchymal spinal cord cystic lesions in small animals. *Neural Regen. Res.* **2021**, *16*, 2293–2298. [PubMed]
32. Barrett, R.T.; Fieler, R.; Anker-Nilssen, T.; Rikardsen, F. Measurements and weight changes of norwegian adult puffins *Fratercula arctica* and kittiwakes *Rissa tridactyla* during the breeding season. *Ring. Migr.* **1985**, *6*, 102–112. [CrossRef]

33. Westmoreland, L.S.; Hadfield, C.A.; Clayton, L.A.; Atkins, R.M.; Hyman, J.A.; Rossi, B.H. Ocular examinations and investigation of intraocular pressure, tear production, central corneal thickness, and corneal touch threshold in a captive flock of Atlantic puffins (*Fratercula arctica*). *J. Avian Med. Surg.* **2020**, *34*, 123–131. [CrossRef]
34. Fernández, M.S.; Archuby, F.; Talevi, M.; Ebner, R. Ichthyosaurian eyes: Paleobiological information content in the sclerotic ring of *Caypullisaurus* (*Ichthyosauria, Ophthalmosauria*). *J. Vertebr. Paleontol.* **2005**, *25*, 330–337. [CrossRef]
35. Hall, M.I. The anatomical relationships between the avian eye, orbit and sclerotic ring: Implications for inferring activity patterns in extinct birds. *J. Anat.* **2008**, *212*, 781–794. [CrossRef]
36. Salgüero, R.; Johnson, V.; Williams, D.; Hartley, C.; Holmes, M.; Dennis, R.; Herrtage, M. CT dimensions, volumes and densities of normal canine eyes. *Vet. Rec.* **2015**, *176*, 386. [CrossRef]
37. Boroffka, S.A.E.B.; Voorhout, G. Direct and reconstructed multiplanar computed tomography of the orbits of healthy dogs. *Am. J Vet. Res.* **1999**, *60*, 1500–1507.
38. Sreenivasan, S.; Linden, A.; Owen, M.; Pemberton, S.; Pinard, C.; Matsuyama, A.; Poirier, V.J. Computed tomography measurements of intraocular structures of the feline eye. *Vet. Rec.* **2019**, *184*, 651–656.
39. Fumero-Hernández, M.; Encinoso Quintana, M.; Ramírez, A.S.; Morales Fariña, I.; Calabuig, P.; Jaber, J.R. Morphometric Study of the Eyeball of the Loggerhead Turtle (*Caretta caretta*) Using Computed Tomography (CT). *Animals* **2023**, *13*, 1016. [CrossRef] [PubMed]
40. Andrade, S.B.D.; Araujo, N.L.L.C.D.; Raposo, A.C.S.; Muramoto, C.; Oriá, A.P. Morphometric descriptive report of scleral ossicle rings, by ultrasound and computed tomography, in three Testudines specimens. *Ciência Rural.* **2023**, *53*, e20210423. [CrossRef]
41. Kuszyk, B.S.; Heath, D.G.; Bliss, D.F.; Fishman, E.K. Skeletal 3-D CT: Advantages of Volume Rendering over Surface Rendering *Skelet. Radiol.* **1996**, *25*, 207–214. [CrossRef]
42. Pérez, S.; Encinoso, M.; Morales, M.; Arencibia, A.; Suárez-Bonnet, A.; González-Rodríguez, E.; Jaber, J.R. Comparative evaluation of the Komodo dragon (*Varanus komodoensis*) and the Green iguana (*Iguana iguana*) skull by three dimensional computed tomographic reconstruction. *Slov. Vet. Res.* **2021**, *58*, 111–116. [CrossRef]
43. Schultz, A.H. The size of the orbit and of the eye in primates. *Am. J. Phys. Anthropol.* **1940**, *26*, 389–408. [CrossRef]
44. Kay, R.F.; Cartmill, M. Cranial morphology and adaptations of *Palaechthon nacimienti* and other paromomyidae (plesiadapoidea,? Primates), with a description of a new genus and species. *J. Hum. Evol.* **1977**, *6*, 19–35. [CrossRef]
45. Kay, R.F.; Kirk, E.C. Osteological evidence for the evolution of activity pattern and visual acuity in primates. *Am. J. Phys. Anthropol.* **2000**, *113*, 235–262. [CrossRef] [PubMed]
46. Heesy, C.P.; Ross, C.F. Evolution of activity patterns and chromatic vision in primates: Morphometrics, genetics and cladistics *J. Hum. Evol.* **2001**, *40*, 111–149. [CrossRef]
47. Metscher, B.D. MicroCT for comparative morphology: Simple staining methods allow high-contrast 3D imaging of diverse non-mineralized animal tissues. *BMC Physiol.* **2009**, *9*, 1–14. [CrossRef]

animals

MDPI

Article

Gross Anatomy of the Female Reproductive System of Sugar Gliders (*Petaurus breviceps*)

María del Mar Yllera [1,†], Diana Alonso-Peñarando [2,3] and Matilde Lombardero [1,*]

1 Unit of Veterinary Anatomy and Embryology, Department of Anatomy, Animal Production and Clinical Veterinary Sciences, Faculty of Veterinary Sciences, University of Santiago de Compostela–Campus of Lugo, 27002 Lugo, Spain; mar.yllera@usc.es

2 Department of Animal Pathology, Faculty of Veterinary Sciences, University of Santiago de Compostela–Campus of Lugo, 27002 Lugo, Spain; diana.alonso.penarando@rai.usc.es

3 DVM at Veterinary Clinic Madivet, Calle Comercio, 5, Bargas, 45593 Toledo, Spain

* Correspondence: matilde.lombardero@usc.es; Tel.: +34-982-822-333

† Retired.

Simple Summary: Sugar gliders are small marsupials that are becoming quite common as pets. As patients, they are challenging for clinical veterinarians. References about exotic animals are difficult to find, especially regarding the internal anatomy of *Petaurus breviceps*, knowledge of which is essential to reach a correct diagnosis of pathologies. Since the reproduction of marsupials is peculiar, we focused on the female genital apparatus of this species, very different from that of the usual mammals in veterinary practices. Sugar gliders have duplicates of all the genital organs: they have two ovaries, two oviducts, two uteri and two vaginas. In addition, in pregnant females, a new conduit arises: the birth canal, through which the offspring will leave the genital tract to go outside. In this species, the urinary, genital and digestive systems flow into a small cavity called the cloaca, which receives urine, the foetus and faeces, and communicates with the outside through a single orifice. Thus, the female genital apparatus of the sugar glider fits the general schema of marsupials.

Abstract: We dissected carcasses of eight mature females, both parous and non-parous specimens, to study the macroscopic anatomy of the female reproductive system in the sugar glider. The genital system includes double organs, namely the right and left ones, which are completely separated. It includes two ovaries, two oviducts, two *uteri* and a vaginal complex. The *uteri* are fusiform-shaped and lack horns. The vaginal complex includes two lateral *vaginae* and a median vagina, also called the 'birth canal'. The cranial end of both lateral *vaginae* partially fuses, forming an expansion named the vaginal sinus, which is divided into two parts by a longitudinal septum, one for each vagina, where the ipsilateral uterine cervix opens. The caudal end of the lateral *vaginae* opens into a medial and impar duct: the urogenital sinus that serves as a common passage for the reproductive and urinary systems. In non-pregnant females, only the lateral *vaginae* are present. In pregnant and recently parous females, a short median vagina extends from the caudal wall of the vaginal sinus to the cranial end of the urogenital sinus. In the ventral wall of this sinus, next to its caudal opening, there is a forked clitoris.

Keywords: sugar glider; female genital anatomy; ovary; uterine tuba; uterus; vaginal complex; clitoris

1. Introduction

Exotic animal clinical practice is challenging for veterinarians because of the widely varying anatomy, peculiar physiology and specific needs of the new companion animals. Very often, it is hard to find appropriate references in the literature, and even sometimes, the available information is contradictory, and erroneous extrapolations from domestic species are applied. Knowledge of anatomy is basic to reach a correct diagnosis of pathologies and to interpret the images obtained with modern diagnostic imaging methods. The purpose of this article is to provide a practical approach to the female genital anatomy of the sugar glider.

Sugar gliders (*Petaurus breviceps*) are becoming quite popular as pets. They are originally from Australia and New Guinea, but were introduced into Tasmania. They arrived as pets to the USA in the 1990s [1] from Western New Guinea (Indonesia) [2], and much later to Europe. They are small, nocturnal and omnivorous marsupials that live in arboreal habitats and have a large gliding membrane named the *patagium* between the front and hind limbs. Therefore, they can glide long distances, up to 50 metres, between trees [3]. Sugar gliders are a social species that builds communitarian nests in holes in eucalypts. In the wild, they feed on plants, insects and arthropods' exudates, with a natural diet depending on the season and habitat [4,5]. Furthermore, some Tasmanian populations have carnivorous preferences, as they feed on eggs and young birds [6].

The body organization of the three great groups of mammals, eutherians, and marsupials and monotremes, is very similar, but they differ in their reproductive performance, which leads to differences in the morphology of their genital apparatus. In marsupials, the urinary, genital and digestive systems end in a common space, the cloaca, which communicates with the exterior. The newborns, named joeys, are in an early stage of development and after a short period of gestation (15–17 days in the sugar glider) [7,8], a highly altricial young emerges through the cloacal opening and climbs by its own into a pouch called the marsupium, where its development will be completed.

Although there is abundant literature on marsupials, there are practically no articles on the anatomy of *Petaurus breviceps*, especially regarding its internal organs. The purpose of this study is to contribute to a better understanding of the female genital apparatus of this species.

2. Materials and Methods

Eight females were investigated, all of whom were adults and intact specimens. They died from different pathologies, and their carcasses were donated by their owners to carry out this study. Seven animals had given birth to joeys at least once, including three females with joeys in the pouch, one with two joeys recently weaned, and three individuals that had not been pregnant for some years; one additional female was non-parous.

Seven specimens were kept frozen and one was preserved in alcohol 70° until the time of the study. The dissections were performed by accessing the body cavities through the ventral midline. In two cases, after the macroscopic study, the entire urogenital apparatus except the kidneys was removed and preserved in formalin for a second and more detailed study.

All the dissections were carried out under a Nikon AFX-II magnifying device (magnification from 0.66 up to 4.0, with oculars of 10×), and photographs were taken with a Nikon Coolpix S6 camera (Zoom Nikkor ED 5.8–1.7 mm 1:3.0–5.4) equipped with a macro lens.

3. Results

Female sugar gliders have a pouch in the ventral part of the abdomen. Its opening is placed in the midline, caudal to the *umbilicus*. The pouch follows the longitudinal axis of the animal; it has two lateral pockets and contains four *mamae* (nipples or teats): two in each pocket, one cranial and one caudal. The mucosa that lines the inner wall of the pouch has secretory glands, which are macroscopically visible (Figure 1).

The female genitalia are in the caudal half of the abdominal cavity, and they are related dorsally to the rectum and ventrally to the urethra and the urinary bladder. The sugar glider female reproductive organs include paired ovaries, paired uterine tubes, paired uteri and a vaginal complex that includes paired lateral *vaginae* (Figures 2 and 3a).

The ovaries are close and dorsolateral to their ipsilateral uterus, near the junction between the uterus and the uterine tube. They have a brownish colour and ellipsoidal form and are flattened dorsoventrally. In this study, all the females were pubescent, so the gonad's surface was irregular due to the projection of follicles and/or corpora lutea. Thus, all functional ovarian structures are visible. Each ovary is attached to the abdominal roof by a long ligament known as the *mesovarium*. The cranial free border of the *mesovarium* is a strong band called the suspensory ligament of the ovary (*ligamentum suspensorium ovarii*) (Figure 4a). The *mesovarium* also supports the proper ligament of the ovary (*ligamentum ovarii propium*), a band of connective tissue covered with peritoneum, which extends from the caudal pole of the gonad (*extremitas uterina ovarii*) to the adjacent tip of the uterus. As the peritoneal folds do not cover the dorsal surface of the ovary, an *ovarian bursa* cannot be described in this species (Figure 4b).

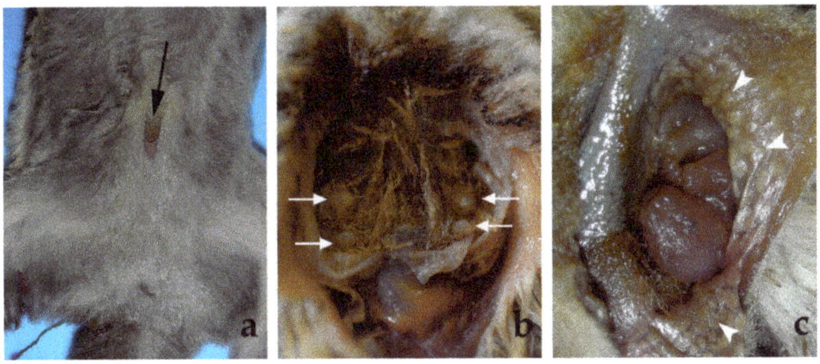

Figure 1. Ventral view of a female sugar glider's abdominal wall. (**a**) The black arrow points to the opening of the pouch (*marsupium*). (**b**) *Marsupium*: the pouch was opened to show the four teats (*mamae*) (⇨). (**c**) *Marsupium*: the left part of the pouch was opened to show the secretory glands (⊳). A joey can be seen deep inside the pouch.

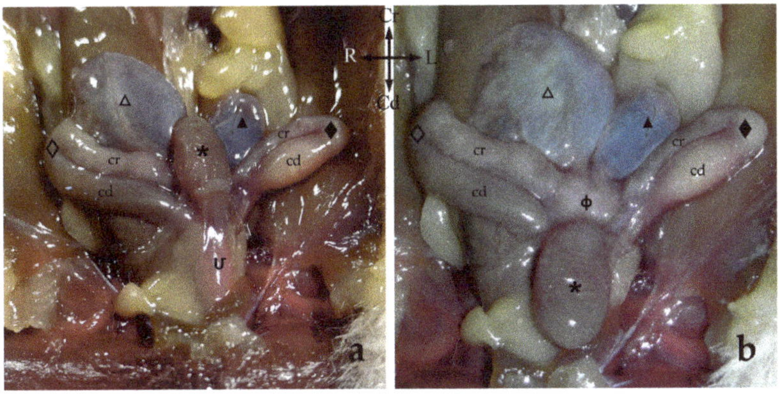

Figure 2. Ventral view of the female sugar glider's apparatus urogenitalis in situ (**a**) and with the urinary bladder displaced caudally (**b**). Fresh specimen: (**Ʊ**) *sinus urogenitalis;* (**Φ**) *sinus vaginalis;* (**★**) urinary bladder. Right-side structures are shown with empty patterns whereas left-side structures are marked with filled patterns: (◊, ♦) lateral *vaginae* with cranial loop (cr) and caudal loop (cd); (△, ▲) *uteri*.

Uterine tubes or oviducts (*tubae uterinae*) are long and flexuous. They begin next to the corresponding ovary and lead to the uterus on the same side. The free cranial extremity of the tube is funnel-shaped and has a very thin wall. It is called the infundibulum (*infundibulum tubae uterinae*) and is close to the ipsilateral ovary. The rest of the oviduct is tubular and very convoluted. It is divided into two parts: the proximal and narrower one, usually called the *ampulla* (*ampulla tubae uterinae*), and the most coiled part, which is the *isthmus* (*isthmus tubae uterinae*). Each tuba is supported by a serous fold called the *mesosalpinx*, a lateral fold released from the *mesovarium*. The *mesosalpinx* and *mesovarium* enclose a pouch that contains the tuba but not the ovary. This fact, together with the proximity of the ovary and the uterus, sometimes makes it difficult to visualize the tubes (Figure 4b).

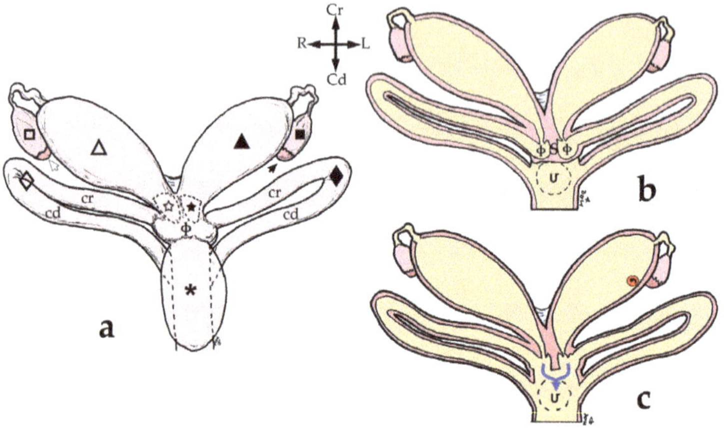

Figure 3. Schematic drawing of the genital system of the female glider: (**a**) intact; (**b**,**c**) opened to expose the lumen; the urogenital sinus is marked with a dotted line. (**a**) The (★) urinary bladder is reflected caudally. (**b**) Schema of the female genitals of a non-pregnant sugar glider, with two independent vaginal sinuses. (**c**) Schema of the female genitals of a pregnant sugar glider showing a single median vagina (confluent blue arrows). Right structures are represented by an empty pattern and left structures with a filled pattern: (⇨, ➜) ovaries; (□, ■) *infundibula* of the uterine tubes; (△, ▲) bodies of the *uteri*; (✩, ★) *cervices*; (◇, ◆) Lateral *vaginae* with cranial loop (cr) and caudal loop (cd); (Φ) aginal sinus; (Ʊ) *Sinus urogenitalis*.

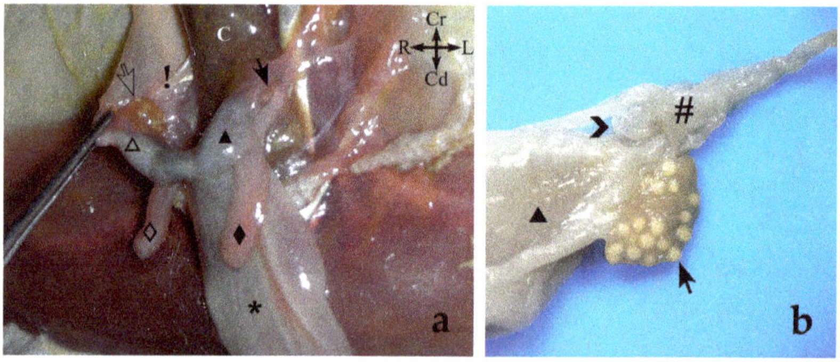

Figure 4. (**a**) Ventrolateral view of the sugar glider's *apparatus urogenitalis* in situ. Fresh specimen: The urinary bladder is reflected caudally to expose the *uterii* and *vaginae*. The (△) right uterus has been displaced ventrally to display the right ovary and its peritoneal folds: (⇨) right ovary; (!) suspensory ligament of the ovary; (★) urinary bladder. Right structures are represented by an empty pattern and left structures with a filled pattern: (◇, ◆) lateral *vaginae*; (△, ▲) *uteri*; (➜) left ovary; (**C**) colon. (**b**) Dorsal view of the female sugar glider's *apparatus genitalis* ex situ: ovary and uterine tube. Specimen is fixed in formalin. The oviduct is included in a pouch formed by the peritoneal folds (*mesovarium* and *mesosalpinx*); (▲) left uterus; (>) uterine tube; (#) pouch; (➜) left ovary.

The paired uteri are elongated and fusiform-shaped, with their major axis running obliquely to the corporal axis. Sugar glider uteri lack horns (*cornua uteri*), but they do have a body (*corpus uteri*) and a neck (*cervix uteri*). A ligament joins the medial surfaces of the caudal part of both uterine bodies. Although they are independent structures, both uteri appear to be fused caudally since they are surrounded by connective tissue (Figure 4). The caudal part of each uterus narrows and they run parallel to each other on each side of the median line forming the uterine neck or cervix, which projects onto a papilla into the vaginal complex. The cervix lumen is almost occluded by mucosal folds, almost hiding the orifice that communicates the interior of the vaginal complex with each uterus (*ostium uteri externum*) (Figures 2 and 5).

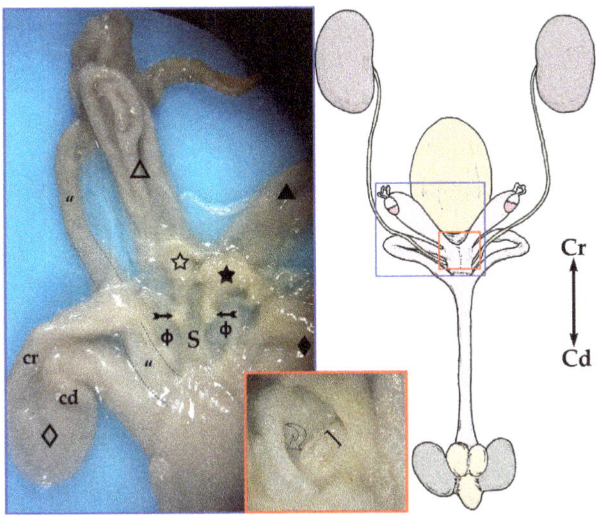

Figure 5. On the right, a general sketch of the female sugar glider urogenital system ex situ (dorsal view) with the areas' detailed photographs. Blue inset: ventral view of the female sugar glider's *apparatus genitalis* ex situ. The caudal parts of the uteri have been dissected to expose the cervix. The vaginal sinus was opened ventrally to show the uterine *papillae* and *septum*. The last section of the right ureter is marked with a dotted line. Specimen is fixed in 76° alcohol. Right structures are represented by an empty pattern and left structures with a filled pattern: (△, ▲) bodies of the uteri; (✩, ★) *cervices*; (➻) *papillae uteri*; (**S**) *septum*; (Φ) vaginal sinus; (◇, ♦) lateral *vaginae* with cranial loop (cr) and caudal loop (cd); (") ureter. Red inset: inner dorsal mucosa of the urinary bladder showing the (]) paired ureteral openings (*ostium ureteris*), close to the neck, and the curved arrow pointing to the internal urethral orifice (*ostium urethrae internum*).

The vaginal apparatus or vaginal complex includes two lateral *vaginae* and a median vagina, also called the 'birth canal'. Lateral *vaginae* are long and curved in a U-shape. They are placed caudal to the uteri and form two loops located on either side of the midline. Each loop has two ends, one cranial and another caudal, very close to each other and located on the same plane and joined by a peritoneal fold. The cranial end of both lateral *vaginae* partially fuse, forming an expansion named the vaginal sinus (formerly called the 'anterior vaginal expansion' or 'cul-de-sac') (Figures 2, 3b and 5). It is divided by a longitudinal *septum* into two parts, one for each vagina, where the corresponding uterine cervix opens. This separation is complete in the non-parous specimen but incomplete in parous ones (Figure 5). The caudal end of the lateral *vaginae* opens into a medial and impar duct: the urogenital sinus that serves as a common passage for the reproductive and urinary systems (Figure 6). The median vagina, very short in the sugar glider, extends from the caudal wall of the vaginal sinus to the cranial end of the urogenital sinus in recently parous females but does not exist in non-pregnant animals (Figures 3c and 7).

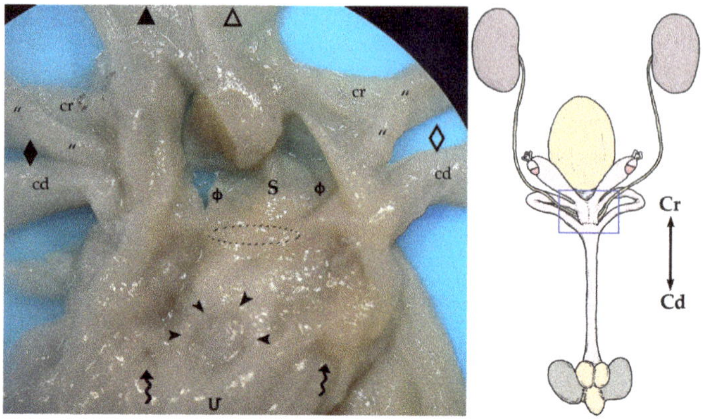

Figure 6. On the right, a general sketch of the female sugar glider urogenital system ex situ (dorsal view) with the area's detailed photograph. On the left, view of the floor of the cranial part of the (**Ʊ**) urogenital sinus that has been opened by cutting its roof in the midline and moving each half laterally. The vaginal sinus (**ɸ**) has also been opened dorsally. In this specimen, the central vagina (marked with a discontinuous oval shape) connects the vaginal and urogenital sinuses. Specimen is fixed in formalin. (⤳) Outlet orifices of the lateral *vaginae*; (➤) external urethral orifice; (″) ureters; (**S**) *septum*. Right structures are represented by an empty pattern and left structures with a filled pattern: (△, ▲) bodies of the *uteri*; (◇, ◆) lateral *vaginae* with cranial loop (cr) and caudal loop (cd).

The urinary bladder is ventral to the median vagina and urogenital sinus (Figure 2a). A large external urethral orifice is located on the ventral wall of the urogenital sinus, the ending point of a short urethra. The ureters, originating in the ipsilateral kidney, run dorsolaterally to the uterus. Each ureter passes dorsally to the cranial branch of the ipsilateral lateral vagina and is incorporated into the peritoneum fold between the two branches. Their course is parallel to the caudal branch until they end in the vicinity of the neck of the urinary bladder, in a dorsal position. The two ureters terminate very close to each other by means of two elongated orifices located near to the internal urethral orifice, where they empty their content (Figures 5 and 6).

The female sugar glider has neither a vulva nor an anus. Instead, there is a single opening, ventral to the beginning of the tail that communicates with a small internal cavity, the cloaca, a common organ of the digestive system and urogenital apparatus.

The urogenital sinus is a long tube that begins next to the caudal wall of the vaginal sinus and runs ventrally to the final part of the digestive system (large intestine) (*intestinum crassum*) to the vicinity of the cloacal orifice (Figure 7). It opens into the floor, close to the rectum that empties into the cranial part of the cloacal cavity.

The genital and urinary ducts empty into the cranial part of the urogenital sinus. The caudal loop of the lateral *vaginae* and the external urethral orifice (*ostium urethrae externum*) flow at the same level, but the lateral *vaginae* open into the dorsolateral wall, whereas the external urethral orifice is on the ventral one (Figure 6).

In the caudal part of the urogenital sinus, next to its cloacal opening, there is a ventral fossa (*fossa clitoridis*) where the clitoris is located. This pit is covered by two folds of the lateral mucosa, one right and one left, which practically hide the fossa. The clitoris is forked. Each of its halves is elongated, with a constant thickness except at its terminal end, near the cloaca, where it tapers to a conical point. Given its great length, at rest, its caudal part must be folded inside the fossa (Figure 8).

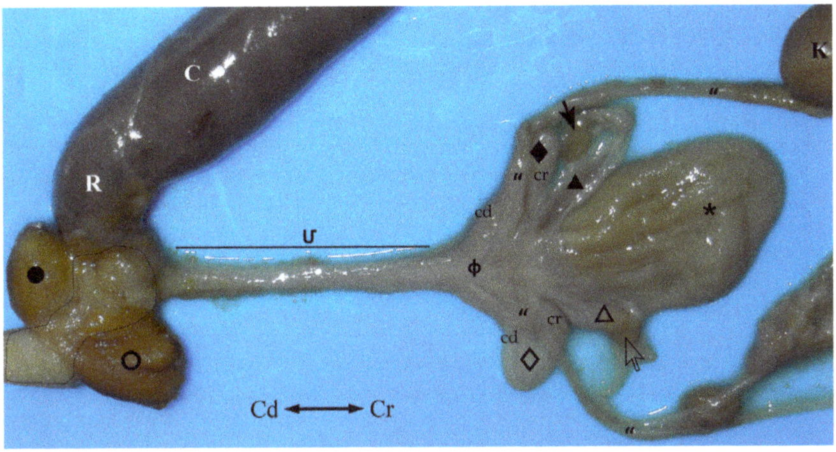

Figure 7. Dorsal view of the female sugar glider's *apparatus genitalis* ex situ. The large intestine (colon and rectum) has been moved laterally to observe the urogenital sinus. Specimen is fixed in 76° alcohol. (**K**) Left kidney; (") ureters; (★) urinary bladder. Right structures are represented by an empty pattern and left structures with a filled pattern: (⇨, ➔) ovaries; (△, ▲) bodies of the uteri; (◇, ◆) lateral *vaginae* with cranial loop (cr) and caudal loop (cd); (**ϕ**) vaginal sinus; (**Ʊ**) *sinus urogenitalis*; (**O**, ●) paracloacal glands; (**C**) colon; (**R**) rectum; (black line) cloaca.

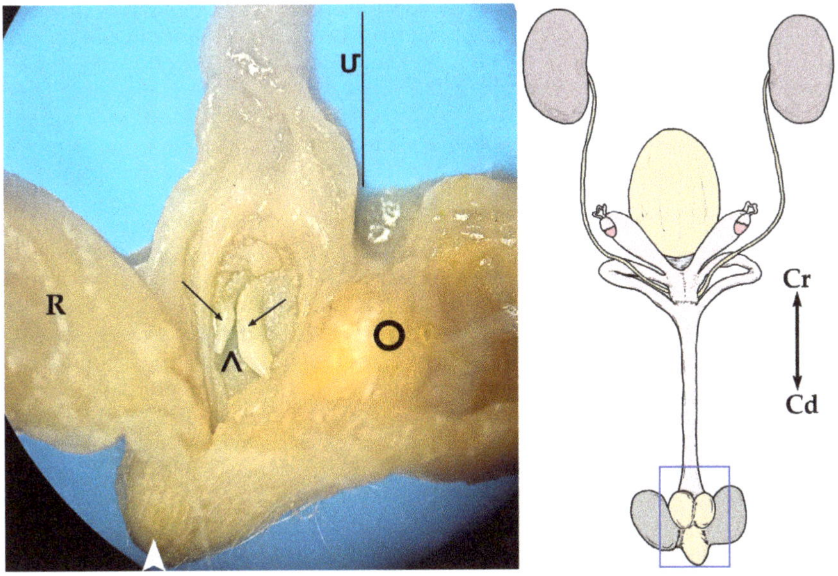

Figure 8. On the right, a general sketch of the female sugar glider urogenital system ex situ, (dorsal view) with the area's detailed photograph. On the left, magnified image of the cloacal area; view of the floor of the caudal part of the urogenital sinus (**Ʊ**). The sinus has been opened by cutting its roof in the midline and moving each half laterally. (**R**) Rectum; (ᐅ) cloacal orifice; (→) clitoris; (>) *fossa clitoridis*; (**O**) right paracloacal gland.

4. Discussion

The name 'marsupials' designates a group of mammals whose females have a pouch, or marsupium, where the young (joeys) finish their development after their birth. The marsupium is usually attached to the ventral wall of the abdomen, but there are considerable variations between species. Sometimes it opens cranially, sometimes caudally or centrally [9]. There are even marsupials that lack a pouch [9,10]. Russell (1982) [11] described six types of marsupium. In type 1, the mammary area is not covered by the skin although lateral ridges of skin develop during the breeding season; in type 2, the mammary area is partially covered; in types 3, 4, 5 and 6, the mammary area is covered by a fold of skin forming a real pouch; they differ in the position of their openings: central (type 3), cranial or anterior (types 4 and 5) and caudal (type 6). In addition, in type 4, the teats are located in two pockets, right and left, projecting forward from the cranial margin of the skin fold. Since the mammary area is completely covered by a skin fold and the opening of the bursa is located in its cranial part, the sugar glider marsupium is included in type 5, according to Russell's (1982) [11] classification.

The morphology of the pouch in the sugar glider matches the description made by Smith in 1973 [7]. It opens near its cranial end and has two pockets extending onto the flanks of the female, where the *mamae* are. According to Tyndale-Biscoe and Renfree (1987) [12], Johnson-Delaney (2021) [13] and our own data, the female sugar glider has four *mamae* (nipples) inside the marsupium, two on the right and two on the left. Nevertheless, other authors point out that they are only two teats in the pouch [7,8,14,15]. Variations between individuals could be possible, as indicated by Smith (1973) [7] and Girling (2013) [16].

We observed macroscopic secretory glands in the mucosa lining the inner surface of the pouch. The presence of apocrine glands in the pouch skin has previously been described in the sugar glider by Smith (1973) [7] and in other marsupials such as the tammar wallaby [17] and red kangaroo [18]. Various functions have been suggested for the secretion of these glands. The role of the chemical protection of the young by the secretions are not well known, but they could have antimicrobial activity as found in the tammar wallaby, whose pouch skin expresses genes for the antimicrobial peptide cathelicidin [19]. The maternal glandular secretions could also provide moisture and their smell could play an important role in guiding the joey on its journey from the cloaca to the pouch after birth [18]. When the young marsupial is born, many physiological processes are still immature for a time after parturition, among them pulmonary respiration. The lungs are immature, so gas exchange takes place through the skin for some days. Inside the pouch, the humid environment may aid this integumental gas exchange [20].

Interestingly, although marsupials have a dual hormonal control of genital organ development, pouch development is not under this control [21]. The initial development of the pouch, as well as of the mammary glands and the scrotum in the male, is evident before the differentiation of the gonads and depends on the expression of a gene or genes on the X-chromosome [22]. In fact, it is the number of X-chromosomes that determines whether a marsupium or a scrotum will form: animals with one X-chromosome will develop a scrotum while those with two X-chromosomes will form a pouch [22,23].

The female genital apparatus of eutherian mammals basically consists of paired organs responsible for producing gametes (ovary) and a system of ducts that communicate with the outside. Each part of the genital tract is specialized to perform one or more functions. Thus, the ovary produces gametes and reproductive hormones, fertilization takes place in the oviduct (pair), and the uterus (unpaired, except in lagomorphs and most rodents which have duplex uteri) provides the environment and adequate nourishment for the embryo development until its birth. In eutherian mammals, the vagina (unpaired) serves both as copulatory organ and as birth canal. The vestibule continues the vagina to the vulva, which opens externally. The genital anatomy of the female sugar glider is similar to that of other marsupials and quite different from that of eutherian mammals. They have double reproductive organs: two ovaries, two oviducts, two uteri and two lateral *vaginae*. The origin of the double reproductive tract is the result of a difference between eutherian and marsupial mammals in the embryonic development of the genital and urinary ducts. In eutherians, when the ureters migrate from the roof of the celomic cavity ventrally to connect with the bladder, they are placed laterally to the paramesonephric (Müllerian) ducts. Therefore, these can fuse with each other to form the body of the uterus and the cranial portion of the vagina. Conversely, in marsupials the ureters migrate medially to the genital ducts, preventing their fusion [10].

As the ovaries were not enclosed in any bursa in the female sugar gliders that we studied, we suggest that there is no true ovarian bursa in this species unlike those described in some South American marsupials such as *Didelphis* sp. [24] and in *Dendrolagus* [25]. Johnson-Delany (2002) [14] described the ovaries as lying against the medioventral side of the uterus, near the junction of the uterus and the oviduct. However, in all the specimens we observed, the gonads were located dorsolateral to the uterus and could not be seen in a ventral view of the genitalia.

Uterine tubes vary in length and their degree of convolution depends on the marsupial species [26]. There are three parts of the mammalian uterine tubes: *infundibulum, ampulla* and *isthmus*. However, unlike in eutherian mammals, in the sugar glider, the widest tubular part is the *isthmus*, as seen in the anatomy of other marsupials referred to by Taggart (1994) [26] and Hughes (2000) [27].

Although the uterine function is similar in marsupial and eutherian mammals, there are important differences in uterine morphology. In marsupials, there are no uterine horns, only a body and neck (*cervix*) are present. The uterine neck length varies between species [26], as does the uterine body shape: usually, it is fusiform, as we have observed in the sugar glider, in accordance to that previously described by Johnson-Delany and Lennox (2017) [28], but was abnormally enlarged in the extinct Tasmanian tiger (*Thylacinus cynocephalus*) [27] and is flattened dorsoventrally in the Tasmanian devil (*Sarcophilus satanicus*) [29]. However, in most species, the caudal ends of both uteri (*cervices*) are surrounded by a common sheath of connective tissue [14,28], as we found in *Petaurus breviceps*.

The uterine papilla formed by the opening of each cervix into the vaginal sinus in the *Petaurus breviceps* was previously described by Johnson-Delany (2002) [14]. This author also reported an *os uteri* ventrolaterally on the papilla that, in contrast, we could not identify in our macroscopic study. This papilla seems to be quite common in marsupials since it has also been observed in *Sarcophilus satanicus* [29], in *Thylacinus cynecephalus* [27] and in *Dendrolagus* [25].

The function of the lateral *vaginae* is to facilitate the arrival of sperm to the *uteri* [30]. The male marsupial deposits semen in the upper part of the urogenital sinus, where the caudal ends of the lateral *vaginae* empty. From here, spermatozoa progress up through these ducts and penetrate the vaginal sinus, near the *cervices* [31,32]. Since there is no direct communication between the urogenital sinus and the *uteri*, this is the only possible pathway in all non-parous and most parous female marsupials. However, in macropods, the birth canal (median vagina) is permanent, so it allows direct access from the urogenital sinus to the cervices in parous females [33].

The shape of the lateral *vaginae* varies among marsupials. In *Petaurus breviceps*, they are long and U-shaped in accordance with our study, but they can become long and convoluted as in *Caenolestes obscurus* or short and straight such as in *Trichosurus vulpecula* and *Tarsupes rostratus*. Additionally, in some species, the cranial part of the lateral vagina expands to form a chamber that serves as a seminal receptacle [12].

In pregnant females, a median vagina is formed in the midline, lying between the lateral *vaginae* [7,16,22]. In most marsupials, this forms a transient birth canal in the medial connective tissue between the vaginal and urogenital sinuses. It opens before each birth and closes soon after the passage of the foetus, and reforms in each subsequent parturition [9]. However, in most *Macropodidae* (kangaroos), the median vagina becomes a permanent structure after the first birth [12]. Judah and Nutall (2008) [8] point out that it is a temporary structure in *Petaurus breviceps*, which coincides with our own macroscopical observations. However, the birth canal could persist even if it is blocked by tissue separating the vaginal and urogenital sinuses. Histological study would be required to confirm that it disappears completely after parturition, as seems to be the case in the Tasmanian tiger [27].

According to our studies, in the sugar glider, the medial vagina is very short whereas the lateral ones are long and thin. Thus, our results are in accordance with those previously described by Johnson-Delaney and Lennox (2017) [28]. However, it is not always so in marsupials. For instance, in opossums and macropods, the medial and lateral *vaginae* are nearly equal in length [28]. In *Petaurus breviceps*, both structures, the vaginal and the urogenital sinuses, are very close, but it could be the same as in other marsupials such as *phalangeridae, dasyuridae* or *vombatidae* [30]: they are contiguous, but there is no communication between them except in pregnant or recently calved females.

We observed an incomplete septum in the midline of the vaginal sinus, separating the beginning of the two cranial loops of the lateral *vaginae*, in parous females (six of seven specimens). This was previously described by Johnson-Delany (2002) [14] and Johnson-Delany and Lennox (2017) [28]. According to Rodger (2020) [32], in non-parous marsupial females this median vaginal septum separates the left and right parts of the genital tract, but after the first parturition, in most marsupials this septum is perforated and the right and left sides of the vaginal sinus communicate. Our observations support this statement in the sugar glider. In some species such as *Dendrolagus matschiei*, the division of the sinus vaginalis is only partial and due to a longitudinal fold, whose height decreases as it advances in the caudal direction until it disappears. It is only attached to the ventral wall but not to the dorsal one, only dividing the upper half of the sinus [25]. Differences in the septum between species (complete, incomplete, or absent) have previously been described by Tyndale-Biscoe and Renfree (1987) [12].

Because of the close relationship of the ureters and vagina in marsupials, unlike eutherian mammals, neutering of female sugar gliders is not a routine procedure [34]. Given the technical complexity of the surgery, the animal's small size [34], and the fact that it has not been confirmed to have a favourable impact on their health [28], to avoid unwanted gestations in this species, only males are usually neutered [34]. However, due to the presence of the cloaca, which connects the urogenital and digestive tracts, this species is relatively prone to ascending infections from the cloaca, which can cause vaginitis and metritis [35]. Sometimes, these pathologies can be resolved conservatively [35]. However, when removal of the female genitalia (ovary-vagina-hysterectomy) is necessary for medical reasons, careful consideration should be given to the course of the ureters, isolating their last section from the ipsilateral lateral vagina. In obese animals it may not be possible to separate these ducts, in which case, only the ovaries and uteri could be removed (ovariohysterectomy) [14].

As stated by Pearson (1945) [30], the long urogenital sinus of the sugar glider seems to be characteristic of marsupials. He also described a clitoris in *Petaurus breviceps*, located on the ventral wall of the urogenital sinus, near its posterior extremity. According to Pavlicev et al. (2022) [36], there seems to be a relationship between the anatomy of the penis and that of the clitoris in marsupials: when the *glans penis* is bifurcated, the clitoris is also forked. Thus, a bifurcated clitoris was found in female short-tailed opossums [14], the Tasmanian devil (*Sarcophilus satanicus*) [29], the Tasmanian tiger (*Thylacinus cynecephalus*) [27] and the sugar glider (*Petarurus breviceps*). In contrast, it is unforked in macropodids [37]. Matthews (1947) [25] described the clitoris in *Dendrolagus ursinus* as a conical structure, flattened on its dorsal side, convex on its ventral side and lying in a pocket, so it did not project into the urogenital sinus lumen. In addition, Gonçalves et al. (2009) [24] reported the existence of erectile tissue in the ventral wall of the urogenital sinus, next to its end, forming a genital tubercle that resembles a clitoris.

The short urethra of the sugar glider follows the general rule of marsupials indicated by Pearson (1945) [30]. According to his studies, the only exceptions were the genders/genere *Parameles* and *Bettongia*, due to the cranial displacement of the urinary bladder. The urethral orifice (*ostium urethrae externum*), placed on the ventral floor of the urogenital sinus, was previously identified by Girling (2013) [16] at the same level of the lateral *vaginae* ends and had also been described in other species (Hughes, 2000) [27], although it does not seem to be a constant rule in marsupials. Thus, in some South American species, such as

Didelphis albiventris [38], the external urethral orifice is located cranially at the end of the lateral *vaginae*, but caudally in *Dendrolagus ursinus* [25].

Although we did not observe it in the sugar glider, a papilla has been described at the end of the urethra (external urethral orifice) in some marsupial species [25].

The existence of a reduced cloaca, such as the sugar glider one, seems to be a common feature of marsupials [39], although in some South American species, such as *Didelphis albiventris* [38], *Chinorectes minimus* [40] and *Monodelphis* [36], the urogenital sinus and the anus open separately on the surface of the body.

5. Conclusions

Sugar gliders are lively, playful and intelligent marsupials that have gained recent popularity as exotic pets. They are a small, lovely and social species, and bond closely with their owners if given the socialisation they require. Like other pets, they are susceptible to various illnesses and they arrive at veterinary clinics. Although dystocias are not frequent because the joeys are very small when they are born, as mammals they can suffer from uterine inertia, genital neoplasias, malpositioned foetus, malformation of the reproductive tract and so on. Occasionally, it may be necessary to perform an ovariohysterectomy in cases of pyometra. On these and other occasions, clinicians need extensive knowledge of the female genital tract to reach a correct diagnosis when using imaging methods and performing surgery, if necessary. According to our study, the female genital system of the *Petaurus breviceps* has duplicate genital organs as it corresponds to a marsupial with a vaginal complex that opens into a long urogenital sinus, carrying both the foetus and urine to a cloaca, which also receives the faeces from the digestive system. Our results provide useful information for clinicians, especially for the correct interpretation of modern imaging techniques, such as magnetic resonance imaging (MRI) and computed tomography (CT), as well as to perform successful surgery in the female genital tract when needed.

Author Contributions: Conceptualization, M.d.M.Y., D.A.-P. and M.L.; Writing—Original Draft Preparation, M.d.M.Y., D.A.-P. and M.L.; Disecctions, M.d.M.Y., D.A.-P. and M.L.; Writing—Review and Editing, M.d.M.Y., D.A.-P. and M.L. All authors have read and agreed to the published version of the manuscript.

Funding: This research received no external funding.

Institutional Review Board Statement: The Institutional Review Board statement was not required as no live animals were used to carry out this study.

Informed Consent Statement: Not applicable.

Data Availability Statement: Data sharing is not applicable to this article as all data associated is available in the text.

Acknowledgments: The authors are thankful to Nuria Pérez for donating the corpses of some specimens used in this study.

Conflicts of Interest: The authors declare no conflict of interest.

References

1. Brust, D. Sugar Gliders. *Exot. DVM* **2009**, *11*, 32–41.
2. Campbell, C.D.; Pecon-Slattery, J.; Pollak, R.; Joseph, L.; Holleley, C.E. The origin of exotic pet Sugar Gliders (*Petaurus breviceps*) kept in the United States of America. *PeerJ* **2019**, *7*, e6180. [CrossRef]
3. Raftery, A. Sugar Gliders (*Petaurus breviceps*). *Companion Anim.* **2015**, *20*, 422–426. [CrossRef]
4. Dierenfeld, E.S.; Thomas, D.; Ives, R. Comparison of commonly used diets on intake, digestion, growth, and health in captive Sugar Gliders (*Petaurus breviceps*). *J. Exot. Pet Med.* **2006**, *15*, 218–224. [CrossRef]
5. Dierenfeld, E.S. Feeding behavior and nutrition of the Sugar Glider (*Petaurus breviceps*). *Vet. Clin. N. Am. Exot. Anim. Pract.* **2009**, *12*, 209–215. [CrossRef]
6. Stojanovic, D.; Webb, M.H.; Alderman, R.; Porfirio, L.L.; Heinsohn, R. Discovery of a novel predator reveals extreme but highly variable mortality for an endangered migratory bird. *Divers. Distrib.* **2014**, *20*, 1200–1207. [CrossRef]
7. Smith, M.J. *Petaurus breviceps*. *Mamm. Species* **1973**, *1*, 1–5. [CrossRef]

8. Judah, V.; Nuttal, K. Exotic animal care & management. In *Exotic Animal Care & Management*; Thomson Delmar Learning: Scarborough, ON, Canada, 2008; pp. 118–131.
9. Shaw, G. Reproduction. In *Marsupials*; Armati, P.J., Dickman, C.R., Hume, I.D., Eds.; Cambridge University Press: Cambridge, UK, 2006; pp. 83–107, ISBN 9780511541889.
10. Renfree, M.; Shaw, G. Reproduction in Monotremes and Marsupials. *eLS* **2001**. [CrossRef]
11. Russell, E.M. Patterns of parental care and parental investment in Marsupials. *Biol. Rev.* **1982**, *57*, 423–486. [CrossRef]
12. Tyndale-Biscoe, H.; Renfree, M. *Reproductive Physiology of Marsupials*; Cambridge University Press: Cambridge, UK, 1987; ISBN 9780521337922.
13. Johnson-Delaney, C. Sugar Gliders. In *Ferrets, Rabbits, and Rodents: Clinical Medicine and Surgery*; Quesenberry, K.E., Carpenter, J.W., Eds.; Elsevier: St. Louis, MO, USA, 2021; pp. 385–400, ISBN 978-0-323-48435-0.
14. Johnson-Delaney, C.A. Reproductive medicine of companion Marsupials. *Vet. Clin. N. Am. Exot. Anim. Pract.* **2002**, *5*, 537–553. [CrossRef]
15. Banks, R.E.; Sharp, J.M.; Doss, S.D.; Vanderford, D. Exotic small mammal care and husbandry. In *Exotic Small Mammal Care and Husbandry*; Wiley: Blackwell, OK, USA, 2010; pp. 157–168, ISBN 9780-8138-1022-5.
16. Girling, S.J. Basic small mammal anatomy and physiology. In *Veterinary Nursing of Exotics Pets*; Blackwell Publishing Company: Iowa, IA, USA, 2013; pp. 21–23, ISBN 9780470659175.
17. Wang, J.; Wong, E.S.W.; Whitley, J.C.; Li, J.; Stringer, J.M.; Short, K.R.; Renfree, M.B.; Belov, K.; Cocks, B.G. Ancient antimicrobial peptides kill antibiotic-resistant pathogens: Australian mammals provide new options. *PLoS ONE* **2011**, *6*, e24030. [CrossRef] [PubMed]
18. Kubota, K.; Shimizu, T.; Shibanai, S.; Nagae, K.; Nagata, S. Histological properties and biological significance of pouch in red Kangaroo, Macropus Rufus. *Anat. Anz.* **1989**, *168*, 169–179.
19. Ambatipudi, K.; Joss, J.; Raftery, M.; Deane, E. A Proteomic approach to analysis of antimicrobial activity in marsupial pouch Secretions. *Dev. Comp. Immunol.* **2008**, *32*, 108–120. [CrossRef] [PubMed]
20. Edwards, M.J.; Deakin, J.E. The Marsupial pouch: Implications for reproductive success and mammalian evolution. *Aust. J. Zool.* **2013**, *61*, 41–47. [CrossRef]
21. Shaw, G.; Renfree, M.B.; Short, R.V.; O, W.S. Experimental manipulation of sexual differentiation in Wallaby pouch young treated with exogenous steroids. *Development* **1988**, *104*, 689–701. [CrossRef]
22. Renfree, M.B.; Chew, K.Y.; Shaw, G. Hormone-independent pathways of sexual differentiation. *Sex. Dev.* **2014**, *8*, 327–336. [CrossRef]
23. Renfree, M.B.; Short, R.V. Sex Determination in Marsupials: Evidence for a Marsupial-Eutherian dichotomy. *Philos. Trans. R. Soc. Lond. B Biol. Sci.* **1988**, *322*, 41–53. [CrossRef]
24. Gonçalves, N.N.; Maçanares, C.A.F.; Miglino, M.A.; Samoto, V.Y.; Martins, D.D.S.; Ambrósio, C.E.; Ferraz, R.H.; Dos, S.; de Carvalho, A.F. Aspectos morfológicos sos órgãos genitais femininos do Gambá (*Didelphis* sp.). *Braz. J. Vet. Res. Anim. Sci.* **2009**, *46*, 332. [CrossRef]
25. Matthews, L.H. A note on the female reproductive tract in the tree Kangaroos (Dendrolagus). *Proc. Zool. Soc. Lond.* **1947**, *117*, 313–333. [CrossRef]
26. Taggart, D.A. A Comparison of sperm and embryo transport in the female reproductive tract of Marsupial and Eutherian mammals. *Reprod. Fertil. Dev.* **1994**, *6*, 451–472. [CrossRef]
27. Hughes, R. Structure of the female reproductive tract of an adult parous Tasmanian Tiger, Thylacinus Cynocephalus. *Aust. J. Zool.* **2000**, *48*, 487–499. [CrossRef]
28. Johnson-Delaney, C.A.; Lennox, A.M. Reproductive disorders of Marsupials. *Vet. Clin. N. Am. Exot. Anim. Pract.* **2017**, *20*, 539–553. [CrossRef]
29. Flynn, T.T. Contributions to a knowledge of the anatomy and development of the Marsupialia. No. i. The Genitalia of Sarcophilus Satanicus (Female). *Proc. Linn. Soc. N. S. Wales* **1911**, *35*, 873–887. [CrossRef]
30. Pearson, J. The Female Urogenital System of the Marsupialia with Special Reference to the Vaginal Complex. *Pap. Proc. R. Soc. Tasman.* 1944, pp. 71–98. Available online: https://eprints.utas.edu.au/13473/ (accessed on 6 June 2023).
31. Rodger, J.C. Fertilization of Marsupials. In *A Comparative Overview of Mammalian Fertilization*; Dunbar, B.S., O'Rand, M.G., Eds.; Springer: Boston, MA, USA, 1991; pp. 117–135.
32. Rodger, J.C. Marsupials the alternative Therians—From gametes to birth. *Theriogenology* **2020**, *150*, 405–411. [CrossRef]
33. Rodger, J.C.; Paris, D.B.B.P.; Czarny, N.A.; Harris, M.S.; Molinia, F.C.; Taggart, D.A.; Allen, C.D.; Johnston, S.D. Artificial insemination in Marsupials. *Theriogenology* **2009**, *71*, 176–189. [CrossRef]
34. Kubiac, M. Sugar Glider. In *Handbook of Exotic Pet Medicine*; Kubiac, M., Ed.; John Wiley & Sons Ltd.: Hoboken, NJ, USA, 2021; pp. 125–140, ISBN 9781119389958.
35. Bassan, T.; Cobos, A.; Mallol, C.; Abarca, M.L.; Martorell, J. Reproductive tract infection caused by *Kocuria kristinae* in an entire female sugar glider (*Petaurus breviceps*). *Vet. Rec. Case Rep.* **2022**, *10*, e507. [CrossRef]
36. Pavlicev, M.; Herdina, A.N.; Wagner, G. Female genital variation far exceeds that of male genitalia: A Review of comparative anatomy of clitoris and the female lower reproductive tract in Theria. *Integr. Comp. Biol.* **2022**, *62*, 581–601. [CrossRef] [PubMed]
37. Renfree, M.B.; Chew, K.Y.; Shaw, G. Inducing sex reversal of the urogenital system of Marsupials. *Differentiation* **2014**, *87*, 23–31. [CrossRef] [PubMed]

. Schimming, B.C.; Cesario, M.D.; Matheus, S.M.M. Morphology of the vaginal complex in the White-Eared Opossum (*Didelphis albiventris*): Gross anatomy and light microscopy. *J. Vet. Med. Ser. C Anat. Histol. Embryol.* **2018**, *47*, 566–572. [CrossRef]
. Kardong, K.V. The Urogenital system. In *Vertebrates: Comparative Anatomy, Function, Evolution*, 6th ed.; Evolution; Mc Graw Hill: New York, NY, USA, 2012; p. 581. ISBN 978-0-07-352423-8.
. Nogueira, J.C.; Castro, A.C.S.; Câmara, E.V.C.; Câmara, B.G.O. Morphology of the male genital system of Chironectes minimus and comparison to other didelphid marsupials. *J. Mammal.* **2004**, *85*, 834–841. [CrossRef]

animals

MDPI

Article

Geometric and Morphometric Analysis of the Auditory Ossicles in the Red Fox (*Vulpes vulpes*)

Nedžad Hadžiomerović [1,*], Ozan Gundemir [2], Faruk Tandir [1], Rizah Avdić [1] and Muhamed Katica [3]

[1] Department of Basic Sciences of Veterinary Medicine, Veterinary Faculty, University of Sarajevo, 71000 Sarajevo, Bosnia and Herzegovina
[2] Department of Anatomy, Faculty of Veterinary Medicine, Istanbul University-Cerrahpasa, Istanbul 34500, Turkey
[3] Department of Clinical Sciences of Veterinary Medicine, Veterinary Faculty, University of Sarajevo, 71000 Sarajevo, Bosnia and Herzegovina
* Correspondence: nedzad.hadziomerovic@vfs.unsa.ba; Tel.: +387-61-914-924

Simple Summary: The aim of the study was to determine the morphological and morphometrical features of the auditory ossicles in the red fox. Moreover, for the first time, shape analysis was performed on all the middle ear bones by geometric morphometry. The auditory ossicles play a vital role in transmitting sound waves through the middle ear. The malleus was considered to be the largest bone, with three distinctive processes. The overall shape of the incus was found to be similar to that in other mammals. The presence of a prominent muscular process was noted on the head of the stapes and the convex base which closes the window of the inner ear.

Abstract: The use of carnivores as experimental models in auditory biology has led to a significant improvement regarding our understanding of the structure and function of the ear. Considering that data regarding the anatomy of the middle ear in the red fox are rare, this study aimed to describe the morphological and morphometrical features of the auditory ossicles in the red fox, as well as to provide their shape characteristics by geometric analysis. Nine adult red foxes were used in the study. The malleus, incus and stapes were extracted from the middle ear, prepared, photographed and measured using the software. For the geometric analysis, 19 landmarks were used. Following Principal Component Analysis (PCA), PC1 was found to explain approximately half of all variance (incus: 49.97%; malleus: 49.93%; stapes: 58.49%). The study demonstrated the similar anatomical organization of the auditory ossicles in line with important morphometric and basic geometric data, which can contribute to this field and add a useful perspective to the literature.

Keywords: ear ossicles; geometric morphometrics; morphometry; red fox

1. Introduction

The extremely adaptable and most numerous representative of the *Vulpes* genus is the red fox (*Vulpes vulpes* Linnaeus, 1758). It is present in the steppe biomes and mixed forests of Eurasia within the temperate climate zone [1,2]. The red fox has adapted to life in different types of habitats and climates and can be seen in urban and suburban areas, which has led to partial changes in behavior and in its manner and type of nutrition [1,3]. *Vulpes vulpes* belongs to the order of Carnivora; however, unlike dogs and cats, it is actually an undomesticated canid, between a large cat and a small dog in size [4]. However, as the largest species of the genus *Vulpes*, its body size ranges between 3 and 14 kg [5]. The ear (organum vestibulocochleare) is a complex sense organ that includes both the organs of balance and of hearing. As the organ is subdivided into three parts, sound waves are received by the external ear (Auris externa), transmitted through the middle ear (Auris media), and transformed into electrical signals by the inner ear (Auris interna), where the vestibular organ is located [6]. The middle ear extends from the tympanum (membrana tympani) to the oval vestibular window (fenestra vestibuli) and houses three auditory ossicles joined in the form of a chain. In many mammals, the tympanic cavity extends ventrally from the skull base and forms the bulla tympanica [7]. The middle ear cavity is connected with the pharynx by way of the auditory or Eustachian tube. The ossicles mediate transmission of vibrations through the tympanic cavity to the inner ear [8,9]. The most lateral and largest bone is the malleus, which consists of a head, neck and handle, embedded in the tympanic membrane. The incus is divided into a body with two limbs and a lenticular process. The stapes consist of a head, two limbs and a base [5,10]. One of the best studied species in auditory biology is the domestic cat (*Felis sylvestris catus*). The use of this species as an experimental model has led to a great deal being known about the structure and function of the ear [4,11–13]. The auditory biology and morphometry of the ears of dogs (*Canis lupus familiaris*) have also been relatively well studied [4,14]. The anatomical features of the middle ear and its parameters are often used to predict the hearing sensitivity of the species. Although the red fox has similar anatomical structures of the middle ear to dogs and cats, the morphological aspect of ear structures should be taken with caution when considering the hearing attributes of related species [4]. Furthermore, the anatomical and morphological characteristics of the hearing apparatus of the red fox have been described, and it was determined that the middle ear is very similar to that of the dog; however, the cochlear structures of red foxes are very similar to those of cats [4]. Detailed descriptions of the morphological features and morphometric parameters of the bulla tympanica, tympanic membrane and middle ear ossicles have been composed for the red fox [15]. The study revealed values for several parameters of the auditory ossicles, as well as weight parameters and area measurements for the tympanic membrane, and the oval and round window. Similar morphological studies on the middle ear structures were performed on carnivores, such as wolves [10], dogs [16,17], badgers [18], red foxes, dogs and cats [4], on domestic and wild carnivores [19], on some herbivores, such as goats [20], sheep [21], horses [22], donkeys [23,24], on ruminants [25], on rodents, such as mice [26], hamsters [27], rabbits [9,28], chinchillas [29,30], and on some birds, such as ostriches [31] and chickens [32]. Geometric morphometry provides more precise information about small anatomical structures and focuses on data analysis from the position of homologous landmarks [33,34]. The process of analysis excludes factors such as position and size, revealing only shape-related differences [35]. To the best of the authors' knowledge, no data were found about the detailed morphometry of the auditory ossicles and their anatomical parts in the red fox. The present study aims to describe detailed morphological and morphometrical data regarding the auditory ossicles of the red fox, as well as to reveal their basic shape characteristics using the geometric morphometric method.

2. Materials and Methods

2.1. Study Sample

Nine red foxes (seven females and two males) were used in this study. All the animals were adults according to their teeth formula, aged between 6–12 months. Additional methods for age determination were the fusion of the cranial sutures and cementum layers on the longitudinal section of the canine tooth. The crania were all obtained from the Department of Pathology after autopsies were performed. The muscles and the fascia were dissected, boiled and macerated to remove fat, and finally the skulls were bleached with 3% hydrogen peroxide. After the cleaning procedure was complete, the ossicula auditus were extracted from the tympanic cavity via the external auditory channel.

2.2. Morphometric–Analysis

The samples were prepared and digital images were collected using a stereomicroscope (ZEISS Stemi 508) under 0.63 magnification with a camera (ZEISS Axiocam 208, Jena, Germany). The images were further processed using ZEISS ZEN 3.0 software (blue edition) and measured using ImageJ® software after the calibration procedure. The samples were measured in order to determine the differences between the auditory ossicles on the right and left sides.

The parameters for measurement of the auditory ossicles were selected according to previous studies [10,18,29].

The following measurements were performed (Figure 1):

- LM—length of the malleus
- WHM—width of the head of the malleus
- LHM—length of the head of the malleus
- LhM—length of the handle of the malleus
- LI—length of the incus
- LLC—length of long crus of the incus
- LSC—length of short crus of the incus
- HBI—height of the body of the incus
- WBI—width of the body of the incus
- LS—length of the stapes
- LCC—length of the caudal crus of the stapes
- LRC—length of the rostral crus of the stapes
- WHS—width of the head of the stapes
- WBS—width of the base of the stapes

The anatomical structures of the auditory ossicles were identified and named according to the NAV [36] and other relevant literature [18,19].

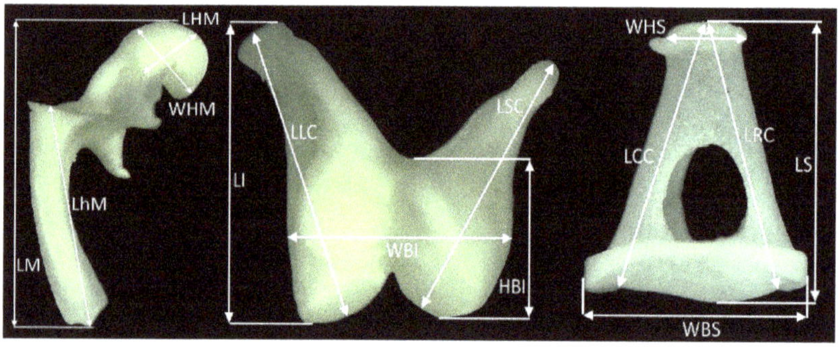

Figure 1. Measurements of the auditory ossicles. Abbreviations—see text.

2.3. Geometric Analysis

For the geometric morphometric analysis, digital images were taken with the same magnification using the stereomicroscope. The files were converted into "tps" format using the tpsUtil (version 1.74) program [37]. The landmarks used in the study are shown in Figure 2. TpsDig2 (version 2.32) was used for landmarking.

Incus, 1—leftmost peak of the crus longum; 2—rightmost peak of the crus longum; 3—angle of the crus breve with the crus longum; 4—highest point of the crus breve; 5—left corner point of corpus incudis; 6—middle point of the articulation incundomalleolaris;

Malleus, 7—connecting point of the caput mallei and collum mallei; 8—top of processus rostralis; 9—top of the processus muscularis; 10—connecting of the manubrium mallei; 11—terminal point of the manubrium mallei; 12—upper segment of the processus lateralis; 13—lower segment of the processus lateralis;

Stapes, 14—highest connecting point of the caput and crus caudale; 15—highest connecting point of the caput and crus rostrale; 16—connecting point of the crus caudale and basis stapedis; 17—caudal point of the basis stapedis; 18—connecting point of the crus rostrale and basis stapedis: 19—rostral point of the basis stapedis.

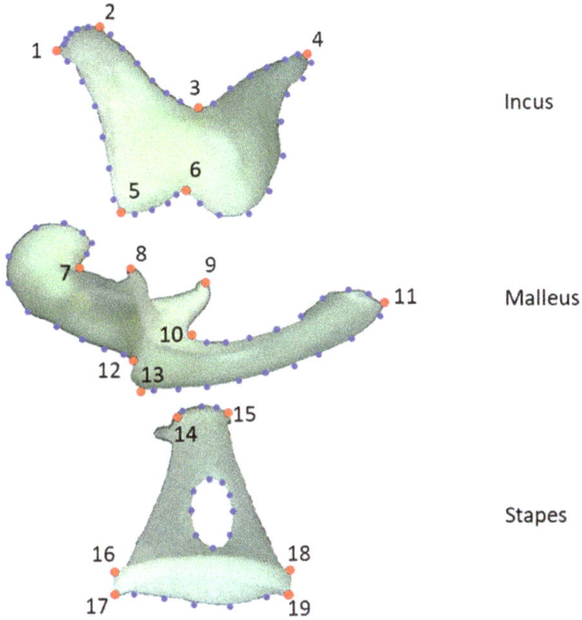

Figure 2. Landmarks (red) and Semilandmarks (blue).

2.4. Statistical Analysis

MorphoJ (version 1.07a) software was used for the statistical part of the geometric morphometric analysis. The landmark file was imported into MorphoJ, and "Procrustes fit" was applied first. A Generalized Procrustes Analysis was applied to the imported landmark data before analysis. Principal Component Analysis (PCA) was performed to determine the shape variations among the bones.

3. Results

The auditory ossicles of the red fox are composed of three bones: the malleus, incus and stapes. The malleus, as the most external lateral part of the ossicles chain, is connected to the tympanic membrane and articulates with the middle ossicle, the incus, followed by the most internal auditory ossicle, the stapes, which closes the fenestra vestibuli. The largest bone is the malleus (Figure 3), consisting of three distinctive anatomical parts: the head of the malleus (caput mallei), the neck of the malleus (collum mallei) and the handle (manubrium mallei). The head of the malleus is slightly ovoidal in shape and connected with the neck. The articular surface, in the form of a saddle-like impression, is located on the caudo-medial aspect of the head. The head (caput) is laterally divided from the neck and osseous lamina by a small notch. These three processes were identified on the malleus of the red fox. The anterior process (processus rostralis) is located near the head and is triangular in shape, oriented laterally. The longer muscular process (processus muscularis) originates near the base of the manubrium and has a sharp ending. The muscular process extends ventrally in the same direction as the manubrium and attaches to the m. tensor tympani muscle. Opposite the muscular process is the short, lateral process (processus lateralis). The handle of the malleus represents around 70% of the total malleus length. The total malleus length (LM) measures 7.21 ± 0.53 mm on the right and 7.18 ± 0.53 mm on the left side, while the manubrium (LhM) measures on the right 5.19 ± 0.53 mm and left 5.27 ± 0.42 mm, respectively (Table 1). At the beginning, the manubrium has three sides in cross section; however, further along distally, it continues as a double-sided structure. The terminal part of the manubrium is embedded in the tympanic membrane.

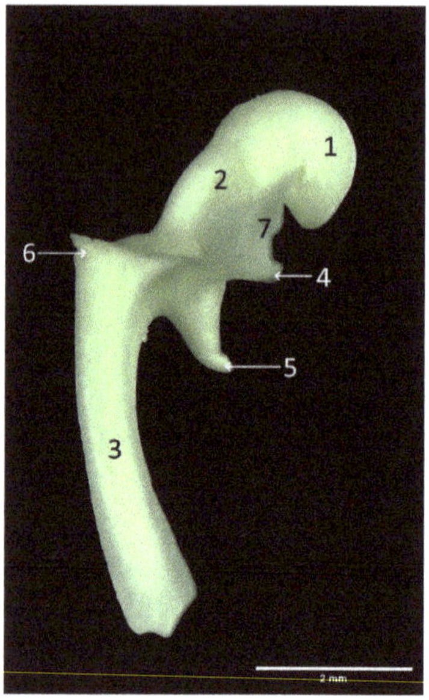

Figure 3. Malleus. 1. Caput mallei; 2. Collum mallei; 3. Manubrium mallei; 4. Processus rostralis; 5. Processus muscularis; 6. Processus lateralis; 7. Lamina.

Table 1. Morphometric values of the auditory ossicles (mm).

Measurement	No	Right Side Mean ± SD	Left Side Mean ± SD	Red Fox (Malkemper, 2014) [15]	Badger (Martonos et al., 2022) [18]	Wolf—Rs (Gürbüz et al., 2019) [10]
Malleus						
LM	7	7.21 ± 0.53	7.18 ± 0.53		8.14 ± 0.14	9.35 ± 0.14
WHM	7	1.75 ± 0.07	1.73 ± 0.10		2.26 ± 0.014	2.16 ± 0.14
LHM	7	1.15 ± 0.10	1.19 ± 0.10			1.60 ± 0.12
LhM	7	5.19 ± 0.53	5.27 ± 0.42	3.22 ± 0.38 * 7.04 ± 0.31 **	4.85 ± 0.15	6.73 ± 0.67
Incus						
LI	7	2.34 ± 0.38	2.25 ± 0.09		2.72 ± 0.12	3.01 ± 0.32
LLC	7	2.44 ± 0.47	2.33 ± 0.11	2.06 ± 0.12	2.69 ± 0.21	3.09 ± 0.23
LSC	7	2.39 ± 0.54	2.23 ± 0.17	2.14 ± 0.19	1.99 ± 0.1	2.73 ± 0.65
HBI	7	1.31 ± 0.27	1.22 ± 0.08		1.48	1.73 ± 0.25
WBI	7	2.19 ± 0.41	2.01 ± 0.10		2.07 ± 0.14	2.26 ± 0.14
Stapes						
LS	3	2.11 ± 0.10	2.11 ± 0.10	2.11 ± 0.15	2.3 ± 0.08	2.57 ± 0.12
LCC	3	2.05 ± 0.14	2.07 ± 0.11		2.09 ± 0.2	2.51 ± 0.03
LRC	3	2.11 ± 0.10	2.11 ± 0.10		1.95 ± 0.11	2.77 ± 0.09
WHS	3	0.61 ± 0.05	0.61± 0.07		0.79 ± 0.13	0.49 ± 0.12
WBS	3	1.88 ± 0.10	1.86 ± 0.09		2.16	2.01 ± 0.21

* Measured as the distance from the presumed axis of rotation. ** Measured from the pivotal point of the malleo-incudal joint. LM—length of the malleus; WHM—width of the head of the malleus; LHM—length of the head of the malleus; LhM—length of the handle of the malleus; LI—length of the incus; LLC—length of long crus of the incus; LSC—length of short crus of the incus; HBI—height of the body of the incus; WBI—width of the body of the incus; LS—length of the stapes; LCC—length of the caudal crus of the stapes; LRC—length of the rostral crus of the stapes; WHS—width of the head of the stapes; WBS—width of the base of the stapes.

The incus, the second auditory ossicle, is significantly smaller than the malleus, measuring 2.34 ± 0.38 mm on the right side and 2.25 ± 0.09 mm on the left side. The overall shape of the incus resembles a biradicular molar tooth (Figure 4). Situated in the middle, it is connected with both auditory ossicles. With its body (corpus incudis) it articulates with the malleus, while its longer crus (crus longum) at its terminal part forms an articular surface with the lenticular process. This process ossifies with the long crus in later life, while juvenile animals have a separate additional bone, the lenticular bone (os lenticulare).

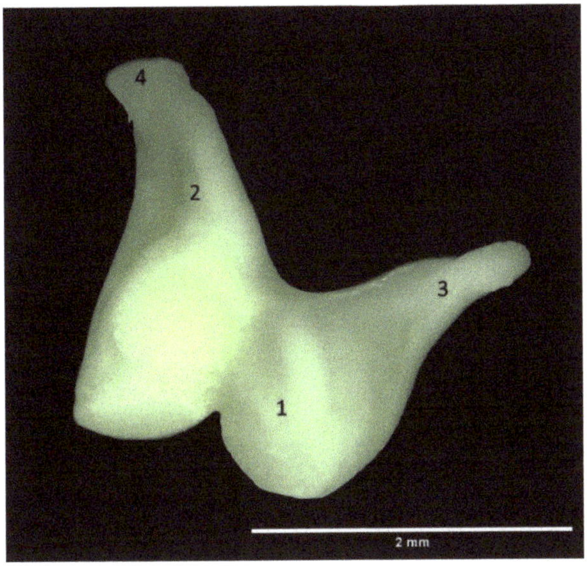

Figure 4. Incus. 1. Corpus incudis; 2. Crus longum; 3. Crus breve; 4. Processus lenticularis.

The articular surface of the corpus is divided by a notch into two smaller parts. The short crus (crus breve) continues in the dorso-caudal direction, ending in a pointed shape. The size of the two limbs is similar, but the right side is slightly larger (Table 1).

The stapes is the third and smallest auditory ossicle (Figure 5). As the innermost ossicle, it is interposed between the incus and the vestibular window, completing the transmission of vibrations to the perilymph in the inner ear. The total length of the stapes is the same on both sides, 2.11 ± 0.10 mm. The stapes is triangular in shape, consisting of the head (caput stapedis), two limbs, the rostral (crus rostrale), the posterior (crus caudale) and the base (basis stapedis). The head is a spherical structure that forms the articulation with the lenticular process. The small prominent part of the head represents the muscular process (tuberositas m. stapedius), used as the insertion point of the stapedial muscle (m. stapedius). This muscle, together with the malleus-attached muscle (m. tensor tympani), has an important role in the vibration enhancement mechanism in the inner ear. The head is distally connected with the base by the rostral and caudal crus, which are quite similar in size (Table 1). The small opening in the form of a tunnel (foramen intercrurale) between the limbs and base is closed by a stapedial membrane (membrana stapedis). The base of the stapes has an elliptical shape with central convexity which closes the window (fenestra vestibuli). This connection is secured with the annular ligament.

PCA was performed to reveal the inter-individual shape variations of the auditory ossicles. PC results for each bone are given in Table 2. For each bone, PC1 was found to explain approximately half of all variance (incus: 49.97%; malleus: 49.93%; stapes: 58.49%). PCA results and shape variations are given in Figures 6–8.

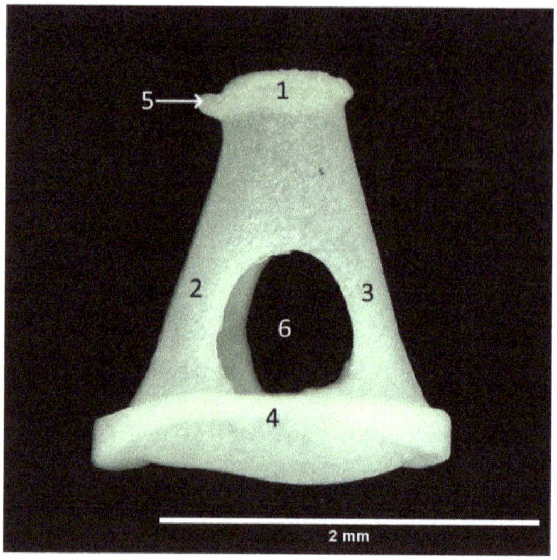

Figure 5. Stapes. 1. Caput stapedis; 2. Crus caudale; 3. Crus rostrale; 4. Basis stapedis; 5. Tuberositas m. stapedius; 6. Foramen intercrurale.

Table 2. Results of PCA (Principal Component Analysis) for the auditory ossicles.

PC	Incus			Malleus			Stapes		
	E	% V	C%	E	% V	C%	E	% V	C%
PC1	0.00362179	49.946	49.946	0.00373181	49.931	49.931	0.00198681	58.485	58.485
PC2	0.00204002	28.133	78.078	0.00172400	23.067	72.998	0.00141034	41.515	100.000
PC3	0.00085646	11.811	89.889	0.00100988	13.512	86.510	-	-	-
PC4	0.00050179	6.920	96.809	0.00049064	6.565	93.075	-	-	-
PC5	0.00023140	3.191	100.000	0.00033112	4.430	97.505	-	-	-
PC6	-	-	-	0.00018645	2.495	100.000	-	-	-

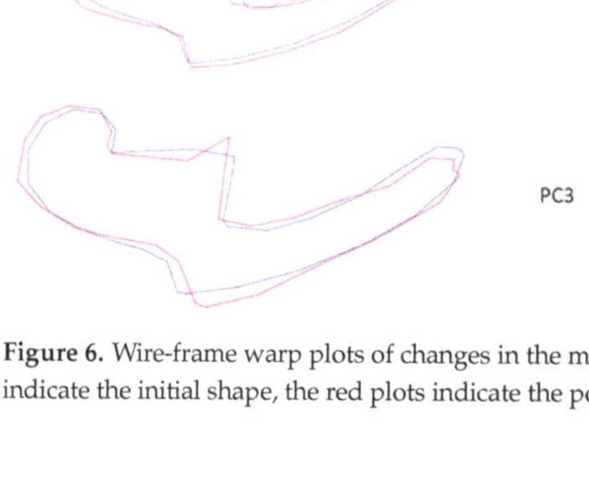

Figure 6. Wire-frame warp plots of changes in the malleus shape in PC1, PC2 and PC3. The blue plots indicate the initial shape, the red plots indicate the positive limit of the principal component's scores.

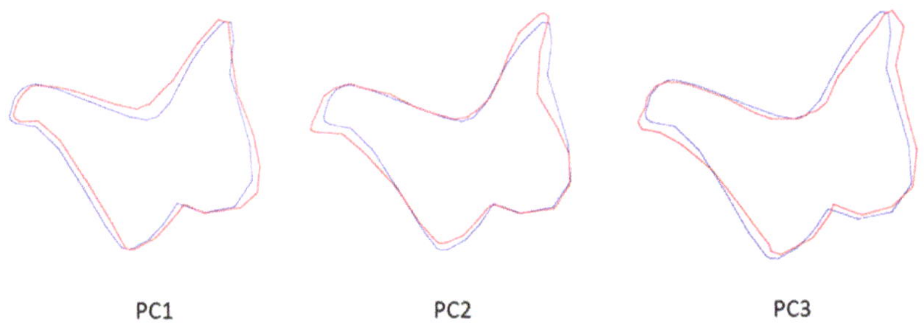

Figure 7. Wire-frame warp plots of changes in incus shape in PC1, PC2 and PC3. The blue plots indicate the initial shape; the red plots indicate the positive limit of the principal component's scores.

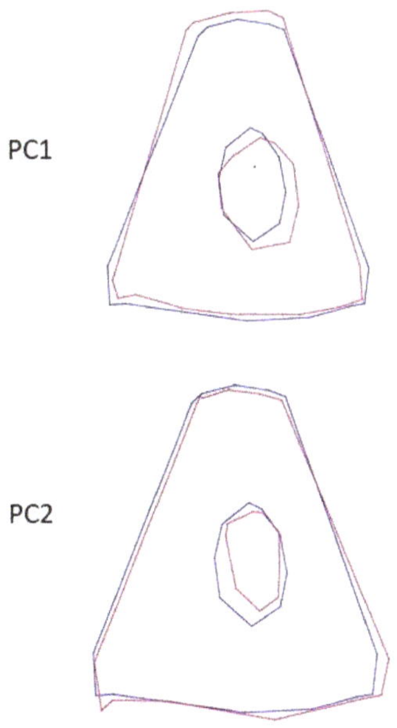

Figure 8. Wire-frame warp plots of changes in shape of the stapes in PC1 and PC2. The blue plots indicate the initial shape; the red plots indicate the positive limit of the principal component's scores.

Wire-frame warp plots of changes in the malleus shape in PC1, PC2 and PC3 are shown in Figure 5. Compared to PC1, the most variation in the shape of the malleus was seen in the collum mallei. The higher PC2 value represents the larger caput mallei. In PC3, the positive value represents the longer processus muscularis. Shape variations in the incus were generally in the middle part of the corpus incudis. There were also inter-individual shape variations in the processus lenticularis. In the stapes, inter-individual shape variations were generally seen in the foramen intercrurale (Figure 8). There were differences in shape between individuals at the borders of the foramen intercrurale.

The mean shape for female and male is given in Figure 9. In incus, the highest point of the crus breve in males was more oval in shape than in females. In addition, the angle of the crus breve with the crus longum in males was narrower than in females. The tip of the male manubrium mallei was wider in shape than that of the females. The connecting point of the caput mallei and collum mallei was deeper in females. Additionally, the intercrural foramen in the stapes was wider in shape in females.

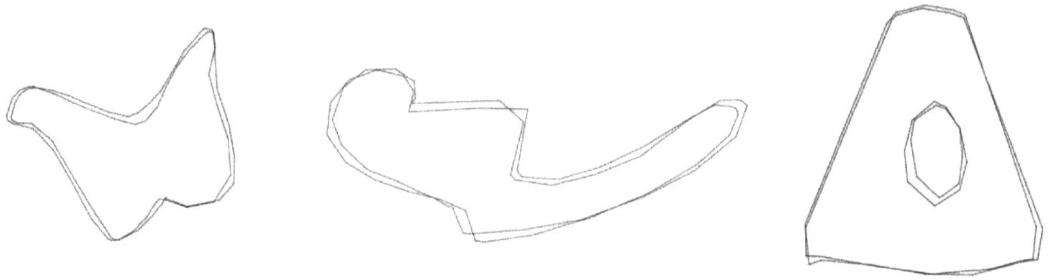

Figure 9. Main shapes for female (blue) and male (red).

4. Discussion

The morphological characteristics of the auditory structures are closely associated with different environmental and behavioral patterns [13,38]. Some reports have suggested, at least in principle, that hearing ability can be predicted from the anatomical structures [39]. However, despite the significant difference in body and tympanic membrane size, behavioral audiograms are similar in various breeds of dog [40]. The red fox audiogram showed similar hearing characteristics with both dogs and cats [41]. The middle ear structures have been relatively well studied in domestic carnivores [12–14]. The middle ear apparatus of the fox, including the auditory ossicles, is, as expected, very similar to that of cats and dogs [4]. The present study revealed a detailed morphometric analysis of the auditory ossicles of the red fox, including shape analysis, using geometric morphometry.

The three auditory ossicles are located in the middle ear of the red fox, as in most mammals: the malleus, incus and stapes. The macroanatomical features of the malleus revealed a similar shape to that in wolves [10]. The overall size of the malleus in the fox was slightly smaller than that of the badger and goat but larger than that of the cat, hamster and macaque monkey [18,19,27,42]. The ovoidal shape of the head of the malleus in the fox was similar to that found in previous studies in carnivores [10,18,19]. The muscular and rostral processes on the malleus were longer than the lateral process in the wolf [10], which is similar to our findings in the fox. The size of the anterior and shorter lateral processes of the malleus was 2.25 ± 0.25 mm and 1.49 ± 0.11 mm, respectively [15]. The robust and curved manubrium was typical for carnivores [4], while its triangular shape in cross-section was in accordance with the results reported from the badger and goat [18,20].

Previous reports described the shape of the incus as similar to a biradicular molar tooth and also as an anvil with a saddle [4,9,15,18,20]. The two divergent crura differ greatly in size in the majority of carnivores and other mammals [9,18,29]. However, our results reported very small differences between the longer and short crus of the incus (Table 1). Slightly smaller values of both limbs of the incus were reported in a previous study in the red fox [15]. Similar sizes of the two incus limbs were reported in goats and water buffaloes [20,43].

The smallest auditory ossicle, the stapes, has a unique triangular shape, similar in many species, and it is composed of the head, two crura and the base of the stapes [9,10,29]. The specific rectangular shape of the stapes has been noted in ruminants [25]. The red fox stapes was recorded as most fragile of the three ossicles with a close resemblance to a stirrup [15]. The small head on the top carries the prominent muscular process for the stapedial muscle, which was previously described in rabbits, dogs and goats [9,18,20]. The overall length of the stapes was smaller than in the wolf [10], badger [18] and goat [20]. A previous report on the length of the stapes in the red fox showed identical results to ours, and the study was performed on a larger number of individuals (n = 26) compared to our study [15]. The rostral crus of the stapes was slightly longer than the caudal crus; however, no differences were found between the right and left middle ears. A longer caudal crus of the stapes has been described in goats, hamsters and chinchilla [20,27,29]. The base of the stapes was represented by a little oval plate which was connected to the vestibular window through an annular ligament [25]. The study measured the weight parameters for ear ossicles in red fox; the mean weight of the malleus was 10.64 ± 1.18 mg, for incus 5.19 ± 0.63 mg and for stapes 0.63 ± 0.13 mg [15].

Recent studies have used geometric morphometric methods for various species to contribute to the taxonomic classification of animals and to determine sex dimorphism [35,44–47]. One study examined the shape variations of the incus and the differences between the donkey and the horse. The results reported that the corpus incudis edges (right, left, and bottom) were flatter in donkeys [48]. The study revealed that taxonomy was possible for ear ossicles using the geometric analysis method. Sex discrimination within the same species may be possible using geometric shape analysis [44]. The differentiation was more evident in the dorsal aspect of the quail skull. In our results, shape variations between the male and female were close to each other. Some differences were noted in part of the collum of the malleus and the intercrural foramen in the stapes. The study was carried out with a small sample for sexual dimorphism, so further advances in this field depend very much on the availability of specimens of both sexes and different age groups.

5. Conclusions

The study presented a morphological description of the auditory ossicles in the red fox, alongside detailed morphometrical data of the particular parts of the ossicles. The results showed that the anatomical organization of the auditory ossicles was similar to that of other carnivores, in terms of morphometric parameters, which may contribute to the study of the comparative morphology of the middle ear. The study revealed the further potential of shape analysis of the auditory bones using the geometric morphometry method, which should provide a significant addition to traditional morphometric studies. However, some limitations should be considered regarding this study and its results. Geometric analysis should aim to include a larger sample size in order to obtain more details about the shape differences between males and females. Moreover, it would be useful to use more than one aspect of the bones, as the previous study showed. Overall, we believe that further geometric morphometric studies of the ear ossicles would be very useful and would contribute to the morphology of these bones.

Author Contributions: Conceptualization, N.H. and O.G.; methodology, N.H., O.G. and M.K.; software, F.T.; validation, R.A.; formal analysis, N.H.; investigation, F.T.; resources, M.K.; data curation, O.G.; writing—original draft preparation, N.H.; writing—review and editing, N.H. and M.K.; visualization, O.G.; supervision, R.A.; project administration, N.H.; funding acquisition, R.A. All authors have read and agreed to the published version of the manuscript.

Funding: This research received no external funding.

Institutional Review Board Statement: Ethical or animal welfare approval for the research was not required since the bones were obtained from the Department of Pathology after autopsy was performed.

Informed Consent Statement: Not applicable.

Data Availability Statement: The data presented in this study are available on request from the corresponding author.

Acknowledgments: The authors thank Darinka Klarić Soldo for her help with the stereomicroscope measurements.

Conflicts of Interest: The authors declare no conflict of interest.

References

1. Porobić, J.M. Geometric-Morphometric Analyses of Golden Jackal (*Canis aureus*) and Red Fox (*Vulpes vulpes*) Skulls from the Territory of Serbia: Biogeographical Aspects of Morphological Variability. Ph.D. Thesis, University of Belgrade, Belgrade, Serbia, 2017.
2. Lloyd, H.G. *The Red Fox*; Batsford: London, UK, 1980.
3. Gloor, S.; Bontadina, F.; Hegglin, D.; Deplazes, P.; Breitenmoser, U. The rise of urban fox populations in Switzerland. *Mamm. Biol.* **2001**, *66*, 155–164.
4. Malkemper, E.P.; Mason, M.J.; Burda, H. Functional anatomy of the middle and inner ears of the red fox, in comparison to domestic dogs and cats. *J. Anat.* **2020**, *236*, 980. [CrossRef] [PubMed]

5. Nowak, R.M. *Walker's Mammals of the World*, 6th ed.; The Johns Hopkins University Press: Baltimore, MD, USA, 1999.
6. König, H.E.; Liebich, H.G. *Veterinary Anatomy of Domestic Mammals: Textbook and Color Atlas*, 3rd ed.; Schattauer Co.: Stuttgart, Germany, 2007; pp. 595–601.
7. Keen, J.A.; Grobbelaar, C.S. The comparative anatomy of the tympanic bulla and auditory ossicles, with a note suggesting their function. *Trans. R. Soc. S. Afr.* **1941**, *28*, 307–329. [CrossRef]
8. Miller, M.E. *Anatomy of the Dog*; W.B. and Sounders: Philadelphia, PA, USA, 1964; pp. 853–858.
9. Kurtul, I.; Cevik, A.; Bozkurt, E.U.; Dursun, N. A detailed subgross morphometric study on the auditory ossicles of the New Zealand rabbit. *Anat. Histol. Embryol.* **2003**, *32*, 249–252. [CrossRef] [PubMed]
10. Gürbüz, I.; Demiraslan, Y.; Dayan, M.O.; Aslan, K. Morphometric and macroanatomic examination of auditory ossicles in male wolves (*Canis lupus*). *Folia Morphol.* **2019**, *78*, 600–605. [CrossRef]
11. Decraemer, W.F.; Khanna, S.M. Measurement, Visualization and Quantitative Analysis of Complete Three-Dimensional Kinematical Data Sets of Human and Cat Middle Ear. In *Middle Ear Mechanics in Research and Otology*; World Scientific: Singapore, 2004; pp. 3–10.
12. Puria, S.; Allen, J.B. Measurements and model of the cat middle ear: Evidence of tympanic membrane acoustic delay. *J. Acoust. Soc. Am.* **1998**, *104*, 3463–3481. [CrossRef]
13. Huang, G.; Rosowski, J.; Peake, W. Relating middle-ear acoustic performance to body size in the cat family: Measurements and models. *J. Comp. Physiol.* **2000**, *186*, 447–465. [CrossRef]
14. Cole, L.K. Anatomy and physiology of the canine ear. *Vet. Dermatol.* **2009**, *20*, 412–421. [CrossRef]
15. Malkemper, E.P. The Sensory Biology of the Red Fox—Hearing, Vision, Magnetoreception. Ph.D. Thesis, University of Duisburg Essen, Duisburg, Germany, 2014.
16. Ortug, G.; Ortug, C.; Türkmenoğlu, I. Scanning electron microscopic (SEM) examination of the incudomalleal (IM) joint in dogs. *Turk. J. Vet. Anim. Sci.* **2005**, *29*, 345–351.
17. Berghes, C.; Parvu, M.; Cucoanes, M.; Cuca, D. Anatomic Considerations on the Middle Ear in Dog. *Sci. Pap. Anim. Sci. Biotechnol.* **2010**, *43*, 450–452.
18. Martonos, C.; Gudea, A.; Latiu, C.; Blagojevic, M.; Stan, F. Morphological and Morphometrical Aspects of the Auditory Ossicles in the European Badger (Meles Meles). *Vet. Sci.* **2022**, *9*, 483. [CrossRef]
19. Wible, J.R.; Spaulding, M.A. Reexamination of the Carnivora Malleus (Mammalia, Placentalia). *PLoS ONE* **2012**, *7*, e50485. [CrossRef]
20. Martonos, C.; Gudea, A.; Damian, A.; Lacatus, R.; Purdoiu, R.; Cocan, D.; Stan, F.G. Morphological and morphometrical aspects of the auditory ossicles in goat (*Capra hircus*). *Anat. Histol. Embryol.* **2021**, *50*, 184–191. [CrossRef]
21. Dalga, S.; Aslan, K. A Macroanatomic and Morphometric Study on Ossicula Auditus in Male Hemshin Sheep. *Atatürk Üniv. Vet. Bil. Derg.* **2019**, *14*, 114–118. [CrossRef]
22. Blanke, A.; Aupperle, H.; Seeger, J.; Kubick, C.; Schusser, G.F. Histological study of the external, middle and inner ear of horses. *Anat. Histol. Embryol.* **2015**, *44*, 401–409. [CrossRef]
23. Demiraslan, Y.; Gurbuz, I.; Aslan, K. A Macroanatomic and morphometric study on Auditory ossicles in Donkey (*Equus asinus*). *Istanbul Üniv. Vet. Fak. Derg.* **2015**, *41*, 151–154.
24. Nazih, A.M. Anatomical Study on the Middle Ear of Donkey (*Equus acinus*). *Int. J. Adv. Res. Biol. Sci.* **2017**, *4*, 110–121. [CrossRef]
25. Botti, M.; Secci, F.; Ragionieri, L.; Dessole, A.A.; Acone, F. Auditory ossicles in the ruminants: Comparative morphological analysis with the analogues formations of horse. *Ann. Fac. Medic. Vet. Parm.* **2006**, *XXVI*, 91–96.
26. Mohammadpour, A.A. Morphological study of auditory ossicles in the mouse. *J. Appl. Anim. Res.* **2010**, *37*, 269–271. [CrossRef]
27. Mohammadpour, A.A. Morphology and morphometrical study of hamster middle ear bones. *Int. J. Virtual Real.* **2011**, *12*, 121–126.
28. Guan, M.; Zhang, J.; Jia, Y.; Cao, X.; Lou, X.; Li, Y.; Gao, X. Middle ear structure and transcanal approach appropriate for middle ear surgery in rabbits. *Exp. Ther. Med.* **2019**, *17*, 1248–1255. [CrossRef]
29. Martonos, C.; Damian, A.; Gudea, A.I.; Bud, I.; Stan, G.F. Morphological and morphometrical study of the auditory ossicles in chinchilla. *Anat. Histol. Embryol.* **2019**, *48*, 340–345. [CrossRef] [PubMed]
30. Wang, X.; Gan, R.Z. 3D finite element model of the chinchilla ear for characterizing middle ear functions. *Biomech. Model. Mechanobiol.* **2016**, *15*, 1263–1277. [CrossRef] [PubMed]
31. Beşoluk, K.; Dayan, M.O.; Eken, E.; Turgut, N.; Aydogdu, S. Macroanatomic structure and morphometric analysis of middle ear in ostrich (*Struthio camelus*). *Arch. Vet. Sci. Med.* **2019**, *2*, 8–16. [CrossRef]
32. Claes, R.; Muyshondt, P.G.G.; Van Hoorebeke, L.; Dhaene, J.; Dirckx, J.J.J.; Aerts, P. The effect of craniokinesis on the middle ear of domestic chickens (*Gallus gallus domesticus*). *J. Anat.* **2017**, *230*, 414–423. [CrossRef]
33. Slice, D.E. Geometric morphometrics. *Annu. Rev. Anthropol.* **2007**, *36*, 261–281. [CrossRef]
34. Gündemir, O.; Hadžiomerović, N.; Pazvan, G.; Erdikmen, D.O. Radiometric and geometric morphometric analysis of the carpal joint area in 2-year-old thoroughbred horses. *Veterinaria* **2021**, *70*, 209–217. [CrossRef]
35. Demircioglu, I.; Demiraslan, Y.; Gurbuz, I.; Dayan, M.O. Geometric Morphometric Analysis of Skull and Mandible in Awassi Ewe and Ram. *Kafkas Üniv. Vet. Fakültesi Derg.* **2021**, *27*, 43–49.
36. International Committee on Veterinary Gross Anatomical Nomenclature. *Nomina Anatómica Veterinaria*, 6th ed.; World Association of Veterinary Anatomists, Ed.; World Association of Veterinary Anatomists: Hannover, Germany, 2017; ISBN 0-9600444-7-7.
37. Rohlf, F.J. The tps series of software. *Hystrix* **2015**, *26*, 9–12.

8. Mason, M.J. Structure and function of the mammalian middle ear. I: Large middle ears in small desert mammals. *J. Anat.* **2015**, *228*, 284–299. [CrossRef]
9. Rosowski, J. Hearing in Transitional Mammals: Predictions from the middle-ear anatomy and hearing capabilities of extant mammals. In *The Evolutionary Biology of Hearing*; Springer: New York, NY, USA, 1992; pp. 615–631.
0. Heffner, H.E. Hearing in large and small dogs: Absolute thresholds and size of the tympanic membrane. *Behav. Neurosci.* **1983**, *97*, 310–318. [CrossRef]
1. Malkemper, E.P.; Topinka, V.; Burda, H. A behavioral audiogram of the red fox (*Vulpes vulpes*). *Hear. Res.* **2015**, *320*, 30–37. [CrossRef]
2. Wysocki, J. Topographical anatomy and morphometry of the temporal bone of the macaque. *Folia Morphol.* **2009**, *68*, 13–22.
3. Nourinezhad, J.; Abedini, M.; Shamsi, M.M.; Dabbaghi, A.; Janeczek, M. Evaluation of the middle ear in water buffaloes (*Bubalus bubalis*) by gross anatomy and cone-beam computed tomography. *Folia Morphol.* **2021**, *80*, 177–185. [CrossRef]
4. Szara, T.; Duro, S.; Gündemir, O.; Demircioglu, I. Sex Determination in Japanese Quails (*Coturnix japonica*) Using Geometric Morphometrics of the Skull. *Animals* **2022**, *12*, 302. [CrossRef]
5. Gürbüz, I.; Aytek, A.I.; Demiraslan, Y.; Onar, V.; Özgel, Ö. Geometric morphometric analysis of cranium of wolf (*Canis lupus*) and German shepherd dog (*Canis lupus* familiaris). *Kafkas Üniv. Vet. Fakültesi Derg.* **2020**, *26*, 525–532. [CrossRef]
6. Hadžiomerović, N.; Gündemir, O.; Kovačević, S. Mandible size and shape of the red fox (*Vulpes vulpes*) and golden jackal (*Canis aureus*). *Adv. Anim. Vet. Sci.* **2022**, *10*, 364–368. [CrossRef]
7. Gündemir, O.; Özkan, E.; Dayan, M.O.; Aydogdu, S. Sexual analysis in turkey (*Meleagris gallopavo*) neurocranium using geometric morphometric methods. *Turk. J. Vet. Anim. Sci.* **2020**, *44*, 681–687. [CrossRef]
8. Gürbüz, I.; Demiraslan, Y. Geometric morphometric investigation of incus in horse (*Equus ferus caballus*) and donkey (*Equus Asinus*). *BSJ. Agric.* **2023**, *6*, 26–31. [CrossRef]

animals

Article

Morphometric Study of the Eyeball of the Loggerhead Turtle (*Caretta caretta*) Using Computed Tomography (CT)

Marcos Fumero-Hernández [1], Mario Encinoso Quintana [1,*], Ana S. Ramírez [2], Inmaculada Morales Fariña [1], Pascual Calabuig [3] and José Raduan Jaber [4,*]

[1] Veterinary Clinical Hospital, Faculty of Veterinary Medicine, University of Las Palmas de Gran Canaria, 35413 Las Palmas, Spain
[2] Department of Pathology and Food Technology, Faculty of Veterinary Medicine, Universidad de Las Palmas de Gran Canaria, 35413 Las Palmas, Spain
[3] Tafira Wildlife Rehabilitation Center (Cabildo de Gran Canaria), 35017 Las Palmas, Spain
[4] Department of Morphology, Faculty of Veterinary Medicine, University of Las Palmas de Gran Canaria, 35413 Las Palmas, Spain
* Correspondence: mencinoso@gmail.com (M.E.Q.); joseraduan.jaber@ulpgc.es (J.R.J.)

Simple Summary: There are few publications describing the use of advanced diagnostic imaging techniques in turtle ophthalmology. However, adequate knowledge of the anatomy of the eye and its associated structures, given their adaptations to be able to develop in the aquatic environment and also in the terrestrial one, could be an important tool to deepen the ophthalmological anatomy, lifestyle, and pathology of these animals.

Abstract: The short bibliography referring to the anatomy and pathology of the eyeball of turtles poses a challenge for veterinarians and conservationists given the increasing presence of this type of turtle in veterinary and wildlife centres. Although they nest on land, these animals spend a large part of their lives in the ocean, which entails a series of eye adaptations such as well-developed nictitating membranes, palpebral scales, highly sensitive corneas, or sclerotic rings to protect the eye. In our study, we performed a morphometric analysis of the loggerhead turtle (*Caretta caretta*) eyeball and its internal structures using advanced imaging techniques such as computed tomography (CT). To the best of the authors' knowledge, there have been no studies published that describe the CT intraocular measurements of presumed normal loggerhead turtle eyes.

Keywords: computed tomography; morphometry; eyeball; sclerotic ring; anatomy; loggerhead turtle

1. Introduction

The loggerhead turtle (*Caretta caretta*) is a species of sea turtle that shows a wide distribution, mainly in all the world's temperate and subtropical regions of the Atlantic, Pacific, and Indian oceans, as well as the Mediterranean Sea [1]. In the wintertime, they also carry out large migrations to tropical and subtropical waters to deposit their eggs on sandy beaches, where the hatchlings move to the open sea to feed. Loggerhead sea turtles are primarily carnivorous, but they can also eat other types of food, making them omnivorous. The loggerhead head is relatively large. It has a roughly triangular shape, wide posteriorly, and the snout tapers anteriorly to the orbits. Moreover, the head presents deep parietal notches as happens in other sea turtle species such as the olive ridley or the Kemp's ridley turtles [2,3]. Compared to other sea turtles, loggerheads have a relatively long secondary palate that separates or partially separates food and air passages. The considerable development of their jaws, which are robust and shaped in a wide V, serve as an effective tool for dismantling their pray, allowing them to eat hard-shelled prey and other invertebrates, such as sponges, jellyfish, cephalopods, shrimp, insects, and sea urchins, including different fish species [2]. There are slight variations in the diet of each life stage, but loggerhead sea turtles are generalists throughout life [3]. The loggerhead sea turtle is the world's largest hard-shelled turtle. Adult individuals of this species have an approximate weight range of 80 to 200 kg, averaging around 135 kg [2,3], and their shells can exceed a meter in length. The head and carapace range from a yellow-orange to a reddish-brown colour that can be covered with barnacles and algae, whereas the plastron is typically pale yellow. Sexual dimorphism of the loggerhead sea turtle can be observed in adults. In the juvenile stage, male specimens have increasing levels of testosterone as they approach maturity, triggering wider carapaces, longer curved claws on each forelimb [3,4], and longer, more robust tails [5,6], whereas females produce oestrogen and small amounts of testosterone, but externally just grow larger [4,7].

The loggerhead turtle is currently on the International Union for Conservation of Nature (IUCN) list of threatened species, which is considered a vulnerable species since it is in decline [8]. Among the multiple causes of decline that have included this animal on the list, we report those associated with human activity and disturbances (72%), entanglement in plastic or fishing nets (50.81%), and the ingestion of hooks (11.88%) [9]. Different studies have highlighted sense of sight as an important element for interaction with the environment and maintaining their survival [10]. Some reports have described that artificial lighting near beaches may confuse emerging hatchlings, causing them to move away from beaches and into hazardous urban areas [4,11]. Therefore, sea turtle conservation is an important concern around the world; conservation efforts include an increase in knowledge of the anatomy and physiology of stranded animals, in part because the behaviour of sea turtles has an anatomic basis [2,4]. Hence, veterinarians, biologists, and conservationists need to know the turtle's ocular structures and their functioning that could affect the greater or lesser incidence of strandings.

Turtles present a series of ophthalmological peculiarities, some common with other groups of vertebrates such as birds. This is the case of the sclerotic rings, made up of a variable number of imbricated ossicles and which, together with the scleral cartilages, form the so-called ocular or scleral skeleton [12,13]. Its function has been scarcely described, but it is believed that it protects the eye against mechanical pressure in those species with diving capacities, support to preserve the shape, and the accommodation mechanism, allowing for the insertion of the ciliary muscle at the edge of the cornea [13–16]. Despite the significance of the sclerotic ring and the applicability of imaging modalities for anatomical evaluations in live animals, most of the studies on the biometric characteristics of the ocular skeleton concerning the eyeball have been conducted in fossils such as Caypullisaurus [17], or non-passerine bird species [18]. Only a few imaging morphometric studies have been carried out in alive animals [19,20] and, interestingly, just a description has been performed on turtles [21]. Nevertheless, specific studies concerning the ocular morphometry of these last animals are essential since they may provide information on the activity patterns of the animals, such as their diurnal or nocturnal behaviour [13,17]. Therefore, this study aimed to measure the eyeball and its structures in the loggerhead turtle (*Caretta caretta)* using non-invasive examinations such as computed tomography.

2. Materials and Methods

2.1. Animals

To perform this study, we used 10 loggerhead turtles from the Tafira Wildlife Rehabilitation Center (Gran Canaria, Canary Islands, Spain), which weighed between 19 and 30 kg and were classified as young and subadult turtles since age is not easily quantifiable in these animals. As physical examination did not provide sufficient information to allow for a characterisation of the exact cause of stranding; the rehabilitation centre requested modern methods of non-invasive diagnostic imaging to evaluate their appendicular skeleton and metabolic bone diseases and rule out foreign bodies and lesions of the internal organs. Therefore, the turtles were enrolled in this study with the prior consent of the person in charge of the rehabilitation centre. Before performing the CT study, we completed an ophthalmologic examination of all turtles and we did not observe signs of disease in any of these animals.

2.2. CT Technique

CT studies of the skull of these individuals were performed with the animals immobilised in plastic containers. These containers, provided by the wildlife center, were the place where they carried out different medical procedures. Therefore, these animals were used to spending time in containers. In addition, the room temperature (ranging between 16 and 20 °C) can produce lethargy, which is helpful when performing the imaging study without sedation. Sequential slices were obtained using a 16-slice helical CT scanner (Toshiba Astelion, Canon Medical System®, Tokyo, Japan). The animals were positioned symmetrically in the prone position on the stretcher, with craniocaudal entry, and a standard clinical protocol was used (120 kVp, 80 mA, 512 × 512 acquisition matrix, 1809 × 858 field of view, pitch of 0.94, and a gantry rotation of 1.5 s), to acquire images of a 0.6 mm thickness. Different CT images were obtained in dorsal, transverse, and sagittal planes with bone and soft tissue windows. All these images were uploaded to an image viewer (OsiriX MD, Apple, Cupertino, CA, USA) to perform data manipulation and measurements of the loggerhead turtle eyeball.

2.3. Measurements

We measured the carapace longitudinal length (CCL), including the midline between the cranial and caudal portions of this structure. In addition, the different measurements of both eyes (n = 20) were analysed by two observers using oblique sagittal, transverse, and dorsal CT images of ten skulls, with a soft tissue attenuation window. The parameters and measures taken are described below:

(A) Transverse plane respect to the eyeball

- Lens diameter, understood as the maximum distance between the lateral and medial edges of the lens (equatorial diameter) (Figure 1A).
- Internal diameter of the sclerotic ring, or maximum distance between the inner lateromedial edges of the ring, close to the cornea (Figure 1A).
- External diameter of the sclerotic ring or maximum distance between the outer lateromedial edges of the ring, close to the sclera (Figure 1A).

(B) Transverse plane concerning the turtle's body

- Dorsal and ventral arch length: the thickness of the upper and lower regions of the scleral ring, with the corneal and scleral margins (Figure 2B).
- Width of the dorsal and ventral arches of the scleral ring, that is, of the upper and lower parts of the lateral portion of the ring (Figure 2A).
- Height of the eyeball or distance between the most dorsal and the most ventral portion of the eye (Figure 3B).
- Attenuation of the sclerotic ring and lens, taken in the dorsal area and vitreous humour expressed in Hounsfield Units (HU) (Figure 4).

(C) Dorsal plane

- Width of the eyeball, which is the lateromedial length of the globe (Figure 3A).
- Length of the eyeball or distance between the most rostral and caudal portion of this structure (Figure 3A).

2.4. Statistical Analysis

We performed the statistical analysis with commercially available software (SPSS 19, Statistical Package for the Social Sciences, Chicago, IL, USA). The mean, median, range, and standard deviation (sd) were calculated for each measurement. The Mann–Whitney U test was used to compare the measurements between the right and left eyes, and the dorsal and ventral arch. To compare differences between the right and left eyes, a two-way analysis of variance (ANOVA) test was used. A linear regression analysis was performed assessing the correlation of all variables to the carapace length, as well as to analyse the quantitative variables. The statistical significance was set at $p < 0.05$.

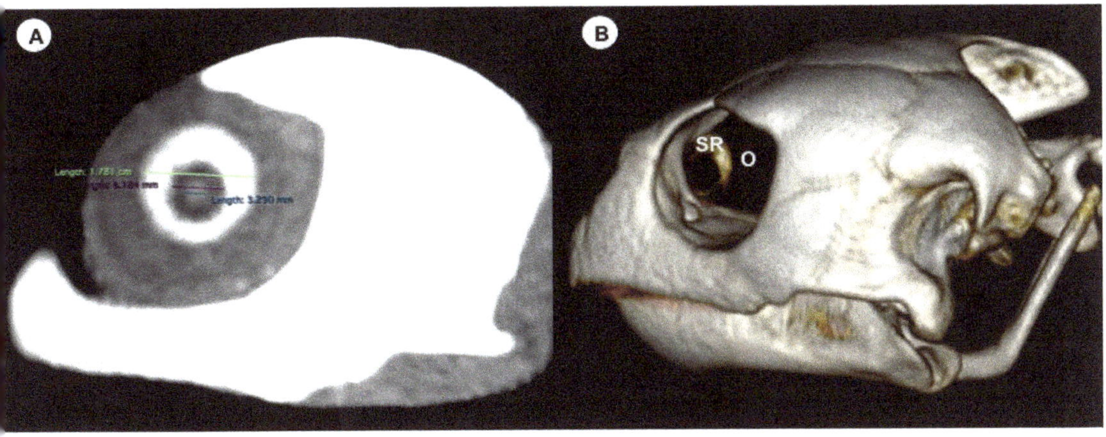

Figure 1. (**A**) Transverse multiplanar reconstruction (MPR) related to the eyeball of the *C. caretta* with measurements of the lens and the internal and external diameters of the sclerotic ring. (**B**) Volume rendering image of the *C. caretta* skull with the sclerotic ring (SR) and orbit (O).

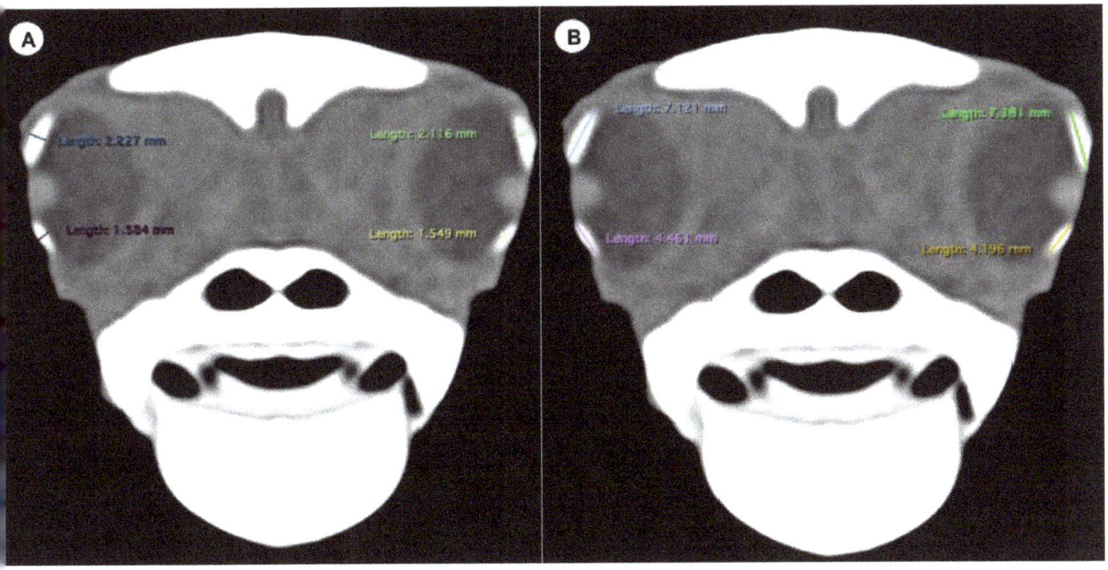

Figure 2. Transverse CT image in soft tissue window of the *C. caretta* head with the width (**A**) and height (**B**) of the dorsal and ventral arches of the sclerotic ring.

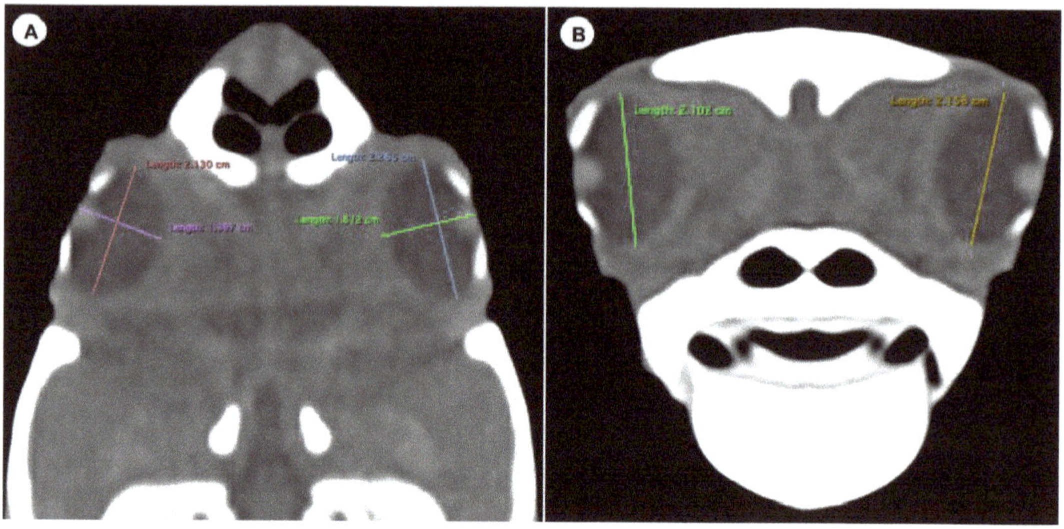

Figure 3. Dorsal multiplanar reconstruction (MPR) and transverse CT images in soft tissue windows of the head with measurements of the length, width (**A**), and height (**B**) of the eyeball of the *C. caretta* head.

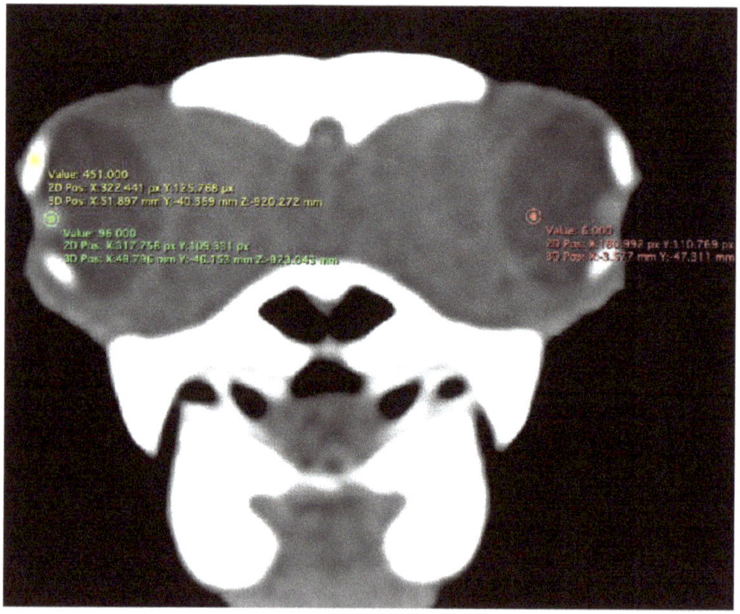

Figure 4. Transverse CT image in soft tissue window of the *C. caretta* head. Regions of interest (ROI) indicating the areas used to measure mean attenuation in Hounsfield Units (HU) of the lens (green circle), sclerotic ring (yellow circle), and vitreous humour (red circle).

3. Results

The dorsal and transverse CT images obtained in this study were helpful to identify and measure the structures of the eyeball (Figures 1–4). From these images, we obtained the different measurements corresponding to the right and left eyes (Table 1).

Table 1. Measurements of the right and left eye of the loggerhead turtle.

	Right Eye				Left Eye				Both Eyes					
	Mean	Median	Range	SD	Mean	Median	Range	SD	Mean	Median	Range	SD	r^2	p
Lens diameter (mm)	4.26	4.34	2.8–5.11	0.70	4.40	4.29	2.6–5.36	0.88	4.33	4.29	2.6–5.6	0.78	0.40	**0.00290**
Internal diameter of the sclerotic ring (mm)	10.18	10.40	6.9–13.8	1.90	10.45	10.20	7.4–14.7	2.45	10.32	10.35	6.9–14.7	2.14	0.67	**0.00001**
External diameter of the sclerotic ring (mm)	16.63	16.50	13–22	2.52	17.29	16.35	13.4–22	2.97	16.96	16.35	13–22	2.70	0.71	**<0.00001**
Lens attenuation (HU)	99.10	99.50	81–116	10.13	98.60	97.50	88–108	6.57	98.85	98.00	81–116	8.31	0.11	0.15480
Vitreous humor attenuation (HU)	18.10	17.50	7–28	5.49	17.50	18.50	9–23	4.20	17.80	18.00	7–28	4.76	0.00	0.80290
Sclerotic ring attenuation (dorsal arch) (HU)	368.00	360.50	292–500	60.23	339.70	359.50	203–417	65.31	353.85	360.50	203–500	62.84	0.05	0.32200
Dorsal arch height (mm)	6.88	6.69	4.6–10.0	1.67	6.76	6.81	4.44–10	1.80	6.82	6.81	4.44–10.1	1.69	0.65	**0.00002**
Ventral arch height (mm)	4.95	4.67	3.7–7.12	1.09	4.97	4.69	3.2–7.51	1.35	4.96	4.69	3.2–7.51	1.19	0.62	**0.00004**
Dorsal arch width (mm)	2.84	2.21	2–6.04	1.59	2.79	2.66	1.89–5.5	1.50	2.81	2.36	1.89–6.71	1.23	0.43	**0.00303**
Ventral arch width (mm)	2.81	2.20	1.98–6.5	1.71	2.65	2.29	1.8–5.48	1.52	2.57	2.15	1.8–5.59	1.08	0.32	**0.01687**
Eyeball height (mm)	22.28	23.05	17–25	2.44	22.17	23.35	15.9–27	3.79	22.23	23.10	15.9–27	0.31	0.36	**0.00557**
Eyeball width (mm)	15.68	15.80	12–19.10	1.88	15.86	15.75	12–19.2	2.27	15.77	15.75	12–19.2	2.00	0.31	**0.01112**
Eyeball length (mm)	22.95	22.80	16–27.9	3.37	22.71	22.40	18–27.1	2.83	22.83	22.40	16–27.9	3.00	0.61	**0.00005**

On the CT image, the eyeball appeared nearly spherical, with a soft tissue attenuating rim in all loggerhead turtles, differing in its sagittal, transverse, and vertical diameters. This border probably represents part of the sclera. Other eyeball structures that were distinguishable were the lens, aqueous and vitreous chambers, and the sclerotic rings (Figure 1). This last chamber was identified as a fluid-attenuating region that accounted subjectively for most of the eyeball volume. In addition, the lens appeared as a spherical hyperattenuated structure suspended in the eyeball and contacting with the vitreous body. The scleral ring, embedded in the rostral portion of the ocular sclera, showed a circular, continuous, and hyperattenuated appearance.

Table 1 shows the measurements made in the right and left eye and in the different planes of all the animals studied. A total of 10 loggerhead turtles were studied, with all 20 eyes measured. The mean carapace length was 41.53 cm (range= 25–71.7), the median was 39.25 cm, and the standard deviation was 13.11 cm.

The mean lens diameter was 4.33 mm for all eyes (range = 2.6–5.6 mm). The mean internal diameter of the sclerotic ring was 10.32 mm (range = 6.9–14.7), while the mean external diameter of the sclerotic ring was 16.96 mm (range = 13–22).

The means arch heights were 6.82 mm (range = 4.44–10.1) and 4.96 (range = 3.2–7.51) for the dorsal and ventral measurements, respectively. The difference between the dorsal and ventral arch height was statistically significant (p = 0.0006). At the same time, the means of the dorsal and ventral arch width were 2.81 mm (range = 1.89–6.71) and 2.57 mm (range = 1.8–5.56), respectively, with no significant differences.

Last, the eyeball presented a mean height of 22.23 mm (range = 15.9–27), a mean width of 15.77 mm (range = 12–19.2), and a mean length of 22.83 mm (range = 16–27.9).

Concerning attenuation, the three circles (yellow, green, and red) indicate the areas used to describe the mean attenuation of different structures of the eyeball. Therefore, the sclerotic ring (yellow circle) was hyperattenuated (mean 353.85 HU: range = 203–500), and the lens (green circle) showed median attenuation (mean 98.85 HU; range = 81–116). Nonetheless, the vitreous chamber (red circle) was hyperattenuated (mean 17.80 HU; range = 7–28) with the adjacent structures (Figure 4).

The Mann–Whitney U test revealed no significant differences when the measurements from the right and left eyes were analysed. Additionally, no statistical differences between the right and left eyes were found when all the variables measured were analysed using a two-way ANOVA. When linear regression analysis was conducted to evaluate the correlation between the measurements and the carapace length, statistical significance was found with the lens diameter (R^2 = 0.40; p = 0.00290), the internal diameter of the sclerotic ring (R^2 = 0.67; p = 0.00001), the external diameter of the sclerotic ring (R^2 = 0.71; p = <0.00001), ventral arch height (R^2 = 0.62; p = 0.00004), eyeball height (R^2 = 0.36; p = 0.00557), and eyeball length (R^2 = 0.61; p = 0.00005).

4. Discussion

Modern diagnostic imaging techniques such as computed tomography and magnetic resonance imaging have revolutionised clinical diagnosis in reptile medicine since they are valuable tools that demonstrate appreciable advantages over conventional imaging techniques. Therefore, they make it possible to obtain views of body sections from various tomographic planes, providing images with an adequate anatomic resolution without the superimposition of the tissues, a high contrast between different structures, and excellent tissue differentiation [22–24]. Thus, advanced imaging technologies have improved the quality of anatomical imaging and diagnosis, enabling an exceptional evaluation of different anatomical regions and a better detection of different diseases such as metabolic bone disease, skull fractures, abscesses, and neoplasia [25]. To the best of the authors' knowledge, this is the first study reporting measurements on loggerhead turtle healthy eyes from CT scans. Hence, the different CT images of the loggerhead turtle eyeball have proven useful in this study since they provided the imaging of orbital soft tissues and ocular adnexa with a high spatial and moderate contrast resolution. We should highlight that the evaluation of anatomic structures within the heads of loggerhead sea turtles and the examination of soft tissues is arduous because of the turtle's anatomic difficulty [26]. Non-invasive imaging exams have already been performed to describe some structures of the eye in reptiles, such as the Komodo dragon [23,24], the green iguana, the common tegu, and the bearded dragon [25], and different species of turtles inhabiting marine, freshwater, and terrestrial environments such as the green turtle, the black-bellied slider, the loggerhead turtle, and the red-footed tortoise [21,26]. Different studies have used contrast media to enhance the densities of both humours [19,27]. However, we did not use a radiocontrast since the sclerotic ring and associated ocular structures, such as the lens or the vitreous chamber, could be adequately visualised. Therefore, we identified most of the eyeball structures already seen in other studies performed in sea and terrestrial turtles [21,26] and domestic mammals such as dogs [19] and cats [27].

The anatomic knowledge of the eyeball of the loggerhead turtle is helpful for the ophthalmology, lifestyle, and visual ability of marine animals already described in other reports [11,12,21,28]. Our CT evaluation revealed that eyeball size and sclerotic ring width were correlated. This positive correlation between these two parameters could be associated with their level of visual acuity. Nonetheless, further studies with a larger number of animals should be done to confirm this statement. Moreover, the width and length of the dorsal arches of the sclerotic rings in all turtles were greater than those of the ventral ones. Similar findings were observed in other marine turtle species such as *Chelonia mydas*, where the sclerotic ring presented a close form, and was wider in the dorsal region than in its ventral aspect [21]. As stated in other reports, sea turtles usually dive and spend most of their lives in the marine environment. Therefore, this greater thickness of the dorsal arches of the ring could be related to the influence of water pressure on the eyes of these animals during diving [21]. Nevertheless, studies with this scope are needed to directly assess the influence of water pressure on the thickness of the ossicle rings.

Some authors have proposed scleral ossification as a possible growth marker. Nonetheless, it is important to highlight that other studies reported that scleral ossicles would not be an alternative for other sea turtle species since other sea turtles such as leatherbacks show some advantage by reaching a large size quicker, which is perhaps related to the thermoregulatory capacity that allows them to exploit a foraging niche not available to other sea turtle species [21,29].

Concerning attenuation variation, we recorded different values for the scleral ring, lens, and vitreous humour. Although bone densities have been reported for reptiles, such as turtles, snakes, and lizards, there are still no attenuation reference values for the eyeball structures of sea turtles [21]. Changes in some of these densities could be considered an important reason to perform a further ocular examination to determine the presence of haemorrhage, inflammation, degeneration, and neoplasia [19]. We also highlight that we did not observe any mineral attenuation structure that we could identify as scleral cartilage, unlike what has been reported by other authors in different species [21,30,31]. However, in agreement with other researchers [13,21], we did not appreciate the junction of this ring with the skull.

In this study, we present a relatively low number of specimens of different sizes and weights. This limited number in our study was due to these animals being free-ranging turtles and due to this fact, it is not easy to obtain a large number of animals as happens in domestic mammals. Therefore, we only access them when they appear stranded on our beaches.

In conclusion, the CT images obtained in this study provided an adequate interpretation of the anatomy of the eyeball and the sclerotic ring in live loggerhead turtles. Our reference values included eyeball and lens presumed normal diameters, as well as different measurements of the scleral ring, which could be related to the vision capability and habits of these animals. However, further studies are necessary since a limited number of animals of different sizes and weights in this study prevents us from attributing this finding to other sea turtles. Moreover, as another limitation, it is important to consider the inherent examiner error related to the manual contouring of computed tomography images.

Author Contributions: Conceptualization, J.R.J., M.F.-H. and M.E.Q.; methodology, M.E.Q., P.C. and M.F.-H.; investigation, J.R.J., M.F.-H., I.M.F., A.S.R. and M.E.Q.; resources, P.C. writing—original draft preparation, J.R.J. and M.F.-H.; writing—review and editing, J.R.J., M.F.-H., A.S.R., I.M.F., P.C. and M.E.Q. All authors have read and agreed to the published version of the manuscript.

Funding: This research received no external funding.

Institutional Review Board Statement: In this study, we did not need ethical approval since the turtles were coming to our hospital to check their health status and rule out by CT any cause of stranding. From these studies, we obtained the information presented in our paper. As happens in domestic mammals, just informed consent from the owner, allowed us to carry out this study. Therefore, Tafira Wildlife Rehabilitation Center was informed that all animal identity information obtained from this study was treated as confidential to the extent permitted by law, and only used for research or teaching purposes.

Informed Consent Statement: Written informed consent has been obtained from Tafira Wildlife Rehabilitation Center to publish this paper.

Data Availability Statement: The data supporting reported results can be found at "accedacris.ulpgc.es".

Acknowledgments: In loving memory of Alvaro Domingo and Honorio Rodriguez Garcia. We also thank Ayesh Mohamad, Carmen Mingot, Marisa Mohamad, Jamal Jaber, and Fatima Alswikie for their support and constructive comments, and Tafira Wildlife Rehabilitation Center for providing the animals of this study.

Conflicts of Interest: The authors declare no conflict of interest.

References

1. Monzón-Argüello, C.; Rico, C.; Carreras, C.; Calabuig, P.; Marco, A.; López-Jurado, L.F. Variation in spatial distribution of juvenile loggerhead turtles in the eastern Atlantic and western Mediterranean Sea. *J. Exp. Mar. Biol. Ecol.* **2009**, *373*, 79–86. [CrossRef]
2. Ernst, C.H.; Lovich, J.E. *Turtles of the United States and Canada*; Smithsonian Institution Press: Washington, DC, USA; London UK, 1994.
3. Spotila, J. *Sea Turtles: A Complete Guide to Their Biology, Behavior, and Conservation*; The Johns Hopkins University Press: Baltimore MD, USA, 2004.
4. Duermit, L. "*Caretta caretta*" (Online), Animal Diversity Web. 2007. Available online: https://animaldiversity.org/accounts/Caretta_caretta/ (accessed on 9 February 2023).
5. Enciclopedia Virtual de los Vertebrados Españoles. Available online: http://www.vertebradosibericos.org/ (accessed on 25 April 2022).
6. Jones, T.T.; Seminoff, J.A. Feeding biology. Advances from Field-Based Observations, Physiological Studies, and Molecular Techniques. In *The Biology of Sea Turtles*, 1st ed.; Wyneken, J., Lohmann, K.J., Musick, J.A., Eds.; CRC Press: Boca Raton, FL, USA, 2013; Volume 3, pp. 211–247.
7. Crespo, J.L.; García-Párraga, D.; Giménez, I.; Rubio-Guerri, C.; Melero, M.; Sánchez-Vizcaíno, J.M.; Marco, A.; Cuesta, J.A.; Muñoz, M.J. Description of two cases of pseudohermaphroditism in loggerhead sea turtles (*Caretta caretta*). *Dis. Aquat. Org.* **2013**, *105*, 183–191. [CrossRef]
8. IUCN. The IUCN Red List of Threatened Species. Version 2022-2. Available online: https://www.iucnredlist.org (accessed on 30 May 2022).
9. Marco, A.; Abella, E.; Carreras, C.; López-Jurado, L.F. Caretta caretta (Linnaeus, 1758). In *Fauna Ibérica*, 2nd ed.; de los Angeles, M., Sánchez, R., Consejo Superior de Investigaciones Científicas (España), Museo Nacional de Ciencias Naturales (España), Eds.; CSIC: Madrid, Spain, 2014; Volume 10, pp. 61–88.
10. Orós, J.; Montesdeoca, N.; Camacho, M.; Arencibia, A.; Calabuig, P. Causes of Stranding and Mortality, and Final Disposition of Loggerhead Sea Turtles (*Caretta caretta*) Admitted to a Wildlife Rehabilitation Center in Gran Canaria Island, Spain (1998–2014): A Long-Term Retrospective Study. *PLoS ONE* **2016**, *11*, e0149398. [CrossRef]
11. "NOAA Fisheries" (Online). Loggerhead Turtle (*Caretta caretta*). 2006. Available online: http://www.nmfs.noaa.gov/pr/species/turtles/loggerhead.htm (accessed on 7 October 2006).
12. Bartol, S.M.; Musick, J.A.; Ochs, A.L. Visual acuity thresholds of juvenile loggerhead sea turtles (*Caretta caretta*): An electrophysiological approach. *J. Comp. Physiol. Neuroethol. Sens. Neural Behav. Physiol.* **2002**, *187*, 953–960. [CrossRef]
13. Franz-Odendaal, T.A. Skeletons of the Eye: An Evolutionary and Developmental Perspective. *Anat. Rec.* **2020**, *303*, 100–109. [CrossRef]

4. Zhang, G.; Boyle, D.L.; Zhang, Y.; Rogers, A.R.; Conrad, G.W. Development and mineralization of embryonic avian scleral ossicles. *Mol. Visión* **2012**, *18*, 348–361.
5. Bayón, A.; Brotóns, N.J.; Albert, A.; Talavera, J. Patología ocular en reptiles. *Clínica Vet. Pequeños Anim. (Avepa)* **1999**, *19*, 227–242.
6. Palumbo, C.; Cavani, F.; Sena, P.; Benincasa, M.; Ferretti, M. Osteocyte Apoptosis and Absence of Bone Remodeling in Human Auditory Ossicles and Scleral Ossicles of Lower Vertebrates: A Mere Coincidence or Linked Processes? *Calcif. Tissue Int.* **2012**, *90*, 211–218. [CrossRef]
7. Fernández, M.S.; Archuby, F.; Talevi, M.; Ebner, R. Ichthyosaurian eyes: Paleobiological information content in the sclerotic ring of *Caypullisaurus* (Ichthyosauria, Ophthalmosauria). *J. Vertebr. Paleontol.* **2005**, *25*, 330–337. [CrossRef]
8. Hall, M.I. The anatomical relationships between the avian eye, orbit and sclerotic ring: Implications for inferring activity patterns in extinct birds. *J. Anat.* **2008**, *212*, 781–794. [CrossRef]
9. Salgüero, R.; Johnson, V.; Williams, D.; Hartley, C.; Holmes, M.; Dennis, R.; Herrtage, M. CT dimensions, volumes and densities of normal canine eyes. *Vet. Rec.* **2015**, *176*, 386. [CrossRef] [PubMed]
0. Boroffka, S.A.E.B.; Voorhout, G. Direct and reconstructed multiplanar computed tomography of the orbits of healthy dogs. *Am. J. Vet. Res.* **1999**, *60*, 1500–1507. [PubMed]
1. Andrade, S.B.D.; Araujo, N.L.L.C.D.; Raposo, A.C.S.; Muramoto, C.; Oriá, A.P. Morphometric descriptive report of scleral ossicle rings, by ultrasound and computed tomography, in three Testudines specimens. *Ciência Rural.* **2023**, *53*, 1–10. [CrossRef]
2. Kuszyk, B.S.; Heath, D.G.; Bliss, D.F.; Fishman, E.K. Skeletal 3-D CT: Advantages of Volume Rendering over Surface Rendering. *Skelet. Radiol.* **1996**, *25*, 207–214. [CrossRef] [PubMed]
3. Pérez, S.; Encinoso, M.; Corbera, J.A.; Morales, M.; Arencibia, A.; González-rodríguez, E.; Déniz, S.; Melián, C.; Suárez-bonnet, A.; Jaber, J.R. Cranial Structure of *Varanus Komodoensis* as Revealed by Computed-tomographic Imaging. *Animals* **2021**, *11*, 1078. [CrossRef]
4. Pérez, S.; Encinoso, M.; Morales, M.; Arencibia, A.; Suárez-Bonnet, A.; González-Rodríguez, E.; Jaber, J.R. Comparative evaluation of the Komodo dragon (*Varanus komodoensis*) and the Green iguana (*Iguana iguana*) skull by three dimensional computed tomographic reconstruction. *Slov. Vet. Res.* **2021**, *58*, 111–116. [CrossRef]
5. Banzato, T.; Selleri, P.; Veladiano, I.A.; Martin, A.; Zanetti, E.; Zotti, A. Comparative evaluation of the cadaveric, radiographic and computed tomographic anatomy of the heads of green iguana (Iguana iguana), common tegu (*Tupinambis merianae*) and bearded dragon (*Pogona vitticeps*). *BMC Vet. Res.* **2012**, *11*, 53. [CrossRef]
6. Mohamad, J.R.J.; González-Rodríguez, E.; Arencibia, A.; Déniz, S.; Carrascosa, C.; Encinoso, M. Anatomical Description of Loggerhead Turtle (*Caretta caretta*) and Green Iguana (Iguana iguana) Skull by Three-Dimensional Computed Tomography Reconstruction and Maximum Intensity Projection Images. *Animals* **2023**, *13*, 621. [CrossRef]
7. Sreenivasan, S.; Linden, A.; Owen, M.; Pemberton, S.; Pinard, C.; Matsuyama, A.; Poirier, V.J. Computed tomography measurements of intraocular structures of the feline eye. *Vet. Rec.* **2019**, *184*, 651–656.
8. De Oliveira, R.L.; Hoffman, J.; Hingst-Zaher, E.; Majluf, P.; Muelbert MM, C.; Amos, W.; Morgante, J.S. Morphological and genetic evidence for two Evolutionary Significant Units (ESUs) in the South American fur seal Arctocephalus australis. *Conserv. Genet.* **2008**, *9*, 1451–1466. [CrossRef]
9. Avens, L.; Taylor, J.C.; Goshe, L.R.; Jones, T.T.; Hastings, M. Use of skeletochronological analysis to estimate the age of leatherback sea turtles *Dermochelys coriacea* in the western North Atlantic. *Endanger. Species Res.* **2009**, *8*, 165–177. [CrossRef]
0. Franz-Odendaal, T.A. Scleral ossicles of Teleostei: Evolutionary and developmental trends. *Anat. Rec.* **2008**, *291*, 161–168. [CrossRef] [PubMed]
1. Franz-Odendaal, T.A.; Hall, B.K. Skeletal elements within teleost eyes and a discussion of their homology. *J. Morphol.* **2006**, *267*, 1326–1337. [CrossRef] [PubMed]

animals

Article

Anatomical Description of Rhinoceros Iguana (*Cyclura cornuta cornuta*) Head by Computed Tomography, Magnetic Resonance Imaging and Gross-Sections

Eligia González Rodríguez [1], Mario Encinoso Quintana [1,*], Daniel Morales Bordon [2], José Guerra Garcés [3], Himar Artiles Nuez [4] and José Raduan Jaber [5,*]

1 Hospital Clínico Veterinario, Facultad de Veterinaria, Universidad de Las Palmas de Gran Canaria, Trasmontaña, Arucas, 35413 Las Palmas, Spain
2 Departamento de Patología Animal, Producción Animal, Bromatología y Tecnología de los Alimentos, Facultad de Veterinaria, Universidad de Las Palmas de Gran Canaria, Trasmontaña, Arucas, 35413 Las Palmas, Spain
3 Rancho Texas Lanzarote Park, Lanzarote, 35510 Islas Canarias, Spain
4 IVC Evidensia Los Tarahales, 35013 Islas Canarias, Spain
5 Departamento de Morfología, Facultad de Veterinaria, Universidad de Las Palmas de Gran Canaria, Trasmontaña, Arucas, 35413 Las Palmas, Spain
* Correspondence: mencinoso@gmail.com (M.E.Q.); joseraduan.jaber@ulpgc.es (J.R.J.)

Simple Summary: The rhinoceros iguana (*Cyclura cornuta cornuta*), like many other iguana species, has become severely endangered as a result of human activities, earning the highest level of protection under the CITES convention for protected species. Moreover, it has been classified as vulnerable by the International Union for Conservation of Nature (IUCN). The rhinoceros iguana belongs to the family Iguanidae and is native to the island of Hispaniola. Since its population has declined, the Canary Island Government has promoted an initiative for its recovery in a completely natural environment. The limited literature regarding the anatomy of this species motivated us to investigate its cranial structures by using modern imaging techniques such as computed tomography (CT) and magnetic resonance imaging (MRI) combined with anatomical dissection, to obtain helpful information on the structures that form the rhinoceros iguana's head.

Abstract: In this paper, we attempted to elaborate on an atlas of the head of the rhinoceros iguana, applying modern imaging techniques such as CT and MRI. Furthermore, by combining the images acquired through these techniques with macroscopic anatomical sections, we obtained an adequate description of the relevant structures that form the head of this species. This anatomical information could provide a valuable diagnostic tool for the clinical evaluation of different pathological processes in iguanas such as abscesses and osteodystrophy secondary to nutrient imbalances, skull malformations, fractures, and neoplasia.

Keywords: computed tomography; magnetic resonance imaging; gross-sections; reptiles; anatomy; head; rhinoceros iguana

1. Introduction

The rhinoceros iguana (*Cyclura cornuta cornuta*), belonging to the Iguanidae family, is an endemic species of the island of Hispaniola, in the Caribbean Sea [1]. Its skin has rough epidermal scales, and its greyish-brown or olive coloring camouflages very well with the environment. The name comes from the bony protuberances or pseudo-horns dorsal to the snout. Males have larger body dimensions than females, and their horns are also larger. This iguana has a polygamous mating system and is oviparous (5–20 eggs per clutch), reaching maturity between 5 and 9 years of age [2]. This species is extremely territorial. To intimidate conspecifics or predators, these animals move their heads and neck, while rotating their bodies. These movements are also used in mating rituals. Like all iguanas, this species is heliothermic, which means it must organize its activities to use solar radiation to regulate its body temperature. This also means that it can only survive in tropical/subtropical climates. Its diet is herbivorous, although it may occasionally feed on insects, land crabs or carrion. Due to human encroachment and destruction of the environment, it has been severely displaced from its habitat (near the coast but with low rainfall). Limestone mining, predation, as well as deforestation by the wood industry and forest fires have forced this species to migrate or even become extinct, as is the case with the closely related subspecies *Cyclura cornuta onchiopsis*, native to Navassa Island [3].

The great interspecific anatomical complexity between mammals and reptiles and the growing interest in reptiles as companion animals present a challenge to veterinary clinicians in diagnostic imaging studies interpretation [4]. Diagnostic imaging has brought a radical change in their clinical practice due to the facility to obtain information on the internal structures of the animal body [5]. Thanks to technological developments in this field, morphological and anatomical information can be obtained in a less invasive and considerably faster way, and this technique has become a fundamental tool for this practice, using conventional radiology [6] and ultrasound [7], as well as more advanced imaging techniques such as CT [8] or MRI [9], which offer certain advantages over conventional ones, such as overlapping structures avoidance, fast image acquisition and high contrast resolution, among others [10,11]. All this has meant that the veterinary clinician has to become acquainted not only with these techniques but also with the anatomy and physiology of reptiles [12].

Some relevant literature on the anatomical, physiological and pathological study of these species is already widely available [8–15]. As far as we know, the anatomy of different reptile species has already been thoroughly described employing diagnostic imaging techniques [13–15]. Authors have reported atlases of green, loggerhead and leatherback sea turtles [15–18], komodo dragons [19,20] and green iguanas [12,15,21,22], snakes and lizards [21]. Concerning the rhinoceros iguana, most of the studies are focused on population genetics [23], evolution and historical biogeography [24], and on some pathologic descriptions such as osteopetrosis-like conditions [25]. However, to the authors' knowledge, none have performed anatomic investigations on the head of the rhinoceros iguana. Therefore, this study aimed to describe the normal anatomy of the head of this species, using CT, MRI and those anatomic sections that are the more informative regions of gross anatomy and can help to understand all the structures that form the head of the rhinoceros iguana. The application of CT, MRI and macroscopic anatomical sections could provide essential information for anatomic descriptions in teaching and clinical practice.

2. Materials and Methods

2.1. Animals

Two carcasses of adult female rhinoceros iguanas (*Cyclura cornuta cornuta*) from the zoological park "Rancho Texas Lanzarote Park" (Lanzarote, Canary Islands, Spain) were collected. One female measured 94 cm, and the other was 91 cm long from snout to tail. They weighed 5 and 4.2 kg, respectively. The animals died due to natural causes. No abnormalities were found on physical examination.

2.2. Anatomic Evaluation

We performed anatomical gross-sections to facilitate the identification of structures observed in the CT and MRI images. Immediately after the scanning procedures, these specimens were placed in a plastic isolation holder in ventral recumbency and successively stored in a freezer (−80 °C) until completely frozen. Subsequently, the two frozen carcasses were sectioned using an electric band saw to obtain sequential anatomical gross-sections. Contiguous 1 cm transverse slices were obtained starting at the snout and extending to the first cervical vertebra region. These slices were thicker than those for CT and MRI to preserve integrity and position of the anatomic structures in the sections. These sections were cleaned with water, numbered and photographed on the cranial and caudal surfaces.

Afterwards, we selected those anatomic sections that better matched the CT and MRI images to identify the relevant structures of the rhinoceros iguana head. To help in this process, we also used anatomical texts and relevant references describing this species [4,21,26,27].

2.3. CT Technique

Transverse CT images were obtained at the Veterinary Hospital of Las Palmas de Gran Canaria University using a 16-slice helical CT scanner (Toshiba Astelion, Canon Medical System, Tokyo, Japan). The animals were placed in ventral recumbency as symmetrically as possible on the CT couch. A standard clinical protocol (100 kVp, 80 mA, 512 × 512 acquisition matrix, 1809 × 858 field of view, a spiral pitch factor of 0.94 and a gantry rotation of 1.5 s) was used to obtain sequential transverse CT images (1 mm thickness). The original transverse data were recorded and transferred to the CT workstation. No CT density or anatomic variations were detected in the head of the reptiles used in the investigation. In this study, we applied two CT windows by adjusting the window widths (WW) and window levels (WL) to appreciate the CT appearance of the head structures: a bone window setting (WW = 1500; WL = 300) and a soft tissue window setting (WW = 350; WL = 40). Moreover, dorsal and sagittal multiplanar reconstructed (MPR) images were also obtained to better visualize other iguana head structures.

2.4. MRI Technique

MRI images were obtained with a Canon Vantage Elan 1.5 T equipment, using T1W sequences in a transversal plane (TR: 634 ms, TE: 10, FOV: 1809 × 829, thickness 2 mm, matrix 192 × 160), T2W sequences in a transversal plane (TR: 4769 ms, TE: 120, FOV: 1809 × 829, thickness 2 mm, matrix 192 × 224), T2W sequences in the dorsal plane (TR: 5271 ms, TE: 120, FOV: 1809 × 829, thickness 2.5 mm, matrix 240 × 192), T2W sequences in the sagittal plane (TR: 4450 ms, TE: 120, FOV: 1809 × 829, thickness 2.9 mm, matrix 224 × 224); enhanced spin-echo sequences were performed in the dorsal, transverse and sagittal planes. The MRI images were obtained with a thickness of 2.7–3.5 mm.

3. Results

Figure 1 corresponds to a bone-window CT sagittal plane, in which each line and number (I–V) represents approximately the level of the following anatomical, CT and MRI transverse planes. Transversal sections revealing the relevant anatomical structures of the rhinoceros iguana head are presented (Figures 2–6). Figures 2–6 are composed of three images: (A) macroscopic, (B) bone window CT and (C) MRI. The images are presented in a rostrocaudal progression from the maxillary bone (Figure 2) to the brainstem levels (Figure 6). Figures 7 and 8 are composed of two images: (A) bone-window CT and (B) MRI in T2W, in a dorsal and sagittal plane, respectively. Figure 9 is a T2W sequence sagittal image showing the angulation concerning the horizontal axis of the myelencephalon (40°).

3.1. Anatomical Sections

Different structures belonging to the central nervous system were visualized by anatomical gross-sections. Therefore, we identified the brain (telencephalon) and the two telencephalic hemispheres (Figures 4A and 5A) separated by the *fissura longitudinalis cerebri* (Figure 5A), the diencephalon (thalamus) (Figures 4A and 5A), the dorsal part of the mesencephalon with the two caudal colliculus (Figure 6A) and the ventral part of the cerebellum with the nodule, covering part of the fourth ventricle, as well as the ventral surface of the brainstem (myelencephalon) (Figure 6A). These sections were also helpful for the observation of the olfactory bulb that showed an extracranial location, ventral to the frontal bones, and coursed between the two eyeballs (Figure 3). In addition, these sections allowed the description of structures of the eyeball, identifying the cornea, the sclera, the retina, the vitreous humor, the lens, as well as associated structures such as the interorbital septum and the orbital sinus (Figures 3A and 4A). In addition, these transverse sections allowed the identification of structures belonging to the oral cavity such as the tongue (Figure 2A) and other structures such as the larynx, with the corniculate tubercles of the arytenoid cartilage, the thyroid cartilage, and the laryngeal ventricle (Figure 3A), as well as of different structures of the nasal cavity such as the nasal glands that filled almost the entire cavity, the conchal grooves and the stammteil located laterally to the nasal septum (Figure 2A). Adjacent structures such as the trachea and the nasopharynx were also well identified (Figures 4A, 5A and 6A). Furthermore, most bony structures that form the neurocranium were observed, such as the pterygoid, frontal, postfrontal-postorbital, parietal, supraoccipital, basioccipital, exoccipital, otoccipital, parabasisphenoid and sphenoid bones (Figures 3A, 4A, 5A and 6A), as well as those that form the splanchnocranium such as the nasal, vomer, palatine and maxillary bones (Figures 2A and 3A) and also the medial horn located dorsal to the nasal bone (Figure 2A), the mandible, with the dentary bone (Figure 3A), and the hyoid apparatus, visualizing the central body of the hyoid arch, between the lateral branches of the dentary bone (Figure 3A). Rostromedially to the two dentary bones, we identified different muscles related to the hyoid apparatus such as the *musculus genihyioideus, hyoglossus and intermandibularis* (Figures 2A and 3A). In the following sections, we also observed the muscle groups corresponding to the mandibular musculature (we were not able to dissect them, and therefore they were treated as a group), including the *pterygoideus, omohyoideus, sternohyoideus, ceratohyoideus, adductor mandibulae externus medialis* and its homonym *superficialis* (Figures 4A and 5A).

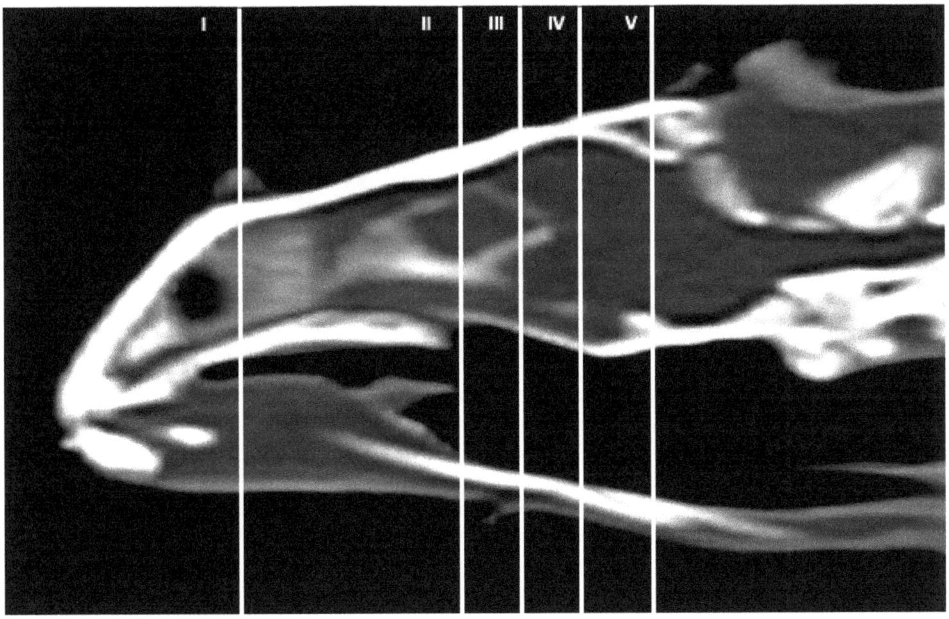

Figure 1. Sagittal CT bone window image of the head of rhinoceros iguana that approximately representing the level of the slices of this study. Segments I–V correspond to Figures 2–6.

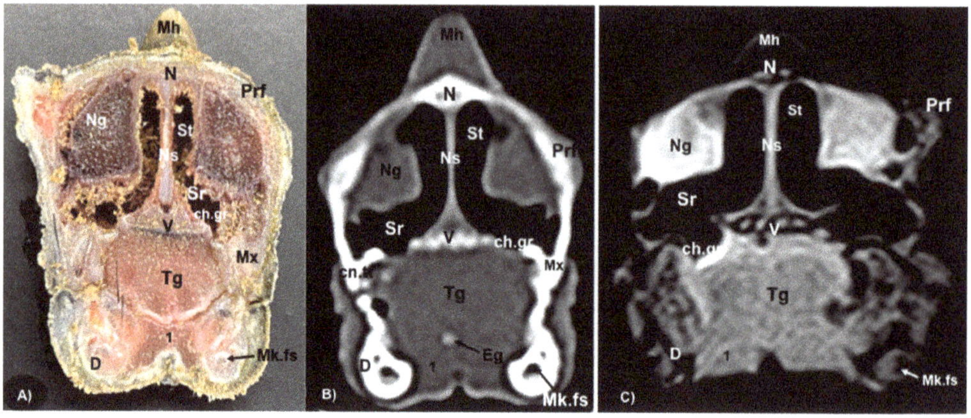

Figure 2. Transversal gross-section (**A**), CT bone window (**B**), and MRI in T1W (**C**), images of the head of rhinoceros iguana at the level of the nasal cavity corresponding to line I in Figure 1. Mh: Medial horn. N: Nasal bone. Prf: Prefrontal bone. Ns: Nasal septum. Ng: Nasal glands. Sr: Subconchal recess. St: Stammteil. ch.gr: Choanal groove. V: Vomer. Mx: Maxillary bone. Tg: Tongue. Eg: Entoglossal process of the basihyal bone. cn.tt: Canine tooth. 1: *Musculus geniohyoideus* + *Musculus hyoglossus* + *intermandibularis*. D: Dentary bone. Mk.fs: Meckelian fossa.

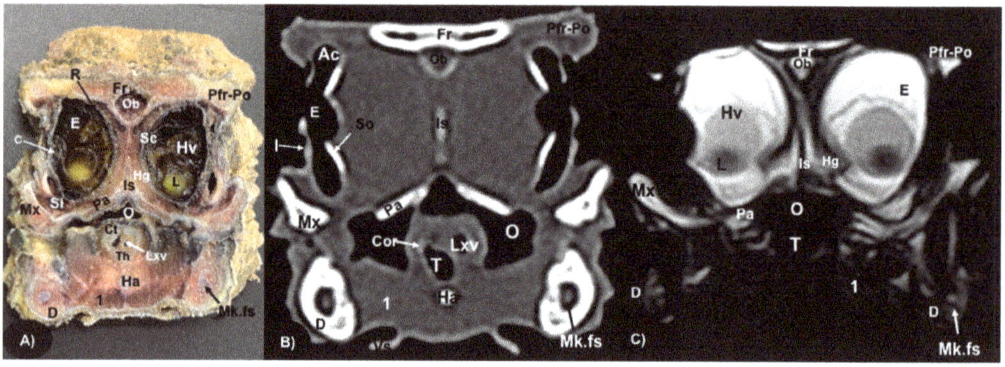

Figure 3. Transversal gross-section (**A**), CT bone window (**B**), and MRI in T2W (**C**), images of the head of rhinoceros iguana in the orbital region corresponding to line II in Figure 1. Fr: Frontal bone. Pfr-Po: Postfrontal-postorbital bone. Ob: Olfactory bulb. Ac: Anterior chamber. R: Retina. So: Scleral ossicles. I: Iris. E: Eyeball. C: Cornea. Sc: Sclera. L: Lens. Hv: Vitreous chamber. Is: Interorbital septum. Hg: Harderian gland. Si: Infraorbital sinus. Pa: Palatine bone. Mx: Maxillary bone. O: Oral cavity. Cor: Corniculated tubercles of arytenoid cartilage. Lxv: Laryngeal ventricle. Th: Thyroid cartilage. D: Dentary bone. Mk.fs: Meckelian fossa. Ha: Hyoid apparatus. T: Trachea. 1: *Musculus geniohyoideus* + *Musculus hyoglossus* + *Musculus intermandibularis*. Vs: Ventral spines.

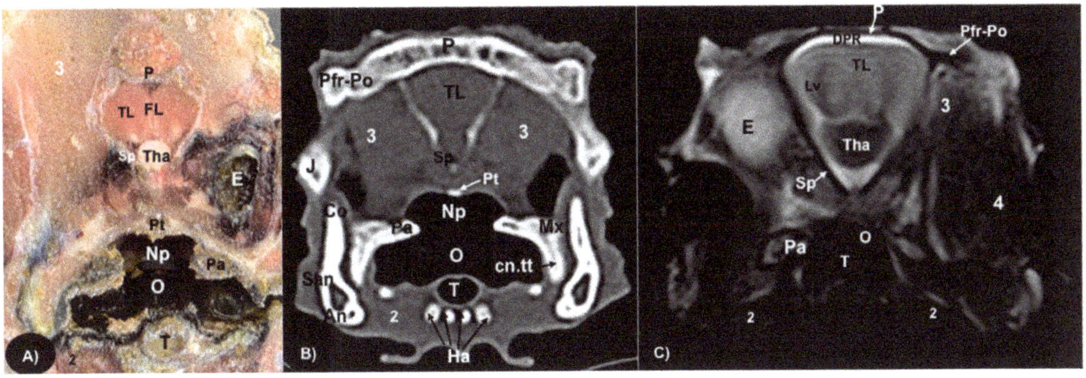

Figure 4. Transversal gross-section (**A**), CT bone window (**B**), and MRI in T2W (**C**), images of the head of rhinoceros iguana at the level of the frontal bone corresponding to line III in Figure 1. 3: *Musculus adductor mandibulae externus medialis*. P: Parietal bone. Pfr-Po: Postfrontal-Postorbital bone. TL: Telencephalon. FL: *Fissura longitudinalis cerebri*. DPR: Dorsal Pallium Rostral Part. Tha: Thalamus. E: Eyeball. Sp: Sphenoid bone. Pt: Pterygoid bone. J: Jugal bone. Pa: Palatine bone. O: Oral cavity. 4: *Musculus adductor mandibulae externus Pars superficialis*. Mx: Maxillary bone. cn.tt: Caniniform tooth. Np: Nasopharyngeal duct. Co: Coronoid. San: Surangular bone. An: Angular bone. 2: *Musculus intermandibularis* + *Musculus geniohyoideus* + *Musculus hyoglossus* + *Musculus pterygoideus* + *Musculus omohyoideus* + *Musculus sterohyoideus* + *Musculus ceratohyoideus*. T: Trachea. Ha: Hyoid apparatus.

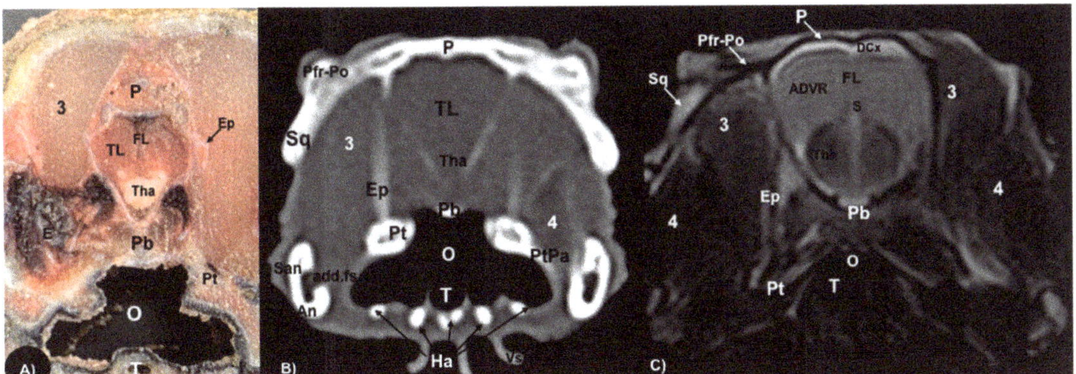

Figure 5. Transversal gross-section (**A**), CT bone window (**B**) and MRI in T2W (**C**), images of the head of rhinoceros iguana at the level of the anterior dorsal ventricular ridge (ADVR) corresponding to line IV in Figure 1. P: Parietal bone. Pfr-Po: Postfrontal-Postorbital bone. Sq: Squamousal bone. E: Eyeball. TL: Telencephalon. FL: *Fissura longitudinalis cerebri*. Tha: Thalamus. ADVR: Anterior Dorsal Ventricular Ridge. DCx: Dorsal cortex. 3: *Musculus adductor mandibulae externus medialis*. 4: *Musculus adductor mandibulae externus superficialis*. Pb: Parabasisphenoid bone. Pt: Pterygoid bone. Ep: Epipterygoid bone. PtPa: Pterygoid-palatine processes. add.fs: Adductor fossa. An: Angular bone. San: Surangular bone. Ha: Hyoid apparatus. O: Oral cavity. T: Trachea. Vs: Ventral spines.

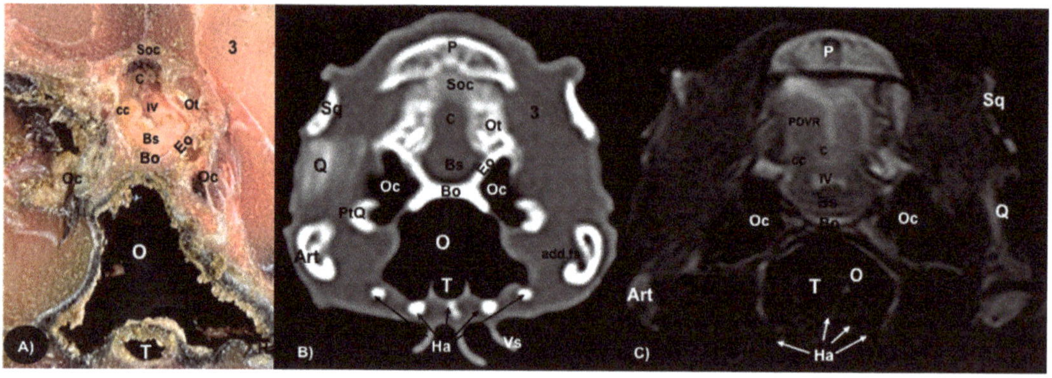

Figure 6. Transversal gross-section (**A**), CT bone window (**B**) and MRI in T2W (**C**), images of the head of rhinoceros iguana at the level of the occipital region corresponding to line V in Figure 1. P: Parietal bone. Soc: Supraoccipital bone. Sq: Squamosal bone. 3: *Musculus adductor mandibulae externus superficialis*. C: Cerebellum. PDVR: Posterior Dorsal Ventricular Ridge. Bs: Brain stem. cc: Caudal colliculus. IV: Fourth ventricle. Ot: Otoccipital bone. Eo: Exoccipital bone. Bo: Basioccipital bone. Oc: Otic cavity. Q: Quadrate bone. PtQ: Pterygoid-quadrate processes. Art: Articular bone. add.fs: Adductor fossa. O: Oral cavity. T: Trachea. Ha: Hyoid apparatus. Vs: Ventral spines.

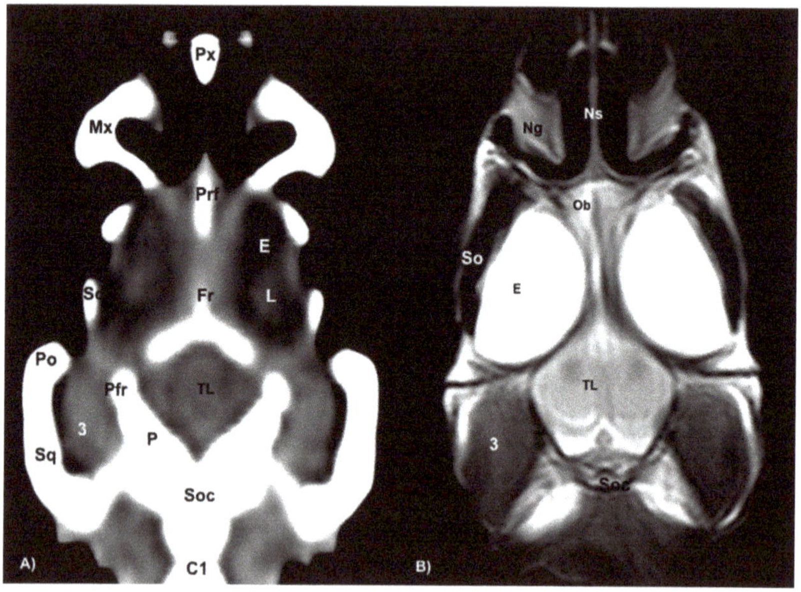

Figure 7. (**A**) Dorsal CT image in the brain window. (**B**) Dorsal MRI in the T2W sequence of the head of rhinoceros iguana telencephalon and at the olfactory bulb level. Px: Premaxillary bone. Ns: Nasal septum. Mx: Maxillary bone. Ng: Nasal glands. Ob: Olfactory bulbs. Prf: Prefrontal bone. E: Eyeball. L: Lens. So: Scleral ossicles. Fr: Frontal bone. TL: Telencephalon. 3: *Musculus adductor mandibulae externus medialis*. Pfr: Postfrontal bone. Po: Postorbital bone. P: Parietal bone. Soc: Supraoccipital bone. Sq: Squamosal bone. C1: First cervical vertebra.

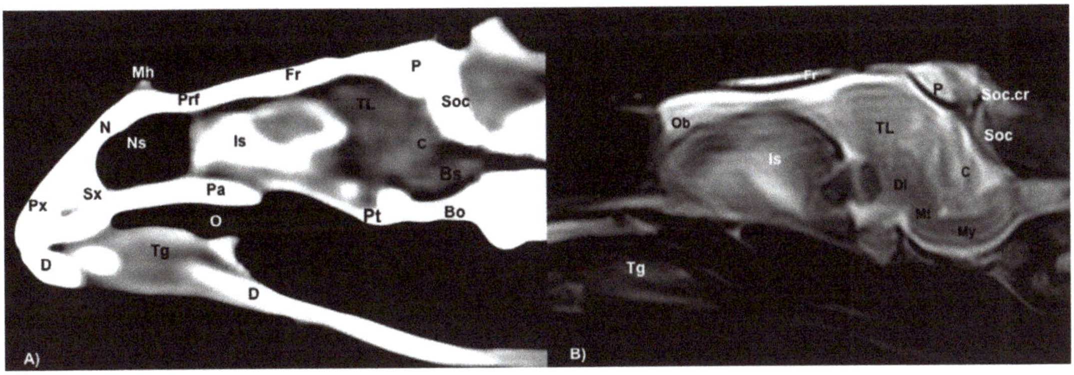

Figure 8. (**A**) Mid-sagittal CT image in the brain window. (**B**) Sagittal MRI in the T2W sequence image of the head of rhinoceros iguana. Px: Premaxillary bone. Sx: Septomaxilla. D: Dentary bone. N: Nasal bone. Mh: Medial horn. Prf: Prefrontal bone. Pa: Palatine bone. O: Oral cavity. Tg: Tongue. Fr: Frontal bone. Is: Interorbital sinus. P: Parietal bone. Soc: Supraoccipital bone. Soc.cr: Supraoccipital crest. Pt: Pterygoid bone. Bo: Basioccipital bone. Ob: Olfactory bulbs. TL: Telencephalon. C: Cerebellum. Bs: Brainstem. Di: Diencephalon. Mt: Metencephalon. My: Myelencephalon.

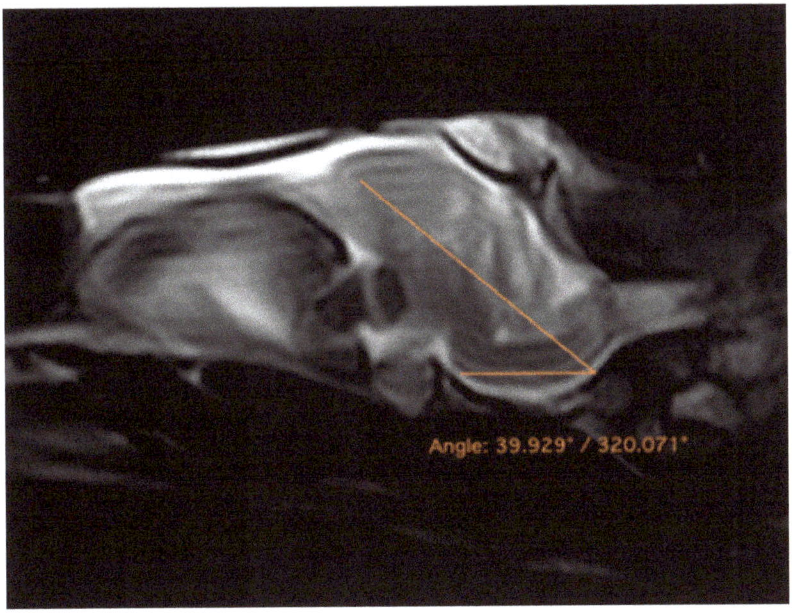

Figure 9. Rhinoceros iguana encephalic angulation, sagittal T2W image with respect to the horizontal axis of the myelencephalon (40°).

3.2. Computed Tomography (CT)

Regarding the neurocranium, the CT images allowed us to distinguish bone structures such as the prefrontal, frontal, postfrontal-postorbital, parietal, squamosal, quadrate, epipterygoid, pterygoid, basioccipital, exoccipital, otoccipital and parabasisphenoid bones (Figures 3B, 4B, 5B and 6B); related to the splanchnocranium, we observed the nasal, premaxilla, maxilla, septomaxilla, vomer, jugal and palatine bones, and the nasal septum (Figures 2B, 3B, 4B, 7B and 8B); the mandible structures such as the dentary, angular, surangular, coronoid and articular bones (Figures 2B, 3B, 4B, 5B and 6B) and the hyoid apparatus

(Figures 2B, 3B, 4B, 5B and 6B). CT scanning and post-processing transverse images showed the relation between the different bones that form the head of the rhinoceros iguana, the junction of the nasal and the prefrontal bone, as well as that of the parietal bone with the postfrontal-postorbital bone (Figures 2B and 4B), and the palatine and quadrate processes of pterygoid bone (Figures 5B and 6B respectively). The prominent medial horn was identified dorsal to the nasal bone with soft-tissue attenuation and a thin, lamellar-shaped mineral structure bordering it regularly on its most external aspect (Figure 2B).

Concerning the nasal cavity, the transverse CT image showed the nasal glands as symmetrical bilateral structures, with regular and well-defined margins, located on both sides of the nasal cavity and with soft tissue attenuation (Figure 2B). Moreover, those structures with intraluminal gas content such as the nasal conchal recess (Figure 2B), oral cavity, nasopharyngeal duct, trachea, adductor fossa (Figure 4B) and the otic cavity (Figure 6B) were identified with this technique, appearing with a vacuum effect. In addition, there were areas of soft tissue attenuation medial to the mandible and bilateral to the hyoid apparatus, compatible with the *intermandibularis*, *genihyioideus* and *hyoglossus* muscles (1 in Figures 2B and 3B), the *pterygoideus*, *omohyoideus*, *sternohyoideus* and *ceratohyoideus* muscles (2 in Figure 4B), the *adductor mandibulae externus medialis* and *superficialis* muscles located, respectively, dorsomedially and ventrolaterally to the adductor fossa (3 and 4 in Figures 4B and 5B).

In addition, we distinguished with adequate resolution different structures of the central nervous system such as the olfactory bulb, the brain (telencephalon and diencephalon), the cerebellum and the brainstem (Figures 3B, 4B, 5B and 6B).

3.3. *Magnetic Resonance Imaging (MRI)*

The soft structures of the iguana's head, such as the central nervous system as well as the eyeball's structures (vitreous humor and lens), the oral cavity with the tongue and the masticatory muscles, showed an accurate visualization using MRI (Figures 2C, 3C, 4C, 5C and 6C). Therefore, an increased volume of both eyeballs in proportion to the size of the head was seen in all sequences (Figures 3C and 7B). As in CT, the structures with gas content (Figures 2C and 6C) appeared with a vacuum effect, being hypointense in all sequences. The nasal glands were bilaterally symmetric, with regular and well-defined margins, located on both sides of the nasal cavity, being iso/hyperintense in T1W and T2W sequences, compared to the encephalic grey matter (Figures 2C and 4C). In contrast, the medial horn appeared hypo/isointense on T1W and T2W sequences concerning the white matter and with mild differentiation of the external bony cortex in relation to the white matter (Figure 2C).

In contrast to the CT images, the bone junctions were not distinguishable on MRI, but those bones that formed the neurocranium, such as the frontal, postfrontal-postorbitary, parietal and supraoccipital bones were identified (Figures 3C, 4C, 5C, 6C and 8B). The cranial musculature was found isointense concerning the thalamus in T2W. This technique enabled a better resolution to identify the muscle groups already mentioned (Figures 2C, 3C, 4C and 5C). In the transverse planes of the encephalon, the cerebral cortex was observed slightly more hyperintense than the white matter, which was more hypointense in T2W sequences (Figure 4C). The diencephalic region (Figures 4C, 5C and 8B) was hypointense (T2W) compared to the cerebral cortex (Figure 5C), showing the thalamus and hypothalamus (Figure 8B). The brainstem appeared hypo/isointense in T2W compared to the cerebral cortex, as well as presenting a markedly tortuous horizontal alignment (Figure 8B). Moreover, the caudal colliculus and the fourth ventricle were also displayed in excellent detail. In the rostral aspect of the telencephalon, we distinguished the dorsal pallium rostral part with its lateral and medial portions (Figure 4C). Interestingly, the dorsal MRI image was quite helpful to identify the olfactory bulb located extracranially, which extended rostromedially between the eyeballs (Figures 3C, 7B and 8B).

4. Discussion

Technological developments in imaging techniques have improved the anatomical knowledge and the diagnosis of several pathologies. From conventional imaging methods, such as radiography and ultrasound, to advanced ones such as CT and MRI, the level of resolution, the rapid acquisition of images and the absence of superimposition have meant innovation in research, daily clinical practice and academic purposes [11,12,28–30].

According to other descriptive studies of different reptile species [1,3,12,14,16,17,20], the images obtained by CT, MRI and gross-sections were adequate to study the rhinoceros iguana head. Therefore, the gross-sections provided accurate anatomic characteristics of the head structures, mainly those related to the brain, the eyeball and the larynx. To the authors' knowledge, a unique description using gross-section and conventional CT equipment was performed of the green iguana head [22], but only a few brain and laryngeal structures were labelled.

Considering the green iguana [26], some differences were found between the bony structures of these species. Thus, we identified the presence of three horns on the dorsal aspect of the nasal region, which are absent in the green iguana. Moreover, the evaluation of the transverse CT images obtained in the bone window and the post-processing in MPR showed a lower prominence of the occipital ridge in the rhinoceros iguana compared to the green iguana. In addition, as described by other authors [22], the CT images of the rhinoceros iguana head displayed in excellent detail the bony structures compared to the anatomic gross-section. With MRI, the palatine, frontal, postfrontal-postorbital, parietal, supraoccipital bones and crest were distinguishable due to a gross-section and with both imaging techniques. In addition, the eyes were easily displayed with all the techniques used. Nonetheless, specific eyeball structures, such as the iris and the scleral ossifications, were better identified by CT and were hardly visible in anatomic gross-sections, where the lens, the retina, the sclera and the vitreous chamber could be visualized. Some of these structures are of scientific and morphological interest to perform further studies on the dimensions of the eyeball and vision ability [29,30].

The use of MRI was valuable for visualizing organs located in the head. Thus, transverse MR images T2W facilitated the accurate identification of the main components of the brain, such as the telencephalon, diencephalon, mesencephalon, metencephalon and myelencephalon. Therefore, we identified specific structures such as the lateral ventricles, the anterior dorsal and posterior dorsal ventricular ridges, the cerebellum, the fourth ventricle, the dorsal pallium and the brain stem with the caudal colliculus. Interestingly, the dorsal MR T2W image showed the olfactory bulb isointense compared to the telencephalon, located in the rostral portion of the encephalon but in an extracranial situation, remaining in the rostromedial aspect of the eyeballs (extracranial structures). In contrast to other articles that excluded the olfactory bulb description, we present the shape, location and intensity of this structure [31]. This extracranial location was already observed in previous studies on the tawny dragon [31]. However, this finding had not been reported in other species of reptiles, such as the rhinoceros iguana. The gross-section images confirmed the presence of these structures and their rostral extension. Interestingly, in other species, such as galliform birds, the olfactory bulbs are separated from the telencephalic hemispheres but located intracranially, protected by the frontal bone [32].

In addition, the MRI and CT findings showed a greater angulation concerning the horizontal axis of the encephalon, with an angulation of 40° between the central horizontal axis of the skull and the central axis of the encephalon (dorsal displacement concerning the encephalon of the dog) (Figure 9), differing from other studies where an angulation of 28° was described [31]. In galliform birds, the brain has a similar angulation to the skull axis [32]. More similar aspects to the latter were found, such as the large size of the eyeballs, which were almost as large as the whole encephalon [32]. Nonetheless, further studies with a large number of animals should be performed to confirm these findings.

Finally, several studies carried out in reptiles have been performed with micro-CT equipment, as it has a higher resolution, allowing a better distinction of anatomical formations [33–35]. However, in our work, the combined application of different advanced techniques (CT, MRI) and macroscopic anatomical sections provided helpful information for anatomic descriptions in teaching and clinical practice.

5. Conclusions

In conclusion, CT and MRI appeared to be adequate tools, providing anatomical references of the different bone and soft tissue structures comprising the head of the rhinoceros iguana. Therefore, the findings obtained in this study are helpful for evaluating numerous processes involving the head of these animals, such as abscesses, metabolic bone diseases, fractures and neoplasia. Moreover, these two imaging techniques could contribute to veterinary anatomy learning by our students as these techniques allow the visualization of structures without overlapping, eliminating the difficulties of visualizing the extent of different types of lesions.

Author Contributions: Conceptualization, J.R.J., E.G.R. and M.E.Q.; methodology, M.E.Q., E.G.R. and H.A.N.; investigation, J.R.J., E.G.R. and M.E.Q.; resources, D.M.B. and J.G.G.; writing—original draft preparation, J.R.J. and E.G.R.; writing—review and editing, J.R.J., M.E.Q., E.G.R., J.G.G., D.M.B. and H.A.N. All authors have read and agreed to the published version of the manuscript.

Funding: This research received no external funding.

Institutional Review Board Statement: Rancho Texas Lanzarote Park is in conformed that all animal identifies information obtained from this study will be treated as confidential to the extent permitted by law, and only used for research or teaching purposes.

Informed Consent Statement: Informed consent was obtained from all subjects involved in the study.

Data Availability Statement: Data is unavailable due to privacy or ethical restrictions.

Acknowledgments: In loving memory of Alvaro Domingo and Honorio Rodriguez Garcia. We also thank Carmen Mingot, Ayesh Mohamad, Marisa Mohamad, Jamal Jaber and Fatima Alswikie for their support and constructive comments and the Rancho Texas Lanzarote Park (Lanzarote, Canary Islands, Spain) for providing the animals of this study.

Conflicts of Interest: The authors declare no conflict of interest.

References

1. Alberts, A.C. *West Indian Iguanas. Status Survey and Conservation Action Plan*; IUCN: Gland, Switzerland; Cambridge, UK, 2000 ISBN 2831704561.
2. Glor, R.E.; Powell, R.; Parmerlee, J.S., Jr. Cyclura Cornuta. *Cat. Am. Amphib. Reptil.* **2000**, *8235*, 1–6.
3. Egnatios-Beene, J. Cyclura Cornuta. Available online: https://animaldiversity.org/accounts/Cyclura_cornuta/ (accessed on 13 February 2023).
4. Banzato, T.; Russo, E.; Di Toma, A.; Palmisano, G.; Zotti, A. Evaluation of Radiographic, Computed Tomographic, and Cadaveric Anatomy of the Head of Boa Constrictors. *Am. J. Vet. Res.* **2011**, *72*, 1–8. [CrossRef]
5. Lauridsen, H.; Hansen, K.; Wang, T.; Agger, P.; Andersen, J.L.; Knudsen, P.S.; Rasmussen, A.S.; Uhrenholt, L.; Pedersen, M. Inside out: Modern Imaging Techniques to Reveal Animal Anatomy. *PLoS One* **2011**, *6*, e17879. [CrossRef] [PubMed]
6. Knotek, Z. Reptile Medicine and Surgery in Clinical Practice. *Reptil. Med. Surg.* **2006**, 471–489. [CrossRef]
7. Urbanová, D.; Halán, M. The Use of Ultrasonography in Diagnostic Imaging of Reptiles. *Folia Vet.* **2016**, *60*, 51–57. [CrossRef]
8. Bannas, P.; Adam, G.; Derlin, T. Modern Imaging Techniques in Patients with Multiple Myeloma. *RoFo* **2013**, *185*, 26–33.
9. Behroozi, M.; Billings, B.K.; Helluy, X.; Manger, P.R.; Güntürkün, O.; Ströckens, F. Functional MRI in the Nile Crocodile: A New Avenue for Evolutionary Neurobiology. *Proc. R. Soc. B Biol. Sci.* **2018**, *285*, 20180178. [CrossRef] [PubMed]
10. Knipe, M.F. Principles of Neurological Imaging of Exotic Animal Species. *Vet. Clin. North Am.—Exot. Anim. Pract.* **2007**, *10*, 893–907. [CrossRef]
11. Banzato, T.; Hellebuyck, T.; Van Caelenberg, A.; Saunders, J.H.; Zotti, A. A Review of Diagnostic Imaging of Snakes and Lizards *Vet. Rec.* **2013**, *173*, 43–49. [CrossRef]
12. Pérez, S.; Encinoso, M.; Morales, M.; Arencibia, A.; Suárez-Bonnet, A.; González-Rodríguez, E.; Jaber, J.R. Comparative Evaluation of the Komodo Dragon (Varanus Komodoensis) and the Green Iguana (Iguana Iguana) Skull by Three-Dimensional Computed Tomographic Reconstruction. *Slov. Vet. Res.* **2021**, *58*, 111–116. [CrossRef]

3. Hernández Morales, C.; Peloso, P.L.V.; Bolívar García, W.; Daza, J.D. Skull Morphology of the Lizard Ptychoglossus Vallensis (Squamata: Alopoglossidae) with Comments on the Variation Within Gymnophthalmoidea. *Anat. Rec.* **2019**, *302*, 1074–1092. [CrossRef]
4. Sterli, J.; Müller, J.; Anquetin, J.; Hilger, A. The Parabasisphenoid Complex in Mesozoic Turtles and the Evolution of the Testudinate Basicranium. *Can. J. Earth Sci.* **2010**, *47*, 1337–1346. [CrossRef]
5. Jaber, J.R.; González-Rodríguez, E.; Arencibia, A.; Deniz, S.; Carrascosa, C.; Encinoso, M. Anatomical Description of Loggerhead Turtle (Caretta Caretta) and Green Iguana (Iguana Iguana) Skull by Three-Dimensional Computed Tomography Reconstruction and Maximum Intensity Projection Images. *Animals* **2023**, *13*, 621. [CrossRef]
6. Arencibia, A.; Melián, A.; Orós, J. Anatomic Interactive Atlas of the Loggerhead Sea Turtle (Caretta Caretta) Head. *Animals* **2021**, *11*, 198. [CrossRef] [PubMed]
7. Arencibia, A.; Rivero, M.A.; De Miguel, I.; Contreras, S.; Cabrero, A.; Orós, J. Computed Tomographic Anatomy of the Head of the Loggerhead Sea Turtle (Caretta Caretta). *Res. Vet. Sci.* **2006**, *81*, 165–169. [CrossRef]
8. Yamaguchi, Y.; Kitayama, C.; Tanaka, S.; Kondo, S.; Miyazaki, A.; Okamoto, K.; Yanagawa, M.; Kondoh, D. Computed Tomographic Analysis of Internal Structures within the Nasal Cavities of Green, Loggerhead and Leatherback Sea Turtles. *Anat. Rec.* **2021**, *304*, 584–590. [CrossRef]
9. Pérez, S.; Encinoso, M.; Corbera, J.A.; Morales, M.; Arencibia, A.; González-rodríguez, E.; Déniz, S.; Melián, C.; Suárez-bonnet, A.; Jaber, J.R. Cranial Structure of Varanus Komodoensis as Revealed by Computed-tomographic Imaging. *Animals* **2021**, *11*, 1078. [CrossRef]
0. Moreno, K.; Wroe, S.; Clausen, P.; McHenry, C.; D'Amore, D.C.; Rayfield, E.J.; Cunningham, E. Cranial Performance in the Komodo Dragon (Varanus Komodoensis) as Revealed by High-Resolution 3-D Finite Element Analysis. *J. Anat.* **2008**, *212*, 736–746. [CrossRef]
1. Banzato, T.; Selleri, P.; Veladiano, I.A.; Zotti, A. Comparative Evaluation of the Cadaveric and Computed Tomographic Features of the Coelomic Cavity in the Green Iguana (Iguana Iguana), Black and White Tegu (Tupinambis Merianae) and Bearded Dragon (Pogona Vitticeps). *J. Vet. Med. Ser. C Anat. Histol. Embryol.* **2013**, *42*, 453–460. [CrossRef]
2. Banzato, T.; Selleri, P.; Veladiano, I.A.; Martin, A.; Zanetti, E.; Zotti, A. Comparative Evaluation of the Cadaveric, Radiographic and Computed Tomographic Anatomy of the Heads of Green Iguana (Iguana Iguana), Common Tegu (Tupinambis Merianae) and Bearded Dragon (Pogona Vitticeps). *BMC Vet. Res.* **2012**, *8*, 1. [CrossRef]
3. Pasachnik, S.A.; Colosimo, G.; Carreras-De León, R.; Gerber, G. Genetic Structure of Rhinoceros Rock Iguanas, Cyclura Cornuta, in the Dominican Republic, with Insights into the Impact of Captive Facilities and the Taxonomic Status of Cyclura on Mona Island. *Conserv. Genet.* **2020**, *21*, 837–851. [CrossRef]
4. Reynolds, R.G.; Miller, A.H.; Pasachnik, S.A.; Knapp, C.R.; Welch, M.E.; Colosimo, G.; Gerber, G.P.; Drawert, B.; Iverson, J.B. Phylogenomics and Historical Biogeography of West Indian Rock Iguanas (Genus Cyclura). *Mol. Phylogenet. Evol.* **2022**, *174*, 107548. [CrossRef] [PubMed]
5. Reece, R.L.; Dickson, D.B.; Butler, R. An Osteopetrosis-like Condition in a Juvenile Rhinocerus Iguana (Cyclura Cornuta). *Aust. Vet. J.* **1986**, *63*, 343–344. [CrossRef]
6. Evans, S.E. The Skull of Lizards and Tuatara. *Biol. Reptil.* **2008**, *20*, 2–227.
7. Arencibia, A.; Hidalgo, M.R.; Vázquez, J.M.; Contreras, S.; Ramírez, G.; Orós, J. Sectional Anatomic and Magnetic Resonance Imaging Features of the Head of Juvenile Loggerhead Sea Turtles (Caretta Caretta). *Am. J. Vet. Res.* **2012**, *73*, 1119–1127. [CrossRef] [PubMed]
8. Spadola, F.; Barillaro, G.; Morici, M.; Nocera, A.; Knotek, Z. The Practical Use of Computed Tomography in Evaluation of Shell Lesions in Six Loggerhead Turtles (Caretta Caretta). *Vet. Med.* **2016**, *61*, 394–398. [CrossRef]
9. Christman, J.; Devau, M.; Wilson-Robles, H.; Hoppes, S.; Rech, R.; Russell, K.E.; Heatley, J.J. Oncology of Reptiles: Diseases, Diagnosis, and Treatment. *Vet. Clin. North Am.—Exot. Anim. Pract.* **2017**, *20*, 87–110. [CrossRef]
0. Avens, L.; Taylor, J.C.; Goshe, L.R.; Jones, T.T.; Hastings, M. Use of Skeletochronological Analysis to Estimate the Age of Leatherback Sea Turtles Dermochelys Coriacea in the Western North Atlantic. *Endanger. Species Res.* **2009**, *8*, 165–177. [CrossRef]
1. De Queiroz, K. The Scleral Ossicles of Sceloporine Iguanids: A Reexamination with Comments on Their Phylogenetic Significance. *Herpetologica* **2023**, *38*, 302–311.
2. Hoops, D.; Desfilis, E.; Ullmann, J.F.P.; Janke, A.L.; Stait-Gardner, T.; Devenyi, G.A.; Price, W.S.; Medina, L.; Whiting, M.J.; Keogh, J.S. *A 3D MRI-Based Atlas of a Lizard Brain*; Wiley Periodicals, Inc: Hoboken, NJ, USA, 2018; Volume 526, ISBN 3415253201.
3. Getty, R. *The Anatomy of the Domestic Animals*, 5th ed.; Macmillan Company of India Limited: Mumbai, India, 2004.
4. du Plessis, A.; Broeckhoven, C.; Guelpa, A.; le Roux, S.G. Laboratory X-Ray Micro-Computed Tomography: A User Guideline for Biological Samples. *Gigascience* **2017**, *6*, 1–11. [CrossRef]
5. Broeckhoven, C.; Du Plessis, A. X-Ray Microtomography in Herpetological Research: A Review. *Amphib. Reptil.* **2018**, *39*, 377–401. [CrossRef]

animals

Article

Comparison of the Rostral Epidural Rete Mirabile and the Patterns of Its Blood Supply in Selected Suiformes and Hippopotamuses

Maciej Zdun [1,2]

[1] Department of Basic and Preclinical Sciences, Nicolaus Copernicus University in Torun, Lwowska 1, 87-100 Torun, Poland; maciejzdun@umk.pl; Tel.: +48-56-611-30-93
[2] Department of Animal Anatomy, Poznan University of Life Sciences, Wojska Polskiego 71C, 60-625 Poznań, Poland

Simple Summary: The rostral epidural rete mirabile is built of considerable small arterial vessels that anastomose with each other at the base of the cranial cavity. This structure is formed as a result of the division of a large arterial vessel into many small arteries. Furthermore, on the other side of the rete mirabile, these small arteries are joined together to form one large artery, through which blood flows out of the rete mirabile. This system participates in decreasing the temperature of the brain and thus protects the body from thermal stress. Moreover, it influences body–water balance and has a crucial role in the retrograde transfer of neuropeptides. According to old systematics, pigs and hippopotamuses are in one suborder. Hence it was decided to compare their structures and blood supplies. It was found that the shape of the analyzed structure and the inflows differed in these groups.

Abstract: The rostral epidural rete mirabile (*rete mirabile epidurale rostrale*) is built of considerable small arterial vessels that anastomose with each other. This structure is formed as a result of the division of a large arterial vessel into many small arteries. Furthermore, on the other side of the rete mirabile, these small arteries are joined together to form one large artery, through which blood flows out of the rete mirabile. This system participates in decreasing the temperature of the brain and thus protects the body from thermal stress. Moreover, it influences body–water balance and has a crucial role in the retrograde transfer of neuropeptides. The goal of this study was to describe the rostral epidural rete mirabile and pathways that provide blood to this structure as well as compare it in selected Suiformes and hippopotamuses. The study was performed on desert warthogs (*Phacochoerus aethiopicus*), Eurasian wild boars (*Sus scrofa*), collared peccaries (*Pecari tajacu*), pygmy hippopotamuses (*Choeropsis liberiensis*), and common hippopotamuses (*Hippopotamus amphibius*). Preparations were made using the latex method and corrosion cast. An elongated shape characterizes its anatomy with a much wider rostral part than caudal part in the Eurasian wild boars, desert warthogs, and collared peccaries. The main source of blood was the branch to the rostral epidural rete mirabile branched off from the internal carotid artery. Moreover, blood enters the rete by the caudal branch and rostral branch to the rostral epidural rete mirabile. In hippopotamuses, the major source of blood was the rostral branches to the rostral epidural rete mirabile.

Keywords: anatomy; angiology; artery; encephalon; RERM; vascularization

1. Introduction

The rostral epidural rete mirabile (*rete mirabile epidurale rostrale*) is an arterial system built of considerable small vessels that anastomose with each other. It is situated at the base of the cranial cavity [1–5]. The rostral epidural rete mirabile is found in the cerebral base of Artiodactyla. It is a characteristic structure for this order [6]. The rostral epidural rete mirabile has not been found in representatives of other orders. Mentions of the occurrence of this rete mirabile in humans appear in the works of Leonardo da Vinci. In 1664, Thomas Willis denied these claims and stated that there is no rostral epidural rete mirabile in humans and horses [7]. A detailed analysis of the rete mirabile was made in the dromedary and llama from the Tylopoda suborder. In this species, the rostral epidural rete mirabile has a spongy appearance and is composed of a tight network of anastomosed arteries occupying the whole cavity of the cavernous sinus. It is H-shaped and composed of two lobes on each side of the hypophysis. Each lobe is divided into rostral and caudal parts to the hypophysis. The rostral part represents a major part of the rete mirabile. The shape of the unilateral structure was described as an elongated oval [4,8]. Similar observations have been made in some representatives of Bovidae and Cervidae from the Ruminantia suborder. In these animals, the rete mirabile assumes an analogous shape and is asymmetrical in the rostral part. The caudal part is acute. Bilateral structures are connected to each other rostrally and caudally to the hypophyseal gland [2,9]. In histological studies conducted on sheep, it was found that the vessels of the rete mirabile and the cavernous sinus share a common adventitia [10]. In addition, there are fenestrations in the arterial vessels of the rete. The diameter of these fenestrations depends on the phase of the cycle [11]. A detailed list of species in which this rete has been found, and information about the vessels that connect to it are included in the discussion section. In addition to the rostral epidural rete mirabile, other rete mirabiles have also been described. In the limbs of the northern tamandua (*Tamandua mexicana*), the rete mirabile was found. The superficial brachial (*arteria brachialis superficialis*), cubital transverse (*arteria transversa cubiti*), and cranial and caudal interosseous arteries (*arteria interossea cranialis et caudalis*) have a structure of rete mirabile sending vessels to the muscles at the craniolateral part of the forearm. The medial and lateral branches of the superficial antebrachial artery (*arteria antebrachialis superficialis*), which arise as an extension of the superficial brachial artery, also form the rete mirabile. This structure is flat and wide. The major vessels of the forearm in the southern tamandua (*T. tetradactyla*), Cyclopes, and sloths also form the rete mirabile [12,13]. The anatomy of the rostral epidural rete mirabile is closely determining its function. This structure participates in the selective decreasing of the brains' temperature and thus protects the body from thermal stress [14,15]. Moreover, it influences body–water balance and plays an important role in the retrograde transfer of neuropeptides such as progesterone and β-endorphin [16]. This mechanism is made possible by the location of the rostral epidural rete mirabile in the cavernous sinus (*sinus cavernosus*). Arterial vessels contain blood at a higher temperature. The blood returning from the nasal cavity with lower temperature contributes to cooling arterial blood located in the rostral epidural rete mirabile. Due to the close proximity of the vessels of the rete mirabile and the cavernous sinus, there is a transfer of heat from the arterial blood to the cooler venous blood. Cooled arterial blood follows to the vessels supplying the brain [14]. As a consequence, it affects the welfare of an animal. Because of the physiological implications of the anatomical structure and location of the rostral epidural rete mirabile, it seems essential and necessary to describe its characteristics.

Three suborders were separated according to old systematics based on morphological features in the Artiodactyla order. These were Suiformes, Ruminantia, and Tylopoda [17,18]. According to more recent studies using genetic studies, Whippomorpha were further separated. This is because hippopotamuses have been shown to be more closely related to Cetacea than pigs [19,20]. According to old systematics, hippopotamuses belonged to the Suiformes, while more recent studies classified them as Whippomorpha. Among the analyzed species were those occurring quite commonly as well as relatively rare.

The information provided in this study, which focuses on anatomy, can be helpful to scientists, especially physiologists. In addition, information about this topic may be helpful to veterinarians who care for animals in zoos or animal rescue centers. Blood supply pathways to the brain are essential during clinical procedures only in cases of congestion or syncope due to poor brain perfusion. The brain is vulnerable to hypoxia and cannot accumulate glucose or glycogen, so an adequate blood supply is critical [21,22].

The aim of this study is to describe the rostral epidural rete mirabile and pathways that provide blood to the encephalon through the rete mirabile in selected Suiformes and hippopotamuses. A further purpose is also to compare the form of this structure and arterial patterns of this rete mirabile between the mentioned groups.

2. Materials and Methods

2.1. Animals

The study was performed on five desert warthogs (*Phacochoerus aethiopicus*), ten Eurasian wild boars (*Sus scrofa*) from the Suidae family, six collared peccaries (*Pecari tajacu*) from the Tayassuidae family from the Suiformes, two pygmy hippopotamuses (*Choeropsis liberiensis*), and three common hippopotamuses (*Hippopotamus amphibius*). The analyzed cadavers were obtained as post-mortem material from private breeders and zoos in Poland or obtained through hunting. The animals had not been euthanized or hunted for an experimental purpose. Of the cadavers obtained from hunting, only animals without damage to the head and neck were included in the study. Both genders were used in the study. Approval from research ethics committees was not needed to accomplish the aims of this study because the experimental work was conducted only on dead specimens. All procedures involving cadavers in accordance with the law on 15 January 2015 for the protection of animals used for scientific or educational purposes did not require the approval of the local ethics committee (Journal of Laws 2015, item 266).

2.2. Methods

In the study, the classical techniques used in cardiovascular anatomy research were used. The specimens were prepared in two ways and assigned to one of the methods in random order. In twenty-one specimens (Table 1), injection of both common carotid arteries with a red solution of the chemo-setting acrylic material Duracryl® Plus (SpofaDental, Jicín, Czech Republic) was performed. In addition, in four specimens (Table 1), a blue solution of the chemo-setting acrylic material Duracryl® Plus (SpofaDental, Jicín, Czech Republic) was injected into the bilateral external jugular vein (*vena jugularis externa*). When the injected solution hardened, the specimens were moved to a special container with a washing powder (Persil, Germany) solution for enzymatic maceration. The temperature used in this process was 40 °C. The maceration lasted 45 days, during which the specimens were controlled every seven days. An arterial casting of the head and cranial cavity on a skeletal scaffold were obtained as a final step in the described process. The second method used for five specimens (Table 1) consisted of injecting red liquid LBS 3060 latex into both the right and left common carotid arteries. Furthermore, in two specimens (Table 1), a blue liquid LBS 3060 latex was injected into the bilateral external jugular vein. Later, the specimens were submerged in 5% formalin solution for 10 days, and the specimens were rinsed in water for 48 h prior to the next step to flush out excess formalin solution for safety reasons. Additional protection was provided with a ventilation system in a room that was used for the preparation and protection of the preparation person using latex gloves, a filtered mask, and safety googles. The ventilation system was set for 15 air changes per hour. The skull bone was cut watchfully using an oscillating saw and then opened to visualize the soft tissue and enable preparation. In the next step, the blood vessels were manually prepared using surgical and anatomical instruments. Then, the excess connective tissue was cleaned from the prepared arterial vessels for better visualization. Using this method, the visible blood vessels patterns on the animal's soft tissues were obtained. The diameter of the casts of blood vessels providing blood to the rete mirabile before the vessel entered the rete mirabile was measured using calipers. Subsequently, all the diameters of a given individual were added together, taking this value as 100%. In the next step, the proportion that a given vessel represents to the sum of all vessels was calculated, and this value was given as a percentage. The average value for a given vessel was calculated from the analyzed specimens in a given family. This presentation of the data was intended to show which vessel has the largest diameter in a given group of animals proportionally.

Table 1. The number of specimens examined in this study.

Species	Liquid into the Common Carotid Artery		Liquid into the External Jugular Vein		Liquid into the Common Carotid Artery and External Jugular Vein	
	Duracryl® Plus (Red)	LBS 3060 Latex (Red)	Duracryl® Plus (Blue)	LBS 3060 Latex (Blue)	Duracryl® Plus (Red + Blue)	LBS 3060 Latex (Red + Blue)
desert warthogs (*Phacochoerus aethiopicus*)	3	1	-	-	1	-
Eurasian wild boars (*Sus scrofa*)	6	1	1	-	1	1
collared peccaries (*Pecari tajacu*)	4	-	-	-	1	1
pygmy hippopotamuses (*Choeropsis liberiensis*)	2	-	-	-	-	-
common hippopotamuses (*Hippopotamus amphibius*)	2	1	-	-	-	-

The names of the anatomical structures were standardized according to the *Nomina Anatomica Veterinaria* [23].

3. Results

3.1. Suidae Family

The rostral epidural rete mirabile is supplied through several ways. They all originate from the common carotid artery (*arteria carotis communis*). This vessel is divided into the external carotid artery (*arteria carotis externa*) and the internal carotid artery (*arteria carotis interna*). The internal carotid artery exists only as an initial segment. It branches off from the common carotid artery together with the condylar artery (*arteria condylaris*) and occipital artery (*arteria occipitalis*). The rest of the extracranial segment of the internal carotid artery is obliterated. The initial part of this artery becomes the ramus to the rostral epidural rete mirabile (*ramus ad rete mirabile epidurale rostrale*). This is the primary blood source to the rostral epidural rete mirabile. In the representatives of this family, the diameter of this vessel is 84.5% of the total diameter of the arteries supplying blood to the rostral epidural rete mirabile (Figure 1). In the Eurasian wild boars and desert warthogs, the caudal epidural rete mirabile (*rete mirabile epidurale caudale*) is located caudally to the ramus to the rostral epidural rete mirabile. The condylar artery and occipital artery merge with the caudal epidural rete mirabile. This is a paired structure formed from a few vessels resembling a fishing net.

Another source of blood for the rostral epidural rete mirabile is branches of the maxillary artery (*arteria maxillaris*). The maxillary artery is a continuation of the external carotid artery. From the maxillary artery, the middle meningeal artery branches off (*arteria meningea media*). From this vessel, the caudal branch to the rostral epidural rete mirabile (*ramus caudalis ad rete mirabele epidurale rostrale)* (Figure 2) branches off and joins the rete mirabile. In this specimen, this vessel is 7% of the total diameter of the arteries supplying blood to the rete mirabile. Next, the external ophthalmic artery (*arteria ophthalmica externa*) branches off from the maxillary artery. From this, the vessel branches off the rostral branch to the rostral epidural rete mirabile (*ramus rostralis ad rete mirabele epidurale rostrale*). It is a single vessel in the analyzed representatives of the Suiformes suborder. In one Eurasian wild boar, it was double bilaterally. The diameter of this vessel is 8.5% of the total diameter of the arteries supplying the rostral epidural rete mirabile.

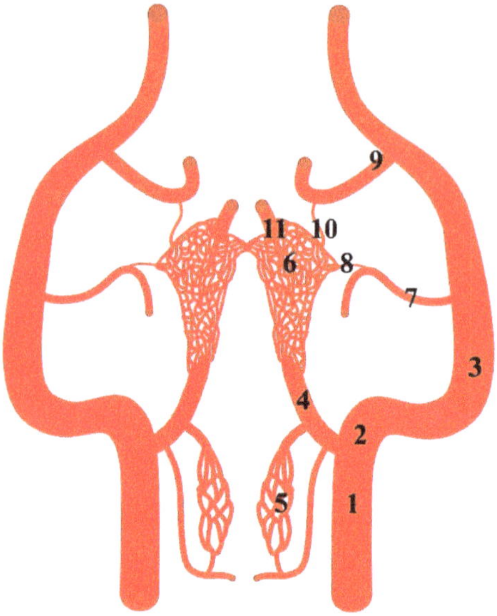

Figure 1. Diagram of the arterial pattern in the Eurasian wild boar (*Sus scrofa*). 1—common carotid artery; 2—external carotid artery; 3—maxillary artery; 4—branch to the rostral epidural rete mirabile; 5—caudal epidural rete mirabile; 6—rostral epidural rete mirabile; 7—middle meningeal artery; 8—caudal branch to the rostral epidural rete mirabile; 9—external ophthalmic artery; 10—rostral branch to the rostral epidural rete mirabile; 11—intracranial segment of the internal carotid artery.

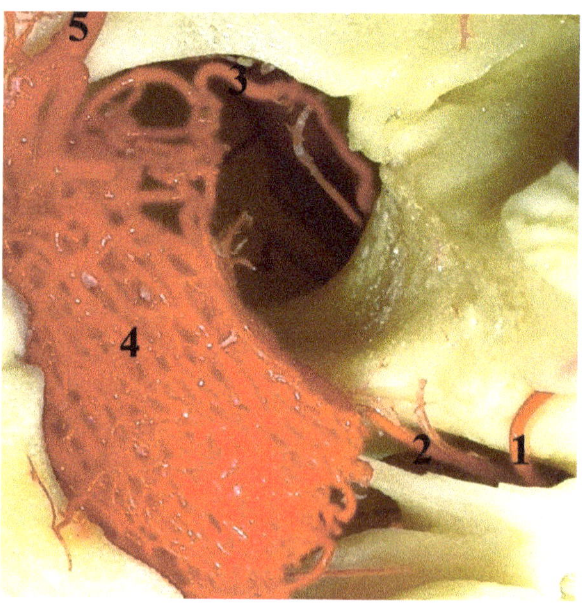

Figure 2. The rostral epidural rete mirabile in the Eurasian wild boar (*Sus scrofa*). A dorsal view and corrosion cast. 1—caudal branch to the rostral epidural rete mirabile; 2—middle meningeal artery; 3—rostral branch to the rostral epidural rete mirabile; 4—rostral epidural rete mirabile; 5—intracranial part of the internal carotid artery.

The rostral epidural rete mirabile is located at the base of the cranial cavity on the basisphenoid bone laterally surrounding the hypophyseal fossa. This is a well-developed, paired structure composed of numerous tiny anastomosing arteries (Figure 2). An elongated shape characterizes its form with a much wider rostral part than caudal. Bilateral structures are connected by several vessels in the rostral part. In a cross-sectional view, the rostral epidural rete mirabile has an oval outline. All arteries in the rete mirabile are similar in diameter (Figure 3). In the center of the rete mirabile, there is no single vessel with a significantly larger lumen than the others. There was no direct connection between the rostral and caudal epidural retia mirabilia. The intracranial segment of the internal carotid artery emerges from the rostral epidural rete mirabile. It forms the cerebral arterial circle (*circulus arteriosus cerebri*), whose branches supply the encephalon.

Figure 3. The cross-sectional view of the rostral epidural rete mirabile in the desert warthogs (*Phacochoerus aethiopicus*). A latex cast. Red—arteries of the rete mirabile; blue—cavernous sinus.

Veins at the base of the brain are expressed in the form of venous sinuses. The most important for functional reasons is the cavernous sinus (Figure 4). This sinus has a porous structure due to the numerous small spaces filled with vessels of the arterial rostral epidural rete mirabile. In the central part of this sinus, there is a fragment connecting the bilateral sinuses. The shape of the cavernous sinus is that of a rete mirabile. It is the intercavernous sinus (*sinus intercavernosus*), situated rostrally from the sella turcica.

3.2. *Tayassuidae Family*

In this family, all ways that provide blood to the rostral epidural rete mirabile also take their origin from the common carotid artery. The internal carotid artery branches off from the common carotid artery together with the condylar artery and occipital artery (Figure 5). The main extracranial part of this vessel is obliterated. Blood enters the rete mirabile by the ramus to the rostral epidural rete mirabile. This is the primary blood source to the rostral epidural rete mirabile. The diameter of this vessel is 66% of the total diameter of the arteries supplying blood to the rostral epidural rete mirabile. In the collared peccaries, the caudal epidural rete mirabile does not occur.

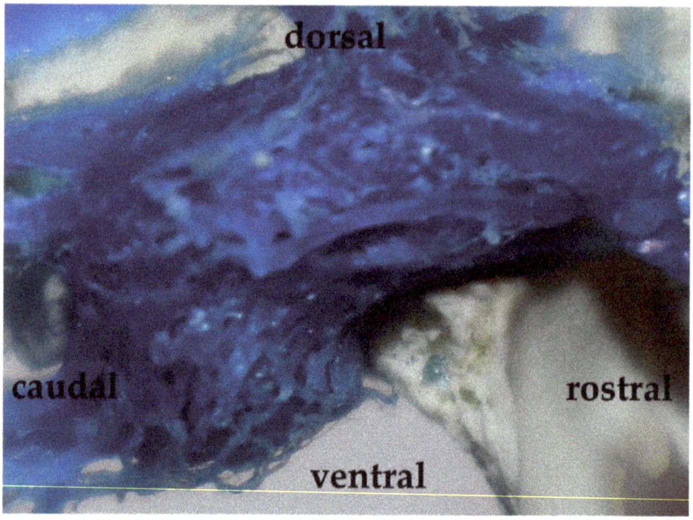

Figure 4. The cavernous sinus in the Eurasian wild boar (*Sus scrofa*). A lateral view and corrosion cast.

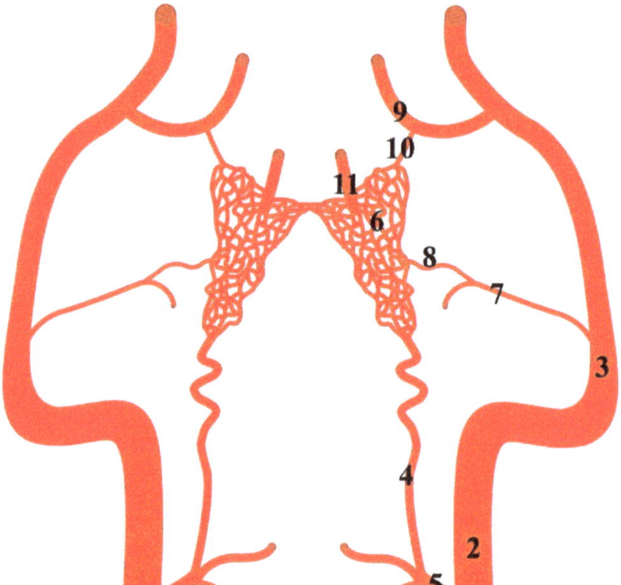

Figure 5. Diagram of the arterial pattern in the collared peccary (*Pecari tajacu*). 1—common carotid artery; 2—external carotid artery; 3—maxillary artery; 4—branch to the rostral epidural rete mirabile; 5—initial part of the internal carotid artery; 6—rostral epidural rete mirabile; 7—middle meningeal artery; 8—caudal branch to the rostral epidural rete mirabile; 9—external ophthalmic artery; 10—rostral branch to the rostral epidural rete mirabile; 11—intracranial segment of the internal carotid artery.

Next, the caudal branch to the rostral epidural rete mirabile branches off from the middle meningeal artery. The diameter of this vessel is 17% of the total diameter of the arteries supplying blood to the rostral epidural rete mirabile. More rostrally, the rostral branch to the rostral epidural rete mirabile arises (Figure 6). It is a branch of the external ophthalmic artery. It is a single vessel. In two collared peccaries, it was double or triple bilaterally. The diameter of this vessel is 17% of the total diameter of the arteries supplying the rostral epidural rete mirabile.

Figure 6. The eye socket of the collared peccary (*Pecari tajacu*). A caudolateral view and corrosion cast. 1—external ophthalmic artery; 2—rostral branch to the rostral epidural rete mirabile; 3—rostral epidural rete mirabile.

The position of the rete mirabile is analogous to that in the representatives of the Suidae family (Figure 7). In this species, it is also a well-developed, paired structure composed of numerous tiny anastomosing arteries. An elongated shape characterizes its form with a much wider rostral part than caudal. The rostro-lateral part is well developed, which makes the bilateral retes resemble the appearance of a butterfly. Bilateral structures are connected by several vessels in the rostral part. In a cross-section view, the rete mirabile looks like one in the Suidae family. The rostral epidural rete mirabile is also located in the cavernous sinus. The structure of this sinus did not differ from that described in the Suidae family. In shape, it corresponded to the rete mirabile, closely adhering to the vessels of the rete.

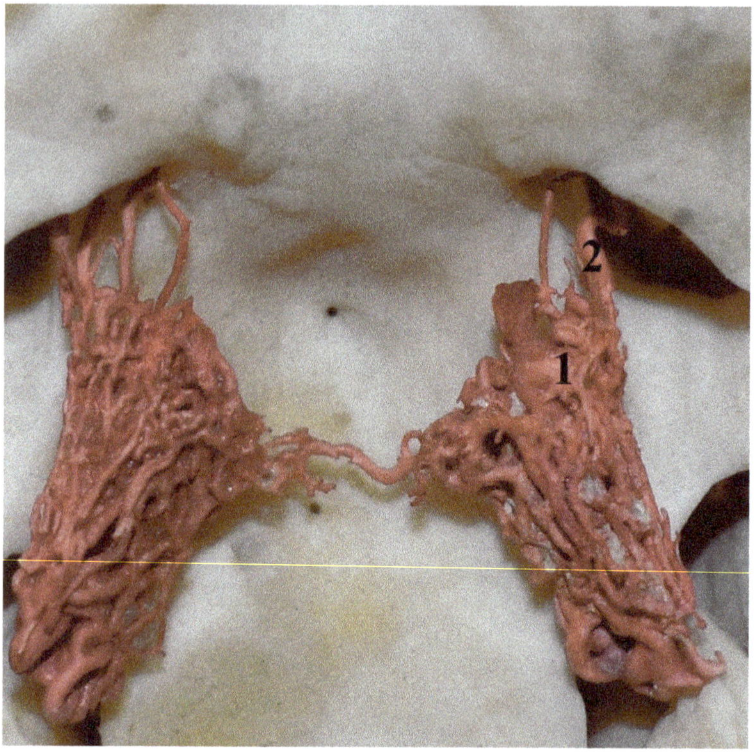

Figure 7. Rostral epidural rete mirabile in the collared peccary (*Pecari tajacu*). A dorsal view and corrosion cast. 1—rostral epidural rete mirabile; 2—rostral branches to the rostral epidural rete mirabile.

3.3. Hippopotamuses

The internal carotid artery arises from the common carotid artery. From the initial part of this vessel, the ramus to the rostral epidural rete mirabile arises (Figure 8). This is a double vessel. Close to the point where it enters the rostral epidural rete mirabile, it sometimes occurs as a single vessel. The diameter of this vessel is 29% of the total diameter of the arteries supplying blood to the rostral epidural rete mirabile. Another feature that occurs in hippopotamuses is the presence of the caudal epidural rete mirabile along the course of this vessel. It is a paired structure formed from few vessels.

From the maxillary artery, the middle meningeal artery arises, from which the caudal branch to the rostral epidural rete mirabile branches off (Figure 9). The diameter of this vessel is 10% of the total diameter of the arteries supplying blood to the rostral epidural rete mirabile. Next, the external ophthalmic artery arises from the maxillary artery. From this vessel, the rostral branches to the rostral epidural rete mirabile branch off. The rostral branches to the rostral epidural rete mirabile are present in numbers three to five. In three cases, bilaterally and one unilaterally, it is a triple vessel. In one case, bilaterally, it is quadruple. In one case, unilaterally, this vessel is in the number of five. The total diameter of these branches is 61% of the total diameter of the arteries supplying blood to the rostral epidural rete mirabile. Most blood in this species enters the rostral epidural rete mirabile through these vessels.

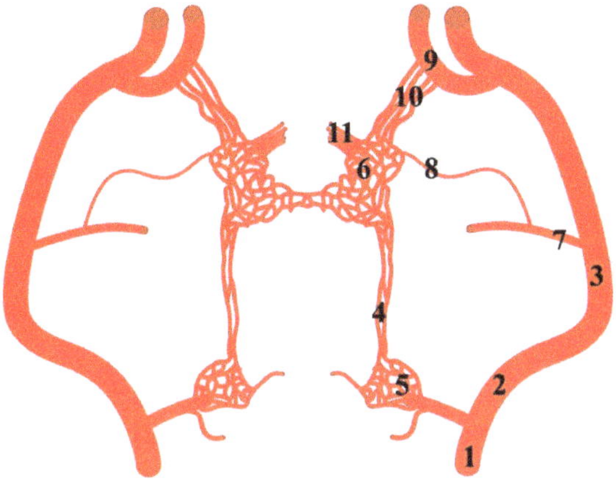

Figure 8. Diagram of the arterial pattern in the common hippopotamus. 1—common carotid artery; 2—external carotid artery; 3—maxillary artery; 4—branch to the rostral epidural rete mirabile; 5—caudal epidural rete mirabile; 6—rostral epidural rete mirabile; 7—middle meningeal artery; 8—caudal branch to the rostral epidural rete mirabile; 9—external ophthalmic artery; 10—rostral branches to the rostral epidural rete mirabile; 11—intracranial segment of the internal carotid artery.

Figure 9. Rostral epidural rete mirabile in the common hippopotamus (*Hippopotamus amphibius*). A dorsolateral view and corrosion cast. 1—branch to the rostral epidural rete mirabile; 2—caudal branch to the rostral epidural rete mirabile; 3—rostral branches to the rostral epidural rete mirabile; 4—rostral epidural rete mirabile; 5—connection between the right and left rete mirabile; 6—intracranial part of the internal carotid artery.

The rostral epidural rete mirabile is a well-developed, paired structure composed of numerous tiny anastomosing arteries. No widening is observed in the rostral part, but it has a similar diameter over its entire length. Bilateral structures are connected by several vessels in the caudal part.

4. Discussion

The presence of the rostral epidural rete mirabile is a feature of almost all animals from the Artiodactyla order. Its occurrence has been found in the Tylopoda suborder, which includes the camels of the Old and New Worlds, specifically the dromedary (*Camelus dromedarius*) and Bactrian camel (*Camelus bactrianus*) as well as the llama (*Lama glama*) and guanaco (*Lama guanicoe*) [4,24–26]. In the Ruminantia suborder, this structure has been found in all described animals from infraorder Pecora, e.g., giraffe (*Giraffa camelopardalis*), sitatunga (*Tragelaphus spekei*), nyala (*Tragelaphus angasi*), greater kudu (*Tragelaphus strepsiceros*), common eland (*Taurotragus oryx*), nilgai (*Boselaphus tragocamelus*), springbuck (*Antidorcas marsupialis*), blackbuck (*Antilope cervicapra*), dik-dik (*Madoqua kirkii*), saiga (*Saiga tatarica*), oribi (*Ourebia ourebi*), domestic cattle (*Bos taurus taurus*), goat (*Capra hircus*), domestic sheep (*Ovis aries*), banteng (*Bos javanicus*), yak (*Bos mutus f. grunniens*), European bison (*Bison bonasus*), American bison (*Bison bison*), common wildebeest (*Connochaetes taurinus*), mouflon (*Ovis aries musimon*), chital (*Axis axis*), European red deer (*Cervus elaphus*), wapiti (*Cervus elaphus Canadensis*), milu (*Elaphurus davidianus*), Eld's deer (*Rucervus eldii*), sika deer (*Cervus nippon*), fallow deer (*Dama dama*), moose (*Alces alces*), roe deer (*Capreolus capreolus*), reindeer (*Rangifer tarandus*), and Chinese muntjac (*Muntiacus reevesi*) [27–38]. In contrast, in Tragulidae of the Tragulina infraorder, the internal carotid artery anastomoses directly to the cerebral arterial circle due to the lack of the rostral epidural rete mirabile. This group of animals survived without significant changes from the Miocene when they separated from the stem of the phylogenetic tree of the Artiodactyla order. It is not known whether the lack of the rostral epidural rete mirabile is an apomorphic or plesiomorphic trait, that is, whether it was lost as a result of evolution or never developed in this group of animals. This is the only group of animals among the Artiodactyla in which the occurrence of this structure has not been described [39,40]. In Suiformes, a detailed analysis of the anatomy of this structure has been described in the domestic pig (*Sus scrofa f. domestica*) [3,41,42].

During the analysis of the anatomy of the rostral epidural rete mirabile in the cross-section, it can be seen that two types of rete mirabiles can be identified. A rete mirabile formed from vessels of similar diameter has been identified as a bipolar type [1]. The same type of rete mirabile has been found in the analyzed species i.e., in the desert warthogs, Eurasian wild boars, collared peccaries, pygmy hippopotamuses, common hippopotamuses, and in the common wildebeests, sheep and goats [2,5]. In domestic cattle, European bison, red deer, roe deer, and fallow deer, one vessel has a much larger diameter than the other vessels in the rete mirabile [2]. According to Simoens' [1] classification, this rete mirabile was called the fascicular type.

In all the analyzed representatives of Suiformes, the rostral branch to the rostral epidural rete mirabile is a single, weak vessel. In the hippopotamuses, the vessels are more numerous (from three to five). They are also the primary source of blood for the rete mirabile (61% of the total diameter of all vessels which provide blood to the rete compared to 17% in Suidae and 8.5% in collared peccaries). In the described Artiodactyla, those branches are developed in various ways. In the Tylopoda, these branches are very well developed. They occur in several numbers, and, close to the point of origin, they divide secondarily into a more significant number of small vessels, thus contributing to the expansion of the volume of the rete mirabile on the rostral side [8,24]. The rostral portion of the rete mirabile is built similarly in the common wildebeest from the Ruminantia [5]. In some representatives of the Ruminantia, such as domestic goats, domestic sheep, cattle, European bison, roe deer, red deer, and fallow deer, those branches were described in a number of two to four. However, they were not such strongly developed as in the Tylopoda or common wildebeest. Moreover, in the Tylopoda and Ruminantia, rostral branches to the rostral epidural rete mirabile branch off mainly from the maxillary artery and less often from the external ophthalmic artery [2,5,8,24]. In the analyzed species, all rostral branches to the rostral epidural rete mirabile branch off from the external ophthalmic artery. The caudal branch to the rostral epidural rete mirabile in all analyzed species in this study is a vessel with a very small lumen. It branches off from the middle meningeal artery. In the llama, this vessel is also described as an artery with a small lumen but branches off directly from the maxillary artery [8]. In the camel, it has a larger diameter [24]. In the Bovini and Cervidae, the diameter is comparable to the diameter of each rostral branch to the rete mirabile. A much larger diameter is found in the domestic goat, domestic sheep, and common wildebeest [2,5]. The phenomenon of obliteration of the extracranial part of the internal carotid artery has been described in the domestic pig and Eurasian wild boar [43]. According to this information, the vessel connecting to the rete mirabile from its caudal side has named the branch to the rostral epidural rete mirabile. This vessel that originated from the initial part of the internal carotid artery is the artery with the largest diameter in the desert warthogs and Eurasian wild boars. Such observations have also been made in the domestic pig [3,27,43].

The caudal epidural rete mirabile as the paired structure is found in desert warthogs, Eurasian wild boars, and hippopotamuses. This structure is also described in domestic pigs [3,27]. Unpaired caudal epidural rete mirabile has been noticed in the representatives of the Bovini tribe, such as domestic cattle, banteng, yak, American bison, and European bison [36]. In contrast to animals of the Suiformes, this rete mirabile in the members of the Bovini tribe is directly connected to the rostral epidural rete mirabile. The connection between these structures is also found in hippopotamuses in this research.

Physiologically, the encephalon exhibits a relatively large temperature fluctuation of 2–4 °C, which is associated with the naturally varied activities of life. In addition to the observed physiological hyperthermia, some situations cause the temperature of the encephalon to rise above acceptable norms. This can cause numerous abnormalities leading to negative effects affecting neural activity and function [44]. For this reason, in most mammals, we observe various adaptations in the course and structure of blood vessels to protect the brain from the negative effects of high temperatures [45]. An example of such adaptations is the presence of a well-developed rostral epidural rete mirabile, as observed in the Artiodactyla. The first description of the functional significance of the presence of a rostral epidural rete mirabile located close to the base of the brain and the relationship of its structure to its role in selective brain cooling was described in 1969 by Hayward and Baker [46]. This mechanism allows the animal's brain temperature to be maintained at 40.5 °C during muscular exertion. In the analyzed species, the rete mirabile is a well-developed structure located in the cavernous sinus. There were several experiments conducted on different species that showed the temperature of the encephalon is lower than the temperature of the blood in the common carotid artery. The difference ranged from 1.3 °C in domestic cattle up to 3.98 °C in camels [47–49]. Therefore, it can be assumed that it is an adaptation to this function. The well-developed structure has a large contact area with the cavernous sinus, which may result in a more efficient heat exchange between the arterial blood in the rete mirabile and the venous blood in the cavernous sinus. The rostral epidural rete mirabile as an evolutionary achievement has provided Artiodactyla with a better opportunity to adapt in terms of physiology to changing environmental conditions throughout history [14,39]. According to Strauss et al. [14], the occurrence of this structure in hippopotamuses was questionable due to the lack of data in this area. The role of this phenomenon is not limited to protecting the encephalon from overheating. It also affects the body's water balance. It represents an adaptation to extremely hot and dry environments [15]. This is because it allows for a reduction in evaporation. In Dorper sheep, it was found to possibly save even more than 2.5 liters of water per day in an individual weighing about 50 kg [50]. Another role of the described structure can be found in the literature. The mechanism of neuropeptide transport (Lh-Rh, progesterone, and β-endorphin) in the cavernous sinus and the rostral epidural rete mirabile was described in 1992 by Krzymowski et al [16]. In the cavernous sinus, the concurrent exchange of oxytocin, gonadoliberin [51,52], and dopamine takes place [53]. The limited mass of the hormones mentioned above, which varies from 0.19 kDa to 3.4 kDa, enables this process to be performed. Other substances such as prolactin and luteinizing hormone have no possibility to exchange this way, most likely due to the higher mass of these hormones, i.e., 23 kDa prolactin and 30 kDa luteinizing hormone, [54]. Another substance physiologically important for this area is carbon monoxide (CO). The light rays due to the phototransduction mechanism enable its synthesis on the retina [55]. CO may influence various areas of the brain and its activity through the concurrent exchange from ophthalmic region circulation via the cavernous sinus with CO from the nearby veins to the rostral epidural rete mirabile.

The function of the rete mirabile mentioned above indicates how multi-dimensional the physiological importance of the anatomical structure is for Artiodactyla species' biology.

5. Conclusions

This study provides information about the anatomy of the rostral epidural rete mirabile and the patterns of its blood supply in desert warthogs, Eurasian wild boars, collared peccaries, pygmy hippopotamuses, and common hippopotamuses. It is an elongated shape with a much wider rostral part in Suiformes. In hippopotamuses, no such widening was found. The caudal epidural rete mirabile is absent only in the collared peccaries. In all examined species, the rete mirabile is a bipolar type. The main blood vessel supporting the rete mirabile is the branch to the rostral epidural rete mirabile from the internal carotid artery. Other sources of blood contributed negligibly to its supply in Suiformes. In hippopotamuses, the main source of blood is the rostral branches to the rostral epidural rete mirabile. The rostral epidural rete mirabile as an evolutionary achievement has provided Artiodactyla with a better opportunity to adapt in terms of physiology to changing environmental conditions throughout history. The above research adds to the state of knowledge on the occurrence of this structure in hippopotamuses, in which it was not clear until now whether this structure exists. This could be useful information in the discussion of the evolution of this structure in Artiodactyla.

Funding: This research received no external funding.

Institutional Review Board Statement: No approval of research ethics committees was required to accomplish the aims of this study because experimental work was conducted only on cadavers. All procedures involving cadavers in accordance with the law of 15 January 2015 on the protection of animals used for scientific or educational purposes do not require the approval of the local ethics committee (Journal of Laws 2015, item 266).

Informed Consent Statement: Not applicable.

Data Availability Statement: The data presented in this study are available on request from the corresponding author.

Conflicts of Interest: The author declares no conflict of interest.

References

1. Simoens, P.; Lauwers, H.; De Geest, J.; De Schaepdrijver, L. Functional morphology of the cranial retia mirabilia in the domestic mammals. *Schw. Arch. Fur Tierh.* **1987**, *129*, 295–307.
2. Graczyk, S.; Zdun, M. The structure of the rostral epidural rete mirabile in selected representatives of the Cervidae and Bovidae families. *Acta Zool.* **2021**, *102*, 496–501. [CrossRef]
3. Graczyk, S.; Zdun, M. The structure of the rostral epidural rete mirabile and caudal epidural rete mirabile of the domestic pig. *Folia Morph.* 2022; *in press*.
4. Jerbi, H.; Khalid, S.; Perez, W. Morphometric study of the rostral epidural rete mirabile in the dromedary (*Camelus dromedarius* Linnaeus 1758). *Int. J. Morph.* **2016**, *34*, 1429–1435. [CrossRef]
5. Zdun, M.; Melnyk, O.O.; Ruszkowski, J.J.; Hetman, M. Arterial circle of the brain in the common wildebeest (*Connochaetes taurinus*). *Anat. Rec.* **2022**, *in press*. [CrossRef] [PubMed]
6. Frąkowiak, H. Magistrale tętnicze głowy u niektórych rzędów ssaków. *Rocz. AR Poz.* **2003**, *336*, 5–80.

Castrillo, G. The carotid rete mirabile: Its history and its function. *An. Real. Acad. Farm.* **2016**, *82*, 372–407.

Zdun, M.; Grzeczka, A.; Zawadzki, M.; Frąckowiak, H. The rostral epidural rete mirabile of the llama as a place of retrograde transport of various substances–anatomical basics. *Med. J. Cell Biol.* **2021**, *9*, 105–109. [CrossRef]

Ocal, M.; Asian, K. A quantitative study on the retial arteries in the bovine fetus. *Ann. Anat.* **1994**, *176*, 151–153.

9. Khamas, W.A.; Ghoshal, N.G.; Bal, H.S. Histomorphologic structure of the carotid rete cavernous sinus complex and its functional importance in sheep (*Ovis aries*). *Am. J. Vet. Res.* **1984**, *45*, 156–158.

. Khamas, W.A.; Ghoshal, N.G. Gross and scanning electron microscopy of the carotid rete-cavernous sinus complex of the sheep (*Ovis aries*). *Anat. Anz.* **1985**, *159*, 173–179.

2. Polania-Guzmán, P.V.; Vélez-García, J.F. Gross anatomical adaptations of the craniolateral forearm muscles in Tamandua mexicana (Xenarthra: Myrmecophagidae): Development of accessory muscles and rete mirabile for its arterial supply. *Heliyon* **2019**, *5*, e02179. [CrossRef]

3. Scholander, P.F. Evolution of climatic adaptation in homeotherms. *Evol* **1955**, *9*, 15–26. [CrossRef]

4. Strauss, W.; Hetem, R.; Mitchell, D.; Maloney, S.; O'Brien, H.; Meyer, L.; Fuller, A. Body water conservation through selective brain cooling by the carotid rete: A physiological feature for surviving climate change? *Conserv. Physiol.* **2017**, *5*, cow078. [CrossRef] [PubMed]

5. Hetem, R.S.; Strauss, W.M.; Fick, L.G.; Maloney, S.K.; Meyer, L.C.R.; Fuller, A.; Shobrak, M.; Mitchell, D. Selective brain cooling in Arabian oryx (Oryx leucoryx): A physiological mechanism for coping with aridity? *J. Exp. Biol.* **2012**, *215*, 3917–3924. [CrossRef] [PubMed]

6. Krzymowski, T.; Skipor, J.; Grzegorzewski, W. Cavernous sinus and carotid rete of sheep and sows as a possible place for countercurrent exchange of some neuropeptides and steroid hormones. *Anim. Reprod. Sci.* **1992**, *29*, 225–240. [CrossRef]

7. McKenna, M.; Bell, S. *Classification of Mammals Above the Species Level*; Columbia University Press: New York, NY, USA, 1997.

8. Simpson, G. The principles of classification and classification of mammals. *Bull. Am. Mus. Nat. Hist.* **1945**, *85*, 1–350.

9. Spaulding, M.; O'Leary, M.A.; Gatesy, J. Relationships of Cetacea (Artiodactyla) among mammals: Increased taxon sampling alters interpretations of key fossils and character evolution. *PLoS ONE* **2009**, *4*, e7062. [CrossRef]

0. Asher, R.J.; Helgen, K.M. Nomenclature and placental mammal phylogeny. *BMC Evol. Biol.* **2010**, *10*, 102. [CrossRef] [PubMed]

1. Kid, S.; Doughty, C.; Goldhaber, S. Syncope (fainting). *Circul* **2016**, *133*, 600–602. [CrossRef]

2. Seymour, R.; Angove, S.; Snelling, E.; Cassey, P. Scaling of cerebral blood perfusion in primates and marsupials. *J. Exper. Biol.* **2015**, *218*, 2631–2640. [CrossRef]

3. International Committee on Veterinary Gross Anatomical Nomenclature. *Nomina Anatomica Veterinaria*, 6th ed.; Editorial Committee: Hanover, Germany, 2017; pp. 73–147.

4. Al Aiyan, A.; Menon, P.; AlDarwich, A.; Almuhairi, F.; Alnuaimi, S.; Bulshawareb, A.; Qablan, M.; Shehab, S. Descriptive analysis of cerebral arterial vascular architecture in dromedary camel (*Camelus dromedarius*). *Front. Neuroanat.* **2019**, *13*, 67. [CrossRef]

5. Kiełtyka-Kurc, A.; Frąckowiak, H.; Nabzdyk, M.; Kowalczyk, K.; Zdun, M.; Tołkacz, M. The arteries on the base of the brain in the camelids. *It. J. Zool.* **2014**, *81*, 215–220.

6. Ocal, M.; Erden, H.; Ogut, I.; Kara, M. A quantitative study on the retial arteries in one-humped camels. *Ann. Anat.* **1998**, *180*, 369–371. [PubMed]

7. Nickiel, R.; Schwarz, R. Vergleichende Betrachtung der Kopfarterien der Haussaugetiere (Katze, Hund, Schwein, Rind, Schaf, Ziege, Pferd). Comparative analysis of the head arteries of domestic mammals (cat, dog, pig, cattle, sheep, goat, horse). *Zent. Für Veterinärmed. Reihe A* **1963**, *10*, 89–120. (In German) [CrossRef]

8. Frąckowiak, H.; Jakubowski, H. Arterial vascularization in the giraffe brain. *Ann. Zool. Fenn.* **2008**, *45*, 353–359.

9. Frąckowiak, H.; Dębiński, D.; Komosa, M.; Zdun, M. The arterial circle of the brain, its branches and connections in selected representatives of the Antiolopinae. *J. Morp.* **2015**, *276*, 766–771. [CrossRef]

0. Frąckowiak, H.; Zdun, M.; Kowalczyk, K.; Komosa, M.; Kiełtyka-Kurc, A. Comparison of cerebral base arteries in antelopes of Tragulus, Taurotragus and Boselaphus genera. *Zoomorph* **2014**, *133*, 351–357. [CrossRef]

1. Frąckowiak, H.; Kulawik, M. Przebieg i zmienność tętnic na podstawie mózgowia u gatunków z podrodziny Caprinae. *Rocz. AR. Poz.* **2001**, *344*, 37–48. (In Polish)

2. Kiełtyka-Kurc, A.; Frąckowiak, H.; Brudnicki, W. The arteries of brain base in species of the Cervid family. *Anat. Rec.* **2015**, *298*, 735–740. [CrossRef]

3. Godynicki, S.; Wiland, C. Tętnice podstawy mózgowia u sarny. *Rocz. AR. Poz.* **1971**, *54*, 47–54. (In Polish)

4. Godynicki, S.; Wiland, C. Tętnice podstawy mózgowia u jelenia. *Rocz. AR. Poz.* **1970**, *49*, 45–51. (In Polish)

5. Zdun, M.; Jabłoński, R.; Dębiński, D.; Frąckowiak, H. The Eurasian elk's (*Alces alces*) brain base arteries in view of vascular variation. *Anat. Rec.* **2019**, *302*, 339–345. [CrossRef]

6. Zdun, M.; Frackowiak, H.; Kiełtyka-Kurc, A.; Kowalczyk, K.; Nabzdyk, M.; Timm, A. The arteries of brain base in species of Bovini tribe. *Anat. Rec.* **2013**, *296*, 1677–1682. [CrossRef]

7. Carlton, C.; McKean, T. The carotid and orbital retia of the pronghorn, deer and elk. *Anat Rec.* **1977**, *189*, 91–107. [CrossRef] [PubMed]

8. Ding, Y.; Shao, B.; Wang, J. The arterial supply to the brain of the yak (*Bos grunniens*). *Ann. Anat.* **2007**, *189*, 31–38. [CrossRef]

9. Fukuta, K.; Kudo, H.; Sasaki, M.; Kimura, J.; Ismail, D.; Endo, H. Absence of carotid rete mirabile in small tropical ruminants: Implications for the evolution of the arterial system in artiodactyls. *J. Anat.* **2007**, *210*, 112–116. [CrossRef] [PubMed]

40. O'Brien, H. Cranial arterial pattern of the Sri Lankan spotted chevrotain, Moschiola memmina, and comparative basicrania osteology of the Tragulidae. *PeerJ* **2015**, *3*, e1451. [CrossRef] [PubMed]
41. Daniel, P.M.; Dawes, J.D.K.; Prichard, M.M.L. Studies of the Carotid Rete and Its Associated Arteries. *Philos. Trans. R. Soc. Londor Ser. B Biol. Sci.* **1953**, *237*, 173–208.
42. McGrath, P. Observations on the intra-cranial carotid rete and the hypophysis in the mature female pig and sheep. *J. Anat.* **197** *124*, 689–699.
43. Graczyk, S.; Zdun, M.; Frąckowiak, H. The internal carotid artery of the domestic pig and Eurasian wild boar in ontogenesis *Aniam. Sci. Gn.* **2022**, *18*, 1–10.
44. Kijatkin, E.A. Brain temperature and its role in physiology and pathophysiology: Lessons from 20 years of thermorecording *Temp. (Austin)* **2019**, *6*, 271–333.
45. Jessen, C. Selective brain cooling in mammals and birds. *Jpn. J. Physiol.* **2001**, *51*, 291–301. [CrossRef] [PubMed]
46. Hayward, J.; Baker, M. A comparative study of the role of the cerebral blood in the regulation of brain temperature in five mammals. *Brain Res.* **1969**, *16*, 417–440. [CrossRef] [PubMed]
47. Taylor, C.R.; Lyman, C.P. Heat storage in running antelopes: Independence of brain and body temperatures. *Am. J. Phys.* **1972** *222*, 114–117. [CrossRef]
48. Caputa, M. *Mechanizmy Obrony Mózgu Przed Przegrzaniem u Człowieka i Niektórych Innych Gatunków Ssaków*; Roczniki UMK: Toruń Poland, 1982.
49. Cain, J.; Krausman, P.; Rosenstock, S.; Turner, J. Mechanisms of thermoregulation and water balance in desert ungulates. *Wild Soc. Bull.* **2006**, *34*, 570–581. [CrossRef]
50. Strauss, A.M.; Hetem, R.S.; Mitchell, D.; Maloney, S.K.; Meyer, L.C.R.; Fuller, A. Selective Brain Cooling Reduces Water Turnover in Dehydrated Sheep. *PLoS ONE* **2015**, *10*, e0115514. [CrossRef]
51. Grzegorzewski, W.; Skipor, J.; Wasowska, B.; Krzymowski, T. Counter Current Transfer of Oxytocin from the Venous Blood of the Perihypophyseal Cavernous Sinus to the Arterial Blood of Carotid Rete Supplying the Hypophysis and Brain Depends on the Phase of the Estrous Cycle in Pigs. *Biol. Reprod.* **1995**, *52*, 139–144. [CrossRef]
52. Grzegorzewski, W.; Skipor, J.; Wasowska, B.; Krzymowski, T. Counter current transfer of 125I-LHRH in the perihypophyseal cavernous sinus-carotid rete vascular complex, demonstrated on isolated pig heads perfused with autologous blood. *Domest Anim. Endocrinol.* **1997**, *14*, 149–160. [CrossRef]
53. Skipor, J.; Wasowska, B.; Grzegorzewski, W.; Zezula-Szpyra, A.; Stefanczyk-Krzymowska, S.; Thiery, J. Transfer of dopamine by counter-current mechanism in the ewes changes with endocrine stage. *Biog. Amines.* **2001**, *16*, 431–445.
54. Skipor, J.; Grzegorzewski, W.; Wasowska, B.; Krzymowski, T. Luteinizing hormone and prolactin are not retrogradetransferred in perihypophyseal vascular complex in ewes. *Reprod. Biol.* **2004**, *4*, 195–201.
55. Oren, D. Humoral Phototransduction: Blood Is a Messenger. *Neuroscientist* **1996**, *2*, 207–210. [CrossRef]

animals

rticle

Iistology and Ultrastructure of the Esophagus in European Beaver (*Castor fiber*) Displays Features Adapted to Seasonal Changes in Diet

amila Martyniuk †, Natalia Ziółkowska †, Maria Hanuszewska-Dominiak, Natalia Szyryńska nd Bogdan Lewczuk *

Department of Histology and Embryology, Faculty of Veterinary Medicine, University of Warmia and Mazury in Olsztyn, Oczapowskiego 13, 10-719 Olsztyn, Poland
* Correspondence: lewczukb@uwm.edu.pl
† These authors contributed equally to this work.

Simple Summary: The European beaver is the biggest rodent in Eurasia and its diet is closely related to the season of the year. The gastrointestinal tract adapts its structure and physiology to ensure resistance to mechanical factors and proper digestion. Among mammals, the esophagus structure is highly variable. We aimed to perform morphological analysis of the esophagus in the European beaver, with a special attention to differences in the thickness of the epithelium in organs collected in spring, summer, and winter. Our results reveal that the mucosa is lined with stratified squamous keratinized epithelium with a structure similar to that of the skin epidermis. The process of epithelial keratinization occurs in both adult and fetal animals, suggesting that it is genetically programmed. The keratinized layer is much thicker in esophagi collected in winter than in spring and summer, while the living cell layer thickness remains unchanged regardless of the season. Moreover, immunohistochemical staining shows increased proliferation of epithelial cells in winter compared to that in spring and summer. Ultrastructural studies reveal the presence of multiple lamellar and non-lamellar bodies in granular cells, whose morphology and location gradually change while reaching the upper epithelial layers. The morphology of the esophagus in the European beaver displays features adapted to seasonal changes in diet. Future studies could employ biochemical characterization of keratinization process in the epithelium and composition of keratinized layer.

Abstract: The European beaver is a herbivorous rodent whose diet changes seasonally, and in winter consists of large quantities of woody plants. It is distinguished among other mammals by a unique organization of the stomach that comprises the cardiogastric gland and by the unusual process of mucus formation in the gastric mucosa. The aim of study was to (i) characterize the structure of the beaver esophagus with particular attention to the mucosal epithelium; (ii) compare the histological structure of the esophagi collected in spring, summer, and winter; (iii) provide preliminary data on the structure of the esophagus in beaver fetuses. The study was conducted on esophagi of 18 adult beavers captured in Poland in April, August, and December, and on 3 fetal organs. The results obtained in adults show that the mucosa is lined with thick stratified squamous keratinized epithelium with a structure similar to that of the skin epidermis. Ultrastructural studies reveal the presence of multiple lamellar and non-lamellar bodies in granular cells, whose morphology and location gradually change while reaching the upper epithelial layers. The muscularis mucosa comprises a layer of longitudinally oriented bundles of smooth muscle cells. Both mucosa and submucosa do not comprise any glands. The thick muscularis externa consists mainly of internal circular and external longitudinal layers of striated muscle fibers. The keratinized layer of mucosa epithelium was 2-3-fold thicker in esophagi collected in winter than in those collected in spring and summer, while the epithelial cell layer thickness remained unchanged regardless of the season. Immunolabeling for proliferating cell nuclear antigen shows a higher index of epithelium proliferation in esophagi collected in winter than in spring and summer. No seasonal differences were noted in other layers of the esophagus. Fetal organs have epithelium covered with a keratinized layer, thinner than in adults, and the muscularis externa comprises both striated and smooth muscle cells.

Keywords: epithelium; keratinization; histology; ultrastructure; PCNA; rodent; wildlife

1. Introduction

The structure of a mammalian esophagus shows prominent interspecies differences, which concern the keratinization of mucous epithelium, the arrangement of muscularis mucosa, the presence of mucus-secreting glands, and the organization of muscularis externa [1–6]. The differences are closely related to physicochemical properties of food and the digestive physiology, however, several aspects of these relationships remain unknown.

The European beaver (*Castor fiber*) is the largest wild rodent species in Eurasia, belonging to the suborder *Castorimorpha* and the family *Castoridae* (together with the second member of this family, *Castor canadensis*). As a herbivorous animal, it feeds mainly on plants, and the type of food it consumes is closely related to the season of the year. In summer, beavers eat leaves and young sprouts, while in winter, their basic food consists of large quantities of woody plants [7–9]. Recent research conducted on the European beaver revealed uncommon features of its stomach including the presence of the cardiogastric gland, which has been described in less than 20 mammalian species, and the occurrence of distinctive mucus layer covering the mucosa [10]. This gland is located close to the entrance of the esophagus to the stomach and comprises mainly parietal cells and chief cells [10]. The formation of mucus covering the stomach mucosa is also a species-specific process and includes the secretory granule accumulation in cytoplasm of pit cells, the granule aggregation inside cells, and the incorporation of degenerating cells into the mucus [10]. The very thick layer of this mucous is obviously a highly effective, protective barrier against hard food particles and gastric juice [10,11]. The analysis of unique features of the stomach leads to the question if the beaver's esophagus has some special protective barriers.

Taking into account the distinctness of the European beaver's lifestyle and several adaptive structural features of its gastrointestinal tract due to the large uptake of woody parts [10,12], the goal of the present study was to (i) characterize the structure of the beaver esophagus with particular attention to histology and ultrastructure of the mucosal epithelium; (ii) compare the histological structure of the esophagi collected in spring, summer, and winter; (iii) provide preliminary data on the structure of the esophagus in beaver fetuses.

2. Materials and Methods

2.1. Ethics and Legal Statement

The study was performed in strict accordance with the Polish and EU laws of animal welfare and protection. The animal capture and euthanasia were approved by the Regional Directorate for Environmental Protection in Olsztyn, a government institution responsible for the management of wildlife in the Warmia and Mazury Voivodship of Poland. Animal captures were allowed due to beaver overpopulation in this area and were conducted by a specialized team from the Polish Hunting Association. The experimental protocol including animal euthanasia was approved by the Local Ethical Commission for Experiments on Animals at the University of Warmia and Mazury in Olsztyn (Permit Numbers 22/2007, 33/2009) and the III Local Ethical Commission for Experiments on Animals at Warsaw University of Life Sciences (Permit Number 11/2010). All possible efforts were made to minimize animal suffering.

2.2. Animals

The study was conducted on 18 adult European beavers captured in their natural habitat in northeastern Poland in April (n = 6), August (n = 6), and December (n = 6). Beavers were euthanized by exsanguination under deep anesthesia induced by intramuscular administration of 60 mg of xylazine (Sedazin, Biowet Puławy, Puławy, Poland) and 300 mg of ketamine (VetaKetam, VetAgro, Lublin, Poland). Esophagi were removed immediately after the heart stopped and used for histological, immunohistochemical, and ultrastructural studies. During experiment, it was discovered that two females were pregnant, thus, three fetuses in the last prenatal development stage were collected for histological and immunohistochemical studies.

2.3. Histology and Immunohistochemistry

Tissue samples were taken from the anterior, middle, and posterior parts of esophagi, then fixed in 4% paraformaldehyde in 0.1 M phosphate buffer (pH 7.4) for 48 h, dehydrated in ethanol (TP 1020, Leica, Wetzlar, Germany), and embedded in paraffin (EG1150, Leica, Wetzlar, Germany). The 4 µm thick sections were prepared (RM2125, Leica, Wetzlar, Germany) and stained with hematoxylin and eosin (HE) using an automated multi-stainer ST 5020 with cover-slipper CV5030 (Leica, Wetzlar, Germany). The slides were digitalized using Pannoramic 250 Flash scanner (3DHistech, Budapest, Hungary).

Immunohistochemistry was performed on paraffin sections (after wet-heat-induced epitope retrieval in Tris-EDTA buffer, pH 9) using primary antibodies against proliferating cell nuclear antigen (PCNA, diluted 1:200, clone PC10, M0879, Dako Agilent Technologies, Santa Clara, CA, USA), human smooth muscle actin (SMA, diluted 1:50, clone 1A4, Dako, Glostrup, Denmark), and human desmin (diluted 1:50, clone D33, Dako, Glostrup, Denmark). The reaction products were visualized using the immunoperoxidase method with 3,3-diaminobenzidine (DAB) or 3-amino-9-ethylcarbazole (AEC) as the substrates. The specimens were counterstained with Mayer's hematoxylin and cover slipped using Mounting Glycergel Medium (Dako, Glostrup, Denmark) or Canadian balsam. For negative controls, the primary antibody was replaced with mouse IgG1 or IgG2a (Dako, Glostrup, Denmark) at the appropriate dilution.

2.4. Morphometric Evaluation

Morphometry was performed on HE-stained cross-sections from the anterior (3rd cm from the pharynx), middle, and posterior (3rd cm from the stomach) parts of esophagi of adult animals collected in April, August, and December. From each part of the organ, ten sections (separated each other by 150 µm) were used in the measurements of the thicknesses of whole mucosa, the keratinized layer, the epithelial cell layer, the muscularis mucosa, the submucosa, the muscularis externa, and the whole esophagus wall. Linear measurements were repeated 10 times per section for each parameter. The analyses were performed using the Panoramic Viewer 1.15 software program (3D-Histech, Budapest, Hungary).

2.5. Degree of Cellular Proliferation

PCNA index was determined to evaluate the degree of cellular proliferation of epithelial cells in the middle part of esophagi collected in April, August, and December from adult beavers. For comparison, the fetal esophagi were also included in this analysis. PCNA index was calculated as the percentage of PCNA-positive nuclei in the total number of nuclei [(PCNA-positive nuclei count/total nuclei count) $\times$ 100%]. The nuclei were manually counted in four rectangle 100 µm $\times$ 400 µm located over the epithelial cell layer in each section. Five sections per esophagus were analyzed.

2.6. Ultrastructural Studies

The samples of mucous membranes were collected for electron microscopy within 3 min after the heart stopped. Tissue pieces were fixed (2 h, 4 °C) in a mixture of 1% paraformaldehyde and 2.5% glutaraldehyde in 0.2 M phosphate buffer (pH 7.4). Samples were then washed, post-fixed in 2% osmium tetroxide (two hours at room temperature), and embedded in Epon 812. Semi-thin sections were cut from each block of tissue, stained with toluidine blue, and examined to identify regions for electron microscopy. Ultrathin sections were contrasted with uranyl acetate and lead citrate and then examined using a Tecnai 12 Spirit G2 BioTwin transmission electron microscope (FEI, Hillsboro, OR, USA) equipped with two digital cameras: Veleta (Olympus, Tokyo, Japan) and Eage 4k (FEI, Hillsboro, OR, USA).

2.7. Statistical Analysis

The data were analyzed by two-way analysis of variance with the season and the region of the organ as factors. The Duncan test was used a post-hoc procedure. Statistical significance was set at $p < 0.05$. The analyses were performed using Statistica 13 software (TIBCO Software, Inc., Palo Alto, CA, USA).

3. Results

3.1. Histology

The mucosa and submucosa form loosely arranged longitudinal folds (Figure 1A,D). In adult beavers, the esophageal mucosa is lined with stratified squamous epithelium composed of an epithelial cell layer covered by a very thick keratinized layer. In the epithelial cell layer, three sublayers comprising morphologically distinct epithelial cells could be distinguished (Figure 1B). The first layer (basal layer) of cells lying on the basement membrane is composed of vertically arranged columnar cells containing an elongated nucleus. The next layer (spinous layer) is composed of a few sheets of cells of various shapes, equipped with a round nucleus containing a plain nucleolus. The last sublayer (granular layer) is composed of 3–5 sheets of cells filled with multiple basophilic granules (Figure 1C). The keratinized layer consists of flattened cells usually without a nucleus (Figure 1B). The flattened cells detach from the surface of the keratinized layer, presenting signs of desquamation (Figure 1A).

The lamina propria is formed by loose connective tissue and contained numerous blood vessels (Figure 1A). The muscularis mucosa comprises a discontinuous layer of longitudinally oriented bundles of smooth muscle cells. The number of these bundles increases towards the stomach. The submucosa has typical structure. We do not observe any kind of glands in submucosa. The thick muscularis externa consists mainly of internal circular and external longitudinal layers of striated muscle fibers. However, a few flat or oval bundles of longitudinally oriented striated muscle fibers are observed on the internal surface of the circular layer. The myenteric plexus, located between the longitudinal and circular layers, is well-developed. The tunica adventitia is composed of loose connective tissue with blood vessels surrounding the esophagus. The tunica serosa is present behind the esophageal hiatus.

The general structure of the fetal esophagus is similar to that observed in adult beavers (Figure 1D). The epithelial cell layer is much thinner than in adult animals and consists of one sheet of basal cells, 2–3 sheets of spinous cells, and 1–2 sheets of granular cells (Figure 1E,F). The stratified epithelium is covered with a layer of keratinized tissue, which is much thinner than in adult animals. In fetal esophagi, the muscularis externa comprises both striated muscle fibers and smooth muscle cells, which is clearly demonstrated by immunochemistry with antibodies against smooth muscle actin and desmin (Figure 2).

3.2. Morphometry

Morphometric analysis shows significant differences in thickness of the keratinized layer depending on the part of the esophagus and the season (Figure 3A). The thickness of this layer increases from the cranial to the caudal part of the organ, irrespective of the time of year. Seasonal differences are the most prominent in the cranial part of the esophagus, where the thickness of keratinized layer is approximately 3-fold higher in December than in August and 1.5-fold higher in December than in April. Similar, but slightly lower, seasonal variations are found in the middle and cranial part of the esophagus. In contrast to the keratinized layer, the epithelial cell layer shows no significant differences of its thickness between parts of organ and seasons (Figure 3B). The thickness of muscularis mucosa increases significantly towards the stomach (Figure 3C). The seasonal and regional variations are also found in the measurement of the whole mucosa (Figure 3D). The thickness of muscularis externa is significantly higher in the caudal part than in cranial and middle parts of the esophagus (Figure 3E). Similar differences are found in the thickness of the whole esophagus wall (Figure 3F).

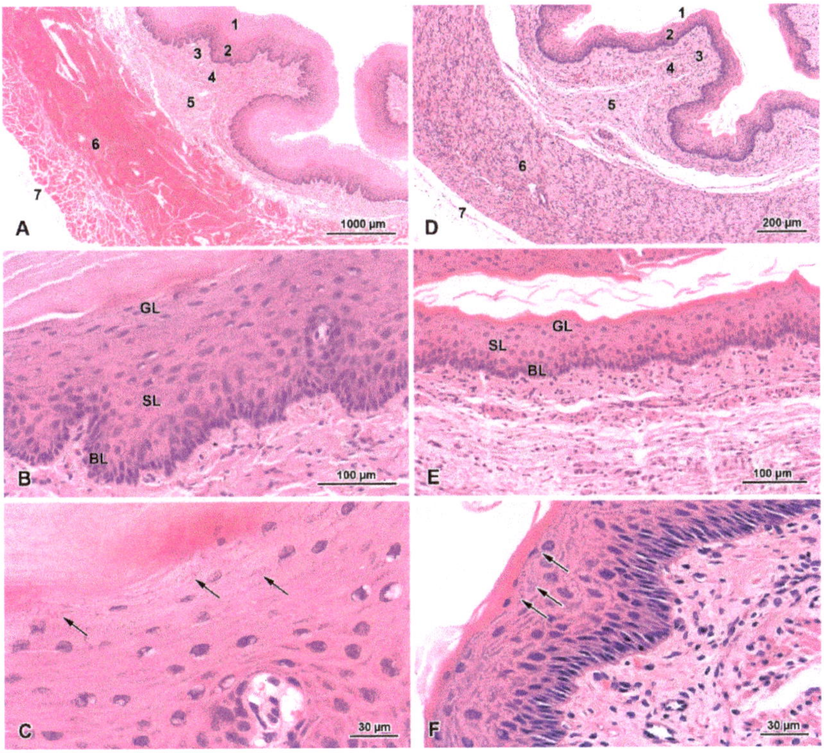

Figure 1. Histological structure of the adult and fetal beaver esophagus. Hematoxylin and eosin staining (HE). (**A**) Histological structure of the esophagus in adult European beaver: 1—thick keratinized layer; 2—epithelial cell layer; 3—lamina propria; 4—muscularis mucosa; 5—submucosa; 6—muscularis externa; 7—adventitia. (**B**) Histological structure of epithelial cell layer in adult beaver esophagus: BL—basal layer; SL—spinous layer; GL—granular layer. (**C**) Magnification of granular layer. Note the presence of basophilic inclusions in the cytoplasm of the granular cell (arrows). (**D**) Histological structure of the esophagus in fetal European beaver: 1—thin keratinized layer; 2—epithelial cell layer; 3—lamina propria; 4—muscularis mucosa; 5—submucosa; 6—muscularis externa; 7—adventitia. (**E**) Histological structure of epithelial cell layer in fetal beaver esophagus: BL—basal layer; SL—spinous layer; GL—granular layer. (**F**) Magnification of granular layer. Note the presence of basophilic inclusions in the cytoplasm of the granular cell (arrows).

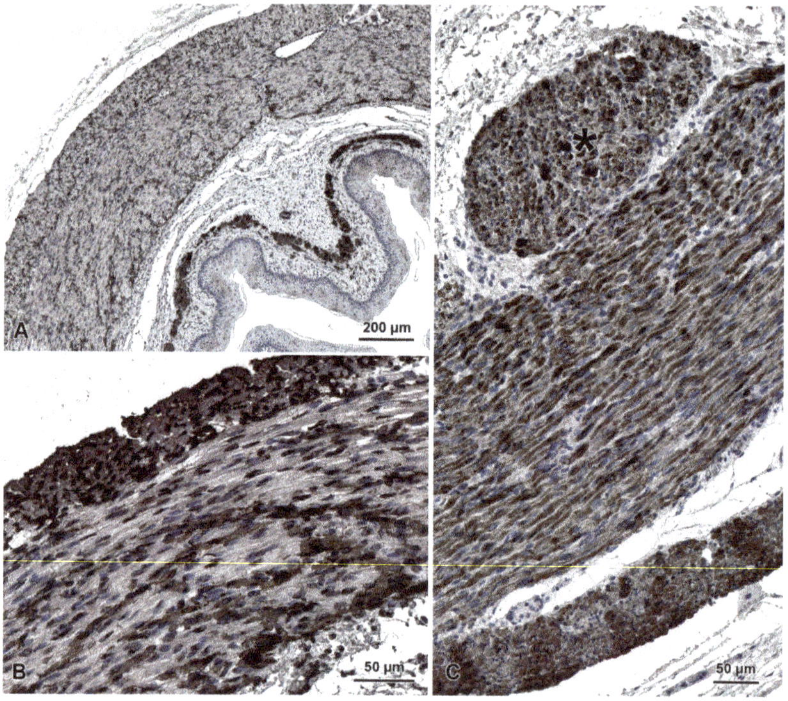

Figure 2. Distribution of smooth muscle and striated muscle cells in the esophagus of beaver fetuses demonstrated by immunohistochemical staining with antibodies anti-SMA (**A**,**B**) and anti-desmin (**C**). SMA-positive smooth muscle cells are visible both in the muscularis mucosa and the muscularis externa (**A**). Higher magnification of muscularis externa shows both types of muscle cells (**B**). Note an oval bundle of longitudinally oriented muscle cells on the internal surface of the circular layer (star in **C**).

3.3. PCNA Immunohistochemistry

Immunocytochemistry for PCNA shows a positive reaction in cells of the basal and spinous layers of the epithelium in both fetuses and adults esophagi (Figure 4A). The percentage of PCNA-positive cells is significantly higher in organs collected in December than in April and August, and in April than in August (Figure 4B). The highest PCNA index (92.1%) is found in the esophagi of fetuses.

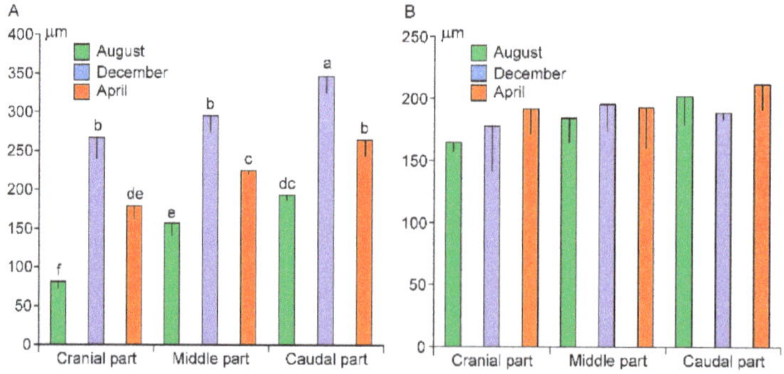

Figure 3. *Cont.*

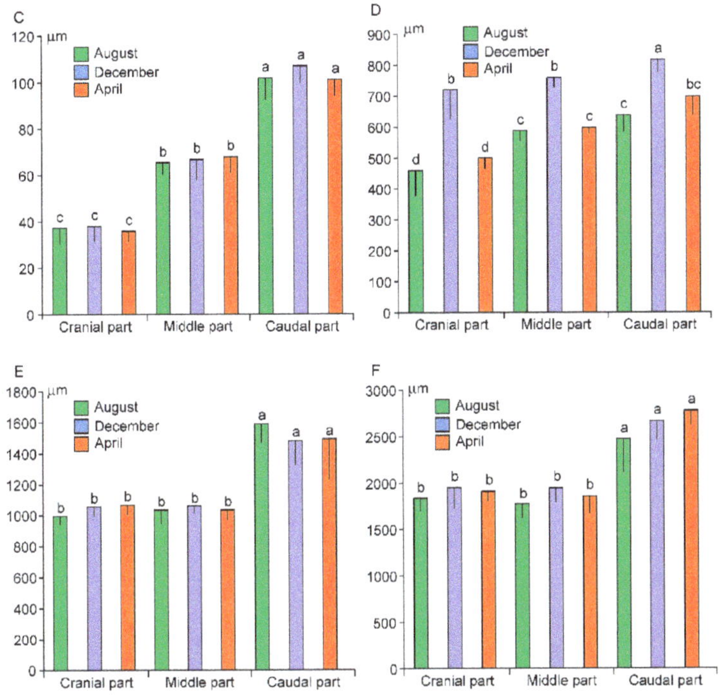

Figure 3. The thickness (means ± SD) of the keratinized layer (**A**), the epithelial cell layer (**B**), the muscularis mucosa (**C**), the mucosa (**D**), the muscularis externa (**E**), and the whole wall (**F**) in the esophagi of adult European beavers collected in August, December, and April. The same letters indicate means that are not significantly different.

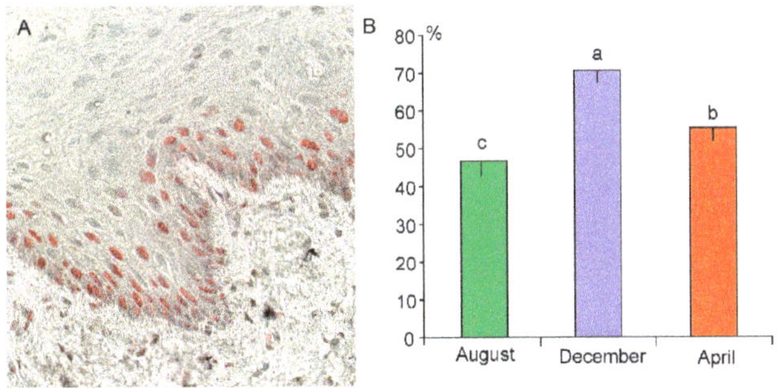

Figure 4. (**A**) PCNA-positive cells in the basal and spinous layers of the epithelium of adult beaver esophagi collected in April. (**B**) Percentage (means ± SD) of PCNA positive cells in the epithelial cell layers in esophagi of collected in August, December, and April. Different letters indicate significant differences between means.

3.4. Ultrastructure of Stratified Squamous Epithelium

The basal layer consists of one layer of columnar cells that has an oval, euchromatin-rich nucleus with deep nuclear envelope invaginations and cytoplasm with rough endoplasmic reticulum, free ribosomes, and mitochondria (Figure 5A). Mitotic figures are occasionally observed in this layer. Basal cells are connected to the basal lamina of a typical structure by hemidesmosomes (Figure 5A).

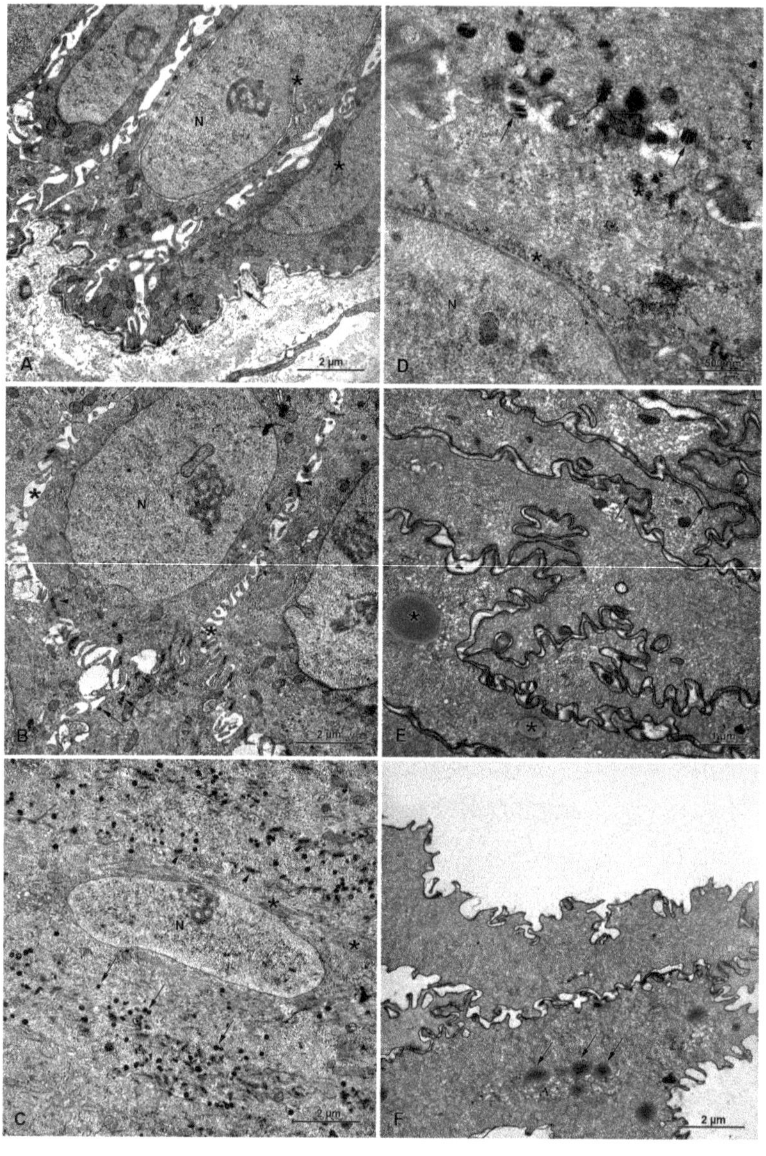

Figure 5. Ultrastructure of stratified squamous epithelium of the adult beaver esophagus. (**A**) Basal cells containing large nucleus (N) rich in euchromatin and nuclear envelope with deep invaginations (stars). Note that the attachment of these cells to the basement membrane (arrow) occurs via hemidesmosomes (arrowhead). (**B**) Spinous cells reveal multiple cytoplasmic processes (stars) and desmosomes (arrowhead). Note the presence of single, electron-dense bodies in the cytoplasm near the cell membrane (arrows). (**C**) Granular cells located near the spinous layer are filled with bundles of intermediate filaments (star) located around the nucleus (N). Numerous electron-dense bodies are present in cytoplasm, near the cell membrane (arrows). Note that granular cells are tightly connected

to each other via desmosomes (arrowhead). (**D**) Granular cells located near the keratinized layer. Note that lamellar bodies spread out from granular cells (arrows) and occur in the extracellular space. Irregular areas of granular bodies are located in the cytoplasm (stars). Letter N shows nucleus (**E**) Keratinized layer contains lipid droplets (star), intermediate filaments, and the remains of granular and lamellar bodies. Note lamellar deposits in the intracellular space along interdigitated cell membrane (arrows). (**F**) Keratinized layer with cell desquamation visible as detaching from the surface. Note the presence of lipid droplets (arrow).

The spinous layer consists of 4–6 layers of cells with dilated cisterns of rough endoplasmic reticulum filled with electron-dense content. The most prominent feature of this layer is the presence of numerous cell processes and desmosomes (Figure 5B). The bundles of intermediate filaments are present in the cytoplasm of spinous cells. Non-lamellar electron-dense bodies are occasionally observed in the cytoplasm of these cells (Figure 5B).

The granular layer is composed of 2–3 layers of flattened cells, connected to each other by desmosomes (Figure 5C). The prominent feature of the granular layer is the presence of numerous electron-dense bodies and accumulation of intermediate filaments (Figure 5C). These bodies are mostly round in shape and located close to the cell membrane. Two types of bodies can be distinguished: (1) non-lamellar bodies with electron-dense granular content and electron-lucent clefts (Figure 6A); (2) lamellar bodies with electron-dense discs arranged horizontally and/or perpendicularly to the long axis of those bodies (Figure 6B). In the superficial region of the granular layer, the localization and morphology of these bodies changes (Figure 5D). Lamellar bodies are present very close to the cell membrane and in the extracellular space. In the cytoplasm, we could distinguish areas of small, electron-dense bodies that form irregular clusters (Figure 5D).

The keratinized layer consists of flattened cells without nuclei, filled with intermediate filaments. The ultrastructure of this layer shows the presence of the remains of non-lamellar bodies and multiple droplets of lipids (Figure 5E,F and Figure 6C). The cell membranes of keratinized cells are interdigitated. Lamellar structures and electron-dense material are present between these cells in the intracellular space (Figure 5E,F).

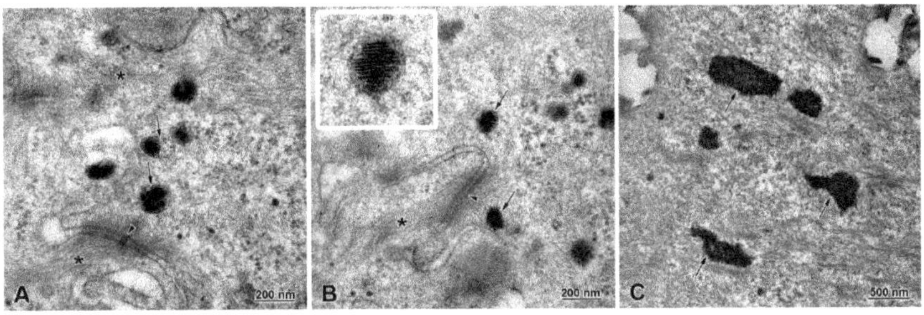

Figure 6. Ultrastructure of lamellar and non-lamellar bodies in the stratified squamous epithelium of the adult beaver esophagus. (**A**) Non-lamellar bodies present in granular cells are covered with membrane and have granular, electron-dense content and lucent clefts (arrows). Note long bundles of intermediate filaments (star) and desmosomes (arrowhead). (**B**) Lamellar bodies present in granular cells are round and ellipsoidal in shape (arrows). Insert shows electron-dense discs inside these bodies. Note bundles of intermediate filaments (star) and desmosomes (arrowhead). (**C**) Non-lamellar bodies present in keratinized layer consist of small electron-dense granules that form irregular shapes and are not covered with membranes (arrows).

4. Discussion

The present study provides a description of morphological features of the esophagus of the European beaver at light and electron microscopy levels, paying a special attention to the differences between organs collected in spring, summer and winter. It also provides some data on the esophagus structure in fetuses.

The obtained data show a typical histological structure with well-separated layers of the European beaver esophagus, both in adults and fetuses. The epithelium covering mucosa is characterized by heavy keratinization. The keratinized layer is very thick and the epithelial cell layer comprises the prominent granular layer with numerous basophilic granules. It should be noted that the histological organization of the esophageal epithelium is similar to the epidermis covering the skin [11]. The absence of nuclei in the cornified layer and the presence of a granular layer allow us to establish that it is an orthokeratinization and not a parakeratinization [13,14]. Keratinization is typical for the esophageal epithelia exposed to mechanical irritation by food compounds and is especially intensive in species with an abrasive diet, for example in rodents such as mice [15], rats [16,17], capybaras [18], guinea pigs [19], Syrian hamsters [20], grasscutters [21], and agoutis [22]; in domestic ruminants, such as cows [23], goats [24–28], and sheep [23,28]; in wild animals such as European roe deer [27], and in camels [1]. In contrast, a lack of keratinized layer in the esophageal epithelium has been described in mammals whose diet is based on a soft type of food, that is, humans [12], cats and dogs [12], buffalo calves [29], and rabbits [30]. In these animals, the superficial cells of the epithelium possess nuclei and only occasionally contain a few keratohyalin granules [31].

In the present study, morphometric analysis of the epithelial thickness in adult beavers reveals that the keratinized layer is 2–3-fold thicker in esophagi collected in winter than in those collected in summer. In spring, the keratinized layer thickness is between the values found in August and December. Such a difference is likely due to availability of different types of food in these seasons and confirms the adaptation of the epithelium to its protective function. In spring and summer, beavers eat green plant parts, and in winter, their diet is based on a large quantity of woody plants [7–9], which are much more mechanically irritating to the digestive tract. Interestingly, we also show that the keratinized layer is already present in the fetal esophagus, which suggests that the process of keratinization is genetically programmed. In contrast, in mice, the squamous epithelium becomes keratinized at least 8 days after birth [32,33].

Although a substantial difference in the thickness of the keratinized layer between adult esophagi collected in spring, summer, and winter is detected, no differences in the thickness of the epithelial cell layer between esophagi collected in those seasons are recorded. This observation leads to the assumption that in winter, due to increased epithelial keratinization and desquamation, the process of epithelial cell proliferation is more intense than in other seasons, ensuring the unaffected thickness of the epithelial cell layer. This hypothesis is strongly supported by PCNA immunolabeling, which shows a higher PCNA index in December than in April and August. The highest PCNA index is observed in fetal esophagi and is likely related to massive mitoses that occur during embryonic development.

Ultrastructural analysis shows that the amount and distribution of intermediate filaments in the esophagus epithelium is typical for the process of intensive keratinization, which reflects beaver's food preferences. Intermediate filaments are involved in defense against strong mechanical stress from heterogeneous nutrition types, microorganisms, and toxins during food passage [16,34–38]. Special attention should be paid to the presence of variable dense bodies in the granular and keratinized layers. We distinguished two types of bodies in the granular cell layer: non-lamellar and lamellar bodies. The non-lamellar bodies are covered with a membrane and filled with electron-dense granular content and lucent clefts. In contrast, lamellar bodies are also covered with membrane, but contain electron-dense discs oriented perpendicularly or longitudinally to the long axis of those bodies. An interesting phenomenon is the rearrangement of lamellar bodies in the cytoplasm, while granular cells reach the upper epithelial layers on the border with the keratinized layer. The lamellar bodies migrate through the cell membrane to the extracellular space and form lamellar deposits. Moreover, granular bodies are arranged in irregular areas near the cell membrane, similar to lamellar bodies also found in the intercellular space. The majority of current knowledge on dense bodies in epithelial cells derives from the study performed on the epidermis. Epithelial lamellar bodies play a major role in skin barrier formation [39–44], but are also responsible for proper desquamation of epidermis and skin health [45–47]. Secretion of granule content to the intracellular space in the epidermis protects the keratinized layer and lower layers of cells from physical and chemical factors. Moreover, cell bodies may affect the flow between keratinized cells, regulating the permeability of the epithelial barrier [48–52]. In the present study, we also demonstrate that lipid droplets are present in keratinized layer. Although we are not able to identify their biochemical composition, it is possible that these droplets contain sphingolipids and glucosylceramides, as it was shown in other squamous epithelia in previous studies [29,51].

The muscularis mucosa is moderately developed and shows the organization typical for a majority of mammalian species [27,53–56]. Although the thickness and the density of muscle bundles increase in a craniocaudal direction, the muscularis mucosa does not form a continuous layer even at the entrance of the esophagus to the stomach.

The submucosa shows a typical structure, however, no esophageal glands, which are important for food translocation and mucosa protection, are observed in any region of the beavers' esophagi. The presence and localization of glands differ distinctly between species. Mucus glands are abundant in the proximal part of the esophagus in humans and Japanese macaque [5,6]. In some carnivores, such as dogs, racoons, Japanese martens, and masked palm civets, glands are present along the entire length of the organ [5]. In turn, mucus glands are poorly developed in cats, rabbits, and rats [5]. Similar to our results, submucosal glands were not observed in guinea pigs [3]. Mucus glands play an important role in maintaining the mucinous pre-epithelial barrier of the squamous epithelium [6,57]. Taking together, our results suggest that keratinized layer of epithelium is the main protective barrier in the European beaver esophagus.

The muscularis externa in esophagi of adult beavers is formed by skeletal muscle tissue. Similarly, in other rodents, such as mice, rats, guinea pigs, and hamsters, the muscularis externa is also composed of skeletal muscle fibres [3,5,58,59]. In contrast, human and cat esophagi have both skeletal and smooth muscle components located zonally: striated muscle fibres in the upper part, smooth muscle cells in the lower part, and both types of muscle tissue in the middle part of the organ [5,6,58,59]. In ruminants and dogs, muscularis externa is composed largely of skeletal muscle tissue [5]. Structural interspecies variations in esophageal walls could be related to animal feeding patterns, but the evolutionary significance of these differences remains unknown [5,60]. Interestingly, the muscularis externa in the esophagi of fetuses of the European beaver at the end of pregnancy comprises both striated and smooth muscle tissue. Our data agree with other studies showing that the developing rodent esophagi is initially composed of smooth muscle tissue that is later gradually replaced by striated muscle in a craniocaudal progression [58,61].

5. Conclusions

The present study shows that the structure of esophagus epithelium in the European beaver is similar to that observed in the skin epidermis. The process of epithelial keratinization occurs in both the adult and fetal animals, suggesting that it is genetically programmed. The thickness of the keratinized layer is season-dependent, and in winter it is 2–3-fold greater than that in summer. Interestingly, the epithelial cell layer thickness remains unchanged regardless of the season. Higher index of proliferation is observed in esophagi collected in winter compared to those collected in spring and summer. Ultrastructural studies reveal that epithelial cells contain lamellar and non-lamellar bodies, which gradually change their morphology and location while reaching the upper epithelial layers. It would be interesting in future studies to investigate cell differentiation using immunohistochemistry for different cytokeratins and for other molecules such as filaggrin, the expression of the genes responsible for keratinization of the epithelium in esophagi of the European beaver, and determine biochemical composition of lipids that are present in keratinized layer. The esophagus of the European beaver does not contain mucous glands, so the keratinized layer is a sole protective barrier against the injury by food components and gastric juice.

Author Contributions: Conceptualization, K.M., N.Z. and B.L.; methodology, K.M., M.H.-D. and N.S.; measurements, K.M. and M.H.-D.; writing—original draft preparation, K.M. and N.Z.; writing—review and editing, B.L.; visualization, N.S. and B.L.; supervision, B.L. All authors have read and agreed to the published version of the manuscript.

Funding: Project financially supported by the Minister of Education and Science under the program entitled "Regional Initiative of Excellence" for the years 2019–2023, Project No. 010/RID/2018/19, amount of funding 12.000.000 PLN.

Institutional Review Board Statement: The animal study protocol was approved by the local Ethical Commission for Experiments on Animals at the University of Warmia and Mazury in Olsztyn (Permit Numbers 22/2007, 33/2009), and the III Local Ethical Commission for Experiments on Animals at Warsaw University of Life Sciences (Permit Number 11/2010).

Informed Consent Statement: Not applicable.

Data Availability Statement: The data presented in this study are available on request from the corresponding author.

Acknowledgments: The authors are grateful to Zygmunt Giżejewski for enabling the sample collection. We also thank Jan Gozdziewski from the Polish Hunting Association in Suwalki for the capturing and delivery of animals.

Conflicts of Interest: The authors declare no conflict of interest.

References

1. Jamdar, M.N.; Ema, A.N. The submucosal glands and the orientation of the musculature in the oesophagus of the camel. *J. Anat* **1982**, *135*, 165–171.
2. Windoffer, R.; Beil, M.; Magin, T.M.; Leube, R.E. Cytoskeleton in motion: The dynamics of keratin intermediate filaments in epithelia. *J. Cell. Biol.* **2011**, *194*, 669–678. [CrossRef]
3. Berghes, C.; Tanase, P.; Parvu, M.; Dinu, C.; Cuca, D. Contributions to the study of the esophagus and stomach morphology in guinea pig. *Sci. Papers Anim. Sci. Biotech.* **2011**, *44*, 150–154.
4. Meyer, W.; Schoennagel, B.; Kacza, J.; Busche, R.; Hornickel, I.N.; Hewicker-Trautwein, M.; Schnapper, A. Keratinization of the esophageal epithelium of domesticated mammals. *Acta Histochem.* **2014**, *116*, 235–242. [CrossRef] [PubMed]
5. Shiina, T.; Shimizu, Y.; Izumi, N.; Suzuki, Y.; Asano, M.; Atoji, Y.; Nikami, H.; Takewaki, T. A comperative histological study of the distribution of striated and smooth muscles and glands in the esophagus of wild birds and mammals. *J. Vet. Med. Sci.* **2005** *67*, 115–117. [CrossRef] [PubMed]
6. Zhang, X.; Patil, D.; Odze, R.D.; Zhao, L.; Lisovsky, M.; Guindi, M.; Riddel, R.; Bellizzi, A.; Yantiss, R.K.; Nalbantoglu, I.; et al. The microscopic anatomy of the esophagus including the individual layers, specialized tissues, and unique components and their responses to injury. *Ann. N.Y. Acad. Sci.* **2018**, *1434*, 304–318. [CrossRef] [PubMed]
7. Doucet, C.M.; Fryxell, J.M. The effect of nutritional quality on forage preference by beavers. *OIKOS* **1993**, *67*, 201–208. [CrossRef]
8. Nolet, B.A.; Rosell, F. Comeback of the beaver Castor fiber: An overview of old and new conservation problems. *Biol. Conserv.* **1998**, *83*, 165–173. [CrossRef]
9. Haarberg, O.; Rosell, F. Selective foraging on woody plant species by the Eurasian beaver (Castor fiber) in Telemark, Norway. *J. Zool.* **2006**, *270*, 201–208. [CrossRef]
10. Ziółkowska, N.; Lewczuk, B.; Petryński, W.; Palkowska, K.; Prusik, M.; Targońska, K.; Giżejewski, Z.; Przybylska-Gornowicz, B Light and electron microscopy of the European beaver (Castor fiber) stomach reveal unique morphological features with possible general biological significance. *PLoS ONE* **2014**, *9*, e945902014. [CrossRef]
11. Squier, C.A.; Kremer, M.J. Biology of oral mucosa and esophagus. *J. Natl. Cancer Inst. Monogr.* **2001**, *29*, 7–15. [CrossRef] [PubMed]
12. Pratama, R.; Schneider, D.; Böer, T.; Daniel, R. First Insights Into Bacterial Gastrointestinal Tract Communities of the Eurasian Beaver (Castor fiber). *Front. Microbiol.* **2019**, *10*, 1646. [CrossRef]
13. Alibardi, L.; Maderson, P.F.A. Observations on the histochemistry and ultrastructure of the epidermis of the tuatara, Sphenodon punctatus (Sphenodontida, Lepidosauria, Reptilia): A contribution to an understanding of the Lepidosaurian epidermal generation and the evolutionary origin of the squamate shedding complex. *J. Morphol.* **2003**, *256*, 111–133. [CrossRef] [PubMed]
14. Krmpotic, C.M.; Carlini, A.A.; Galliari, F.C.; Favaronc, P.; Miglinoc, M.A.; Scarano, A.C.; Barbeitoa, C.G. Ontogenetic variation in the stratum granulosum of the epidermis of Chaetophractus vellerosus (Xenarthra, Dasypodidae) in relation to the development of cornified scales. *Zoology* **2014**, *117*, 392–397. [CrossRef] [PubMed]

5. Parakkal, P.F. An electron microscopic study of esophageal epithelium in the newborn and adult mouse. *Am. J. Anat.* **1967**, *121*, 175–195. [CrossRef]
6. Marques-Pererira, J.P.; Leblond, C.P. Mitosis and differentiation in the stratified squamous epithelium of the rat esophagus. *Am. J. Anat.* **1965**, *117*, 73–90. [CrossRef]
7. Borghesi, J.; Mario, L.C.; Carvalho, R.C.; Rodrigues, M.N.; Favaron, P.O.; Miglino, M.A. Morphology of the digestive apparatus in Oligoryzomys nigripes (Rodentia, Sigmodontinae). *Open J. Anim. Sci.* **2015**, *5*, 132–141. [CrossRef]
8. Carrascal Velásquez, J.C.; Ortiz Bedoya, S.A.; Petro Hernández, V.G. Microscopic characterization of esophageal regions of a group of Capybara (Hydrochoerus hydrochaeris) free in Brazil. *Rev. CES Med. Zootec.* **2016**, *11*, 73–81. [CrossRef]
9. Philipsen, H.P.; Fejerskov, O. Normal histology and the effect of acute mechanical stress on the esophagus epithelium in the guinea pig. *Acta Odontol. Scand.* **1973**, *31*, 201–210. [CrossRef]
0. Chu, E.W.; Malmgren, R.A. An inhibitory effect of vitamin A on the induction of tumors of forestomach and cervix in the Syrian hamster by carcinogenic polycyclic hydrocarbons. *AACR* **1965**, *25*, 884–895.
1. Alogninouwa, T.; Agba, K.C.; Agossou, E.; Kpodekon, M. Anatomical, histological and functional specificities of the digestive tract in the male grasscutter (Thryonomys swinderianus, Temminck 1827). *Anat. Histol. Embryol.* **1996**, *25*, 15–21. [CrossRef]
2. Garcia, G.W.; Baptiste, Q.S.; Adogwa, A.O.; Kakuni, M.; Arishima, K.; Makita, T. The digestive system of the agouti (Dasyprocta leporina)-gross anatomy and histology. *Japanese J. Zoo Wildl. Med.* **2000**, *5*, 55–66. [CrossRef]
3. Goetsch, E. The structure of mammalian oesophagus. *Am. J. Anat.* **1910**, *10*, 1–40. [CrossRef]
4. Islam, M.S.; Awal, M.A.; Quasem, M.; Asaduzzaman, M.; Das, S.K. Morphology of esophagus of Black Bengal goat. *Bangl. J. Vet. Med.* **2008**, *6*, 223–225. [CrossRef]
5. Islam, M.S.; Quasem, M.A.; Awal, M.A.; Das, S.K. Histology of esophagus of Black Bengal goat. *Bangl. J. Vet. Med.* **2005**, *3*, 152–154. [CrossRef]
6. Kumar, P.; Mahesh, R.; Kumar, P. Histological architecture of esophagus of goat (Capra hircus). *Haryana Vet.* **2009**, *48*, 29–32.
7. Sokołowska, J.; Urbańska, K.; Matusiak, J.; Wiśniewski, J. New aspects of the esophageal histology of the domestic goat (Capra hircus) and European roe deer (Capreolus capreolus). *Vet. Med. Sci.* **2021**, *7*, 1743–1756. [CrossRef]
8. Hameed, B.K.; Ebraheem, A.H.; Hussein, F.A. Histological structure of the cervical segment oesophagus in goats and sheep (Comparison study). *Tikrit J. Pure Sci.* **2018**, *23*, 55–60. [CrossRef]
9. Gupta, S.K.; Sharma, D.N. Regional histology of the oesophagus of buffalo calves. *Indian J. Anim. Sci.* **1991**, *61*, 722–724.
0. Tseng, S.C.; Hatchell, D.; Tierney, N.; Huang, A.J.; Sun, T.T. Expression of specific keratin markers by rabbit corneal, conjunctival, and esophageal epithelia during vitamin A deficiency. *J. Cell. Biol.* **1984**, *99*, 2279–2286. [CrossRef]
1. Rao, R.S.; Patil, S.; Ganavi, B.S. Oral cytokeratins in health and disease. *J. Contemp. Dent. Pract.* **2014**, *15*, 127–136. [CrossRef]
2. Raymond, C.; Anne, V.; Millane, G. Development of esophageal epithelium in the fetal and neonatal mouse. *Anat. Rec.* **1991**, *230*, 225–234. [CrossRef] [PubMed]
3. Yu, W.Y.; Slack, J.M.W.; Tosh, D. Conversion of columnar to stratified squamous epithelium in the developing mouse oesophagus. *Dev. Biol.* **2005**, *284*, 157–170. [CrossRef]
4. Meyer, W.; Kacza, J.; Schnapper, A.; Verspohl, J.; Hornickel, I.N.; Seeger, J. A first report on the microbial colonisation of the equine oesophagus. *Ann. Anat.* **2010**, *192*, 42–51. [CrossRef] [PubMed]
5. Meyer, W.; Kacza, J.; Hornickel, I.; Schoennagel, B. Immunolocalization of succinate dehydrogenase in the esophagus epithelium of domesticated mammals. *Eur. J. Histochem.* **2013**, *57*, e182013. [CrossRef] [PubMed]
6. Schweizer, J.; Bowden, P.E.; Coulombe, P.A.; Langbein, L.; Lane, E.B.; Magin, T.M.; Maltais, L.; Omary, M.B.; Parry, D.A.D.; Rogers, M.A.; et al. New consensus nomenclature for mammalian keratins. *J. Cell. Biol.* **2006**, *174*, 169–174. [CrossRef] [PubMed]
7. Moll, R.; Divo, M.; Langbein, L. The human keratins: Biology and pathology. *Histochem. Cell. Biol.* **2008**, *129*, 705–733. [CrossRef]
8. Bragulla, H.H.; Homberger, D.G. Structure and functions of keratin proteins in simple, stratified, keratinized and cornified epithelia. *J. Anat.* **2009**, *214*, 516–559. [CrossRef]
9. Grayson, S.; Johnson-Winegar, A.D.; Elias, P.M. Isolation of lamellar bodies from neonatal mouse epidermis by selective sequential filtration. *Science* **1983**, *221*, 962–964. [CrossRef]
0. Freinkel, R.K.; Traczyk, T.N. Lipid composition and acid hydrolase content of lamellar granules of fetal rat epidermis. *J. Investig. Dermatol.* **1985**, *85*, 295–298. [CrossRef]
1. Rassner, U.A.; Crumrine, D.A.; Nau, P.; Elias, P.M. Microwave incubation improves lipolytic enzyme preservation for ultrastructural cytochemistry. *Histochem. J.* **1997**, *29*, 387–392. [CrossRef] [PubMed]
2. Ishida-Yamamoto, A.; Deraison, C.; Bonnart, C.; Bitoun, E.; Robinson, R.; O'Brien, T.J.; Wakamatsu, K.; Ohtsubo, S.; Takahashi, H.; Hashimoto, Y.; et al. LEKTI is localized in lamellar granules separated from KLK5 and KLK7, and is secreted in the extracellular spaces of the superficial stratum granulosum. *J. Investig. Dermatol.* **2005**, *124*, 360–366. [CrossRef]
3. Ishida-Yamamoto, A.; Simon, M.; Kishibe, M.; Miyauchi, Y.; Takahashi, H.; Yoshida, S.; O'Brien, T.J.; Serre, G.; Iizuka, H. Epidermal lamellar granules transport different cargoes as distinct aggregates. *J. Investig. Dermatol.* **2004**, *122*, 1137–1144. [CrossRef]
4. Elias, P.M.; Feingold, K.R. *Skin Barrier*; CRC Press: New York, NY, USA, 2005.
5. Dale, B.A.; Holbrook, K.A.; Fleckman, P.; Kimball, J.R.; Brumbaugh, S.; Sybert, V.P. Heterogeneity in harlequin ichthyosis an inborn error of epidermal keratinization: Variable morphology and structural protein expression and a defect in lamellar granules. *J. Investig. Dermatol.* **1990**, *94*, 6–18. [CrossRef]

46. Milner, M.E.; O'Guin, W.M.; Holbrook, K.A.; Dale, B.A. Abnormal lamellar granules in harlequin ichthyosis. *J. Investig. Dermatol.* **1992**, *99*, 824–829. [CrossRef] [PubMed]
47. Stone, S.J.; Myers, H.M.; Watkins, S.M.; Brown, B.E.; Feingold, K.R.; Elias, P.M.; Farese, R.V., Jr. Lipopenia and skin barrier abnormalities in DGAT2-deficient mice. *J. Biol. Chem.* **2004**, *279*, 11767–11776. [CrossRef] [PubMed]
48. Matoltsy, A.G.; Parakkal, P.F. Membrane-coating granules of keratinizing epithelia. *J. Cell. Biol.* **1965**, *24*, 297–307. [CrossRef] [PubMed]
49. Akiyama, M.; Sawamura, D.; Nomura, Y.; Sugawara, M.; Shimizu, H. Truncation of CGI-58 protein causes malformation of lamellar granules resulting in ichthyosis in Dorfman-Chanarin syndrome. *J. Investig. Dermatol.* **2003**, *121*, 1029–1034. [CrossRef] [PubMed]
50. Demerjian, M.; Crumrine, D.A.; Milstone, L.M.; Williams, M.L.; Elias, P.M. Barrier dysfunction and pathogenesis of neutral lipid storage disease with ichthyosis (Chanarin-Dorfman syndrome). *J. Investig. Dermatol.* **2006**, *126*, 2032–2038. [CrossRef]
51. Houben, E.; Hachem, J.P.; De Paepe, K.; Rogiers, V. Epidermal ceramidase activity regulates epidermal desquamation via stratum corneum acidification. *Skin Pharmacol. Physiol.* **2008**, *21*, 111–118. [CrossRef]
52. Menon, G.K.; Orso, E.; Aslanidis, C.; Crumrine, D.; Schmitz, G.; Elias, P.M. Ultrastructure of skin from Refsum disease with emphasis on epidermal lamellar bodies and stratum corneum barrier lipid organization. *Arch. Dermatol. Res.* **2014**, *306*, 731–737. [CrossRef]
53. DeNardi, F.G.; Riddell, R.H. The normal esophagus. *Am. J. Surg. Pathol.* **1991**, *15*, 296–309. [CrossRef] [PubMed]
54. Nagai, K.; Noguchi, T.; Hashimoto, T.; Uchida, Y.; Shimada, T. The organization of the lamina muscularis mucosae in the human esophagus. *Arch. Histol. Cytol.* **2003**, *66*, 281–288. [CrossRef]
55. Christensen, J.; Rick, G.A.; Soll, D.J. Intramural nerves and interstitial cells revealed by the Champy Maillet stain in the opossum esophagus. *J. Auton. Nerv. Syst.* **1987**, *19*, 137–151. [CrossRef]
56. Uchida, K.; Kamikawa, Y. Muscularis mucosae—The forgotten sibling. *J. Smooth Muscle Res.* **2007**, *43*, 157–177. [CrossRef] [PubMed]
57. Sarosiek, J. Does the healing of the esophageal mucosa improve the function of the esophageal submucosal and salivary glands? *Ann. N. Y. Acad. Sci.* **2016**, *1380*, 155–161. [CrossRef] [PubMed]
58. Su, P.H.; Wang, T.C.; Wong, Z.R.; Huang, B.M.; Yang, H.M. The expression of nestin delineates skeletal muscle differentiation in the developing rat esophagus. *J. Anat.* **2011**, *218*, 311–323. [CrossRef]
59. Fullington, N.M.; Potanos, K.M.; Cauley, R.P.; Purcell, P.; Zurakowski, D.; Fishman, S.J.; Vakili, K.; Kim, H.B. Strain induced esophageal growth in a novel rodent model. *J. Pediatr. Surg.* **2016**, *51*, 1273–1278. [CrossRef]
60. Krauss, R.S.; Chihara, D.; Romer, A.I. Embracing change: Striated-for-smooth muscle replacement in esophagus development. *Skelet. Muscle* **2016**, *6*, 27. [CrossRef]
61. Wörl, J.; Neuhuber, W.L. Ultrastructural analysis of the smooth-to-striated transition zone in the developing mouse esophagus: Emphasis on apoptosis of smooth and origin and differentiation of striated muscle cells. *Dev. Dyn.* **2005**, *233*, 964–982. [CrossRef]

animals

MDPI

Article

Anatomical Description of Loggerhead Turtle (*Caretta caretta*) and Green Iguana (*Iguana iguana*) Skull by Three-Dimensional Computed Tomography Reconstruction and Maximum Intensity Projection Images

Jose Raduan Jaber Mohamad [1,*], Eligia González-Rodríguez [2], Alberto Arencibia [1], Soraya Déniz [2], Conrado Carrascosa [1] and Mario Encinoso [2]

1 Department of Morphology, Veterinary Faculty, University of Las Palmas de Gran Canaria, Trasmontaña, Arucas, 35413 Las Palmas, Spain
2 Veterinary Teaching Hospital, University of Las Palmas de Gran Canaria, Trasmontaña, Arucas, 35413 Las Palmas, Spain
* Correspondence: joseraduan.jaber@ulpgc.es

Simple Summary: Due to the growing interest in reptiles and the fact that variation in reptile skull anatomy remains poorly documented, we compared the heads of two reptile species using computed tomography reconstructive procedures, such as maximum intensity projections (MIP) and volume rendering (VR). The resulting images demonstrated that these procedures are suitable for performing comparative anatomical studies as they provide adequate information about the different bones that comprise the head while avoiding the overlapping of other structures. A detailed description of the morphology of the head and associated structures could expand the knowledge of clinicians for the diagnosis of traumatic lesions with small or large fractures, skull malformations, and osteodystrophy secondary to nutrition imbalances or neoplasms.

Abstract: The growing interest in reptiles has posed a challenge to veterinary clinicians due to the lack of a standardized system to perform anatomical studies similar to those used for dogs and cats. In this paper, we have attempted to describe, employing computed tomography and subsequent three-dimensional reconstructions, the normal anatomical features that comprise the skulls of two species of reptiles: the loggerhead turtle (*Caretta caretta*) and the green iguana (*Iguana iguana*). Computed tomography (CT) and subsequent image processing allowed the identification of the bony structures that comprise the head of these species. As a result, and based on previous articles, we propose the most significant anatomical differences and similarities between these species.

Keywords: computed tomography; maximum intensity projection; volume rendering; reptiles; anatomy; skull; lizard; turtle

1. Introduction

Modern diagnostic imaging techniques allow for the creation of three-dimensional representations that can be of considerable value in the dissemination of clinical diagnoses and anatomical studies [1]. Thus, the development of such advanced imaging technologies has improved the quality of anatomical imaging, which enables better evaluation and treatment of different animal diseases.

Therefore, these imaging modalities provide highly reliable diagnostic images that have enabled advancements in the fields of both human and veterinary medicine by reconstructing axial images [2,3]. *Reconstruction* is widely misused as it refers to a very specific procedure that converts the raw data into an axial image, however, *reformatting* displays the images produced from the original reconstruction process in a different orientation than that which was initially produced [4]. Among these, we highlight multiplanar reconstructions (MPR), maximum and minimum-intensity projections (MIP and MinIP), and volume rendering (VR) [5–9].

MPR produces a reconstruction along the three axes of space. The X-axis constitutes the right-to-left view, the Y-axis corresponds to the craniocaudal view, whereas the Z-axis is the dorsoventral view of the patient under study, resulting in sagittal and coronal views from cross-sectional images [6].

The most commonly used approaches are the MIP and MinIP. MIP is a procedure that uses only the maximum relative value detected along the ray trajectories retained by the computer, showing preferentially osseous structures and high-contrast regions, while other structures of lower attenuation are not well visualized. Due to its ability to retain attenuation information despite thresholds, it is adequate for the identification of simple structures. Nevertheless, it is conditioned by the poor differentiation of overlapping structures, pulse motion artifact, and breathing [7]. On the other hand, the MinIP detects the minimum values along the ray paths in each view, showing better image quality in those structures with low attenuation [8]. In the literature, these techniques have been reported in human medicine to diagnose different disorders, especially neoplastic diseases [10]. In contrast, only a few articles in veterinary medicine have used these imaging procedures [11,12], although this is expected to increase as they are very useful for the visualization of internal structures [13,14].

Another tool is volume rendering (VR), where the voxels that comprise the image can be programmed to be visible or not, to have different colors, and also to have different levels of opacity, which is called "transparency". VR additionally provides a very useful tool: "spherical clipping volume" or "shutter tool", which establishes borderlines between voxels that need to remain visible (tissues inside the sphere) and voxels that need to remain invisible (all adjacent structures). Therefore, the volume data can be resampled based on a series of software algorithms, and manually edited to depict the region of interest from surrounding structures. This procedure has been applied in both human and veterinary medicine, as it is a valuable tool, e.g., for surgical planning [15] and the anatomical study of many species, allowing an acute visualization of bone and soft structures [9].

Reptiles have gained increasing interest in veterinary medicine. It has motivated veterinary clinicians to understand physiology and anatomy to improve diagnostics. However, to the authors' knowledge, most of the studies performed on these species have been conducted using dedicated micro-CT scanners, and only a few descriptions have been reported using conventional CT equipment, highlighting those carried out on different species of reptiles, such as the Komodo dragon, the loggerhead turtle, the common tegu, the green iguana, the boa constrictor, and the bearded dragon [16–22]. Therefore, this paper aims to provide further insight into the comparative anatomical study of the heads of the green iguana and the loggerhead turtle using conventional CT equipment and 3-D reconstructed images.

2. Materials and Methods

2.1. Animals

In this study, we studied five adult female loggerhead turtles (*Caretta caretta*), which weighed between 82 and 106 kg and had a carapace length ranging between 78 and 101 cm. This data was essential for confirming that they were adult specimens. Additionally, we included four adult male green iguanas (*Iguana iguana*), which weighed between 3.8 and 4.1 kg and had a length from snout-to-vent ranging between 28 and 45 cm. All these carcasses were collected from the Rancho Texas Lanzarote Park (Lanzarote, Spain). The animals were scanned at the Veterinary Hospital of Las Palmas de Gran Canaria University. No pathological findings were observed during the clinical examination of the head. We performed anatomical measurements of the skulls of both species (Figure 1) Therefore, we measured three head characters: head length (HL), from the tip of the snout to the posterior border of the collar, head width (HW) at the widest point of the head, and head height

(HH) at the highest point of the head. In the loggerhead turtle's head, the HL mean was 11.03 ± 2.1 cm, the HW mean was 7.94 ± 1.02 cm, and the HH mean was 5.29 ± 0.7 cm. Whereas in the green iguana head, the HL mean was 8.08 ± 0.28 cm, the HW mean was 6.83 ± 0.09 cm, and the HH mean was 4.89 ± 0.07 cm.

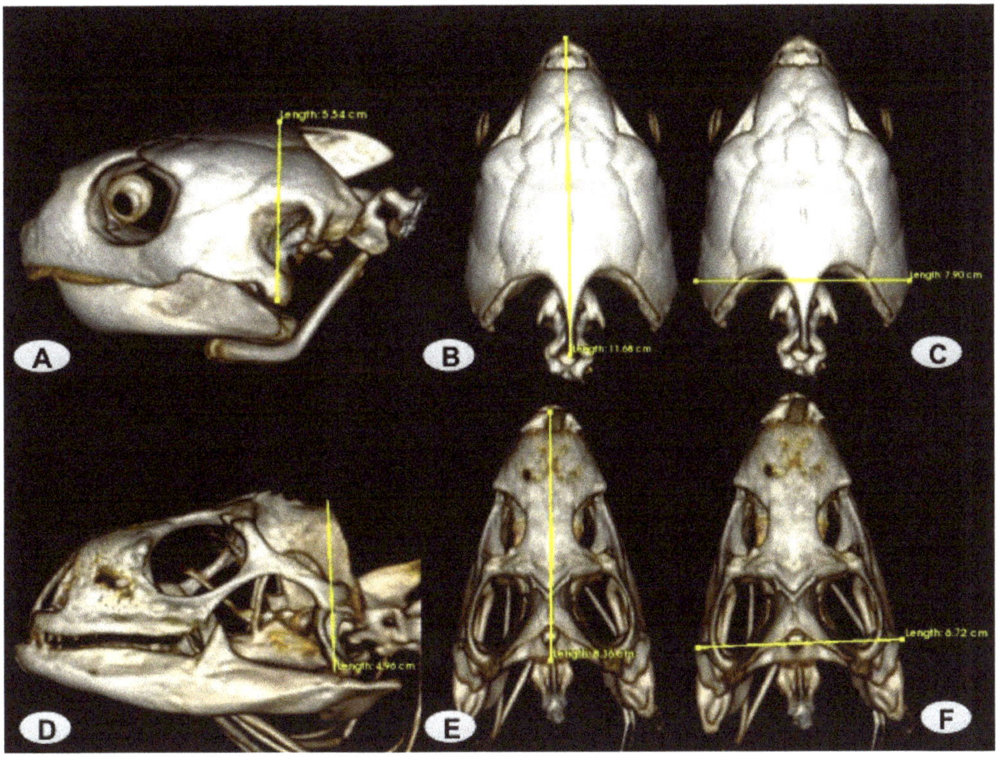

Figure 1. Measurements of the loggerhead turtle and green iguana Head. (**A**) Example of HH in a loggerhead turtle lateral VR image. (**B**,**C**) Examples of HL and HW in loggerhead dorsal VR images. (**D**) Example of HH in a green iguana lateral VR image. (**E**,**F**) Examples of HL and HW in green iguana dorsal VR images.

2.2. CT Technique

Transverse CT images were obtained using a 16–slice helical CT scanner (Toshiba Astelion, Toshiba Medical System, Tokyo, Japan). The reptiles were placed symmetrically in ventral recumbency on the CT couch. A standard clinical protocol (120 kVp, 80 mA, 512 × 512 acquisition matrix, 1809 × 858 field of view, a spiral pitch factor of 0.94, and a gantry rotation of 1.5 s) was used to obtain sequential transverse CT images (1 mm thickness). The original transverse data were recorded and transferred to the CT workstation. No CT density or anatomic variations were detected in the heads of the reptiles used in the investigation. In this study, we applied two CT windows by adjusting the window widths (WW) and window levels (WL) to appreciate the CT appearance of the head structures: a bone window setting (WW = 1500; WL = 300) and a soft tissue window setting (WW = 350; WL = 40). The original data were used to generate head VR and MIP reconstructed images after manual editing of the transverse CT images to remove soft tissues using a standard Dicom 3D format (OsiriX MD, Geneva, Switzerland).

3. Results

Different rostral, lateral, ventral, dorsal, and caudal VR and MIP images were obtained for better visualization and description of the different bones comprising the head (dermatocranium, neurocranium, and mandibula) of the turtle and the iguana (Figures 2–11). We selected images that better matched each other, providing valuable information.

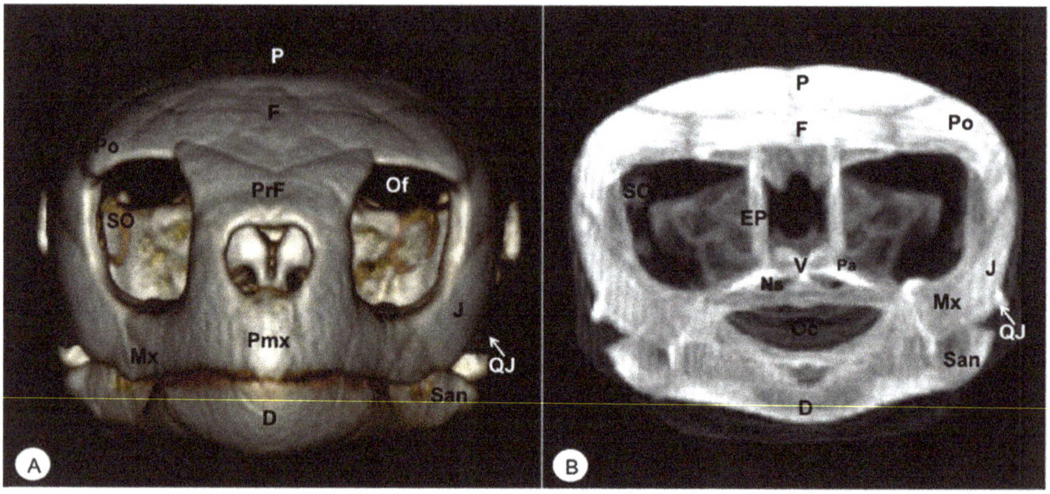

Figure 2. Head of *Caretta caretta*. (**A**) Rostral VR image. (**B**) Rostrotransversal MIP image. Pmx: Premaxillary bone. PrF: Prefrontal bone. F: Frontal bone. Po: Postorbital bone. O: Orbit. Of: Orbital fossa. J: Jugal bone. QJ: Quadratojugal. Mx: Maxillary bone. SO: Scleral ossifications. P: Parietal bone. EPt: Epipterygoid bone. V: Vomer. Pa: Palatine bone. Ns: Nasal sinuses. Oc: Oral cavity. D: Dentary. San: Surangular.

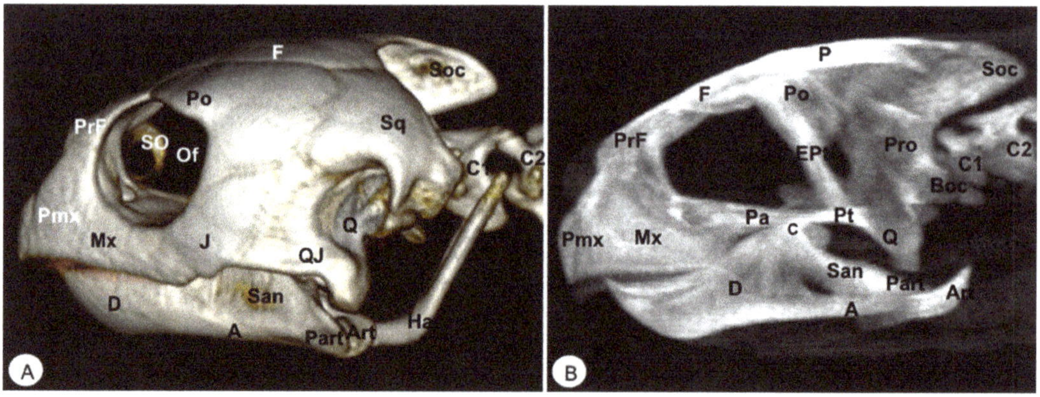

Figure 3. Head of *Caretta caretta*. (**A**) Lateral VR image. (**B**) Medial MIP image. Pmx: Premaxillary bone. Mx: Maxillary bone. PrF. Prefrontal bone. F: Frontal. Po: Postorbital. J: Jugal. SO: Scleral ossifications. QJ: Quadratojugal. Q: Quadrate. Sq: Squamosal. Boc: Basioccipital. Pt: Pterygoid. EPt: Epipterygoid. Pa: Palatine. P: Parietal. Soc: Supraoccipital. Pro: Prootic. C: Coronoid. O: Orbit. Of: Orbital fossa. D: Dentary. A: Angular. San: Surangular. Part: Prearticular. Art: Articular. Ha: Hyoid apparatus. C1: First cervical vertebra. C2: Second cervical vertebra.

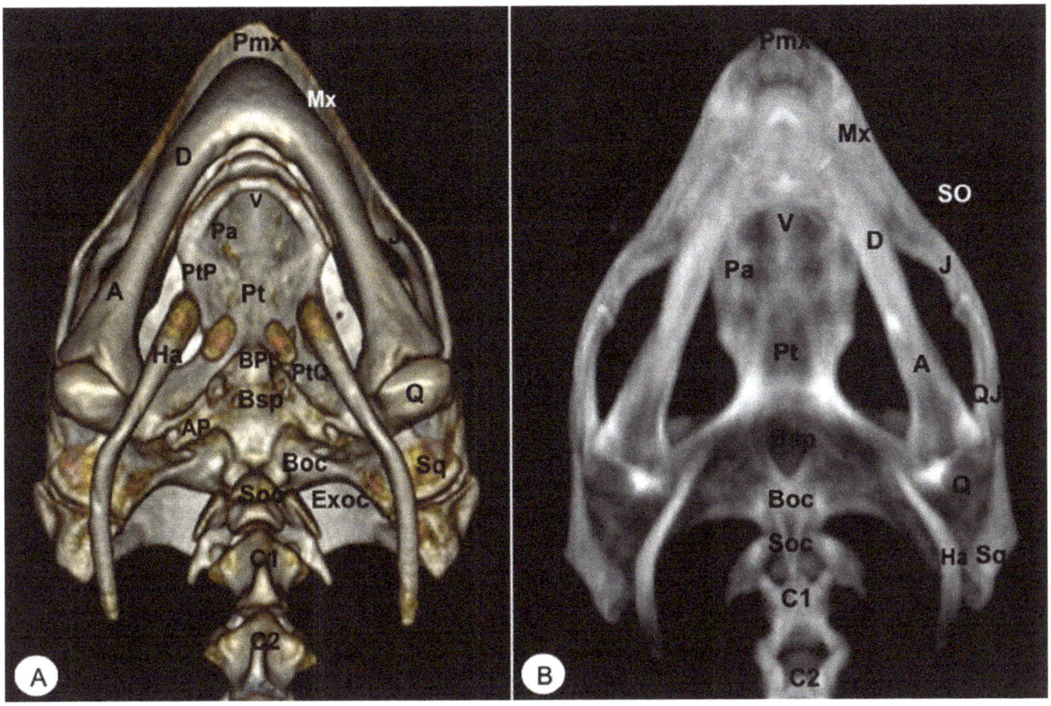

Figure 4. Head of *Caretta caretta*. (**A**) Ventral VR image. (**B**) Ventral MIP image. Pmx: Premaxillary bone. Mx: Maxillary bone. J: Jugal. V: Vomer. Pa: Palatine. PtP: Pterygoid palatine process. BPt: Basipterygoid. Pt: Pterygoid. PtQ: Pterygoid quadrate process. SO: Scleral ossifications. Sq: Squamosal. Q: Quadrate. QJ: Quadratojugal. Bsp: Basisphenoid. AP: Articular process of Bsp. Boc: Basioccipital. Soc: Supraoccipital. Exoc: Exoccipital. M: Mandible. D: Dentary. A: Angular. Ha: Hyoid apparatus. C1: First cervical vertebra. C2: Second cervical vertebra.

3.1. Dermatocranium

3.1.1. PREMAXILLA (Os Premaxillare)

The premaxillary bones are narrow, showing a V shape. These bones are the most rostral bony structures of the turtle head (Figures 2A, 3, 4 and 5A) and caudoventrally with the anterior end of the vomer (Figure 4)). Its rostral VR image showed the absence of dorsomedial processes in these bones, allowing the joining of the nostrils (*nares*) at the midline. This paired bone articulates laterally with the maxillary bone (Figures 2A, 3, 4 and 5A) and caudoventrally with the anterior end of the vomer (Figure 4). We distinguished a similar configuration in the iguana, where it comprises a tooth-bearing alveolar plate. This bone contacts the nasal bone dorsally (Figures 7, 8 and 10), the maxilla caudolaterally (Figures 7, 8 and 10), and the vomer caudomedially (Figure 10B).

3.1.2. MAXILLA (Os Maxillare)

The maxilla is a pair of bones that meet each other at the midline under the palatal bones. Therefore, it is located along the lateral edge of the vomer and palatine bones, surrounding the ventral border of the eye orbit (Figure 2). In the rostral and lateral MIP images of the loggerhead turtle (Figures 2B and 3B), this bone meets rostromedially with the premaxillary bone, ventromedially with the vomer and palatine bones, and caudally along the ventral border of the jugal bone (Figures 2 and 3A). Concerning the iguana, this bone arises lateral and ventral to the nasal bone on each side of the skull, contacting it rostrally with the premaxilla (Figures 7 and 8), the lacrimal and nasal bones dorsally, and the jugal caudolaterally (Figure 8).

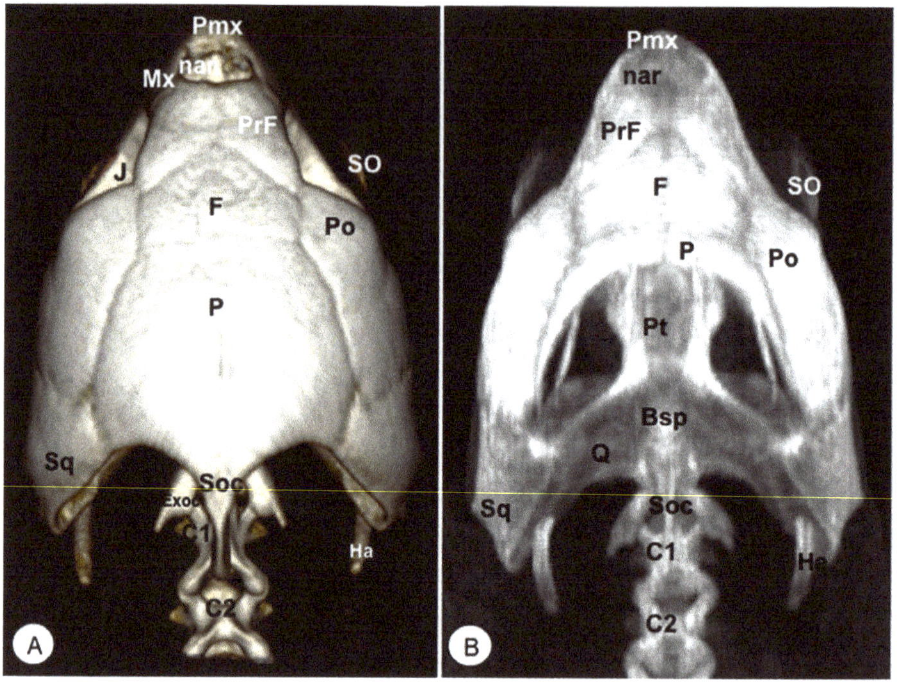

Figure 5. Head of *Caretta caretta*. (**A**) Dorsal VR image. (**B**) Dorsal MIP. Pmx: Premaxillary bone. Mx: Maxillary bone. nar: Nares. PrF: Prefrontal. F: Frontal. Po: Postorbital. J: Jugal. O: Orbit. SO: Scleral ossifications. Sq: Squamosal. Pt: Pterygoid. Bsp: Basisphenoid. P: Parietal. Exoc: Exoccipital. Soc: Supraoccipital. Ha: Hyoid apparatus. C1: First cervical vertebra. C2: Second cervical vertebra.

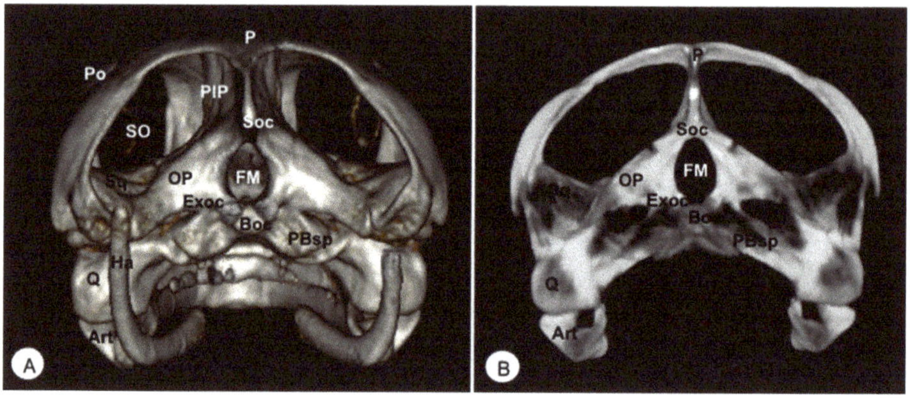

Figure 6. Head of *Caretta caretta*. (**A**) Caudal VR image. (**B**) Caudal MIP image. P: Parietal bone. PIP: Procesus inferior parietalis. Po: Postorbital. Sq: Squamosal bone. Soc: Supraoccipital bone. Exoc: Exoccipital bone. Boc: Basioccipital bone. OP: Opisthotic bone. FM: Foramen magnum. PBsp: Parabasisphenoid bone. SO: Scleral ossifications. Q: Quadrate. Art: Articular. Ha: Hyoid apparatus.

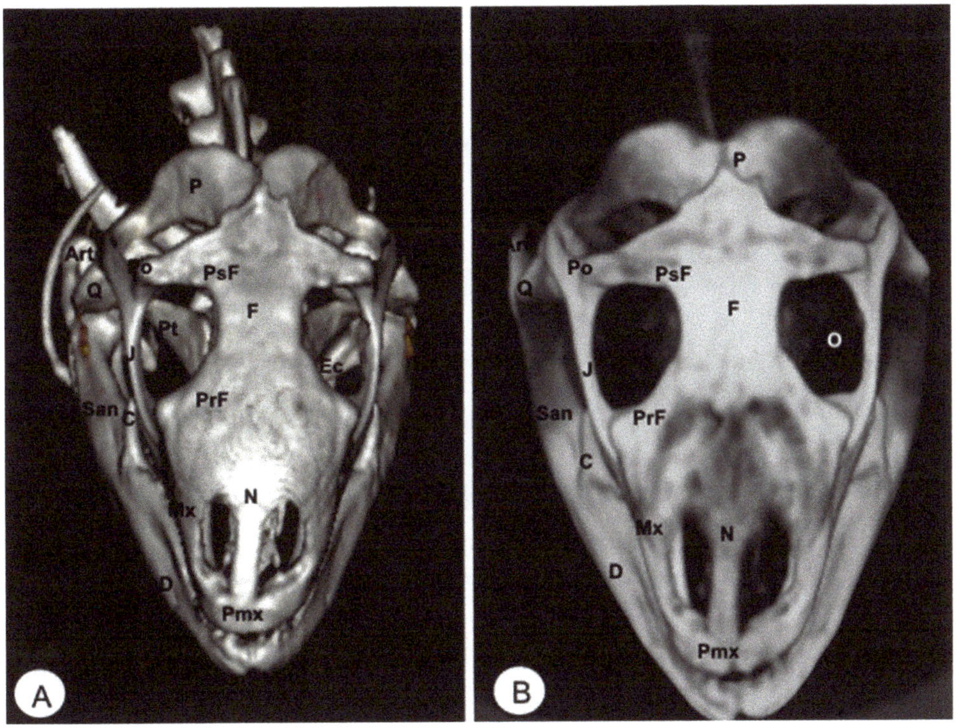

Figure 7. Head of *Iguana iguana*. (**A**) Rostral VR image. (**B**) Rotral MIP image. N: Nasal. Pmx: Premaxillary bone. Mx: Maxillary bone. PrF: Prefrontal. F: Frontal. PsF: Postfrontal. Po: Postorbital. J: Jugal. Q: Quadrate. P: Parietal. O: Orbit. Pt: Pterygoid. Ec: Ectopterygoid. Art: Articular. C: Coronoid. D: Dentary. San: Surangular.

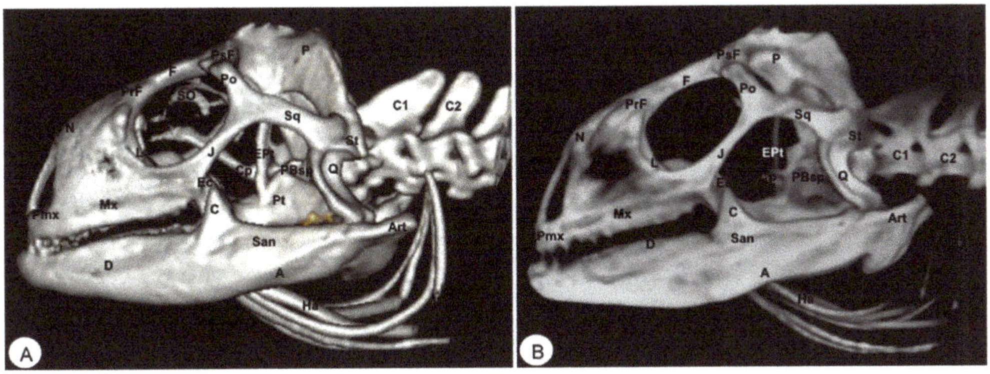

Figure 8. Head of *Iguana iguana*. (**A**) Lateral VR image. (**B**) Lateral MIP image. Pmx: Premaxillary. Mx: Maxillary bone. N: Nasal. PrF: Prefrontal. L: Lacrimal. F: Frontal. PsF: Postfrontal. Po: Postorbital. SO: Scleral ossifications. J: Jugal. Q: Quadrate. Sq: Squamosal. Ec: Ectopterygoid. Pt: Pterygoid. EPt: Epipterygoid. Cp: Cultriform process. PBsp: Parabasisphenoid. P: Parietal. St: Supratemporal. D: Dentary. A: Angular. San: Surangular. Art: Articular. C: Coronoid. Ha: Hyoid apparatus. C1: First cervical vertebra. C2: Second cervical vertebra.

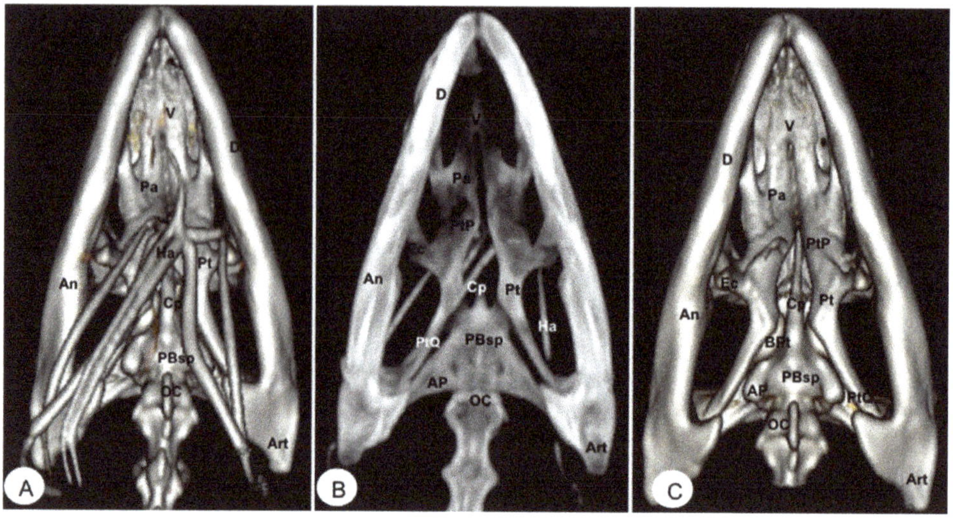

Figure 9. Head of *Iguana iguana*. (**A**) Ventral VR image. (**B**) Ventral MIP image. (**C**) Ventral VR image without Hyoid apparatus. D: Dentary bone. An: Angular bone. Art: Articular bone. V Vomer. Pa: Palatine bone. Pt: Pterygoid bone. Ec: Ectopterygoid. PtP: Pterygoid palatine process. PtQ: Pterygoid quadrate process. BPt: Basipterygoid process. PBsp: Parabasisphenoid bone. Cp: Cultriform process. OC: Occipital condyle. Ha: Hyoid apparatus. AP: Alar process.

3.1.3. NASAL (Os Nasale)

In the iguana, the nasal bones are broad and slightly convex dorsally. The VR and MIP images showed the nasal bone junction in the mid-sagittal plane and how the nasal bones were distinctly longer than wide (Figure 8), contacting with the neighboring bones such as the lacrimal bone caudolaterally, the maxilla rostrolaterally, the prefrontal bone caudodorsolaterally, and the frontal bone caudodorsomedially (Figures 7, 8 and 10).

3.1.4. VOMER

The vomer is an unpaired laminar bone divided into a dorsal and a ventral portion. This last is roughly rectangular and could be distinguished in ventral VR and MIP reconstructions. The vomer is located in the anteromedial palatal region, separating the two external nostril openings (*Apertura narium externa*). The anterodorsal section of this bone forms part of the floor of the nasal passages (*fossa nasalis*), which is well seen with both VR and MIP ventral reconstructions. In the turtle, this bone contacts rostrally with the premaxilla and the maxilla, laterally with the palatine, and caudally with the pterygoid (Figure 4). Referring to the iguana, the vomer is a paired bone with a triangular shape that articulates with the maxilla (rostrally), the palatine bones (caudally) (Figure 9), and medially with one another. Interestingly, this bone shows little or no contact with the premaxilla and none with the pterygoid bone.

3.1.5. PALATINE (Os Palatinum)

The palatine bone overlays the dorsomedial surface of the maxillary bone (Figure 2B). It is a paired laminar bone between the vomer and pterygoid bones. It shows a ventral concave shape that narrows posteriorly, contacting laterally with the jugal bone and medially with the pterygoid bone. The loggerhead turtle and green iguana showed the same configuration, which was easily seen by the VR and MIP ventral reconstructed images (Figures 3B, 4 and 9).

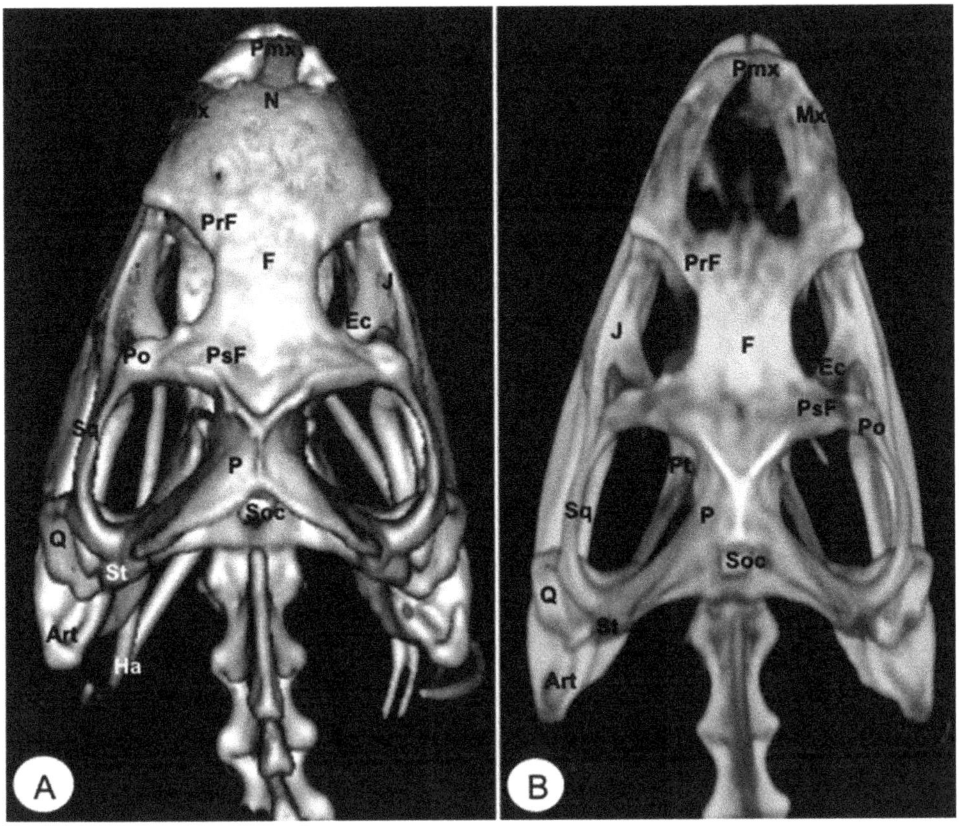

Figure 10. Head of *Iguana iguana*. (**A**) Dorsal VR view. (**B**) Dorsal MIP view. Pmx: Premaxillary bone. Mx: Maxilla. N: Nasal. PrF: Prefrontal. F: Frontal. PsF: Postfrontal. Po: Postorbital. J: Jugal. Pt: Pterygoid. Ec: Ectopterygoid. Q: Quadrate. Sq: Squamousal. St: Supratemporal. P: Parietal. Soc: Supraoccipital. Art: Articular. Ha: Hyoid apparatus.

3.1.6. PREFRONTAL-FRONTAL

In these species, the frontal bone shows a flat shape and lies in the anterior part of the skull (Figures 2, 3, 5, 7, 8 and 10). The rostral VR turtle reconstruction showed a clear view of the prefrontal and frontal bone junctions (caudomedially). Therefore, the frontal bone contacts the prefrontal bone (rostrally), the postorbital bone (laterally), and the parietal bone (caudally) (Figures 2, 3 and 5). As our animals were adults, we could observe how this bone was excluded from the orbit. The VR and MIP images of the green iguana showed that the prefrontal bone contacts rostromedially with the nasal bone, rostrolaterally with the lacrimal bone (paired small bones located in the lateral region of the orbit observed only in the iguana), and caudomedially with the frontal bone (Figures 7, 8 and 10). The frontal bone is broad and forms a major component of the skull roof, contributing to the dorsal border of the orbit. It meets the prefrontal bone rostrolaterally, the postfrontal bone caudolaterally, and the parietal bone caudally (Figures 7, 8 and 10). Laterally, it is differentiated from the postorbital by the postfrontal bone.

3.1.7. POSTFRONTAL-POSTORBITAL

The postorbital bone forms part of the temporal arch, and most of the posterior margin of the orbit. This bone contacts dorsocaudally with the parietal bone, anterodorsally with the prefrontal bone, and ventrally with the jugal bone (Figures 2, 3 and 5), excluding, therefore, the frontal bone from the orbit. It reaches the squamosal caudally (Figures 3A and 5) and contacts the prefrontal bone rostrally (Figures 2A, 3A and 5). Regarding the iguana, those bones were broad and could be well identified in VR and MIP rostral, dorsal, and lateral images as two different bones. The postfrontal bone arises in the caudodorsal part of the orbit and meets the frontal bone (rostromedially), the parietal bone (caudomedially), and the postorbital bone (caudolaterally) (Figures 7, 8 and 10). In contrast, the postorbital bone meets the postfrontal bone dorsally, the squamosal bone caudolaterally, and the jugal bone rostrolaterally (Figures 8 and 10).

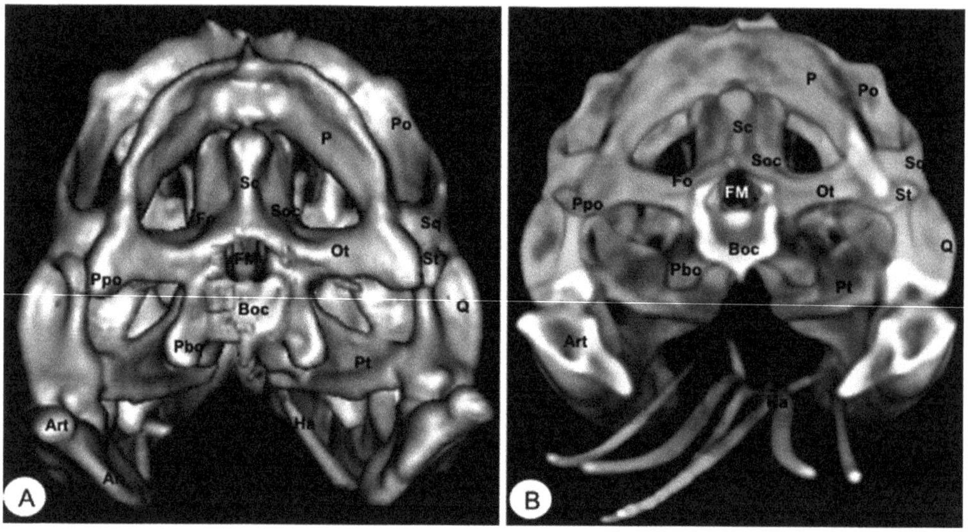

Figure 11. Head of *Iguana iguana*. (**A**) Caudal VR image without hyoid apparatus. (**B**) Caudal MIP image. Sc: Occipital sagital crest. Soc: Supraoccipital. P: Parietal. Ot: Otoccipital. FM: Foramen magnum. Pt: Pterygoid. Po: Postorbital. Sq: Squamosal. St: Supratemporal. Boc: Basioccipital. Ppo: Paroccipital process. Pbo: Parabasioccipital process. Fo: Otoccipital facet. An: Angular. Art: Articular. Ha: Hyoid apparatus.

3.1.8. PARIETAL (Os Parietale)

The parietal bone in loggerhead turtles is wide and relatively flat. This bone is the largest component of the skull roof (Figure 5A) and is formed by a horizontal dorsal plate comprising part of the temporal skull roof and a vertical ventral plate (*processus inferior parietalis*) parasagittally (Figures 2, 3, 5 and 6). Moreover, this bone separated the cranial cavity (*cavum cranii*) from the temporal fossa (*fossa temporalis) (*observed in caudal VR view). It contacts the postorbital bone (laterorostrally) (Figures 2, 3B and 5), the frontal bone (mid-rostrally) (Figures 2, 3B and 5), the supraoccipital bone (caudally), and the squamosal bone (caudolaterally) (Figures 3B and 5A). Similarly, the parietal bone in the green iguana is a short, plate-shaped bone with a descending portion that forms part of the temporal fossa (Figures 7, 8 and 10). It was clearly seen using both MIP and VR tools. This bone meets the postfrontal bone (rostrolaterally) (Figures 7, 8 and 10), the squamosal and the supratemporal (laterally) (Figure 8), and the supraoccipital (caudomedially) (Figures 10 and 11).

3.1.9. JUGAL (Os Jugale)

The jugal bone is one of the main elements of the zygomatic arch under the orbit in both species, showing an "L" shaped configuration in VR and MIP lateral images. Rostral and lateral VR images were quite useful in distinguishing the arrangement of this bone at the caudoventral border of the orbit. Moreover, the rostral, lateral, and ventral VR and MIP images of the turtle show how this bone contacts dorsocaudally with the postorbital bone, rostrally with the maxilla, and posteriorly with the quadratojugal bone (resulting in the zygomatic arch). In contrast, the jugal bone in the iguana is observed as a curved bone (seen by rostral, lateral, and dorsal VR and MIP images) that runs forward along the orbital border to make contact with the lacrimal (dorsally) and maxillary bones rostrally. Caudodorsally, it angles slightly upwards to meet with the postorbital and the squamosal bones. Ventrally, it contacts the ectopterygoid bone.

3.1.10. QUADRATE (Os Quadratum)

The quadrate is a conch-shaped bone and is part of the craniomandibular joint, forming the *fossa temporalis* in both species. The quadrate bone of the turtle contacts dorsally with squamosal (Figures 3 and 4), rostrally is sutured to the pterygoid by the quadratopterygoid process (Figure 4A), and rostrolaterally to the jugal bone by the quadratojugal bone (Figure 3A). In the iguana, this bone contacts dorsally with the supratemporal and the caudal edge of the squamosal bone (Figures 8 and 10). Ventrally, it reaches the articular bone forming the craniomandibular joint (Figure 8).

3.1.11. QUADRATOJUGAL (Os Quadratojugale)

The quadratojugal bone was only observed in the turtle's head and forms part of the anteroventral border of the cavum tympani and half of the cheek emargination. They are well seen in the lateral VR image. This roughly square bone joins the jugal anteriorly, the postorbital anterodorsally, the squamosal caudodorsally, and the quadrate bone caudally (Figure 3A).

3.1.12. SQUAMOSAL (Os Squamosum)

The squamosal bone is located posterodorsally above the quadrate and the quadratojugal, and both form part of the antrum postoticum, where the quadrate bone comprises the base and the squamosal the roof in these species. This bone, in the turtle, is attached to the caudal region of the quadrate bone (Figure 3A), dorsally joins the parietal bone (Figure 5A), caudomedially contacts with the paraoccipital process of the opisthotic bone (Figure 6), rostrolaterally with the quadratojugal (Figure 3A), and ventrally with the basioccipital bone (Figure 4). In the iguana, the squamosal bone is a roughly L-shaped bone that shares the same configuration as in the turtle and is well visualized using both VR and MIP techniques in lateral (Figure 8), dorsal (Figure 10), or caudal images (Figure 11). Nonetheless, in the iguana, we could distinguish the presence of a large ventral process of the squamosal that is offset from the main axis of the bone.

3.1.13. PTERYGOID (Os Pterygoideum)

This flat bone lies between the vomer and palatine bones (rostrally) and the parabasisphenoid (caudally) bones (Figure 4) in the loggerhead turtle, reaching the quadrate bone caudally and the palatine bone rostrally (Figures 3B and 4). Interestingly, this bone is fussed with the turtle. In contrast, the iguana was completely separated (Figure 9B,C). This bone has a dorsal projection called the epipterygoid that contacts the parietal bone (Figures 3B and 8) in both species. Nonetheless, we observed that the iguana showed two rostral projections called ectopterygoid bones (Figures 8 and 9), which contacted rostromedially with the pterygoid palatine process (Figure 9B,C), and laterally, it contacts with the jugal and the maxillary bones (Figures 8 and 10).

3.1.14. EPIPTERYGOID

The epipterygoid bone in the turtle consists of a tubular-shaped bone that extends vertically from the pterygoid bone and contacts the parietal (Figure 3B). A similar configuration was observed in the iguana, where this bone is a thin road that touches dorsally the parietal bone (Figure 8).

3.2. *Neurocranium*

3.2.1. PARABASISPHENOID-BASISPHENOID (Os Basisphenoidale)

The parabasisphenoid is a roughly trapezoidal bone located between the pterygoid (rostrally), and the basioccipital (caudally) bones (Figures 4, 5B, 6, 8 and 9). This bone forms part of the ventral surface of the skull and has two lateral processes that contact the pterygoid bone, called basipterygoid processes, which are short and thick, and well identified in the iguana (Figure 9C). Between these processes, we observe a thin and tubular structure named the cultriform process (Figures 8 and 9), which is well developed compared to other iguana species. This latter also shows two alar processes in the laterocaudal direction (Figure 9B,C).

3.2.2. BASIOCCIPITAL (Os Basioccipitale)

The basioccipital bone in the turtle is a rectangular-shaped bone located in the posterior half of the planum basale, contacting the basisphenoid (rostrally) and exoccipital (caudally) bones, forming part of the floor of the *cavum cranii* (Figures 3B, 4 and 6). In the iguana, it roughly reaches the parabasisphenoid and caudally the occipital bone (Figure 11), forming part of the occipital condyle.

3.2.3. SUPRAOCCIPITAL (Os Supraoccipitale)

The supraoccipital bone is an unpaired midline element located in the posterodorsal portion of the skull (Figures 3–6, 10 and 11). Both turtles and iguanas have a similar configuration. This bone has a broad contact with the posteromedial margins of the parietals (Figures 3B, 5A, 10 and 11). Posteroventrally, it contacts the opisthotic and otoocipital, forming part of the medial part of the occipital condyle, near the foramen magnum (Figures 6 and 11). These features were observed by VR and MIP reconstructions in the caudal view of both species.

3.2.4. EXOCCIPITAL (Os Exoccipitale)

This bone is a paired bone that lies lateral to the foramen magnum and comprises part of the occipital condyle. Dorsally, it is in contact with the supraoccipital, anterodorsolaterally with the opisthotic bone, and ventromedially with the basioccipital bone (Figures 4A, 5A and 6). The exoccipital forms the posterior border of the metotic fissure. The ossification extends ventrally to the dorsolateral aspect of the occipital condyles, where it joins the basioccipital bone (Figure 4A).

3.3. *Mandible*

In both species, the mandible is very robust and shows two pieces that are fused cranially at the intermandibular symphysis.

3.3.1. DENTARY (Os Dentale)

The dentary bone is the major bone of the mandible in both species. Its lateral surface appears pierced by small foramina, where mandibular nerve branches and blood vessels run. In the iguana, it carries pleurodont teeth. This bone was better visualized in the rostral, lateral, and ventral VR and MIP images. Either in the turtle or the iguana, it joins the coronoid and the surangular caudodorsally (Figures 2–4, 8 and 9), the angular caudoventrally (Figures 3, 4, 8 and 9), and the splenial bone medially.

3.3.2. ANGULAR (Os Angulare)

The angular bone is a laminar bone located in the caudal aspect of the mandible from the rostromedial to the caudolateral aspect, forming the ventrocaudal margin of the mandible. This bone was well visualized in the lateral MIP and VR images of these species (Figures 3 and 8). The connections of the angular bone in both species are the dentary bone rostrally, the splenial and prearticular bones dorsomedially, and the articular bone in the caudal direction.

3.3.3. SURANGULAR (Os Surangulare)

The surangular is a widely flat bone located below the coronoid along the dorsal border of the mandible and forming most of the caudolateral surface of the mandible. In both species, it is located posteriorly and slightly dorsal to the dentary, connected to the coronoid rostrodorsally, to the angular bone caudoventrally, and to the articular bone medially (Figures 3 and 8).

3.3.4. CORONOID (Os Coronoideum)

The coronoid is a broadly triangular bony structure located approximately medial to each branch on the caudodorsal aspect of the mandible. This process is the highest part of the mandible in both species. This bone was better visualized in the lateral MIP view of the iguana than in the turtle, where we could distinguish how the coronoid tends to be slightly higher and more dorsally pointed than in the loggerhead turtle. Both species have the same configuration where the coronoid meets the dentary rostrolaterally, the surangular caudolaterally, the articular caudoventromedially, and the splenial bone ventromedially (Figures 3B, 7 and 8).

3.3.5. PREARTICULAR (Os Prearticulare)

The prearticular bone is a wide, flat lamella overlying a large area of the caudomedial surface of the lower jaw, forming a long, narrow process whose dorsal border constitutes the medial border of the adductor fossa. This bone reaches the coronoid rostrodorsally, the articular bone caudomedially, and the angular ventrally (Figure 3). If the splenial bone is present, this bone is sutured to the rostral edge of the prearticular bone.

3.3.6. ARTICULAR (Os Articulare)

The articular bone is an irregularly block-shaped bone found at the caudal edge of the mandible in the turtle and the iguana. It was observed using both VR and MIP techniques (Figures 3 and 6–11) and had the same configuration in both species. Thus, it joins with the surangular laterally and the angular ventrally, forming the retroarticular process, and dorsally with the coronoid bone.

3.3.7. HYOID APPARATUS (Os Hyoideum)

The hyoid apparatus is located in the intermandibular space. It provides support to the tongue, pharynx, as well as floor of the mouth. This apparatus consists of a pentagonal piece, which has three structures: a central (basihyal) structure, an anterior process, and two caudolateral processes (Figures 3A, 4, 5, 6A, 8, 9, 10A and 11).

4. Discussion

In this study, we have used different imaging techniques to evaluate the skulls of two species of reptiles. The results of this investigation have demonstrated that MIP and VR images of the turtle and iguana skulls provided adequate information and understanding of the different bones that comprise the skull. Nonetheless, although a direct comparison between VR and MIP images was not possible, a remarkable correlation in the matched images was obtained. As in other reptiles, the heads of the loggerhead turtle and the green iguana are highly complex structures composed of different bones with significant differences [17,18]. Our results have demonstrated that the imaging reconstruction techniques used in our study were valuable in visualizing these differences. Compared to other studies performed on reptiles, such as the boa constrictors [21], the green iguana, the common tegu, and the bearded dragon that used conventional radiography [22], the acquisition of reconstructed images of the skull reduced the superimposition of the bilateral structures of the snout and the neurocranium, providing better visualization of the configuration of the bones that make up the skull of these species.

MIP is a volume rendering procedure that projects a volume of interest onto a 3-dimensional viewing plane based on high-intensity structures on CT [9]. Thus, the ability of MIP images to preserve attenuation information makes them widely applicable for visualizing areas of interest or specific anatomic structures. Hence, the images obtained in our study allowed excellent visualization of the external and internal structures that concern the skull of both species. Therefore, we identified specific bones of the dermatocranium as the premaxilla, the maxilla, the nasal, the vomer, and the palatine. In the neurocranium, we distinguished dorsal and lateral bones as the frontal, the parietal, and the different parts of the pterygoid, as well as their configuration and relation with other bones, such as the epipterygoid and the parietal bones. However, we could not identify other head bones, such as the basisphenoid, the supraoccipital, or the dentary, which have already been identified using micro-computed tomography [23]. This specific equipment provides higher spatial resolution and thinner slice thickness than that obtained with conventional CT scanners; therefore, micro-CT scanners provide standard images and three-dimensional renderings of extraordinary quality and a superior level of detail compared with standard CT units [24,25]. Nonetheless, micro-CT units are not usually available in veterinary hospitals because of the limited range of species that can fit into the gantry [26].

Three-dimensional volume rendering (VR) is a powerful imaging tool that reconstructs 2D CT data into a 3D model with high fidelity [9]. The main advantage is that it facilitates a three-dimensional visualization, avoiding the superimposition of surrounding structures and therefore allowing for the identification of internal structures [9,16,19]. The VR images allowed us to see how the head of the loggerhead turtle was wider in the posterior part and narrower before the orbits, and the jaw was robust with a fused symphysis. These features have been associated with greater biting performance than those of other sea turtle species [27]. Despite these findings, the skull of *Caretta caretta* shares similar features to those of other sea turtles, such as large orbits and a short and blunt rostrum [23,28]. Concerning the orbit, VR and MPI were quite helpful in visualizing how the frontals were excluded from the border of the orbit by the prefrontal bones. A similar finding was identified in other studies using micro-CT [23]. This orbital disposition is not observed in other species of sea turtles, such as the Kemp's ridley. The skull of the iguana is convex with a short preorbital region but a similar V-shaped jaw configuration. Concerning internal bones, VR ventral images allowed the depiction of different configurations of the pterygoid bone in both species. As stated in other studies performed on sea turtles [23,28], the pterygoid bone of the loggerhead turtle shows prominent lateral processes. In contrast, in the iguana, this bone was separated from the bone on the other side. Some authors have reported an age-dependent degree of ossification of the pterygoid bone, especially at the pterygoid-epipterygoid junction, tending to fuse with age [29]. However, with both techniques, we could observe that this fusion did not occur in the iguana. Other valuable structures of the head identified by VR were the scleral ossifications, which have been studied by several researchers as a possible growth marker, although scleral ossicles would not be an alternative for loggerhead sea turtle skeletochronology. However, they could be helpful for other sea turtle species [30]. Additionally, this technique allowed us to ascertain the differences in the quadrate bone in both species. Therefore, VR lateral images served to visualize the relation between other bones and the quadratojugal bone and to confirm the absence of this last bone in the iguana head. Similar findings were also described in other studies [31–33], which confirmed that the presence of quadratojugal is a misinterpretation. In contrast, VR and MIP lateral images were helpful in identifying the adductor fossa in the iguana's head. The presence of this chamber in the iguana has been proposed as indicative of the development of jaw adductor muscles in this species [34].

In conclusion, these two techniques were quite helpful, providing anatomical references for the bone structures that comprise the heads of these species. Compared to the excellent resolution provided by micro-CT, VR combined with MPI provides adequate detail of the bone structures that compose the head. Therefore, the results obtained in this study can be adequate to evaluate numerous processes, such as traumatic lesions with small or large fractures involving the skull that can produce brain damage, skull malformations, and osteodystrophy secondary to nutrient imbalances or neoplasms. Moreover, these two reconstruction techniques can facilitate teaching veterinary anatomy to students by realistically allowing the view of structures without superimposition of other structures, eliminating the difficulties of visualizing the extension of different types of lesions.

Author Contributions: Conceptualization, J.R.J.M., A.A. and M.E.; methodology, M.E. and E.G.-R.; investigation, J.R.J.M., E.G.-R. and M.E.; resources, S.D. and C.C.; writing—original draft preparation, J.R.J.M., E.G.-R., A.A., S.D., C.C. and M.E.; writing—review and editing, J.R.J.M. All authors have read and agreed to the published version of the manuscript.

Funding: This research received no external funding.

Institutional Review Board Statement: Not applicable.

Informed Consent Statement: Not applicable.

Data Availability Statement: Not applicable.

Acknowledgments: In loving memory of Alvaro Domingo Rodriguez Garcia. We also thank Marisa Mohamad, Carmen Mingot, Ayesh Mohamad and Jamal Jaber for their support and constructive comments, and Rancho Texas Lanzarote Park (Lanzarote, Canary Islands, Spain) for providing the animals for this study.

Conflicts of Interest: The authors declare no conflict of interest.

References

1. Kuszyk, B.S.; Heath, D.G.; Bliss, D.F.; Fishman, E.K. Skeletal 3-D CT: Advantages of Volume Rendering over Surface Rendering. *Skeletal Radiol.* **1996**, *25*, 207–214. [CrossRef] [PubMed]
2. Hughes, J.R.; Johnson, V.S.; Genain, M.A. CT Characteristics of Primary Splenic Torsion in Eight Dogs. *Vet. Radiol. Ultrasound* **2020**, *61*, 261–268. [CrossRef] [PubMed]
3. Rowe, S.P.; Fishman, E.K. Three-Dimensional Computed Tomography Cinematic Rendering of Mandibular Odontogenic Myxofibroma. *Oral Surg. Oral Med. Oral Pathol. Oral Radiol.* **2019**, *128*, e122–e125. [CrossRef]
4. Dianna, D.; Cody, P. AAPM/RSNA Physics Tutorial for Residents: Topics in CT. *Radiographics* **2002**, *22*, 14. [CrossRef]
5. Fishman, E.K.; Ney, D.R.; Heath, D.G.; Corl, F.M.; Horton, K.M.; Johnson, P.T. Volume Rendering versus Maximum Intensity Projection in CT Angiography: What Works Best, When, and Why. *Radiographics* **2006**, *26*, 905–922. [CrossRef] [PubMed]
6. Calhoun, P.S.; Kuszyk, B.S.; Heath, D.G.; Carley, J.C.; Fishman, E.K. Three-Dimensional Volume Rendering of Spiral CT Data: Theory and Method. *Radiographics* **1999**, *19*, 745–764. [CrossRef]
7. Prokop, M.; Shin, H.O.; Schanz, A.; Schaefer-Prokop, C.M. Use of Maximum Intensity Projections in CT Angiography: A Basic Review. *Radiographics* **1997**, *17*, 433–451. [CrossRef]
8. Thrall, D. *Textbook of Veterinary Diagnostic Radiology*; Elsevier: St. Louis, MO, USA, 2012; ISBN 9780323482479.
9. Ho, J.L.; Konda, A.; Rahman, J.; Harris, E.; Korn, R.; Sabir, A.; Bawany, B.; Gulati, R.; Harris, G.J.; Boswell, W.D.; et al. Comparative Analysis of Three-Dimensional Volume Rendering and Maximum Intensity Projection for Preoperative Planning in Liver Cancer. *Eur. J. Radiol. Open* **2020**, *7*, 100259. [CrossRef]
10. Li, W.J.; Chu, Z.G.; Zhang, Y.; Li, Q.; Zheng, Y.N.; Lv, F.J. Effect of Slab Thickness on the Detection of Pulmonary Nodules by Use of CT Maximum and Minimum Intensity Projection. *Am. J. Roentgenol.* **2019**, *213*, 562–567. [CrossRef]
11. Bertolini, G.; Rolla, E.C.; Zotti, A.; Caldin, M. Three-Dimensional Multislice Helical Computed Tomography Techniques for Canine Extra-Hepatic Portosystemic Shunt Assessment. *Vet. Radiol. Ultrasound* **2006**, *47*, 439–443. [CrossRef]
12. Secrest, S.; Bugbee, A.; Waller, K.; Jiménez, D.A. Comparison of Transverse Computed Tomographic Excretory Urography Images and Maximum Intensity Projection Images for Diagnosing Ectopic Ureters in Dogs. *Vet. Radiol. Ultrasound* **2017**, *58*, 163–168. [CrossRef] [PubMed]
13. Ribas, L.M.; Massad, M.R.R.; Pinto, A.C.B.C.F.; Heng, H.G.; Tremori, T.M.; Reis, S.T.J.; Baroni, C.O.; Massad, E.; Rocha, N.S. Post-Mortem CT vs Necropsy in Feline Medicine. *J. Feline Med. Surg.* **2020**, *22*, 1206–1213. [CrossRef] [PubMed]
14. Jeong, Y.; Lim, C.; Oh, S.; Jung, J.; Chang, J.; Yoon, J.; Choi, M. Three-Dimensional CT Angiography of the Canine Hepatic Vasculature. *J. Vet. Sci.* **2008**, *9*, 407–413. [CrossRef]
15. Jha, D.K.; Khera, P.; Bhaskar, S.; Garg, M. Three-Dimensional Volume Rendering: An Underutilized Tool in Neurosurgery. *World Neurosurg.* **2019**, *130*, 485–492. [CrossRef] [PubMed]
16. Pérez, S.; Encinoso, M.; Corbera, J.A.; Morales, M.; Arencibia, A.; González-Rodríguez, E.; Déniz, S.; Melián, C.; Suárez-Bonnet, A.; Jaber, J.R. Cranial Structure of *Varanus komodoensis* as Revealed by Computed-Tomographic Imaging. *Animals* **2021**, *11*, 1078. [CrossRef] [PubMed]
17. Arencibia, A.; Melián, A.; Orós, J. Anatomic Interactive Atlas of the Loggerhead Sea Turtle (Caretta Caretta) Head. *Animals* **2021**, *11*, 1–13. [CrossRef]

18. Pérez, S.; Encinoso, M.; Morales, M.; Arencibia, A.; Suárez-Bonnet, A.; González-Rodríguez, E.; Jaber, J.R. Comparative evaluation of the Komodo dragon (*Varanus komodoensis*) and the Green iguana (*Iguana iguana*) skull by three dimensional computed tomographic reconstruction. *Slov. Vet. Res.* **2021**, *58*, 111–116. [CrossRef]
19. Hernández Morales, C.; Peloso, P.L.V.; Bolívar García, W.; Daza, J.D. Skull Morphology of the Lizard Ptychoglossus Vallensis (Squamata: Alopoglossidae) With Comments on the Variation Within Gymnophthalmoidea. *Anat. Rec.* **2019**, *302*, 1074–1092. [CrossRef]
20. Roscito, J.G.; Rodrigues, M.T. Comparative Cranial Osteology of Fossorial Lizards from the Tribe Gymnophthalmini (Squamata: Gymnophthalmidae). *J. Morphol.* **2010**, *271*, 1352–1365. [CrossRef]
21. Banzato, T.; Russo, E.; Di Toma, A.; Palmisano, G.; Zotti, A. Anatomic imaging of the Boa constrictor head: A comparison between radiography, computed tomography and cadaver anatomy. *Am. J. Vet. Res.* **2011**, *72*, 1592–1599. [CrossRef]
22. Banzato, T.; Selleri, P.; Veladiano, I.A.; Martin, A.; Zanetti, E.; Zotti, A. Comparative evaluation of the cadaveric, radiographic and computed tomographic anatomy of the heads of green iguana (Iguana iguana), common tegu (Tupinambis merianae) and bearded dragon (Pogona vitticeps). *BMC Vet. Res.* **2012**, *11*, 53. [CrossRef] [PubMed]
23. Jones, M.E.H.; Werneburg, I.; Curtis, N.; Penrose, R.; O'Higginns, P.; Fagan, M.; Evans, S. The head and neck anatomy of sea turtles (Cryptodira: Chelonioidea) and skull shape in Testudines. *PLoS ONE* **2012**, *7*, e47852. [CrossRef] [PubMed]
24. Sasai, H.; Iwai, H.; Fujita, D.; Seto, E.; Izumi, Y. The use of micro-computed tomography in the diagnosis of dental and oral diseases in rabbits. *BMC Vet. Res.* **2014**, *10*, 209. [CrossRef] [PubMed]
25. Spoor, F.; Jeffery, N.; Zonneveld, F. Using diagnostic radiology in human evolutionary studies. *J. Anat.* **2000**, *197*, 61–76. [CrossRef]
26. Capello, V. Disorders of the oral cavity. *Vet. Clin. N. Am. Exot. Anim. Pract.* **2016**, *19*, 669–998. [CrossRef]
27. Herrel, A.; O'Reilly, J.C.; Richmond, A.M. Evolution of bite performance in turtles. *J. Evol. Biol.* **2002**, *15*, 1083–1094. [CrossRef]
28. Chatterji, R.M.; Hutchinson, M.N.; Jones, M.E.H. Redescription of the skull of the Australian flatback sea turtle, Natator depressus provides new morphological evidence for phylogenetic relationships among sea turtles (Chelonioidea). *Zool. J. Linn. Soc.* **2021**, *191*, 1090–1113. [CrossRef]
29. Wernerburg, I.; Maier, W. Diverging Development of Akinetic Skulls in Cryptodire and Pleurodire Turtles: An Ontogenetic and Phylogenetic Study Ingmar. *Vertebr. Zool.* **2019**, *69*, 113–143.
30. Avens, L.; Taylor, J.C.; Goshe, L.R.; Jones, T.T.; Hastings, M. Use of Skeletochronological Analysis to Estimate the Age of Leatherback Sea Turtles Dermochelys Coriacea in the Western North Atlantic. *Endanger. Species Res.* **2009**, *8*, 165–177. [CrossRef]
31. Abdala, F.; Lobo, F.; Scrocchi, G. Patterns of ossification in the skeleton of *Liolaemus quilmes* (Iguania: Tropiduridae). *Amphibia-Reptilia* **1997**, *18*, 75–83.
32. Lions, M.L.; Álvarez, B.B. Desarrollo del esqueleto de *Tropidurus etheridgei* (Iguania: Tropiduridae). *Rev. Esp. Herpetol.* **1998**, *12*, 7–18.
33. Evans, S.E. The skull of lizards and tuatara. In *The Skull of Lepidosauria*; Biology of the Reptilia; Gans, C., Ed.; Society for the Study of Amphibians and Reptiles: Ithaca, NY, USA, 2008; Volume 20, pp. 1–347.
34. Holliday, C.; Witmer, L.M. Archosaur Adductor Chamber Evolution: Integration of Musculoskeletal and Topological Criteria in Jaw Muscle Homology. *J. Morphol.* **2007**, *268*, 457–484. [CrossRef] [PubMed]

animals

MDPI

Article

Origin and Distribution of the Brachial Plexus in Two Procyonids (*Procyon cancrivorus* and *Nasua nasua*, Carnivora)

Juan Fernando Vélez García [1,2,*], Roseãmely Angélica de Carvalho Barros [3] and Maria Angélica Miglino [2]

1 Department of Animal Health, Faculty of Veterinary Medicine and Zootechnics, Universidad del Tolima, Ibague 730006299, Colombia
2 Department of Surgery, Faculty of Veterinary Medicine and Animal Science, Universidade de São Paulo, Butantã CEP 05508-270, Brazil
3 Comparative Animal Laboratory, Biological Sciences Department, Biotechnology Institute, Universidade Federal de Catalão, Catalão 75704-020, Brazil
* Correspondence: jfvelezg@ut.edu.co

Simple Summary: The brachial plexus is a nervous network from which nerves originate to innervate the thoracic limb and its adjacent parts. It has been found with different origins and distributions in the thoracic limb between carnivoran species (interspecific) and even in the same species (intraspecific). Anatomical knowledge of the brachial plexus allows us to understand the differences and similarities between species to be applied in evolutionary biology. *Procyon cancrivorus* and *Nasua nasua* are two neotropical carnivoran species that have different evolutionary adaptations in their thoracic limbs. The aim of this study was to describe the anatomical arrangement of the brachial plexus of these species to review intra- and interspecific anatomical variations. The brachial plexus may originate in both species from the last four cervical spinal nerves (C5, C6, C7 and C8) and the first two thoracic spinal nerves (T1 and T2). The contribution from C5 was higher in *Nasua nasua*, while the contribution from T2 was higher in *Procyon cancrivorus*. Interestingly, both species develop ansa axillaris and ansa pectoralis, which is a primitive anatomical arrangement that is shared with other members of arctoid carnivorans. Therefore, this study confirmed that both structures can be present in some wild carnivorans.

Abstract: *Procyon cancrivorus* and *Nasua nasua* are two procyonids with different evolutionary adaptations to use their thoracic limbs. Therefore, this study aimed to characterize the differences in the brachial plexus between both species. Five *P. cancrivorus* and five *N. nasua* cadavers were used to perform this investigation with the permission of the bioethics committee and environmental license. Gross dissections were performed on the cervical, pectoral, and thoracic limb regions to find the origin and distribution of the brachial plexus. The brachial plexus of both species originated in a variant manner from C5-T1, C5-T2, C6-T1, or C6-T2. All brachial plexus nerves were observed and, interestingly, the musculocutaneous sent a communicating branch to the median nerve medially to the axillary artery, forming an ansa axillaris in both species. An ansa pectoralis was also observed medially to the axillary artery. Additionally, in *P. cancrivorus*, the musculocutaneous nerve innervates the pronator teres and flexor carpi radialis muscles and communicates with the median nerve at the elbow level to continue as a common trunk at the antebrachium. The brachial plexus has differences between both procyonids, although in both species, it could conserve a primitive arrangement present within the infraorder Arctoidea.

Keywords: anatomy; ansa axillaris; Caniformia; communicating branches; nerve; neurology

1. Introduction

The crab-eating raccoon (*Procyon cancrivorus*) and the coati (*Nasua nasua*) are two mammal species that belong to the order Carnivora, suborder Caniformia, and family Procyonidae [1]. *P. cancrivorus* is geographically distributed in Central and South America [2], while *N. nasua* is only distributed in South America [3]. The genera *Procyon* and *Nasua* had a common ancestor within the family Procyonidae [1,4], which diverged approximately 17 million years ago [1]. *P. cancrivorus* has crepuscular, nocturnal, and solitary habits [5,6], while *N. nasua* is gregarious with diurnal habits, although adult males have mainly solitary habits [7].

The *P. cancrivorus* diet is based on fruits, crabs, mollusks, arthropods, and small vertebrates [8,9]. The diet of *N. nasua* is mainly composed of fruits and ground invertebrates [7,10,11], although it is also an opportunistic hunter of small vertebrates and sometimes eats eggs [7,10,12,13]. Both species are mainly terrestrial species [7]; however, *N. nasua* sometimes has arboreal preferences [11] and uses trees to rest, escape predators, or nest [11,14,15]. Both species are good climbers [16], although *N. nasua* uses its hands mainly to excavate (fossorial abilities) and shred dead logs [5,11,17]. *Procyon*, due to the well-developed sense of touch in its hands, uses them to locate and handle animal prey [5,17], manipulating the food to wash it and put it in the mouth [16,18]. In addition, *Procyon* is also a good swimmer and fisher [16].

Procyon has long thoracic limbs adapted to semidigitigrade and cursorial locomotion, while *Nasua* has short thoracic limbs adapted to palmigrade locomotion with long claws [7,17]. *Nasua* has an enlarged sensory cortex region for receiving the afferent projections from the vibrissae of the nose and tactile carpal hairs, and *Procyon* has a well-developed cerebral sensorimotor cortex, which is focused in its hands due to the high density of cutaneous nerves [5]. Therefore, the brachial plexus and its nerves could have evolved in different forms in both procyonid species, since they are responsible for the sensitivity and movements of the thoracic limb. In addition, specific knowledge about the brachial plexus could be used for medical and surgical procedures and locoregional anesthesia in wild carnivorans [19].

Some brachial plexus studies in wild carnivorans only reported the origin of the nerves [20–22], while others reported the origin and distribution of the nerves up to the antebrachium [23–27] and others up to the manus [24,28–31]. However, among these latter studies, only two report innervation of the intrinsic muscles of the manus [24,30]. In most carnivoran species, the brachial plexus originates from the ventral branches of the last three cervical spinal nerves and the first thoracic spinal nerve (C6-T1). The contributions of the fifth cervical spinal nerve (C5) and the second thoracic spinal nerve (T2) have been found in few carnivorans. However, C5 has been found in a high proportion in some felids [23,32], and T2 in some musteloids [30,31]. Therefore, the aim of this study was to describe the origin and distribution of the brachial plexus along the thoracic limb in *P. cancrivorus* and *N. nasua*.

2. Materials and Methods

2.1. Specimens

Five cadavers of *P. cancrivorus* (three adult males and two juvenile females) (Table A1) and five of *N. nasua* (two adult males, one juvenile male, one infant male, and one adult female) were used to perform this investigation (Table A2). Two *P. cancrivorus* (one male and one female) and one *N. nasua* were donated by CORPOCALDAS (Environmental authority of Caldas, Colombia) to the Universidad del Tolima. One *P. cancrivorus* was donated by CORTOLIMA (Environmental authority of Tolima, Colombia). Four *N. nasua* and one *P. cancrivorus* were collected from roads to the Universidade Federal de Catalão with a license from the Ministry of Environment of Brazil (SISBIO Number 37072-2). Three *P. cancrivorus* and four *N. nasua* had died from being run over on roads, and the other specimens died of natural causes. The bioethics committees of the Universidad del Tolima (agreement number 2.3-059), Universidade de São Paulo (CEUAx agreement number 3928240820), and Universidade de Catalão (CEUA-UFCAT agreement number 01/22) approved the use of these specimens in this investigation.

2.2. Fixation and Vascular Repletion of Specimens

The specimens were fixed with a solution of 10% formaldehyde via the femoral artery and were posteriorly immersed and conserved in closed containers with 5% formaldehyde. One *P. cancrivorus* and all *N. nasua* specimens were repleted with natural latex tinctured with red vinyl via the femoral artery. One *P. cancrivorus* was repleted via the axillary artery due to a previous necropsy. The other specimens were not replete.

2.3. Dissection and Documentation

Gross dissections were performed at the Amphitheater of Veterinary Anatomy of the Universidad del Tolima (Tolima, Colombia) and Comparative Anatomy of Wild Animals Laboratory of the Universidade Federal de Catalão (Goias, Brazil). First, the skin was removed from the dorsal regions of the neck and trunk to the distal extreme of the thoracic limbs. The fascia around the muscles, nerves, and vessels was removed to review the distribution of the nerves. The pectoral muscles were disinserted from the sternum and the ventral part of the thoracic wall was removed to review the vascular relationships of the brachial plexus. Finally, the vessels, viscera, and ventral muscles of the neck and thorax were removed until the emergence of the ventral branches of the cervical and thoracic spinal nerves was found. The anatomical descriptions were performed according to the terminology of the *Nomina Anatomica Veterinaria -NAV-* [33]. However, other terms that do not appear in the *NAV* were used due to the presence of other nerves and muscles in carnivorans, such as n. brachiocephalicus [34], n. teres major [30,35], m. pectoralis abdominalis [36,37], m. anconeus medialis [38], and m. palmaris longus [39]. Photographs of the dissections were taken with a Canon EOS 6D 20.2 MP fitted with a 100 mm Canon macro lens and a Canon T5i 18 MP camera fitted with a 50 mm Canon macro lens. The origin of each brachial plexus nerve was expressed in percentage values.

3. Results

3.1. Origin of the Brachial Plexus in Procyon cancrivorus and Nasua nasua

The brachial plexus in four specimens of *P. cancrivorus* originates bilaterally from the ventral branches of the cervical spinal nerves C6, C7, and C8, and the thoracic spinal nerves T1 and T2 (C6-T2). In one specimen, it originates from C5-T1 bilaterally (Pc5). In *N. nasua*, the brachial plexus originates from C5-T2 in six limbs, C5-T1 in two limbs, and C6-T1 in two limbs.

The branch of C5 to the brachial plexus is a small branch that extends parallel to the longus capitis muscle until reaching the ventral branch of C6. The ventral branch of C6 emerges between the longus capitis and the scalenus muscles, while the ventral branches of C7 to T1 emerge between the longus colli and the scalenus muscles. T2 joins with T1 in the medial surface of the first rib before it emerges toward the axillary space, which is a small branch. In one *P. cancrivorus* (Pc2), T2 is a large branch bilaterally. The ventral branches of C5, C6, C7, and C8 have a ventral relationship with the external jugular vein and the superficial cervical artery, while the common trunk of T1 and T2 has a ventral relationship with the axillary artery. C6 crosses ventrally to the scalenus muscle and sends two large branches (cranial and caudal), with the cranial directed laterally to form the suprascapular and brachiocephalic nerves. The caudal branch extends parallel to the scalenus muscle, forms the cranial subscapular nerve, and joins the ventral branch of C7. When C5 contributes to the brachial plexus, it joins C6 and forms the brachiocephalicus, suprascapular, and cranial subscapular nerves in *N. nasua*, and only the two former nerves in *P. cancrivorus* (Figure 1).

The ventral branch of C7 forms two large branches, one dorsal and one ventral, where both receive branches from the caudal branch of C6 and form two trunks, dorsal and ventral, respectively. The ventral cranial trunk forms the musculocutaneous and cranial pectoral nerves. In two *N. nasua* (Nn1 and Nn5), it also forms an ansa pectoralis, which is a communicating branch with the caudal pectoral nerves (Figure 2). In the other three *N. nasua*, C7 directly forms the ansa pectoralis, and in one case, it also communicates with the lateral thoracic nerve (Nn5). In the case of *P. cancrivorus*, the ansa axillaris forms the ansa pectoralis (Figure 1).

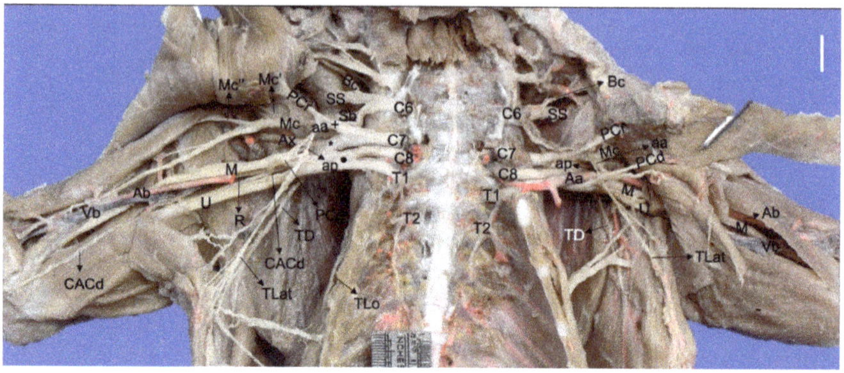

Figure 1. Ventral photographic view of the brachial plexus in a *Procyon cancrivorus*. (Aa) axillary artery, (aa) ansa axillaris (proximal communicating branch to the median and caudal pectoral nerves), (Ab) brachial artery, (ap) ansa pectoralis (communicating branch to the nn. Pectorales caudales), (Ax) n. axillaris, (Bc) n. brachiocephalicus, (C6–8) cervical spinal nerves, (CACd) n. cutaneus antebrachii caudalis, (M) n. medianus, (Mc) n. musculocutaneus, (Mc′) branch to the m. coracobrachialis, (Mc″) ramus muscularis proximalis, (PCr) nn. Pectorales craniales, (PCd) nn. Pectorales caudales, (R) n. radialis, (Sb) nn. subscapulares, (SS) n. suprascapularis, (T1–T2) thoracic spinal nerves, (TD) n. thoracodorsalis, (TLat) n. thoracicus lateralis, (TLo) n. thoracicus longus, (U) n. ulnaris, (Vb) brachial vein, (+) cranial ventral trunk, (*) cranial dorsal trunk, (black circle) caudal ventral trunk. White bar: 10 mm.

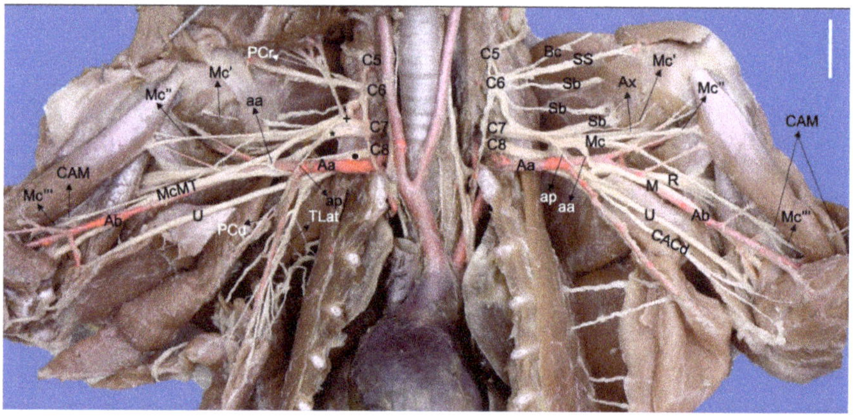

Figure 2. Ventral photographic view of the brachial plexus with arterial relationships in a *Nasua nasua*. (Aa) axillary artery, (aa) ansa axillaris (proximal communicating branch to the n. medianus), (Ab) brachial artery, (ap) ansa pectoralis (communicating branch to the nn. pectorales caudales), (Ax) n. axillaris, (Bc) n. brachiocephalicus, (Bv) brachial vein, (C5–8) cervical spinal nerves, (CACd) n. cutaneus antebrachii caudalis, (CAM) n. cutaneus antebrachii medialis, (M) n. medianus, (Mc) n. musculocutaneus, (Mc′) branch to coracobrachialis, (Mc″) ramus muscularis proximalis, (Mc‴) ramus muscularis distalis, (PCr) nn. pectorales craniales, (PCd) nn. pectorales caudales, (R) n. radialis, (Sb) nn. subscapulares, (SS) n. suprascapularis, (TD) n. thoracodorsalis, (TLat) n. thoracicus lateralis, (TLo) n. thoracicus longus, (McMT) Mc and M common trunk, (U) n. ulnaris, (Vb) brachial vein, (+) cranial ventral trunk, (*) cranial dorsal trunk, (black circle) caudal ventral trunk. White bar: 10 mm.

The dorsal cranial trunk forms the caudal subscapular and axillary nerves. It also forms an independent nerve directed to the m. teres major (*n. teres major*) and the m. subscapularis in all cases of *N. nasua* and one case bilaterally in *P. cancrivorus* (Nn5). The ventral branches of C8, T1, and T2 form two common trunks just lateral to the m. scalenus, the ventral caudal and dorsal caudal trunks. The former forms the common trunk of the median and ulnar nerves and a common trunk to the lateral thoracic and caudal pectoral nerves (Figure 1). The dorsal caudal trunk joins with a branch from the dorsal cranial trunk and forms the radial and thoracodorsal nerves (Figure 3).

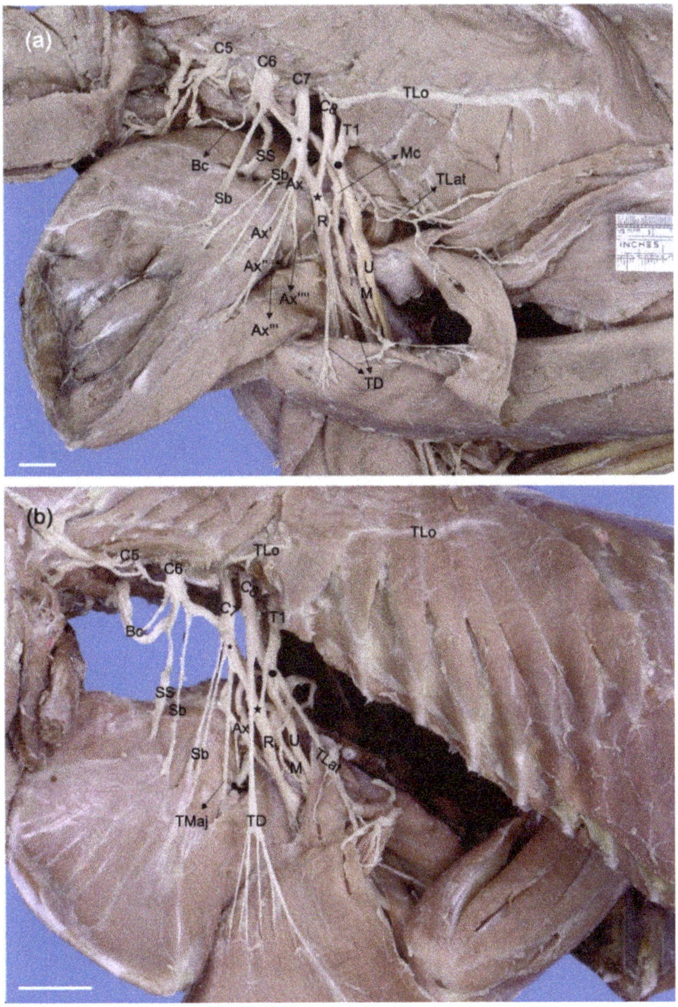

Figure 3. Dorsolateral photographic views of left brachial plexuses of a (**a**) *Procyon cancrivorus* and (**b**) *Nasua nasua*. (Ax) n. axillaris, (Ax′) branch to the m. subscapularis, (Ax″) branch to the teres major and subscapularis muscles, (Ax‴) branch to the m. teres minor, (Ax⁗) branch to the deltoideus muscles and n. cutaneus brachii cranialis, (Bc) n. brachiocephalicus, (C6–8) cervical spinal nerves, (M) n. medianus, (Mc) n. musculocutaneus, (R) n. radialis, (Sb) nn. subscapulares, (SS) n. suprascapularis, (T1) first thoracic spinal nerve, (TD) n. thoracodorsalis, (TLat) n. thoracicus lateralis, (TLo) n. thoracicus longus, (TMaj) n. teres major, (U) n. ulnaris, (*) cranial dorsal trunk, (black star) caudal dorsal trunk, (black circle) caudal ventral trunk. White bars: 10 mm.

3.2. Origin and Distribution of the Brachial Plexus Nerves in Procyon cancrivorus and Nasua nasua

The origin and distribution of each nerve are summarized in Table 1.

Table 1. Origin and distribution of the brachial plexus nerves in *Procyon cancrivorus* and *Nasua nasua*.

Nerve	Origin	*Procyon cancrivorus* (%)	*Nasua nasua* (%)	Muscle Innervation	Skin Innervation
Suprascapularis	C5-C6 C5-C7 C6 C6-C7	10 10 70 10	80 0 10 10	Supraspinatus and infraspinatus	
Subscapularis cranialis	C5-C6 C6 C6-C7	0 70 30	60 30 10	Subscapularis	
Subscapularis medium	absent C6 C6-C7	90 10 0	20 30 50		
Subscapularis caudalis	C6-C7	100	100		
Teres major	C6-C7	20	100	Teres major and subscapularis	
Axillaris	C6-C7	100	100	Subscapularis, teres, minor, deltoideus (Pars scapularis, pars acromialis, pars clavicularis -m. Cleidobrachialis-), teres major (only in *P. cancrivorus*).	Cranial, lateral, and medial surfaces of the distal half of the brachium and proximal extreme of the antebrachium.
Musculocutaneus	C6-C7 C7	90 10	100 0	Coracobrachialis, biceps brachii, brachialis, pronator teres (only in *P. cancrivorus*), and flexor carpi radialis (only in *P. cancrivorus*).	Medial surfaces of the antebrachium and manus. Abaxial and axial dorsal surfaces of the digit I, and abaxial dorsal surface of the digit II (via communicating branches with the radial nerve). Abaxial and axial palmar surfaces of the digits I and II, and abaxial palmar surface of the digit III (via common trunk with the median nerve in *P. cancrivorus*) Abaxial palmar surface of the digit I (*N. nasua*)
Radialis	C6-C8 C6-T1 C6-T2 C7-T1	20 20 60 0	0 60 0 40	Tensor fasciae antebrachii, triceps brachii (Caput longum, caput mediale, caput accessorium, and caput laterale), anconeus lateralis (anconeus), brachioradialis, extensor carpi radialis, extensor digitorum communis, extensor digitorum lateralis, extensor carpi ulnaris, supinator, abductor digiti I longus, and extensor digiti I et II.	Cranial and lateral surfaces of the antebrachium, and dorsal and medial surfaces of the manus Axial and abaxial dorsal surfaces of the digits I to III. Abaxial dorsal surface of the digit IV and axial dorsal surface of the digit V (100% in *P. cancrivorus* and 40% in *N. nasua*)

Table 1. *Cont.*

Nerve	Origin	*Procyon cancrivorus* (%)	*Nasua nasua* (%)	Muscle Innervation	Skin Innervation
Medianus	C6-T1 C6-T2 C7-T2	20 50 30	60 40 0	Pronator teres (only in *N. nasua*), flexor carpi radialis, flexor digitorum superficialis, palmaris longus, flexor digitorum profundus (caput humerale and caput radiale), interflexorii, pronator quadratus, abductor digiti I brevis, flexor digiti I brevis, and the most medial lumbricalis.	Axial and abaxial palmar surfaces of the digits II to III. Abaxial palmar surface of the digit I (*P. cancrivorus*) Abaxial palmar surface of the digit IV and axial palmar surface of the digit V (10% of *N. nasua*)
Ulnaris	C8-T1 C8-T2	20 80	60 40	Anconeus medialis (Anconeus epitrochlearis), flexor carpi ulnaris (Caput humerale and caput ulnare), flexor digitorum profundus (Caput humerale and caput ulnare), flexor digitorum brevis, adductor digiti I, adductor digiti II, abductor digiti V, flexor digiti V, adductor digiti V, interossei, and the three lateral lumbricales.	Caudal surface of the antebrachium. Carpal tactile hairs and distal extreme of the antebrachium. Lateral dorsal and palmar surfaces of the manus. Axial and abaxial palmar surfaces of the digit V. Abaxial palmar and dorsal surfaces of the digit IV. Axial and abaxial dorsal surfaces of the digit V.
Cutaneus antebrachii caudalis	C8-T1 C8-T2 T1 T1-T2	0 60 20 20	10 20 70 0		Caudal surface of the antebrachium
Brachiocephalicus	C5-C6 C6 C6-C7	60 30 10	80 20 0	Supraspinatus (only 10% in *P. cancrivorus*)	Cranial surface of the shoulder and proximal half of the cranial surface of the brachium
Thoracicus longus	C7 C7-C8	100 0	90 10	Serratus ventralis thoracis	
Thoracodorsalis	C6-C8 C6-T1 C6-T2 C7-C8 C7-T2	20 20 20 20 20	10 10 0 80 0	Latissimus dorsi	
Pectorales craniales	C6-C7 C7 C7-T2	90 0 10	40 60 0	Pectoralis descendens and pectoralis transversus	
Pectorales caudales	C6-T1 C6-T2 C7-T1 C7-T2	20 50 0 30	20 0 80 0	Pectoralis profundus and pectoralis transversus	
Thoracicus lateralis	C6-T2 C7-T1 C8-T1 C8-T2	40 0 20 40	0 10 90 0	Cutaneus trunci and pectoralis abdominalis	

The brachiocephalic (*N. brachiocephalicus*) and suprascapular nerves (*N. suprascapularis*) are formed from a single trunk in all *P. cancrivorus* and one limb of *N. nasua* (Nn1), while both originate independently in most limbs of *N. nasua* (Figures 1–3). The brachiocephalic nerve extends laterally toward the space between the omotransversarius and cleidocephalicus muscles, passing superficially to the m. cleidobrachialis. In two *P. cancrivorus* (Pc3 and Pc4), the nerve receives communication from the ventral branch of C5 bilaterally just before passing through the space toward the skin. It innervates the cranial regions of the shoulder and the proximal half of the brachium and does not send a muscle branch to the m. cleidobrachialis. In one *P. cancrivorus*, it also communicates bilaterally with the cranial lateral brachial cutaneous nerve (Pc4). In one limb of *P. cancrivorus* (Pc5), the n. brachiocephalicus perforates the m. supraspinatus, innervates it, and continues laterally along its normal course. The suprascapular nerve passes between the subscapularis and supraspinatus muscles, cranially to the scapular incisure and laterally to the scapular neck until the m. infraspinatus. Two subscapular nerves (*Nn. subscapulares*) are formed in *P. cancrivorus*, and three are formed in *N. nasua*. However, three nerves are formed in one limb of *P. cancrivorus* (Pc4) and two nerves bilaterally in *N. nasua* (Nn4) (Figure 4).

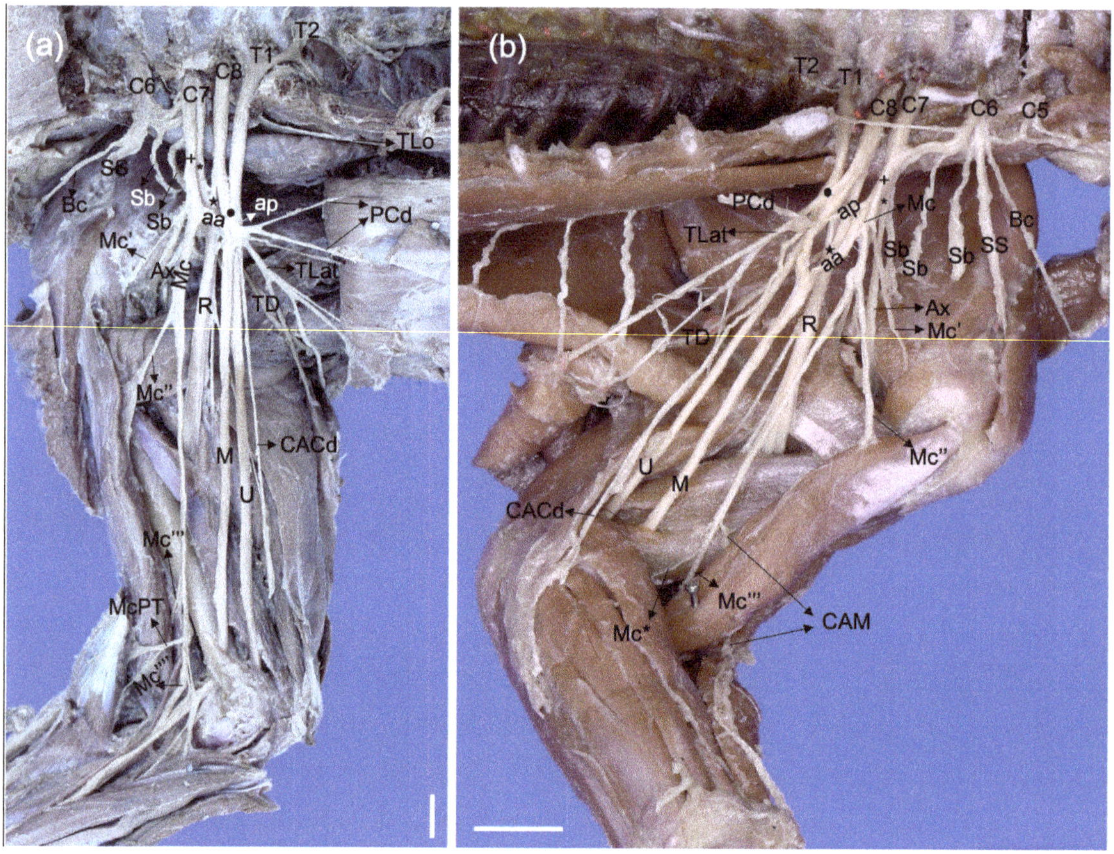

Figure 4. Medial photographic views of a right brachial plexus of a *Procyon cancrivorus* (**a**) and a left brachial plexus of a *N. nasua* (**b**). (aa) ansa axillaris (proximal communicating branch of the Mc to M), (ap) ansa pectoralis, (Ax) n. axillaris, (Bc) n. brachiocephalicus, (C6–8) cervical spinal nerves, (CACd) n. cutaneus antebrachii caudalis, (CAM) n. cutaneus antebrachii medialis, (M) n. medianus, (Mc) n. musculocutaneus, (Mc′) branch to coracobrachialis, (Mc″) ramus muscularis proximalis, (Mc‴) ramus muscularis distalis, (Mc⁗) distal communicating branch with the median nerve, (Mc*) branch to the joint capsule of the elbow, (McPT) branch to the m. pronator teres, (PCr) nn. pectorales craniales, (PCd) nn. pectorales caudales, (R) n. radialis, (Sb) nn. subscapulares, (SS) n. suprascapularis, (T1-2) thoracic spinal nerves, (TD) n. thoracodorsalis, (TLat) n. thoracicus lateralis, (TLo) n. thoracicus longus, (U) n. ulnaris, (+) cranial ventral trunk, (*) cranial dorsal trunk, (black star) caudal dorsal trunk, (black circle) caudal ventral trunk. White bars: 10 mm.

The axillary nerve (*N. axillaris*) first sends one branch to the most caudal belly of the m. subscapularis and, in most cases, to the m. teres major in *P. cancrivorus*. In all *N. nasua* and one *P. cancrivorus* (Pc5), the nerve to the teres major muscle (*N. teres major*) originates directly from the dorsal cranial trunk and sends a small branch to the m. subscapularis (Figure 3). The axillary nerve passes between the subscapular and teres major muscles to continue laterally deeply to the m. deltoideus, sending branches to the m. teres minor, and acromial and scapular parts of the m. deltoideus and m. cleidobrachialis. Deep in the m. deltoideus, it forms the cranial lateral brachial cutaneous nerve (*N. cutaneus brachii lateralis cranialis*), which extends between the m. brachialis and acromial part of the m. deltoideus to innervate the distal half of the brachium and elbow. It communicates with the medial branch of the superficial branch of the radial nerve in *N. nasua* and directly with the superficial branch of the radial nerve in *P. cancrivorus*. However, in one limb of *N. nasua* (Nn1), it also communicates with the lateral branch of the superficial radial nerve.

The radial nerve (N. *radialis*) extends laterally to the axillary and brachial arteries. Medial to the common tendon of the latissimus dorsi and teres major muscles, it forms branches to the m. tensor fasciae antebrachii, capita longum, mediale, and accessorium of the m. triceps brachii. The nerve passes laterally between the capita accessorium and mediale of the m. triceps brachii at the proximal third of the brachium (Figure 4). It continues laterally at the brachium between the m. brachialis and the caput laterale of m. triceps brachii, sending a branch to this caput and another branch to the m. anconeus. At this level, the nerve divides into deep and superficial branches, where the former innervates the craniolateral antebrachial muscles, and the latter innervates the craniolateral surface of the antebrachium and dorsum of the manus. The deep branch (*Ramus profundus*) first innervates the brachioradialis and extensor carpi radialis muscles and passes deeply to the m. supinator to innervate the other muscles. It also innervates the m. brachialis unilaterally in two cases in *P. cancrivorus* (Pc2 and Pc5).

The branches of the superficial branch of the radial nerve (*Ramus superficialis*) are relatively larger in *P. cancrivorus* than in *N. nasua*. In *N. nasua*, it divides into two branches, lateral (*Ramus lateralis*) and medial (*Ramus medialis*), which extend laterally and medially to the cephalic vein, respectively. In *P. cancrivorus*, the superficial branch at the proximal fourth of the antebrachium is a large trunk and distally divides into three branches: lateral, medial, and accessory medial. The accessory medial branch forms at the proximal third of the antebrachium and joins the medial antebrachial cutaneous nerve, and both form the abaxial dorsal digital nerve I and the dorsal common digital nerve I. The accessory medial branch is not present in *N. nasua;* thus, the latter two nerves are formed by the medial branch of the superficial radial nerve. However, there are communicating branches from the medial antebrachial cutaneous nerve at the level of the antebrachium in *N. nasua*. The medial and lateral branches form the common dorsal nerves II and III due to communication between them proximal to the manus and the dorsum of the manus in both species. In *P. cancrivorus*, the lateral branch of the radial nerve and the dorsal branch of the ulnar nerve form the common dorsal digital nerve IV, although the radial nerve contributes mainly to forming the abaxial proper dorsal digital nerve IV, and the ulnar nerve mainly forms the axial proper dorsal digital nerve V. In *N. nasua*, the dorsal branch of the ulnar nerve forms the common dorsal digital nerve IV, although it receives weak communicating branches from the lateral branch of the radial nerve in three cases (Nn2, Nn3, and Nn4). The superficial branch of the radial nerve forms the lateral antebrachial cutaneous nerve (*n. cutaneus antebrachii lateralis*) in *P. cancrivorus*, while it forms from the lateral branch of the superficial branch of the radial nerve in *N. nasua*. However, in one limb of *N. nasua*, it formed similarly to *P. cancrivorus* (Nn4) (Figures 5 and 6).

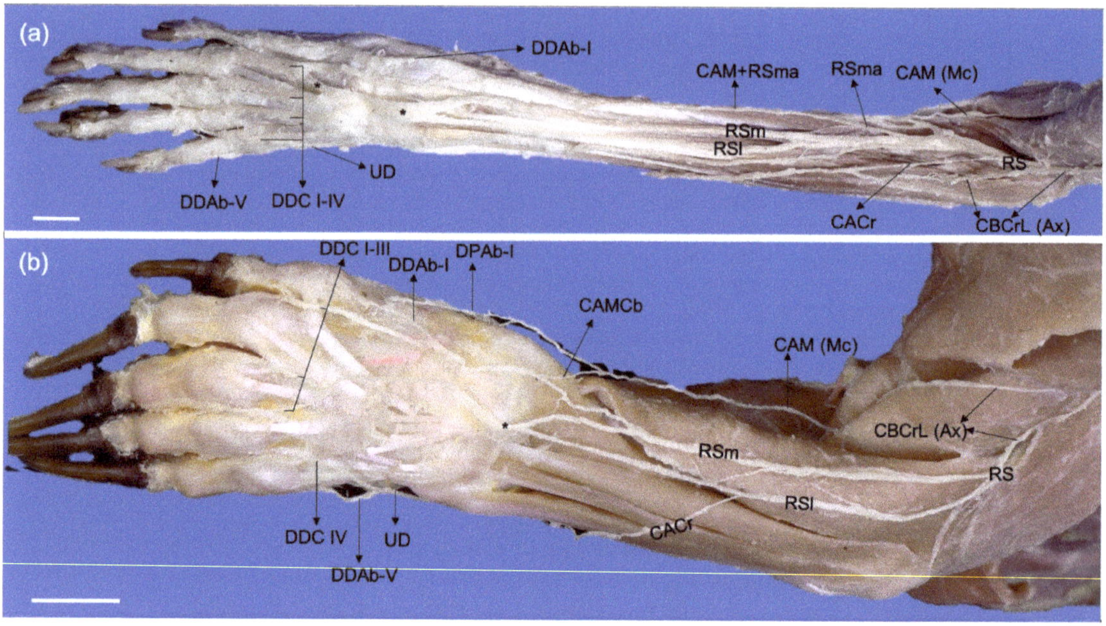

Figure 5. Cranial photographic views of the left antebrachia of a *P. cancrivorus* (**a**) and a *N. nasua* (**b**). (Ax) n. axillaris, (CACr) n. cutaneus antebrachii cranialis, (CAM) n. cutaneus antebrachii medialis, (CAMCb) communicating branch from CAM to RSm, (CBCrL) n. cutaneus brachii cranialis lateralis, (DDAb-I) n. digitalis dorsalis I abaxialis, (DDAb-V) n. digitalis dorsalis V abaxialis, (DDC I-IV) nn. digitales dorsales communes I-IV, (DPAb-I) n. digitalis palmaris I abaxialis, (Mc) n. musculocutaneus, (RS) superficial branch of the n. radialis, (RSl) lateral branch, (RSm) medial branch, (RSma) accessory medial branch, (UD) dorsal branch of the n. ulnaris, (*) communications between the RSm and RSl. White bars: 10 mm.

The musculocutaneous nerve (*N. musculocutaneus*) extends cranially to the axillary and brachial arteries, which form several branches. At the axillary level in *P. cancrivorus*, it forms the branch to the m. coracobrachialis and an ansa axillaris. This latter is named ansa axillaris because it passes medially to the axillary artery and forms two communicating branches, one proximal to the caudal pectoral nerves (*ansa pectoralis*) and one distal to the median nerve (Figure 1). In one specimen bilaterally (Pc4), C7 directly forms the ansa axillaris and not the musculocutaneous nerve. In one specimen unilaterally (Pc3), the ansa axillaris forms a branch to the m. pectorlais profundus. The branch to m. coracobrachialis forms proximally to the ansa axillaris in two specimens (Pc2 and Pc5), while in three specimens, it forms distally. In two specimens bilaterally (Pc1 and Pc2), the ansa pectoralis also communicates with the lateral thoracic nerve (Figures 1 and 3).

At the axillary level in *N. nasua*, the ansa axillaris originates from the musculocutaneous nerve and joins the median nerve at the level of the transition between the axillary and brachial arteries (more distal than *P. cancrivorus*). The ansa pectoralis does not form from the musculocutaneous nerve. The origin of the branch to the m. coracobrachialis is variant; therefore, it forms proximally to the ansa axillaris in two limbs (Nn3 bilaterally), from the ansa axillaris in four limbs (two unilaterally—Nn4 and Nn5—and one bilaterally—Nn2), from one branch of the ansa axillaris and another branch proximal to the ansa axillaris in one limb (Nn4), and distally to the ansa axillaris in three limbs (one bilaterally—Nn1—and one unilaterally—Nn5) (Figure 7).

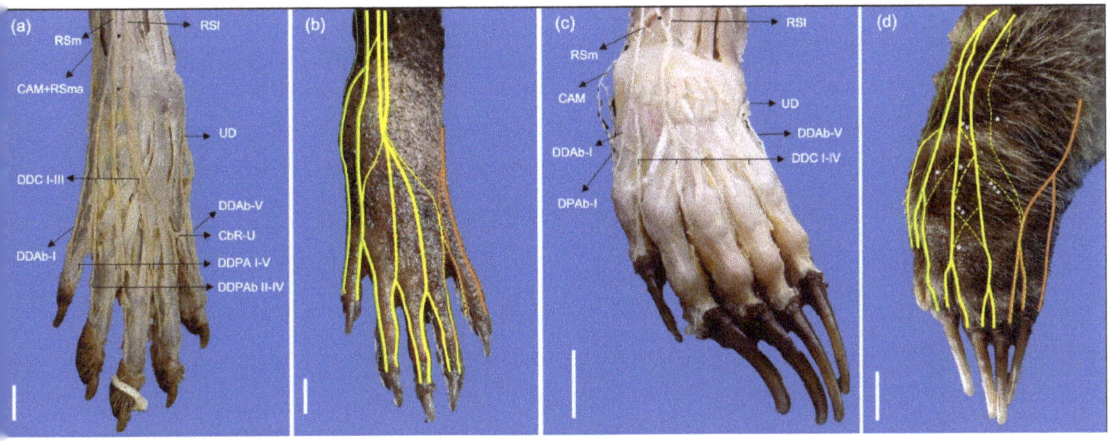

Figure 6. Dorsal photographic views of the dorsal digital nerves: (**a**) Dissection a left manus of a *P. cancrivorus*; (**b**) Scheme of the nerve distribution in a left manus of a *P. cancrivorus*; (**c**) Dissection a left manus of a *N. nasua*; (**d**) Scheme of the nerve distribution in a left manus of a *N. nasua*. (CAM) n. cutaneus antebrachii medialis, (CAM-RSma) common trunk of CAM and RSma, (CbR-U) communicating branches between DDAbIV and DDAV, (DDAb-I) n. digitalis dorsalis I abaxialis, (DDAb-V) n. digitalis dorsalis V abaxialis, (DDAV) n. digitalis dorsalis V axialis, (DDC I-IV) nn. digitales dorsales communes I-IV, (DDPA I-IV) nn. digitales dorsales propii I-IV axiales, (DDPAb II-IV) nn. digitales dorsalis propii II-IV abaxiales, (RS) superficial branch of the n. radialis, (RSl) lateral branch, (RSm) medial branch, (RSma) accessory medial branch, (black asterisks) communications between RSm and RSl, (UD) dorsal branch of the n. ulnaris, (yellow lines) distribution of the superficial branch of the radial nerve, (orange lines) ulnar nerve, (green lines) musculocutaneus nerve, (yellow lines) median nerve. The dashed lines indicate the variant communicating branches. White asterisks indicate the number of times that the communicating branch is present. White bars: 10 mm.

In both species, the proximal muscular branch (*Ramus muscularis proximalis*) of the musculocutaneous nerve forms at the axillary level, and the main trunk of the musculocutaneous nerve continues distally between the brachial artery and the m. biceps brachii. At the distal third of the brachium, the nerve forms three branches: the medial antebrachial cutaneous nerve (*N. cutaneus antebrachii medialis*), the distal muscular branch (*Ramus muscularis distalis*), and a branch to the elbow joint capsule. In *P. cancrivorus*, the main trunk of the nerve continues distally toward the antebrachium, which, at the elbow level, sends branches to the pronator teres and flexor carpi radialis muscles and communicates with the median nerve distally to the supracondylar foramen (*Ramus communicans cum n. mediano*), forming a common trunk toward the manus (Figure 4). In both species, the medial cutaneous antebrachial nerve passes between the brachialis and biceps brachii muscles and extends medially to the m. brachioradialis until the medial side of the manus. Therefore, it sends branches to the medial side of the palmar pad, and in the case of *N. nasua*, it forms the abaxial palmar nerve of digit I. Proximal to the radial styloid process, it sends a communicating branch to the medial branch of the superficial branch of the radial nerve. A communicating branch is also sent to the abaxial dorsal digital nerve I (Figure 7).

On the right brachial plexus of one *N. nasua* (Nn1), the musculocutaneous has a different arrangement at the axillary and brachial regions. The ansa axillaris receives communicating branches from the median nerve, forming an ansa axillaris that is more proximal and medial to the axillary artery. From there, the musculocutaneous and median nerves continue as a common trunk medial to the brachial artery, which receives a communicating branch from the proximal muscular branch of the musculocutaneous nerve. The distal muscular branch and medial antebrachial cutaneous nerve form from a branch of the common trunk at the level of the middle of the brachium (see the right brachial plexus of Figure 2 and the schemas of Figure 7).

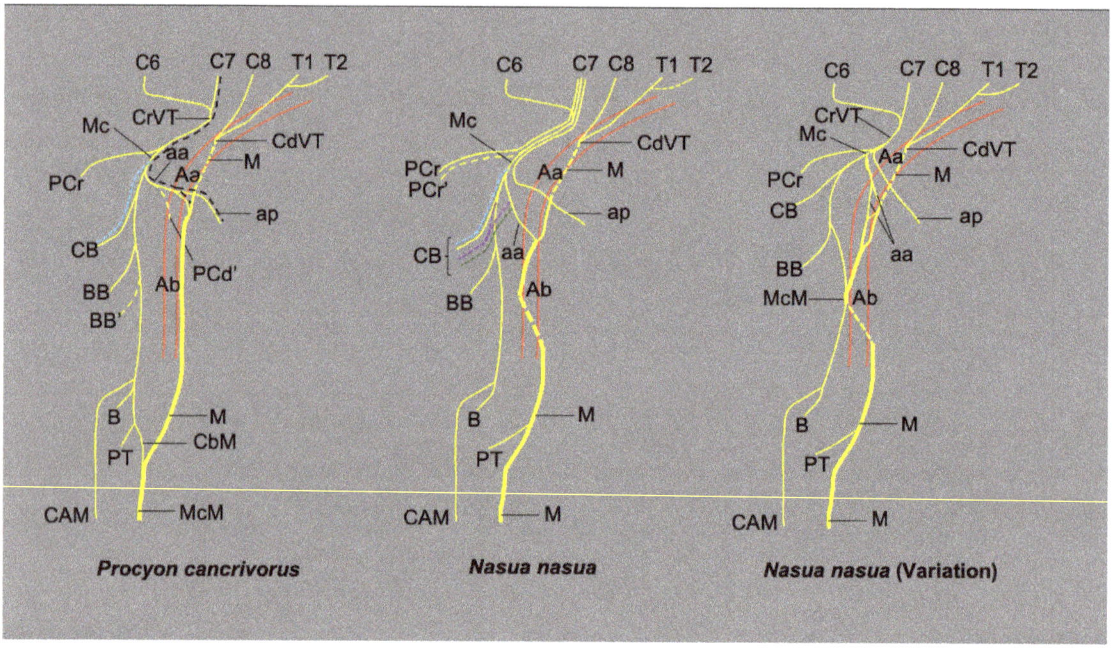

Figure 7. Schemas of the median and musculocutaneous nerves in *P. cancrivorus* and *N. nasua*. (Aa) arteria axillaris, (Ab) arteria brachialis, (aa) ansa axillaris, (ap) ansa pectoralis, (BB) branch to the m. biceps brachii, (BB′) variant branch to BB, (CAM) N. cutaneus antebrachii medialis, (CB) branches to the m. coracobrachialis, (CbM) ramus communicans cum n. mediano, (CdVT) caudal ventral trunk, (CrVT) cranial ventral trunk, (M) n. medianus, (Mc) n. musculocutaneus, (McM) common trunk of Mc and M, (PCd′) variant branch to m. pectoralis profundus, (PCr′) variant origin of the PCr from the cranial ventral trunk, (PT) branch to the m. pronator teres, (black dashed line) anatomical variation where the ansa axillaris is only formed by C7, (blue dashed line) origin of CB proximal to aa, (green dashed line) origin of CB from aa, (purple dashed line) origin of CB from Mc and aa. *Nasua nasua* (Variation): Variation present in one limb (10%) where branches of Mc and M form the ansa axillaris and a common trunk at the level of the brachium.

The median nerve (*N. medianus*) extends medially to the brachial artery at the proximal half of the brachium and at the distal half lateral to the brachial artery to pass alone through the supracondylar foramen (without vessels). At the level of the antebrachium, the median nerve sends branches to the caudomedial antebrachial muscles but not the flexor carpi ulnaris and capita ulnaris of the flexor digitorum profundus. In *P. cancrivorus*, it does not send branches to m. pronator teres and communicates with the musculocutaneous nerve. The median nerve extends together with the median artery between the flexor carpi radialis and the capita radialis of the m. flexor digitorum profundus to reach the manus passing through the carpal canal. Distal to the carpus, it forms common palmar nerves I to III. It also forms the common palmar nerve IV in one *N. nasua* bilaterally (Nn1), which receives a communicating branch from the superficial palmar branch of the ulnar nerve) (Figure 8).

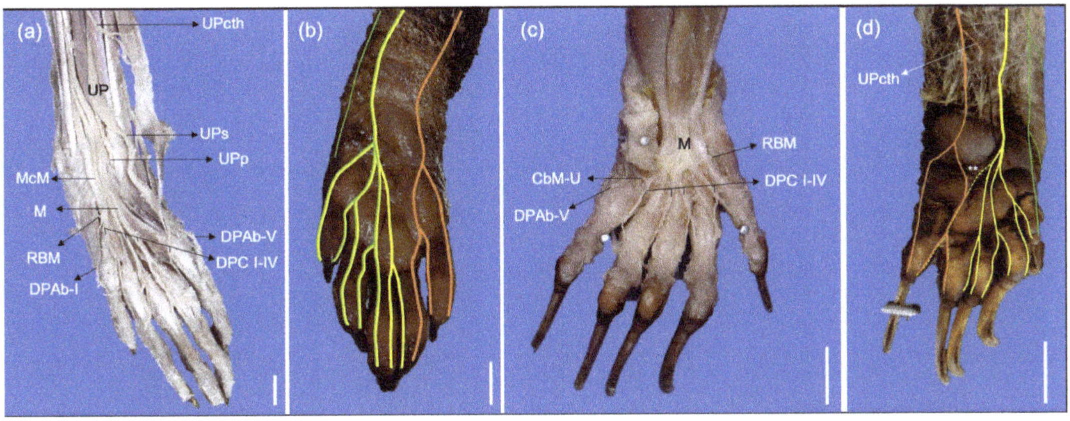

Figure 8. Palmar photographic views of the manus with deep dissections and skin. (**a**) Palmar photographic view of a deep dissection of a right manus of a *P. cancrivorus;* (**b**) Scheme of the palmar nerves in a right manus of a *P. cancrivorus;* (**c**) Palmar photographic view of a deep dissection of a left manus of a *N. nasua;* (**d**) Scheme of the palmar nerves in a left manus of *N. nasua.* (CbM-U) communicating branch from UPs to DPC IV, (DPC I-IV) nn. digitales palmares communes I-IV, (DPAb-I) n. digitalis palmaris I abaxialis, (DPAb-V) n. digitalis palmaris V abaxialis, (M) n. medianus, (MCM) common trunk of median and musculocutaneous nerves, (RBM) recurrent branch of M, (UP) n. ulnaris ramus palmaris, (UPs) ramus superficialis, (UPp) ramus profundus, (UPcth) branch to the distal extreme of the antebrachium and carpal tactile hairs, (yellow lines) distribution of the superficial branch of the radial nerve, (orange lines) ulnar, (green lines) musculocutaneous and (yellow lines) median nerves, (dashed lines) variant communicating branches, (*) number of times that the communicating branch is present. White bars: 10 mm.

The ulnar nerve (*N. ulnaris*) extends caudal to the brachial artery and cranial to the brachial vein and is covered by the m. tensor fasciae antebrachii at the level of the brachium. It passes deeply to the m. anconeus medialis just proximally and caudally to the medial epicondyle to continue toward the antebrachium between the m. flexor carpi ulnaris and caput ulnare of the m. flexor digitorum profundus. The nerve divides into dorsal and palmar branches (*Ramus dorsalis* and *Ramus palmaris,* respectively) at the proximal third of the antebrachium in *N. nasua* and at the middle third in *P. cancrivorus.* The dorsal branch sends branches to the distal third of the caudolateral surface of the antebrachium and continues distally to pass superficially between the tendon of the m. flexor carpi ulnaris and the styloid process of the ulna. It passes laterally to the accessory carpal bone toward the dorsum of the manus and forms the common dorsal digital nerve IV and the abaxial dorsal digital nerve V. The palmar branch at the distal fourth of the antebrachium sends a branch to the skin, which extends to the venous sinuses of the tactile carpal hairs and the distal extreme of the antebrachium. This branch passes between the palmaris longus and flexor digitorum superficialis muscles in *N. nasua,* while in *P. cancrivorus,* it perforates the m. palmaris longus. At the carpal level, the palmar branch continues deeply to the flexor retinaculum and medial to the accessory carpal bone, where it forms the superficial and deep branches. The superficial branch forms the abaxial palmar digital nerve V and the common palmar digital nerve IV, and the deep branch forms the metacarpal nerves and innervates most of the intrinsic muscles of the manus (Table 1) (Figures 7 and 8).

The caudal antebrachial cutaneous nerve (*N. cutaneus antebrachii caudalis*) is an independent nerve to the ulnar nerve, which originates directly from T1 or T1-T2. It passes medially to the ulnar nerve at the axillary level, while at the brachial level, it passes medially to the brachial artery and cranially to the m. tensor fasciae antebrachii, reaching the caudal surface of the antebrachium (Figures 1, 2 and 4). The nerve communicates with the branch to the carpal tactile hairs bilaterally in one *N. nasua* (Nn4).

4. Discussion

4.1. Comparative Origin of the Brachial Plexus in Carnivorans

The origins of the brachial plexus in *P. cancrivorus* and *N. nasua* from C5-T1, C5-T2, and C6-T2 have been found in few carnivorans. C5-T1 origin has only been reported in other caniformes, such as *Meles meles, Nyctereutes procyonoides* [31], and *Canis lupus familiaris* [40], and feliformes, such as *Leopardus geoffroyi* [32], *Puma yagouaroundi* [23], and *Felis catus* [41]. C5-T2 origin has been found in *Potos flavus* [30], *Ailuropoda melanoleuca* [35], *Vulpes vulpes* [22], and *C. l. familiaris* [40]. C6-T2 was reported in *P. flavus* [30], *Bassariscus astutus* [35], *M. meles, N. procyonoides, V. vulpes* [31], *C. l. familiaris* [40], and *F. catus* [29]. In contrast, the origin from C6-T1 is the most frequent in most carnivorans (Table 2), which agrees with the brachial plexus evolution in carnivorans, where the ventral branches of C6, C7, C8, and T1 are the largest branches to form the brachial plexus, whereas the branches of C5 and T2 are weak branches [22,30,34,42], as was found in *P. cancrivorus* and *N. nasua*.

Table 2. Comparative origins of the brachial plexus in carnivorans.

Suborder	Infraorder	Superfamily	Family	Species	Number of Limbs (Number of Specimens)	Brachial Plexus Origin (%)
		Musteloidea	*Procyonidae*	*Procyon cancrivorus*	10 (2 females and 3 males)	C5-T1 (20%) C6-T2 (80%)
				Nasua nasua	10 (1 female and 4 males)	C5-T1 (20%) C5-T2 (60%) C6-T1 (20%)
	Arctoidea			*Bassariscus astutus* [35]	2 (1 unreported sex)	C6-T2 (100%)
				Potos flavus [30]	10 (3 females and 2 males)	C5-T2 (30%) C6-T2 (70%)
				Martes foina [43]	12 (3 females and 3 males)	C6-T1 (100%)
Caniformia				*Martes martes* [31]	16 (6 female and 2 male)	C6-T1 (100%)
			Mustelidae	*Neovison vison* [31]	60 (18 female and 12 male)	C6-T1 (26.67%) C6-T2 (73.33%)
				Neovison vison [44]	52 (26 unreported sexes)	C6-T1 (100%)
				Meles meles [31]	50 (15 females and 10 males)	C5-T1 (10%) C6-T1 (40%) C6-T2 (50%)
		Ursoidea	*Ursidae*	*Ailuropoda melanoleuca* [35]	2 (1 male)	C5-T2 (100%)
				Ursus americanus [35]	2 (1 unreported sex)	C6-T1 (100%)
				Nyctereutes procyonoides [31]	40 (10 female and 10 males)	C5-T1 (12.5%) C6-T1 (25%) C6-T2 (62.5%)
		Canoidea	*Canidae*	*Vulpes vulpes* [31]	52 (10 females and 16 male)	C5-T1 (7.69%) C6-T1 (61.54%) C6-T2 (30.77%)
				Vulpes vulpes [22]	12 (6 males)	C5-T2 (33.33%) C6-T1 (66.67%)
				Canis lupus familiaris [40]	58 (unreported specimens and sexes)	C6-T1 (58.62%) C6-T1 (20.69%) C5-T1 (17.24%) C5-T2 (0.034%)
				Atelocynus microtis [25]	1 female	C6-T1 (100%)
				Cerdocyon thous [45]	20 (8 females and 2 males)	C6-T1 (100%)
				Cerdocyon thous [20]	6 (3 males)	C6-T1 (100%)
				Lycalopex gymnocercus [46]	20 (3 females and 7 males)	C6-T1 (100%)
		Pinnipedia	*Otaridae*	*Arctocephalus australis* [26]	4 (1 female and 1 male)	C6-T1 (100%)
	Aeluroidea	Feloidea		*Felis catus* [47]	12 (6 unreported sexes)	C6-T1 (100%)
				Felis catus [28]	10 (5 females)	C6-T1 (100%)
				Panthera onca [47]	4 (2 unreported sexes)	C6-T1 (100%)
				Puma concolor [47]	6 (3 unreported sexes)	C6-T1 (100%)
Feliformia			*Felidae*	*Puma concolor* [27]	2 (1 female)	C6-T1 (100%)
				Puma yagouaroundi [23]	14 (5 males and 2 females)	C5-T1 (57%) C6-T1 (43%)
				Leopardus geoffroyi [32]	6 (4 females)	C5-T1 (66.66%) C6-T1 (33.33%)
				Leopardus pardalis [21]	4 (1 male and 1 female)	C6-T1 (100%)

Most origins of the brachial plexus nerves in *P. cancrivorus* and *N. nasua* in the present study have also been found in other carnivorans. Some nerve origins of both species have not been found in other carnivorans, although the most of these were found in very low proportions. However, the origin of the median nerve from C6-T1 in *N. nasua* was present in most limbs (60%). The predominant origins of the subscapular, musculocutaneous, axillary, ulnar, long thoracic, lateral thoracic, and cranial pectoral nerves of both procyonids were similar to those of most carnivorans (Table 3). The other nerves had several inter- and intraspecific variations, mainly due to the contribution of C5 and T2 and the presence of the ansa axillaris and ansa pectoralis to the median and caudal pectoral nerves, respectively.

4.2. Comparative Distribution of the Brachial Plexus Nerves in Carnivorans and Evolutionary Comments

The brachial plexus nerves conserve a similar distribution among carnivorans mainly to the innervation of the thoracic limb muscles, although some differences have been found. Among them, the suprascapular nerve also innervates other muscles, such as the m. deltoideus in *Martes foina* [43], m. serratus ventralis cervicis in *Puma concolor* [27], and m. trapezius in *F. catus* [28]. One of the subscapular nerves innervates the m. teres major in *M. foina* [43], *A. australis* [26], and *L. geofffroyi* [32]. An independent nerve from the brachial plexus innervates the m. teres major in *P. flavus* [30], *B. astutus,* and *U. americanus* [35], which differs from *P. concolor* [27] and *N. nasua,* where that nerve also innervates the subscapular muscle. The radial nerve also innervates the m. teres minor in *M. foina* [43]. The musculocutaneous nerve may innervate the m. pronator teres as an anatomical variant in *C. thous* [49], which was found in the common pattern of *P. cancrivorus* together with the innervation of the m. flexor carpi radialis. The m. latissimus dorsi may also be innervated by the lateral thoracic nerve in *P. concolor* [27] and *F. catus* [50]. In *M. foina,* the caudal pectoral nerves innervate not only m. pectoralis profundus but also m. cutaneus trunci [43].

Some nerves develop other branches when there are extra muscles. Therefore, the musculocutaneous nerve innervates the caput breve of the m. biceps brachii and the m. coracobrachialis longus in *P. flavus* [30] and *A. melanoleuca* [35]. The median nerve innervates the palmaris longus muscles in *P. flavus* and *P. cancrivorus* [39], as was found in the present study in the latter species and *N. nasua*. The ulnar nerve innervates the m. anconeus medialis (m. anconeus epitrochlearis) in *F. catus* [38], *P. concolor* [27], *A. melanoleuca* [35], *P. flavus* [30], *P. cancrivorus,* and *N. nasua*.

The name n. brachiocephalicus is not in the *NAV* [33], but it has been used by several authors because it innervates m. cleidobrachialis [23,30–32,34,45,46], which is a part of m. brachiocephalicus [33]. Other authors named it n. supraclavicularis in *F. catus* [29,51] or n. subclavius in *M. foina* [43] and *Leopardus pardalis* [21]. On the other hand, in *P. concolor,* it was considered a branch of n. suprascapularis to m. cleidobrachialis [27]. In *F. catus,* it was also considered only as a cutaneous branch of the n. suprascapularis, which extends to the cranial skin to the shoulder and brachium [50,52]. In the case of *P. cancrivorus* and a variant manner in *N. nasua,* the latter arrangement was similar since the nerve formed from a common trunk with the n. suprascapularis and did not innervate the m. cleidobrachialis. In another procyonid, *P. flavus,* the brachiocephalic nerve innervated m. cleidobrachialis in only 50% of cases [30]. However, it is mainly innervated by the axillary nerve in all cases of *P. flavus* [30], *N. nasua,* and *P. cancrivorus* [37], which was confirmed in all specimens of the present study. The innervation of m. cleidobrachialis by the axillary nerve has also been described in other caniformes, such as *A. melanoleuca* [35], *A. microtis* [25], and *C. l. familiaris* [42,53], and feliforms, such as *F. catus* [29,50,51] and *P. concolor* [27]. Based on Singh et al. [42], m. cleidobrachialis splits from a primordium different from m. cleidocephalicus in the embryo since the former retains the appropriate innervation by the axillary nerve. Even innervation of m. cleidobrachialis is only reported by the axillary nerve in all domestic mammals [42,53]. Therefore, the present study confirms previous studies [30,37] in which it was concluded that m. cleidobrachialis was evolutionarily derived from m. deltoideus and is its corresponding clavicular part (*Pars clavicularis*). On the other hand, n. brachiocephalicus has developed mainly to innervate the cranial and lateral skin of the shoulder and brachium in carnivorans, which was found in both procyonids.

The communication of the cranial lateral cutaneous brachial nerve of the axillary nerve with the medial branch of the superficial branch of the radial nerve was found in the common pattern of *N. nasua* and *P. cancrivorus*, such as occurs in most carnivorans [24,54,55]. Communication with the lateral branch of the superficial branch of the radial nerve found in lesser proportion in *N. nasua* has been described in *A. melanoleuca* [35]. Therefore, some afferent fibers of the axillary nerve could come from the manus in these species.

The communicating branch from the musculocutaneous nerve to the median nerve in *P. cancrivorus* and *N. nasua* is medial to the axillary artery, as was found in *P. flavus* [30] and other caniformes [35]. Therefore, the communicating branch was called the "ansa axillaris" in the present study, based on the description written by several authors, who have described that this anatomical structure is a communication from the musculocutaneous nerve to the median nerve around the axillary artery [42,44,53,56,57]. The ansa axillaris of *N. nasua* has also been reported previously by other authors in the same species [48,58]. However, it is called a communicating branch of the musculocutaneous nerve with the median nerve at the axillary level [48] or axillary loop [58]. In another study, it was reported as ansa mediana in procyonids, such as *Procyon lotor* and *P. flavus*, while it was absent in the procyonid *B. astutus* [35]. Thus, the ansa axillaris is present in procyonids of the genera *Procyon*, *Nasua*, and *Potos*. However, in the case of *P. cancrivorus*, the ansa axillaris is formed more proximally than in other procyonid species. Other names were given to the ansa axillaris, such as axillary loop by Arlamowska-Palider [58], ansa mediana by Davis [35], and communicating branch at the axillary level by Grzeczka and Zdun [31]. Therefore, the ansa axillaris is also present in mustelids such as *Neovison vison* [31], *M. martes* [31,58], *M. foina* [58], *M. meles* [31,58], *Mustela lutreola*, *Mustela putorius*, and *Mustela nivalis* [58]; ursids such as *Ailuropoda melanoleuca*, *Ursus americanus* [35], and *Ursus arctos* [58]; the ailurid *Ailurus fulgens* [35]; and the canid *N. procyonoides* [31]. In the case of *M. meles*, the ansa axillaris is formed by two branches [31], but both communicate with the median nerve, while in *P. cancrivorus*, one of the branches communicates with the caudal pectoral nerves. Therefore, the presence of an ansa axillaris in arctoid carnivorans (infraorder Arctoidea, where the Musteloidea and Ursoidea superfamilies are included [1,4]) can be normal and differs from other authors who described that carnivorans do not have ansa axillaris [44]. Even its presence in a canid such as *N. procyonoides* [31] allows us to suggest that the presence of an ansa axillaris in carnivorans is not only due to its phylogenetic relationship from the last common ancestor of Caniformes, but also due to the anatomical shape of its body. *N. procyonoides* has a body shape very similar to procyonids of the genus *Procyon* but is phylogenetically a canid [1,4]. However, it may be absent as an anatomical variant in *N. procyonoides* since it was not reported in one specimen dissected by Davis [35], though the absence is the common pattern to most canids [31,34,35,45,46,58]. Therefore, when an ansa axillaris is present in some canids different from *N. procyonoides*, it could be considered an aberrant arrangement of the n. musculocutaneous since, as was reported in one *C. l. familiaris*, the formation of an ansa axillaris could obstruct the blood flow of the axillary artery [56].

A common trunk of the musculocutaneous and median nerves at the brachial level has been found in Mephitis and some specimens of M. putorius [58], similar to that which occurred in one case of *N. nasua*. This common trunk of both nerves is a normal arrangement in ungulates [44,53,58] and pangolins [59]. Therefore, it is a primitive arrangement that could have been present in the last common ancestor of Carnivora, Ungulata, and Pholidota [60]. Consequently, the common trunk of both nerves may be present as an anatomical variant in some carnivorans.

Table 3. Comparative origins of the brachial plexus nerves in carnivorans.

Nerves	Origin	Species
Suprascapularis	C5-C6	*P. cancrivorus* (10%), *N. nasua* (80%), *P. flavus* (20%), *L. geofffroyi* (50%), and *C. l. familiaris.*
	C5-C7	*P. cancrivorus* (10%), *P. flavus* (10%), *V. vulpes** (33.33%), *M. meles, N. procyonoides, V. vulpes*, and *C. l. familiaris.*
	C6	*P. cancrivorus* (70%), *N. nasua* (10%), *C. thous* (25%), *L. gymnocercus* (20%), *C. l. familiaris, F. catus** (80%), *P. concolor**, *P. yagouaroundi* (85.7%), and *L. geofffroyi* (50%).
	C6-C7	*P. cancrivorus* (10%), *N. nasua* (10%), *P. flavus* (70%), *M. meles, M. foina, M. martes, N. vison, N. procyonoides, C. l. familiaris, A. microtis, C. thous* (75%), *C. thous*, L. gymnocercus* (75%), *A. australis, V. vulpes** (33.33%), *V. vulpes, F. catus** (20%), *P. yagouaroundi* (14.3%), and *L. pardalis* (100%).
Subscapulares	C5-C7	*V. vulpes* (16.67%) and *N. nasua* (60%).
	C6-C7	*P. cancrivorus* (100%), *N. nasua* (40%), *P. flavus* (100%), *N. vison, M. meles, C. l. familiaris, C. thous* (45%), *C. thous*, L. gymnocercus* (90%), *V. vulpes** (83.33%), *V. vulpes, P. yagouaroundi* (78.7%), *L. geofffroyi* (100%), *P. concolor**, and *L. pardalis* (100%).
Musculocutaneus	C6-C7	*P. cancrivorus* (90%), *N. nasua* (100%), *P. flavus* (100%), *N. vison, M. martes, M. meles, V. vulpes, N. procyonoides, C. l. familiaris, A. microtis, C. thous* (30%), *C. thous*, L. gymnocercus* (20%), *F. catus** (100%), *F. catus, P. onca, P. concolor, P. concolor*, P. yagouaroundi* (57.3%), *L. geofffroyi* (50%), and *L. pardalis* (100%).
	C7	*P. cancrivorus* (10%), *N. nasua** (100%).
Axillaris	C6-C7	*P. cancrivorus* (100%), *N. nasua* (100%), *P. flavus* (80%), *A. microtis, C. thous* (20%), *V. vulpes, F. catus** (100%), *F. catus, P. onca, P concolor, P concolor*, P. yagouaroundi* (57.3%), and *L. geofffroyi* (50%)
Radialis	C6-C8	*P. cancrivorus* (20%) and *C. thous* (5%).
	C6-T1	*P. cancrivorus* (20%), *N. nasua* (60%), and *P. yagouaroundi* (14.3%).
	C6-T2	*P. cancrivorus* (60%), and *P. flavus* (90%).
	C7-T1	*N. nasua* (40%), *N. vison, M. martes, M. meles, C. l. familiaris, C. thous* (70%), *C. thous*, L. gymnocercus* (70%), *V. vulpes, N. procyonoides, F. catus, P. onca, P. concolor, P. concolor*, L. geoffroyi* (66.67%), and *P. yagouaroundi* (64.3%), and *L. pardalis* (100%).
Medianus	C6-T1	*N. nasua* (60%) and *P. cancrivorus* (20%).
	C6-T2	*P. cancrivorus* (50%), *N. nasua* (40%), and *P. flavus* (100%).
	C7-T2	*P. cancrivorus* (30%), *N. vison*, and *C. l. familiaris.*
Ulnaris	C8-T1	*P. cancrivorus* (20%), *N. nasua* (60%), *N. vison, M. meles, M. martes, N. procyonoides, C. l. familiaris, A. microtis, C. thous* (80%), *C. thous*, L. gymnocercus* (70%), *V. vulpes, A. australis, F. catus** (100%), *F. catus, P. onca, P concolor, P. yagouaroundi* (85.7%), *L. geoffroyi* (100%), and *L. pardalis* (100%).
	C8-T2	*P. cancrivorus* (80%), *N. nasua* (40%), *P. flavus* (100%), *N. vison, M. meles, N. procyonoides*, and *C. l. familiaris.*
Brachiocephalicus	C5-C6	*P. cancrivorus* (60%), *N. nasua* (80%), *P. flavus* (30%), *M. meles, N. procyonoides*, and *C. l. familiaris.*
	C5-C7	*P. cancrivorus* (10%).
	C6	*P. cancrivorus* (30%), *N. nasua* (20%), *P. flavus* (60%), *N. vison, M. martes, M. foina, C. l. familiaris, C. thous* (100%), *L. gymnocercus* (100%), *V. vulpes, P. yagouaroundi* (29.6%), and *L. geoffroyi* (33.34%).
	C6-C7	*P. cancrivorus* (10%) and *L. pardalis* (100%).
Thoracicus longus	C7	*P. cancrivorus* (100%), *N. nasua* (90%), *P. flavus* (100%), *M. meles, C. l. familiaris, C. thous* (100%), *L. gymnocercus* (100%), *V. vulpes, F. catus** (100%), *P. concolor*, P. yagouaroundi* (92.9%), and *L. geoffroyi* (100%).
	C7-C8	*N. nasua* (10%), *N. vison, M. martes, A. australis*, and *V. vulpes** (83.33%).
Thoracodorsalis	C6-C8	*P. cancrivorus* (20%), *N. nasua* (10%), and *P. flavus* (10%).
	C6-T1	*P. cancrivorus* (20%) and *N. nasua* (10%).
	C6-T2	*P. cancrivorus* (20%) and *P. flavus* (60%).
	C7-C8	*P. cancrivorus* (20%), *N. nasua* (80%), *M. foina, M. martes, M. meles, C. l. familiaris, C. thous* (5%), *L. gymnocercus* (50%), *V. vulpes, F. catus* *(80%), *P. concolor*, P. yagouaroundi* (7.1%), and *L. geoffroyi* (83.33%).
Thoracicus lateralis	C6-T2	*P. cancrivorus* (40%) and *P. flavus* (10%).
	C7-T1	*N. nasua* (10%), *C. thous* (25%), *L. gymnocercus* (35%), and *P. yagouaroundi* (14.2%).
	C8-T1	*P. cancrivorus* (20%), *N. nasua* (90%), *M. martes, N. procyonoides, C. l. familiaris, C. thous* (40%), *C. thous*, L. gymnocercus* (45%), *F. catus** (100%), *P. yagouaroundi* (50%), and *L. geoffroyi* (16.67%).
	C8-T2	*P. cancrivorus* (40%), *P. flavus* (90%), *N. vison, N. procyonoides.*
Pectorales craniales	C6-C7	*P. cancrivorus* (90%), *N. nasua* (40%), *P. flavus* (60%), *C. thous* (25%), *L. gymnocercus* (5%), *V. vulpes** (33.33%), *V. vulpes, P. yagouaroundi* (21.4%), and *L. geoffroyi* (16.67%).
	C7	*N. nasua* (60%), *P. flavus* (10%), *N. vison, M. meles, M. martes, M. foina, N. procyonoides, C. thous* (25%), *L. gymnocercus* (40%), *F. catus** (100%), *P. concolor*, P. yagouaroundi* (35.7%), and *L. geoffroyi* (50%).
	C7-T2	*P. cancrivorus* (10%).
Pectorales caudales	C6-T1	*P. cancrivorus* (10%) and *N. nasua* (10%).
	C6-T2	*P. cancrivorus* (50%) and *P. flavus* (20%).
	C7-T1	*N. nasua* (80%), *M. martes, N. vison, C. thous* (20%), and *L. gymnocercus* (35%).
	C7-T2	*P. cancrivorus* (30%).

References: *Procyonidae: P. cancrivorus* (present study), *N. nasua* (present study), *N. nasua** [48], and *P. flavus* [30]. *Mustelidae: M. foina* [43], *M. martes, N. vison*, and *M. meles* [31]. *Canidae: N. procyonoides, V. vulpes* [31], *V. vulpes** [22], *C. l. familiaris* [34], *A. microtis* [25], *C. thous* [45], *C. thous** [20], and *L. gymnocercus* [46]. *Otaridae: A. australis* [26]. *Felidae: F. catus, P. onca, P. concolor* [24,47], *F. catus** [28], *P. concolor** [27], *P. yagouaroundi* [23], *L. geoffroyi* [32], and *L. pardalis* [21].

The communicating branch of the musculocutaneous nerve with the median nerve at the elbow level was absent in all cases of *N. nasua*, which confirms other studies performed in the same species [48,58]. In another procyonid, *P. flavus*, it may be present but as a weak branch [30]. In *P. cancrivorus*, it is a large branch that is more a continuation of the musculocutaneous nerve together with the median nerve than a communicating branch. In one case of *C. l. familiaris*, it was reported that the musculocutaneous nerve did not join the median nerve but did not continue distally at the antebrachial region, replacing the median nerve to form common digital nerves I to III [58]. In other carnivorans, the unique communication at the elbow level is normal, as has been reported in canids [31,34,45,46,49,58], an otariid such as *A. australis* [26], and felids [23,24,32]. However, this communicating branch may be absent as a variation in felids [27–29,58]. Therefore, the main nerve of the brachial plexus that presents changes in carnivorans is the musculocutaneous nerve, mainly in the presence of communicating branches with the median nerve. In the case of *P. cancrivorus*, the musculocutaneous nerve was a large nerve with two communicating branches (*ansa axillaris* and *ramus comunicans*), which differs from most carnivorans, who only have one communicating branch.

The median nerve passes through the supracondylar foramen within vessels in *A. melanoleuca* [35], *P. cancrivorus*, and *N. nasua*, while it passes with the brachial artery in mustelids [31] and F. catus [42,61] and even with the brachial vein in *P. flavus* [30], *P. concolor* [24,27], and *P. onca* [24]. In *A. melanoleuca*, it has been reported that the median nerve also innervates the m. flexor carpi ulnaris [35]. However, when we observe the figures of the antebrachial muscles of *A. melanoleuca* presented by Davis [35], the caput humerale of the m. flexor carpi ulnaris he identified could actually be the palmaris longus lateralis found in *P. flavus* [39] or the palmaris longus of *P. cancrivorus* and *N. nasua*. It has a similar location in procyonids and is innervated by the median nerve. Therefore, palmaris longus or palmaris longus lateralis could have evolutionarily derived from m. flexor digitorum superficialis, not only from the last common ancestor of musteloids but also within arctoid carnivorans (based on the phylogenetic three of carnivorans [1]).

The ulnar nerve always innervates the m. anconeus medialis in both procyonids, similar to other carnivorans that present it [27,30,49,51,62]. This corroborates the fact that the muscle was evolutionarily derived from the m. flexor carpi ulnaris, as has been found in most mammals [63]. It may only be derived from the triceps brachii and flexor carpi ulnaris muscles in other phylogenetically distant mammals, such as xenarthrans, since in some of them, it is innervated by the radial and ulnar nerves [64]. Thus, it is not a head derived from the m. triceps brachii in carnivorans.

The branch from the ulnar nerve to innervate the carpal tactile hairs has only been reported in *F. catus* [29]. In *N. nasua*, the carpal tactile hairs were highly developed, while in *P. cancrivorus*, they were very small and short. The carpal tactile hairs in *N. nasua* should help to locate prey underground while digging, since these hairs may collect information about substrate vibration [65,66]. The carpal tactile hairs also collect information about surface properties related to substrate diameter, which is necessary to influence the body and thoracic limb postures while walking in several substrates [65]. However, the homologous branch from the ulnar nerve to the carpal tactile hairs is present in both species, similar to that reported in *P. flavus* [39]. Thus, although the carpal tactile hairs are more developed in *Nasua*, the distal fourth of the antebrachium remains innervated by the ulnar nerve in procyonids.

The caudal antebrachial cutaneous nerve originated independently from T1 or T1-T2, such as in *P. cancrivorus* and *N. nasua*, and has also been described in *P. flavus* [30], *F. catus* [29,51], and *A. melanoleuca* [35], while in other carnivoran species, it is a branch of the ulnar nerve. However, in a study that stimulated the dorsal roots of C5 to T2 in *C. l. familiaris*, the caudal antebrachial cutaneous nerve originated predominantly from T1 and T2 [67]. Therefore, the caudal surface of the antebrachium is predominantly innervated by the thoracic spinal nerves in these species of carnivorans.

The radial nerve innervated the m. brachialis in two cases of *P. cancrivorus*, similar to *A. melanoleuca* [35]; however, its main innervation is from the musculocutaneous nerve. Therefore, in these cases, the m. brachialis could embryologically have derived not only from the myotomes that migrate with the musculocutaneous nerve but also from the radial nerve. This could be a primitive phylogenetic arrangement within a last common ancestor of reptiles and mammals since, in some reptiles, a part of the radial nerve goes within a common trunk with the musculocutaneous nerve [68,69]. The branches to the craniolateral antebrachial muscles in carnivorans seem consistent [68,70–72], although the course of the deep branch of the radial nerve with respect to the m. supinator is variable. It passes superficially to the m. supinator in *P. flavus* [30] and *F. catus* [29,51] and passes deeply to the m. supinator in *C. l. familiaris* [55], *A. melanoleuca* [35], *P. cancrivorus,* and *N. nasua.*

The presence of an ansa pectoralis in *P. cancrivorus* and *N. nasua* has also been found in other procyonids, such as *P. flavus* [30] and *B. astutus* [35], and even in other arctoids, such as *A. fulgens, A. melanoleuca,* and *U. americanus* [35]. It could be a primitive arrangement due to the pectoralis superficialis and profundus muscles of mammals evolutionarily derived from an undivided pectoral muscle of reptiles [73]. Thus, the pectoral nerve to the reptilian pectoral muscle was divided into cranial and caudal pectoral nerves during the transition to mammals when the muscle divided into superficial and deep pectoral muscles; thus, communicating branches between both nerves may be present in carnivorans. The innervation of the m. pectoralis transversus from the caudal pectoral nerves has only been found in procyonids [30,37], which was further strengthened by findings in the present study with more specimens. This indicates that the m. pectoralis transversus in procyonids was derived not only from the m. pectoralis superficialis but also from the m. pectoralis profundus of the last common ancestor of procyonids [37].

The lateral thoracic nerve innervates m. cutaneus trunci in all carnivorans, although in some carnivorans, it also innervates m. pectoralis profundus, such as in *M. martes, V. vulpes* [31], *C. thous* [45], *L. gymnocercus* [46], *F. catus* [28], *P. yagouaroundi* [23], and *L. geoffroyi* [32]. In *P. cancrivorus* and *N. nasua,* it innervates the m. pectoralis abdominalis, which has been considered a part of the m. pectoralis profundus by some authors. However, in a recent study, it was hypothesized that the m. pectoralis abdominalis of carnivorans is an evolutionary derivation from m. cutaneus trunci of the last common ancestor of placental and marsupial mammals, but not from m. pectoralis profundus [37]. Even in a monotreme such as the platypus (*Ornithorhynchus anatinus*), both nerves also originate from a common trunk [74]. Therefore, the common trunk of the caudal pectoral and lateral thoracic nerves supports the hypothesis that the m. cutaneus trunci of mammals is derived from the pectoralis muscle of reptiles [73] and the m. pectoralis abdominalis from the m. cutaneus trunci in carnivorans [37].

4.3. Comparative Nerve Branching on the Manus of Carnivorans

The branching of the medial and lateral branches of the superficial radial nerve at the dorsum of the manus in *P. cancrivorus* and *N. nasua* is highly variable due to several communications between them. However, the communications between them in *P. cancrivorus* are by large branches of almost the same thickness as the main branch. In other carnivorans, both branches have a common pattern. Therefore, the medial branch forms the abaxial dorsal digital nerve I and the common dorsal digital nerve I, and the lateral branch forms the common dorsal digital nerves II to IV in canids, mustelids, and ursids [31,34,35,43,54]. In *P. cancrivorus* and *N.* nasua, the common dorsal digital nerves II and III originate from contributions of the lateral and medial branches of the superficial radial nerve. The medial cutaneous antebrachial nerve communicates with the medial branch of the radial nerve in felids [24,54] and procyonids, such as *N. narica* [54], *N. nasua,* and *P. cancrivorus.* Therefore, the axial and abaxial dorsal surfaces of digit I are innervated by the musculocutaneous and radial nerves in these species.

In *P. cancrivorus,* the most lateral branch forms the abaxial dorsal digital nerve IV, and this branch communicates with the axial dorsal digital nerve V, having a similar arrangement to that found in *P. flavus* [30] and *A. melanoleuca* [35]. In a feliform such as *F. catus*, the lateral branch only forms the dorsal common digital nerves II to III, while the dorsal branch of the ulnar nerve forms the dorsal common digital nerve IV [28,29,52]. However, a communicating branch from the common dorsal digital nerve III (radial nerve) to the common dorsal digital nerve IV (ulnar nerve) has been found in felids [29,54]. Similarly, it also occurs in mustelids [43,54], ursids, and procyonids such as *N. narica* [54] and *N. nasua*. In *P. onca*, the radial nerve does not innervate digit IV and is, thus, supplied axially and abaxially by the dorsal branch of the ulnar nerve [24]. In *P. concolor*, dorsal common digital nerve III may be formed by communication between the lateral branch of the radial nerve and the dorsal branch of the ulnar nerve [54].

The medial cutaneous antebrachial nerve forms the abaxial palmar digital nerve I in *P. flavus* [30], similar to that found in *N. nasua*. The abaxial palmar digital nerve I in *P. cancrivorus* is formed by the median nerve, which occurs normally in other carnivorans [24,28,29,31,34,43].

The ulnar nerve forms the palmar common digital nerve IV in canids [31,34], mustelids [31,43], procyonids such as *P. cancrivorus* and *N. nasua*, and a felid such as *F. catus* [24,28,29]. It is formed by the median nerve in a ursid as *A. melanoleuca* [35], in a procyonid as *P. flavus* [30], and as an anatomical variant in *N. nasua*. However, in *A. melanoleuca* [35] and *N. nasua*, it may receive a communicating branch from the ulnar nerve. In felids such as *P. concolor* and *P. onca*, the median nerve does not innervate digit IV since the axial and abaxial palmar surfaces are innervated by the ulnar nerve [24].

The branches of the ulnar and median nerves to the intrinsic muscles of the manus in *N. nasua* and *P. cancrivorus* were similar to those reported in *P. flavus* [30], *C. l. familiaris* [34], and felids [24]. However, the innervation by the median nerve to the most medial lumbrical muscle in both procyonids was not described in those species, although in *C. l. familiaris*, it is normal because the fourth lumbrical muscle is absent [55].

The large size of the palmar branches to the digits from the median, musculocutaneous, and ulnar nerves together with the large dorsal branches of the superficial branch of the radial nerve in *P. cancrivorus* corroborates the anatomical adaptation of the nervous peripheral system of the genus *Procyon* to allow a higher touch sensitivity than other carnivorans. Even in *P. lotor*, a high innervation has been found with the presence of all types of sensory receptors in similar proportions compared with other mammals, such as primates [75,76].

5. Conclusions

The brachial plexus of both procyonid species originates in a variant manner from C5-T1, C5-T2, C6-T1, or C6-T2, from which form the thoracic limb nerves having a similar distribution to the muscles and skin. The major nerves, such as the radial and median nerves, originated from more spinal nerves than other nerves in both procyonids, which could be an anatomical adaptation to increase the function of the antebrachial muscles and skin sensitivity of the antebrachium and manus. However, some specific differences could be associated with the locomotor behavior in each species. The higher contribution from C5 in *N. nasua* to the suprascapular and cranial subscapular nerves could be associated with a higher use of the scapular muscles to move the shoulder during digging. In addition, the innervation from the ulnar nerve to the sinusoids of the carpal tactile hairs should help to feel invertebrates underground and coordinate the movements in several terrestrial and arboreal substrates. On the other hand, *P. cancrivorus* has a higher contribution from T2, two communications from the musculocutaneous nerve to the median nerve, and relatively larger nervous branches to the skin. This arrangement could indicate that the thoracic limb nerves in *P. cancrivorus* have been more adapted to touch and greater manual skills than those in *N. nasua*, since *P. cancrivorus* uses its manus to bring the food to the mouth and even to search for food at night and underwater.

In general, through a comparative study with other species, the distribution and branching of the brachial plexus nerves also depends on the evolutive derivation of the muscles. The main nerve of the brachial plexus that presents changes in carnivorans is the musculocutaneous nerve, mainly due to the presence of communicating branches with the median nerve. The presence of ansa axillaris and ansa pectoralis in both procyonids mainly has a phylogenetic relationship within the infraorder Arctoidea since both structures can be in ursoid and musteloid species. The communicating branches between nerves also have a phylogenetic relationship not only with the common ancestor of carnivorans but also with other mammals and even reptiles. The dorsal sensory innervation of the manus is highly variable in most carnivorans due to the dorsal communicating branches and does not depend on the taxonomic family. However, in canids, it seems constant.

Author Contributions: Conceptualization, J.F.V.G., M.A.M. and R.A.d.C.B.; methodology, J.F.V.G., and M.A.M.; formal analysis, J.F.V.G.; investigation, J.F.V.G.; resources, J.F.V.G. and R.A.d.C.B.; data curation, J.F.V.G.; writing—original draft preparation, J.F.V.G.; writing—review and editing, J.F.V.G., M.A.M. and R.A.d.C.B.; visualization, J.F.V.G.; supervision, M.A.M. and R.A.d.C.B.; project administration, J.F.V.G., M.A.M. and R.A.d.C.B.; funding acquisition, J.F.V.G. and M.A.M. All authors have read and agreed to the published version of the manuscript.

Funding: This research was funded by the Universidad del Tolima (funding number 390116).

Institutional Review Board Statement: All experimental protocols were approved by the bioethics committee of the Universidad del Tolima (agreement number 2.3-059), Universidade de São Paulo (CEUAx agreement number 3928240820), and Universidade de Catalão (CEUA-UFCAT agreement number 01/22).

Data Availability Statement: The data presented in this study are available on request from the corresponding author. The data are not publicly available due to the high quantity of photos.

Acknowledgments: We want to thank the Universidad del Tolima, Universidade de São Paulo, Universidade Federal de Catalão, and CAPES (Coordenação de Aperfeiçoamento de Pessoal de Nível Superior, Bolsa No. 88887.685526/2022-00) for supporting this research. Thanks to CORTOLIMA and CORPOCALDAS for donating the cadavers, and Stephanie Delgado Puentes for the support during the first dissections of the project.

Conflicts of Interest: The authors declare no conflict of interest.

Appendix A

Table A1. Origins of the brachial plexuses in five specimens of *P. cancrivorus*.

Nerve	Pc1 (JF)		Pc2 (JF)		Pc3 (AM)		Pc4 (AM)		Pc5 (AM)	
	Right	Left	Right	Left	Right	Left	Right	Left	Right	Left
Brachial plexus	C6-T2	C6-T2	C6-T2	C6-T2	C6-T2	C6-T2	C6-T2	C6-T2	C5-T1	C5-T1
Suprascapularis	C6	C6	C6-C7	C6	C6	C6	C6	C6	C5-C7	C5-C6
Subscapularis cranialis	C6	C6	C6-C7	C6	C6	C6	C6	C6	C6-C7	C6-C7
Subscapularis medium	—	—	—	—	—	—	—	C6	—	—
Subscapularis caudalis	C6-C7	C6-C7	C6-C7	C6-C7	C6-C7	C6-C7	C6-C7	C6-C7	C6-C7	C6-C7
Musculocutaneus	C6-C7	C6-C7	C7	C6-C7	C6-C7	C6-C7	C6-C7	C6-C7	C6-C7	C6-C7
Axilaris	C6-C7	C6-C7	C6-C7	C6-C7	C6-C7	C6-C7	C6-C7	C6-C7	C6-C7	C6-C7
Radialis	C6-T2	C6-T2	C6-T2	C6-T2	C6-C8	C6-C8	C6-T2	C6-T2	C6-T1	C6-T1
Medianus	C6-T2	C6-T2	C7-T2	C6-T2	C6-T2	C6-T2	C7-T2	C7-T2	C6-T1	C6-T1
Ulnaris	C8-T2	C8-T2	C8-T2	C8-T2	C8-T2	C8-T2	C8-T2	C8-T2	C8-T1	C8-T1
Cutaneus antebrachii caudalis	C8-T2	C8-T2	C8-T2	C8-T2	C8-T2	C8-T2	T1-T2	T1-T2	T1	T1
Brachiocephalicus	C6	C6	C6-C7	C6	C5-C6	C5-C6	C5-C6	C5-C6	C5-C6	C5-C6
Thoracicus longus	C7	C7	C7	C7	C7	C7	C7	C7	C7	C7
Thoracicus lateralis	C6- T2	C6- T2	C6-T2	C6- T2	C8-T2	C8-T2	C8-T2	C8-T2	C8-T1	C8-T1
Thoracodorsalis	C7-T2	C7-T2	C7-C8	C7-C8	C6-C8	C6-C8	C6-T2	C6-T2	C6-T1	C6-T1
Pectorales craniales	C6-C7	C6-C7	C7-T2	C6-C7	C6-C7	C6-C7	C6-C7	C6-C7	C6-C7	C6-C7
Pectorales caudales	C6-T2	C6-T2	C7-T2	C6-T2	C6-T2	C6-T2	C7-T2	C7-T2	C6-T1	C6-T1

AM: adult male, JF: juvenile female, Pc#: number of the specimen, —: absent.

Table A2. Origins of the brachial plexuses in five specimens of *N. nasua*.

Nerve	Nn1 (IM)		Nn2 (JM)		Nn3 (AF)		Nn4 (AM)		Nn5 (AM)	
	Right	Left	Right	Left	Right	Left	Right	Left	Right	Left
Brachial plexus	C5-T2	C5-T2	C5-T2	C5-T1	C5-T1	C5-T2	C6-T1	C6-T1	C5-T2	C5-T2
Suprascapularis	C5-C6	C5-C6	C5-C6	C5-C6	C5-C6	C5-C6	C6-C7	C6	C5-C6	C5-C6
Subscapularis cranialis	C5-C6	C5-C6	C6	C6	C5-C6	C5-C6	C6-C7	C6	C5-C6	C5-C6
Subscapularis medium	C6	C6	C6-C7	C6-C7	C6-C7	C6-C7	—	—	C6	C6-C7
Subscapularis caudalis	C6-C7	C6-C7	C6-C7	C6-C7	C6-C7	C6-C7	C6-C7	C6-C7	C6-C7	C6-C7
Musculocutaneus	C6-C7	C6-C7	C6-C7	C6-C7	C6-C7	C6-C7	C6-C7	C6-C7	C6-C7	C6-C7
Axilaris	C6-C7	C6-C7	C6-C7	C6-C7	C6-C7	C6-C7	C6-C7	C6-C7	C6-C7	C6-C7
Radialis	C6-T1	C6-T1	C7-T1	C7-T1	C7-T1	C7-T1	C6-T1	C6-T1	C6-T1	C6-T1
Medianus	C6-T2	C6-T2	C6-T2	C6-T1	C6-T1	C6-T2	C6-T1	C6-T1	C6-T1	C6-T1
Ulnaris	C8-T2	C8-T2	C8-T1	C8-T2	C8-T1	C8-T2	C8-T1	C8-T1	C8-T1	C8-T1
Cutaneus antebrachii caudalis	C8-T2	C8-T2	T1	C8-T1	T1	T1	T1	T1	T1	T1
Brachiocephalicus	C5-C6	C5-C6	C5-C6	C5-C6	C5-C6	C5-C6	C6	C6	C5-C6	C5-C6
Thoracicus longus	C7	C7-C8	C7	C7	C7	C7	C7	C7	C7	C7
Thoracicus lateralis	C8-T1	C8-T1	C8-T1	C8-T1	C7-T1	C8-T1	C8-T1	C8-T1	C8-T1	C8-T1
Thoracodorsalis	C6-T1	C6-C8	C7-C8	C7-C8	C7-C8	C7-C8	C7-C8	C7-C8	C7-C8	C7-C8
Pectorales craniales	C6-C7	C6-C7	C7	C7	C7	C7	C7	C7	C6-C7	C6-C7
Pectorales caudales	C6-T1	C6-T1	C7-T1	C7-T1	C7-T1	C7-T1	C7-T1	C7-T1	C7-T1	C7-T1

AM: adult male, AF: Adult female, IM: infant male JM: juvenile male, Nn#: number of the specimen, —: absent.

References

1. Nyakatura, K.; Bininda-Emonds, O.R. Updating the Evolutionary History of Carnivora (Mammalia): A New Species-Level Supertree Complete with Divergence Time Estimates. *BMC Biol.* **2012**, *10*, 12. [CrossRef] [PubMed]
2. Reid, F.; Helgen, K.; González-Maya, J.F. Procyon Cancrivorus. Available online: https://www.iucnredlist.org/species/41685/45216426 (accessed on 19 December 2022).
3. Emmons, L.; Helgen, K. Nasua nasuai. Available online: https://www.iucnredlist.org/species/41684/45216227 (accessed on 19 December 2022).
4. Hassanin, A.; Veron, G.; Ropiquet, A.; Jansen van Vuuren, B.; Lécu, A.; Goodman, S.M.; Haider, J.; Nguyen, T.T. Evolutionary History of Carnivora (Mammalia, Laurasiatheria) Inferred from Mitochondrial Genomes. *PLoS ONE* **2021**, *16*, e0240770. [CrossRef]
5. Whiteside, D.P. Nutrition and Behavior of Coatis and Raccoons. *Vet. Clin. N. Am. Exot. Anim. Pract.* **2009**, *12*, 187–195. [CrossRef]
6. Arispe, R.; Venegas, C.; Rumiz, D. Abundancia y Patrones de Actividad Del Mapache (*Procyon cancrivorus*) En Un Bosque Chiquitano de Bolivia. *Mastozool. Neotrop.* **2008**, *15*, 323–333.
7. Hirsch, B. *Nasua nasua* (Ring-Tailed Coati). *CABI Compend.* **2022**. [CrossRef]
8. Gatti, A.; Bianchi, R.; Rosa, C.R.X.; Mendes, S.L. Diet of Two Sympatric Carnivores, *Cerdocyon thous* and *Procyon cancrivorus*, in a Restinga Area of Espirito Santo State, Brazil. *J. Trop. Ecol.* **2006**, *22*, 227–230. [CrossRef]
9. Quintela, F.M.; Iob, G.; Artioli, L.G.S. Diet of Procyon Cancrivorus (Carnivora, Procyonidae) in Restinga and Estuarine Environments of Southern Brazil. *Iheringia. Série Zool.* **2014**, *104*, 143–149. [CrossRef]
10. Alves-Costa, C.P.; Fonseca, G.; Christófaro, C. Variation in the Diet of the Brown-Nosed Coati (*Nasua nasua*) in Southeastern Brazil. *J. Mammal.* **2004**, *85*, 478–482. [CrossRef]
11. Beisiegel, B.M. Notes on the Coati, *Nasua nasua* (Carnivora: Procyonidae) in an Atlantic Forest Area. *Braz. J. Biol.* **2001**, *61*, 689–692. [CrossRef]
12. Hirsch, B.T. Seasonal Variation in the Diet of Ring-Tailed Coatis (*Nasua nasua*) in Iguazu, Argentina. *J. Mammal.* **2009**, *90*, 136–143. [CrossRef]
13. Campos, Z.; Mourão, G. Camera Traps Capture Images of Predators of Caiman Crocodilus Yacare Eggs (Reptilia: Crocodylia) in Brazil's Pantanal Wetlands. *J. Nat. Hist.* **2015**, *49*, 977–982. [CrossRef]
14. Desbiez, A.L.; Borges, P.A. Density, Habitat Selection and Observations of South American Coati Nasua Nasua in the Central Region of the Brazilian Pantanal Wetland. *Small Carniv. Conserv.* **2010**, *42*, 14–18.
15. Olifiers, N.; de Bianchi, R.C.; de Mourão, G.M.; Gompper, M.E. Construction of Arboreal Nests by Brown-Nosed Coatis, *Nasua nasua* (Carnivora: Procyonidae) in the Brazilian Pantanal. *Zoologia* **2009**, *26*, 571–574. [CrossRef]
16. Nowak, R.M. *Walker's Carnivores of the World*; The John Hopkins University Presss: Baltimore, MD, USA, 2005.
17. McClearn, D. Locomotion, Posture, and Feeding Behavior of Kinkajous, Coatis, and Raccoons. *J. Mammal.* **1992**, *73*, 245–261. [CrossRef]
18. Santos, C.M.D.; Santos, S.M.D.; Pizzutto, C.S.; Custódio, A.E.I. Enriquecimento Ambiental Para Guaxinim, *Procyon cancrivorus* (Cuvier, 1798). *Biosci. J.* **2015**, *31*, 275–282. [CrossRef]

9. Demczuk Thomas, L.; Piccoli, R.J.; Quintana Bernardi, P.E.; Sinotti, J.F.; Andrade Silva, V.; Fucks de Souza, C.; Bono Fukushima, F. Braquial Plexus Block and Lumbosacral Epidural in a South American Coati (*Nasua nasua*). *Acta Sci. Vet.* **2021**, *49* (Suppl. S1), 651. [CrossRef]
0. Pinheiro, L.L.; Branco, É.; Souza, D.C.; Pereira, L.H.C.; Lima, A.R. Descrição Do Plexo Braquial Do Cachorro-Do-Mato (*Cerdocyon thous* Linnaeus, 1766). *Ciência Anim. Bras.* **2014**, *15*, 213–219. [CrossRef]
1. Chagas, K.L.S.; Moura Fé, L.C.; Pereira, L.C.; Lima, A.R.; Branco, É. Descrição Morfológica Do Plexo Braquial de Jaguatirica (*Leopardus pardalis*). *Biotemas* **2014**, *27*, 171. [CrossRef]
2. Haligur, A.; Ozkadif, S. Macroanatomical Investigation of the Plexus Brachialis in the Red Fox (*Vulpes vulpes*). *Pak. J. Zool.* **2021**, *53*, 1617–1622. [CrossRef]
3. Souza Junior, P.; da Carvalho, N.C.; Medeiros-do-Nascimento, R.; de Dantas, P.O.; Bernardes, F.C.S.; Abidu-Figueiredo, M. Brachial Plexus Formation in Jaguarundi (*Puma yagouaroundi*). *Anat. Histol. Embryol.* **2022**, *51*, 746–755. [CrossRef]
4. Sánchez, H.L.; Silva, L.B.; Rafasquino, M.E.; Mateo, A.G.; Zuccolilli, G.O.; Portiansky, E.L.; Alonso, C.R. Anatomical Study of the Forearm and Hand Nerves of the Domestic Cat (*Felis catus*), Puma (*Puma concolor*) and Jaguar (*Panthera onca*). *Anat. Histol. Embryol.* **2013**, *42*, 99–104. [CrossRef] [PubMed]
5. Pinheiro, L.L.; Branco, É.R.; De Souza, D.C.; De Souza, A.C.B.; Pereira, L.C.; Lima, A.R. De Descrição Do Plexo Braquial Do Cachorro-Do-Mato-de-Orelhas-Curtas (*Atelocynus microtis*—Sclater, 1882): Relato de Caso. *Biotemas* **2013**, *26*, 203–209. [CrossRef]
6. De Souza, D.A.S.; de Castro, T.F.; Franceschi, R.D.C.; Silva Filho, R.P.; Pereira, M.A.M. Formação Do Plexo Braquial e Sistematização Dos Territórios Nervosos Em Membros Torácicos de Lobos-Marinhos *Arctocephalus australis*. *Braz. J. Vet. Res. Anim. Sci.* **2010**, *47*, 168–174. [CrossRef]
7. Barreto-Mejía, R.; Ceballos, C.P.; Tamayo-Arango, L.J. Anatomical Description of the Origin and Distribution of the Brachial Plexus to the Antebrachium in One Puma (*Puma concolor*) (Linnaeus, 1771). *Anat. Histol. Embryol.* **2022**, *51*, 153–159. [CrossRef] [PubMed]
8. Hakkı Nur, İ.; Keleş, H.; Pérez, W. Origin and Distribution of the Brachial Plexus of the Van Cats. *Anat. Histol. Embryol.* **2020**, *49*, 251–259. [CrossRef]
9. Roos, H.; Vollmerhaus, B. Konstruktionsprinzipien an Der Vorder- Und Hinterpfote Der Hauskatze (*Felis catus*). 4. Mitteilung: Muskelinnervation Und Bewegungsanalyse+. *Anat. Histol. Embryol.* **2005**, *34*, 2–14. [CrossRef]
0. Enciso-García, L.M.; Vélez-García, J.F. Origin and Distribution of the Brachial Plexus in Kinkajou (*Potos flavus*—Schreber, 1774). *Anat. Histol. Embryol.* **2022**, *51*, 221–235. [CrossRef]
1. Grzeczka, A.; Zdun, M. The Structure of the Brachial Plexus in Selected Representatives of the Caniformia Suborder. *Animals* **2022**, *12*, 566. [CrossRef]
2. Souza-Junior, P.; Wronski, J.G.; Carvalho, N.C.; Abidu-Figueiredo, M. Brachial Plexus in the *Leopardus geoffroyi*. *Ciência Anim. Bras.* **2018**, *19*, 1–14. [CrossRef]
3. International Committee on Veterinary Gross Anatomical Nomenclature. *Nomina Anatómica Veterinaria*, 6th ed.; World Association of Veterinary Anatomists, Ed.; World Association of Veterinary Anatomists: Hannover, Germany, 2017; ISBN 0-9600444-7-7.
4. Hermanson, J.; Evans, H.; De Lahunta, A. *Miller and Evan's Anatomy of the Dog*, 5th ed.; Hermanson, J.W., Evans, H., De Lahunta, A., Eds.; Elsevier Inc.: St. Louis, MO, USA, 2020; ISBN 978-0-323-54601-0.
5. Davis, D.D. The Giant Panda: A Morphological Study of Evolutionary Mechanisms. *Fieldiana* **1964**, *3*, 1–339.
6. Langworthy, O.R. The Panniculus Carnosus in Cat and Dog and Its Genetical Relation to the Pectoral Musculature. *J. Mammal.* **1924**, *5*, 49. [CrossRef]
7. Vélez García, J.F.; Miglino, M.A. Evolutionary Comparative Analysis of the Extrinsic Thoracic Limb Muscles in Three Procyonids (*Procyon cancrivorus* Cuvier, 1798, *Nasua nasua* Linnaeus, 1766, and *Potos flavus* Schreber, 1774) Based on Their Attachments and Innervation. *Anat. Sci. Int.* **2022**. [CrossRef] [PubMed]
8. Barone, R. *Anatomie Comparée Des Mammifères Domestiques. Tome 2: Arthrologie et Myoulogie*, 4th ed.; Association Centrale D'Entraide Vétérinaire: Paris, France, 2020; ISBN 978-2-9571960-1-2.
9. Perdomo-Cárdenas, V.; Patiño-Holguín, C.; Vélez-García, J.F. Evolutionary and Terminological Analysis of the Flexor Digitorum Superficialis, Interflexorii and Palmaris Longus Muscles in Kinkajou (*Potos flavus*) and Crab-Eating Racoon (Procyon Cancrivorus). *Anat. Histol. Embryol.* **2021**, *50*, 520–533. [CrossRef] [PubMed]
0. Allam, M.W.; Lee, D.G.; Nulsen, F.E.; Fortune, E.A. The Anatomy of the Brachial Plexus of the Dog. *Anat. Rec.* **1952**, *114*, 173–179. [CrossRef] [PubMed]
1. Aubert, L.; Carozzo, C.; Devillaire, A.-C.; Crevier-Denoix, N.; Moissonnier, P. Macro- and Microanatomical Characterization of the Cat Brachial Plexus. *Cells Tissues Organs* **2004**, *176*, 205–210. [CrossRef] [PubMed]
2. Singh, B. *Dyce, Sack and Wensing' Textbook of Veterinary Anatomy*, 5th ed.; Elsevier: Saint Louis, MO, USA, 2018.
3. Demiraslan, Y.; Aykut, M.; Özgel, Ö. Macroanatomical Characteristics of Plexus Brachialis and Its Branches in Martens (*Martes foina*). *Turk. J. Vet. Anim. Sci.* **2015**, *39*, 693–698. [CrossRef]
4. Backus, T.C.; Solounias, N.; Mihlbachler, M.C. The Brachial Plexus of the Sumatran Rhino (Dicerorhinus Sumatrensis) and Application of Brachial Plexus Anatomy Toward Mammal Phylogeny. *J. Mamm. Evol.* **2016**, *23*, 71–79. [CrossRef]
5. Souza-Junior, P.; Carvalho, N.C.; Mattos, K.; Santos, A.L.Q. Origens e Ramificações Do Plexo Braquial No Cachorro-Do-Mato *Cerdocyon thous* (Linnaeus, 1766). *Pesqui. Vet. Bras.* **2014**, *34*, 1011–1023. [CrossRef]

46. Souza-Junior, P.; da Cruz de Carvalho, N.; de Mattos, K.; Abidu Figueiredo, M.; Luiz Quagliatto Santos, A. Brachial Plexus in the Pampas Fox (*Lycalopex gymnocercus*): A Descriptive and Comparative Analysis. *Anat. Rec.* **2017**, *300*, 537–548. [CrossRef]
47. Silva, L.B.; Sánchez, H.L. La Inervación Del Miembro Torácico En Felinos. *AnAlectA Vet.* **2013**, *33*, 10–17.
48. Felipe, R.; Silva, F.; França, G.L.; Silva, E.M.; Leonel, L.C.; Carvalho-Barros, R.A.; Silva, D.C.; Silva, Z. Anatomia Descritiva Do Nervo Musculocutâneo Em Quatis (*Nasua nasua*, Linnaeus, 1766). *Enciclopédia Biosf.* **2014**, *10*, 92–98.
49. Vélez-García, J.F.; Patiño-Holguín, C.; Duque-Parra, J.E. Anatomical Variations of the Caudomedial Antebrachial Muscles in the Crab-Eating Fox (*Cerdocyon thous*). *Int. J. Morphol.* **2018**, *36*, 1193–1196. [CrossRef]
50. Reighard, J.; Jennings, H.S. *Anatomy of the Cat*; Henry Holt and Company: New York, NY, USA, 1901.
51. König, H.E. *Anatomie Der Katze: Mit Hinweisen Für Die Tierärztliche Praxis*; Gustav Fischer: Portland, OR, USA, 1992.
52. Hudson, L.C.; Hamilton, W.C. *Atlas of Feline Anatomy for Veterinarians*, 2nd ed.; Teton NewMedia: Jackson, MI, USA, 2010.
53. König, H.; Mülling, C.; Seeger, J.; Liebich, H. Nervous System (Systema Nervosum). In *Veterinary Anatomy of Domestic Animals Textbook and Colour Atlas*; König, H.E., Liebich, H.G., Eds.; Thieme: Stuttgart, Germany, 2020; pp. 515–586.
54. Arłamowska-Palider, A. Morphological Studies on the Main Branches of the Radial Nerve in Mammals. *Acta Theriol.* **1970**, *15*, 185–197. [CrossRef]
55. Hermanson, J. The Muscular System. In *Miller's Anatomy of the Dog*; Hermanson, J.W., Evans, H., De Lahunta, A., Eds.; Elsevier Inc.: St. Louis, MO, USA, 2020; pp. 444–658. ISBN 978-0-323-54601-0.
56. Kamali, Y. Aberrant Arrangement of the Musculocutaneous and Median Nerves in the Thoracic Limbs of a Mixed-breed Dog Cadaver. *Anat. Histol. Embryol.* **2022**, *51*, 419–423. [CrossRef]
57. Vélez García, J.F.; Ospina Orozco, A.; Duque Parra, J.E. Origen de Los Nervios Del Plexo Braquial Del Venado Coliblanco (*Odocoileus virginianus*) En Comparación Con Otros Rumiantes. *Rev. Investig. Vet. Del Perú* **2018**, *29*, 713–722. [CrossRef]
58. Arłamowska-Palider, A. Comparative Anatomical Studies of Nervus Musculocutaneus in Mammals. *Acta Theriol.* **1970**, *15*, 343–356. [CrossRef]
59. Kawashima, T.; Thorington, R.W.; Bohaska, P.W.; Chen, Y.-J.; Sato, F. Anatomy of Shoulder Girdle Muscle Modifications and Walking Adaptation in the Scaly Chinese Pangolin (*Manis pentadactylapPentadactyla*: Pholidota) Compared with the Partially Osteoderm-Clad Armadillos (Dasypodidae). *Anat. Rec.* **2015**, *298*, 1217–1236. [CrossRef]
60. Upham, N.S.; Esselstyn, J.A.; Jetz, W. Inferring the Mammal Tree: Species-Level Sets of Phylogenies for Questions in Ecology, Evolution, and Conservation. *PLoS Biol.* **2019**, *17*, e3000494. [CrossRef]
61. Liebich, H.G.; Maierl, J.; König, H.E. Forelimbs or Thoracic Limbs (Membra Thoracica). In *Veterinary Anatomy of Domestic Animals Textbook and Colour Atlas*; König, H.E., Liebich, H.G., Eds.; Georg Thieme Verlag: Stuttgart, Germany, 2020; pp. 171–242. ISBN 978-3-13-242933-8.
62. Windle, B.; Parsons, F. Myology of the Terrestrial Carnivora. Part I. Muscles of the Head, Neck, and Fore-Limb. *Proc. Zool. Soc Lond.* **1897**, *65*, 370–409. [CrossRef]
63. Abdala, V.; Diogo, R. Comparative Anatomy, Homologies and Evolution of the Pectoral and Forelimb Musculature of Tetrapods with Special Attention to Extant Limbed Amphibians and Reptiles. *J. Anat.* **2010**, *217*, 536–573. [CrossRef]
64. Vélez-García, J.F.; Arbeláez-Quiñones, A.C.; Montealegre-Hurtado, K.D. Evolutionary Adaptations in the Flexor Digitorum Profundus Muscle in *Tamandua mexicana* (Xenarthra, Myrmecophagidae). *Anat. Rec.* **2021**, *304*, 758–770. [CrossRef] [PubMed]
65. Niederschuh, S.J.; van Beesel, J.; Schmidt, M. The Role of Sensory Feedback from Carpal Sinus Hairs in Locomotor Kinematics of Rats (*Rattus norvegicus*, Rodentia) during Walking on Narrow Substrates. *Zoology* **2022**, *155*, 126055. [CrossRef] [PubMed]
66. Fundin, B.T.; Arvidsson, J.; Rice, F.L. Innervation of Nonmystacial Vibrissae in the Adult Rat. *J. Comp. Neurol.* **1995**, *357*, 501–512 [CrossRef] [PubMed]
67. Bailey, C.S.; Kitchell, R.L.; Johnson, R.D. Spinal Nerve Root Origins of the Cutaneous Nerves Arising from the Canine Brachial Plexus. *Am. J. Vet. Res.* **1982**, *43*, 820–825. [PubMed]
68. Numata, N.; Kida, M.Y.; Kudoh, H. Ramification Patterns of the Nerves Innervating the Forearm Extensors in Mammals and Reptiles. *Okajimas Folia Anat. Jpn.* **1996**, *73*, 55–63. [CrossRef] [PubMed]
69. Koizumi, M. Comparative Anatomy of the Subscapularis, Teres Major and Latissimus Dorsi Muscles from Salamanders to Mammals with Special Reference to Their Innervations from the Brachial Plexus. *Anat. Sci. Int.* **2022**, *97*, 124–137. [CrossRef]
70. Vélez-García, J.F.; Chunganá-Caicedo, D.; Saavedra-Montealegre, S. Gross Anatomy of the Craniolateral Antebrachial Muscles in Kinkajou (*Potos flavus*, Carnivora): Intra- and Interspecific Variants within the Family Procyonidae. *Anat. Histol. Embryol.* **2022**, *51*, 308–313. [CrossRef]
71. Vélez-García, J.F.; Marín-González, L.; Monroy-Cendales, M.J.; Miglino, M.A. Craniolateral Forearm Muscles of the Crab-Eating Raccoon (*Procyon cancrivorus*) and a Comparative Review with Other Carnivorans. *Iheringia. Série Zool.* **2022**, *112*, e2022012. [CrossRef]
72. Echeverry, J.; Vélez, J.; Sánchez, C. Descripción Anatómica de Los Músculos Cráneo-Laterales Superficiales Del Antebrazo Del Zorro Perruno (*Cerdocyon thous*). *Rev. Colomb. Cienc. Anim.* **2015**, *8*, 44–51.
73. Diogo, R.; Abdala, V. *Muscles of Vertebrates: Comparative Anatomy, Evolution, Homologies and Development*, 1st ed.; CRC Press: Enfield, NH, USA, 2010; ISBN 9781578086825.
74. Koizumi, M.; Sakai, T. On the Morphology of the Brachial Plexus of the Platypus (*Ornithorhynchus anatinus*) and the Echidna (*Tachyglossus aculeatus*). *J. Anat.* **1997**, *190*, 447–455. [CrossRef]

3. Turnbull, B.G.; Rasmusson, D.D. Sensory Innervation of the Raccoon Forepaw: 1. Receptor Types in Glabrous and Hairy Skin and Deep Tissue. *Somatosens. Res.* **1986**, *4*, 43–62. [CrossRef] [PubMed]
›. Munger, B.L.; Pubols, L.M. The Sensorineural Organization of the Digital Skin of the Raccoon (Part 1 of 2). *Brain. Behav. Evol.* **1972**, *5*, 367–379. [CrossRef] [PubMed]

animals

MDPI

Article

The Respiratory System of the *Arctocephalus australis* in Comparison to the Dog as a Land-Carnivore: Are There Adaptations to Marine Life?

Ignacio Molpeceres-Diego [1], Rosario Martín-Orti [2], Juan-Pablo Loureiro [1], Carlos Tostado-Marcos [1], Enrique Tendillo-Domínguez [1], Inmaculada Santos-Álvarez [2], Pilar Pérez-Lloret [2] and Juncal González-Soriano [2,*]

[1] Fundación Mundo Marino, Av. X 157, San Clemente del Tuyú B7105, Provincia de Buenos Aires, Argentina
[2] Departamento de Anatomía y Embriología, Sección Departamental de Anatomía y Embriología Veterinaria, Facultad de Veterinaria, Universidad Complutense, Avenida Puerta de Hierro s/n, 28040 Madrid, Spain
* Correspondence: juncalgs@vet.ucm.es

Simple Summary: It is normally recognized that anatomy is crucial for an improved knowledge of many physiological adaptations. In the case of marine mammals, their ability to dive for long periods of time is especially striking. They spend most of their lives in water, although in general terms, they behave and have characteristics very similar to those of land mammals. Our results demonstrate that, due to its capacity to stay under water, the respiratory apparatus of the South American fur seal shows specific characteristics. Yet, in general terms, being a carnivore seems to be the main characteristic of this species, and the adaptations to the aquatic environment being less important.

Abstract: Marine mammals are divided into three groups, with similar adaptations resulting from their aquatic lifestyle: sirenians, pinnipeds, and cetaceans. The present work focused on the South American fur seal, or *Arctocephalus australis*, a carnivore included in the pinnipeds group. We assessed whether the anatomical features of the *Arctocephalus australis'* respiratory system are comparable to those of other land-carnivores or whether these individuals show anatomical adaptations related to their ability to dive or their breath-holding capacities. We studied 11 cadavers of *Arctocephalus australis*, which included adult ($n = 2$) and juvenile ($n = 9$) individuals, by anatomically dissecting their isolated entire respiratory system. Although it is generally similar to that in land-carnivores, we demonstrated that the *Arctocephalus australis*'s respiratory apparatus shows several specific characteristics. Therefore, our results are of great importance for clinical diagnostic and wildlife conservation purposes.

Keywords: *Arctocephalus australis*; anatomy; respiratory system; land-carnivore; dog

1. Introduction

Marine mammals have evolved from common ancestors that changed from a terrestrial life to an aquatic environment. Yet, aquatic life presents numerous challenges for those mammals originally adapted for life on land. Most of them develop their life, at least in part, in an aquatic environment; therefore, it is supposed that some of their morphological and physiological characteristics are linked to an organized evolution to marine-life adaptation.

However, despite all these changes, marine mammals still share some basic characteristics with terrestrial mammals, such as breathing through lungs, being warm-blooded, having fur (periodically in life), and producing milk to nurse their young [1–3]. As marine mammals do not breathe in water, modifications to the respiratory tract are needed to protect a system designed to function on land. Many of these adaptations have already been described, for example, valvular nostrils that exclude water, or an intranarial larynx in odontoceti that, while swallowing, further protects the respiratory tract from water inundation [4]. In addition, diving presents additional challenges, as ambient pressure rises with depth, potentially causing the lungs to collapse [5].

The present study focused on the South American fur seal or *Arctocephalus australis*, an otariid included in the taxonomic family of pinnipeds (Order Carnivora), such as the dog or the cat [6,7]. Surprisingly, although it is generally accepted that anatomy is crucial to a better understanding of different physiological adaptations [2], and although studies on the physiology of diving are numerous [1,8–19], only a small number of them analyze the anatomical characteristics of marine mammals [5,20–24], and even fewer are focused on the anatomy of the respiratory system [2,3,25–27].

In the South American fur seal, its capability to remain submerged for some time is well known. However, previous studies on the anatomy of the South American fur seal are scarce. Thus, the possible anatomical adaptations of its respiratory system are yet to be described. In other words, the question remains whether the respiratory system of the *Arctocephalus australis* is similar to that in other carnivores or if there are significant anatomical adaptations that allow the South American fur seal to dive.

In this work, we tried to answer this question. Thus, the objective of the present paper was to perform an anatomical, systematic, and descriptive study of the South American fur seal's respiratory system and then compare it to the respiratory apparatus of another carnivore model, the dog.

2. Materials and Methods

The study was performed on 11 cadavers of the species *Arctocephalus australis*, including adult (n = 2) and juvenile (n = 9) individuals. All animals were found stranded on the coast and taken to the Rescue and Rehabilitation Center Fundación Mundo Marino (FMM), where they died. It is important to highlight that none of the individuals were euthanized. They were all frozen until their dissection. After thawing for a period of 24/48 h, animals were placed in dorsal recumbency and dissected by veterinarians.

As a first step, the skin and part of the blubber and muscle were removed. With a scalpel blade, we made a first incision under the chin and continued down the midline (*linea mediana ventralis*) to the xiphoid cartilage (*cartilago xiphoidea*). To make the internal tissues visible, we extended the cut laterally along the costal arch.

Before cutting the ribs, we perforated the diaphragm. To open the thoracic cavity, we cut through each thoracic rib mid-articulation (*genu costae*), which is the cartilaginous flex point, and removed the rib cage..

For exposing the trachea and esophagus, the superficial musculature of the neck area (*regio collis ventralis*) was removed. Then, we made a triangular incision along the inner sides of the mandible body. The hyoid apparatus was cut, and the tongue was extracted through the intermandibular space. The respiratory system was then progressively separated.

As the present study included both juvenile and adult individuals, each with a different degree of development, the corresponding arithmetic means of the data obtained were calculated separately.

Pictures were taken with a Canon EOS 500D (Tokyo, Japan) camera equipped with an 18–55 mm lens.

As a complementary study, we prepared additional material:

- Sample 1. An exhaustive washout period of the trachea, bronchi and lungs was performed by leaking liquid for several days. Once cleaned, the empty ducts were filled by injecting compressed air.
- Sample 2. To obtain a template of the whole bronchial tree, expansive polyurethane was inserted through the trachea. Once polyurethane was expanded and cured, the sample was incubated with flies and larvae for tissue digestion to obtain the polymeric template of the lower respiratory tract.

3. Results

3.1. Nose and Nares

The nose is completely coated with hairy skin except the apex, where nares are placed and joined with the upper lip to give a higher functional mobility. The bones of the dorsal and lateral nasal walls are formed by the facial bones, except in the rostral region. This rostral region contains the dorsal and ventral lateral nasal cartilages. They are the bilateral extension of the dorsal and ventral borders of the cartilaginous nasal septum. The nares are placed in the nose apex. They are "boomerang"-shaped openings (Figure 1). The nose apex, the nasal mimetic muscles, and the nares are fixed by a cartilaginous skeleton. The surrounding skin, as well as the skin between the nares, is hairless and pigmented. It forms the nasal plane, which presents small superficial furrows. The middle furrow or philtrum extends all along the nasal plane, dividing it partially and not reaching the upper lip (Figure 1).

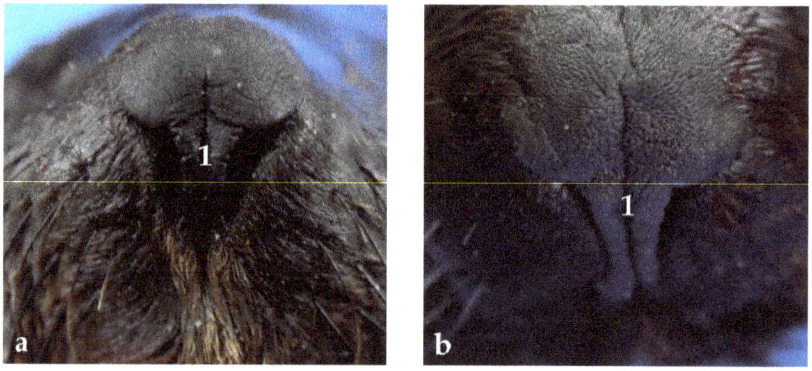

Figure 1. Frontal view of the nares: opened (**a**) and slightly closed (**b**). Note the presence of different furrows, (1) middle furrow or philtrum.

3.2. Nasal Cavity

The nasal cavity of the South American fur seal is narrow and long. It is divided into two halves (right and left) by the nasal septum. In both halves, the ventral part is bony, whereas the dorsal and rostral parts are cartilaginous. The perpendicular plate of the ethmoid bone and the vomer forms the bony portion. Caudally, it does not reach the lower part of the nasal cavity; thus, both nasal cavity halves are communicated. The cartilaginous segment expands from the perpendicular plate of the ethmoid bone to the incisive bones.

Caudally to the nasal cavity, the choanae connects the nasal cavity to the nasopharynx. Each nasal cavity is mainly filled by the nasal conchae (dorsal, middle, and ventral conchae), extending from the walls until nearly the nasal septum. The dorsal nasal conchae are relatively long and narrow, with a smooth surface, whereas the ventral is shorter and wider and forms little parallel folds (*plicae parallelis*). The medial nasal conchae are placed in between the other two.

The conchae divide each nasal cavity into subsequent parts called meatus. The dorsal nasal meatus is narrow and short. It is placed between the dorsal nasal conchae and the roof of the nasal cavity. The middle nasal meatus is located between the dorsal and ventral nasal conchae. It is also short and narrow. The ventral nasal meatus is longer than the other two. It extends from the bottom of the nasal cavity to the nasal ventral conchae and leads into the nasopharynx. Therefore, the ventral meatus represents a direct passage of air into the larynx and lower respiratory tract. The common nasal meatus is the narrow space between the nasal septum and the conchae, and it continues laterally with the other meatuses.

In fact, the choanae represent a direct path from the nasal cavity into the nasopharynx. They are bounded rostrally by the horizontal portion of the palatine bone, laterally by the pterygoid bone and the perpendicular plate of the palatine bone, and caudally by the vomer bone. Interestingly, there are no paranasal sinuses in the *Arctocephalus australis* (Figure 2).

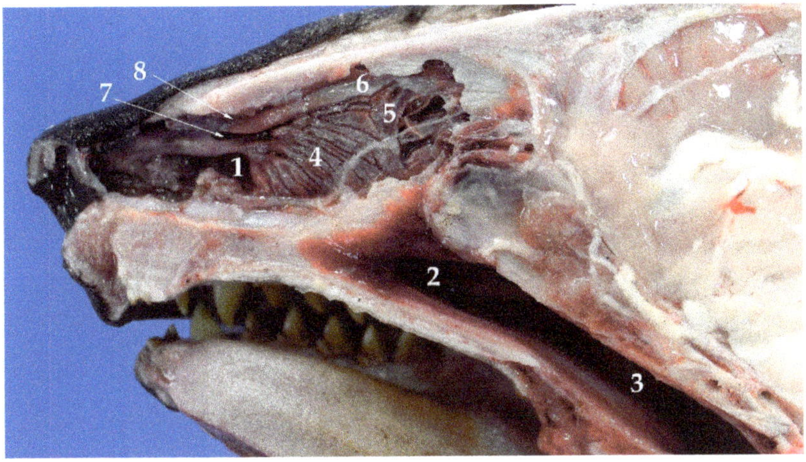

Figure 2. Lateral view of the left nasal cavity: (1) ventral nasal meatus; (2) choana; (3) nasopharynx; (4) ventral nasal conchae with parallel folds (*plicae parallelis*); (5) middle nasal conchae; (6) dorsal nasal conchae; (7) middle nasal meatus; (8) dorsal nasal meatus.

3.3. Pharynx

The pharynx is divided into the oropharynx, nasopharynx, and laryngopharynx. The most noteworthy detail is the length of the soft palate, with a significant amount of peripharyngeal soft tissue.

3.4. Larynx

The larynx is relatively short and wide (Figure 3). It is located underneath the skin, ventral to the CI and CII vertebral body. It is dorsally related to the pharynx and the esophagus, ventrally to the sternohyoid muscle, and laterally to the sternothyroid muscle and the mandibular gland. The cartilages are cricoid, thyroid, arytenoid, and epiglottis (Figure 4).

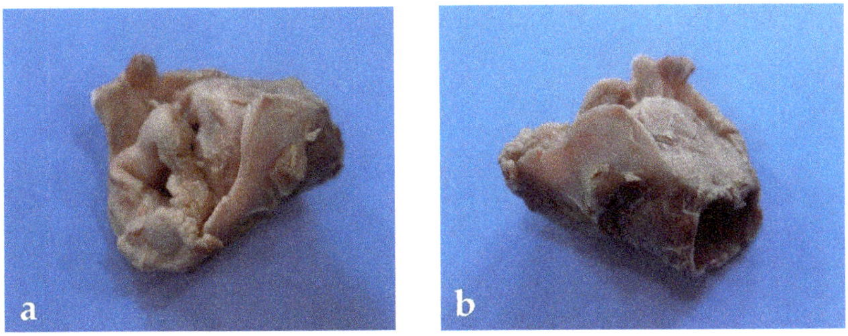

Figure 3. Larynx of *Arctocephalus australis*: (**a**) oblique view of the rostral larynx; (**b**) oblique view of the caudal larynx.

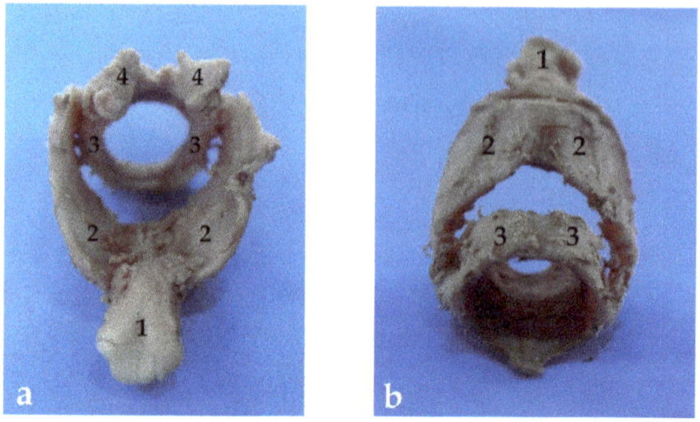

Figure 4. Larynx cartilaginous structure of *Arctocephalus australis*: (**a**) rostral view, (**b**) caudal-ventral view, (1) epiglottis, (2) thyroid cartilage, (3) cricoid cartilages, (4) arytenoid cartilages.

3.5. Trachea

The trachea is a membranous and cartilaginous flexible duct that forms the proximal part of the lower respiratory airways. It extends from the larynx to the tracheal bifurcation (*bifurcate tracheae*). In young individuals, the trachea reaches the area below the sternal manubrium. On the other hand, it does not go that far in adults. Additionally, in adults, the trachea splits into two main bronchi (left and right) immediately cranial to the tip of the manubrium (Figure 5a). In the South American fur seal, the trachea shows only a cervical portion, whereas it is not possible to identify a thoracic portion. The tracheal length average is around 8 cm in juveniles and 13 cm in adults, whereas the width in juveniles has a value of approximately 2 cm and 3 cm in adults (Table 1).

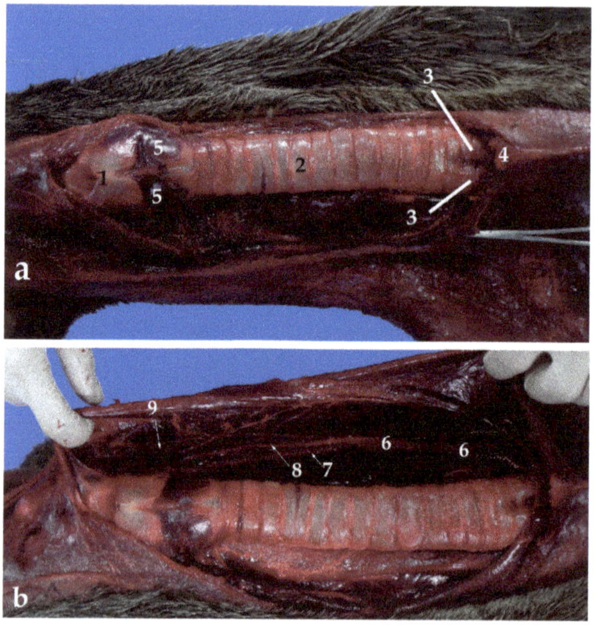

Figure 5. Dorsal view of the neck: (**a**) superficial view, (**b**) deeper view with the carotid sheath in detail, (1) larynx; (2) trachea; (3) bronchi; (4) sternal manubrium; (5) thyroid gland; (6) carotid sheath (opened); (7) common carotid artery; (8) vagosympathetic trunk; (9) internal jugular vein.

Table 1. Morphometric data of the trachea from the 11 individuals of *Arctocephalus australis* investigated in the present study. A: adult; J: juvenile.

ID	Age Group	Sex	Length (cm)	Width (cm)	Lumen Tracheal Diameter (cm)	Tracheal Circumference (cm)	N° Cartilaginous Rings
M9719	J	F	6.9	1.6	1.2	5.4	12 + 1
M7419	J	M	7.3	1.7	1.3	5.6	12 + 1
M7019	J	M	7.8	1.9	1.4	5.9	13 + 1
M8419	J	F	8	1.8	1.4	6	14 + 1
M8319	J	M	8	2	1.5	6	13 + 1
M10019	J	F	8.1	2	1.5	6.1	12 + 1
M7919	J	M	8.3	2	1.5	6.2	12 + 1
M8219	J	F	8.7	2.1	1.6	6.2	12 + 1
M10419	J	M	9.1	2.2	1.7	6.2	12 + 1
M7319	A	F	12.8	2.7	2.4	6.5	13 + 1
M9919	A	F	14	3	2.5	6.7	12 + 1
Mean	J	-	8	1.9	1.5	6	12–14
Mean	A	-	13.4	2.9	2.5	6.6	12–14

As mentioned before, the trachea of the *Arctocephalus australis* has only a cervical portion, practically in the midline (Figures 5 and 6). The common carotid artery, the internal jugular vein, and the vagosympathetic trunk are included in the carotid sheath. The recurrent laryngeal nerve is located on the dorsolateral portions of the trachea, either close to its dorsolateral side, in the cranial half of the neck, or close to its lateral side, in the caudal half of the neck. (Figure 5b). The lobes of the thyroid gland and the small parathyroid glands are related to the lateral aspects of the cranial part of the trachea. The trachea is also related to different lymphatic structures and to the external jugular vein, at the thoracic entrance.

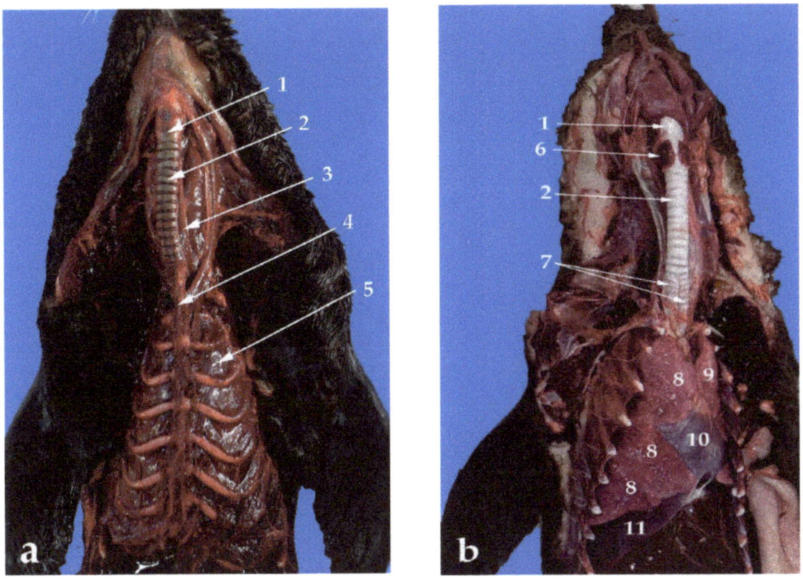

Figure 6. Ventral view of the neck and the thorax (**a**) and the neck and thoracic and abdominal cavities (**b**): (1) larynx; (2) trachea; (3) esophagus; (4) sternal manubrium; (5) thorax; (6) thyroid gland; (7) main bronchi; (8) right lung; (9) left lung; (10) heart; (11) diaphragm.

There are 12–14 incomplete cartilaginous rings, U-shaped, that constitute the trachea of the South American fur seal, plus one more as a transition between the trachea and the bronchi (Table 1). Their dorsal part is free but attached by tracheal muscles that form the dorsal part of the tracheal wall. It is known as the membranous portion (Figure 7).

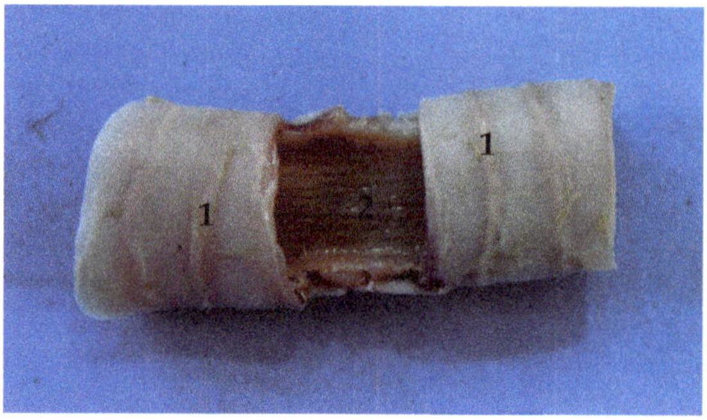

Figure 7. Trachea section, ventral view: (1) cartilaginous rings; (2) membranous part.

The diameter of the tracheal lumen is 1.5 cm in juveniles and 2.5 cm in adults, whereas the tracheal circumference is about 6 cm in the former and 6.8 cm in the latter (Table 1).

3.6. Bronchi

As previously described, the trachea of the South American fur seal early bifurcates into the two main bronchi at the level of the caudal third of the neck, practically at the sternal manubrium. The right and left main bronchi are in close contact with some of the great thoracic vessels, such as the cranial vena cava or the aortic arch. The first branch is the right main bronchus (Figure 8).

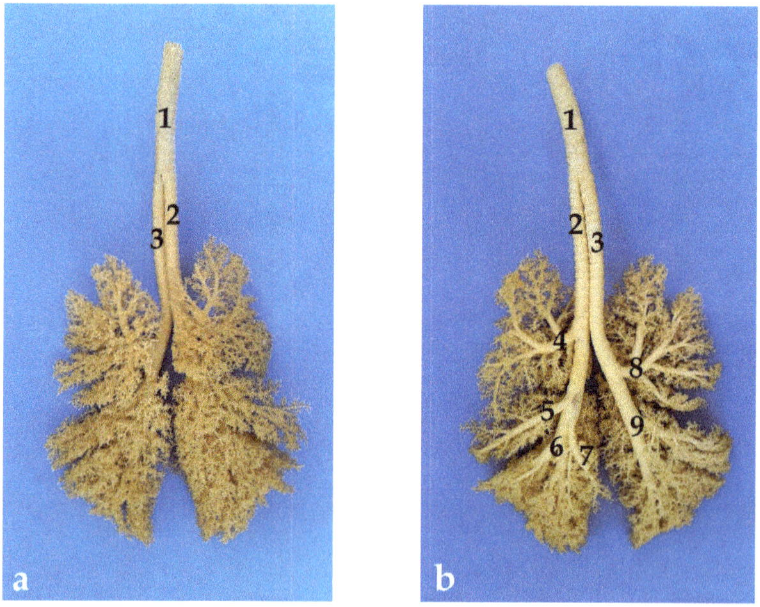

Figure 8. Polyurethane bronchial tree: (**a**) dorsal view; (**b**) ventral view; (1) trachea; (2) right main bronchus; (3) left main bronchus; (4) right cranial lobar bronchus; (5) right mid lobar bronchus; (6) right caudal lobar bronchus; (7) accessory lobar bronchus; (8) left cranial lobar bronchus; (9) left caudal lobar bronchus.

After traveling about 8 cm in juveniles and about 14 cm in adults, a second bronchial branch emerges. It is the right cranial lobar bronchus, which ventilates the cranial lobe of the right lung. This right main bronchus continues traveling about 5 cm more, inside the thoracic cavity, until the lung hilus, where the other three branches of the right main bronchus arise almost simultaneously.

The first of these three branches is the so-called right middle lobar bronchus, which ventilates the middle lobe of the right lung. This bronchus extends throughout the entire middle lobe, branching into several collateral bronchi (*bronchi segmentalis*). Immediately after the right middle lobar bronchus, the caudal right lobar bronchus emerges, which ventilates the caudal right lobe. The right accessory lobar bronchus appears practically at the same level, which enters the accessory lobe of the right lung.

The left main bronchus gives rise only to two branches, both located at the level of the pulmonary hilus. This second left bronchial generation arises to about 11 cm from the tracheal bifurcation in juveniles and approximately 16 cm in adults, cranially to the triple branching of the right main bronchus. The left main bronchus branches into two secondary bronchi: the left cranial bronchus and the left caudal bronchus. The left cranial bronchus ventilates the left cranial lobe, while the left caudal bronchus ventilates the left caudal lobe. Each of these secondary branches divide further into tertiary bronchi or segmental bronchi, which form the bronchioles (arbor alveolaris), ending in the alveolar sacs (Figure 8).

As mentioned before, the right main bronchus is slightly longer than the left one (Figure 9), which means that it has more cartilaginous rings, between 16 and 18 until the exit of the first branch of the right cranial bronchus, and another 7 or 8 more rings before the triple branching described previously (Table 2). The left main bronchus has between 19 and 22 cartilaginous rings up to the region where it divides into the two left secondary bronchi (Table 2). From this point, the diameter and width of the bronchi decrease accordingly until the alveolar sacs (Table 2). The cross-sectional area of each bronchus is approximately half of the tracheal area. Thus, the total cross-sectional area of the two primary bronchi is practically equal to the cross-sectional area of the trachea.

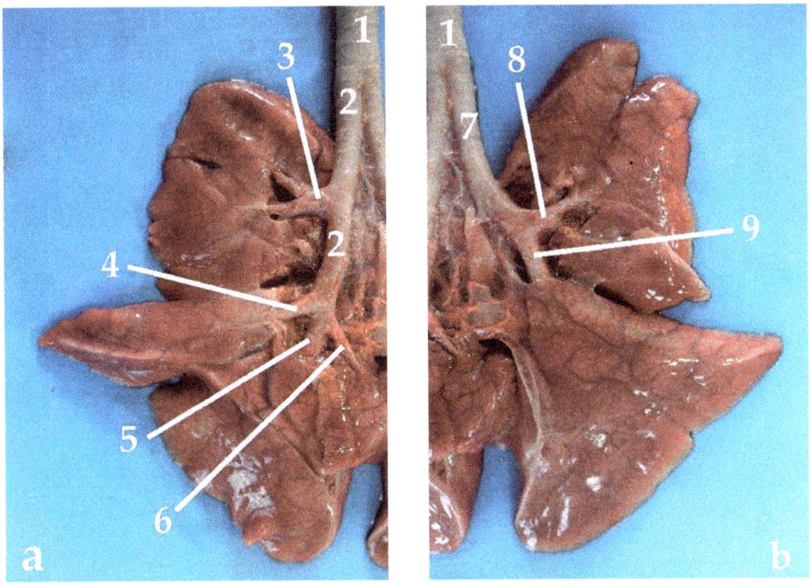

Figure 9. Ventral view of the lung: (**a**) right lung; (**b**) left lung; (1) trachea; (2) right main bronchus; (3) right cranial lobar bronchus; (4) right mid lobar bronchus; (5) right caudal lobar bronchus; (6) accessory lobar bronchus; (7) left main bronchus; (8) left cranial lobar bronchus; (9) left caudal lobar bronchus.

Table 2. The morphometric data of the bronchi from the 11 individuals of *Arctocephalus australis* investigated in the present study. A: adult; J: juvenile.

ID	Age Group	Sex	Length (cm)		Width (cm)	Lumen Bronchus Diameter (cm)	Bronchus Circumference (cm)	N° Cartilaginous Rings	
			Right	Left				Right	Left
M9719	J	F	6.4 + 4.7	9.2	1.1	0.9	3.5–3.2	17 + 8	20
M7419	J	M	6.8 + 5	9.5	1.1	1	3.7–3.5	18 + 8	22
M7019	J	M	7 + 5	9.8	1.1	1	3.7–3.5	18 + 7	21
M8419	J	F	7 + 5	10	1.1	1	3.7–3.5	16 + 7	19
M8319	J	M	7.4 + 5.1	10.2	1.1	1	3.7–3.5	18 + 7	21
M10019	J	F	8.4 + 5.2	10.6	1.1	1	3.8–3.6	17 + 8	20
M7919	J	M	9 + 5.2	11	1.2	1.1	3.8–3.6	18 + 8	21
M8219	J	F	9.3 + 5.4	11.8	1.2	1.1	3.9–3.7	16 + 7	20
M10419	J	M	9.8 + 5.4	12.4	1.3	1.2	3.9–3.7	17 + 7	20
M7319	A	F	13 + 6	15.9	1.4	1.3	4.1–3.9	17 + 8	21
M9919	A	F	14 + 6.4	17.6	1.6	1.5	4.3–4	18 + 8	22
Mean	J	-	7.9 + 5	10.5	1.1	1	3.7	16–18 + 7/8	20–22
Mean	A	-	14 + 6.2	16.7	1.5	1.4	4.2	16–18 + 7/8	20–22

Width, lumen bronchus diameter, and bronchus circumference are equal in both bronchi, right and left.

3.7. Lungs

There are two lungs, one right and one left (Figure 10), that occupy the largest part of the thoracic cavity (Figure 11). In the case of the South American fur seal, they approach the level of the 9th rib. Each lung is covered by the pulmonary pleura and invaginated in the ipsilateral pleural sac, to which it is attached by the pulmonary ligament. Inside this pleural sac, the lung can move freely.

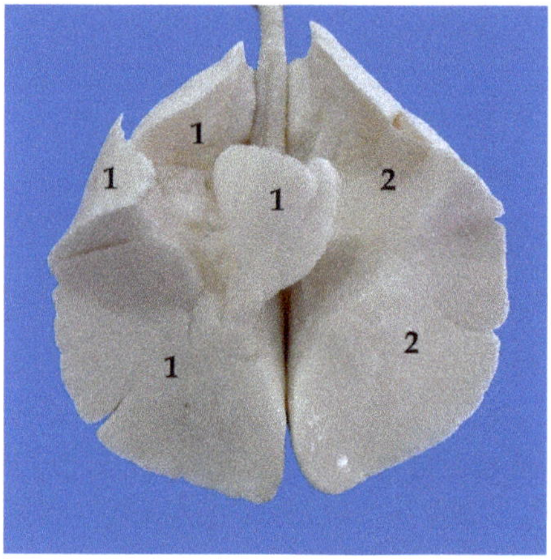

Figure 10. Ventral view of the lungs, washed with water and filled with compressed air: (1) right lung; (2) left lung.

The lungs of the *Arctocephalus australis* are divided into lobes by deep interlobar fissures. In other words, these lungs show a well-defined lobulation, the patter of which is as follows (Figure 12):

- Right lung: It has four differentiated lobes, the first being the cranial lobe. Next, and separated from the cranial lobe by a deep interlobar fissure, is the middle lobe, with an elongated and pyramidal shape. Caudally to the middle lobe is the caudal lobe, with an expanded fan-like shape. The *Arctocephalus australis* has a fourth lobe in the right lung, the accessory lobe. It is located between both lungs, specifically in the caudal mediastinum (Figure 12). Ventrally, this accessory lobe has a groove for the passage of the caudal vena cava and the right phrenic nerve (Figure 13).

- Left lung: The left lung shows one pair of lobes. The first is the cranial lobe, which has an irregular surface. It is divided into two parts by a short fissure. The caudal lobe is large and has scarce subdivisions. Interestingly, its shape is very similar to that of the caudal right lobe (Figure 12).

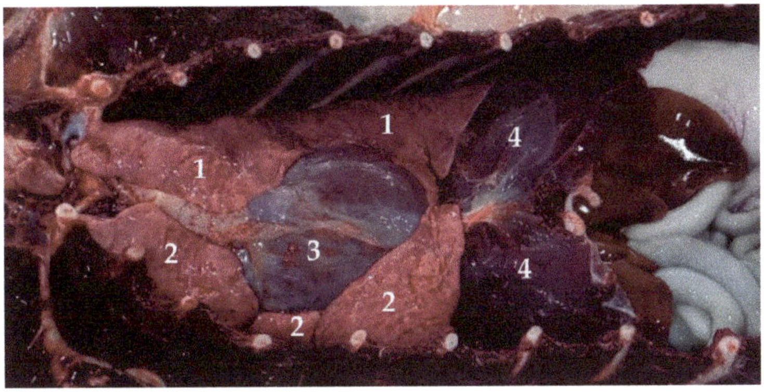

Figure 11. Ventral view of the thoracic cavity of *Arctocephalus australis*: (1) left lung; (2) right lung; (3) heart; (4) diaphragm.

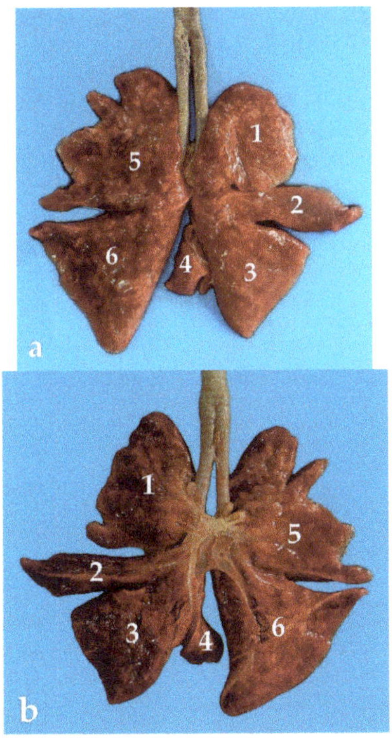

Figure 12. Lungs of *Arctocephalus australis*: (**a**) dorsal view; (**b**) ventral view; (1) right cranial lobe; (2) right middle lobe; (3) right caudal lobe; (4) right accessory lobe; (5) left cranial lobe; (6) left caudal lobe.

Figure 13. Detail of the groove formed by the accessory lobe for the passage of the caudal vena cava and the right phrenic nerve.

The different parts of both lungs can be described as follows:

1. Base, or diaphragmatic surface of the lung: It is concave as it overlaps the convex thoracic surface of the diaphragm. It is limited by the basal border.
2. Apex: It is free, sharpened, and laterally flattened. It fills the space of the pleural dome.
3. Two surfaces: (A) The costal surface: Lateral, larger, smooth, and convex. It is in contact with the inner surface of the lateral thoracic wall. (B) The medial surface: It is smaller than the costal and is related to the mediastinum and the mediastinal structures. It presents a deep depression formed by the heart and its pericardium, the cardiac impression. The cardiac impression is deeper in the left lung (Figure 14).
4. Three distinct borders (ventral, dorsal, and basal): (A) Ventral border: It is sharp and irregular. It occupies the costomediastinic sinus and presents the cardiac notch. The ventral border of the left lung is slightly greater than the one of the right lung (Table 3). (B) Basal border: The basal border separates the base (or diaphragmatic surface) from the costal and medial surfaces. As in the case of the ventral border, the basal border of the left lung has a greater extension than in the right lung (Table 3). (C) Dorsal border: It is thick and rounded. It forms the dorsal boundary between the costal surface and the medial surface. Contrary to what happens in the ventral and basal borders, the dorsal border of the right lung is the one with a greater length (Table 3).

Table 3. Morphometric data of the lungs from the 11 individuals of *Arctocephalus australis* investigated in the present study. A: adult; J: juvenile.

ID	Age Group	Sex	Length (cm)						Maximum Width (cm)	
			Dorsal Border		Ventral Border		Basal Border			
			Right	Left	Right	Left	Right	Left	Right	Left
M9719	J	F	20.1	19.3	12.8	13.9	10.7	13.6	12.6	11.3
M7419	J	M	20.5	20	13	14.3	11	14	12.9	11.8
M7019	J	M	21.6	21	13.4	15	11.3	15.3	13.2	12.1
M8419	J	F	22.6	21.3	13.7	15.5	11.6	15.5	14	12.8
M8319	J	M	23.3	22	14.2	16.3	11.9	15.7	14.3	13
M10019	J	F	24	22.9	14.7	16.5	12.1	16.3	14.5	13.4
M7919	J	M	25.8	24.2	15	16.7	12.6	16.7	15.1	13.8
M8219	J	F	27.7	26	16.8	18.2	13.7	17.7	16.4	15.2
M10419	J	M	28	26.2	17.2	19	14.1	18	18	16.7
M7319	A	F	33	30.5	23.5	25.7	19.6	24.5	24	22.6
M9919	A	F	35	32.7	24.5	26.3	20.1	25	24.7	23
Mean	J	-	23.7	22.5	14.5	16.1	12.1	15.8	14.5	13.3
Mean	A	-	34	31.6	24	26	19.8	24.7	24.3	22.8

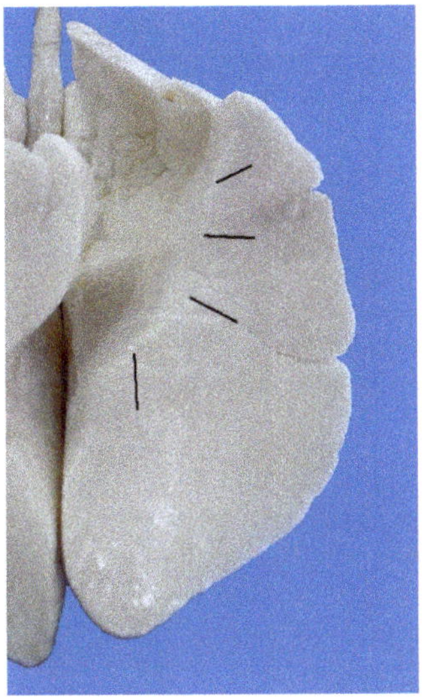

Figure 14. Detail of the cardiac impression in the left lung.

The maximum width of both lungs is also different, and the values of the right side are slightly lower (Table 3).

4. Discussion

Marine mammals, which live most of their lives in the ocean or in an aquatic environment, have typical characteristics of all mammals. Thus, in the case of their respiratory system, it also consists of the same parts, starting with the nose and nares and ending with the lungs.

It is important to remark that, in general terms, knowledge of anatomy helps to understand any physiological phenomenon. Surprisingly, and although they could provide key insights into physiological responses to dive, in the case of the *Arctocephalus australis*, previous reports on the anatomy of its respiratory system are scarce [2]. For example, the depths to which these animals can dive and the length of time they remain submerged have intrigued researchers for decades. Unlike what happens with land mammals, *Arctocephalus australis* has enough oxygen to perform long dives and is able to avoid pressure-related diseases from repetitive and deep diving. The question that arises is if there are anatomical differences between its respiratory apparatus and that of dogs, as a typical model of land-mammal, or if on the contrary, this species is able to live in the water environment without special anatomical adaptations.

As mentioned before, the respiratory apparatus of the South American fur seal starts with the nose and the nares. The nares have a greater mobility compared to other mammal species with similar nose physiognomy. They can intentionally be totally closed or even expand to double their size as compared to resting conditions [5]. Thus, the South American fur seal has the ability to open or close the nares as if they were valves to avoid a massive water uptake through the airways while diving. Additionally, they can increase the air expelled per unit time when approaching the surface [5]. In addition, contrary to what happens in other mammals such as the dog, the philtrum divides the nasal plane partially, as it does not reach the upper lip [6]. The nasal septum is also significantly shorter than in dogs or cats [6], which facilitates different pathologies affecting the upper respiratory tracts [28]. On the contrary, the nasal cartilage is similar to that of the dog [6].

There are four groups of bone cavities connected to the nasal cavity in land mammals: the paranasal sinuses. These sinuses act synergistically with the nasal cavity to warm and humidify the inhaled air during breathing. Pinnipeds, other marine mammals, have lost these paranasal sinuses due to the high pressures they face under water. These structures are too fragile to resist these high pressures. Thus, in the case that these structures remained in pinnipeds, they would fracture easily [18]. As in phocids, the soft palate of the *Arctocephalus australis* is much longer than that of the dog, with a significant amount of peripharyngeal soft tissue. This anatomical characteristic is of great clinical interest, taking into consideration the difficulties these marine species have for intubation [29]. In contrast, the larynx of the South American fur seal is practically like that of other carnivores [6]. Thus, the laryngeal cartilages are similar in shape and size to those of the dog.

As shown by other authors, tracheal length is directly related to the need of determined speed air flows. When analyzing the functional implications of this fact, it is important to compare the tracheal width to its length, since the ability to exchange large air volumes per unit time is directly related to the absolute diameter/length ratio [25]. This seems to be the reason why cetaceans have short and wide tracheas with a large pulmonary ventilations volume (80–90%) in shorter time, which is the opposite of what happens in land-carnivores, as these species do not have this need [2]. The *Arctocephalus australis* trachea is longer than in cetaceans, with an early bifurcation, but it is shorter than that in dogs and cats. This anatomical difference is probably related to its diving adaptation and its distinct ventilation need in comparison to other marine species [30].

Mammals show variations in the bronchial branching pattern, which is of great importance when analyzing their functional implications. Thus, their bronchial tree consists of a sequence of branches, whose number, generation, and distribution vary according to the different mammalian species. In summary, the branching pattern is similar to that in the dog [31,32].

As previously shown in these species, the cross-sectional area of each bronchus is approximately half of the tracheal area. Thus, the total cross-sectional area of the two primary bronchi is practically equal to the cross-sectional area of the trachea [2]. According to our results, this is also the case for the *Arctocephalus australis*, which, due to the early bifurcation of its trachea, has extremely long bronchi in comparison to other species of land mammals [6,7].

Interestingly, the *Arctocephalus australis* bronchial tree shows differences compared to other marine mammals, such as certain dolphin species. These animals have developed a third bronchus or tracheal bronchus. This bronchus ventilates the most cranial part of the right lung, independently of the main right bronchus entry [26,27].

The explanation could be in the South American fur seal lifestyle, as this species alternates land and water environments. They do not need rapid ventilation, as is the case in cetaceans, and neither live constantly on land. This could also be the reason why the *Arctocephalus australis* has a short trachea and long bronchi, especially when compared to land mammals, whose ventilation demands are different [25].

Concerning the lungs, our results showed that the lobulation pattern is very similar to that in other species of land mammals, such as the dog or the cat [6,7]. Surprisingly, the lungs of the South American fur seal are different from those of cetaceans, which show unilobed lungs [25,26,33]. This could be related to the ability of these species to mobilize large volumes of air at high speeds in short intervals of time; meanwhile, pinnipeds do not have that need as they spend more time on land [33].

As it has been previously shown, the lung lobulation of land mammals, with a greater absorption surface, is able to better bear internal impacts during terrestrial locomotion [26]. We think that this could be the case for the South American fur seal. However, as in other species of marine mammals, the lung microanatomy of this species has thicker and firmer connective tissue partitions in comparison to land mammals, which is a clear adaptation to the aquatic environment. This greater development of the septa forms a more rigid and resistant lung support stroma, giving the parenchyma a greater resistance to lung collapse that could occur during diving [2].

In summary, the lungs of the South American fur seal show their own differential characteristics, as they are a mixture of those of other mammal species, either marine or terrestrial

5. Conclusions

The anatomical characteristics of the South American fur seal's respiratory apparatus are very similar to those observed in other land-carnivores, such as the dog.

Nevertheless, this species exhibits several differences with other land-carnivores, of course resulting from their diving ability. Examples of this are its capacity to open or close the nares and the early bifurcation of the trachea, which does not have a thoracic portion.

There are also interesting differences between the lungs of the South American fur seal and other marine mammals, which is probably related to their different lifestyle and ventilation demands.

The conclusion is that, besides these differences, being a carnivore seems to be the critical characteristic defining the respiratory apparatus of the South American fur seal, even though there are clear differential anatomical details to facilitate its adaptation to the aquatic environment.

Author Contributions: Conceptualization, I.M.-D., R.M.-O. and J.G.-S.; methodology, J.G.-S. and R.M.-O.; data collection and validation, I.M.-D., C.T.-M., J.-P.L. and E.T.-D.; formal analysis, I.M.-D., R.M.-O. and J.G.-S.; writing—original draft preparation, I.M.-D., I.S.-Á. and P.P.-L.; writing—review and editing, J.G.-S., R.M.-O. and I.S.-Á.; supervision, I.M.-D., R.M.-O. and J.G.-S.; project administration, J.-P.L., I.M.-D., C.T.-M. and E.T.-D.; funding acquisition, J.-P.L. All authors have read and agreed to the published version of the manuscript.

Funding: This research received no external funding.

Institutional Review Board Statement: Ethical review and approval were waived for this study by Fundación Mundo Marino, authorizing the use of several cadavers for an anatomical study.

Informed Consent Statement: Not applicable.

Data Availability Statement: Not applicable.

Acknowledgments: Authors thank Fundación Mundo Marino for the use of its facilities and equipment to carry out this work.

Conflicts of Interest: The authors declare no conflict of interest.

eferences

Leith, D. Adaptations to deep breath-hold diving: Respiratory and circulatory mechanics. *Undersea Biomed. Res.* **1989**, *16*, 345–354.

Crespo, F.A. Morfología Funcional en el Sistema Respiratorio de Mamíferos Marinos. Ph.D. Thesis, Facultad de Ciencias Exactas y Naturales, Universidad de Buenos Aires, Viamonte, Argentina, 1998.

Wislocki, G. On the structure of the lungs of the porpoise (*Tursiops truncatus*). *Am. J. Anat.* **1929**, *44*, 47–77. [CrossRef]

Reidenberg, J.S.; Laitman, J.T. Position of the larynx in Odontoceti (toothed whales). *Anat. Rec.* **1987**, *218*, 98–106. [CrossRef] [PubMed]

Reidenberg, J.S. Anatomical adaptations of aquatic mammals. *Anat. Rec.* **2007**, *290*, 507–513. [CrossRef] [PubMed]

Getty, R.; Sisson, S.; Grossman, J.D. *Anatomy of Domestic Animals*, 5th ed.; Salvat Editores: Barcelona, Spain, 1982; Volume 2, pp. 1710–1727.

Climent, S.; Sarasa, M.; Muniesa, P.; Latorre, R. *Manual de Anatomía y Embriología de los Animales Domésticos. Cabeza, Aparato Respiratorio, Aparato Digestive, Aparato Urogenital*; Editorial Acribia S.A.: Zaragoza, Spain, 2005; pp. 13–60.

Kooyman, G.L. Physiology without restraint in diving mammals. *Mar. Mam. Sci.* **1985**, *1*, 166–178. [CrossRef]

Kooyman, G.L. *Diverse Divers: Physiology and Behavior*, 1st ed.; Springer: Berlin, Germany, 1989; pp. 53–65.

0. Zapol, W.M.; Liggins, G.C.; Schneider, R.C.; Qvist, J.; Snider, M.T.; Creasy, R.K.; Hochachka, P.W. Regional blood flow during simulated diving in the conscious Weddell seal. *J. Appl. Physiol.* **1979**, *47*, 968–973. [CrossRef]

1. Sinnet, E.E.; Kooyman, G.L.; Wahrenbrock, E.A. Pulmonary circulation of the harbor seal. *J. Appl. Physiol.* **1978**, *45*, 718–727. [CrossRef] [PubMed]

2. Elsner, R.; Franklin, D.L.; Van Citters, R.L.; Kenney, D.W. Cardiovascular defense against asphyxia. *Science* **1966**, *153*, 941–949. [CrossRef]

13. Kooyman, G.L.; Castellini, M.A.; Davis, R.W. Physiology of diving in marine mammals. *Annu. Rev. Physiol.* **1981**, *43*, 343–356. [CrossRef]
14. Williams, T.M.; Haun, J.E.; Friedl, W.A. The diving physiology of bottlenose dolphins (*Tursiops truncatus*) I. Balancing the demands of exercise for energy conservation at depth. *J. Exp. Biol.* **1999**, *202*, 2739–2748. [CrossRef]
15. Kooyman, G.L.; Campbell, W.B. Heart rates in freely diving Weddell Seals (*Leptonychotes weddelli*). *Comp. Biochem. Physiol.* **1972**, *43*, 31–36. [CrossRef] [PubMed]
16. Murphy, B.; Zapol, W.M.; Hochachka, P.W. Metabolic activities of heart, lung and brain during diving and recovery in the Wedel seal. *J. Appl. Physiol. Respir. Environ. Exerc. Physiol.* **1980**, *48*, 596–605. [CrossRef] [PubMed]
17. Hochachka, P.W.; Storey, K.B. Metabolic consequences of diving in animals and man. *Science* **1975**, *187*, 613–621. [CrossRef] [PubMed]
18. Ponganis, P. *Diving Physiology of Marine Mammals and Seabirds*; Cambridge University Press: Cambridge, UK, 2015; pp. 162–170. [CrossRef]
19. Noren, S.R.; Williams, T.M.; Pabst, D.A.; McLellan, W.A.; Dearolf, J.L. The development of diving in marine endotherms: Preparing the skeletal muscles of dolphins, penguins, and seals for activity during submergence. *J. Comp. Physiol. B* **2001**, *171*, 127–134. [CrossRef] [PubMed]
20. Katz, H.; Pérez, W.; Bielli, A.; Chavez, R. Histomorphology of prepuberal ovaries in the South American fur seal (*Arctocephalus australis*, Zimmerman, 1783). *Folia Morphol.* **2009**, *68*, 277–286.
21. Drabeck, C.M. Some anatomical aspects of the cardiovascular system of antarctic seals and their possible functional significance in diving. *J. Morphol.* **1975**, *145*, 85–105. [CrossRef] [PubMed]
22. Pérez, W.; Katz, H.; Lima, M. Gross heart anatomy of *Arctocephalus australis* (Zimmerman, 1783). *Anat. Sci. Int.* **2008**, *83*, 6–10. [CrossRef]
23. Stewardson, C.L.; Hemsley, S.; Meyer, M.A.; Canfield, P.J.; Maindonald, J.H. Gross and microscopic visceral anatomy of the male Cape fur seal, Arctocephalus *pusillus pusillus* (Pinnipedia: Otariidae), with reference to organ size and growth. *J. Anat.* **1999**, *195*, 235–255. [CrossRef]
24. Martín-Orti, R.; Tostado-Marcos, C.; Loureiro, J.P.; Molpeceres-Diego, I.; Tendillo-Domínguez, E.; Santos-Álvarez, I.; Pérez-Lloret, P.; González-Soriano, J. The digestive system of the *Arctocephalus australis* in comparison to the dog as a land-carnivore model. *Animals* **2022**, *12*, 1634. [CrossRef]
25. Tarasoff, F.J.; Kooyman, G.L. Observations on the anatomy of the respiratory system of the river otter, sea otter and harp seal. II The trachea and bronchial tree. *Can. J. Zool.* **1973**, *51*, 171–177. [CrossRef]
26. Fanning, J.C.; Harrison, R.J. The strucuture of the trachea and lungs of the south australian bottle-nosed dolphin. In *Functional Anatomy of Marine Mammals*, 2nd ed.; Harrison, R.J., Ed.; Academic Press: London, UK, 1974; pp. 231–252.
27. Yamasaki, F.; Takahashi, K.; Kamiya, T. Lungs of Franciscana (*Pontoporia blainvillei*), with special references to their external aspects, weihts and bronchial ramifications. *Okajimas Folia Anat. Jpn.* **1977**, *53*, 337–358. [CrossRef] [PubMed]
28. Field, C.L. Bacterial Diseases of Marine Mammals. 2022. Available online: https://www.msdvetmanual.com/exotic-and-laboratory-animals/marine-mammals/bacterial-diseases-of-marine-mammals (accessed on 1 October 2022).
29. Higgins, J.L.; Hendrickson, D.A. Surgical procedures in pinniped and cetacean species. *J. Zoo Wildl Med.* **2013**, *44*, 817–836. [CrossRef] [PubMed]
30. Repenning, C.A. Adaptative evolution of sea lions and walruses. *Syst. Zool.* **1976**, *25*, 375–390. [CrossRef]
31. Horsefield, K.; Cumming, G. Morphology of the bronchial tree in man. *J. Appl. Physiol.* **1968**, *24*, 373–383. [CrossRef] [PubMed]
32. Schlesinger, R.B.; McFadden, L.A. Comparative morphometry of the upper bronchial tree in six mammalian species. *Anat. Rec.* **1981**, *199*, 99–108. [CrossRef] [PubMed]
33. Piscitelli, M.A.; Raverty, S.A.; Lillie, M.A.; Shadwick, R.E. A review of cetacean lung morphology and mechanics. *J. Morphol.* **2013**, *274*, 1425–1440. [CrossRef] [PubMed]

animals MDPI

rticle

ʒody Length and Craniometrics of Non-Native Raccoons in ʿwo Regions in Middle Japan during Early Invasion Stages

akuya Kato [1,*,†], Fumiaki Yamasaki [1,†], Kandai Doi [1,2], Mieko Kawamichi [3] and Shin-ichi Hayama [1]

1 Faculty of Veterinary Science, Nippon Veterinary and Life Science University, Tokyo 180-8602, Japan
2 Japan Society for Promotion of Science Research Fellow, Forestry and Forest Products Research Institute, Ibaraki 305-8687, Japan
3 Kansai Wildlife Research Association, Kyoto 605-0981, Japan
* Correspondence: tkato@nvlu.ac.jp
† These authors contributed equally to this work.

Simple Summary: Morphological characteristics related to sex-age class in introduced species during the early invasion stages, before further environmental adaptation, may indicate important information regarding their adaptability. For raccoons introduced to different regions in middle Japan, the results of multiple linear regression on body length showed that males were larger than females and body length increased with age class, while there was no difference between regions. The cranial size, which indicated different allometry in each sex, regressed positively according to increasing age class, body length, and body mass index in both sexes; however, it only differed between regions for females. Therefore, our results supported the hypothesis that older individuals with larger body sizes have an advantage in intrasexual selection and competition for food resources in raccoons. Considering the fact that multiple subspecies are sympatric in North America and the morphometrics of introduced raccoons in Japan were inconsistent with any of them, it is likely that hybridization occurred prior to introduction or naturalization. Further studies are necessary to demonstrate subspecific hybridization and its impact on morphological and ecological changes.

Abstract: Although the genetic distribution of introduced raccoons (*Procyon lotor*) in recent years is well known, few studies have examined their morphometrics, especially the relationships between sex and age in the introduced populations. The aim of this study was to describe the morphological characteristics of raccoons from parts of eastern and western Japan during their early invasion stages, focusing on the relationships between body length and the principal component of craniometrics, with region, sex, age class, body length, and body mass index using a regression model. The body length increased more in males than females and in the older age class, supporting the association with intrasexual selection and competition for food resources. Positive relationships for body length and body mass index were found in craniometric analyses, particularly regarding cranial size components, in addition to age class for both sexes, while cranial size also differed between regions for females. The relationship between body length and craniometrics was inconsistent with that of subspecies originating in North America. Given the sympatric distribution of haplotypes of multiple subspecies without reproductive-isolating barriers in North America and in several introduced areas, hybridization must have occurred prior to the introduction or naturalization of this species.

Keywords: body length; cranial size; morphometrics; *Procyon lotor*; raccoon

1. Introduction

Globally, several carnivore species have been introduced to non-native areas, and they have managed to naturalize as invasive alien species owing to their high adaptability to various factors, including productivity and food availability, in each environment [1–4]. Following adaptation, the non-native population of a species may become ecologically and morphologically different from the population inhabiting the native region [5,6]. Compared to the indigenous animals, introduced raccoon dog (*Nyctereutes procyonoides*) populations in Europe demonstrate different craniometrics and non-metrical features [5,6]. Korablev and Szuma [6] proposed that the main factors influencing morphological variation in the raccoon dog could be related to the net primary production of ecosystems, adaptability to ecologically different regions, stochastic factors (such as the founder effect), and isolation of populations due to human activity. However, some omnivorous mammals seem to have rapid morphological changes in their new distribution [7–10]. Retrospective studies on the morphological characteristics during the early invasion stage in introduced species may provide insights regarding their adaptability in the future.

The Northern raccoon (*Procyon lotor*) is a medium-sized carnivore that is native to and widespread in North and Central America and has been introduced to Alaska [11], Russia [12], several countries in Europe [4,13], and Japan [1]. The raccoon has tolerance to a wide variety of bioclimatic conditions that results in its adaptation to various regions, and it is predicted that they will expand to new favorable areas north of the current favorable areas by 2050, due to climatic change [14]. This expansion may negatively impact the new areas because of the predation of native species of, for instance, amphibians, insects, and crustaceans [1,15]. The first raccoon naturalization in Japan was reported when 12 raccoons escaped from a zoo in Inuyama City, Aichi Prefecture in 1962 [16]; however, the raccoon had not yet been recognized as an invasive alien species. Miyashita [17] described the importation of over 20,000 raccoons to Japan as pets, following a famous television animation show in 1977. In Kamakura City, Kanagawa Prefecture, located in east central Japan, raccoon naturalization was confirmed in 1988 [18]; subsequently, the presence of naturalized raccoons was confirmed in most prefectures by 2006 [19]. The number of 5 km^2 grids containing confirmed raccoon inhabitants increased from 1388 to 3862 between 2006 and 2018, which indicated an increase in distribution [20]. The growing invasive raccoon populations in Japan have become considerably problematic to the native ecosystem, agriculture, and public health [1].

Ecological characteristics, including productivity, home range, seasonal activity, feeding habits, and mortality, vary in different regions for North American raccoons, as described by Gehrt [21]. In addition, previous studies regarding the native raccoon have also clarified both sexual size dimorphism (SSD) and geographical variation in body size based on craniometrics. The SSD of the raccoon is known to involve larger body sizes for males than for females in each regional distribution [22,23]. Furthermore, the largest native population was observed in the northwestern region and the smallest in the southeastern region with a geographical gradient for intraspecific variation [24,25]. Geographic variation was also found in litter size, linked to the variation pattern of female raccoon body size [26]. Considering the widespread raccoon distribution in the northern and southern regions, development of ecological and morphological characteristics might be one of the most important factors explaining their adaptation to various environments.

Based on a study of mitochondrial DNA haplotypes of native raccoon distributions [27], introduced populations in the eastern part of middle Japan, including Chiba Prefecture [28] and Kanagawa Prefecture [29], are considered to originate from the central part of North America. Additionally, different haplotypes were found in several introduced populations in Japan [30,31], which suggests that independent introductions occurred and that the populations did not interact. Some reports on non-native raccoon populations demonstrated ecological characteristics regarding reproduction [32,33] and feeding habits [34–36] that suggest a high adaptability to new distribution areas. Although body size in some introduced carnivores may be related to their ecology [6–8], little is known about morphological characteristics, such as body length and craniometrics, for the introduced raccoons of Japan. The morphological characteristics of raccoons in the early stages of introduction are expected to provide important information for understanding its subsequent environmental adaptation as an invasive alien species.

The aim of this study was to describe the morphological characteristics of raccoons from parts of eastern and western Japan during their early invasion stages, focusing on the relationships between body length and the principal component of craniometrics with region, sex, age-class, body length, and body mass index using a regression model. We predict that: (1) SSD and age-related body length differences would occur even in non-native populations; (2) body length and craniometrics would differ between regions; and (3) the craniometrics and cranial growth pattern in each sex would be related to age class, body length, and nutritional condition which implicates an ecological background.

2. Materials and Methods

2.1. Study Site

The study was performed in Kamakura City, Kanagawa Prefecture and Maizuru City, Kyoto Prefecture; these sites were chosen to represent the eastern and western parts of middle Japan, and the distance between the two areas is approximately 500 km (Figure S1). Although Kamakura is located on the Pacific side and Maizuru on the side of the Sea of Japan, both regions have temperate humid climates. The mean annual temperatures and annual precipitation in Kamakura and Maizuru are 16.2 °C and 1730.8 mm and 14.8 °C and 1941.2 mm, respectively, which indicate that there is little difference between the climates of the two regions.

2.2. Samples and Measurement

Raccoon carcasses were collected from 2005 to 2007 in Kamakura City, and from 2009 to 2014 in Maizuru City. Considering the timing of first reports and the progress of the invasion in Kamakura and Maizuru, each raccoon collection period in this study can be assumed as the early post-introduction stage. All raccoons were captured by trappers permitted by municipalities using box traps. The raccoons were then euthanized by CO_2 inhalation according to the Guidelines for the Management of Invasive Alien Species in Japan [37].

Postmortem examination in each carcass identified the gender by genital observation, and tooth eruption was assessed to exclude individuals under five months of age when pre-weaned [38,39]. Body weight (in g) and length (in mm: nose to tail, subtracting the tail length) were then measured. To analyze the body condition and nutritional status of individuals, the body mass index (BMI) was estimated using the formula W/L^2, where W is the body weight expressed in kg and L is the body length expressed in meters [40].

The skinned head was removed, and the skull was processed to estimate age class and for further craniometrics. Based on the size of the canine root apical foramen [41] and cranial suture obliteration [42], all individuals were classified into three age groups: 5–11 months of age, juveniles; 12–23 months of age, yearlings; and $\geq$24 months of age, adults. Consequently, the number of raccoon samples were 143 from Kamakura (15 juveniles, 30 yearlings, and 15 adult females; 16 juveniles, 34 yearlings, and 33 adult males) and 63 from Maizuru (6 juveniles, 10 yearlings, and 10 adult females; 11 juveniles, 16 yearlings, and 10 adult males). Measurements were performed for 13 cranial components using digital calipers with 0.01 mm precision as per previous studies [24,25,43,44] (Figure 1). The dataset of all specimens analyzed in this study is shown in Table S1.

2.3. Data Analysis

2.3.1. Body Length by Region, Sex, and Age Class

The correlation of body length with sex, age class, and region was tested using a multiple regression analysis. The body length was checked for normality and homogeneity, set as the outcome variable without transformation of raw data, whereafter the region, sex, and age class were entered as explanatory variables, and was examined for interactions between sex-age classes. Multicollinearity between the explanatory variables was assessed using the variance inflation factor (VIF) [45]. The best model selected using the Akaike information criteria (AIC) as well as other informative models with ΔAIC < 2 [46] were evaluated in terms of coefficient parameters and significance of explanatory variables.

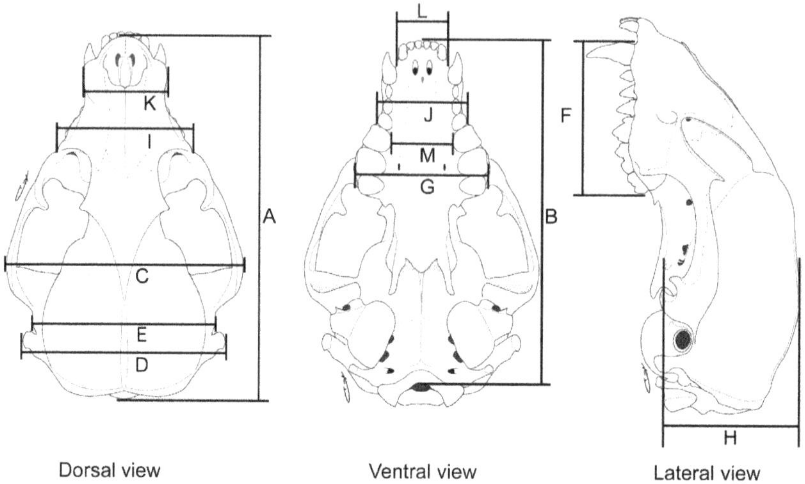

Figure 1. Lateral, ventral, and dorsal craniometrics of the raccoon (*Procyon lotor*) skull. A: greatest length of skull (GLS), B: basal length (BaL), C: zygomatic width (ZW), D: paraoccipital width (PW), E: cranial width (CW), F: length of maxillary tooth row (LMT), G: greatest width across first upper molars (M1s) (GWM), H: depth of skull (DS), I: least width between infraorbital foramen (LWIF), J: greatest width across lower third premolars (P3s) (GWP), K: rostral width (RW), L: width of incisors (WI), M: least distance between lower fourth premolars (P4s) (LDP).

2.3.2. Craniometrics in Each Sex

For raccoons, sexual dimorphism was reported for craniometrics in previous studies [23,43]; therefore, further analyses were conducted separately for female and male raccoons. Since it is possible to summarize the number of variables to a small number of composite data, principal component analysis (PCA) is appropriate for craniometric studies [25,47,48]. In order to derive principal components (PCs), the PCAs for each sex were performed using the correlation matrix on the raw data, consisting of 13 variables of cranial measurements, without log transformation as described by O'Keefe et al. [46].

2.3.3. Relationships between Cranial Size and Basal Factors

To evaluate the basal factors related to craniometrics, significant PCs were considered as summarized cranial characteristics for each sex using the eigenvalue and screen plot, and the relationship between the PCs and basal factors was analyzed using a multiple regression model [47,49]. The basal factors included region, age class, body length, and BMI and were entered as explanatory variables, which confirmed multicollinearity based on the VIF among each other [45], and the multiple regression analyses in each sex were performed for the PCs as outcome variables. The best model and other valid models with ΔAIC < 2 were selected using the Akaike information criteria (AIC) to examine factors related to the cranial characteristics in each sex [46].

2.3.4. Allometry of Craniometrics in Each Sex

To determine cranial growth patterns of raccoon in each sex, we used a multivariate approach based on the generalized allometry equation proposed by Jolicoeur [50]. In this theory, a global multivariate allometry vector uses the first PC value (PC1) calculated from PCA for a variance-covariance matrix of log-transformed variables; if it derives a strong size-factor that obtained high eigenvalue, measurements are positive. Confidence intervals (CI) for each coefficient were estimated from sampling distributions based on

1000 bootstrap iterations. The isometric vector in this study was 0.277 derived by $1/\sqrt{p}$, with p representing the number of variables [50].

2.3.5. Functions for Statistical Analysis

Statistical analyses were performed using R. ver. 4.1.2 [51], using the functions *VIF* in the package "car" for VIF checking [52], and *lm* and *princomp* in the package "Stats" for the regression model and PCA [51], respectively; the function *dredge* in the package "MuMIn" was used for valid model selection using the AIC [53]; and the function *boot* in the package "boot" to calculate the 95% CI of the global allometry vector [54].

3. Results

3.1. Body Length by Region, Sex, and Age Class

For all 206 specimens, the descriptive analysis showed that body length differed between sexes within each age class and region. For adult females and males, the mean ± SD of body lengths was 557.0 ± 28.3 and 569.4 ± 35.6 in Kamakura and 540.4 ± 30.0 and 592.7 ± 28.1 in Maizuru, respectively (Table S2). The violin plots of body length (mean ± SD) per region, sex, and age class showed that males were larger than females in each age class and had increased body lengths with older age class. Moreover, despite the high variance of yearling males, little difference was found between regions (Figure S2). As a result of the regression analysis for body length, the region was excluded from the best model (model 1; Adjusted $R^2 = 0.35$, $F_{(3, 303)} = 37.39$, $p < 0.001$: Table 1), while the body length was significantly correlated with both sex ($p < 0.001$) and age classes ($p < 0.001$). In the best model, male raccoons were larger than the females (coefficient = 31.83, 95% CI = 22.52–41.14) and juveniles were smaller than yearlings (coefficient = 30.24, 95% CI = 18.49–41.99) and adults (coefficient = 49.12, 95% CI = 36.71–61.53). Although, in other valid models with ΔAIC < 2, sex-age interaction (model 2) and region (model 3) were selected in addition to sex and age class, these were not significantly correlated to the body length despite indicating sex and age class strongly affected the body length.

Table 1. Multiple linear regression analysis for body length with region, sex, and age class of feral raccoons in Japan. The best models were selected using the Akaike information criteria (AIC) and valid models within ΔAIC < 2.0.

Category	Variable	Regression Coefficient [1]	Standard Error	95% CI [2]	VIF
Model 1 (Best model): Adjusted $R^2 = 0.35$, $F_{(3, 202)} = 37.39$, $p < 0.001$, AIC = 2035.4					
Sex	Female	Reference			1.00
	Male	31.83 †	4.72	22.52–41.14	
Age class	Juvenile	Reference			1.00
	Yearling	30.24 †	5.50	18.49–41.99	
	Adult	49.12 †	5.96	36.71–61.53	
(Intercept)		496.57 †	5.79	485.73–507.41	
Model 2: Adjusted $R^2 = 0.35$, $F_{(5, 200)} = 23.26$, $p < 0.001$, AIC = 2035.9					
Sex	Female	Reference			2.06
	Male	23.62 *	9.67	4.56–42.69	
Age class	Juvenile	Reference			1.55
	Yearling	20.28 *	8.95	2.63–37.94	
	Adult	49.17 †	9.84	29.77–68.56	
Interaction	Sex–Yearling	17.82	11.96	−5.77–41.41	1.94
	Sex–Adult	0.83	12.78	−24.37–26.03	
(Intercept)		501.19 †	7.25	486.89–515.48	

Table 1. *Cont.*

Category	Variable	Regression Coefficient [1]	Standard Error	95% CI [2]	VIF
	Model 3: Adjusted $R^2 = 0.34$, $F_{(4, 201)} = 27.91$, $p < 0.001$, AIC = 2037.3				
Region	Kamakura	Reference			1.00
	Maizuru	0.77	5.06	−9.22–10.75	
Sex	Female	Reference			1.00
	Male	31.83 *	4.73	22.49–41.16	
Age class	Juvenile	Reference			1.00
	Yearling	30.29 *	5.98	18.50–42.09	
	Adult	49.17 †	6.32	36.71–61.62	
(Intercept)		496.31 †	5.79	484.89–507.72	

[1] *p*-values refer to Ho: Regression coefficient, 0; *: $p < 0.05$; †: $p < 0.001$. [2] 95% confidence interval (CI) of the regression coefficient.

3.2. Craniometrics in Each Sex

Regarding craniometrics, the mean values of males were higher than those of females of all age classes and regions, except for the length of the maxillary tooth row, greatest width across upper molars (M1s), greatest width across lower third premolars (P3s), rostral width, width of incisors, and least distance between lower fourth premolars (P4s) (Table S3). In a PCA using craniometrics, the axes of PC1, PC2, and PC3 indicated 56.79%, 12.73%, and 7.03% for females and 58.19%, 9.95%, and 6.89% for males, respectively (Table 2). The positive and high coefficients of all measurements for PC1 in both sexes indicated large size variations, indicating that these were defined as robust size factors. Contrary to those of PC1, the coefficients of PC2 were a mixture of positive and negative coefficients for both females and males, indicating that there is a shape factor. The individual scores for the PC1 and PC2 axes by region and age class in each sex are plotted in Figure 2, showing that craniometric size varied with age class and region in females, but only with age class in males.

Table 2. Factor loadings of principal component analysis (PCA) in cranial measurements in each principal component (PC) for each sex of introduced raccoons in Kamakura and Maizuru, Japan.

	Female			Male		
	PC1	PC2	PC3	PC1	PC2	PC3
Eigenvalue	7.38	1.65	0.91	7.56	1.29	0.90
Proportion	56.79	12.73	7.03	58.19	9.95	6.89
Cumulative	56.79	69.52	76.55	58.19	68.14	75.03
Measurement loadings [1]						
GLS	0.28	0.44	0.03	0.30	0.36	0.06
BaL	0.30	0.36	0.01	0.31	0.22	0.11
ZW	0.32	−0.07	−0.30	0.32	−0.02	−0.20
PW	0.30	0.14	−0.40	0.30	0.21	−0.11
CW	0.30	0.13	−0.35	0.30	0.28	0.03
LMT	0.20	0.38	0.53	0.23	0.29	0.19
GWM	0.27	−0.40	0.01	0.22	−0.46	0.43
DS	0.30	−0.01	−0.17	0.26	−0.14	−0.51
LWIF	0.26	−0.13	0.41	0.28	0.08	0.29
GWP	0.28	−0.38	0.23	0.27	−0.37	0.14
RW	0.27	−0.01	0.13	0.30	−0.01	−0.25
WI	0.24	−0.07	0.27	0.21	−0.36	−0.45
LDP	0.26	−0.39	−0.08	0.27	−0.34	0.29

[1] Measurement letters corresponded to those of Figure 1. GLS, greatest length of skull. BaL, basal length. ZW, zygomatic width; PW, paraoccipital width; CW, cranial width; LMT, length of maxillary tooth row; GWM, greatest width across first upper molars (M1s); DS, depth of skull; LWIF, least width between infraorbital foramen; GWP, greatest width across lower third premolars (P3s); RW, rostral width; WI, width of incisors; LDP, least distance between lower fourth premolars (P4s).

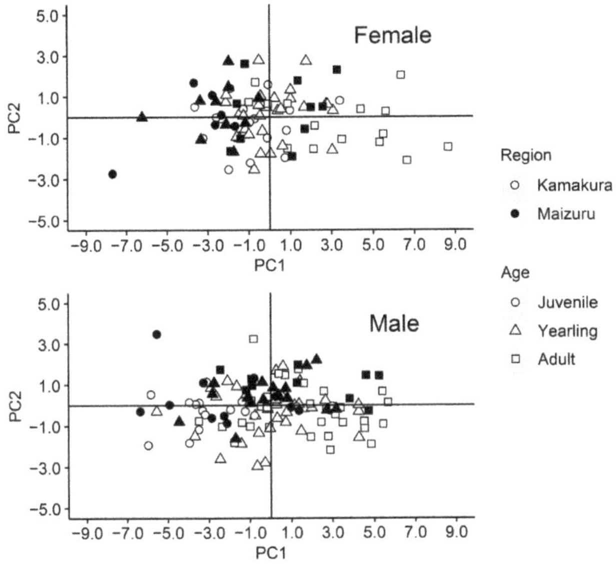

Figure 2. Principal component (PC) analysis of skull measurements in each sex of introduced raccoons in Kamakura and Maizuru, Japan. The white and black colors indicate Kamakura and Maizuru, respectively; and the shape indicate that circle, triangle, and cube are juvenile, yearling, and adult, respectively.

3.3. Relationships between Cranial Size and Basal Factors

Following O'Keefe et al. [46], because of their positive and high coefficients, the PC1s of both sexes were used as size factors in craniometrics; therefore, region, age class, body length, and BMI were used in the regression analysis (Table 3). For the best model in females (Adjusted R^2 = 0.68, $F_{(5, 80)}$ = 36.68, p < 0.001: Table 3), PC1 indicated that the cranial size of raccoons in the Maizuru region was significantly smaller than that in the Kamakura region (coefficient = −2.09, 95% CI = −2.85–−1.33) and for juveniles were smaller than for adults (coefficient = 2.03, 95% CI = 0.94–3.11). The PC1 value of females was also positively correlated with body length (BL) and BMI (BL: coefficient = 0.04, 95% CI = 0.02–0.05; BMI: coefficient = 0.14, 95% CI = 0.01–0.26: Table 3). There were no other valid models with ΔAIC < 2 for the PC1 in females. On the other hand, the PC1 value of males in the best model (Adjusted R^2 = 0.54, $F_{(4, 115)}$ = 35.71, p < 0.001: Table 3) indicated that juveniles were smaller than yearlings (coefficient = 1.30, 95% CI = 0.39–2.31) and adults (coefficient = 1.76, 95% CI = 0.69–2.84). In addition, the PC1 value of males was also positively correlated with BL and BMI (BL: coefficient = 0.03, 95% CI = 0.02–0.04; BMI: coefficient = 0.37, 95% CI = 0.25–0.50) in the best model (Table 3). However, the PC2 value of females was slightly positively correlated with only BL in females for all valid models including the best model (Adjusted R^2 = 0.12, $F_{(4, 81)}$ = 3.94, p < 0.01; BL: coefficient = 0.02, 95% CI = 0.01–0.02; Table 3), whereas the PC2 value of males was high in Maizuru (region: coefficient = 0.85, 95% CI = 0.45–1.25) and positively slight correlated with BL (coefficient = 0.01, 95% CI = 0.004–0.01; best model: Adjusted R^2 = 0.19, $F_{(2, 117)}$ = 15.10, p < 0.001; Table 3).

3.4. Allometry of Cranial Size in Each Sex

The global allometry vector presented by the bootstrap allometric coefficient and 95% CI of cranial size differed in positive or negative patterns against isometry between sexes (Figure 3, Table S4). Variables changing with significant negative allometry included greatest length of skull, basal length, and length of maxillary toothrow; however, positive allometry was only represented by zygomatic width, which indicated that it is associated with sex. Although the cranial width in females and greatest width across first upper molars and greatest width across lower third premolars in male were found to have significant negative allometry, positive allometry was found for paraoccipital width and least distance between lower fourth premolars in males.

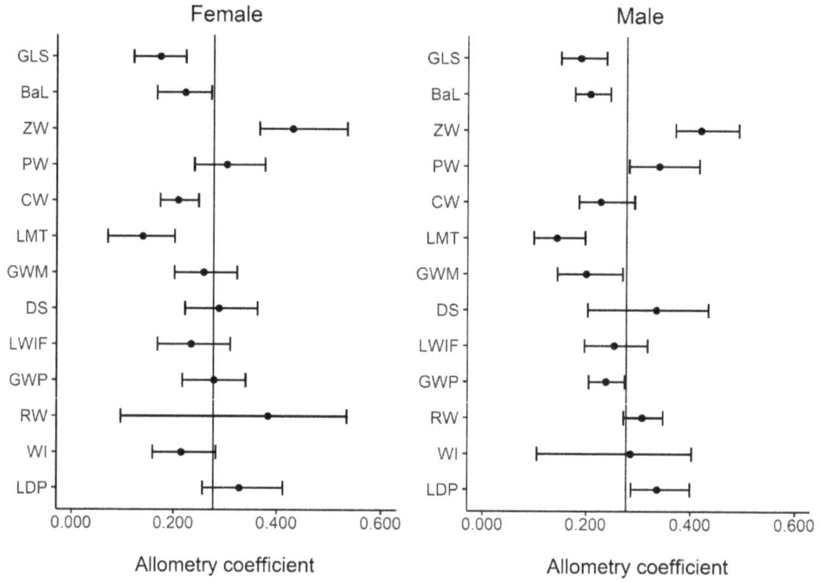

Figure 3. Multivariate allometry vector (the 1st eigenvector derived from the covariance matrix of ln-transformed variables) and 95% confidence interval using the bootstrap resampling procedure or skull morphometrics in each sex of introduced raccoons in Kamakura and Maizuru, Japan. The isometric value is 0.277 according to Jolicoeur [50]. Measurement letters corresponded to those of Figure 1. GLS, greatest length of the skull; BaL, basal length; ZW, zygomatic width; PW, paraoccipital width; CW, cranial width; LMT, length of maxillary tooth row; GWM, greatest width across first upper molars (M1s); DS, depth of skull; LWIF, least width between infraorbital foramen; GWP, greatest width across lower third premolars (P3s); RW, rostral width; WI, width of incisors; LDP, least distance between lower fourth premolars (P4s).

Table 3. Multiple regression analysis for principal component (PC) scores regressed against region, age class, body length (BL), and body mass index (BMI) of female and male introduced raccoons in Japan.

		Regression Coefficient in Modeled Variables						
		Region [1]	**Age Class** [3]		**BL**	**BMI**	**AIC**	**ΔAIC**
			Yearling	**Adult**				
PC1	Female model 1 (Best model)	−2.09 †	0.13	2.03 †	0.04 †	0.14 *	327.5	0.00
	Other valid models with ΔAIC: none							
	Male model 1 (Best model)		1.35 *	1.76 *	0.03 †	0.37 †	498.5	0.00
	model 2	−0.39	1.30 *	1.66 *	0.03 *	0.38 *	499.3	0.86

Table 3. *Cont.*

		Regression Coefficient in Modeled Variables						
		Region [1]	Age Class [3]		BL	BMI	AIC	ΔAIC
			Yearling	Adult				
PC2	Female							
	model 1 (Best model)	0.49	0.09	−0.60	0.02 †		284.1	0.00
	model 2				0.01 *		284.2	0.15
	model 3	0.38			0.01 *		284.5	0.39
	model 4				0.01 *	−0.06	284.6	0.55
	model 5		0.10	−0.50	0.01 *		285.1	0.97
	model 6	0.32			0.01 *	−0.05	285.4	1.34
	model 7	0.44	0.08	−0.58	0.02 †	−0.03	285.6	1.46
	model 8		0.07	−0.48	0.02 †	−0.05	285.9	1.77
	Male							
	model 1 (Best model)	0.85 †			0.01 †		352.0	0.00
	model 2	0.85 †			0.01 †	0.02	353.6	1.63

[1] *p*-values refer to Ho: regression coefficient, 0. *: $p < 0.05$. †: $p < 0.001$. [3] The variable is categorical, which means "Yearling" and "Adult" coefficients against "Juvenile (Reference)", respectively.

4. Discussion

Body length and craniometrics of mammalian species are standard measurements used to understand their morphological characteristics, including those used to test Biogeography theories and identify alien species [6,55]. For native raccoon populations, although the cranial morphology is well-understood [24], only a few reports describe systematic body length details. Some studies have reported the body lengths of introduced raccoons in Germany [43], Spain [44], and Japan [56]; however, it is unclear how the body length differs with sex and age classes. This is the first study to report that the body length and craniometrics of introduced raccoon populations during their early invasion stage are related to sex and age classes (Tables 1 and 3).

The evolution of SSD has been explained by three major hypotheses based on: sexual selection, intersexual food competition, and reproductive role division [57]. In raccoons, SSD has been associated with body length, namely in terms of the larger size of males compared to that of females [22,58]. Since raccoons are typically promiscuous [21], their SSD may be caused by sexual selection. However, if body size increases with age in not only males but also females, resulting in an advantage for intrasexual selection or food competition, the SSD of raccoons could indicate a relatively low value compared to differences in age class. While older males with large body size have been considered to dominate over smaller individuals, larger individuals (regardless of sex) have an advantage, particularly in the competition for food resources [58,59]. Isaac [60] suggested that the pattern of reproductive efforts in the common brushtail possums (*Trichosurus vulpecula*) differed with regard to age in males, and the older, more dominant individuals may invest less time than younger males in finding receptive females owing to a hierarchal social system, resulting in an increase in foraging behavior. In addition, female reproductive characteristics of raccoons, including pregnancy rates and litter sizes, were increased for adults compared to yearlings in both native and introduced areas [21,32,33,61]. Our study demonstrated that both sex and age classes were positively related to the body length (Table 1); the resulting differences in body length by the sex-age class of raccoons indicated an association with intrasexual selection for both sexes and competition for food resources.

Regarding cranial characteristics, the SSD of raccoons in native distributions revealed that males were larger than females [23]. The PC1s for both sexes derived for introduced raccoons in Kamakura and Maizuru could be considered cranial size indicators as described by O'Keefe et al. [47]. Our study demonstrated that the PC1s were high for the older age class and had a week positive correlation with body length and BMI (Table 3), except between juveniles and yearlings in females. In the allometry assessment using PC1 as a size indicator, the zygomatic width of both sexes and the paraoccipital width correlations of males were found to be positive (Figure 3), which are important for masseter muscle attachment. Given that in raccoons, juvenile male dispersal is known to occur earlier than female dispersal [62], the difference between sexes in cranial growth pattern can be considered an important characteristic. Consequently, our results on cranial size also suggest that older individuals exercising dominance in the social group have larger body sizes [59]. In addition, the BMI of female raccoons in Kamakura was reported to increase with their growth until maturity despite seasonal fluctuations [63]. Based on BMI evaluation of the body condition index of raccoons [40,64], individuals with high BMI can be considered to have good physiques. Raccoon cranial width dimensions in particular, indicated an advantageous factor for competition among individuals and was related to body length growth and BMI in the two introduced populations in middle Japan.

The regression model for body length indicated no difference between Kamakura and Maizuru raccoon populations (Table 1). However, slight differences were found in the craniometrics between regions in the sizes (PC1) of females and the shapes (PC2) of males (Table 3). Goldman [22] described that at least three subspecies including *P. lotor lotor*, *P. l. hirtus*, and *P. l. varius* have widespread distribution in North America, and that the body length increases in the order of *P. l. varius* < *P. l. lotor* < *P. l. hirtus* based on the measurement of few specimens of each subspecies in several regions. The body lengths of both sexes of adult raccoons in Kamakura and Maizuru tended to be close to those of *P. l. lotor* in Virginia [22], as shown by the least square mean analysis (females: $F_{(1, 1373)} = 1.64$, $p = 0.22$; males: $F_{(1, 95)} = 0.08$, $p = 0.78$). However, craniometrics for individuals from both regions indicated a similar size compared to *P. l. hirtus*, which was a larger subspecies than *P. l. lotor* [22,24]. Hence, the introduced raccoons in Kamakura and Maizuru were found to be morphologically inconsistent with subspecies in North America.

Genetic structure evaluation of raccoons has indicated an absence of reproductive-isolating barriers due to the degree of lineage mixing across subspecific native distributions [27]. Some of the haplotypes across the subspecies distribution in native regions have been found in raccoons introduced into Europe and Japan [65,66]; this has led to the hypothesis that raccoon populations had already undergone subspecific hybridization before their distribution and introduction into these regions. The raccoons introduced into Kamakura and Maizuru seem to be independent in terms of their genetic background [30,31]; which we consider to be an important reason for the slight morphological differences in body size between these populations. The first raccoon naturalizations in Kamakura City and Maizuru City were reported in 1988 [18] and 2001 [67], respectively. Capture removal of raccoons has subsequently been promoted since 2006 in Kamakura and 2009 in Maizuru by the Raccoon Control Program, which was formulated by the municipalities based on the Invasive Alien Species Act 2005. Considering that raccoon naturalization occurred in the 1980s [18], mating occurs once yearly, and females mature at only 1 year of age [33], this species has not yet been considered as affected by environmental factors because there have been only 20–30 generations since its establishment. Comparisons between morphological variations in introduced and indigenous populations have been reported in some carnivore species. These include the raccoon dog [6], American mink (*Neovison vison*) [7,10], and the small Indian mongoose (*Herpestes javanicus*) [8], in which the size variability of the introduced populations is connected with the invasion processes as well as the history of the introduction site, in terms of factors such as the animal fur industry, adoption as pets, the number of naturalized animals, and the influence of several biotic and abiotic environmental factors.

5. Conclusions

Non-native raccoons introduced to Kamakura and Maizuru present that: (1) SSD and that individuals of the older age class have larger body length compared to the other age classes; (2) no major differences of body length and craniometrics were found between regions, indicating that the distance is not influence their morphologies; (3) the craniometrics and cranial growth pattern for the cranial width dimensions in particular, related to age class, body length, and body mass index implicate an important factor for competition among individuals in each sex. These characteristics may be related to the ecology of the raccoons, including intrasexual selection for both sexes and competition for food resources. Future studies should investigate the morphological differences among raccoon populations in various introduced areas and the morphological changes of the introduced populations over the next few decades, focusing on factors such as genetic differentiation, environmental adaptation, and population density.

Supplementary Materials: The following supporting information can be downloaded at: https://www.mdpi.com/article/10.3390/ani13010055/s1, Figure S1: Map of the study area for raccoon specimen collecting from two introduced regions in Japan; Figure S2: Relationship of body length with sex-age class categories of introduced raccoons in Kamakura and Maizuru, Japan; Table S1: Data set; Table S2: Descriptive analysis of body length with region, sex, and age class categories for introduced raccoons in Kamakura and Maizuru, Japan; Table S3: Descriptive analysis for 13 craniometric parameters with region, sex, and age class categories for introduced raccoons in Kamakura and Maizuru, Japan; Table S4: Allometry coefficient and 95% confidence interval (CI) for cranial size of each sex in introduced raccoons of Kamakura and Maizuru, Japan.

Author Contributions: Conceptualization, T.K. and F.Y.; methodology, T.K.; software, T.K.; validation, T.K., K.D. and S.-i.H.; formal analysis, T.K.; investigation, F.Y.; resources, M.K. and S.-i.H.; data curation, T.K.; writing—original draft preparation, T.K.; writing—review and editing, T.K.; visualization, T.K. and K.D.; supervision, M.K. and S.-i.H.; project administration, T.K. All authors have read and agreed to the published version of the manuscript.

Funding: This research received no external funding.

Institutional Review Board Statement: The animal study protocol was approved by the Bioethics Committee of Nippon Veterinary and Life Science University (protocol code: 2022K-3; approval date: 31 January 2022).

Informed Consent Statement: Not applicable.

Data Availability Statement: The data presented in this study are available in Supplementary Material Table S1.

Acknowledgments: We thank the Kamakura and Maizuru local government offices for their cooperation in constructing the raccoon retrieval system and M. Kaneda, T. Yamamoto, S. Ishihara, and T. Uno for assisting in research and providing information. We are also grateful to the members of the Laboratory of Wildlife Medicine, Nippon Veterinary and Life Science University for their assistance with analyses. This study was supported by the Nippon Veterinary and Life Science University Research Program for Young Scientists.

Conflicts of Interest: The authors declare no conflict of interest.

References

1. Ikeda, T.; Asano, M.; Matoba, Y.; Abe, G. Present status of invasive alien raccoon and its impact in Japan. *Glob. Environ. Res.* **2004**, *8*, 125–131.
2. Long, J.L. Carnivora. In *Introduced Mammals of the World: Their History, Distribution and Influence*; Csiro Publishing: Collingwood, VIC, Australia, 2003; pp. 237–334.
3. Kamler, J.F.; Ballard, W.B. A review of native and nonnative red foxes in North America. *Wildl. Soc. Bull.* **2002**, *30*, 370–379. [CrossRef]
4. Kauhara, K. Introduced carnivores in Europe with special reference to central and northern Europe. *Wildl. Biol.* **1996**, *2*, 197–204.
5. Ansorge, H.; Ranyuk, M.; Kauhala, K.; Kowalczyk, R.; Stier, N. Raccoon dog, *Nyctereutes procyonoides*, populations in the area of origin and in colonised regions—The epigenetic variability of an immigrant. *Ann. Zool. Fenn.* **2009**, *46*, 51–62. [CrossRef]

6. Korablev, N.P.; Szuma, E. Variability of native and invasive raccoon dogs' *Nyctereutes procyonoides* populations: Looking at translocation from a morphological point of view. *Acta Theriol.* **2014**, *59*, 61–79. [CrossRef]
7. Zalewski, A.; Bartoszewicz, M. Phenotypic Variation of an alien species in a new environment: The body size and diet of American mink over time and at local and continental scales. *Biol. J. Linn. Soc.* **2012**, *105*, 681–693. [CrossRef]
8. Simberloff, D.; Dayan, T.; Jones, C.; Ogura, G.O. Character displacement and release in the small Indian mongoose, *Herpestes javanicus*. *Ecology* **2000**, *81*, 2086–2099. [CrossRef]
9. Yom-Tov, Y.; Yom-Tov, S.; Moller, H. Competition, coexistence, and adaptation amongst rodent invaders to Pacific and New Zealand Islands. *J. Biogeogr.* **1999**, *26*, 947–958. [CrossRef]
10. Melero, Y.; Santulli, G.; Gómez, A.; Gosàlbez, J.; Rodriguez-Refojos, C.; Palazón, S. Morphological variation of introduced species: The case of American mink (*Neovison vison*) in Spain. *Mamm. Biol.* **2012**, *77*, 345–350. [CrossRef]
11. MacDonald, S.O.; Cook, J.A. The land mammals fauna of southeast Alaska. *Can. Field-Nat.* **1996**, *110*, 571–598.
12. Aliev, F.F.; Sanderson, G.C. Distribution and status of the raccoon in the Soviet Union. *J. Wildl. Manag.* **1966**, *30*, 497–502 [CrossRef]
13. Salgado, I. Is the Raccoon (*Procyon lotor*) out of control in Europe? *Biodivers. Conserv.* **2018**, *27*, 2243–2256. [CrossRef]
14. Louppe, V.; Leroy, B.; Herrel, A.; Veron, G. Current and future climatic regions favourable for a globally introduced wild carnivore, the raccoon *Procyon lotor*. *Sci. Rep.* **2019**, *9*, 9174. [CrossRef] [PubMed]
15. Boncompagni, L.; Molfini, M.; Ciampelli, P.; Fazzi, P.; Lucchesi, M.; Mori, E.; Petralia, L.; Mazza, G. No country for native crayfish: Importance of crustaceans in the diet of native and alien Northern raccoons. *Ethol. Ecol. Evol.* **2021**, *33*, 576–590. [CrossRef]
16. Agetsuma-Yanagihara, Y. Process of establishing an introduced raccoon (*Procyon lotor*) population in Aichi and Gifu Prefectures, Japan: Policy for Managing Threats Posed by Introduced Raccoons. *Honyurui Kagaku Mamm. Sci.* **2004**, *44*, 147–160. [CrossRef]
17. Miyashita, M. Prevalence of *Baylisascaris procyonis* in raccoons in Japan and experimental infections of the worm to laboratory animals. *Seikatsu Eisei J. Urban Living Health Assoc.* **1993**, *37*, 137–151. [CrossRef]
18. Nakamura, K. A Note on the naturalization of common raccoon in Kanagawa Prefecture. *Nat. Hist. Rep. Kanagawa* **1991**, *12*, 17–19. (In Japanese)
19. Ikeda, T. Problems controlling the invasive raccoon in Japan. *Honyurui Kagaku Mamm. Sci.* **2006**, *46*, 95–97. (In Japanese) [CrossRef]
20. Ministry of Environment Results of Habitat Distribution Survey of Raccoon, Palm Civet, Nutria. Available online: https://www.env.go.jp/press/105902.html (accessed on 14 September 2020).
21. Gehrt, S.D. Raccoon (*Procyon lotor* and Allies). In *Wild mammals of North America: Biology, management, and Conservation*; Feldhamer, G.A., Thompson, B.C., Chapman, J.A., Eds.; Johns Hopkins University Press: Baltimore, MD, USA, 2003; pp. 611–634.
22. Goldman, E.A. Raccoons of north and middle America. *Fish Wildl. Serv.* **1950**, *60*, 1–153.
23. Ritke, M.E. Sexual Dimorphism in the Raccoon (*Procyon lotor*): Morphological evidence for intrasexual selection. *Am. Midl. Nat.* **1990**, *124*, 342–351. [CrossRef]
24. Ritke, M.E.; Kennedy, M.L. Intraspecific morphologic variation in the raccoon (*Procyon lotor*) and its relationship to selected environmental variables. *Southwest. Nat.* **1988**, *33*, 295–314. [CrossRef]
25. Kennedy, M.L.; Lindsay, S.L. Morphologic variation in the raccoon, *Procyon lotor*, and its relationship to genic and environmental variation. *J. Mammal.* **1984**, *65*, 195–205. [CrossRef]
26. Ritke, M.E. Quantitative assessment of variation in litter size of the raccoon *Procyon lotor*. *Am. Midl. Nat.* **1990**, *123*, 390–398. [CrossRef]
27. Cullingham, C.I.; Kyle, C.J.; Pond, B.A.; White, B.N. Genetic structure of raccoons in eastern North America based on MtDNA: Implications for subspecies designation and rabies disease dynamics. *Can. J. Zool.* **2008**, *86*, 947–958. [CrossRef]
28. Yoshida, K.; Hirose, M.; Hasegawa, M.; Inoue, E. Mitochondrial DNA analyses of invasive raccoons (*Procyon lotor*) in the Boso Peninsula, Japan. *Mammal Study* **2020**, *45*, 71–76. [CrossRef]
29. Hamaya, Y.; Tezuka, T.; Kato, T.; Hayama, S. Population structure of raccoon (*Procyon lotor*) based on mitochondrial DNA analysis in Miura Peninsula, Kanagawa Prefecture. 2019; *manuscript in preparation*.
30. Takada-Matsuzaki, Y.; Mukaida, M.; Suzuki, K.; Asano, M.; Makino, T.; Matoba, Y.; Ohtaishi, N.; Imai, T. Geographical distribution of raccoon (*Procyon lotor*) mitochondrial DNA polymorphism. *DNA Polymorph.* **2004**, *12*, 39–44. (In Japanese)
31. Takada, Y.; Murayama, T.; Kanetake, J.; Imai, T.; Suzuki, K. Mitochondrial DNA analysis of *Procyon lotor* in Kyoto. *DNA Polymorph.* **2009**, *17*, 51–54. (In Japanese)
32. Asano, M.; Matoba, Y.; Ikeda, T.; Suzuki, M.; Asakawa, M.; Ohtaishi, N. Reproductive characteristics of the feral raccoon (*Procyon lotor*) in Hokkaido, Japan. *J. Vet. Med. Sci.* **2003**, *65*, 369–373. [CrossRef]
33. Kato, T.; Ichida, Y.; Tei, K.; Asano, M.; Hayama, S. Reproductive characteristics of feral raccoons (*Procyon lotor*) captured by the pest control in Kamakura, Japan. *J. Vet. Med. Sci.* **2009**, *71*, 1473–1478. [CrossRef]
34. Bartoszewicz, M.; Okarma, H.; Zalewski, A.; Szczęsna, J. Ecology of the raccoon (*Procyon lotor*) from western Poland. *Ann. Zool. Fenn.* **2008**, *45*, 291–298. [CrossRef]
35. Matsuo, R.; Ochiai, K. Dietary overlap among two introduced and one native sympatric carnivore species, the raccoon, the masked palm civet, and the raccoon dog, in Chiba Prefecture, Japan. *Mammal Study* **2009**, *34*, 187–194. [CrossRef]
36. Osaki, A.; Sashika, M.; Abe, G.; Shinjo, K.; Fujimoto, A.; Nakai, M.; Shimozuru, M.; Tsubota, T. Comparison of feeding habits and habitat use between invasive raccoons and native raccoon dogs in Hokkaido, Japan. *BMC Ecol.* **2019**, *19*, 35. [CrossRef] [PubMed]

7. Wildlife Committee, Small Animal Clinical Section, Japan Veterinary Medical Association. *Guidelines for the Management of Invasive Alien Species*; Japan Veterinary Medical Association: Tokyo, Japan, 2007; pp. 11–16. Available online: http://nichiju.lin.gr.jp/kousyu/pdf/h19_07_yasei.pdf(In Japanese). (accessed on 19 November 2022). (In Japanese)
8. Montgomery, G.G. Tooth eruption in preweaned raccoons. *J. Wildl. Manag.* **1964**, *28*, 582–584. [CrossRef]
9. Stuewer, F.W. Raccoons: Their habits and management in Michigan. *Ecol. Monogr.* **1943**, *13*, 203–257. [CrossRef]
0. Kato, T.; Uno, T.; Fujioka, Y.; Hayama, S. Indices using external measurements for assessing fat deposition of adult feral raccoons (*Procyon lotor*) in Kanagawa Prefecture, Japan. *J. Vet. Med. Sci.* **2012**, *74*, 181–188. [CrossRef] [PubMed]
1. Grau, G.A.; Sanderson, G.C.; Rogers, J.P. Age determination of raccoons. *J. Wildl. Manag.* **1970**, *34*, 364–372. [CrossRef]
2. Junge, R.; Hoffmeister, D.F. Age determination in raccoons from cranial suture obliteration. *J. Wildl. Manag.* **1980**, *44*, 725–729. [CrossRef]
3. Lutz, W. Occurrence and morphometrics of the raccoon *Procyon lotor* L. in Germany. *Ann. Zool. Fenn.* **1995**, *32*, 15–20.
4. Lassnig, N.; Colomar, V.; Picó, G.; Perelló, E.; Febrer-Serra, M.; Truyols-Henares, F.; Pinya, S. Assessment of the invasion process of the common raccoon *Procyon lotor* (Carnivora: Procyonidae) on a Mediterranean Island a decade after its introduction. *Sci. Total Environ.* **2020**, *699*, 134191. [CrossRef]
5. Tay, R. Correlation, variance inflation and multicollinearity in regression model. *J. East. Asia Soc. Transp. Stud.* **2017**, *12*, 2006–2015. [CrossRef]
6. Flajšman, K.; Jerina, K.; Pokorny, B. Age-related effects of body mass on fertility and litter size in roe deer. *PLoS ONE* **2017**, *12*, e0175579. [CrossRef] [PubMed]
7. O'Keefe, F.R.; Meachen, J.; Fet, E.V.; Brannick, A. Ecological determinants of clinal morphological variation in the cranium of the North American gray wolf. *J. Mammal.* **2013**, *94*, 1223–1236. [CrossRef]
8. Haba, C.; Oshida, T.; Sasaki, M.; Endo, H.; Ichikawa, H.; Masuda, Y. Morphological variation of the japanese raccoon dog: Implications for geographical isolation and environmental adaptation. *J. Zool.* **2008**, *274*, 239–247. [CrossRef]
9. Chong, V.K.; Fung, H.F.; Stinchcombe, J.R. A note on measuring natural selection on principal component scores. *Evol. Lett.* **2018**, *2*, 272–280. [CrossRef] [PubMed]
0. Jolicoeur, P. The multivariate generalization of the allometry equation. *Biometrics* **1963**, *19*, 497–499. [CrossRef]
1. R Core Team. *R: A Language and Environment for Statistical Computing*; Version 4.2.2; Windows; R Foundation for Statistical Computing: Vienna, Austria, 2022. Available online: https://www.R-project.org/ (accessed on 19 November 2022).
2. Fox, J.; Weisberg, S. *An R Companion to Applied Regression*, 3rd ed.; Sage publications, Inc.: Thousand Oaks, CA, USA, 2019; pp. 429–434.
3. Bartoń, K. MuMIn: Multi-Model Inference (R Package Version 1.47.1). 2022. Available online: https://cran.r-project.org/package=MuMIn (accessed on 19 November 2022).
4. Canty, A.; Ripley, B. Boot: Bootstrap R (S-Plus) Functions (R Package Version 1.3-28). 2021. Available online: https://cran.r-project.org/web/packages/boot/ (accessed on 19 November 2022).
5. Mori, E.; Ancillotto, L.; Lovari, S.; Russo, D.; Nerva, L.; Mohamed, W.F.; Motro, Y.; Di Bari, P.; Plebani, M. Skull shape and Bergmann's rule in mammals: Hints from Old world porcupines. *J. Zool.* **2019**, *308*, 47–55. [CrossRef]
6. Asano, M.; Matoba, Y.; Ikeda, T.; Suzuki, M.; Asakawa, M.; Ohtaishi, N. Growth pattern and seasonal weight changes of the feral raccoon (*Procyon lotor*) in Hokkaido, Japan. *Jpn. J. Vet. Res.* **2003**, *50*, 165–173. [PubMed]
7. Hedrick, A.V.; Temeles, E.J. The evolution of sexual dimorphism in animals: Hypotheses and tests. *Trends Ecol. Evol.* **1989**, *4*, 136–138. [CrossRef]
8. Johnson, A.S. *Biology of the Raccoon (*Procyon lotor *Varius Nelson and Goldman) in Alabama*; Agricultural Experiment Station, Auburn University: Auburn, AL, USA, 1970; pp. 30–37.
9. Hauver, S.; Hirsch, B.T.; Prange, S.; Dubach, J.; Gehrt, S.D. Age, but not sex or genetic relatedness, shapes raccoon dominance patterns. *Ethology* **2013**, *119*, 769–778. [CrossRef]
0. Isaac, J.L. Mass Loss during the mating season in male common brushtail possums (*Trichosurus vulpecula*): Implications for age-specific variation in mating strategy. *J. Zool.* **2005**, *266*, 181–185. [CrossRef]
1. Sanderson, G.C.; Nalbandov, A.V. The reproductive cycle of the raccoon in Illinois. *Ill. Nat. Hist. Surv. Bull.* **1973**, *31*, 25–85. [CrossRef]
2. Gehrt, S.D.; Fritzell, E.K. Duration of familial bonds and dispersal patterns for raccoons in South Texas. *J. Mammal.* **1998**, *79*, 859–872. [CrossRef]
3. Kato, T.; Uno, T.; Fujioka, Y.; Kobayashi, K.; Takayanagi, E.; Hayama, S. Difference in age at first conception between early and late litters of feral raccoon (*Procyon lotor*) in Kanagawa Prefecture. *J. Vet. Med. Sci.* **2012**, *74*, 637–640. [CrossRef] [PubMed]
4. Pitt, J.A.; Larivière, S.; Messier, F. Condition indices and bioelectrical impedance analysis to predict body condition of small carnivores. *J. Mammal.* **2006**, *87*, 717–722. [CrossRef]
5. Biedrzycka, A.; Zalewski, A.; Bartoszewicz, M.; Okarma, H.; Jędrzejewska, E. The genetic structure of raccoon introduced in central Europe reflects multiple invasion pathways. *Biol. Invasions* **2014**, *16*, 1611–1625. [CrossRef]

66. Okuyama, M.W.; Shimozuru, M.; Nakai, M.; Yamaguchi, E.; Fujii, K.; Shimada, K.; Ikeda, T.; Tsubota, T. Genetic population structure of invasive raccoons (*Procyon lotor*) in Hokkaido, Japan: Unique Phenomenon Caused by Pet Escape or Abandonment. *Sci. Rep.* **2020**, *10*, 8108. [CrossRef] [PubMed]
67. Kawamichi, M.; Kawamichi, T.; Yamamoto, K.; Yahiro, Y.; Hazama, K.; Kaneda, M.; Kato, T. Raccoon invasion status quo and countermeasures. *Anim.-Husb.* **2013**, *67*, 633–641. (In Japanese)

animals

MDPI

rticle

Arterial Circle of the Brain of the Red-Necked Wallaby (*Notamacropus rufogriseus*)

Maciej Zdun [1,2], Jakub J. Ruszkowski [1,*], Maciej Gogulski [3,4,5], Agata Józefiak [3] and Mateusz Hetman [1]

1 Department of Animal Anatomy, Poznan University of Life Sciences, Wojska Polskiego 71C, 60-625 Poznan, Poland
2 Department of Basic and Preclinical Sciences, Nicolaus Copernicus University in Torun, Lwowska 1, 87-100 Torun, Poland
3 Department of Preclinical Sciences and Infectious Diseases, Poznan University of Life Sciences, Wołynska 35, 60-637 Poznan, Poland
4 University Centre for Veterinary Medicine, Szydłowska 43, 60-656 Poznan, Poland
5 Centre of Biosciences, Institute of Animal Physiology, Soltesovej 4-6, 040-01 Kosice, Slovakia
* Correspondence: jakub.ruszkowski@up.poznan.pl; Tel.: +48-732-192-981

Simple Summary: The red-necked wallaby is a small, herbivorous mammal that is native to Australia. In the study, the anatomy of arterial vessels that transport blood to the brain in this species was described and compared with that of other groups of mammals. The results of this research can contribute to further physiological and pathophysiological studies. This is the first description of this anatomical area that has been carried out in a marsupial species.

Abstract: The red-necked wallaby is a medium-sized marsupial species, which have increasingly been kept as pets around the world. In the study, the arterial blood supply for the brain in this species was described. The study was conducted on 50 specimens with two preparation methods. The main artery supplying the brain was the internal carotid artery. The arterial circle of the brain was closed from the caudal side. The anatomy of the arteries of the described region was compared with other groups of mammals. This is the first description of this anatomical area that has been carried out in a marsupial species. Understanding the anatomy of the circulatory system in the wallaby can be valuable for further physiological and pathophysiological studies.

Keywords: anatomy; vascularization; angiology; marsupials; circle of Willis

1. Introduction

Marsupials are a large and diversified group of mammals. Members of the Macropodoidea, a big marsupial superfamily that contains kangaroos, wallabies, and rat kangaroos, live naturally only in Papua New Guinea and Australia, where they are the main large terrestrial herbivores. The evolution of bipedal locomotion that is typical for macropodoids has affected various morphological and functional adaptations of their cardiovascular, respiratory, and musculoskeletal systems [1–4]. Wallabies, including the red-necked wallaby or Bennet's wallaby (*Notamacropus rufogriseus*), are medium-sized hopping, herbivorous marsupials that are native to eastern Australia.

In captivity, Bennet's wallabies are kept in European zoos and wildlife parks. Keeping them in zoos dates back to the 1930s, according to the Zoological Society of London [5]. In some countries, they are also considered an alternative to farm animals as a source of meat [6]. Currently, the anatomy of cerebral brain arteries has not been described in any marsupial species, including the red-necked wallaby.

Wallaby medicine is a developing branch of veterinary medicine worldwide [7–15]. Among the diseases described in the literature, parasitic infestations [12], bacterial diseases of the oral cavity [13], and cardiomyopathies can be mentioned [8]. Since it is not easy to restrain these species (most of them are kept in extensive enclosures, or are free-roaming), most of the veterinary activities (clinical examination, routine treatment) are carried out with the use of chemical restraint. Furthermore, macropods are exceedingly sensitive to capture myopathy [5]. The increasing number of wallabies kept at zoos, as farm animals, or as pets worldwide, makes it necessary to create diagnostic and treatment protocols for veterinarians. It is important to know the detailed anatomy of every species, in order to acknowledge the pathophysiology of various, often newly discovered diseases. Understanding the anatomy of the brain's vascular system creates new opportunities for further studies on cardiovascular diseases and their complications. Such diseases have also been described in kangaroos, including red-necked wallabies [14]. Hypertrophic cardiomyopathy may lead to thromboembolic events that affect cerebral vasculature in other mammals [15].

In this study we described, for the first time, the arterial blood supply to the brain of a red-necked wallaby.

2. Materials and Methods

2.1. Animals

The study was conducted on 50 adult (21 male and 29 female) red-necked wallabies (*Notamacropus rufogriseus*). The analyzed specimens were delivered as post-mortem material from zoological gardens and private breeders in Poland, where animals are maintained under constant veterinary control and anti-parasitic prophylaxis or vaccinations against bacterial diseases, for example, tetanus. All animals that were included in this study had been euthanized [with xylazine 10 mg/kg (intramuscular; i.m.), ketamine 90 mg/kg (i.m.), and pentobarbital 100 mg/kg (intravenous; i.v.)] for medical reasons other than for neurological or cardiovascular disease. The animals had not been euthanized for an experimental purpose. In the study, only animals lacking head and neck trauma of any source were included. All of the animals were frozen prior to the study.

2.2. Methods

The methods used in the study were based on traditional anatomical techniques used in angiology research; however, new, high-quality advanced imaging techniques were also used.

The specimens were prepared using three methods, with each specimen being randomly assigned to one of the methods. In twenty-eight specimens, injecting both common carotid arteries with a tinged solution of the chemo-setting acrylic material Duracryl® Plus (SpofaDental, Jičín, Czech Republic) was used. When the material hardened, the specimens were subjected to an enzymatic maceration process (using Persil washing powder). The temperature used for the maceration process was 38 °C, and it lasted 40 days. The final effects of the process were arterial casting of the head and cranial cavity on a skeletal scaffold. The second method, used for 20 specimens, consisted of introducing liquid LBS 3060 latex into the bilateral common carotid arteries. After curing in 5% formalin solution, the specimens were rinsed in water for two days to flush out excess formalin solution. Additional protection was provided with a ventilation system in a room that was used for preparation. The system was set for 15 air changes per hour. The skull bone was cut, carefully, using an oscillating saw, and the skull was opened to enable soft tissue preparation. Then, the blood vessels were prepared manually using surgical instruments. Preparation began with removing skin from the entire neck and head. After this, muscle tissue was gently prepared to avoid damage to the surrounding veins and arteries. Then, prepared arteries were cleaned from excess connective tissue. In this way, blood vessels on the animal's soft tissues were obtained.

Two cadavers of both genders were used for angio-CT examination. Before the scans, the bilateral common carotid arteries were injected with barium sulphate (barium sulfuricum 1.0 g/mL, Medana, Sieradz, Poland). The heads were fixed motionless on the tomograph's table. Cone-bean computed tomography (Fidex Animage, Pleasanton,

CA, USA) was performed at the University Centre for Veterinary Medicine in Poznan, Poland, with scanning parameters of 110 kVp, 0.08 mAs pet shot, 20.48 mAs (Total mAs), and a reconstructed slice thickness of 0.29 mm. After the examination scan was reconstructed, the 3D model was created using public available software 3D Slicer (version 5.0.3). All of the excess tissue, except for bone tissue and arteries, was cut out, digitally. The bone tissue and arteries were digitally colored—yellow for bone and red for arteries.

The names of the anatomical structures were standardized according to Nomina Anatomica Veterinaria [16].

For the next process, all of the preparations were photographed and described in detail. The described material was then compared to those of other groups of mammals. A digital camera (Nikon D3200) was used to obtain the photographs. The photographs were saved in JPG format. GIMP v2.10.18, digital image editing software, was used to process the photographs.

3. Results

Blood enters the cranial cavity via the internal carotid artery (*arteria carotis interna*), which is a branch of the common carotid artery (*arteria carotis communis*) (Figure 1).

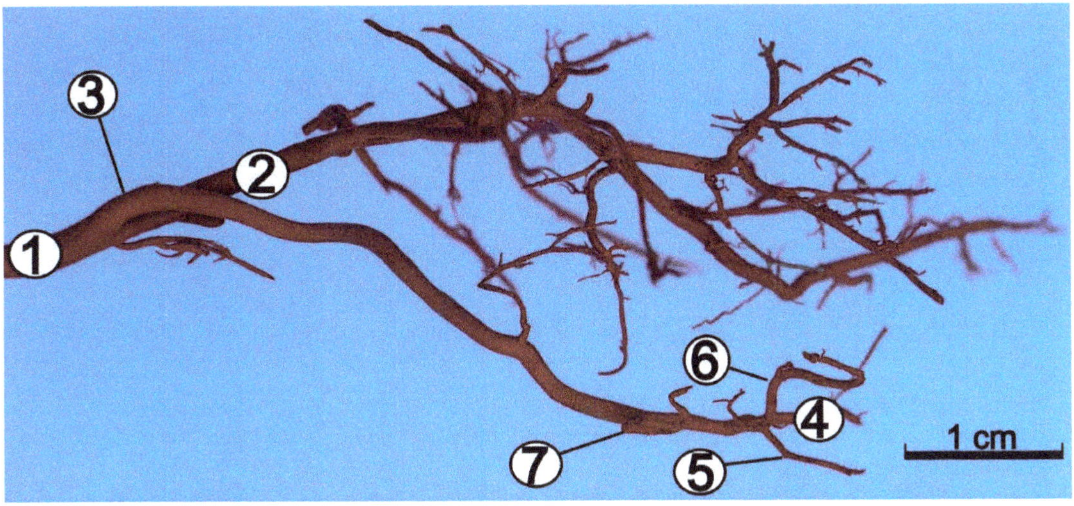

Figure 1. Arteries of the left side of the head of the red-necked wallaby. Corrosion cast. 1—common carotid artery; 2—external carotid artery; 3—internal carotid artery; 4—internal ophthalmic artery; 5—rostral cerebral artery; 6—middle cerebral artery; 7—caudal communicating artery.

There was no carotid sinus (*sinus caroticus*) found in the initial course of the internal carotid artery. The course of the internal carotid artery was straight, and it did not turn sideways. On its extracranial part (from branching off of the common carotid artery to enter the cranial cavity), it was not providing any branches, and was not anastomosing with other arterial vessels. This artery penetrated the cranial cavity and laid on the base of the cerebrum (Figure 2), moving towards the optic canal.

After passing through it, the internal ophthalmic artery (*arteria ophthalmica interna*), was observed to enter the orbit. The internal carotid artery was a strong vessel over its entire length. It divided into the internal ophthalmic artery and the rostral cerebral artery (*arteria cerebri rostralis*) (Figures 1 and 2). At that point, the diameters of these vessels became smaller. The rostral choroidal artery (*arteria choroidea rostralis*), which runs dorsally between the temporal lobe of the brain and the cerebral crus, branched off from the beginning segment of the rostral cerebral artery. The rostral cerebral artery gave way to the middle cerebral artery (*arteria cerebri media*), a vessel with a large diameter (Figure 2). It ran dorsally, and surrounded the piriform lobe from its rostral side, and divided into numerous branches. The middle cerebral artery branched off into four main branches (Figure 3).

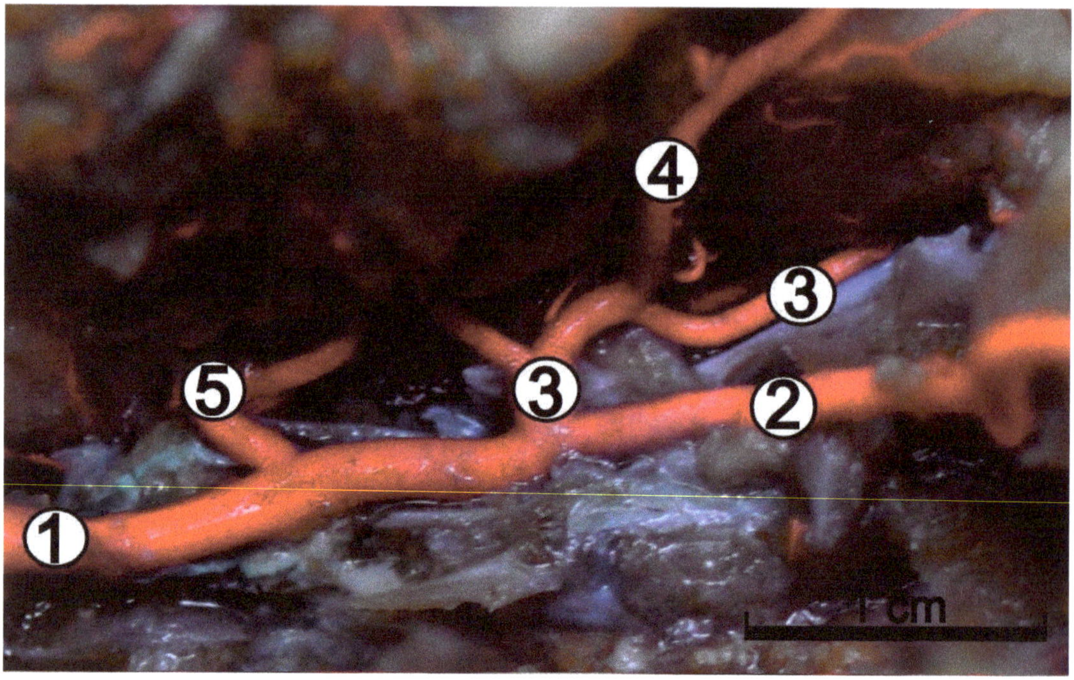

Figure 2. Branches of the internal carotid artery in the cranial cavity between the brain and the base of the skull. Latex preparation. 1—internal carotid artery; 2—internal ophthalmic artery; 3—rostral cerebral artery; 4—middle cerebral artery; 5—caudal communicating artery.

From the beginning section of the middle cerebral artery branched off two vessels: the frontal and the orbital branches. Altogether, those vessels were observed to branch off together as a common trunk. The frontal branch divided further into two smaller vessels once again: the ventral frontal branch and the dorsal frontal branch. The caudal olfactory artery branched off near the point where the above-mentioned common trunk branched off. Next, in the caudal direction, branched off temporal branches (divided secondarily into the middle and ventral temporal branches). Finally, the middle cerebral artery divided into the parietal branch (divided secondarily into the rostral and caudal parietal branches) and the dorsal temporal branch. The dorsal temporal branch was the vessel observed to have the largest diameter. Once the middle cerebral artery branched off, the rostral cerebral artery's lumen reduced significantly. In 39 specimens, the rostral cerebral artery branched off the rostral communicating artery (*arteria communicans rostralis*). This vessel branched off at a short distance from the middle cerebral artery. The presence of the rostral communicating artery provided evidence that the arterial circle of the brain is closed on the rostral side. The rostral cerebral artery was observed to head rostrally, and then change direction, laying in the longitudinal fissure of the brain. Between the olfactory bulbs, the internal ethmoidal artery (*arteria ethmoidalis interna*) branched off, leading to the cribriform plate of the ethmoid bone.

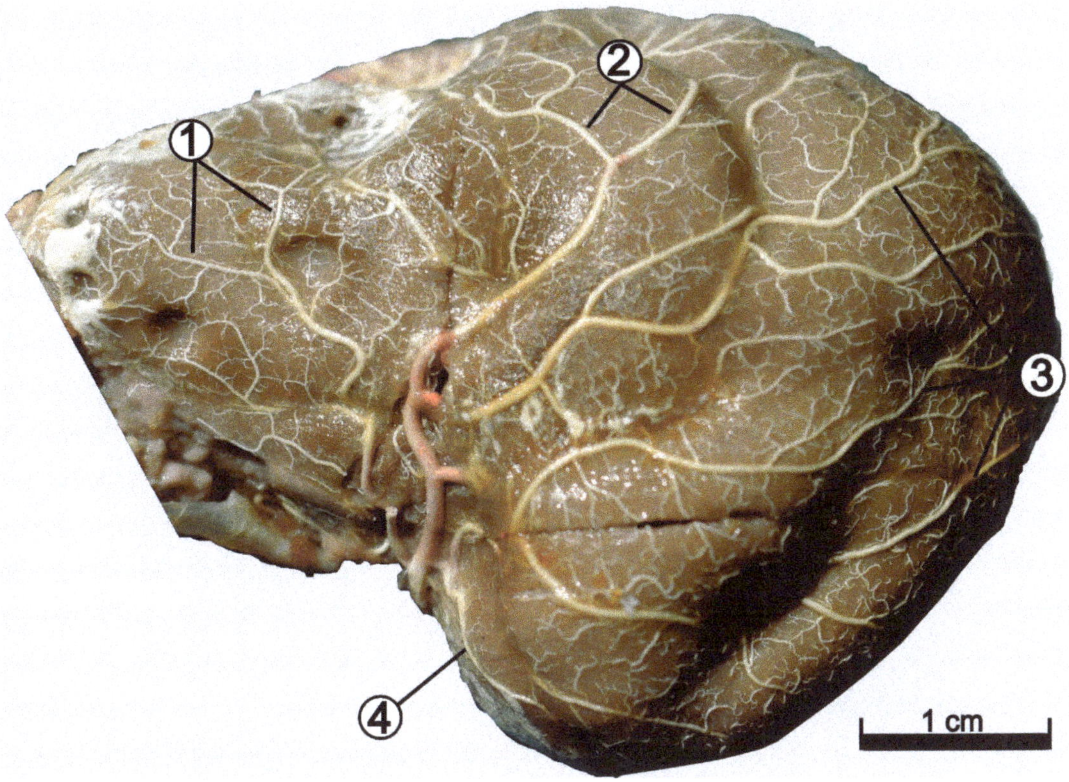

Figure 3. Lateral view of the left cerebral branches of a middle cerebral artery. Latex preparation of red-necked wallaby brain. 1—frontal branches; 2—parietal branches; 3—temporal branches; 4—caudal olfactory artery.

The caudal communicating arteries (*arteriae communicans caudales*), with the basilar artery (*arteria basilaris*) together creates the caudal part of the arterial circle of the brain (Figure 4).

In the vicinity of the origin of the caudal communicating arteries, the internal carotid artery gives way to the caudal cerebral artery (*arteria cerebri caudalis*) (Figure 5).

In 20 specimens, these vessels were found to branch off with a common trunk. The caudal cerebral artery is relatively weak, and supplies a small, caudally located area of the brain. From the caudal communicating arteries branched off the rostral cerebellar artery (*arteria cerebelli rostralis*) (Figure 4), supplying the anterolateral part of the cerebellum. Near the middle of the length of the basilar artery, the caudal cerebellar artery (*arteria cerebelli caudalis*) branched off. This vessel supplies the caudal part of the cerebellum. In addition, small arteries leading to the medulla oblongata were observed to branch off from the basilar artery. The basilar artery is an odd vessel with the same diameter over its entire length (Figure 6), and caudally joins the vertebral arteries (*arteriae vertebrales*).

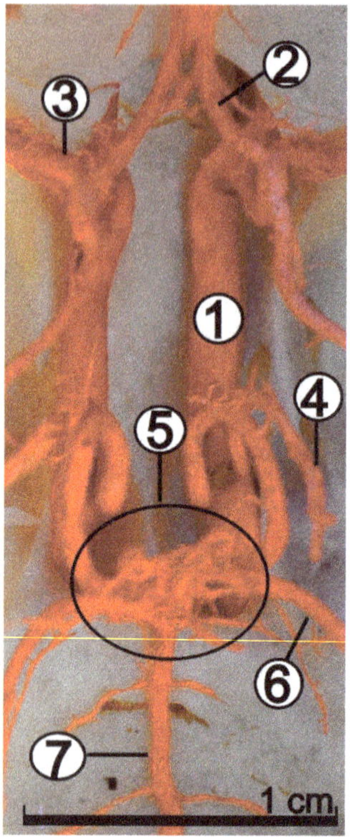

Figure 4. Dorsal view of the arterial circle of the brain. Vessels in the cranial cavity on the skeletal scaffold. Corrosion cast. 1—internal carotid artery; 2—rostral cerebral artery; 3—middle cerebral artery; 4—caudal cerebral artery; 5—caudal communicating artery; 6—rostral cerebellar artery; 7—basilar artery.

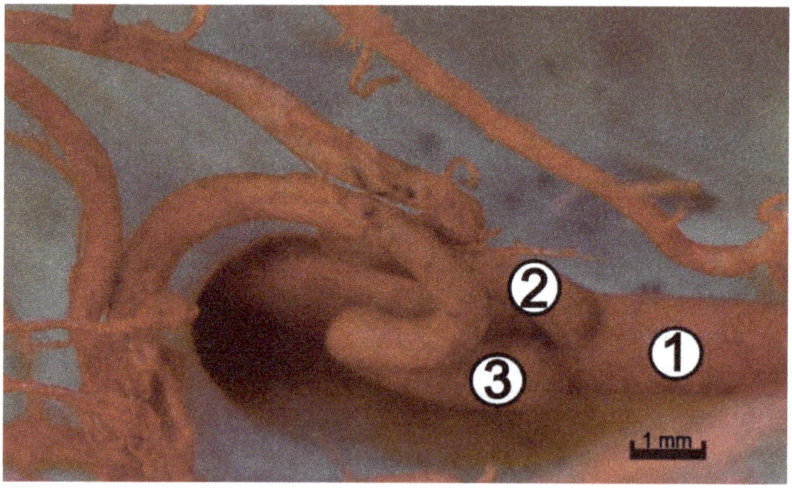

Figure 5. The left side of the caudal part of the arterial circle of the brain. Branches of the internal carotid artery, close up. Independent branching of the caudal cerebral artery and caudal communicating artery from the internal carotid artery. Corrosion cast. 1—internal carotid artery; 2—caudal cerebral artery; 3—caudal communicating artery.

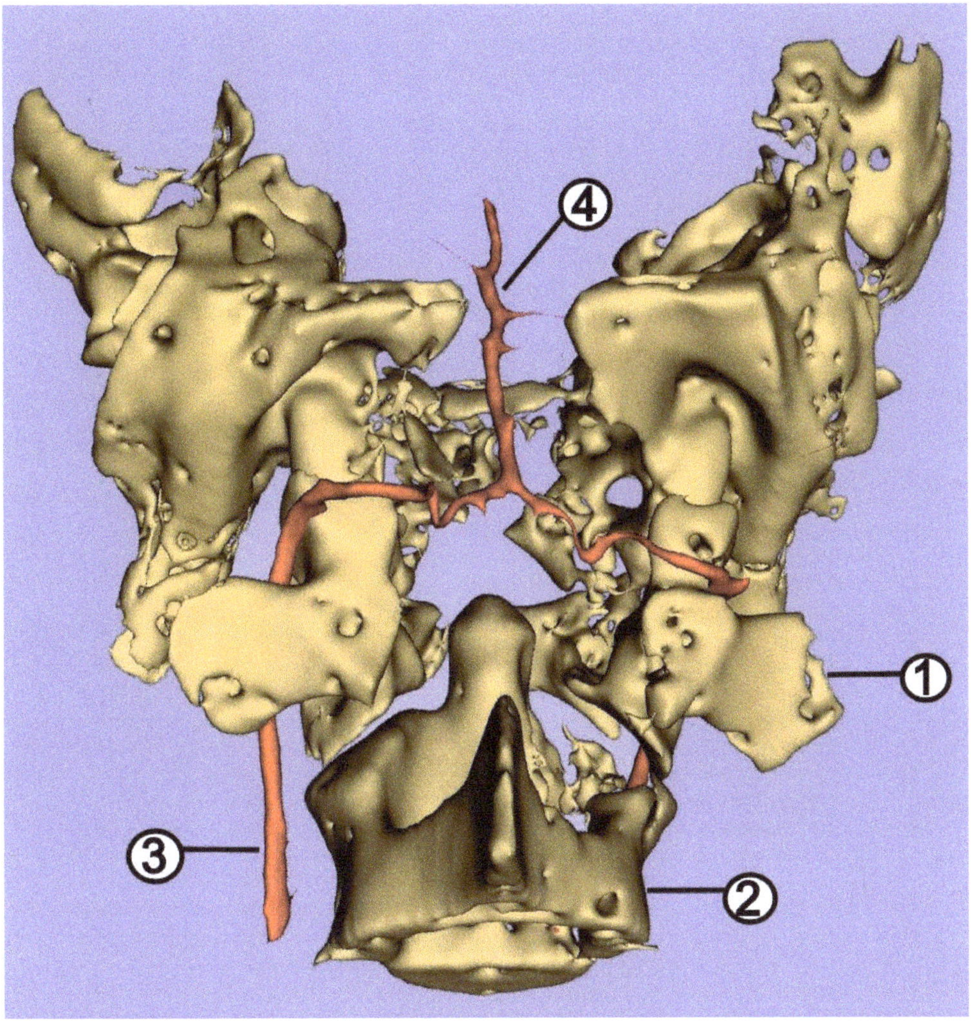

Figure 6. 3D reconstruction of the basal artery on bone scaffolding. 1—first cervical vertebra; 2—second cervical vertebra; 3—vertebral artery; 4—basilar artery.

4. Discussion

The samples collected from fifty red-necked wallabies were investigated for the arterial blood supply to the brain. In the analyzed species, there is a strong internal carotid artery, the main blood source for the brain. Similar observations have been made in the representatives of the Carnivora: domestic dog [17,18], dingo, arctic fox, red fox, pale fox [19], and silver fox [20,21]; *Perissodactyla*: horse [17,22], Przewalski's horse, Grévy's zebra, Grant's zebra, Mongolian khulan, donkey, and South American tapir [22]; lagomorphs: European rabbit [23] and European brown hare [24]; as well as Old and New World camels: Bactrian camel, guanaco, lama [25], and dromedary camel [25–27]. Among rodents, only some of the described species were found to have an internal carotid artery: agouti [28], Mongolian gerbil [29], European beaver [30], Egyptian spiny mouse [31], and rat [32]. In Habermehl's [18] description of this vessel, there is information about the presence of a carotid sinus in the dog. This is a thickening of 3–4 mm in size, at the beginning of this vessel. The presence of this structure has also been described in the cat [33], despite the absence of an internal carotid artery in adults of this species. In other animals with a preserved internal carotid artery, there is no information on the occurrence of the carotid sinus. Similarly, in the analyzed red-necked wallaby, the occurrence of this structure was not observed. The straight course of the internal carotid artery, as observed in the red-necked wallaby, has been previously described in a horse [34]. In camelids and canines, this vessel's course is sinuous. In the dog, it creates a pronounced vascular loop at the height of the rostral foramen of the carotid artery canal, changing its direction 180 degrees [17,25]. In some domestic cats' fetuses and juvenile domestic cats, the connection between the internal carotid artery and the pharyngeal artery has been described. This anastomosis has been observed to be very close to the base of the skull. In adult domestic cats, the internal carotid artery obliterates, and after this, the connection of this vessel to the arteries at the base of the brain no longer exists [33]. The results of the final division of the internal carotid artery are two branches: the rostral cerebral artery, and the internal ophthalmic artery. In some rodent species, the internal ophthalmic artery has been found to have a large diameter [35,36].

The vessels branching off from the arterial circle of the brain show some characteristics of the analyzed species. In the red-necked wallaby, it was confirmed that four main branches branched off from the middle cerebral artery. Skoczylas et al. [37] found three main branches in the polar fox, which he named the rostral middle cerebral artery, dorsal middle cerebral artery, and the caudal middle cerebral artery. The domestic pig and wild boar have shown significant variations in the branching off the vessels from the middle cerebral artery. In some individuals, the division of the middle cerebral artery occurred in the dorsally located point, while in others there was found a double middle cerebral artery [38,39]. The dorsal temporal branch was the strongest in the sheep, polar fox, and silver fox [21,37,40].

In most of the analyzed specimens, the rostral communicating artery was present. In a study that was conducted by König [41] on domestic cattle, the arterial circle of the brain was open on the rostral side in the majority of the cases. An open arterial circle of the brain was also found in representatives of rodents, such as in the guinea pig [36,42], nutria [43], the ground squirrel [44], and the chinchilla [45].

In the described species, a strongly developed internal ophthalmic artery was observed. This vessel was an elongation of the internal carotid artery. Usually, the internal ophthalmic artery is a tiny vessel [17]. The well-developed internal ophthalmic artery has been previously observed in the guinea pig [36]. In this species, the connection of the previously mentioned vessel to the arterial circle of the brain was similar to the pattern found in the red-necked wallaby; however, the vessel originated from the caudal communicating artery near the middle cerebral artery because of the absence of the internal carotid artery. Among Carnivora, the internal ophthalmic artery is also well-developed in the American mink [19].

The caudal cerebral artery in the carnivores, domestic dog [17,18], dingo, arctic fox, red fox, pale fox [19], and silver fox [20,21]; in *Perissodactyla*, horse [17,18], Przewalski's horse, Grévy's zebra, Grant's zebra, Mongolian khulan, donkey, and South American tapir [22]; in lagomorphs, European rabbit [23] and European brown hare [24]; and even-toed ungulates, Bactrian camel, guanaco, lama [25], dromedary camel [25–27], domestic cattle, banteng, yak, American bison, and European bison [46], branched off from the caudal communicating arteries. In the red-necked wallaby, in most cases, this vessel branched off from the internal carotid artery. Similar observations have been made in some rodents with the internal carotid artery. This mode of caudal cerebral artery departure has been described in the Egyptian spiny mouse [31], Mongolian gerbil [29], and only parts of rats [47]. In some rodents with the internal carotid artery, such as the European beaver or agouti, the caudal cerebral artery branched off from the terminal branches of the basilar artery [28,30].

Although the red-necked wallaby's arterial circle of the brain is closed on the caudal side, unlike the Mongolian gerbil, areas that are supplied by the basilar artery and internal carotid artery can be seen [29]. In the Mongolian gerbil, the caudal communicating arteries are not present. This makes it clear that the basilar artery supplies the midbrain and hindbrain, and the internal carotid artery supplies the forebrain and interbrain. In the red-necked wallaby, admittedly, there are the caudal communicating arteries, but their diameters are similar to the basilar artery. These vessels are definitely smaller in diameter than that of the internal carotid artery. This leads us to believe that blood flowing to the midbrain and hindbrain comes from both the internal carotid artery and the basilar artery, and that the basilar artery is not involved in supplying blood to the forebrain and interbrain. The basilar artery is the only source of blood to the cerebellum in some rodents, such as ground squirrels [44]. In the chinchilla, nutria, and the porcupine, the basilar artery creates an arterial circle of the brain alone [35,43,45,48]. In those species, the caudal communicating arteries are strong vessels, and the internal carotid artery does not exist, or is highly reduced, hence not supplying blood to arteries at the base of the brain. Baldwin and Bell [49–51] pointed out the effect of reduced vessel diameter on blood delivery to a given vascular area. Their series of studies on ruminants proved that reducing the lumen of the basilar artery in the caudal segment results in blood flowing from the circle to the basilar artery, and not vice versa. In this group of animals, despite the basilar artery not transporting blood to the brain, the extracranial segment of the internal carotid artery obliterates [46]. Therefore, blood is not transported to the brain from vertebral arteries or the common carotid artery, which is exactly the opposite to the red-necked wallaby. In ruminants, the brain is supplied with blood by arteries that are derivatives of the maxillary artery—caudal branch to the rostral epidural rete mirabile and rostral branches to the rostral epidural rete mirabile [46]. This vascular pattern creates a necessity for the rostral epidural rete mirabile to be present in this group of mammals.

This study complements the anatomical research on brain vascularization among various species with the red-necked wallaby. Differences in the arterial blood flow in the described area may be related to the phylogenetic distance between marsupials and the other above-mentioned mammal species. Additionally, it may be important for anatomists, physiologists, and veterinarians to know how to describe in detail the vascular pattern of the arteries that supply blood to the brain, which differs significantly between various groups of mammals.

5. Conclusions

This study provided detailed information about the brain vascularization of the red-necked wallaby using three methods of specimen preparation, including advanced imaging with cone-bean computed tomography. In the described species, the main blood vessel supporting the brain is an internal carotid artery. Moreover, the arterial circle of the brain is closed on the caudal side. In some of the described preparations, independent branching of the separate vessels from the internal carotid artery was seen: the caudal cerebral artery and the caudal communicating artery. The internal ophthalmic artery is the well-developed elongation of the internal carotid artery. Obtained findings were compared with other mammal species' brain blood flow, described in scientific literature. The results could be useful for further research about wallaby physiology and medicine, due to its increasing popularity among breeders and private holders. Since various cardiovascular diseases have been described before in this species, a detailed description of angio-anatomical structures may contribute to establishing diagnostic protocols, and help in pathophysiological studies.

Author Contributions: Conceptualization, M.Z.; formal analysis, M.Z. and J.J.R.; resources, M.Z., J.J.R., M.H. and M.G.; writing—original draft preparation, M.Z., J.J.R., M.H., M.G. and A.J.; interpretation of data for the work: M.Z., M.G. and A.J.; writing—review and editing, M.Z., J.J.R., M.H., M.G. and A.J.; supervision, M.Z. All authors have read and agreed to the published version of the manuscript.

Funding: This research received no external funding.

Institutional Review Board Statement: Not applicable.

Informed Consent Statement: Not applicable.

Data Availability Statement: The data presented in this study are available on request from the corresponding author.

Acknowledgments: Not applicable.

Conflicts of Interest: The authors declare no conflict of interest.

References

1. Jarman, P.J. Social Behavior and Organization in the Macropodoidea. *Adv. Study Behav.* **1991**, *20*, 1–50.
2. McCarron, H.C.K.; Dawson, T.J. Thermal Relations of Macropodoidea in Hot Environments. *Kangaroos Wallabies Rat-Kangaroos* **1989**, *1*, 255–263.
3. McGowan, C.P.; Skinner, J.; Biewener, A.A. Hind Limb Scaling of Kangaroos and Wallabies (Superfamily Macropodoidea): Implications for Hopping Performance, Safety Factor and Elastic Savings. *J. Anat.* **2008**, *212*, 153–163. [CrossRef] [PubMed]
4. Baudinette, R.V. Locomotion in Macropodoid Marsupials-Gaits, Energetics and Heat-Balance. *Aust. J. Zool.* **1994**, *42*, 103–123. [CrossRef]
5. Bouts, T.; Harrison, N.; Berry, K.; Taylor, P.; Routh, A.; Gasthuys, F. Comparison of three anaesthetic protocols in Bennett's wallabies (*Macropus rufogriseus*). *Vet. Anaesth. Analg.* **2010**, *37*, 207–214. [CrossRef]
6. Basso, W.; Venturini, M.C.; Moré, G.; Quiroga, A.; Bacigalupe, D.; Unzaga, J.M.; Larsen, A.; Laplace, R.; Venturini, L. Toxoplasmosis in captive Bennett's wallabies (*Macropus rufogriseus*) in Argentina. *Vet. Parasitol.* **2007**, *144*, 157–161. [CrossRef]
7. Díaz-Delgado, J.; Kellerman, T.E.; Auckland, L.; Ferro, P.J.; Groch, K.R.; Gomez, G.; Hamer, S.A. Trypanosoma cruzi genotype I and Toxoplasma gondii co-infection in a red-necked wallaby. *J. Comp. Pathol.* **2020**, *179*, 52–58. [CrossRef]
8. Tanner, M.C.; Thomason, J.D.; Rooney, T.; Schneider, S.M.; Lynn, C.R.; Eshar, D. Takotsubo-like cardiomyopathy associated with capture myopathy in a zoo-kept, red-necked wallaby (*Macropus rufogriseus*). *Isr. J. Vet. Med.* **2022**, *77*, 44–48.
9. Kane, L.P.; Langan, J.N.; Adkesson, M.J.; Chinnadurai, S.K.; Nevitt, B.N.; Drees, R. Treatment of mandibular osteomyelitis in two red-necked wallabies (*Macropus rufogriseus*) by means of intensive long-term parenteral drug administration and serial computed tomographic monitoring. *J. Am. Vet. Med. Assoc.* **2017**, *251*, 1070–1077. [CrossRef]
10. Burgdorf-Moisuk, A.; Pye, G.W.; Smith, J.A.; Papendick, R.; Ivy, J.A.; Hamlin-Andrus, C. A Retrospective and prospective study of megaesophagus in the parma wallaby (*Macropus parma*) at the San Diego Zoo, California, USA. *J. Zoo Wildl. Med.* **2012**, *43*, 89–94. [CrossRef]
11. Twomey, D.F.; Boon, J.D.; Hume, M.J.; Schock, A.; Wood, R. Concurrent coccidiosis and listeriosis in a Bennett's wallaby (*Macropus rufogriseus*). *Vet. Rec.* **2008**, *163*, 635. [CrossRef]
12. Bermúdez, R.; Faílde, L.D.; Losada, A.P.; Nieto, J.M.; Quiroga, M.I. Toxoplasmosis in Bennett's Wallabies (*Macropus rufogriseus*) in Spain. *Vet. Parasitol.* **2009**, *160*, 155–158. [CrossRef] [PubMed]
13. Antiabong, J.F.; Boardman, W.; Moore, R.B.; Brown, M.H.; Ball, A.S. The Oral Microbial Community of Gingivitis and Lumpy Jaw in Captive Macropods. *Res. Vet. Sci.* **2013**, *95*, 996–1005. [CrossRef] [PubMed]
14. Stern, A.W.; Smith, S.; Snider, T.A. Hypertrophic cardiomyopathy in two captive Bennett's wallabies (*Macropus rufogriseus rufogriseus*). *J. Vet. Diagnostic Investig.* **2009**, *21*, 889–892. [CrossRef] [PubMed]
15. de Laforcade, A. Diseases Associated with Thrombosis. *Top. Companion Anim. Med.* **2012**, *27*, 59–64. [CrossRef] [PubMed]
16. International Committee on Veterinary Gross Anatomical Nomenclature. *Nomina Anatomica Veterinária*, 6th ed.; Editorial Committee: Hanover, Germany, 2017; pp. 73–147.
17. Nickiel, R.; Schwarz, R. Vergleichende Betrachtung der Kopfarterien der Haussaugetiere (Katze, Hund, Schwein, Rind, Schaf, Ziege, Pferd). Comparative analysis of the head arteries of domestic mammals (cat, dog, pig, cattle, sheep, goat, horse). *Zent. Für Veterinärmedizin Reihe A* **1963**, *10*, 89–120. (In German) [CrossRef]
18. Habermehl, K.H. Zur Topographie der Gehirngefäße des Hundes. *Anat. Histol. Embryol.* **1973**, *2*, 327–353. [CrossRef]
19. Bugge, J. The cephalic arterial system in carnivores, with special reference to the systematic classification. *Acta Anat.* **1978**, *101*, 45–61. [CrossRef] [PubMed]
20. Frąckowiak, H.; Zawidzka, B. Tętnice głowy u lisa srebrzystego. *Rocz. Akad. Rol. W Poznaniu. Zootech.* **1990**, *220*, 27–36.
21. Skoczylas, B.; Brudnicki, W.; Kirkiłło-Stacewicz, K.; Nowicki, W.; Wach, J. Cortical branches of the middle cerebral artery in silver fox (*Vulpes vulpes*). *Pesqui. Veterinária Bras.* **2016**, *36*, 1053–1057. [CrossRef]
22. Frąckowiak, H.; Giejdasz, K. Przebieg i zmienność tętnic na podstawie mózgowia u gatunków z rzędu Perissodactyla. *Rocz. Akad. Rol. W Poznaniu. Zootech.* **1998**, *50*, 109–117.
23. Brudnicki, W.; Nowicki, W.; Skoczylas, B.; Brudnicki, A.; Kirkiłło-Stacewicz, K.; Wach, J. Arteries of the brain in wild European rabbit Oryctolagus cuniculus (Linnaeus, 1758). *Folia Biol. (Kraków)* **2012**, *60*, 189–194. [CrossRef] [PubMed]
24. Brudnicki, W.; Kirkiłło-Stacewicz, K.; Skoczylas, B.; Nowicki, W.; Jabłoński, R.; Brudnicki, A.; Wach, J. The arteries of the brain in hare (*Lepus europaeus* Pallas, 1778). *Anat. Rec.* **2015**, *298*, 1774–1779. [CrossRef] [PubMed]
25. Kiełtyka-Kurc, A.; Frackowiak, H.; Zdun, M.; Nabzdyk, M.; Kowalczyk, K.; Tołkacz, M. The arteries on the base of the brain in the camelids (Camelidae). *Ital. J. Zool.* **2014**, *81*, 215–220. [CrossRef]
26. Al Aiyan, A.; Menon, P.; AlDarwich, A.; Almuhairi, F.; Alnuaimi, S.; Bulshawareb, A.; Qablan, M.; Shehab, S. Descriptive analysis of cerebral arterial vascular architecture in dromedary camel (*Camelus dromedarius*). *Front. Neuroanat.* **2019**, *13*, 67. [CrossRef]

7. Jerbi, H.; Khaldi, S.; Pérez, W. Morphometric Study of the rostral epidural rete mirabile in the dromedary (*Camelus dromedarius*, Linnaeus 1758). *Int. J. Morphol.* **2016**, *34*, 1429–1435. [CrossRef]
8. Da Silva, R.S.B.; De Oliveira, G.B.; Oliveira, C.M.; Bezerra, F.V.F.; Câmara, F.V.; De Oliveira, R.E.M.; De Oliveira, M.F. Arterial vascularization of the brain of the agouti (*Dasyprocta aguti* Linnaeus, 1766). *Semin. Ciências Agrárias* **2016**, *37*, 773–784. [CrossRef]
9. Kuchinka, J.; Nowak, E.; Szczurkowski, A.; Kuder, T. Arteries supplying the base of the brain in the mongolian gerbil (*Meriones unguiculatus*). *Pol. J. Vet. Sci.* **2008**, *11*, 295–299.
0. Frąckowiak, H.; Śmiełowski, J. Cephalic arteries in the european beaver Castor fiber. *Acta Theriol.* **1998**, *43*, 219–224. [CrossRef]
1. Szczurkowski, A.; Kuchinka, J.; Nowak, E.; Kuder, T. Topography of arterial circle of the brain in Egyptian spiny mouse (*Acomys cahirinus*, Desmarest). *Anat. Histol. Embryol.* **2007**, *36*, 147–150. [CrossRef]
2. Komatsu, T.; Ohta, H.; Motegi, H.; Hata, J.; Terawaki, K.; Koizumi, M.; Muta, K.; Okano, H.J.; Iguchi, Y. A novel model of ischemia in rats with middle cerebral artery occlusion using a microcatheter and zirconia ball under fluoroscopy. *Sci. Rep.* **2021**, *11*, 12806. [CrossRef]
3. Ziemak, H.; Frackowiak, H.; Zdun, M. Domestic cat's internal carotid artery in ontogenesis. *Vet. Med.* **2021**, *66*, 292–297. [CrossRef]
4. Du Boulay, G.; Kendall, B.E.; Crockard, A.; Sage, M.; Belloni, G. The Autoregulatory Capability of Galen's Rete Cerebri and Its Connections. *Neuroradiology* **1975**, *9*, 171–181. [CrossRef] [PubMed]
5. Kuchinka, J. Analysis of morphological variation of the internal ophthalmic artery in the chinchilla (*Chinchilla laniger*, Molina). *Vet. Med.* **2015**, *60*, 161–169. [CrossRef]
6. Kuchinka, J. Internal ophthalmic arteries within the brain-base arterial system in guinea pig. *Anat. Rec.* **2018**, *301*, 887–891. [CrossRef]
7. Skoczylas, B.; Brudnicki, W.; Kirkiłło-Stacewicz, K.; Nowicki, W.; Wach, J. Ramifications of the middle cerebral artery in polar fox (*Alopex lagopus*). *Electron. J. Pol. Agric. Univ. Ser. Vet. Med.* **2017**, *20*, 02. [CrossRef]
8. Skoczylas, B.; Wiland, C. Cortical branches of the middle cerebral artery in the wild boar [*Sus scrofa* l.]. *Electron. J. Pol. Agric. Univ. Ser. Vet. Med.* **1999**, *2*, 01.
9. Skoczylas, B. Cortical branches of middle cerebral artery in domestic pig [*Sus scrofa* f. domestica]. *Electron. J. Pol. Agric. Univ. Ser. Vet. Med.* **2000**, *3*, 01.
0. Skoczylas, B.; Brudnicki, W.; Kirkiłło-Stacewicz, K.; Nowicki, W.; Wach, J. Telencephalon vascularity in sheep (Ovis aries). *Pesqui. Veterinária Bras.* **2018**, *38*, 1449–1452. [CrossRef]
1. König, H.E. *Anatomie und Entwicklung der Blutgefässe in der Schädelhöhle der Hauswiederkäuer: (Rind, Schaf und Ziege)*; Ferdinand Enke Verl: Stuttgart, Germany, 1979; pp. 1–215.
2. Majewska-Michalska, E. Vascularization of the brain in guinea pig. I. Gross anatomy of the arteries and veins. *Folia Morphol.* **1994**, *53*, 249–268.
3. de Azambuja, R.C.; Ver Goltz, L.; Campos, R. Systematization of the brain base arteries in nutria (*Myocastor coypus*). *Acta Sci. Vet.* **2018**, *46*, 9. [CrossRef]
4. Aydin, A.; Ozkan, Z.E.; Yilmaz, S.; Ilgun, R. The morphology of the circulus arteriosus cerebri in the ground squirrel (*Spermophilus citellus*). *Vet. Med.* **2009**, *54*, 537–542. [CrossRef]
5. Kuchinka, J. Morphometry and variability of the brain arterial circle in chinchilla (*Chinchilla laniger*, Molina). *Anat. Rec.* **2017**, *300*, 1472–1480. [CrossRef]
6. Zdun, M.; Frackowiak, H.; Kiełtyka-Kurc, A.; Kowalczyk, K.; Nabzdyk, M.; Timm, A. The arteries of brain base in species of Bovini tribe. *Anat. Rec.* **2013**, *296*, 1677–1682. [CrossRef] [PubMed]
7. Brown, J.O. The morphology of circulus arteriosus cerebri in rats. *Anat. Rec.* **1966**, *156*, 99–106. [CrossRef] [PubMed]
8. Aydın, A.; Yılmaz, S.; Dinç, G.; Özdemir, D.; Karan, M. The Morphology of Circulus Arteriosus Cerebri in the Porcupine (Hystrix Cristata). *Vet. Med.* **2005**, *50*, 131–135. [CrossRef]
9. Baldwin, B.A.; Bell, F.R. The contribution of the carotid and cerebral arteries to the blood pressure in sheep. *J. Physiol.* **1960**, *151*, 9–10.
0. Baldwin, B.A.; Bell, F.R. Effect of clamping cartotid and vertebral arteries on the blood pressure in sheep. *J. Physiol.* **1960**, *151*, 39–40.
1. Baldwin, B.A.; Bell, F.R. The anatomy of the cerebral circulation in the sheep and ox. *J. Anat.* **1963**, *97*, 203–215.

animals

MDPI

Article

Morpho-Histological Studies of the Gastrointestinal Tract of the Orange-Rumped Agouti (*Dasyprocta leporina* Linnaeus, 1758), with Special Reference to Morphometry and Histometry

Kegan Romelle Jones *, Roger Edmund John and Venkatesan Sundaram

Department of Basic Veterinary Sciences (DBVS), School of Veterinary Medicine (SVM), Faculty of Medical Science (FMS), University of the West Indies (UWI), St. Augustine Campus, St. Augustine 999183, Trinidad and Tobago
* Correspondence: kegan.jones@sta.uwi.edu

Simple Summary: The agouti (*Dasyprocta leporina*) is a neo-tropical rodent that has the potential to be domesticated. These animals are considered to be omnivores from studies conducted in the wild and in captivity. However, an in-depth morphometric and histometric analysis of the gastrointestinal tract of this animal was never conducted. Thus, the objective of this paper was to give a quantitative analysis of the gastrointestinal tract of the agouti and relate it to the feeding habits of the animals. This study showed that the agouti had an esophagus which was lined with keratinized epithelium and esophageal glands. The stomach was simple, with well-developed gastric glands, and the small intestines accounted for the majority of the animal's digestive tract in length. The long cecum and well-developed colon indicated that this animal had the ability to digest fibrous material in the diet. Based on the analysis of the gastrointestinal tract, these animals can be fed a dry omnivorous diet, with protein sources from either vegetation or animal matter.

Abstract: The morphology of the gastrointestinal tract (GI) is a strong indicator of a species' dietary habits. The objective of this study was to describe and quantitatively analyze the gross and microanatomy of the digestive tract of the orange/red-rumped agouti (*Dasyprocta leporina*) and relate it to the feeding habits of this animal. The digestive tracts of six adult males were used for this study. The results showed that the esophagus was thick (mean thickness of 1023.78 $\pm$ 28.97 μm) and lined by keratinized epithelium with scant esophageal glands. Mucosa-associated lymphocytic infiltration was robust throughout the GI tract. These findings suggest that the esophagus was well adapted to a coarse diet. The simple stomach with well-developed gastric glands in the fundus region (mean thickness of 605.39 $\pm$ 28.68 μm) was indicative of an adaptation to a carnivorous diet. The small intestine constituted approximately 80% of the length of the GI tract. The remarkable development of the jejunum with a greater villus length (mean thickness of 182.50 $\pm$ 27.38 μm) indicated a greater absorptive capacity in frugivorous and carnivorous diets. The long cecum and well-developed colon clearly indicated that the GI tract was well adapted to frugivorous and herbivorous diets. Overall, *D. leporina* showed that it is well adapted to an omnivorous diet. These results suggest that *D. leporina* can be fed a balanced omnivorous dry diet with a high protein content of plant or animal origin that is well suited to the architecture of the GI tract in captivity.

Keywords: agouti; gastrointestinal tract; gross anatomy; histology; histometry; morphometry

1. Introduction

The vertebrate digestive tract is a highly specialized structure through which food enters the organism and is digested, nutrients are absorbed, and waste products are excreted [1]. The morphology of the digestive tract can influence the efficiency of digestion and is closely related to feeding habits. Ecological conditions and feeding habits lead to changes in the basic metabolic rate, which, in turn, lead to changes in the size of the digestive tract [2–4]. Changes in the digestive tract can have a seasonal pattern [5–7] and often reflect the physiological state of an organism [8,9].

Orange-rumped agoutis (*Dasyprocta leporina*) are large rodents belonging to the order Caviomporpha, a rodent order with diverse anatomical characteristics among its members [10]. It is a very popular exotic meat in Trinidad and Tobago [11,12]. An estimated 90,000 agoutis are hunted in Trinidad and Tobago each year during the hunting season [13]. These large forays and loss of habitat can lead to threats to the survival of this species. In recent years, a growing number of breeders [12] in Trinidad and Tobago have attempted to domesticate this animal through breeding to meet the demand for meat. The attempts to domesticate and intensively produce *D. leporina* in Trinidad and Tobago have sparked interest in studying the biology of this animal.

In any intensive livestock production system, knowledge of digestive-system anatomy leads to improvements in feeding and nutrition [14]. *D. leporina* feeds omnivorously and consumes both plant and animal matter [15,16]. A preliminary anatomical study on the digestive tract of agouti was conducted by Garcia et al. [17] that lacks a detailed quantitative and qualitative analysis of the microscopic anatomy of the digestive tract. Therefore, the objective of this work was to perform a comprehensive anatomical and histological study, with special emphasis on morphometric and histometric measurements of the gastrointestinal tract of agouti in order to understand the digestive biology of this animal. Further research with various experimental diets would be a better model to analyze the histological microstructure in depth.

2. Materials and Methods

Six adult male *D. leporina* were collected for this study from the Neotropical Wildlife Unit (Figure 1) at the University of the West Indies Field Station (UFS) in Valsayn, (10°38′15″ N, 61°25′39″ W) Trinidad and Tobago (Figure 1). The study was carried out in the Anatomy Unit of the School of Veterinary Medicine, The University of The West Indies, St. Augustine Campus, Trinidad and Tobago. The Campus Research Ethics Committee, The University of the West Indies, St. Augustine, Trinidad and Tobago (No. CREC-SA 1095/07/2021) approved all animal experiments in accordance with the National Institutes of Health's Guide for the Care and Use of Laboratory Animals [18].

Figure 1. Photograph showing *D. leporina* in a cage housing at the Neotropical Wildlife Unit at University Field station.

Animals were sedated with a solution of xylazine and ketamine hydrochloride at 5 mg/kg body weight and 35 mg/kg body weight, respectively. Live weight was measured and recorded. Animals were then euthanized by intracardiac injection of pentobarbital sodium (release 300 mg/mL). Gross in situ morphological observations were then made and photographed. The entire gastrointestinal tract from the esophagus to the anus was removed, and the length of the different segments was measured. Gross anatomical features were examined in each segment, and specimens were collected and preserved in formalin and embedded in paraffin for histologic examination. The relative size of each segment of the digestive tract was determined by dividing the segment length by the length of the entire digestive tract. Tissue samples collected for histological examination were fixed in 10% buffered neutral formalin and processed for routine paraffin embedding. Sections of 5–8 μm in thickness were cut and stained with routine hematoxylin and eosin (H&E). An Olympus BX51 microscope (Olympus, Tokyo, Japan) with cell Sens imaging software (ver. 1.12) was used to examine histologic features and obtain histometric measurements and photomicrographs.

Histometric measurements were taken by using a micrometric scale for microscopy to measure specimens. The thickness of the mucosa, submucosa, tunica muscularis, and serosa layers was obtained from the esophagus, stomach (cardiac, fundic, and pyloric regions), small intestine (duodenum, jejunum, and ileum), and large intestine (cecum, colon, and rectum). In the small-intestine region, villi length and width and crypt height and width (μm) were measured. The length of the villi was measured from the tip of the villus to the beginning of the crypt. The width of the villi was measured transversely from the widest point of the villi. The depth of the crypt was measured from the bottom of the villi to the base of the crypt. Crypt width is the transverse diameter of the crypt. The average number of goblet cells per 10,000 μm^2 area was also measured. Ten measurements were taken per slide, and three slides were taken for each segment of the gastrointestinal tract; measurements were averaged to determine the value for each animal. Descriptive statistics were calculated by using IBM SPSS Statistics V21 software (IBM, Armonk, NY, USA).

3. Results

3.1. Gross Anatomy of the GI Tract

The mean length of the entire GI tract was 780.5 ± 50.62 cm (Table 1). The esophagus had a mean length of 16.41 ± 1.45 cm, and the width of the esophagus remained fairly uniform throughout its length, merging into a simple rounded stomach (Figures 2 and 3). The stomach was unicameral, had an average length of 12.70 ± 0.71 cm, and was thin-walled. Three regions of the stomach were distinguished: the cardia, the fundus, and the pylorus. The cardia was the junction between the esophagus (Figure 4) and the stomach, and the pylorus formed the dilated posterior end of the stomach. The pylorus was thick-walled and had an average diameter of 3.2 ± 0.03 cm. A thick muscular band, the pyloric sphincter, marked the constricted junction between the stomach and the small intestine.

Table 1. The length (cm) and relative length (%) of various segments of the gastrointestinal tract of adult male *D. leporina*. (n = 6).

Parameter	Length (cm)	Relative Length (%)
Whole digestive tract	780.5 ± 50.62	–
Esophagus	16.41 ± 1.45	2.05
Stomach	12.70 ± 0.71	1.54
Small intestine	603.9 ± 10.15	77.42
Cecum	23.55 ± 0.64	3.02
Colon and rectum	123.55 ± 24.64	15.84

Note: Data are stated as mean ± standard deviation.

The small intestine (SI) consisted of a simple smooth-walled tube with an average length of 603.9 ± 10.15 cm. The small intestine consisted of 77.4% (Figure 5) of the length of the entire gastrointestinal tract (Figure 3). There were no demarcations that distinguished it from the duodenum, jejunum, and ileum. Overall, however, the small intestine accounted for a large proportion of the total length of the digestive tract (Figure 5; Table 1). The duodenum has a sigmoid flexure in which the pancreas is located. The large intestine occupied the caudal end of the digestive tract and was divided into the cecum, colon, and rectum. The most prominent of the three subdivisions of the large intestine (LI) was the large thin-walled greenish cecum. The cranial and caudal portions of the cecum had mean diameters of 4.2 + 0.04 and 2.4 ± 0.02 cm, respectively, making it only a smaller diameter than the stomach. The outer surface of the cecum had dorsal and ventral longitudinal bands and 14 saccular outpouchings. The average length was 23.55 ± 0.64 cm. The ileum and colon united with the cecum to form the ileocolic junction.

The colon and rectum formed a continuous tube without sharp demarcations. The first segment of the colon was enlarged. After the first segment, the rest of the colon was gradually reduced in diameter until it merged into the rectum. The rostral part of the colon had a slightly tortuous spiral consisting of two folds, namely a centripetal fold and a centrifugal fold. The distal end of the colon was bulging due to fecal matter in the lumen. The mean diameter of the colon at the cranial and caudal ends was 2.4 ± 0.02 and 1.3 ± 0.01 cm, respectively. The rectum had a fairly uniform diameter, averaging 1.3 ± 0.01 cm. The mean length of the colon and rectum was 123.55 ± 24.64 cm.

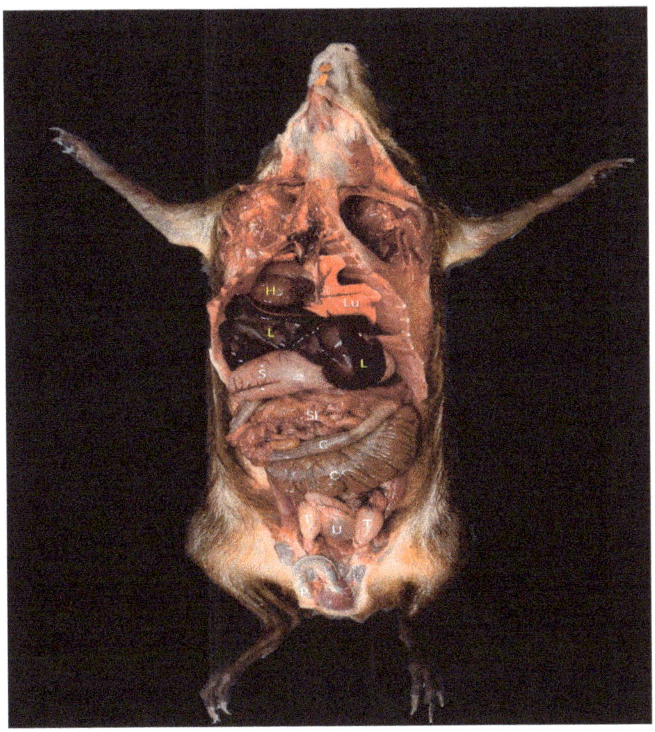

Figure 2. Photograph showing the anatomical view of visceral organs in situ of an adult male *D. leporina*. H—heart; L—liver; Lu—lungs; S—stomach; SI—small intestine; C—colon; T—testis; U—urinary bladder; P—penis.

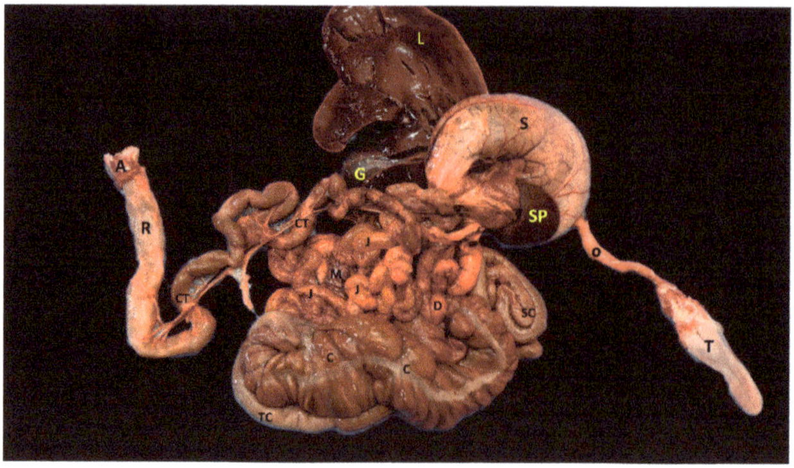

Figure 3. Photograph showing the anatomical view of the gastrointestinal tract of an adult male *D. leporina*. T—tongue; O—esophagus; SP—spleen; L—liver; G—gall bladder; S—stomach; D—duodenum; J—jejunum; C—cecum; SC—spiral colon; TC—transverse colon; CT—terminal colon; R—rectum; M—mesentery; A—anus.

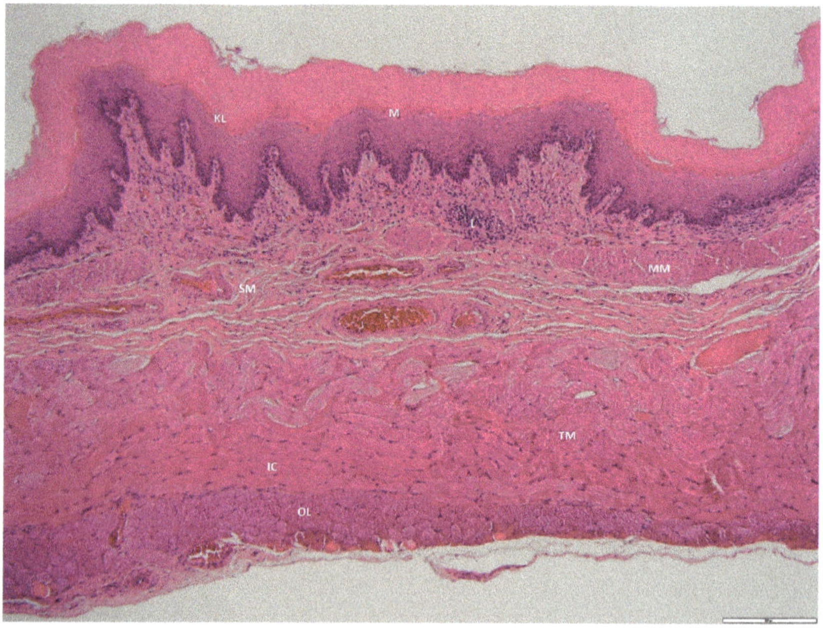

Figure 4. Photomicrograph of a transverse section through the esophagus of an adult male *D. leporina*. H&Ex200. KL—keratinized epithelium; M—mucosa; MM—muscularis mucosa; SM—submucosa; L—lymphocyte infiltration; TM—tunica muscularis; IC—inner circular layer; OL—outer longitudinal layer.

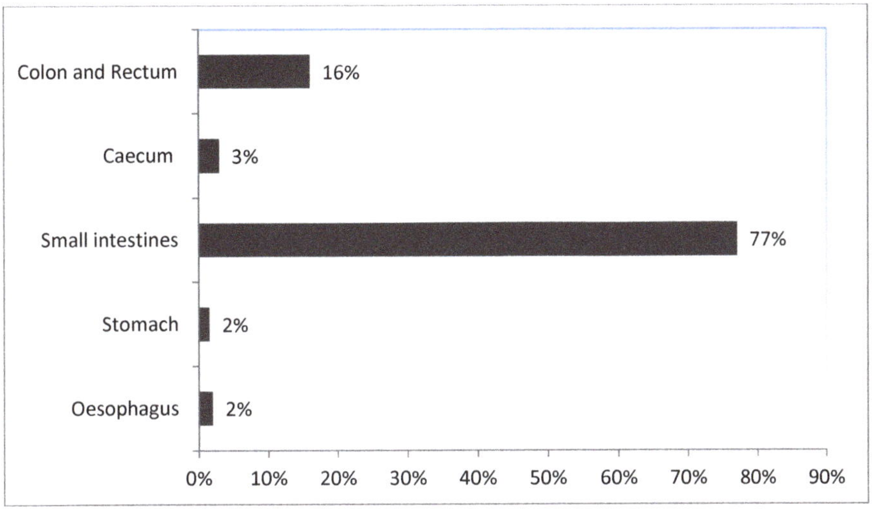

Figure 5. Relative length (%) of the gastrointestinal tract of the agouti.

3.2. Histology of the GI Tract

The mucosa of the esophagus was folded and lined by stratified keratinized squamous epithelium with an average thickness of 1023.78 ± 28.97 μm. The lamina propria consisted of lymphocytic infiltration throughout the esophagus. In addition, scattered lymph nodules were also found. The muscularis mucosae was distinct and consisted of smooth muscle fibers. The submucosa was the thickest layer at 465.67 ± 74.78 μm and consisted of a dense irregular connective tissue layer with scant esophageal glands, lymphocytes, blood vessels, and nerves. The tunica muscularis consisted of outer longitudinal and inner circular muscle fibers. In the present study, the cranial tunica muscularis consisted mainly of skeletal muscle and transitioned to smooth muscle at the cardiac sphincter. The cervical esophagus is covered by adventitia, a loose connective tissue layer containing lymphatic and blood vessels and nerves. The thoracic esophagus is covered by serosa, a loose connective tissue layer with an outer mesothelial covering.

The division of the stomach into three distinct regions is based on the length and composition of the gastric glands: cardiac, fundic, and pyloric. The surface stomach of *D. leporina* was essentially lined and folded with simple columnar epithelium. The gastric pits were characterized by the shape and location of the three regions. The lamina propria consisted of a loose connective tissue layer that was rich in capillaries and lymphoid cells and was completely occupied by gastric glands. The tunica muscularis consisted of two layers of smooth muscle, namely inner circular and outer longitudinal layers of smooth muscle fibers (Figure 5). The outer surface of the stomach was covered by the serosa, which consisted of loose connective tissue, blood vessels, lymphatic vessels, and nerves.

The mucosa in the cardiac region was characterized by densely packed tortuous tubular mucous glands that lay beneath the surface of the simple columnar epithelium. The surface was indented into numerous sharp gastric pits that opened freely into the lumen and were wide and deep. Gastric glands made up most of the mucosa below the pits (Figures 6–8). The glands were relatively short and had few superficial mucosal cells. Some parietal cells were observed at the base of the glands, and many neck cells were identified. There were also two layers of smooth muscle in the tunica muscularis.

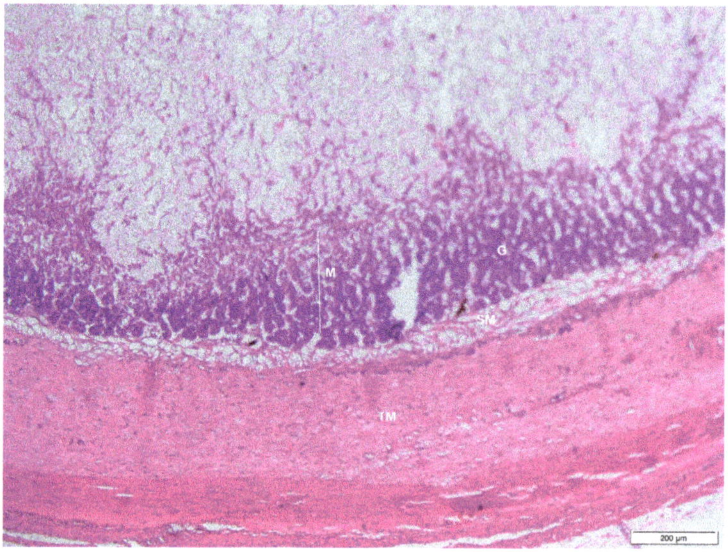

Figure 6. Photomicrograph of a transverse section through the cardiac region of an adult male *D. leporina*. H&Ex200. M—mucosa; G—gastric glands; SM—submucosa; TM—tunica muscularis; S—serosa.

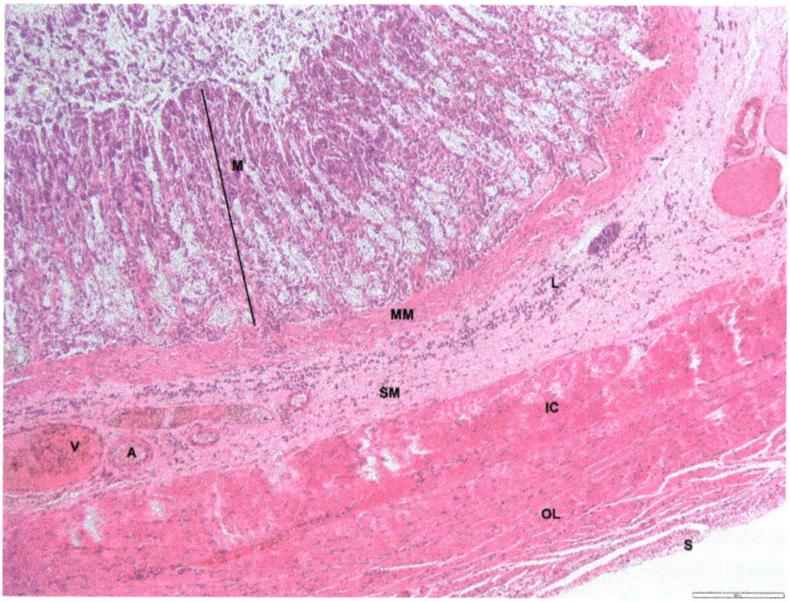

Figure 7. Photomicrograph of a cross-section through the fundic region of an adult male *D. leporina*. H&Ex200. M—mucosa; MM—muscularis mucosa; SM—submucosa; L—lymphocyte infiltration; A—arteriole; V-venule; IC—inner circular layer; OL—outer longitudinal layer; S—Serosa.

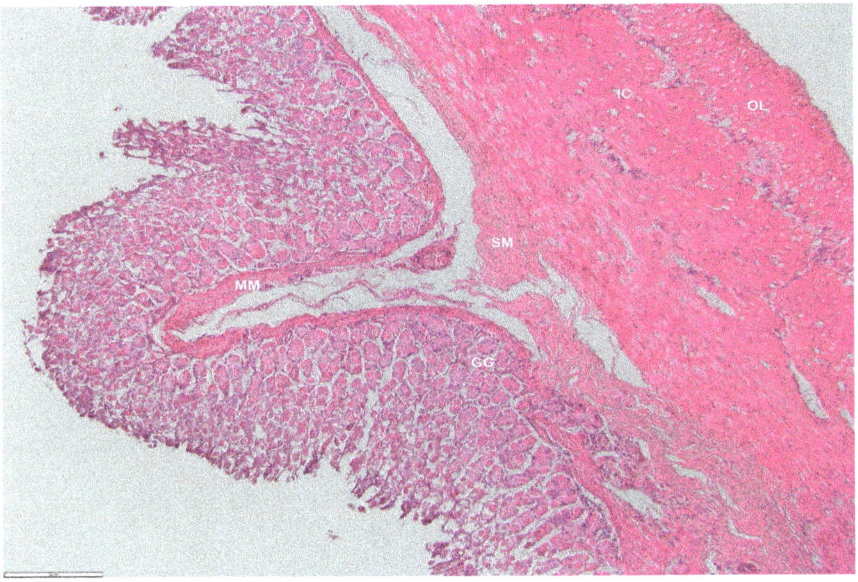

Figure 8. Photomicrograph of a transverse section through the pyloric region of an adult male *D. leporina*. H&Ex200. GG—gastric glands; MM—muscularis mucosa; SM—submucosa; IC—inner circular layer; OL—outer longitudinal layer.

The mucosal folds and gastric pits in the fundus region were shallower in the mucosa, and the mean thickness was measured to be 605.39 ± 28.68 μm, which was higher than in other regions (Table 2). The gastric glands were straight simple tubular glands that ran parallel to each other and extended deep into the mucosa to reach the muscularis mucosae (Figure 8). The surface of the epithelium consisted of fewer mucosal cells. The glands in the fundus were divided into three regions: a basal region, a middle region, and an upper isthmus. The upper isthmus contained both parietal and surface mucosal cells, whereas the parietal and mucosal cells were abundant in the middle region of the glands. The basal region consisted mainly of a few parietal cells and mainly chief cells. There was a much-thickened muscularis mucosae with some striations that extended into the lamina propria between the gastric glands to the gastric folds. The submucosa was composed of dense connective tissue. The tunica muscularis consisted of two layers of smooth muscle, namely the thin outer longitudinal layer and the thick inner circular layer.

The region of the pylorus was similar to the fundus, with the following exceptions: the gastric glands were more branched and shorter, the gastric pits were short, and more principal chief cells and fewer parietal cells were observed. The thick tunica muscularis formed a sphincter at the gastroduodenal junction (Figure 8).

Table 2. The histometric measurements (μm) of esophagus and stomach of adult male *D. leporina* (n = 6).

Parameters	Esophagus	Stomach		
		Cardiac	Fundic	Pyloric
T. mucosa	235.91 ± 19.12	421.71 ± 15.52	605.39 ± 28.68	523.26 ± 10.19
T. submucosa	465.67 ± 74.78	64.20 ± 4.14	77.26 ± 8.46	56.75 ± 16.24
T. muscularis	424.59 ± 11.12	274.66 ± 26.31	315.16 ± 23.14	439.91 ± 35.44
T. serosa	40.34 ± 9.49	13.51 ± 588	17.78 ± 7.43	12.08 ± 1.83
Total thickness	1023.78 ± 28.97	890.34 ± 22.76	978.78 ± 35.65	1034.87 ± 43.56

Note: Data are stated as mean ± standard deviation.

The duodenum was lined by a simple columnar epithelium composed of absorptive cells with enteroendocrine cells and some goblet cells. (Figure 9). Crypts and flattened villi characterized the duodenal mucosa, and the plicae circulares were not prominent. Between the villi were short tubular structures, Lieberkühn crypts. There was extension of the lamina propria between the crypts into the villous core, which had lymphocyte infiltration. The muscularis mucosa was thin, with a large number of duodenal glands beneath. The muscular coat was well developed and consisted of two layers of smooth muscle: a longitudinal outer layer and a circular inner layer.

Finger-like long villi characterized the jejunal mucosa with an average thickness of 182.50 ± 27.38 μm (Table 3), and plicae circulares were observed (Figure 10). The epithelium on the surface had few goblet cells and many absorptive cells. Paneth cells were located at the base of each distinct crypt. Diffuse lymphocyte infiltration was observed in the lamina propria. Two layers of smooth muscle were present in both the tunica muscularis and duodenum. The tunica serosa was typical, and the smooth muscle layers in the tunica muscularis were also thicker than in the duodenum.

Plicae circulares were clearly observed in the ileum and were more prominent than in the duodenum and jejunum (Figure 11). The number of goblet cells was increased numerically with an average of 24.67 ± 2.73 cells (Table 3) compared with the other two regions (duodenum, 2.00 ± 0.63; ileum, 5.50 ± 1.52). In the jejunum, there was similarity with the appearance of the other structures. No Peyer's patches were observed, but diffuse lymphoid infiltration was noted.

There were no villi in the mucosa of the cecum, and there were very well-developed plicae circulares (Figure 12). There were more goblet cells than simple columnar cells. Intestinal crypts and diffuse lymphoid tissue were observed in the mucosa, whereas no glands were found in the submucosa. The tunica serosa was typical, and there was a very thick tunica muscularis.

In the mucosa of the colon, there was a simple columnar epithelial layer with a large number of goblet cells. (Figure 13). The crypts were straight and tubular with absent villi. Lymphocytic infiltration was observed in the lamina propria. Tubular, straight mucous glands were identified in the mucosa, and there were two muscle layers in the muscular layer.

The rectal mucosa was lined with a large number of goblet cells, and the tubular straight glands extended throughout the thickness of the mucosa (Figure 14, Table 4). A markedly thick circular inner area was observed.

Table 3. The histometric measurements (μm) of the small intestine of adult male *D. leporina* ($n = 6$).

Parameters	Small Intestine		
	Duodenum	Jejunum	Ileum
T. mucosa	213.00 ± 5.19	274.51 ± 7.80	232.85 ± 11.35
Villi Length	149.27 ± 22.97	182.50 ± 27.38	172.30 ± 12.51
Villi Width	43.10 ± 3.40	54.49 ± 3.31	35.66 ± 3.48
Crypts Length	22.65 ± 2.36	35.88 ± 3.85	27.40 ± 3.34
Crypts Width	21.13 ± 2.43	37.96 ± 3.96	26.84 ± 2.54
No. Goblet Cells/10,000 μm^2	2.00 ± 0.63	5.50 ± 1.52	24.67 ± 2.73
T. submucosa	95.33 ± 3.33	109.90 ± 4.64	67.94 ± 6.13
T. muscularis	60.42 ± 7.99	141.94 ± 4.83	112.73 ± 4.55
T. serosa	8.70 ± 1.69	7.76 ± 2.70	7.33 ± 1.63
Total Thickness	345.23 ± 19.46	476.89 ± 23.78	405.24 ± 18.45

Note: Data are stated as mean ± standard deviation.

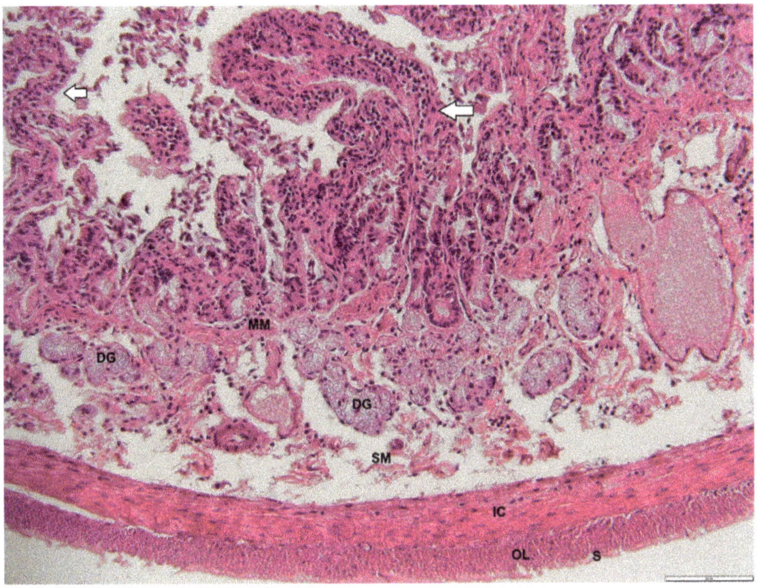

Figure 9. Photomicrograph of a transverse section through the duodenum of an adult male *D. leporina*, showing villi (arrows). H&Ex200. MM—muscularis mucosa; DG—duodenal glands; SM—submucosa; IC—inner circular layer; OL—outer longitudinal layer; S—serosa.

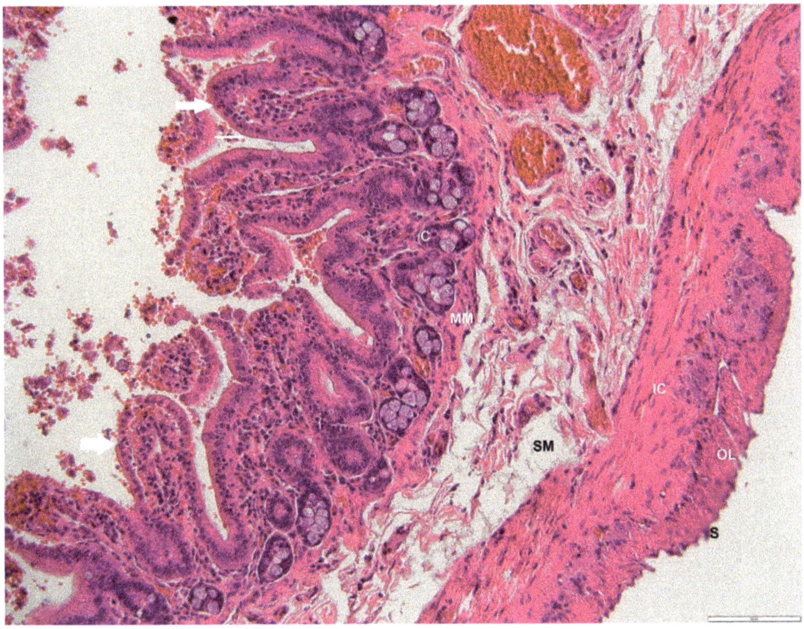

Figure 10. Photomicrograph of a transverse section through the jejunum of an adult male *D. leporina*, showing villi (arrows). H&Ex200. MM—muscularis mucosa; C—intestinal crypts; SM—submucosa; IC—inner circular layer; OL—outer longitudinal layer; S—serosa.

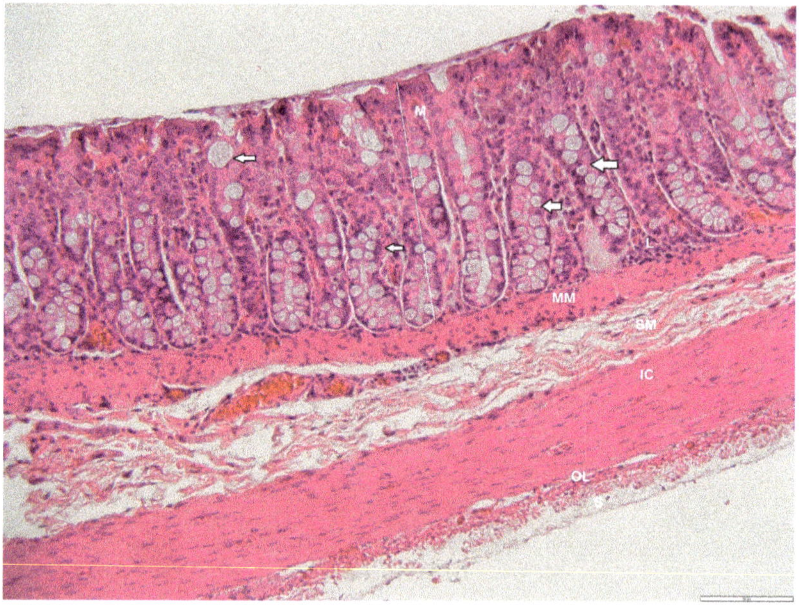

Figure 11. Photomicrograph of a cross-section through the ileum of an adult male *D. leporina*, showing numerous goblet cells in the mucosa (arrows). H&Ex200. M—mucosa; MM—muscularis-mucosa; L—lymphocyte infiltration; SM—submucosa; IC—inner circular layer; OL—outer longitudinal layer; S—serosa.

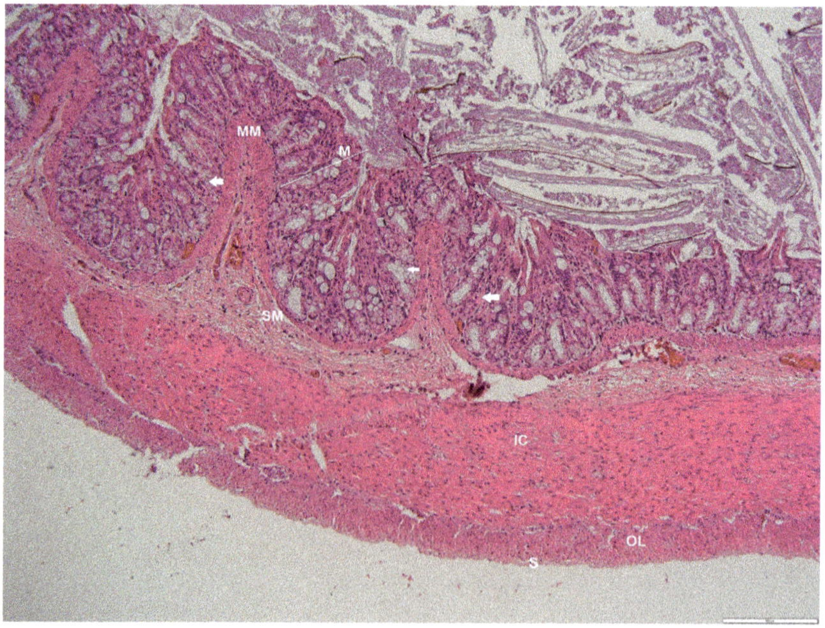

Figure 12. Photomicrograph of a transverse section through the cecum of an adult male *D. leporina*, showing goblet cells (arrows). H&Ex200. M—mucosa; MM—muscularis mucosa; SM—submucosa; IC—inner circular layer; OL—outer longitudinal layer; S—serosa.

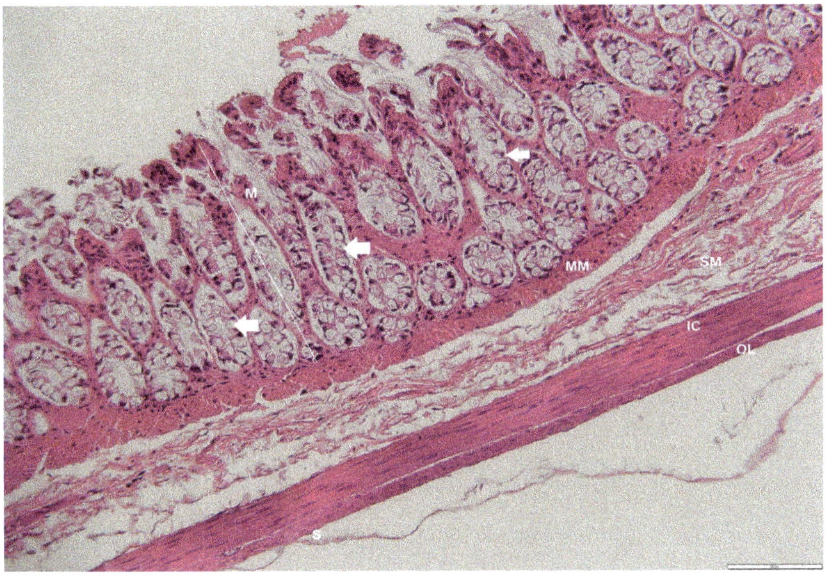

Figure 13. Photomicrograph of a cross-section through the colon of an adult male *D. leporina*, showing goblet cells (arrows). H&Ex200. M—mucosa; MM—muscularis mucosa; SM—submucosa; IC—inner circular layer; OL—outer longitudinal layer; S—serosa.

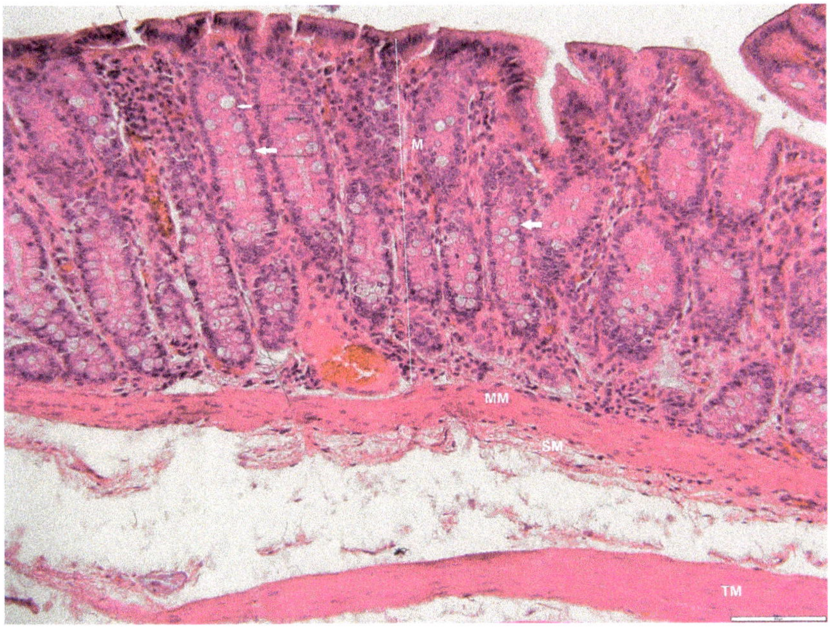

Figure 14. Photomicrograph of a cross-section through the rectum of an adult male *D. leporina*, showing goblet cells (arrows) H&Ex200. M—mucosa; MM—muscularis mucosa; SM—submucosa; TM—tunica muscularis.

Table 4. The histometric measurements (µm) of the large intestine of adult male *D. leporina* (n = 6).

Parameters	Large Intestine		
	Cecum	Colon	Rectum
T. mucosa	281.10 ± 8.89	241.96 ± 9.95	314.02 ± 8.76
No. Goblet Cells/10,000 µm^2	26.00 ± 1.41	38.5 ± 4.32	24.50 ± 5.17
T. submucosa	56.62 ± 3.76	78.19 ± 1.85	121.48 ± 4.25
T. muscularis	328.97 ± 12.26	144.99 ± 34.91	218.36 ± 8.12
T. serosa	7.22 ± 1.17	9.01 ± 2.02	7.09 ± 2.46
Total thickness	680.98 ± 33.89	530.90 ± 22.87	630.54 ± 25.61

Note: Data are stated as mean ± standard deviation.

4. Discussion

In this study, an attempt was made to explore the anatomy and histology of the GI tract of *D. leporina*. The results of the study show that the anatomy is typical of monogastric fermenters, and the histology resembles the tubular organs of mammals with few variations.

The esophagus is a relatively thick muscular tube with all four histological layers, similar to the findings in *D. leporina* [17]. The epithelium is lined with keratinized epithelium, similar to that found in ruminants, horses, and rodents, an adaptation to roughage feeding, whereas in humans, dogs, and cats, a non-keratinized stratified epithelium is found [19]. This suggests that the epithelial tissue renews itself, which is essential for maintaining the mucosal layer and protecting it from pathogen invasion and mechanical abrasion [19]. A scant number of esophageal glands (not shown) indicated that *D. leporina* can process roughage. The increased number of glands is associated with increased buffering capacity and absorption of volatile fatty acids, as reported in *Lama glama* [20]. The tunica muscularis is composed predominantly of striated muscle, similar to that found in dogs; ruminants; and rodents, including mice, rats, and hamsters [21], and transitions into smooth muscle at the cardiac sphincter.

The overall morphology of the mammalian stomach varies widely among species, although there are some basic structural differences. Stomach morphology is strongly influenced by adaptation, type of food, frequency of food intake, duration and need for food storage, body size and shape, etc. The stomach in the present study was unilocular and simple and approximately pear-shaped. The body and fundus of the stomach are not clearly demarcated, as noted by Garcia et al. [17]. Grossly, the stomach of the agouti is similar to that of the lappe [22], but it differs from the capybara, which has a gastric diverticulum [23]. The capybara and lappe are also hystricomorphic rodents, but the capybara is an herbivore, and the lappe is a frugivore. The intermediate, simple morphology and size of the stomach may indicate that *D. leporina* is better adapted to an omnivorous diet.

Histologically, the stomach consists of the same structure as in mammals in terms of layers and cell types. The gastric mucosa is lined by a simple columnar epithelium with gastric glands. The mucosa of the fundus and pyloric glands are much better developed than the cardiac region. However, the parietal cells are the predominant cell type; the chief cells were also increased, indicating that the glandular stomachs are well equipped to digest a coarse diet. The presence of a well-developed muscularis mucosa adapts the stomach to distension and has the function of preventing the deformation of the glandular layer. In addition, the muscularis mucosa may be related to the absorption of easily digestible substrates such as disaccharides and short-chain fatty acids [24].

Schieck and Millar [25] compared the morphology of the digestive tract of 35 rodent species and also found that mass and length were greater in herbivores than in granivores, explaining that the length of the small intestine in small mammals does not reflect the amount of dietary fiber in each species. In the present study, the small intestine was found to be the largest segment of the GI tract, and dietary habits in the wild and in captivity showed that it could process herbivorous food very well. However, no significant differences were found in measurements of the small intestine of the white-footed mouse (*Peromyscus leucopus*), an omnivorous species, and the meadow mouse (*Microtus pennsylvanicus*), an herbivorous species, but measurements of the hindgut showed diet-specific anatomical differences [9]. The intestine of *D. leporina* has a simple histological organization compared to that of rodents. The typical histology of the intestinal tube consists of four layers. The innermost layer is the mucosa with the muscularis mucosa, the submucosa, the tunica muscularis, and the tunica serosa, which is the outermost layer [19]. Mucous cells were found in the intestinal mucosa, and there was a numeric increase (non-significant) in number in the ileum. Although the amount and composition of the mucilaginous substance produced by the gastric and intestinal epithelium and cells of the glands might be mainly related to nutrition and the functional importance of the environment in the stomach and intestine of *D. leporina*, the villi are more developed (based on numeric difference) in the jejunum than in the other areas, suggesting that the jejunum is the most active site of absorption.

The crypts in the present study are mostly unbranched, and some branched crypts have more than one base per opening. They are thicker than the muscularis mucosa. Many authors found a close correlation between the depth of the crypts and the proliferation rate of epithelial cells [26,27]. In the present study, the crypts are well developed, suggesting that the absorbing cells of the intestine proliferate very well, which, in turn, contributes to better absorption. Some studies have shown that increased energy requirements lead to an increase in the length and volume of the small intestine [28,29] and that dietary habits are closely related to intestinal morphology and structure [30,31].

In the present study, the lamina propria throughout the small intestine showed strong lymphocyte infiltration. In particular, lymphocytes were numerous in the ileum, which was not observed in normal Peyer's patches [32]. This suggests that the intestine is well adapted to cope with infectious agents that may be present in the ingested food.The large intestine (LI) was wide and short in length compared to the small intestine in *D. leporina*. The LI was highly developed, suggesting increased adaptation to water absorption from fruits [33,34]. The results suggested that *D. leporina*, like opossums, can consume a large amount of fruit in the wild [35]. The cecum in the present study is large and well-developed, and there is also a large and well-developed colon, suggesting that this animal is a hindgut fermenter and can store and process large amounts of herbivorous food. Overall, *D. leporina* possessed the typical structures of mammalian species in terms of microanatomy. However, as for the SI and LI, they have a considerable amount of lymphoid tissue, indicating that these species have a well-developed immunological response.

5. Conclusions

The anatomy and histology of the GI tract of *D. leporina* were studied. The results showed that the well-folded esophagus lined with keratinized epithelium with scant esophageal glands, lymphatic infiltration, and a thick wall (1023.78 $\pm$ 28.97 μm) indicated adaptation to a large amount of coarse food. The simple stomach with a well-developed gastric gland suggests a carnivorous diet, and the strongly developed small intestine with the remarkable development of villi in the jejunum suggests that they absorb food very well and saturate it for a long time in the intestine. The long cecum and well-developed colon clearly indicate that the GI tract is adapted to frugivorous and herbivorous diets. Overall, *D. leporina* suggests that it is well adapted to its omnivorous feeding behavior. The morphology and histo-architecture of the various regions of the GIT, including the morphometry and histometry of the alimentary canal, justify these adaptations. These results provide qualitative and quantitative background information on the gross morphology and histology, as well as on some functional components of the GIT in *D. leporina*, which are essential for generating data for this animal. However, further detailed studies at the ultrastructural level are needed to confirm these results. Detailed feeding studies have to be performed to obtain the nutritional requirements for this animal, as well as enzymatic studies.

Author Contributions: Conceptualization, K.R.J.; Methodology, K.R.J., R.E.J. and V.S.; Validation, K.R.J., V.S.; Formal Analysis, V.S.; Investigation, K.R.J. and R.E.J.; Data Curation, K.R.J. and R.E.J.; Writing—Original Draft Preparation, K.R.J., R.E.J. and V.S.; Writing—Review and Editing, K.R.J. and V.S.; Supervision, K.R.J. and V.S. All authors have read and agreed to the published version of the manuscript.

Funding: This research received no external funding.

Institutional Review Board Statement: The study was conducted according to the guidelines of the Declaration of Helsinki and approved by the Ethics Committee of The University of the West Indies (No. CREC-SA 1095/07/2021).

Informed Consent Statement: Not applicable.

Data Availability Statement: All data was presented in the manuscript.

Acknowledgments: Special thanks must be given to Gary Garcia for his comments on the draft document. Special thanks are also given to the technicians present in the Anatomy Unit of the Department of Basic Veterinary Sciences for the processing of samples.

Conflicts of Interest: The authors declare no conflict of interest.

References

1. Smith, D.M.; Grasty, R.C.; Theodosiou, N.A.; Tabin, C.J.; Yoder, N.M.N. Evolutionary relationships between the amphibian, avian, and mammalian stomachs. *Evol. Dev.* **2000**, *2*, 348–359. [CrossRef]
2. McNab, B.K. The influence of food habits on the energetics of eutherian mammals. *Ecol. Monogr.* **1986**, *56*, 919. [CrossRef]
3. Gorman, M.L.; Corp, N.; Speakman, J.R. Apparent absorption efficiency and gut morphometry of wood mice, *Apodemus sylvaticus*, from two distinct populations with different diets. *Physiol. Zool.* **1997**, *70*, 610–614.
4. Del Valle, J.C.; Busch, C. Body composition and gut length of *Akodon azarae* (Muridae: Sigmodontinae): Relationship with energetic requirements. *Acta Theriol.* **2003**, *48*, 347–357. [CrossRef]
5. Bozinovic, F.; Novoa, F.F.; Veloso, C. Seasonal changes in energy expenditure and digestive tract of *Abrothrix andinus* (Cricetidae) in the Andes range. *Physiol. Zool.* **1990**, *63*, 1216–1231. [CrossRef]
6. Hammond, K.A. Seasonal changes in gut size of the wild prairie vole (*Microtus ochrogaster*). *Can. J. Zool.* **1993**, *71*, 820–827. [CrossRef]
7. Borkowska, A. Seasonal changes in gut morphology of the striped field mouse (*Apodemus agrarius*). *Can. J. Zool.* **1995**, *73*, 1095–1099. [CrossRef]
8. Weiner, J. Physiological limits to sustainable energy budgets in birds and mammals: Ecological implications. *Trends Ecol. Evol.* **1992**, *7*, 384–388. [CrossRef]
9. Derting, T.L.; Noakes, E.B. Seasonal changes in gut capacity in the white-footed mouse (*Peromyscus leucopus*) and meadow vole (*Microtus pennsylvanicus*). *Can. J.* **1995**, *73*, 243–252. [CrossRef]
10. Eisenberg, J.F.; Redford, K.H. *Mammals of the Neotropics: Ecuador, Bolivia, Brazil*; University of Chicago Press: Chicago, IL, USA, 1989; Volume 3, p. 200.
11. Brown-Uddenberg, R.C.; Garcia, G.W.; Baptiste, Q.S.; Counand, T.; Adogwa, A.O.; Sampson, T. *The Agouti Booklet and Producers Manual*; GWG Publications: Champ Fleurs, Trinidad and Tobago, 2004; p. 100.
12. Mollineau, W.M.; Garcia, G.W.; Samayah, D.; Kissoonsingh, W.; Procope-Garcia, A. The wildlife industry in Trinidad: A case study towards developing a sustainable industry model for a small Twin Island State (Trinidad and Tobago). In Proceedings of the Conference on Managing Space for Sustainable Living in Small Island Development States, St. Augsutine, Trinidad and Tobago, 7–10 March 2000.
13. Roopchand, A. Wildlife and the Oil Industry: Catshill Reserve a Case Study. Masters' Thesis, The University of the West Indies, St Augustine, Trinidad and Tobago, 2002.
14. Leus, K.; Macdonald, A.A. Gastrointestinal anatomy, diet selection and digestion in mammals: A brief overview. In *Research and Captive Propagation*; Gansloßer, U., Hodges, J.K., Kaumanns, W., Eds.; Filander: Furth, Germany, 1995; pp. 100–112.
15. Jones, K.R.; Lall, K.R.; Garcia, G.W. Omnivorous Behaviour of The Agouti *(Dasyprocta Leporina*): A Neotropical Rodent With The Potential For Domestication. *Scientifica* **2019**, *2019*, 3759783. [CrossRef]
16. Lall, K.R.; Jones, K.R.; Garcia, G.W. Nutrition of six selected neo-tropical mammals in Trinidad and Tobago with the potential for domestication. *Vet. Sci.* **2018**, *5*, 52. [CrossRef] [PubMed]
17. Garcia, G.W.; Baptiste, Q.S.; Adogwa, A.O.; Kakuni, M.; Arishima, K.; Makita, T. The digestive system of the agouti *(Dasyprocta leporina*)-gross anatomy and histology. *Jpn. J. Zoo Wildl. Med.* **2000**, *5*, 55–66. [CrossRef]
18. National Research Council (US). Committee for the update of the guide for the care and use of laboratory animals. In *Guide for the Care and Use of Laboratory Animals*, 8th ed.; National Academies Press: Washington, DC, USA, 2011; Available online: https://www.ncbi.nlm.nih.gov/books/NBK54050 (accessed on 1 July 2022). [CrossRef]

9. Dellmann, D.H.; Brown, E.M. Histology of the digestive system. In *Textbook of Veterinary Histology*, 3rd ed.; Lea And Febiger: Philadelphia, PA, USA, 1987; pp. 229–247.

0. Sukon, P.; Timm, K.I.; Valentine, B.A. Esophageal Anatomy of the Llama (*Lama Glama*). *Int. J. Morphol.* **2009**, *27*, 811–817. [CrossRef]

1. Schummer, A.; Nickel, R.; Sack, W.O. The alimentary canal. In *The Viscera of Domestic Mammals*; Springer: New York, NY, USA, 1979; pp. 99–202.

2. Machado, M.R.F.; Leal, L.M.; Sasahara, T.H.C.; Zanetti, A.S.; Cruz, C. Morfologia do estômago e do duodeno da paca (Cuniculus paca Linnaeus, 1766). *Arq. Bras. De Med. Vet. Zootec.* **2015**, *67*, 1616–1624. [CrossRef]

3. Barros Moraes, P.T.; Pacheco, M.R.; Souza, W.M.; Da Silva, A.A.; Neto, P.B.S.; De Figuieriredo Barreto, C.S.; Ribiero, A.A.C.M. Morphological aspects of the capybara stomach (*Hydrochoerus hydrochaeris*): Gross and microscopic structure. *Anat. Histol. Embyol.* **2002**, *31*, 362–366. [CrossRef]

4. Grau, A.; Crespo, S.; Sarasquete, M.C.; Gonzalez De Canales, M.L. The digestive tract of the amberjack *Seriola dumerili*, Risso: A light and scanning electron microscope study. *J. Fish Biol.* **1992**, *41*, 287–303. [CrossRef]

5. Schieck, J.O.; Millar, J.S. Alimentary Tract Measurements as Indicators of Diets of Small Mammals. *Mammalia* **1985**, *49*, 93–104. [CrossRef]

6. Jin, L.; Reynolds, L.P.; Redmer, D.A.; Caton, J.S.; Crenshaw, J.D. Effects of dietary fiber on intestinal growth, cell proliferation, and morphology in growing pigs. *J. Anim. Sci.* **1994**, *72*, 2270–2278. [CrossRef] [PubMed]

7. Van Beers-Schreurs, H.M.G.; Nabuurs, M.J.A.; Vellenga, L.; Kalsbeek-van der Valk, H.J.; Wensing, T.; Breukink, H.J. Weaning and the weanling diet influence the villous height and crypt depth in the small intestine of pigs and alter the concentrations of short-chain fatty acids in the large intestine and blood. *J. Nutr.* **1998**, *128*, 947–953. [CrossRef] [PubMed]

8. Hammond, K.A.; Wunder, B.A. The role of diet quality and energy need in the nutritional ecology of a small herbivore, *Microtus ochrogaster*. *Physiol. Zool.* **1991**, *64*, 541–567. [CrossRef]

9. Gębczyńska, Z.; Gębczyński, M. Length and weight of the alimentary tract of the root vole. *Acta Theriol.* **1971**, *16*, 359–369. [CrossRef]

0. Wilczynska, B.; Przystalski, A. Morphometry and histometry of the alimentary canal in *Bufo orientalis*. *Zool. Pol.* **1998**, *43*, 4.

1. Wilczyńska, B. Anatomical structure and size of large intestinal mucosa in selected species of shrews and rodents. *Acta Theriol.* **1998**, *43*, 363–370. [CrossRef]

2. Banks, W.J. *Applied Veterinary Histology*, 2nd ed.; Williams & Wilkins: Baltimore, MD, USA, 1986; p. 400.

3. Charles-Dominique, P.; Atramentowicz, M.; Charles-Dominique, M.; Gerard, H.; Hladik, A.; Hladik, C.M.; Prévost, M.-F. Les mammiferes frugivores arboricoles nocturnes d'une foret guyanaise: Inter-relations plantes-animaux. *Rev. D'ecol. Terre Vie* **1981**, *35*, 341–436.

4. Santori, R.T.; Astua De Moraes, D.; Cerqueira, R. Comparative gross morphology of the digestive tract in ten Didelphidae marsupial species. *Mammalia* **2004**, *68*, 27–36. [CrossRef]

5. Redford Kent, H.; Einsenberg, J.F. *Mammals of the Neotropics: The Southern Cone*; The University of Chicago Plates: Chicago, IL, USA, 1992; p. 210.

animals

MDPI

Article

The Digestive System of the *Arctocephalus australis* in Comparison to the Dog as a Land-Carnivore Model

Rosario Martín-Orti [1], Carlos Tostado-Marcos [2], Juan-Pablo Loureiro [2], Ignacio Molpeceres-Diego [2], Enrique Tendillo-Domínguez [2], Inmaculada Santos-Álvarez [1], Pilar Pérez-Lloret [1] and Juncal González-Soriano [1,*]

[1] Departamento de Anatomía y Embriología, Sección Departamental de Anatomía y Embriología (Veterinaria), Facultad de Veterinaria, Universidad Complutense, Avenida Puerta de Hierro s/n, 28040 Madrid, Spain; rosamart@ucm.es (R.M.-O.); inmasant@ucm.es (I.S.-Á.); pilper01@ucm.es (P.P.-L.)

[2] Fundación Mundo Marino. Av. X 157, San Clemente del Tuyú B7105, Provincia de Buenos Aires, Argentina; tostadomarcoscarlos@gmail.com (C.T.-M.); juanploureiro@gmail.com (J.-P.L.); ignamolp11@gmail.com (I.M.-D.); enriquetd95@gmail.com (E.T.-D.)

* Correspondence: juncalgs@vet.ucm.es

Simple Summary: Marine mammals are warm-blooded vertebrates that behave in the same way as land mammals do but they spend most or all of their lives in the ocean. There is much previous research on whales, dolphins, or even different types of seals, including their behavior, health, anatomy or perception. Between all these fields, it is commonly accepted that anatomy is considered critical to understanding many physiological adaptations. For example, their ability to dive by holding their breath underwater for long periods of time. During this process they can postpone digestion for several hours. However, and contrary to what might be expected, our results show that being a carnivore seems to be the critical characteristic defining their digestive apparatus, the adaptations to the aquatic environment being less significant.

Abstract: Marine mammals play a critical ecological role as both predator and prey. They are divided into three groups that share similar adaptations to their aquatic life, but that have very different origins and life patterns: sirenians, pinnipeds, cetaceans. The species object of our interest is the South American fur seal or *Arctocephalus australis*, a carnivore classified within the group of pinnipeds. The objective of the present study was to evaluate whether the anatomical characteristics of the *Arctocephalus australis'* digestive system are similar to that of other land-carnivores or if, on the contrary, this species shows anatomical adaptations related to their life in the ocean. The study was carried out on 11 cadavers of the species *Arctocephalus australis*, made up of two adults and nine juveniles, by means of the anatomical dissection of their entire isolated digestive system. We demonstrate that, with several exceptions, the anatomical characteristics of the digestive system of the *Arctocephalus australis* are similar to those in other carnivores. Therefore, our data constitute an important contribution for clinical diagnostic and conservation purposes, for both veterinarians and biologists.

Keywords: *Arctocephalus australis*; anatomy; digestive system; adaptations

1. Introduction

Conducting research into marine mammals is important in order to have a better understanding of their aquatic ecosystems. Pinnipeds, cetaceans and sirenians are mammals that develop their life in the marine aquatic environment, thanks to the large number of anatomical and physiological adaptations they have developed. A part of the life cycle of pinnipeds, such as feeding or exploration, takes place in the water, while another part, usually linked to periods of reproduction, hair shedding and rest, takes place on land. Instead, the life cycle of sirenians and cetaceans is completely aquatic [1–6].

The species object of our interest is the South American fur seal or *Arctocephalus australis*, a carnivore classified within the group of pinnipeds. Previous studies on these pinnipeds are numerous and focused on very different fields [7–15]. Yet, although anatomy is considered critical to understanding fundamental processes such as certain physiological adaptations, bibliography on the anatomy of the South American fur seal is scarce [5,16–18].

One of these adaptations in marine mammals (including the *Arctocephalus australis*) is their ability to postpone digestion for several hours while they are diving. Thanks to this process, it is possible to prevent oxygen consumption and instead, improve the oxygenation of muscles and vital organs, to extend the diving time safely [19–24].

The question that arises is whether this ability or other physiological adaptations modify the anatomical characteristics of the digestive system of the *Arctocephalus australis*, or if, on the other hand, it has the same general characteristics of the digestive system of other carnivores [25–28].

In the present study, we try to answer this question and thus, our main goal is to carry out an anatomical, systematic, and descriptive study of the South American fur seal's digestive system, in comparison with the digestive apparatus of the dog, as this species could be considered as a carnivore model.

2. Materials and Methods

The anatomical study was carried out on 11 cadavers of the species *Arctocephalus australis*, made up of 2 adults and 9 juveniles. In all cases, the animals were found stranded on the coast, and taken to the Centro de Rescate y Rehabilitación Fundación Mundo Marino (FMM), where they died. The individuals were never euthanized. In all cases, they were frozen and preserved until their dissection in the necropsy room. Thawing time was at least 24–48 h. The animals were placed in dorsal recumbency. All dissections were carried out by veterinarians.

As a first step, the skin, blubber, and muscle must first be removed. With a scalpel blade, we made a first incision, under the chin, near the apex of the throat, and continued down the midline to the anus. If the animal was a male, we made the cut slightly left of the penile opening. The cut incised down to the blubber/muscle interface, without penetrating the skeletal muscle. Sometimes, we made perpendicular incisions on either side of the midline incision, creating a series of panels down the length of the animal. The blubber was removed by cutting through the fascia plane and reflecting the sections of skin and blubber away from the body, in a ventral to dorsal manner.

For the internal examination, we removed both front flippers, by pulling the appendage away from the body, and carefully cutting through the tissue that connects the flipper and scapula to the body wall. The same technique was carried out on the hip joints, sectioning the inguinal muscles, and displacing the femurs' heads.

To expose the abdominal organs, we incised the abdominal wall from the last rib mid-ventral to the hip. To expose the internal tissues, we extended the most cranial cut laterally along the thoracic arch, reflecting the abdominal musculature.

Before cutting the ribs, to check the thoracic pressure, we punctured the diaphragm with a scalpel or scissors. To open the thoracic cavity, we removed the rib cage by cutting through each thoracic rib mid-articulation, which is the cartilaginous flex point that allows movement of the rib cage during inhalation and exhalation. Then, the rib cage was removed, and the thoracic cavity exposed.

To keep the esophagus intact, we removed the superficial musculature of the neck area, exposing the trachea and esophagus. Then, we made a triangular incision along the inner sides of the jaws. The hyoid apparatus was cut, and the tongue was extracted through the intermandibular space. Then, the cardiorespiratory system was progressively separated.

To make the esophagus independent, its connections with the respiratory system (esophagus–trachea) were removed. On the other hand, the esophagus was sectioned at the beginning, exactly at the level of the pharynx. To maintain the esophagus–stomach union, an eyelet-shaped cut was made around the esophageal hiatus. In the abdominal cavity, all adhesions with the omentum, as well as the junctions with the uro-genital apparatus, were detached. Finally, the entire digestive system was externalized by sectioning the most caudal part of the rectum, including the liver and the pancreas.

As mentioned before, the present study included different specimens: 9 juveniles and 2 adults, obviously each with a different degree of development. Because of this, the corresponding arithmetic means of the data obtained were calculated separately, to avoid deviations and erroneous data.

The anatomical data obtained from dissection were completed with a radiological analysis of the main digestive viscera. The radiographs were provided by FMM. All proceeded from animals that needed X-rays for diagnostic purposes. The X-ray equipment used for this purpose was a portable brand poskom 40 ma/90 kv. The visualizer was a FujiFilm model devo II.

3. Results

3.1. Oral Cavity

The South American fur seal has an elongated and narrow oral cavity, with a wide mouth opening (Figure 1). The lips are thin and mobile with no tactile hairs. Instead, they have vibrissae in the muzzle, which are functional when diving. The upper lip has a little cleft (not very evident) in the central area, which divides it into two equal parts. In addition, the lateral edges of the lower lip are firm (Figure 1). The hard palate is made up of the palatine, maxillary and incisor bones. It is narrow in the rostral area, and widens caudally. The median palatine raphe divides the palate into two symmetrical halves, with a palatine raphe which is practically inconspicuous. There are also six to seven palatine ridges, curved and poorly marked extending on both sides of the palate. They never reach the median raphe. Caudally to the incisive teeth, there is a rounded shaped papilla: the incisive papilla. The mucosa may be slightly or not at all pigmented.

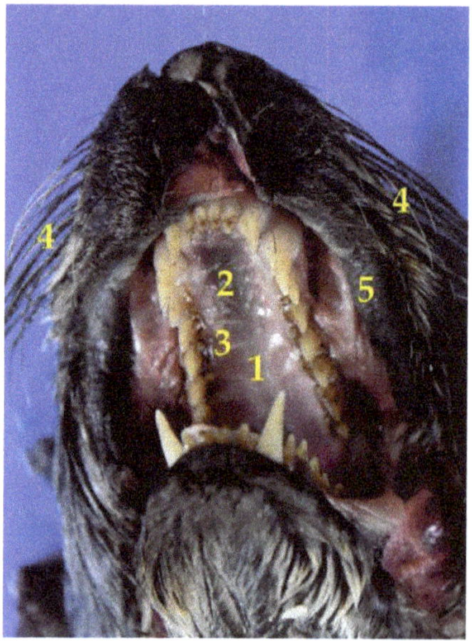

Figure 1. Image of the snout and the oral cavity of the *Arctocephalus australis*. (**1**) Hard palate; (**2**) median raphe; (**3**) curved ridges; (**4**) vibrissae; (**5**) upper lip.

Something that is common and striking in the South American fur seals is the presence of a blackish coloration covering the surface of the teeth and the dorsal surface of the tongue, which is elongated and very mobile. It is wide at the base, although it progressively thins in a rostral direction, until it ends in a blunt and slightly bifid vertex. The tongue consists of three parts: base, body, vertex. No dividing mid-groove is observed on the dorsal surface (Figure 2). Some of the morphometric characteristics of the tongue are reflected in Table 1.

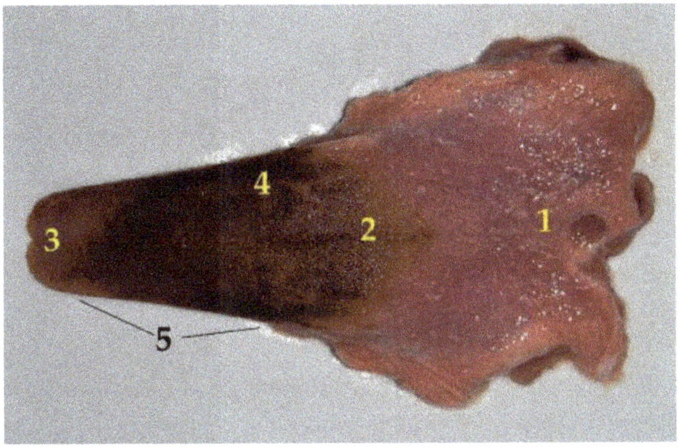

Figure 2. Dorsal surface of the South American fur seal tongue isolated from the oral cavity: (**1**) base; (**2**) body; (**3**) vertex (blunt and slightly bifid); (**4**) black coloration; (**5**) tongue ridge.

Table 1. Morphometric data of the tongue of *Arctocephalus australis*. A: adult; J: juvenile.

	Age		Length	Diameter (cm)			Weight
ID	Group	Sex	(cm)	Base	Body	Tip	(g)
M7319	A	♀	16	14	12	6	135
M99	A	♀	14	13	12	15	130
M104	J	♂	11	8.5	8	4	86
M100	J	♀	9.5	7.3	6.5	3	62.5
M74	J	♂	10	8	6.5	4.5	50
M83	J	♂	10.2	8.5	8	4.5	65
M82	J	♀	9	10	7.8	5	58.4
M70	J	♂	8.7	8	6	3	54
M79	J	♂	10.5	9.2	7	4	71
M97	J	♀	7	9	7.2	5	53.7
M84	J	♀	7.6	7.5	6	4.5	46
Mean	A		15	13.5	12	5.5	132.5
Mean	J		9.2	8.4	7	4.1	60.7

Teeth

An important anatomical feature is the dental formula:

$$\textit{Arctocephalus australis}: 2\ (\mathrm{I}\ 3/2\ \mathrm{C}\ 1/1\ \mathrm{P}\ 4/4\ \mathrm{M}\ 2/1) = 36 \qquad (1)$$

There are characteristics that are worth highlighting. First, the morphological differentiation between premolars and molars in the South American fur seal has disappeared. Thus, the *Arctocephalus australis* has four jaw teeth and two mandibular teeth

It is interesting to remark that the crown of these dental pieces in the *Arctocephalus australis* is tritubercular, with the central prominence clearly larger than the laterals (Figure 3).

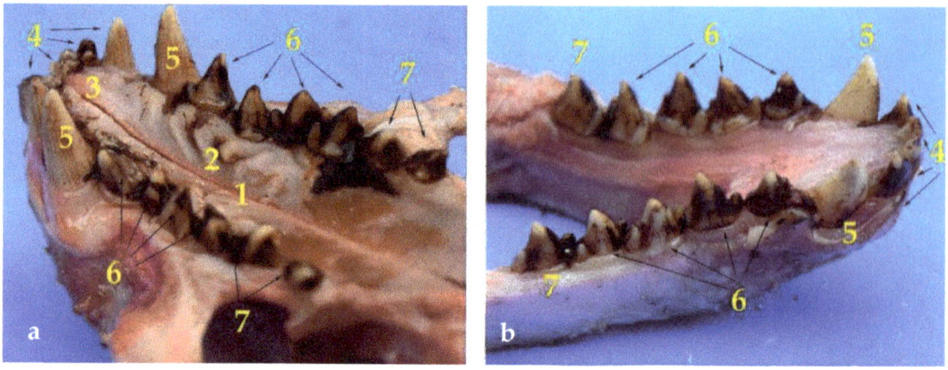

Figure 3. Images of the palate and teeth of the maxilla (**a**) and the jaw (**b**), which are composed of: (**1**) Hard palate; (**2**) palatine ridges; (**3**) incisive papilla; (**4**) incisors; (**5**) canines; (**6**) premolars; (**7**) molars. Additionally, the images show the substantia nigra adhered to the teeth surface.

Concerning the incisors, the upper dental arch has a total of six, three on each side (Figure 3). The four centrals are bitubercular, both prominences being of identical size. The third incisor on each side (more lateral) has a completely different shape, as its size and appearance are similar to that of a canine. The crown has a single tubercle, is conical and curved. It is wide in the root, reducing its size progressively until ending in a sharp vertex. It could resemble a second, somewhat less developed, canine.

The mandibular incisors in the *Arctocephalus australis* number four in total, two on each side (Figure 3). Thus, the South American fur seal has lost two of its lower incisors.

3.2. Pharynx and Esophagus

The oropharynx has the typical characteristics of any carnivore. On the contrary, there are differences that affect the respiratory portion of the pharynx, which are not worth analyzing because the respiratory system is not the object of the present study.

The esophagus looks similar to that of other carnivores (Figure 4), although it is particularly wide and distensible, to avoid obstructions or different potential tissue damages when swallowing big prey. The esophageal width is constant along its course, except when approaching the cardiac sphincter, where a slight widening occurs. It has a great length, as in adults the average is 61 cm (Table 2).

Table 2. Morphometric data of the *Arctocephalus australis* esophagus (11 specimens). A: adult; J: juvenile.

ID	Age Group	Sex	Length (cm)	Diameter (cm)	Weight (g)
M7319	A	♀	63	4.5	210
M99	A	♀	59	4.4	175
M104	J	♂	46	3.5	75
M100	J	♀	34	2.5	45
M74	J	♂	38	2.7	52.6
M83	J	♂	33	2.5	55
M82	J	♀	36	2.4	48
M70	J	♂	38	2.7	43
M79	J	♂	37	2.3	55
M97	J	♀	32	2	39.6
M84	J	♀	31	2.4	40
Mean	A		61	4.4	192.5
Mean	J		36.1	2.5	50.3

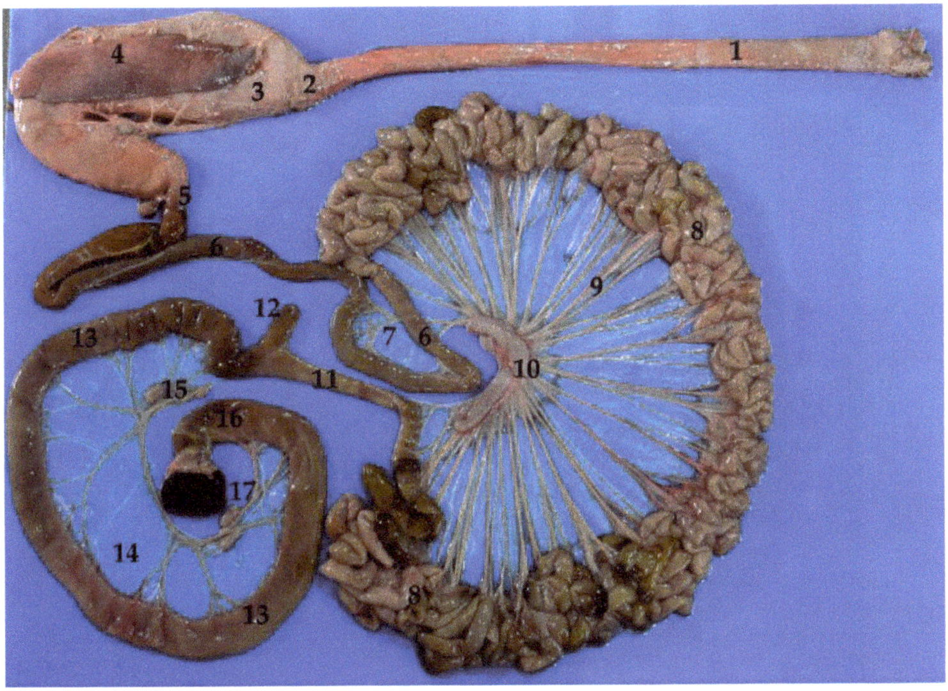

Figure 4. Isolated digestive system of the *Arctocephalus australis*. (**1**) Esophagus; (**2**) cardiac sphincter; (**3**) stomach; (**4**) spleen; (**5**) pylorus; (**6**) duodenum; (**7**) mesoduodenum; (**8**) jejunum; (**9**) mesojejunum and mesenteric vessels; (**10**) small intestine mesenteric lymphnodes; (**11**) ileum; (**12**) cecum; (**13**) colon; (**14**) mesocolon and mesenteric vessels; (**15**) large intestine mesenteric lymphnodes; (**16**) rectum; (**17**) anal canal.

Three esophageal parts are considered: cervical, thoracic and abdominal. The cervical part runs initially dorsal to the trachea (Figure 5). However, at the level of the caudal part of the neck, it becomes lateral and has a dorsolateral position with respect to the trachea, which in the *Arctocephalus australis* bifurcates at the thorax entrance, to become the thoracic esophagus. The esophagus position remains constant in relation to the bronchi (Figure 5), and ventral to the longus colli muscle, up to the third intercostal space and the aortic arch. From this point, it runs slightly tilted to the left, between the pulmonary lobes, until the diaphragm, passing through the esophageal hiatus, joining the stomach to form the abdominal esophagus.

3.3. Stomach

The stomach is a relatively large organ, which is located on the left side of the abdominal cavity. It is related to the liver craniodorsally and to the small intestine caudally (Figure 6), and is intimately related to the spleen (Figure 4). Laterally, it is in contact with the left abdominal wall, while ventrally it reaches the ventral abdominal wall, at the level of the xiphoid cartilage. It has two clearly differentiated parts: the left, comprising the cardiac part, the fundus and the body, and the right part, which is also called pyloric part (Figure 7a).

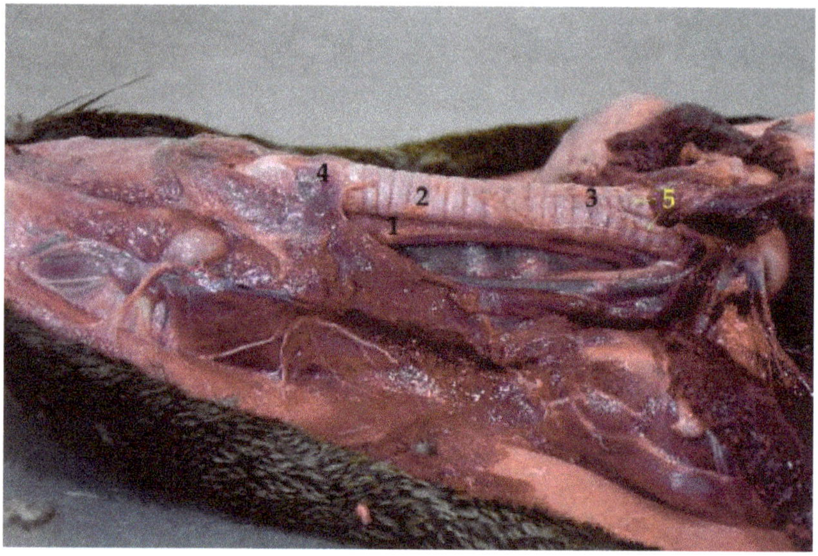

Figure 5. Cervical region with esophagus and trachea in situ (body in dorsal recumbency). Note the dorsal position of the esophagus in relation to the trachea. (**1**) Esophagus; (**2**) trachea; (**3**) tracheal bifurcation in the cervical region; (**4**) larynx; (**5**) bronchi.

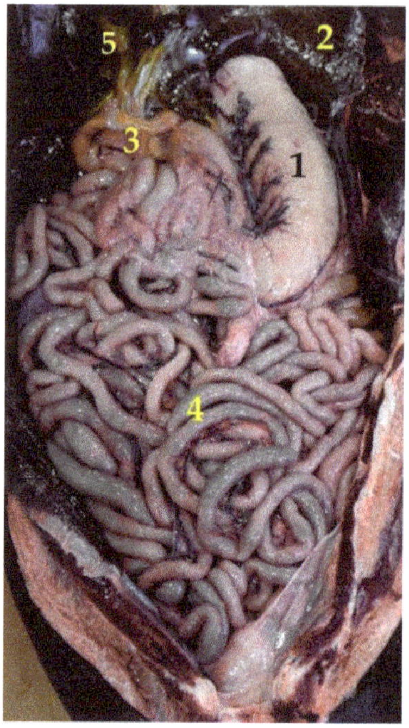

Figure 6. Topographical arrangement of the abdominal cavity (body in dorsal recumbency). The liver has been removed to see the gallbladder. (**1**) Stomach; (**2**) liver; (**3**) duodenum; (**4**) jejunum; (**5**) gallbladder.

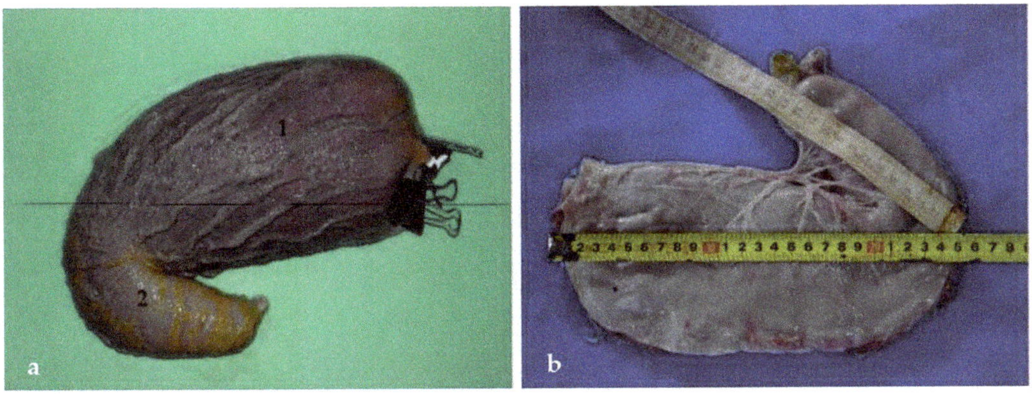

Figure 7. Stomach isolated from the abdominal cavity. (**a**) Anatomical pieces made up of expanded polyurethane (Fundación Mundo Marino) (**1**) Left part, (**2**) right or pyloric part.; (**b**) total length of the stomach or sum of the left and the pyloric parts (37 cm).

One of the relevant pieces of data included in the present study is the total length of the stomach, which is calculated by the addition of the left and right parts (Figure 7). The left is elongated and tubular shaped (Figure 8) and extends from the cardiac sphincter to the end of the stomach body. The right part, or pyloric part, is small and cylindrical in shape. It comprises the length from the end of the left part to the pylorus (Figure 8). According to the data obtained from the specimens of *Arctocephalus australis*, the mean average of the stomach length is 31 cm in juveniles and 53 cm in adults (Table 3). One of the most notable details of the stomach is the appearance of the mucosa, with numerous folds, which can be well appreciated thanks to an anatomical piece of expanded polyurethane, prepared according to the technique of De Sordi et al. [29] (Figures 7a and 9). These anatomical pieces are also very useful for analyzing the stomach variation in size and its capacity of dilation when it is full of content. If the stomach is empty, the left part is strongly contracted but, as it progressively fills up, this left part expands widely. Curiously, the pyloric part (right), is much less affected by the intake and thus, its size is maintained whether the stomach is full or empty.

Table 3. Morphometric data of the stomach proceeding of the 11 specimens of *Arctocephalus australis* used in the present study. A: adult; J: juvenile.

	Age		Length [1]	Curvature (cm)		Diameter (cm)		Weight
ID	Group	Sex	(cm)	Major	Minor	Left Part	Pyloric Part	(g)
M7319	A	♀	60 (38 + 22)	68	42	31	17	380
M99	A	♀	47 (29 + 18)	56	41	29	15	330
M104	J	♂	32 (19 + 13)	39	21	22	10	300
M100	J	♀	27 (17 + 10)	31	20	18.5	7.5	200
M74	J	♂	28 (17 + 11)	35	17	20.5	11	164
M83	J	♂	29 (17 + 12)	32	21	17	10	190
M82	J	♀	33 (21 + 12)	42	21	23.5	13	285
M70	J	♂	32 (19 + 13)	37	22	17	10	191
M79	J	♂	33 (20 + 13)	40	19	21	11	210
M97	J	♀	37 (23 + 14)	46	24	26	14	186
M84	J	♀	32 (21 + 11)	41	21	19	12	190
Mean	A		53.5	62	41.5	30	16	355
Mean	J		31.4	38.1	20.6	20.5	10.9	212.8

[1] Length (addition of the left and the right parts, respectively).

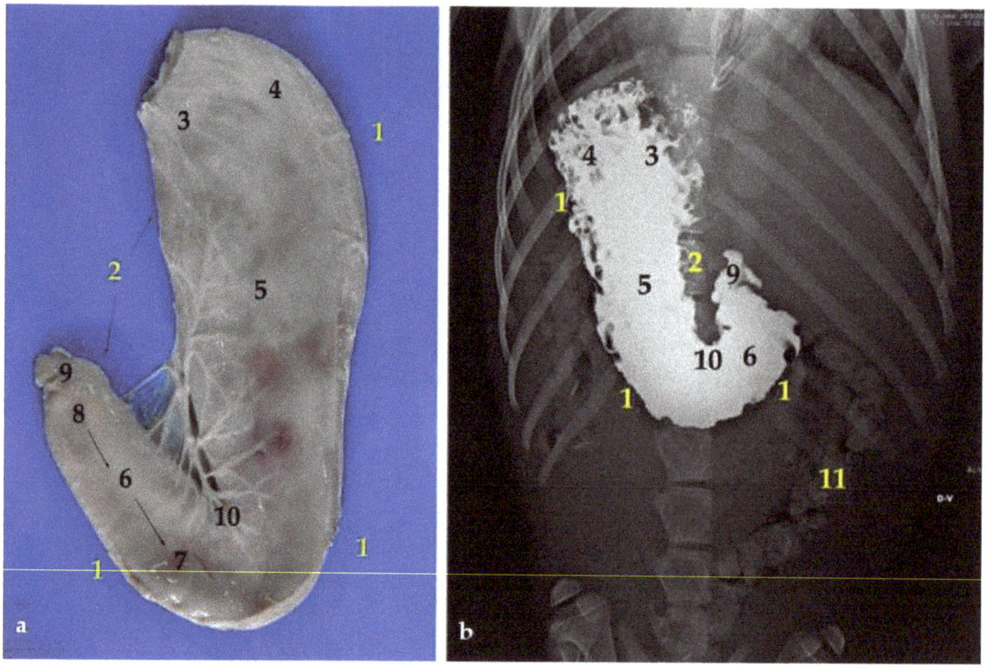

Figure 8. (**a**) Ventral view of the stomach; (**b**) dorsoventral contrast radiography. (**1**) Major curvature; (**2**) minor curvature; (**3**) cardiac part; (**4**) fundus; (**5**) body (**6**) right or pyloric part; (**7**) antrum pylorus; (**8**) pyloric canal; (**9**) pylorus; (**10**) angular incisure or angular notch; (**11**) colon.

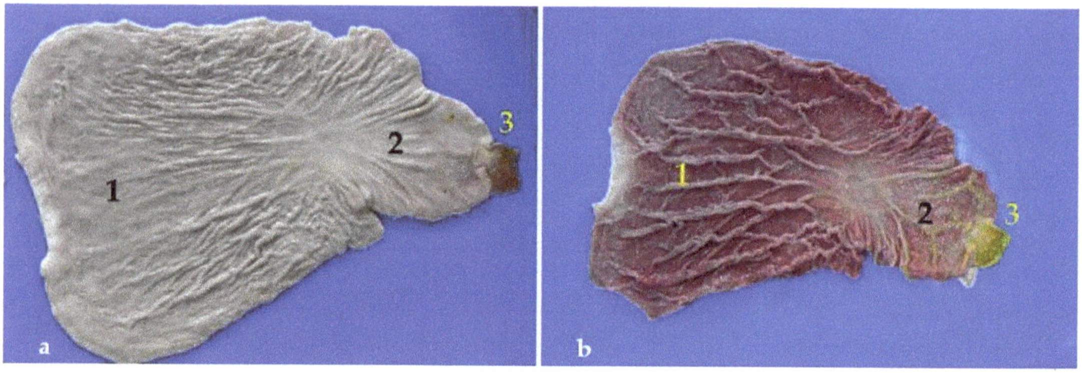

Figure 9. (**a**,**b**) Open stomach. Note the numerous folds of the mucosa. In (**a**) the stomach has been washed with water for days. (**1**) Fundic part; (**2**) pyloric part; (**3**) pylorus.

The stomach shows two curvatures: lesser and greater. First, the lesser curvature is straight and vertical, and forms the angular incisure, just at the bottom. This incisura is the consequence of a sudden change in direction of the pyloric part, which is directed craniodorsally (Figure 8). According to our results, the average length of the lesser curvature in juveniles is 20 cm, just half of that in adults, where it is 41 cm. On the opposite side is the greater curvature, which is considerably larger, with a length of 62 cm in adults and 38 cm in juveniles. Interestingly, in the South American fur seal, the greater curvature is not even double than the lesser (Table 3).

Another noteworthy fact is the stomach capacity, which in adults is of about 2–3 L and in juveniles it is of about 0.6–1 L.

3.4. Small Intestine

The South American fur seal's small intestine is peculiar, due to its significant large length. The shortest section is the ileum followed by the duodenum. The jejunum is the longest part, with an approximate length of 25 m in adults and between 12 and 15 m in juveniles (Table 4). Due to its size, the small intestine occupies a large part of the abdominal cavity (Figures 6 and 10).

Table 4. Morphometric data of the small intestine proceeding of the 11 specimens of *Arctocephalus australis* used in the present study. A: adult; J: juvenile; D: duodenum; J: jejunum; I: ileum.

ID	Age Group	Sex	Length D (cm)	Length J (m)	Length I (cm)	Diameter (cm) D	Diameter (cm) J	Diameter (cm) I	Weight (g) D	Weight (g) J	Weight (g) I
M7319	A	♀	43	20	16	5.5	3.7	4	25	1585	10
M99	A	♀	40	29	14	5	3.2	3.6	21	2015	12
M104	J	♂	26	15	10	4.1	3.4	2.9	16	655	6.6
M100	J	♀	19	13	15	3.5	2.4	3	13	410	8.1
M74	J	♂	22	12	9.5	3.3	2	3.2	12.5	490	6
M83	J	♂	36	13.5	6.1	4.8	2.5	2.7	15	405	5
M82	J	♀	25	14.4	7	4.5	2.6	3	19	680	4.9
M70	J	♂	27	13.5	8	4.2	2.4	2.3	12.8	525	4.3
M79	J	♂	26	14	8	3.7	2.6	3.2	17	615	4.7
M97	J	♀	22	12.5	9	2.3	2.2	2.4	11.5	765	3.8
M84	J	♀	25	12.5	8	3.5	2.5	2.6	10	405	3.8
Mean	A		41.5	24.5	15	5.2	3.4	3.8	23	1800	11
Mean	J		25.3	13.3	8.9	3.8	2.5	2.8	14	550	5.2

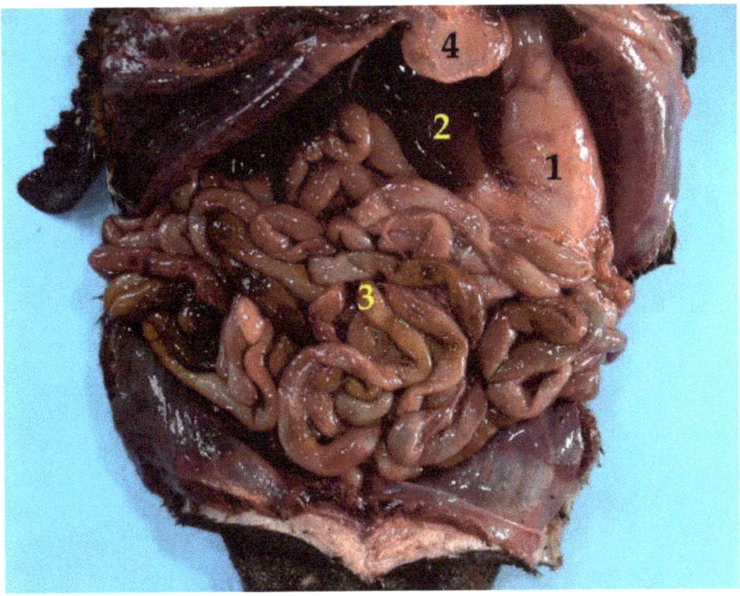

Figure 10. General view of the abdominal cavity (body in dorsal recumbency). (**1**) Stomach; (**2**) liver; (**3**) jejunum; (**4**) xiphoid cartilage.

To recognize the small intestine sections division, we used two criteria. First, the visualization of the mesenteric arteries, especially the last mesenteric artery, which runs isolated from the others and irrigates the ileum (instead of the jejunum), pointing out its beginning (Figure 11). Second, the different layout of the small intestine parts, which is striking. Thus, whereas the jejunum folds on itself and shows noticeable intestinal loops (Figures 4 and 12), the ileum runs straight, while the duodenum is curved, without loops (Figure 12).

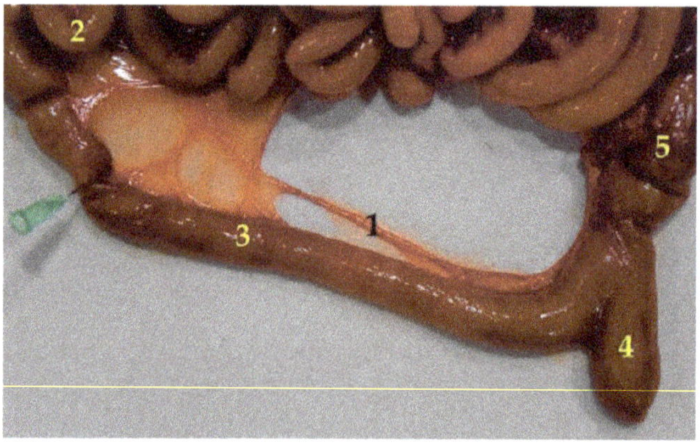

Figure 11. Differentiation of two segments of the small intestine (jejunum and ileum), pointed out by the last mesojejunal vessel, which splits off from the rest to reach the ileum. (**1**) Mesenteric vessel; (**2**) jejunum; (**3**) ileum; (**4**) cecum; (**5**) colon.

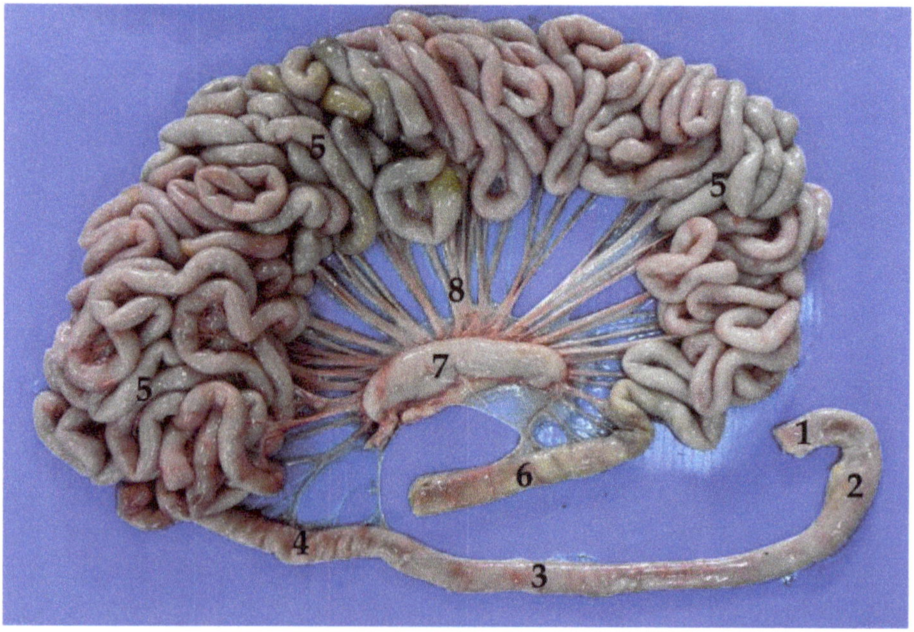

Figure 12. Isolated small intestine. Duodenum: (**1**) cranial part, (**2**) descending part, (**3**) transverse part, (**4**) ascending part; (**5**) jejunum; (**6**) ileum; (**7**) small intestine mesenteric lymph nodes; (**8**) mesenteric vessels.

3.4.1. Duodenum and Pancreas

The duodenum is the first section of the small intestine, beginning at the pyloric sphincter. Initially, it runs caudodorsally, in relation to the visceral surface of the liver. Then, it continues caudally in contact with the right flank, until it turns medially from the right side to the left, at the level of the caudalmost part of the stomach, close to the xiphoid cartilage. Near the pelvis, it turns again medially and goes in relation to the ascending colon and the right kidney. Finally, the duodenum bends ventrally to join the jejunum (Figure 13).

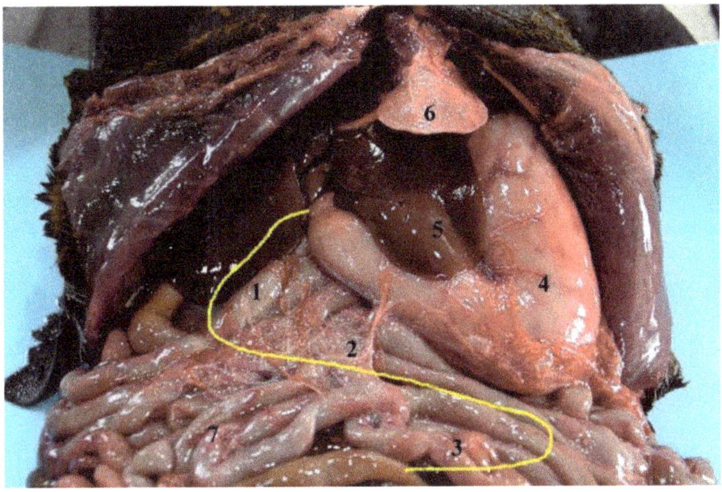

Figure 13. Topography of the duodenum in the abdominal cavity (body in dorsal recumbency). (**1**) Descending duodenum; (**2**) transverse duodenum; (**3**) ascending duodenum; (**4**) stomach; (**5**) liver; (**6**) xiphoid cartilage; (**7**) jejunum.

The pancreas is formed of two lobes, arranged in a V shape (Figure 14). The right lobe is closely related to the cranial part of the duodenum (Figures 15 and 16). The left lobe is located between the stomach and the transverse colon. In the case of the *Arctocephalus australis*, this left lobe does not end in relation to the cranial pole of the left kidney.

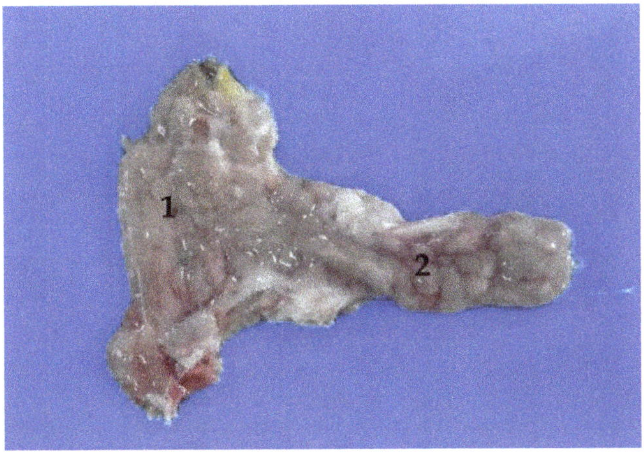

Figure 14. Isolated pancreas. (**1**) Right lobe; (**2**) left lobe.

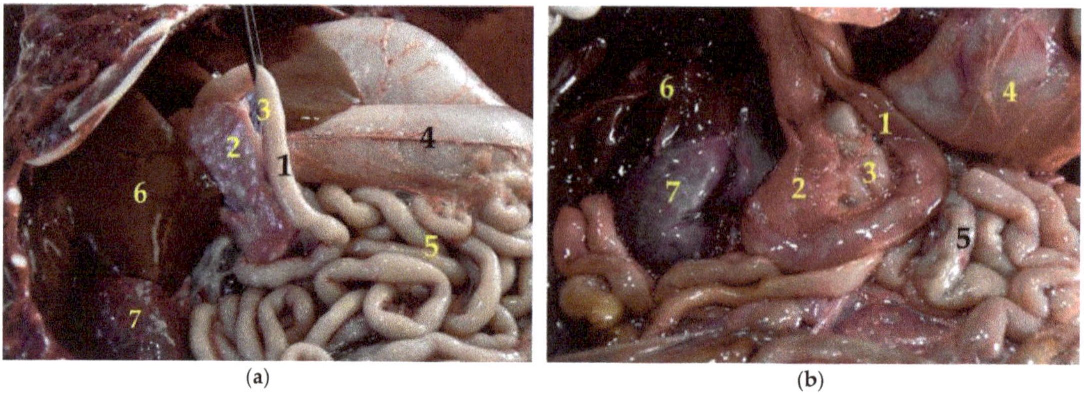

Figure 15. (**a**,**b**) Pancreas and duodenum in situ (body in dorsal recumbency). Note how they are intimately related. In (**b**) the stomach is displaced to visualize both viscera. (**1**) Duodenum; (**2**) pancreas; (**3**) mesoduodenum; (**4**) stomach; (**5**) jejunum; (**6**) liver; (**7**) right kidney.

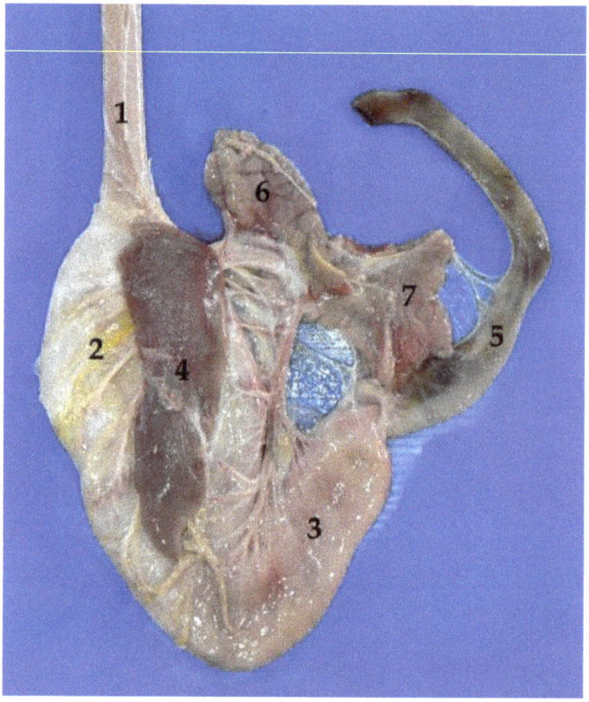

Figure 16. Partial view of the isolated digestive system. (**1**) Esophagus; (**2**) left part of the stomach; (**3**) pyloric part of the stomach; (**4**) spleen; (**5**) duodenum; (**6**) pancreatic left lobe; (**7**) pancreatic right lobe.

3.4.2. Jejunum

As mentioned above, the jejunum is the longest section, with a length of 12–15 m in juvenile animals and 20–30 m in adults (Table 4 and Figure 17). It is located between the stomach, the liver, and the pelvis, occupying a great part of the abdominal cavity. The jejunum forms about 30 intestinal loops and maintains a constant width throughout its length.

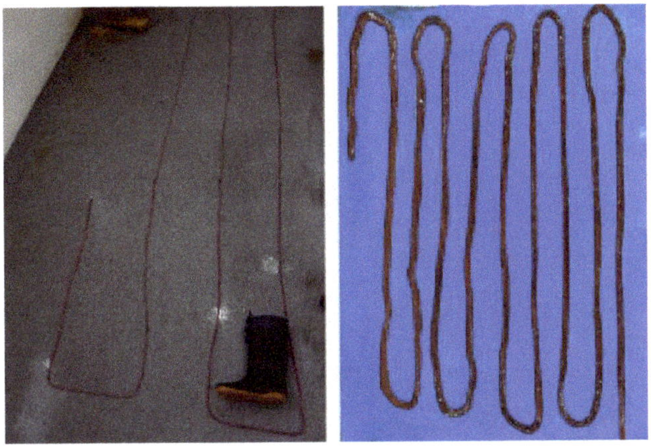

Figure 17. Two views of the extended jejunum after removing the mesojejunum. Note its great length.

3.4.3. Ileum

The ileum is the shortest section and represents the end of the small intestine. It is in the sublumbar region, joining the colon through the so-called ileocolic opening. The connection between ileum and cecum is close. However, it was not the same in all the specimens used in the present study. Thus, there are individual differences in the position, relationship and union between ileum, cecum, and colon, as it is practically parallel in some cases and perpendicular in others. (Figures 11 and 18).

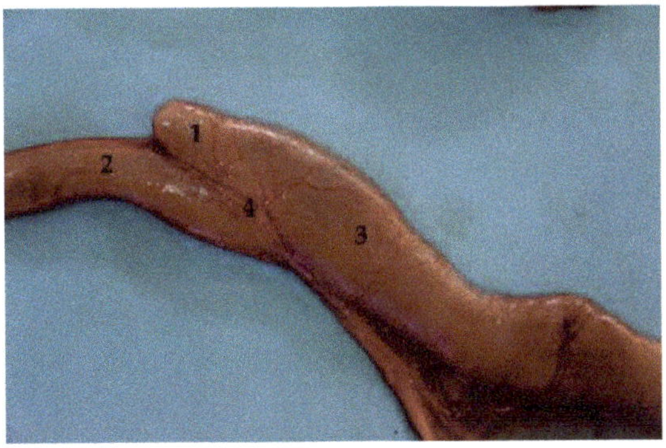

Figure 18. Anatomical relationship between ileum and cecum. (**1**) Cecum (**2**) ileum (final part); (**3**) beginning of the colon; (**4**) ileocecal union.

3.5. Large Intestine

The large intestine is composed of cecum, colon, rectum and anus. It does not show "teniae" or separate longitudinal ribbons of smooth muscle on the outside of the large intestine which give the colon its segmented appearance, or "haustra" (small pouches caused by sacculation). Regarding its total length, the mean of the adult South American fur seal large intestine was of 132.5 cm (Table 5). It is worth remarking that, in the case of the South American fur seal, there is a variation in the size between the small intestine segments (Table 4), and both colon and rectum (Table 5) have a significantly greater width in comparison with that of the small intestine.

Table 5. Morphometric data of the large intestine proceeding of the 11 specimens of *Arctocephalus australis* used in the present study. A: adult; J: juvenile.

	Age		Length (cm)			Diameter (cm)			Weight (g)		
ID	Group	Sex	Cecum	Colon	Rectum	Cecum	Colon	Rectum	Cecum	Colon	Rectum
M7319	A	♀	10	116	10	6	8	6	13	230	20
M99	A	♀	7	113	9	4	7	7	6	235	35
M104	J	♂	1.8	59	8.5	3.5	5.5	5.5	3.1	62	18
M100	J	♀	2.2	67	8	4	5	4.7	2.2	46	11
M74	J	♂	4.3	64	8	4.1	5.2	4.5	2.2	68	16.2
M83	J	♂	3.5	70	5.5	3.2	6	4	1.4	55	15
M82	J	♀	4	83	9	4	6.5	5.5	2	90	14
M70	J	♂	1.6	59	8	3.2	4.7	4.6	1.3	52	14.3
M79	J	♂	3	69	9	4	6	5	1.8	80	9.3
M97	J	♀	2.4	48	8	3.3	4	4.9	2.8	54	11.6
M84	J	♀	2.6	56	7	3	4.5	4.5	1.5	47.5	15.8
Mean	A		8.5	114.5	9.5	5	7.5	6.5	9.5	232.5	27.5
Mean	J		2.8	63.8	7.8	3.5	5.2	4.8	2	61.6	13.9

3.5.1. Cecum

The cecum is located on the right side of the abdominal cavity, ventral to the pancreas and duodenum. One of its ends begins at the origin of the colon, lateral to the ileocolic orifice. The ileum only communicates with the colon, not with the cecum. So, the cecum could be described as a diverticulum in the proximal part of the colon, without any defined function. It communicates with the colon through the cecocolic orifice (Figures 18 and 19).

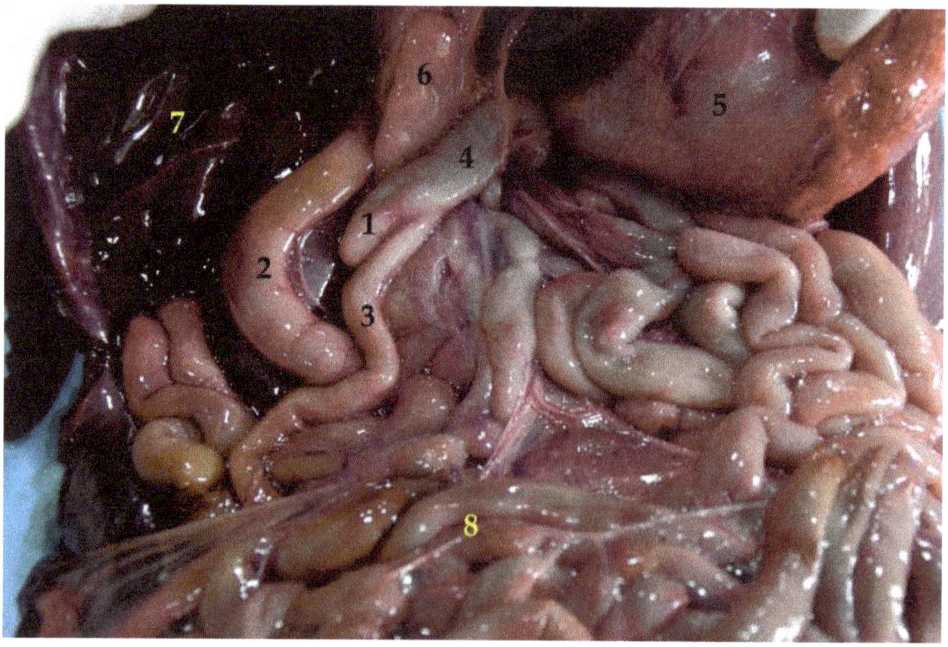

Figure 19. General view of the abdominal cavity (body in dorsal recumbency). The stomach is displaced to visualize the cecum in situ. (**1**) Cecum; (**2**) duodenum; (**3**) ileum; (**4**) colon; (**5**) stomach; (**6**) pancreas; (**7**) liver; (**8**) jejunum.

The cecum of the South American fur seal is relatively small. In adults it measured 8.5 cm on average and 2.8 cm in juveniles (Table 5).

3.5.2. Colon

The colon is composed of three sections that run through the abdominal cavity. The first part corresponds to the ascending colon, followed by the transverse colon and, lastly, the descending colon (Figures 20 and 21). It is attached to the sublumbar region by the mesocolon, which comes from the common mesentery. The ascending colon is short. It passes cranially along the medial surface of the duodenum and the right lobe of the pancreas, until it reaches the pyloric portion of the stomach. Then, the ascending colon changes direction to the left side and crosses the median plane. This section is known as the transverse colon. The last portion is the descending colon, which is located entirely on the left side. It passes caudally to the sublumbar region, along the medial border of the left kidney. Finally, it leans towards the median plane, conforming a "sigmoid" to start the rectum. It is possible to find isolated lymph nodes in both the first part and the final parts of the colon (Figure 20).

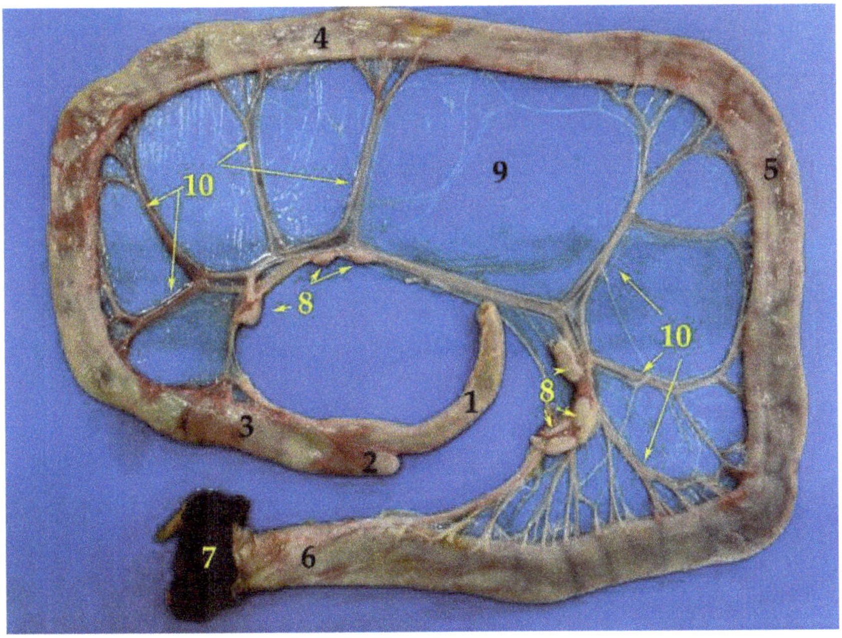

Figure 20. Image of the isolated large intestine. (**1**) Ileum; (**2**) cecum; (**3**) ascending colon; (**4**) transverse colon; (**5**) descending colon; (**6**) rectum; (**7**) anus; (**8**) lymph nodes; (**9**) mesocolon; (**10**) mesenteric vessels.

3.5.3. Rectum

The rectum is the final part of the digestive system (Figure 22). It is intrapelvic in terms of location and practically covered by the peritoneum. No lymph nodes are present in this digestive section.

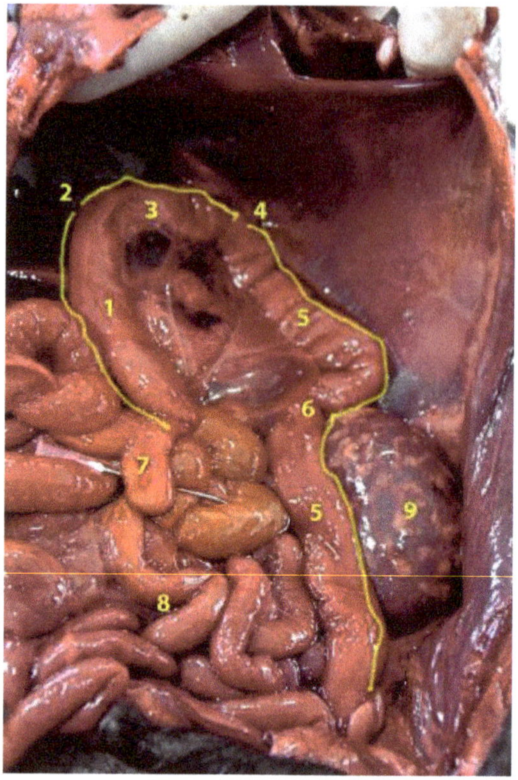

Figure 21. General view of the abdominal cavity (body in dorsal recumbency). Observe the colon with its different parts in situ. (**1**) Ascending colon; (**2**) right flexure of the colon; (**3**) transverse colon; (**4**) left flexure of the colon; (**5**) descending colon; (**6**) sigmoid of the colon; (**7**) cecum; (**8**) jejunum; (**9**) left kidney.

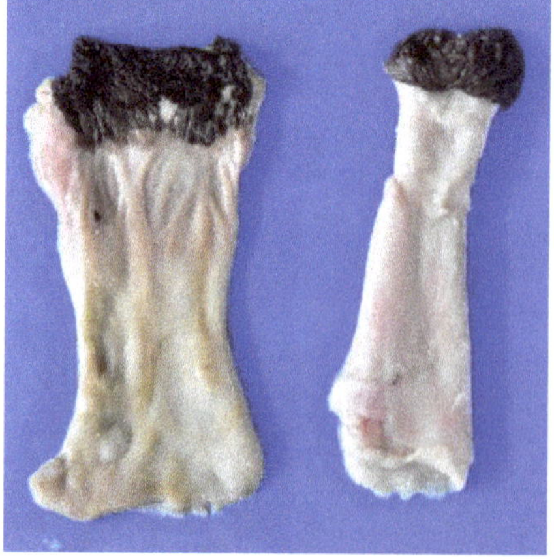

Figure 22. Image of two rectum of different specimens (the one on the left is opened).

3.6. Liver

This is the largest gland of the organism, as it represents 3% of the total weight. For example, in an adult South American fur seal (much bigger in size), the average reached 1000 g.

The liver is located in the intrathoracic portion of the abdominal cavity. It is divided into four main lobes: the quadrate lobe, the caudate lobe, the right lobe and the left lobe (Figures 23 and 24). The right and left lobes are also subdivided, giving rise to a lateral and a medial lobe. On the visceral side, it is possible to observe the caudate lobe, which is made up of the caudate process on the right side, and the papillary process on the left (Figures 25 and 26).

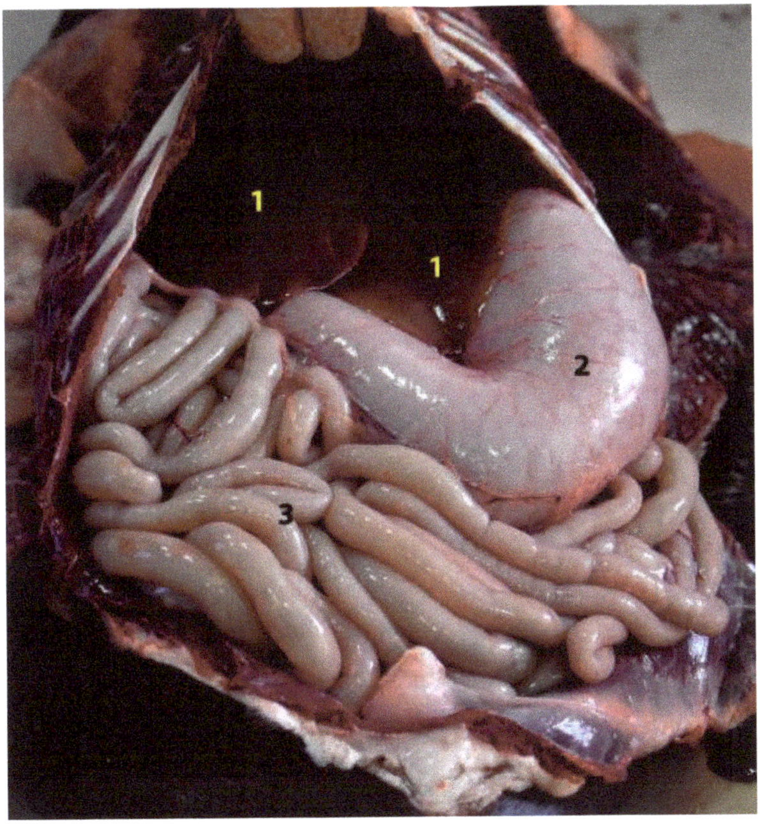

Figure 23. General view of the abdominal cavity (body in dorsal recumbency). Note the liver in situ. (**1**) Parietal surface of the liver; (**2**) stomach; (**3**) small intestine.

On the other hand, the liver is a relatively moldable organ, as it is easily adapted to the surrounding structures. The diaphragmatic or parietal surface is very convex. This surface fits perfectly with the concavity of the diaphragm and the ventral wall of the abdomen. On the visceral side, the hepatic lobes vary their configuration and are arranged irregularly, to have a perfect adaptation to the adjacent viscera. The stomach is located among them, especially because of its changing size, which depends on its degree of repletion.

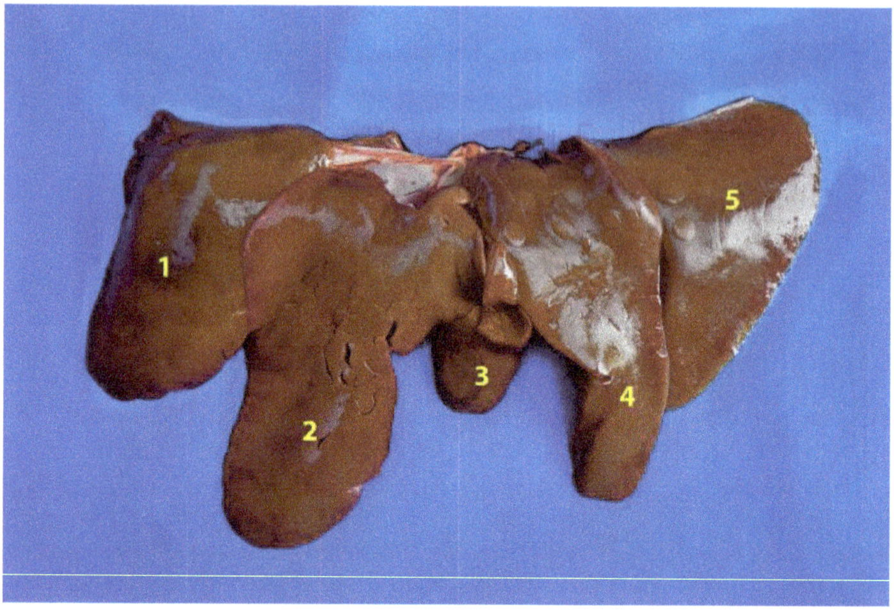

Figure 24. Diaphragmatic face of the liver. (**1**) Right lateral lobe; (**2**) right medial lobe; (**3**) quadrate lobe; (**4**) left medial lobe; (**5**) left lateral lobe.

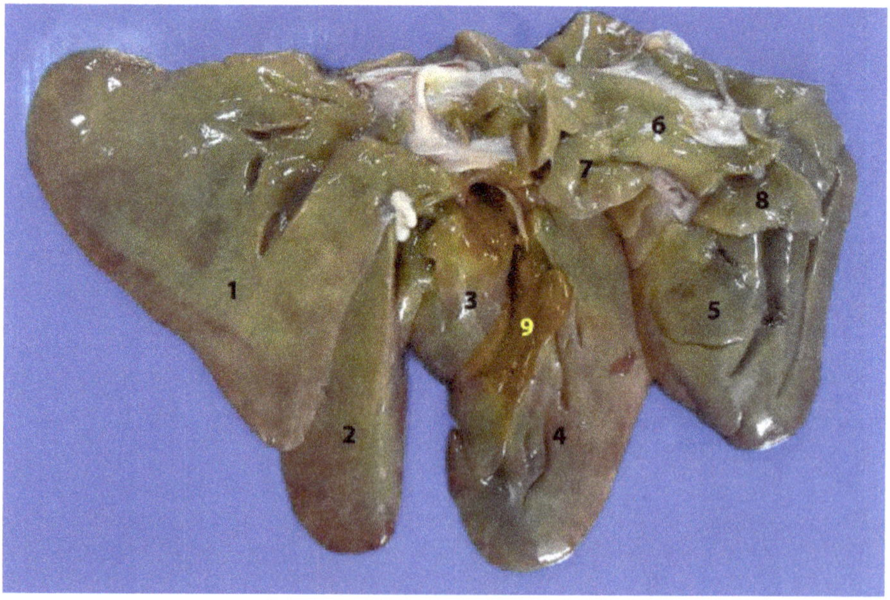

Figure 25. Visceral face of the liver. (**1**) Left lateral lobe; (**2**) left medial lobe; (**3**) quadrate lobe; (**4**) right medial lobe; (**5**) right lateral lobe; (**6**) caudate lobe; (**7**) papillary process; (**8**) caudate process; (**9**) gallbladder fossa.

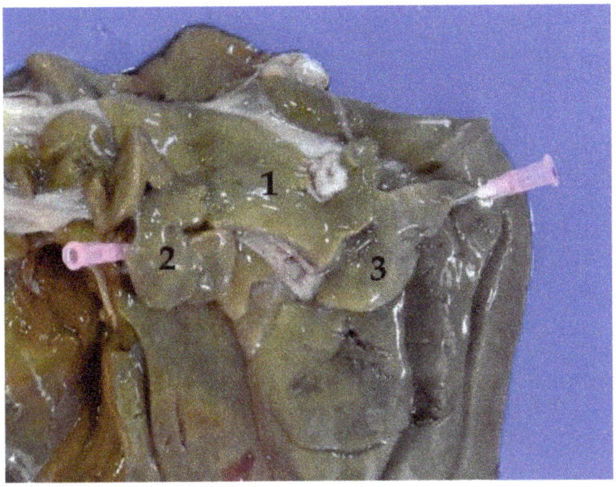

Figure 26. Visceral face of the liver. (**1**) Caudate lobe; (**2**) papillary process; (**3**) caudate process.

3.7. Gallbladder

The gallbladder is located in the gallbladder fossa, on the visceral face of the liver (to which it is firmly attached) and in relation to the quadrate lobe (Figure 27). It has three distinct parts, fundus, body and neck, which continues as a duct. No more anatomical features are worth mentioning.

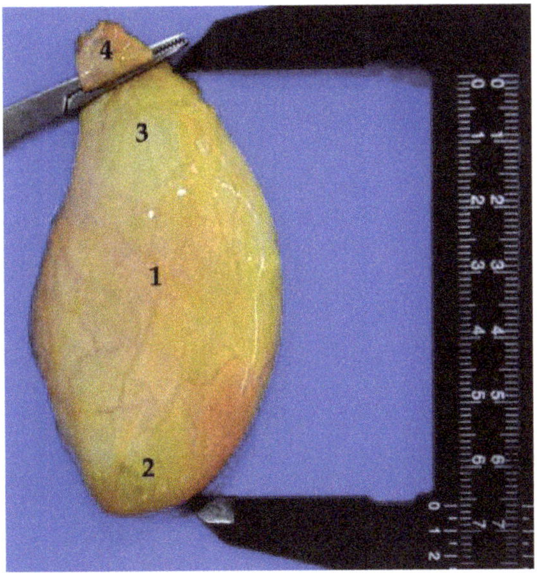

Figure 27. Isolated gallbladder. The total length can be observed, 6.5 cm. (**1**) Body; (**2**) fundus; (**3**) neck; (**4**) begining of the cystic duct.

3.8. Radiology

As mentioned, the anatomical description of the digestive system of the *Arctocephalus australis* was completed with a radiological study. Figure 28 shows the appearance of the main digestive viscera and its topographical relationships.

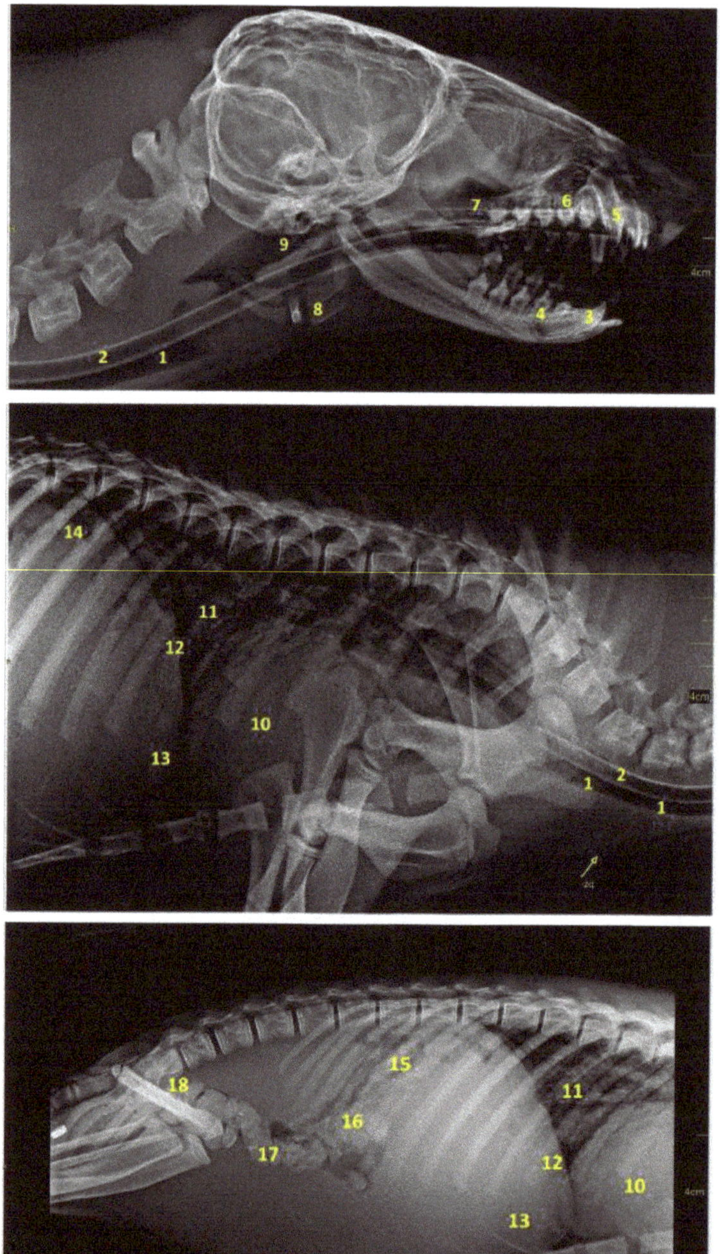

Figure 28. Radiographic images in laterolateral projection of the head, neck and cavities. (**1**) Trachea; (**2**) esophagus; (**3**) lower right canine; (**4**) lower right 2nd molar; (**5**) upper right canine; (**6**) maxillary right 2nd premolar; (**7**) maxillary; (**8**) hyoid; (**9**) pharynx; (**10**) cardiac profile; (**11**) bronchial tree; (**12**) diaphragmatic border; (**13**) liver; (**14**) stomach; (**15**) jejunal loops; (**16**) ascending colon; (**17**) transverse colon; (**18**) descending colon.

4. Discussion

As mentioned before, marine mammals play a critical environmental role, ensuring a balance in the ocean's ecosystem. Although they live most of their lives in or very near the ocean, they have the typical characteristics of all mammals. For example, they breathe air through lungs, they are warm-blooded and they nurse their young. Surprisingly, however, the anatomy of many of these species is not well known. This is the case of the *Arctocephalus australis*, on which there are practically no scientific reports to explain its anatomy in detail. It is important to remark that the animal's anatomy could justify certain physiological phenomena such as its ability to postpone digestion for several hours while they are diving. So, the question is: does the digestive system of *Arctocephalus australis* present adaptations that justify this physiological phenomenon? Or, on the contrary, is its digestive system similar to that of other land-carnivores, and not adapted to aquatic life? To answer this question, in the present paper we describe the digestive apparatus of the South American fur seal, to make a comparison to that of the dog, as a typical model of land-mammal.

As mentioned before in relation to the oral cavity, the upper lip has a little cleft (not very evident) in the central area, which divides it into two equal parts. In other carnivores, this cleft is much more marked, and is called the middle groove or filter [26]. The lateral edges of the lower lip are firm, while in dogs these edges are flaccid and denticulated [26].

In the *Arctocephalus australis*, the palatine raphe is almost inconspicuous, but in dogs, it forms a crest, as the palatine ridges transverse and are much more marked [25]. The incisive papilla is the same in both species [26]. Something curious to mention in the South American fur seal is the presence of a blackish coloration covering the surface of the teeth and the dorsal surface of the tongue. This striking detail is probably linked to their eating habits, and this does not exist in dogs.

The tongue consists of three parts: base, body, vertex. No dividing mid-groove is observed on the dorsal surface, contrary to what has been described in other carnivores [26]. Erdogan et al. [30] have shown that gustatory papillae on the tongue of pinnipeds are reduced in number and simplified in comparison to other carnivores; this and other factors may be responsible for the reduced or absent taste perception in these animals, in comparison with dogs.

In relation to teeth, the *Arctocephalus australis* shows important differential anatomical features in its dental formula, with variations that are worth highlighting. First, the morphological differentiation between premolars and molars in the South American fur seal has disappeared. This is probably because these teeth have currently come to have the same function. They have also decreased in number.

$$\textit{Arctocephalus australis}\text{: } 2\,(\mathrm{I}\ 3/2 - \mathrm{C}\ 1/1 - \mathrm{P}\ 4/4 - \mathrm{M}\ 2/1) = 36$$

$$\text{Dog: } 2\,(\mathrm{I}\ 3/3 - \mathrm{C}\ 1/1 - \mathrm{P}\ 4/4 - \mathrm{M}\ 2/3) = 42$$

The dog has four jaws and six mandibular molars [31], whereas the *Arctocephalus australis* has four jaw teeth and two mandibular teeth, that is, four fewer mandibular teeth. This decrease could be linked to a progressive functionality loss of these dental pieces, as South American fur seals do not chew the food. They simply tear large prey before swallowing. This lack of use has probably caused its disappearance.

Although premolars and molars are similar in size and shape, there are clear differences concerning the number of roots. The first two upper and lower premolars have one root; the rest have two. In the case of dogs, there are three roots in the upper premolar four (carnassial molar tooth) and in the upper molars as well [31]. The crown of these dental pieces in the *Arctocephalus australis* is tritubercular, with the central prominence clearly larger than the laterals.

The mandibular incisors in the *Arctocephalus australis* number four in total, two on each side. Again, there are differences in comparison with the dog. This species presents three mandibular incisors on each side, which makes six in total. Thus, the South American fur seal has lost two of its lower incisors, probably because of its eating habits.

The esophagus looks similar to that of other carnivores. In comparison with dogs, there are two characteristics to remark upon, which could be related to the eating habits of the species. First, the esophageal length, as in adult South American fur seals, the average is 61 cm, which double that of a medium-sized dog (15–25 Kg), where it is about 30 cm [25]. Then, the esophagus of the *Arctocephalus australis* it is particularly wide and distensible, probably to avoid obstructions or different potential tissue damages when swallowing big prey.

The stomach of the South American fur seal is elongated and tubular shaped in contrast to that of other carnivores, such as dogs, where the stomach is piriform-shaped [26]. It is interesting to remark that in the South American fur seal, the greater curvature is not even double that of the lesser (Table 3), whereas in the dog, the greater curvature is almost four times larger than the lesser [26]. Another noteworthy fact is the stomach capacity, which is poor in comparison to other carnivores. Thus, in adults it is about 2–3 L and in juveniles it is about 0.6–1 L; this is much less than in dogs, where it is approximately of 5 L in individuals of an average size of 15–25 kg.

This small stomach of the South American fur seal remembers that in the horse, which is very small in comparison with its body size [25]. This is probably an indication that the capacity of feed storage, one of the stomach main functions, is poor in the Arctocephalus australis in comparison with other carnivores.

The South American fur seal's small intestine is peculiar, due to its significant large length in comparison to other carnivores' species. It would resemble the small intestine of herbivores, instead of that of dogs (33 m in horses or 31 m in small ruminants, whereas in dogs it is 3.6 m) [25]. Herbivores have a longer small intestine than carnivores because they eat plant- and grass-based foods that are high in cellulose, which takes a long time to digest. Although the *Arctocephalus australis* is a carnivore, its small intestine's length could indicate that the nutrients will have more time to be absorbed into the body as they pass through. This is probably a different strategy of extracting food nutrients compared to that of other carnivores, including the dog.

The jejunum is the longest section, with a length of 12–15 m in juvenile animals and 20–30 m in adults (Table 4 and Figure 17). It is interesting to highlight its exaggerated length in comparison with other carnivore species, such as the dog (2.5 m) [25]. The jejunum forms about 30 intestinal loops, much more than in the case of the dog [26] and maintains a constant width throughout its length. As products of digestion (sugars, amino acids, and fatty acids) are absorbed into the bloodstream here, this long jejunal length indicates a great amount of surface area to optimize digestion and nutrient absorption in the *Arctocephalus australis*.

As in dogs, the large intestine of the South American fur seal does not show "teniae" or separate longitudinal ribbons of smooth muscle on the outside of the large intestine which give the colon its segmented appearance, or "haustra" (small pouches caused by sacculation). The mean total length of the adult South American fur seal's large intestine was of 132.5 cm (Table 5), which is double that of the dog (60–75 cm) [26].

When compared, significant differences in the large intestine internal gauge of dogs and *Arctocephalus australis* were observed. In the dog, there is no variation between the small and the large intestines, as the internal diameter remains practically the same in all their parts [26]. On the contrary, in the case of the South American fur seal, there is a variation in size between the small intestine segments (Tables 4 and 5), and both colon and rectum have a significantly greater width in comparison with that of the small intestine. This characteristic could mean a different way to absorb water and salts from the material that has not been digested.

As in the rest of non-marine carnivore species, the ileum only communicates with the colon, not with the cecum. As in the dog, the other end is blind and round-shaped [32].

The cecum of the South American fur seal is relatively small. In adults, it measured 8.5 cm on average, and it was 2.8 cm in juveniles (Table 5). These values are considerably lower than those measured for the dog, in which it can measure up to 15 cm [26]. This is probably due to the nature of their diet. The shape is also different, since in the *Arctocephalus australis* it is straight, while in the dog it is flexuous [26,32].

Concerning the colon, the arrangement is identical to that of other carnivores. As in the case of the dog, it is possible to find isolated lymph nodes in both the first part and the final parts of the colon (Figure 20) [25].

Finally, in relation to the rectum, no lymph nodes are present in this digestive section, contrary to what happens in the dog [32].

5. Conclusions

The anatomical and topographic features of South American fur seal's digestive system are very similar to those presented by other non-marine carnivores, as is the case of the dog. However, the present study shows some specific anatomical differences with other non-marine carnivores. For instance, the changes described on the teeth formula, as a result of the fur seals' adapted diet and their way of ingesting prey, without chewing. When studying adult seal individuals, we find that both the small and the large intestines double their length and weight with respect to their juvenile counterpart. However, surprisingly, in adults fur seals, the lumen of both intestines is only increased by 50%, so these parameters do not seem to be proportional. Compared to the dog, both intestines also show significant differences. The South American fur seal's large intestine is double that of the dog, whereas its small intestine can be up to eight times longer. There is no evidence that could explain the longer length of the small intestine of *Arctocephalus australis*, although we hypothesize that it could be related to their food habits or, as mentioned before, a different strategy to extract food nutrients compared to the dog. Apart from these differences, the conclusion is that being a carnivore seems to be the critical characteristic defining the digestive apparatus of the *Arctocephalus australis*, the adaptations to the aquatic environment being less significant.

Author Contributions: Conceptualization, R.M.-O. and J.G.-S.; methodology, J.G.-S. and R.M.-O.; data collection and validation, C.T.-M., J.-P.L., I.M.-D. and E.T.-D.; formal analysis, C.T.-M., R.M.-O. and J.G.-S.; writing—original draft preparation, C.T.-M., I.S.-Á. and P.P.-L.; writing—review and editing, J.G.-S., R.M.-O. and I.S.-Á.; supervision, R.M.-O. and J.G.-S.; project administration, J.-P.L., C.T.-M., I.M.-D. and E.T.-D.; funding acquisition, J.-P.L. All authors have read and agreed to the published version of the manuscript.

Funding: This research received no external funding.

Institutional Review Board Statement: Ethical review and approval were waived for this study by Fundación Mundo Marino authorizing the use of several cadavers for an anatomical study.

Informed Consent Statement: Not Applicable.

Data Availability Statement: All data supporting reported results are avalaible in Fundación Mundo Marino (by request) https://www.mundomarino.com.ar/.

Acknowledgments: Authors thank Fundación Mundo Marino for the use of its facilities and equipment to carry out this work.

Conflicts of Interest: The authors declare no conflict of interest.

References

1. Rodríguez, S.A.; Karina, C.; Loureiro, J.D.; Ruoppolo, V. *Rescue and Attention of Pinnipeds on the Beach*, 1st ed.; Fundación Mundo Marino: San Clemente, Argentina, 2015; pp. 11–46.
2. English, A.W. Structural correlates of forelimb function in fur seals and sea lions. *J. Morphol.* **1977**, *151*, 325–352. [CrossRef] [PubMed]
3. Noren, S.R.; Williams, T.M.; Pabst, D.A.; McLellan, W.A.; Dearolf, J.L. The development of diving in marine endotherms: Preparing the skeletal muscles of dolphins, penguins, and seals for activity during submergence. *J. Comp. Physiol. B* **2001**, *171*, 127–134. [CrossRef]
4. Dehnhardt, G.; Mauck, B.; Hanke, W.; Bleckmann, H. Hydrodynamic trail-following in harbour seals (*Phoca Vitulina*). *Science* **2001**, *293*, 102–104. [CrossRef]
5. Reidenberg, J.S. Anatomical adaptations of aquatic mammals. *Anat. Rec.* **2007**, *290*, 507–513. [CrossRef] [PubMed]

6. Ortiz, R.M. Osmoregulation in marine mammals. *J. Exp. Biol.* **2001**, *204*, 1831–1844. [CrossRef]
7. Arbiza, J.; Blanc, A.; Castro-Ramos, M.; Katz, H.; Ponce de León, A.; Clara, M. Uruguayan Pinnipeds (*Arctocephalus australis* and *Otaria flavescens*): Evidence of Influenza Virus and Mycobacterium pinnipedii Infections. In *New Approaches to the Study of Marine Mammals*; Romero, A., Keith, E.O., Eds.; InTech: Rijeka, Croatia, 2012; ISBN 9789535108443.
8. Churchill, M.; Clementz, M.T.; Kohno, N. Predictive equations for the estimation of body size in seals and sea lions (Carnivora: Pinnipedia). *J. Anat.* **2014**, *225*, 232–245. [CrossRef] [PubMed]
9. Katz, H.; Pérez, W.; Bielli, A.; Chavez, R. Histomorphology of prepuberal ovaries in the South American fur seal (*Arctocephalus australis* Zimmerman, 1783). *Folia Morphol.* **2009**, *68*, 277–286.
10. Le Bas, A.E. Renal handling of water, urea and electrolytes in wild South America fur seal (*Arctocephalus australis*). *LAJAM* **2003**, *2*, 13–20. [CrossRef]
11. Lord, E. Olfactory Discrimination of Aliphatic 2-Ketones and 1-Alcohols in South African fur Seals (*Arctocephalus pusillus* pusillus). Master's Thesis, Linköpings Universitet, Linköpings, Sweden, 2009.
12. Phillips, A.V. Behavioral cues used in reunions between mother and pup south american fur seals (*Arctocephalus australis*). *J. Mammal.* **2003**, *84*, 524–535. [CrossRef]
13. Seguel, M.; Pavés, H.; Paredes, E.; Schlatter, R. Causes of mortality in South American fur seal pups (*Arctophoca australis gracilis*) at Guafo Island, southern Chile (2004–2008). *Mar. Mam. Sci.* **2013**, *29*, 36–47. [CrossRef]
14. Silva, R.Z.; Pereira, J., Jr.; Cousin, J.C. Histological patterns of the intestinal attachment of *Corynosoma australe* (Acanthocephala: Polymorphidae) in *Arctocephalus australis* (Mammalia: Pinnipedia). *J. Parasit. Dis.* **2013**, *38*, 410–416. [CrossRef] [PubMed]
15. Szteren, D.; Naya, D.E.; Arim, M. Overlap between pinniped summer diet and artisanal fishery catches in Uruguay. *LAJAM* **2004**, *3*, 119–125. [CrossRef]
16. Drabeck, C.M. Some anatomical aspects of the cardiovascular system of antarctic seals and their possible functional significance in diving. *J. Morphol.* **1975**, *145*, 85–105. [CrossRef] [PubMed]
17. Pérez, W.; Katz, H.; Lima, M. Gross heart anatomy of *Arctocephalus australis* (Zimmerman, 1783). *Anat. Sci. Int.* **2008**, *83*, 6–10. [CrossRef]
18. Stewardson, C.L.; Hemsley, S.; Meyer, M.A.; Canfield, P.J.; Maindonald, J.H. Gross and microscopic visceral anatomy of the male Cape fur seal, Arctocephalus *pusillus pusillus* (Pinnipedia: Otariidae), with reference to organ size and growth. *J. Anat.* **1999**, *195*, 235–255. [CrossRef]
19. Ponganis, P. *Diving Physiology of Marine Mammals and Seabirds*; Cambridge University Press: Cambridge, UK, 2015; pp. 162–170. [CrossRef]
20. Pauly, D.; Trites, A.W.; Capuli, E.; Christensen, V. Diet composition and trophic levels of marine mammals. *ICES J. Mar. Sci.* **1998**, *55*, 467–481. [CrossRef]
21. Boyd, I.L.; Arnould, J.P.Y.; Barton, T.; Croxall, J.P. Foraging behaviour of Antarctic fur seals during periods of contrasting prey abundance. *J. Anim. Ecol.* **1994**, *63*, 703–713. [CrossRef]
22. Harvey, J.T. Assesment of errors associated with harbor seal (*Phoca vitulina*) fecal sampling. *J. Zool.* **1989**, *219*, 101–111. [CrossRef]
23. Reid, K.; Arnould, J.P.Y. The diet of Antarctic fur seals Arctocephalus gazella during the breeding season at South Georgia. *Polar Biol.* **1996**, *16*, 105–114. [CrossRef]
24. Naya, D.; Arim, M.; Vargas, R. Diet of South American fur seals (*Arctocephalus australis*) in Isla de Lobos, Uruguay. *Mar. Mamm. Sci.* **2002**, *18*, 734–745. [CrossRef]
25. Climent, S.; Sarasa, M.; Muniesa, P.; Latorre, R. *Manual of Anatomy and Embryology of Domestic Animals*; Acribia: Zaragoza, Spain, 2005; pp. 73–249.
26. Getty, R.; Sisson, S.; Grossman, J.D. *Anatomy of Domestic Animals*, 5th ed.; Salvat Editores: Barcelona, Spain, 1982; Volume 2, pp. 1688–1709.
27. Lavigne, D.M.; Innes, S.; Worthy, G.A.J.; Kovacs, K.M.; Schmitz, O.J.; Hickie, J.P. Metabolic rates of seals and whales. *Can. J. Zool.* **1986**, *64*, 279–284. [CrossRef]
28. Gallivan, G.J.; Ronald, K. Temperature regulation in freely diving harp seals (*Phoca groenlandica*). *Can. J. Zool.* **1979**, *57*, 2256–2263. [CrossRef] [PubMed]
29. De Sordi, N.; Bombardi, C.; Chiocchetti, R.; Clavenzani, P.; Trerè, C.; Canova, M.; Grandis, A. A new method of producing casts for anatomical studies. *Anat. Sci. Int.* **2014**, *89*, 255–265. [CrossRef] [PubMed]
30. Erdogan, S.; Villar Arias, S.; Pérez, W. Morphology of the lingual Surface of South American fur seal (*Arctocephalus australis*) and sea lion (*Otaria flavescens*). *Microsc. Res. Tech.* **2015**, *78*, 140–147. [CrossRef] [PubMed]
31. Colin, E.H.; Peter, P.E. *Small Animal Dentistry*, 1st ed.; Mosby-Year Book: Maryland Heights, MI, USA, 1993; pp. 20–28.
32. Dyce, K.M.; Sack, W.O.; Wensing, C.J.G. *Anatomía Veterinaria*, 2nd ed.; McGraw-Hill Interamericana: Madrid, Spain, 1996; pp. 141–145.

MDPI AG
Grosspeteranlage 5
4052 Basel
Switzerland
Tel.: +41 61 683 77 34

Animals Editorial Office
E-mail: animals@mdpi.com
www.mdpi.com/journal/animals

www.ingramcontent.com/pod-product-compliance
Lightning Source LLC
LaVergne TN
LVHW070502061126
851858LV00011B/278
9783725882311